Thermodynamics in Geochemistry

Thermodynamics in Geochemistry

The Equilibrium Model

GREG M. ANDERSON

DAVID A. CRERAR

New York Oxford
OXFORD UNIVERSITY PRESS
1993

Oxford University Press

Oxford New York Toronto
Delhi Bombay Calcutta Madras Karachi
Kuala Lumpur Singapore Hong Kong Tokyo
Nairobi Dar es Salaam Cape Town
Melbourne Auckland Madrid

and associated companies in
Berlin Ibadan

Published by Oxford University Press, Inc.,
200 Madison Avenue, New York, New York 10016

Library of Congress Cataloging-in-Publication Data
Anderson, G. M. (Gregor Munro), 1932–
Thermodynamics in geochemistry / Greg M. Anderson and David A. Crerar.
p. cm. Includes bibliographical references (p.) and index.
ISBN 0-19-506464-X
1. Geochemistry. 2. Thermodynamics.
I. Crerar, David A. II. Title.
QE515.A6 1993 551.9—dc20
91-18041.

3 5 7 9 8 6 4 2
Printed in the United States of America
on acid-free paper

For Maria Borcsic, steadfast colleague and friend

Aurora Alexandra MacRae Crerar,

the daughter who lights up our lives

Scotia Wendell MacRae, agent, editor, friend and wife

D. A. C.

PREFACE

> Almost all books on thermodynamics contain some errors which are not purely typographical.
>
> Reiss (1965) Preface, p.ix

In the preface to a book on thermodynamics it is customary to offer an excuse for writing yet another. Often the excuse is that no good text is available for this or that specialized subject. Although it is true that we are not completely satisfied with the texts available in the thermodynamics of geochemistry, one would have to be tremendously devoted to teaching for this to be sufficient reason for undertaking the prodigious amount of work involved in writing a book. In our case the reason is probably to be found rather in a long fascination with the subject, an accumulation over the years of topics that we felt were usually badly presented in books, and a desire to see what the final product would look like if we were to put all our convictions on paper. Still, it will have been a sterile effort if there are not at least some students who are helped a little over the hurdles that are encountered in trying to learn this most useful and most difficult of subjects.

Asking "why another book on thermodynamics?" is reminiscent of the story of the man who, asked if he was going to the concert that night to hear one of Beethoven's symphonies, replied no, he'd already heard it. Thermodynamics texts are somewhat like that. Mostly they do not say anything that other books have not said (what is left to say after Gibbs, after all?), but a lot remains in the interpretation, the nuances; authors of thermodynamics texts are like conductors—their worth is shown in the insights of their interpretation, and it seems there will always be room for new books on thermodynamics, just as there is always an audience for more performances of Beethoven.

If there is a difference between this and other books in the same general area, it is not so much in describing the applications of thermodynamics in geochemistry, as in outlining the fundamentals required to understand thermodynamics itself. This is where most of the fascination of the subject is, and where most difficulties arise in learning and teaching the subject. Thermodynamics is of course usually learned in a series of several encounters, in each of which one probes a little deeper into the mysteries of hypothetical systems, unattainable states, unknowable parameters, and impossible processes. We have pondered these mysteries for a long time, and admired the writings of many masters of the subject. We should mention in particular the books by Badger, Callen, and by Reiss, from which we learned a great deal, and whose ideas we in part follow in this book. Students of Gibbs will find other sections where we follow the Master very closely. We sincerely hope that what we have distilled from these writings and combined with our own experience will be found useful in spite of our "errors which are not purely typographical."

Who Is the Book For?

In the sense that we do not skip anything fundamental that is necessary to understanding thermodynamics, the book could be used in a first introduction to the subject. However, we treat these subjects in more detail than is customary in introductory courses, and we do go on to some more advanced topics, so that in that sense the book is for more advanced readers. Then again, we have certain emphases and points of view that are not so much thermodynamic as philosophical, so that we hope that the book will be read with interest by our peers. If there must be a designated intended readership, we could say that it might be people who have some acquaintance with chemical thermodynamics and are eager for more.

Ideally, a student of thermodynamics in the Earth Sciences should have a good grounding in linear algebra because it simplifies and unifies some topics, such as those dealing with components, phases, reactions, and the phase rule, and some of the best modern applications are built along these lines. Unfortunately, we were unable to cram the necessary material into Chapter 2. We can only somewhat lamely recommend to students that they pick up linear algebra somewhere along the way. A similar comment can be made about statistical mechanics and its applications in thermodynamics. We make a gesture in this direction in Chapter 6, but again, serious students of the subject will want to learn more.

Historical Note

Although we have made an occasional foray into biographies and other literature on the development of thermodynamics, and mention bits of history here and there, we are not very knowledgeable in this area. For the most part, we follow traditional attributions of equations and concepts, which may be incorrect. McGlashan (1966) claims that

> The history of thermodynamics is in fact a much more difficult subject than thermodynamics itself, . . .

Truesdell's (1980) Law of Attributions states that

> For arbitrary n, any false attribution outlasts n documented corrections.

Thus it may be that Rankine rather than Clausius discovered entropy, or that Reech preceded Gibbs in the definition of fundamental equations (Truesdell, 1980), but we take no stand on these and other matters. In fact, we are inclined to leave the history of thermodynamics to the specialists, in view of Truesdell's (1980) opinion:

> I feel myself permitted, therefore, to select instead of fields of brilliant success like hydrodynamics, elasticity, and electromagnetism, one accursed by misunderstanding, irrelevance, retreat, and failure. Thus I write of thermodynamics in the nineteenth century.

A Note on Units

Many readers may be dismayed to find that we do not use only SI units. The fact is that in geochemistry at the present time this is not possible without a great deal of difficulty, because at least one very important group of contributors to the data of geochemistry, "Prediction Central" at Berkeley, continues to use calories and related units. Changing all their numbers to SI units simply to preserve SI uniformity in this text would be counter-productive. Students of this subject simply must become accustomed to switching back and forth between calories and Joules, and all their related units. We have tried to include all necessary conversion factors and constants in Appendix A.

ACKNOWLEDGMENTS

The late Professor A. R. Gordon used to teach a course in the chemistry department at Toronto called "Advanced Thermodynamics," which the senior author was obliged to take as a beginning graduate student. Professor Gordon never used lecture notes, but would cover innumerable blackboards in rapid succession in an impressive display twice a week. What is most memorable, however, was his concern for a proper understanding of the fundamentals. The lecture notes from that course became dog-eared from consultation over many years, and several influences and examples from them are incorporated in this book. Prof. Gordon used to say

> "In thermodynamics, you have to get to the point where you understand what any damn fool means no matter what he says."

We hope to help people get to that point.

A great deal of this book was written during a one-year period (1986–87) when we spent a sabbatical year together at the Laboratoire de Géochimie, Université Paul Sabatier, in Toulouse, France. Jacques Schott and the other members of the CNRS lab and the Geology department were extraordinarily hospitable, and the year passed incredibly quickly.

More recently, we have received useful comments on parts of this book from a number of individuals. Hugh Greenwood and Edgar Froese should get medals. Both critically read almost the whole book, and provided detailed criticisms that were always useful. We are especially grateful, too, to Terry Gordon. Terry provided insightful comments on many of the early chapters on fundamentals, and encouraged us to extend the "model" aspect of the presentation, which we have done. The late Al Henry was subjected to preliminary versions of each chapter as a graduate student, and provided much careful and useful criticism. He was an extraordinary student, and would have contributed much to science but for his untimely accidental death. Useful comments on various chapters or groups of chapters were also provided by Paul Barton, Jr., Jim Nicholls, Terry Bowers, Stuart Whittington, Heather Jamieson, Frank Spera, Dick Bailey, Larry Barron and Peter Renders. Christophe Monnin and Susan Brantley helped a great deal on the Pitzer equations. Not all of these people agreed with what we have written, or the way we have written it, or our choice of emphases. We made many changes in response to these comments, but in many other cases we have persisted in our perhaps misguided ways. Either way, the contributions of these people were quite significant, and we thank them all. Thanks also to Susan Lehre for help with the diagrams, and to Domenic Channer for a thorough checking of references and equation numbers.

Toronto G. M. A.
Princeton D. A. C.

Editor's Note

Since the original draft for this book was completed, David A. Crerar has become seriously ill. Were he able to speak for himself, he would express his deep admiration for Greg Anderson, who inspired him to become a geochemist. Originally his mentor, then scientific colleague, Greg took on the formidable task of bringing the manuscript to publication.

David would also have wanted to express his gratitude to the undergraduate and graduate students he worked with at Princeton University, whom he frequently acknowledged as having inspired him by leading him into areas of inquiry he would never have otherwise investigated.

TABLE OF CONTENTS

Thermodynamics in Geochemistry

1

THERMODYNAMICS—A MODEL SUBJECT

> ... the almost certain truth that nobody (authors included) understands thermodynamics completely. The writing of a book therefore becomes a kind of catharsis in which the author exorcises his own demon of incomprehension and prevents it from occupying the soul of another.
>
> Reiss (1965) Preface, p. vii

The teaching of thermodynamics, not to mention the writing of a book on thermodynamics, presents a great challenge for anyone foolhardy enough to undertake it. How to present the subject?—historical or postulational; classical or statistical; macroscopic or microscopic; mathematical or non-mathematical; rigorous or close enough; applications or fundamentals; and so on. Then there is the uncomfortable fact that, as Reiss (1965) points out, one very probably doesn't completely understand the subject. But if teaching thermodynamics presents difficulties, what of the poor student who is subjected to an almost insuperable series of obstacles; everything seems to depend on infinitesimals, on the most unusual and unlikely hypothetical systems, on differentials which may or may not be integrable, and especially on reversible processes which are completely unimaginable and seem to be the key to everything. Somehow out of this mess come concepts and relationships which are very useful in the real world, and it can be argued that in introductory courses the goal should be an ability to use these relationships, not to understand them.

At the same time the subject is so profound and esthetically pleasing to scientists that many feel the urge to rhapsodize on it. Lewis and Randall (1923), who wrote one of the more significant books on the subject after Gibbs, likened thermodynamics to a cathedral inspiring solemnity and awe; Einstein wrote that it was the only physical theory of universal content that he was sure would never be overthrown. The profundity of the subject comes from the seemingly universal applicability of its premises and conclusions to some of the most fundamental questions of science, and the elegance from the simplicity and symmetry of its mathematical operations. For example, it is quite amazing to realize that essentially all the geochemical applications discussed in this book involve a single equation, along with a few direct descendants, derived directly from the first and second laws of thermodynamics. It is a considerable challenge to show the meaning of this equation and how to use it.

In this text we present the subject as we would have liked to have had it presented to us—with a careful and complete explanation of some fundamental points, a partial or complete neglect of others, a combination of historical, postulational, and statistical approaches as seems appropriate at the moment, and a smattering of applications in geochemistry. The choice of topics to present and to neglect is indefensible, except that it is based on our experience in trying to understand the subject and in trying to teach it in a geological context. The reader will find a perhaps unusual emphasis on the concepts of metastability, constraints, fundamental equations, thermodynamic potentials, and some mathematics such as Legendre Transforms, much of which is traceable to books by Reiss (1965) and Callen (1960), and all of which is related to our conviction that thermodynamics is best understood as a model, a system of equations which is quite separate from reality. A central theme of the book is that the reason reversible processes (defined as a continuous succession of equilibrium states) are of such fundamental importance in thermodynamics is not really because of entropy or heat engines, but because reversible processes are the physical equivalent of continuous functions. Reversible processes (strictly speaking reversible and quasistatic processes, an important distinction in our usage) are necessary so that mathematics can be used in connection with processes; they are the only kind of thermodynamic process to which differentiation and integration can be applied. A careful consideration of the equilibrium state reveals that real systems never achieve this state, so that the system of equations describing reversible processes is therefore completely removed from physical reality. These equations have as variables physical properties, and represent a series of surfaces and planes in multidimensional space (thermodynamic state space) which in their symmetry and shining beauty might well be thought of as a cathedral. Any such physical representation probably is inappropriate however, because the surfaces and planes of the model show no signs of dust or dirt or indeed of imperfections of any kind.

We refer to this edifice of surfaces and planes described by the differential equations of thermodynamics in many places throughout the book as "the equilibrium model" or "the thermodynamic model." The properties of real systems (rocks, minerals, magmas, hydrothermal solutions, etc. in our cases) which are (approximately) at equilibrium can be thought of as points (almost) on these surfaces, and all aspects of our development of thermodynamics follows from this situation.

We find that in science it is always essential to keep a clear mental distinction between reality and the model that one develops to describe reality, and this is just as true for thermodynamics as for any other subject. We have found that many of the problems in teaching and learning thermodynamics are made easier by making this distinction between the model, in which mathematical relationships are simple, in which tangents and points slide around surfaces with the greatest of ease; and reality, infinitely more complex, which only approaches the model in states of equilibrium. We hope the reader will agree.

1.1. CHEMICAL REACTIONS

The following is a brief introduction to our usage and notation. Words in italics in this section are defined and discussed more fully in later sections.

Thermodynamics is primarily concerned with *energy* changes, and in chemical thermodynamics, these changes are commonly associated with the rearrangement of atoms from one configuration into another. This rearrangement is called a chemical reaction, and the writing and understanding of chemical reactions is of course central to all of physical chemistry.

Consider the generalized chemical reaction

$$aA + bB + \cdots = mM + nN + \cdots \tag{1.1}$$

The quantities A, B, M, and N are in our usage *constituents* , or, for the moment, chemical formulae representing various combinations of elements in a *system*. The stoichiometric coefficients a, b, m, and n are a set of dimensionless numbers, usually integers, which allow a mass and charge balance in the reaction. There may be any number of *phases* involved, and the constituents represented are normally only a few of the large number in the system. They are chosen to represent a *process* of interest to us. For example, consider a clay mineral (unspecified) on the sea bottom. An essential constituent of clay minerals is SiO_2, and seawater also contains SiO_2 in dissolved form. If we are interested in the exchange of silica between clay minerals and seawater, we might write a reaction such as

$$SiO_2{}^{in\ clay\ mineral} = SiO_2{}^{in\ seawater} \tag{1.2}$$

or, using (s) and (aq) to denote solid and aqueous constituents,

$$SiO_2(s) = SiO_2(aq) \tag{1.3}$$

or, if we are interested in a particular aqueous silica *species*,

$$SiO_2(s) + 2H_2O = H_4SiO_4(aq) \tag{1.4}$$

Note that the clay mineral contains many other constituents, as does the seawater. They affect the properties of the constituents we have chosen to consider, but they do not appear in the reaction. They may also be very important in the way that the reaction proceeds in real systems. In other words, reactions do not always involve only the reactants and products which have been chosen. Thus in reaction (1.1), A and B may not form M and N directly, but may in fact form X, which then changes to M and N. This may or may not be important to the user of the equation, but it makes no difference to the energy balances involved, as long as equilibrium states are compared.

Another fact to note is that the constituents chosen may not actually exist in the real system, only in the model of the system. The commonest example of this in geochemistry is the use of oxygen gas ($O_2(g)$) as a constituent in systems under highly reducing conditions (see §18.5.1).

1.1.1. *Notation*

The reaction as written with the = sign implies little more than a mass and charge balance. The constituents may or may not have reached *equilibrium*, or one side may be *metastable* with respect to the other. For reactions that go *irreversibly* to completion, we will often use $\rightarrow$, as in A $\rightarrow$ B, and for an equilibrium, we can use A $\rightleftharpoons$ B. However, we have probably not been completely consistent in this usage. The = sign covers all cases.

One of the purposes in writing the reaction is to consider the change in some property of the system between the states represented by the left and right sides. For example, in reaction (1.4), we may find that $\Delta_r G^\circ = x$ Joules. In this statement, G° is the property of interest, and Δ means

$$\sum G^\circ_{products} - \sum G^\circ_{reactants} \tag{1.5}$$

To obtain the correct sum on each side, of course, the stoichiometric coefficients must be used. For example, in reaction (1.4),

$$\Delta_r G^\circ = \underbrace{G^\circ_{\mathrm{H_4SiO_4}(aq)}}_{products} - \underbrace{2G^\circ_{\mathrm{H_2O}} - G^\circ_{\mathrm{SiO_2}(s)}}_{reactants} \qquad (1.6)$$

The small r in $\Delta_r G^\circ$ means we are dealing with a reaction. Other letters in this position have more specialized meanings, such as $\Delta_f G^\circ$ (a reaction in which a compound is formed from its elements), or $\Delta_t G^\circ$, a phase transition. The individual constituents usually have molar units (e.g., J mol^{-1}, but by convention, the Δ quantities normally omit the mol^{-1} term, as in $\Delta_r G^\circ = x$ J.

We usually indicate the constituent as a subscript and the phase or other information as a superscript, as in

$$G^{kaolinite}_{\mathrm{SiO_2}}$$

1.2. WHAT'S AHEAD

The relatively simple reactions (1.2), (1.3), and (1.4) are typical of many in geochemistry, and serve to illustrate many of the aspects of thermodynamics that we will be considering. We said that thermodynamics is concerned with the energy changes in such reactions. Apart from the fundamental question of what exactly energy *is*, we have to know exactly what kind of energy to deal with in connection with different problems—there are several. Then how do you measure such energy changes, or where do you look them up? And what good is the energy change once you do know it? What can you do with it? Then there is the problem that each of the constituents in these reactions is part of a solution. How can we break up the energy content or the volume of a solution into energies or volumes of its various parts? The formula SiO_2 is also the formula for several minerals—how do they fit in? How can we consider one constituent of a clay mineral and ignore the others? What happens if the clay mineral or the seawater changes composition? And probably the hardest question of all—how do we know if we have chosen an appropriate equation to consider? Maybe the energy level of SiO_2 in clays is difficult to deal with. Maybe other reactions would be more appropriate, or lead to more useful results. Nevertheless, the questions we have raised are valid, and our answers form the substance of this book.

It is relatively easy to write and balance chemical reactions. It takes years of experience and considerable scientific insight to know which reactions are worth writing and investigating in any given research situation. The problem of the silica balance in the oceans is one of the innumerable ongoing problems in geochemistry, many of which have partial answers. At the present time the reactions we have written above are probably not very useful for several reasons. The point is that thermodynamics cannot decide such questions—you the investigator must decide. In this book we deal with the simpler problem of how to deal with the thermodynamics of given reactions.

2

MATHEMATICAL BACKGROUND

No attempt is made at mathematical rigor, since in physics this is always illusory.

Landau and Lifshitz, *Statistical Physics* quoted in Lewins (1985), p. 168.

2.1. THE MATHEMATICAL SIDE OF THERMODYNAMICS

Thermodynamics, like other sciences, has a theoretical side, expressed in mathematical language, and a practical side, in which experiments are performed to produce the physical data required and interpreted by the theoretical side. The mathematical side of thermodynamics is simple and elegant and is easily derived from first principles. This might lead to the conclusion that thermodynamics is a simple subject, one that can be easily absorbed early in one's education before going on to more challenging and interesting topics. This is true, if by learning thermodynamics one means learning to manipulate its equations and variables and showing their interrelationships. But for most students the subject is actually far from simple, and for professors it is a considerable challenge to present the necessary material intelligibly. The equations and the variables are somehow related to the real world of beakers and solutions, fuels and engines, rocks and minerals, and it is this interface that provides most of the difficulties. What do variables such as entropy and free energy *really* mean, and what physical processes do the equations describe? The difficulty in understanding and using thermodynamics is conceptual, not mathematical. We will attempt to explain the relationship between the mathematical and the physical sides of thermodynamics, but it is advisable first to review the mathematics involved and subsequently to define the terms used in thermodynamics.

The mathematics required for thermodynamics consists for the most part of nothing more complex than differential and integral calculus. However, several aspects of the subject can be presented in various ways that are either more or less mathematically based, and the "best" way for various individuals depends on their mathematical background. The more mathematical treatments are elegant, concise, and satisfying to some people, and too abstract and divorced from reality for others.

In this book we attempt to steer a middle-of-the-road course. We review in the first part of this chapter those aspects of mathematics that are absolutely essential to an understanding of thermodynamics. The chapter closes with mathematical topics that, although not essential, do help in understanding certain aspects of thermodynamics.

We do not strive for the last ounce of mathematical rigor, and the mathematical demonstrations of points in thermodynamics are usually presented in addition to other physical or conceptual ways of looking at the same material. With this approach we hope to convey a sense of the elegance of the subject without restricting the book to mathematically-oriented readers.

2.2. ESSENTIAL MATHEMATICS

2.2.1. *Variables and Functions*

A *variable* is a quantity in mathematical relationships, represented by a symbol, that may take on any value from a given set of values. Variables are related to one another by *functions*. If for every value of the variable x there corresponds at least one value of the variable y, then y is a function of x, written $y = y(x)$ or $y = f(x)$.

Functions may also involve more than two variables, so that if values are assigned to all but one variable (the independent variables), the value of the remaining one (the dependent variable) is fixed, *i.e.*, in $w = w(x, y, z)$, x, y, and z are the independent variables and w the dependent. Usually the choice is arbitrary, that is, we can usually solve for x and find $x = x(w, y, z)$, making x the dependent variable, and so on. In cases where it is difficult or impossible to solve explicitly for the desired dependent variable, we may write an *implicit* function $f(x, y, w, z) = 0$ and treat it as described, for example, by Dence (1975, p. 53).

A function having three variables, x, y, and z, therefore has two independent variables, and can be said to be *divariant*, or to have *two degrees of freedom*. "Divariance" and "two degrees of freedom" refer to the fact that we are free to choose the values of two of the variables (perhaps within certain ranges), the third then being fixed by the functional relationship. For example, for the function

$$x + y + z = 0$$

if we choose $x = 2$, $y = 2$, then z is fixed at -4. If in addition to this function we have another one involving the same variables, such as

$$2x - y + 4z = -19$$

we now have three variables and two functional relationships, and we are only free to choose one of the three variables, the other two then being fixed. This situation can be said to be *univariant* or to have one degree of freedom.

And, of course, if we have a third functional relationship, *e.g.*,

$$-3x + 2y - 7z = 35$$

then we have no choice, x, y and z are fixed at 2, 3 and -5 respectively, and the situation is *invariant*. These equations or functions can also be thought of as *constraints* on the numerical values of the variables.

2.2.2. *Variable Space*

Graphs of functions involve one dependent variable and one or two independent variables. For example, the function $y = x^2$ is shown in Figure 2.1. The two or three variables involved can be said to define a "Variable Space," within which the function forms a line or a surface. Functions with more than three variables cannot, of course, be represented in three dimensional space, but can still be thought of as surfaces in a higher-order space or hyperspace.

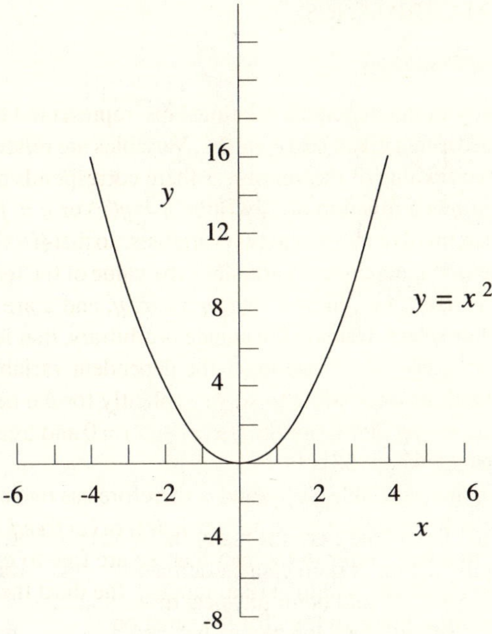

FIG. 2.1. The function $y = x^2$.

2.2.3. *Single-Valued and Continuous Functions*

Functions may have many attributes and may be classified in various ways—see, for example, Perrin (1970, Chap. 1). From our point of view, the two most important attributes that functions may have are to be (a) single-valued and (b) continuous. A single-valued function, as the name suggests, is one for which there is only one value of the dependent variable for given values of the independent variables. Thus $y = x^2$ (Figure 2.1) is single-valued in x (there is only value of y for chosen x-values). However, $y = x^{\frac{1}{2}}$ (Figure 2.2) is not single-valued in x, since for a given value of x there are two values of y.

Functions may also have ranges of the independent variables within which they are single-valued and other ranges in which they are multi-valued.

Continuous functions are ones in which the dependent variable changes smoothly and continuously for smooth and continuous changes of the independent variables. Figures 2.1 and 2.2 represent continuous functions, but Figure 2.3 represents a function which is continuous for $x \neq a$ but shows a discontinuity from $-\infty$ to $+\infty$ at $x = a$. The mathematical definition of continuity is that $f(x)$ is continuous at $x = a$ if $f(a)$ is defined, and if $\lim_{x \to a} f(x) = f(a)$.

Most functions that have discontinuities, such as the one shown, have them only for certain isolated values of the independent variables, and they are continuous for all other values. As long as we stay within the allowed ranges then, the function is

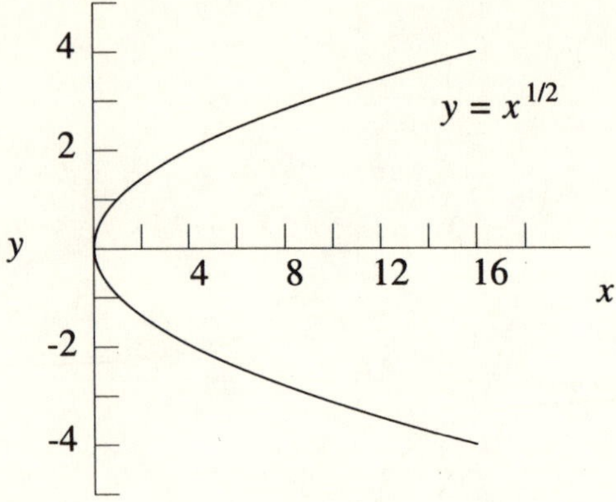

FIG. 2.2. The function $y = x^{\frac{1}{2}}$.

continuous, and in fact this is the case for many thermodynamic functions. That is, they may be used only within certain ranges defined by phase transitions such as boiling points, freezing points, and other physical discontinuities.

Multi-valued and discontinuous functions often present difficulties for differentiation and integration. They are mathematically "not well-behaved." There is always great difficulty, for example, in fitting an equation of state in the vicinity of critical points (critical points being both mathematical and physical discontinuities).

2.2.4. *Differentials and Derivatives*

Thermodynamic relationships are so often presented in differential form that it is essential to have a clear understanding of what differentials are. In Figure 2.4 we illustrate the usual definition of a derivative. For any function $y = y(x)$, the *derivative* of y is a function $y'(x)$ where

$$y'(x) = \lim_{\Delta x \to 0} \left(\frac{y(x + \Delta x) - y(x)}{\Delta x} \right) \tag{2.1}$$

As shown in Figure 2.4, the quantity

$$\left(\frac{y(x + \Delta x) - y(x)}{\Delta x} \right)$$

is the slope of a line that intersects the function $y(x)$ at two points, (x_1, y_1) and (x_0, y_0) and $\Delta x = x_1 - x_0$, $\Delta y = y_1 - y_0$. As Δx gets smaller, x_1 and y_1 approach x_0 and y_0 and in the limit as $\Delta x \to 0$, the line of intersection becomes the tangent to $y(x)$ at (x_0, y_0). The notation $y'(x)$ indicates that the derivative, or the slope of the tangent, is a new function of x, quite distinct from the original function $y(x)$.

If we let $\Delta x = dx$ and define dy, the *differential of* y as

$$dy = y'(x)\, dx \tag{2.2}$$

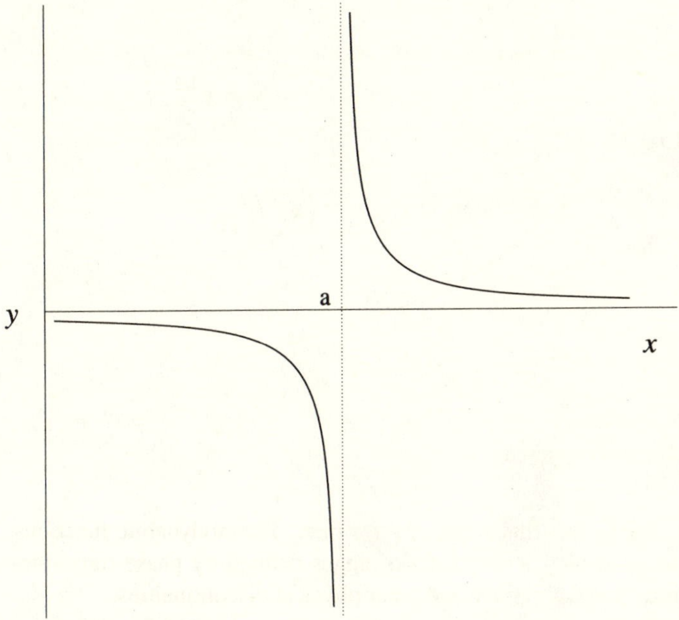

FIG. 2.3. Function showing a discontinuity at $x = a$.

then

$$y'(x) = \frac{dy}{dx} \tag{2.3}$$

and the symbol dy/dx is often used to represent the derivative. dx has already been defined as equal to Δx, i.e., any increment of x, and dy is a linear function of dx as shown in Figure 2.5. Obviously neither dx nor dy is necessarily an infinitesimal. It is an unfortunate fact that because dy/dx is equal to the derivative, many students of thermodynamics get the idea that differentials are infinitesimal quantities, and this is a stumbling block to the intuitive grasp of the many equations involving differentials. During integration, of course, differentials can and do take on infinitesimal values.

Of particular interest in thermodynamics is the extremum value of certain functions, i.e., the maximum or minimum point. According to the calculus, this is the point where the derivative passes through zero, or $dy/dx = 0$. In Figure 2.1, $dy/dx = 2x$ (or $dy = 2x\,dx$), which equals zero at $x = 0$. In differential form, we say the minimum occurs at $dy = 0$. This means that at the minimum, where the tangent is horizontal, y will not change ($dy = 0$) no matter what the size of dx. Keep this in mind when you get to Chapter 5 (§5.4).

2.2.5. *Partial Derivatives and Total Differentials*

A function having several variables may be differentiated with respect to one of the

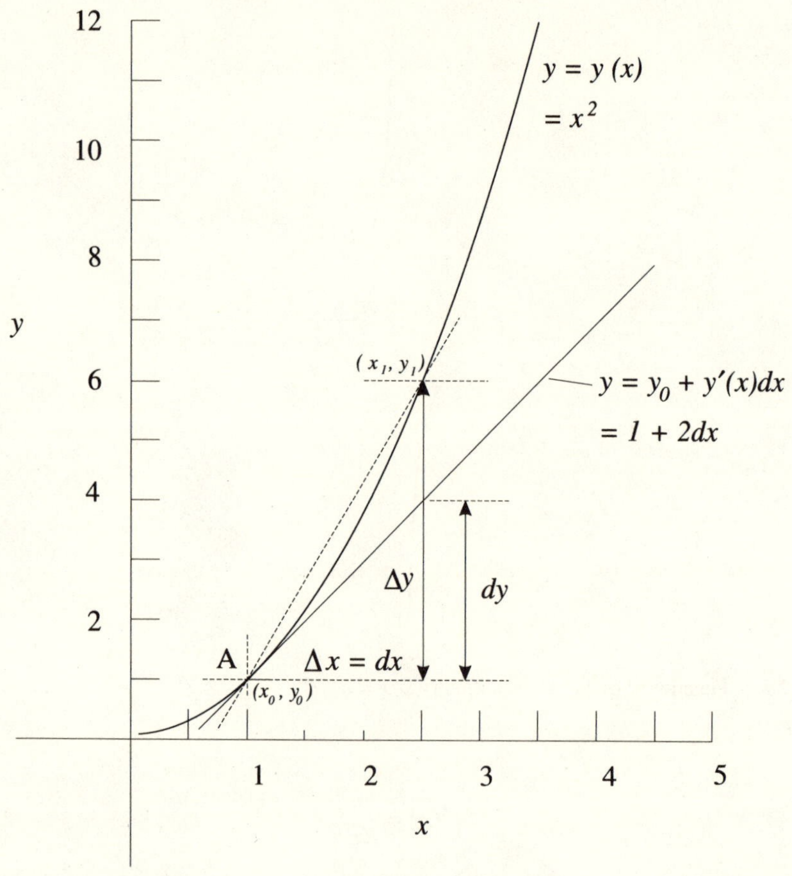

FIG. 2.4. The meaning of dy/dx.

variables, keeping all the others at fixed values. Thus the function

$$z = z(x, y)$$

can be differentiated with respect to x, keeping y constant, thus evaluating $(\partial z/\partial x)_y$, and it can also be differentiated with respect to y keeping x constant, evaluating $(\partial z/\partial y)_x$. These quantities are termed *partial derivatives*. The new shape of the "*d*" symbol is to remind us of the partial nature of the differentiation process and the subscripts remind us which variables are being held constant. In cases where there is no likelihood of confusion, the subscripts are often omitted. For example, if

$$z = 2x^2 + 4y^3$$

then

$$\partial z/\partial x = (\partial z/\partial x)_y = 4x$$

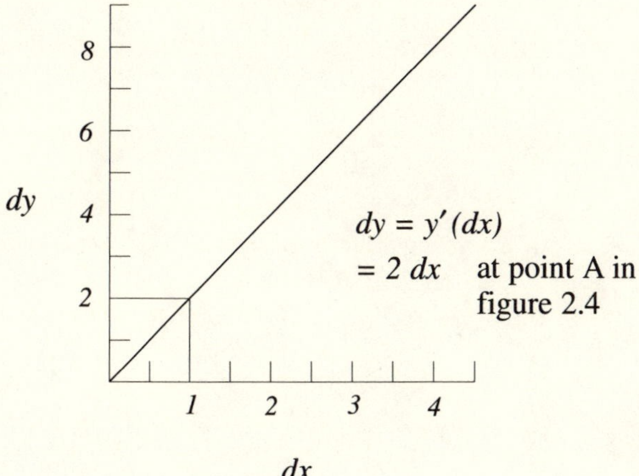

FIG. 2.5. dy is a homogeneous linear function of dx.

and

$$\partial z/\partial y = (\partial z/\partial y)_x = 12y^2$$

The *total differential* of z, dz, is defined as

$$dz = \left(\frac{\partial z}{\partial x}\right)_y dx + \left(\frac{\partial z}{\partial y}\right)_x dy \tag{2.4}$$

For example, if the function $V = V(T, P)$ is

$$V = RT/P \tag{2.5}$$

where R is a constant, then

$$dV = \left(\frac{\partial V}{\partial T}\right)_P dT + \left(\frac{\partial V}{\partial P}\right)_T dP \tag{2.6}$$

$$= (R/P)dT + (-RT/P^2)dP \tag{2.7}$$

Equation (2.4) has a very straightforward geometrical meaning, illustrated in Figure 2.6. Here, PQRS is a surface in x, y, z space representing the function $z = z(x, y)$. PABC is the plane tangent to the surface at point P, and PDEF is a horizontal ($z = constant$) plane through P. Now $(\partial z/\partial x)_y$ is the slope of the tangent plane in the x-direction and $(\partial z/\partial y)_x$ is the slope in the y-direction. Thus $(\partial z/\partial x)_y \cdot dx$ and $(\partial z/\partial y)_x \cdot dy$, where dx and dy are arbitrary increments in x and y starting at P, will give the change in z along the tangent plane (not along the function $z = z(x, y)$) in the x and y directions respectively. It can be seen from the diagram that the sum of these two quantities is the change in z, dz, at the point (P + dx), (P + dy), since the tangent is planar. Thus equation (2.4) is the *equation of the tangent plane* to the surface

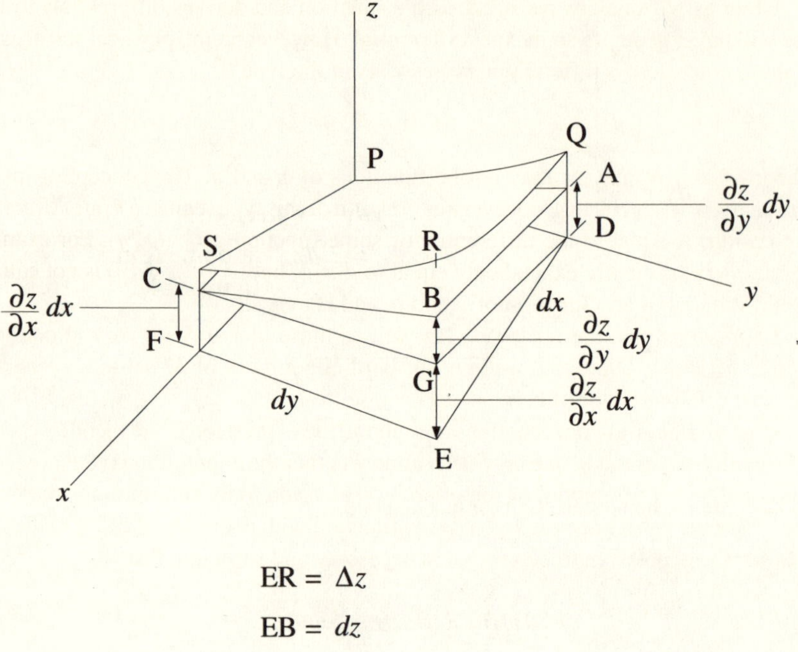

FIG. 2.6. A plane tangent to a curved surface, illustrating the meaning of partial derivatives and total differentials.

$z = z(x, y)$ at an arbitrary point P. We reiterate that dx and dy are not necessarily infinitesimals, and can generally be thought of as macroscopic quantities, except during integration. If dx and dy are small, then dz will be a good approximation to Δz, and this fact is used in some types of calculations. However, rather than thinking of equation (2.4) as a poor representation of the surface $z = z(x, y)$ unless dx and dy are infinitesimals, it is preferable to think of it as an *exact* representation of the tangent plane.

Thermodynamics commonly deals with continuous changes in multivariable systems. For this reason, total differentials are frequently used and it is essential to have a clear idea of their meaning.

2.2.6. *Exact and Inexact Differentials*

Exact differentials are those obtained by differentiating some continuous, single-valued function. Since the function can be obtained again by integration, exact differentials are expressions whose integrals will be the same no matter which path of integration is chosen. Accordingly, the total differentials we have just discussed are also exact differentials. They are the "normal" kind of differential. What then is an inexact differential?

Obviously if one always starts with a function and derives differentials from it, one will never generate an inexact differential. However, in the physical sciences one is apt to come across differential expressions of the type

$$M\,dx + N\,dy \tag{2.8}$$

where M and N may or may not be functions of x and y. The concept of inexact differentials arises because not every expression of this type (called a Pfaff differential expression) is equal to the differential of some function $z = z(x,y)$. For example, $(x\,dy + y\,dx)$ is clearly exact, being equal to $d(xy)$, but $(x\,dy - y\,dx)$ is not equal to the differential of any function of x and y, and is inexact.

Expressions that are not exact, (as well as those that are) are very important in thermodynamics. (Note that we may speak of either dz or of $M\,dx + N\,dy$ as being an exact differential, or of being exact).

If M and/or N are not functions of x and y, the expression is inexact. If M and N are functions of x and y, one may determine whether the expression is exact (whether it is equal to the differential of some function of x and y) by determining whether M and N are the partial derivatives of some function with respect to x and y. If they are, then for the expression to be exact it is necessary and sufficient that

$$\left(\frac{\partial M}{\partial y}\right)_x = \left(\frac{\partial N}{\partial x}\right)_y \tag{2.9}$$

for all pairs of terms in the expression.

Equation (2.9) is known as the *reciprocity relation* or cross-differentiation identity. It follows from the fact that the order of differentiation of our original function $z = z(x,y)$ with respect to x and y is immaterial. Mathematically this is written

$$\left(\frac{\partial}{\partial y}\left(\frac{\partial z}{\partial x}\right)_y\right)_x = \left(\frac{\partial}{\partial x}\left(\frac{\partial z}{\partial y}\right)_x\right)_y \tag{2.10}$$

If, in equation (2.8), $M = (\partial z/\partial x)_y$ and $N = (\partial z/\partial y)_x$ for some function $z = z(x,y)$, then equation (2.9) must follow from (2.10).

As an example of the use of this relation, let us test equation (2.7) for exactness. Applying the reciprocity relation gives

$$\left(\frac{\partial}{\partial T}\left(\frac{-RT}{P^2}\right)\right)_P = \left(\frac{\partial}{\partial P}\left(\frac{R}{P}\right)\right)_T$$
$$= \frac{-R}{P^2}$$

Hence equation (2.6) has been shown to be an exact differential expression, as we knew it must be because we derived it by differentiating the function $V = RT/P$.

Frankly, it is rarely of interest in thermodynamics to test expressions for exactness. The expressions in common use are generally well known to be exact or inexact. However, the reciprocity relationship is useful in deriving relationships between thermodynamic variables.

Getting back to the inexact differentials, suppose we have an expression

$$X\,dx + Y\,dy + Z\,dz \tag{2.11}$$

where X, Y, and Z are functions of x, y, and z but reciprocity is not satisfied,[1] and the expression is not exact. We may assign another variable to represent this expression, e.g.

$$\Phi = X\,dx + Y\,dy + Z\,dz$$

and this variable (Φ) is called an inexact differential. To remind ourselves that it represents a differential expression rather than an ordinary algebraic expression, it is usually combined with some kind of letter "d" such as D or δ or $đ$—any form except d, which is reserved for exact differentials. We use D, thus

$$D\Phi = X\,dx + Y\,dy + Z\,dz$$

Inexact differentials such as $D\Phi$ are thus simply a shorthand notation for a differential expression such as (2.11) and are not the differential of any function. It follows that they cannot be integrated without specifying a path (see *Line Integrals*, §2.2.9).

Of special interest is the simplest of inexact differentials, $y\,dx$. When y and x are both independent variables, $y\,dx$ is clearly inexact, since we may write the equivalent expression

$$y\,dx + 0\,dy$$

and applying the reciprocity relationship,

$$\left(\frac{\partial y}{\partial y}\right)_x = 1 \neq \left(\frac{\partial 0}{\partial x}\right)_y$$

Clearly there is no function the differential of which is equal to $y\,dx$. Similarly, not being exact, the integral of $y\,dx$ is not independent of the path chosen. But expressions like $y\,dx$ are important in thermodynamics and must be integrated. The most common examples are PdV and TdS.

Expressions that are not exact can only be integrated after making them exact, and this is done by choosing a functional relationship between x and y, i.e., by making

$$y = y(x)$$

Now y and x are no longer independent of one another, and the expression

$$y(x)\,dx$$

is exact and may be integrated unambiguously. Reciprocity is satisfied since we may write $y(x)\,dx$ as

$$y(x)\,dx + 0\,dy$$

[1] In the case of three or more terms, the reciprocity relation must hold between any two.

and

$$\left(\frac{\partial y(x)}{\partial y}\right)_x = 0$$
$$= \left(\frac{\partial 0}{\partial x}\right)_y$$

The manner in which y is transformed to $y(x)$ is arbitrary, but is commonly done by applying an "integrating denominator", or an "integrating factor." Thus, if the differential expression

$$D\Phi = M\,dx + N\,dy$$

is not exact, it can frequently be made so by dividing through by some particular $\lambda = \lambda(x, y)$. λ is thus an integrating denominator. (Of course, if there is such a function λ for $M\,dx + N\,dy$, finding out what it is may be another matter.) It happens that Pfaff differential expressions in two variables will always have integrating denominators, while with more than two they may or may not. This topic is discussed in detail by Reiss (1965, Chap. 2), who gives the conditions necessary to have integrating denominators.

Probably the most important example we shall encounter on this topic is the relationship between entropy, heat, and temperature:

$$dS = Dq/T \tag{2.12}$$

As we shall see, S is a function of state, hence dS is exact. Hence, T is an integrating denominator for the Pfaff differential Dq. That is, dividing the inexact differential Dq by T produces the exact differential dS.

2.2.7. *Simple Transformations of Partial Derivatives*

It is often helpful in thermodynamic manipulations to be able to replace one partial derivative with an equivalent but more convenient expression. For example, the reciprocity condition (2.9) may be used to interchange variables in a partial derivative. Like the reciprocity relation, most such transformations derive quite simply from the properties of partial derivatives. As an example, consider equation (2.5) for the volume of an ideal gas. Equation (2.6) may be solved for $(\partial V/\partial T)_P$ by in effect dividing through by dT. For a system at constant volume, $dV = 0$, so

$$dV = 0$$
$$= \left(\frac{\partial V}{\partial P}\right)_T dP + \left(\frac{\partial V}{\partial T}\right)_P dT \tag{2.13}$$
$$= \left(\frac{\partial V}{\partial P}\right)_T \left(\frac{\partial P}{\partial T}\right)_V + \left(\frac{\partial V}{\partial T}\right)_P \tag{2.14}$$

Rearranging (2.14) gives the new relationship

$$\left(\frac{\partial V}{\partial T}\right)_P = -\left(\frac{\partial V}{\partial P}\right)_T \left(\frac{\partial P}{\partial T}\right)_V \tag{2.15}$$

Such derivatives may also be treated, within limits, as simple fractions. Thus

$$\left(\frac{\partial P}{\partial T}\right)_V = \frac{1}{(\partial T/\partial P)_V} \tag{2.16}$$

And therefore,

$$\left(\frac{\partial V}{\partial P}\right)_T = -\left(\frac{\partial V}{\partial T}\right)_P \left(\frac{\partial T}{\partial P}\right)_V \tag{2.17}$$

Note that all three of the variables V, P, and T are interrelated interchangeably by these two expressions. Thus, experimental determination of any two of these partial derivatives would give the third.

A new dependent variable, such as $x(P, T)$ in the above example could be introduced by the relationship

$$\left.\begin{aligned} \left(\frac{\partial V}{\partial T}\right)_P &= \left(\frac{\partial V}{\partial T}\cdot\frac{\partial x}{\partial x}\right)_P \\ &= \left(\frac{\partial V}{\partial x}\cdot\frac{\partial x}{\partial T}\right)_P \\ &= \left(\frac{\partial V}{\partial x}\right)_P \left(\frac{\partial x}{\partial T}\right)_P \end{aligned}\right\} \tag{2.18}$$

Next, we might wish to evaluate the partial derivatives $(\partial V/\partial T)_x$ or $(\partial V/\partial P)_x$. These may be obtained from (2.6) by taking the appropriate total derivative with the restriction that x be held constant:

$$\left.\begin{aligned} \left(\frac{\partial V}{\partial T}\right)_x &= \left(\frac{\partial V}{\partial P}\right)_T \left(\frac{\partial P}{\partial T}\right)_x + \left(\frac{\partial V}{\partial T}\right)_P \\ \left(\frac{\partial V}{\partial P}\right)_x &= \left(\frac{\partial V}{\partial T}\right)_T \left(\frac{\partial T}{\partial P}\right)_x + \left(\frac{\partial V}{\partial P}\right)_T \end{aligned}\right\} \tag{2.19}$$

Finally, the partial derivative $\partial V/\partial x$ may be evaluated from differential expressions such as (2.6) using the chain rule of elementary calculus. From (2.5), V is a function of T and P, or $V = V(T, P)$. Let T and P each be functions of two other variables x and y,

$$T = T(x, y)$$

$$P = P(x, y)$$

Because $V = V(T, P)$, it follows from the chain rule that

$$\left.\begin{aligned} \left(\frac{\partial V}{\partial y}\right)_x &= \left(\frac{\partial V}{\partial P}\right)_T \left(\frac{\partial P}{\partial y}\right)_x + \left(\frac{\partial V}{\partial T}\right)_P \left(\frac{\partial T}{\partial y}\right)_x \\ \left(\frac{\partial V}{\partial x}\right)_y &= \left(\frac{\partial V}{\partial P}\right)_T \left(\frac{\partial P}{\partial x}\right)_y + \left(\frac{\partial V}{\partial T}\right)_p \left(\frac{\partial T}{\partial x}\right)_y \end{aligned}\right\} \tag{2.20}$$

We may conveniently regard equations (2.20) as being derived from (2.6) through division by the differentials dx or dy. As an example, consider a generalized total differential

$$dF = \left(\frac{\partial F}{\partial x}\right)_y dx + \left(\frac{\partial F}{\partial y}\right)_x dy$$

or

$$dF = X\,dx + Y\,dy \tag{2.21}$$

where

$$F = F(x, y)$$

and

$$x = x(u, v)$$
$$y = y(u, v)$$

Then in a similar way, expressions such as (2.20) may be quickly simplified to the forms

$$\left(\frac{\partial F}{\partial u}\right)_v = X\left(\frac{\partial x}{\partial u}\right)_v + Y\left(\frac{\partial y}{\partial u}\right)_v$$

or

$$\left(\frac{\partial F}{\partial v}\right)_u = X\left(\frac{\partial x}{\partial v}\right)_u + Y\left(\frac{\partial y}{\partial v}\right)_u$$

The above techniques are used quite commonly to interrelate and manipulate thermodynamic variables. Later in this chapter we introduce two additional and more general methods of manipulating thermodynamic derivatives called Legendre and Jacobian transformations.

2.2.8. *Integration*

As with differentiation, we do not give a full discussion but only recall to the reader the essentials of integration. Integration is the inverse of differentiation. That is, the problem is to find a function when its rate of change is known. It is performed by summing up (functions of) differentials that are chosen to be very small (infinitesimals). This can be done either in the general case, giving indefinite integrals, or between specified limits, giving definite integrals. For example,

$$\int y'(x)\,dx = y(x) + constant$$

is the general case, since differentiation of $y(x)$ plus any constant will give $y'(x)$, the derivative of a constant being zero.

$$\int_a^b y'(x)\,dx = y(b) - y(a)$$

is the definite integral between the limits a and b, and can be thought of as the area under the curve $y'(x)$ in the x–y plane, between the limits $x = a$ and $x = b$. Both methods of integration have been used in applications of thermodynamics in the Earth sciences, but in this text we prefer the definite integral.

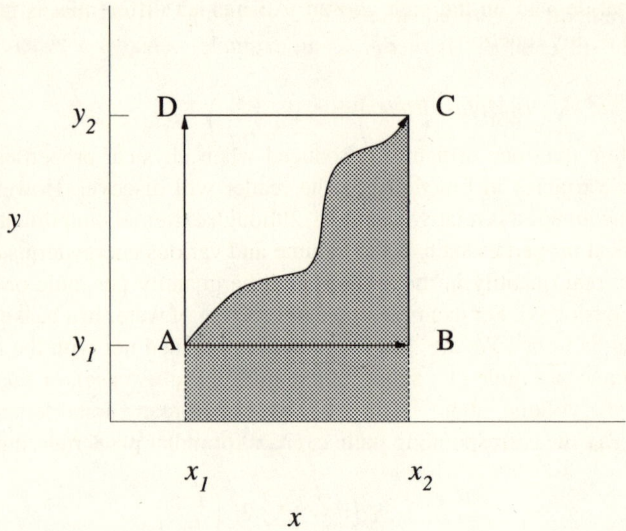

FIG. 2.7. Integration of the function $y\,dx$ illustrating that the functional relationship between y and x must be specified.

2.2.9. *Line Integrals*

The properties of exact versus inexact differentials are well illustrated by integrating them along different paths. An integral whose path is specified is called a *line integral*. Consider the expression $y\,dx$, that we have just seen is inexact. Integration of $y\,dx$ can be viewed as determining the area under some curve in the x–y plane (Figure 2.7), but the obvious question is what curve?

If we are integrating from A to C there are an infinite number of paths to follow, giving an infinite number of areas. For example, the path ABC results in the area $y_1(x_2 - x_1)$, while ADC results in $y_2(x_2 - x_1)$, and intermediate paths such as the one shown give intermediate results. This is clearly also true of the expression $x\,dy$. However, the exact expression

$$dz = x\,dy + y\,dx$$

is different. Integration from (x_1, y_1) to (x_2, y_2) always results in the area

$$z_2 - z_1 = x_2y_2 - x_1y_1$$

no matter which path is followed, illustrating once again that $x\,dy + y\,dx$ is an exact differential.

Thus in summary, exact differentials have coefficients that satisfy the reciprocity relations and have definite integrals that are independent of the path followed during integration. Exact differentials are obtained by differentiating some function. Inexact differentials have coefficients that do not satisfy the reciprocity relations, and have integrals that depend on the path chosen. An inexact differential is not the (total) differential of any function.

2.2.10. *Total versus Molar Properties*

There is more than one difficulty introduced when physical properties of systems are used as variables in functions, as the reader will discover. However, the first difficulty encountered is relatively simple, although extremely important. That is that many physical properties such as the volume and various energy terms come in two forms—the total quantity in the system, and the quantity per mole or per gram of substance considered. For example, if we have 513 g of water in a beaker, its volume (V) is about 513 cm^3. We are, however, often concerned not with the total volume but the volume per mole (V, about 18 cm^3mol^{-1} in this case), or the volume per gram (specific volume, about 1 cm^3g^{-1}). We use Roman capital letters for "total" properties, and the corresponding italic capitals for molar properties, and

$$Z = \mathrm{Z}/\sum_i n_i$$

where Z is any total property, Z is the corresponding molar property, and $\sum_i n_i$ is the total number of moles in the system (see §14.3 for an example of the change from one to the other). We don't have a special type of symbol for *specific* properties (i.e. property per gram rather than per mole) as they are not often used.

The reader is advised to notice whether Roman or Italic capitals are used in the equations to be presented, as it is an aid to clear thinking. In fact, many sections of this text will be quite confusing if the distinction is not made. Generally speaking, any functions or equations that are used in real applications will be in the molar form, while theoretical discussions are often more convenient using the total form. For example, much of the theoretical discussion in Chapters 4 to 6 inclusive will use total properties, but in Chapter 7 we start to discuss measurements and properties with numerical values, and of necessity we use molar properties.

Of course, many equations look much the same with total and molar properties, because ratios are involved. That is, if $(\partial \mathrm{U}/\partial \mathrm{S})_{\mathrm{V}} = T$, then it is also true that $(\partial U/\partial S)_V = T$; or if $(\partial \mathrm{G}/\partial P)_T = \mathrm{V}$, then $(\partial G/\partial P)_T = V$, so that the distinction may seem to be unimportant. However, sometimes it *is* important. For example, understanding the meaning of partial molar properties (Chapter 9) is not really possible without making this distinction clear, and many of the equations in Chapter 14 look quite different in the two forms.

2.3. ADDITIONAL MATHEMATICAL TECHNIQUES

In this section we continue the presentation of the mathematics of thermodynamics, but the concepts here are less essential to the understanding of thermodynamics. This is not to imply that the following material is of little use or that the reader should blithely skip ahead to Chapter 3. On the contrary, the next three topics are extremely powerful and may be used to quickly derive and illustrate most of the theoretical foundation of thermodynamics. Mastery of these somewhat more difficult techniques actually helps reveal the fundamental simplicity of thermodynamics, and unfolds much of its symmetry, elegance, and unity.

2.3.1. *Molar and Partial Molar Properties*

It is possible to subdivide the properties used to describe a thermodynamic system (e.g., $T, P, V, U, \ldots$) into two main classes termed intensive and extensive variables. This distinction is quite important since the two classes of variables are often treated in significantly different fashion. For present purposes, *extensive properties* are defined as those that depend on the mass of the system considered, such as volume and total energy content, indeed all the "total" system properties (Z) mentioned above. On the other hand, *intensive properties* do not depend on the mass of the system, an obvious example being density. For example, the density of two grams of water is the same as that of one gram at the same P, T, though the volume is double. Other common intensive variables include temperature, pressure, concentration, viscosity and all molar (Z) and partial molar ($\bar{Z}$, defined below) quantities.[2]

Partial molar quantities are very commonly used to describe solutions, or systems containing more than one component. Mathematically, a partial molar quantity $\bar{Z}_i$ is defined as the partial derivative

$$\bar{Z}_i = \left(\frac{\partial \mathrm{Z}}{\partial n_i}\right)_{T,P,\hat{n}_i} \tag{2.22}$$

where Z is an extensive or "total" property of a system that contains constituent i and (usually) other constituents as well, and the partial derivative of Z is taken at constant T, P and $\hat{n}_i$, where n_i is the number of moles of constituent i, and $\hat{n}_i$ refers to all constituents *other* than the constituent i being considered.

Equation (2.22) is a good example of the conceptual difficulties peculiar to thermodynamics, which we mentioned at the beginning of this chapter. Mathematically, (2.22) is a simple enough definition, but what does it mean physically? Partial molar quantities are discussed in more detail in Chapter 9; for present purposes, we might use volume, V, as an example of property Z. Equation (2.22) then refers to the change in total volume V of a solution when one mole of component i is added to a quantity of that solution sufficiently large that the concentrations of all other components ($\hat{n}_i$) do not change significantly at constant T and P. In other words, $\bar{V}_i$ is the effective volume of one mole of component i in this solution at the T, P and concentration of interest. Notice also that while V is obviously an extensive property, $\bar{V}_i$ is intensive since by (2.22) it is defined in terms of volume per mole (and this cannot change with the size of a system). In general, all partial molar quantities such as $\bar{V}_i, \bar{G}_i, \bar{H}_i, \bar{S}_i$, etc. are intensive and derived from their extensive equivalents (V,G,H,S) by (2.22).

[2]The word "molar" in this phrase refers to "per mole," and has no connection with the molarity scale of concentration. An alternative name is partial molal quantity, which is synonymous, and has no connection with the molality scale of concentration.

2.3.2. *Homogeneous Functions*

Functions are said to be homogeneous and of degree n if we can multiply every variable in the function by a constant, then factor out the constant. For example, $f(x, y, z)$ is homogeneous in degree n if

$$f(kx, ky, kz) = k^n f(x, y, z) \tag{2.23}$$

Thus, $x^2 + y^2 + z^2$ is homogeneous second degree because

$$(kx)^2 + (ky)^2 + (kz)^2 = k^2 (x^2 + y^2 + z^2)$$

Clearly, homogeneity is concerned with the exponents of the variables in a function. Homogeneity is not always quite so obvious, however. For example

$$\frac{x^2y^3}{z} + xyz^2$$

is homogeneous in the fourth degree.

Some functions are homogeneous in only some of the variables and not in others, which is the usual case in thermodynamics. Thus

$$x^2y^3 + x^2y^4$$

is homogeneous second degree in x, but not in y, and not overall, because

$$(kx)^2y^3 + (kx)^2y^4 = k^2(x^2y^3 + x^2y^4)$$

Consider an ideal gas, for which

$$\mathrm{V} = nRT/P$$

V being the total volume, n the number of moles of gas, R the gas constant, T the absolute temperature, and P the pressure. Clearly V is homogeneous first degree in (n, T, P), homogeneous first degree in n only or in T only, and of degree -1 in P only. The derivative

$$(\partial \mathrm{V}/\partial n)_{T,P} = RT/P$$

which is a *partial molar quantity,* is homogeneous zeroth degree in T, P (or n, T, P) and zeroth degree in n only. Also

$$(\partial \mathrm{V}/\partial T)_{n,P} = nR/P$$

is zeroth degree in T, but still first degree in n and degree -1 in P. Similarly, differentiating V with respect to P produces a function homogeneous in degree -2 in P. Thus, without bothering to derive a formal proof, we see that differentiation of a function with respect to a variable lowers the degree of homogeneity of that variable by one. These relations are illustrated in Figure 2.8 where we see that V is a function of T, P and n, but $(\partial V/\partial T)_{P,n}$ is a function of P and n but not of T, and so on. As we have seen, *extensive properties* are those that depend on the mass of the system considered, such as volume or total energy. Mathematically speaking then, extensive properties are homogeneous in the first degree in the masses of the components. For a two component system for example,

$$\mathrm{V}(T, P, kn_1, kn_2) = k^1\, \mathrm{V}(T, P, n_1, n_2)$$

where n_1 and n_2 are the masses of components 1 and 2, and P, T and V are the pressure, temperature, and total volume of the system.

Intensive properties were defined as those that do not depend on the mass of the system considered. They are thus homogeneous in the zeroth degree in the masses of the components. For example, for the density

$$\rho(T, P, kn_1, kn_2) = k^0\, \rho(T, P, n_1, n_2)$$

where ρ is the density. Temperature and pressure are normally considered to be intensive variables, since intuitively they do not depend on what mass of system is considered. However, care is required in the equating of intensive variables and variables homogeneous first order in the masses of the components, because in

$$P = nRT/\mathrm{V}$$

P is clearly first order in n, and in

$$T = P\mathrm{V}/nR$$

T is homogeneous in degree -1 in n. In other words, we must be clear as to what variables are being held constant when we say something is homogeneous in the masses of its components. Obviously if we stuff twice the mass into a system without changing V or T, the P will rise, but this is not usually the operation considered when T and P are said to be intensive, and that intensive variables are zeroth order in mass. There is no real danger of confusion here since the terms intensive and extensive are usually used in a very general sense, whereas homogeneity is a very clearly defined concept, and hence more useful. Its usefulness arises mainly from Euler's theorem regarding homogeneous functions.

2.3.3. *Euler's Theorem for Homogeneous Functions*

This theorem says that for a function such as $f(x, y, z)$ that is homogeneous in degree n,

$$n \cdot f(x, y, z) = x\left(\frac{\partial f}{\partial x}\right) + y\left(\frac{\partial f}{\partial y}\right) + z\left(\frac{\partial f}{\partial z}\right) \tag{2.24}$$

thus relating a function to its partial derivatives in a way somewhat reminiscent of the relation between the total differential and the partial derivatives. Euler's theorem for homogeneous nth order functions is obtained by differentiating (2.23) with respect to k (using the chain rule on the left hand side):

$$\frac{\partial f}{\partial kx} \cdot \frac{\partial kx}{\partial k} + \frac{\partial f}{\partial ky} \cdot \frac{\partial ky}{\partial k} + \frac{\partial f}{\partial kz} \cdot \frac{\partial kz}{\partial k} = n \cdot k^{n-1} \cdot f(x, y, z)$$

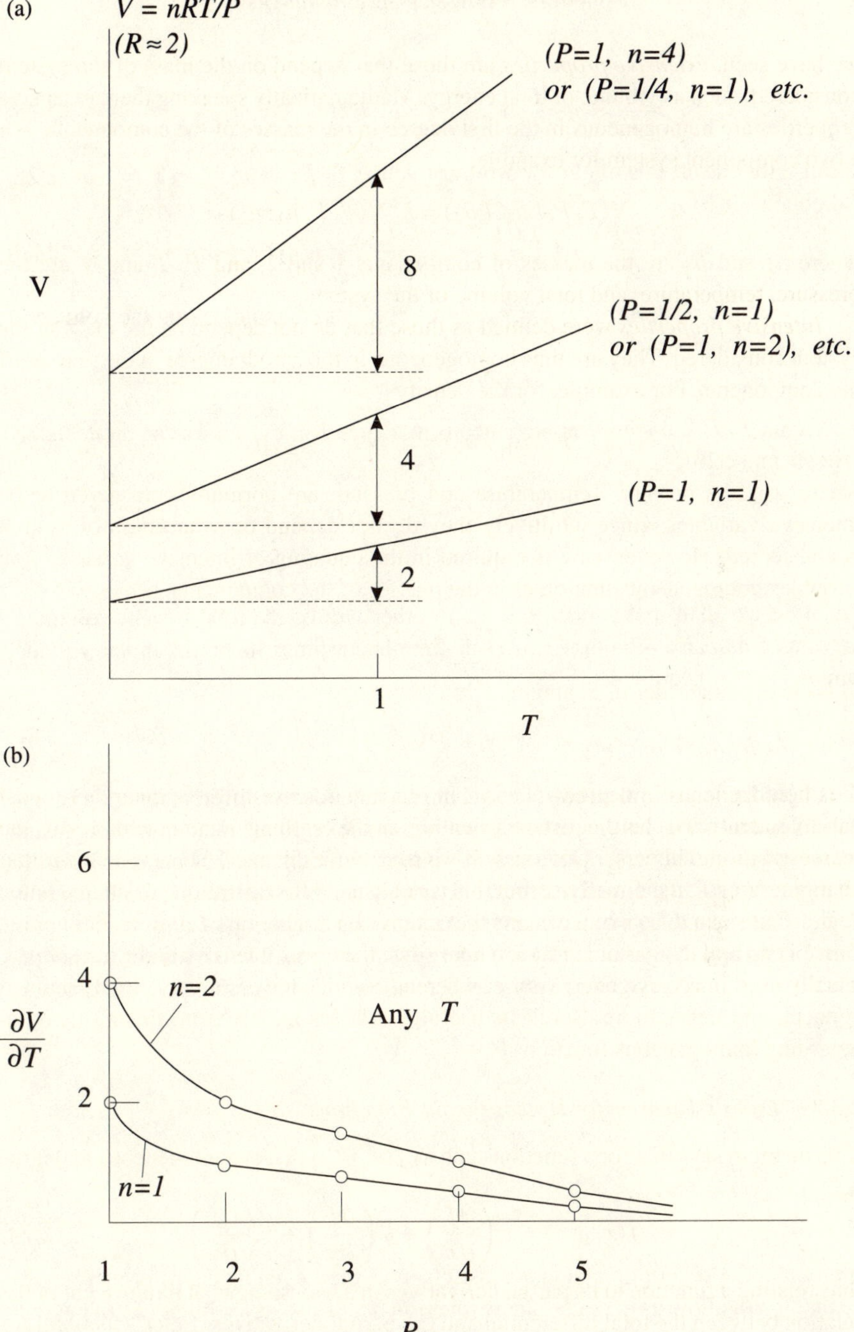

FIG. 2.8. (a) Volume versus temperature for various quantities of an ideal gas at various pressures. R, the gas constant, is taken as equal to 2, and n is the number of moles of the gas. (b) $\partial V/\partial T$ versus pressure at any temperature for one mole (n = 1) and two moles (n = 2) of an ideal gas.

or

$$x\frac{\partial f}{\partial kx} + y\frac{\partial f}{\partial ky} + z\frac{\partial f}{\partial kz} = n \cdot k^{n-1} \cdot f(x, y, z) \tag{2.25}$$

Because this result must hold for arbitrary values of k, we may set $k = 1$ in (2.25) and obtain

$$x\left(\frac{\partial f}{\partial x}\right) + y\left(\frac{\partial f}{\partial y}\right) + z\left(\frac{\partial f}{\partial z}\right) = n \cdot f(x, y, z) \tag{2.26}$$

This simple relationship is extremely useful. For example, taking the volume of a two component system

$$\mathrm{V} = \mathrm{V}(T, P, n_1, n_2)$$

at constant T, P and with V homogeneous in degree 1 in n_1, n_2, we can immediately write, from (2.26),

$$\mathrm{V} = n_1\left(\frac{\partial \mathrm{V}}{\partial n_1}\right)_{T,P,n_2} + n_2\left(\frac{\partial \mathrm{V}}{\partial n_2}\right)_{T,P,n_1} \tag{2.27}$$

a relation we shall make much use of. In other words, the total system volume is the sum of its parts—the moles of each component times its partial molar volume, summed over all components.

2.3.4. *Legendre Transforms*

The Legendre Transform allows one to change a function to a different function having as independent variables the partial derivatives of the original function, without losing any information. This description in words is more difficult than the operation itself. To see its usefulness in thermodynamics, one simply needs to realize that fundamentally the first and second laws of thermodynamics give us a criterion of system stability in terms of entropy (S), volume (V) and energy (U), which we will consider further on. In other words, we have some very useful relationships beginning with the function

$$U = U(S, V) \tag{2.28}$$

If you then realize that

$$T = (\partial U/\partial S)_V$$

and

$$P = -(\partial U/\partial V)_S$$

and look at the description we have just given of the Legendre Transform, you will see that it will allow us to define a new function that is just as useful as (2.28) but that uses T and P as independent variables instead of S and V. As we said before, the development of thermodynamics does not depend on the Legendre Transform—other paths through thermodynamic pedagogic space are available—but it is elegant and concise. It illustrates quite beautifully the underlying unity and symmetry among the thermodynamic state functions and their independent variables.

We derive the transform first using the simple algebraic approach of Boas (1966, p. 159). The total differential of a function $f = f(x, y)$ is written

$$df = \left(\frac{\partial f}{\partial x}\right)_y dx + \left(\frac{\partial f}{\partial y}\right)_x dy \tag{2.29}$$

If we define

$$\left(\frac{\partial f}{\partial x}\right)_y = p \quad \text{and} \quad \left(\frac{\partial f}{\partial y}\right)_x = q$$

then

$$df = p\,dx + q\,dy \tag{2.30}$$

Subtracting the quantity $d(qy)$ from both sides of (2.30), we have

$$df - d(qy) = p\,dx + q\,dy - q\,dy - y\,dq$$

or

$$d(f - qy) = p\,dx - y\,dq \tag{2.31}$$

This is the total differential of a new function,

$$\begin{aligned}\Psi(x, q) &= \Psi(x, \frac{\partial f}{\partial y}) \\ &= f - qy\end{aligned} \tag{2.32}$$

that has the partial derivatives

$$\left(\frac{\partial \Psi}{\partial x}\right)_q = p$$

and

$$\left(\frac{\partial \Psi}{\partial q}\right)_x = -y$$

Notice that the transformation merely replaces the original variable y in $f(x, y)$ by its partial derivative $\partial f/\partial y$ to give the new function $\Psi(x, \partial f/\partial y)$, which is

$$\Psi = f - y\left(\frac{\partial f}{\partial y}\right)$$

For many people, it is helpful to view this solution geometrically (Callen, 1960, p. 90). To state the problem in a different way then, given the function

$$y = y(x_1, x_2, x_3, \cdots, x_n)$$

we want a method whereby the derivatives

$$p_i = \partial y/\partial x_i$$

are used as independent variables in a new function containing the same information

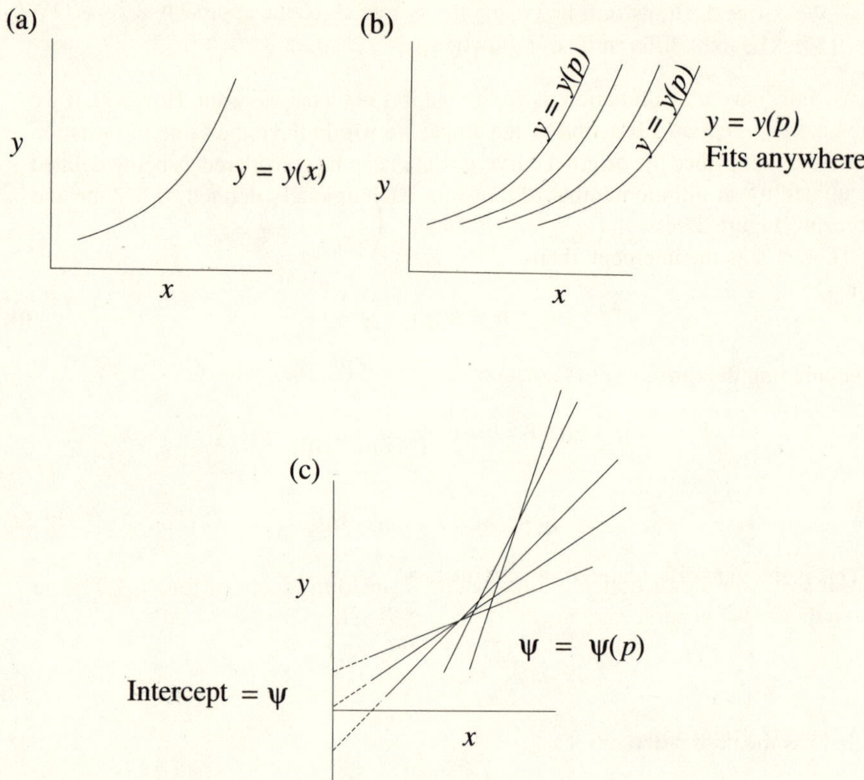

FIG. 2.9. Illustration of why one must know the slope (p) as a function of the y-intercept (Ψ) to have the same information as one has in the function $y = y(x)$. (a) the function $y = y(x)$. y is known for any x. (b) y is known as a function of the slope ($p = dy/dx$) of $y = y(x)$. This does not fix the position of the curve with respect to the x-axis. (c) the slope p as a function of the y-intercept of the slope (Ψ). Defines an infinite set of tangents that outline the original function $y = y(x)$.

as the original. To begin, consider a function of a single independent variable

$$y = y(x)$$

Geometrically, $p = dy/dx$ is a tangent (Figure 2.9a). We might be tempted to simply eliminate x and find $y = y(p)$, but this would lose some information since knowing y as a function of the slope does not give us y as a function of x (Figure 2.9b).

For example, let

$$y = x^3$$

then

$$\begin{aligned} p &= dy/dx \\ &= 3x^2 \end{aligned}$$

and

$$y = (p/3)^{1.5}$$

We now have y as a function of dy/dx but it is not what we want. However, if we knew the *intercept* as a function of the slope, we would have the same information we started with, since the original curve $y = y(x)$ can be considered as being defined or outlined by an infinite number of tangents, each uniquely defined by a slope and intercept. (Figure 2.9c).

Thus, if Ψ is the intercept, then

$$\Psi = \Psi(p)$$

is the relation we want.

Now since

$$p = \frac{y - \Psi}{x - 0} \qquad \text{(Figure 2.10)}$$

then

$$\Psi = y - px$$

and, in case you didn't notice, the Legendre Transform has been found. It can be shown that in the general case

$$y = y(x_1, x_2, x_3, \cdots, x_n)$$

the Legendre Transform is

$$\begin{aligned} \Psi &= y - p_1 x_1 - p_2 x_2 - p_3 x_3 \cdots - p_n x_n \\ &= y - \sum_i p_i x_i \end{aligned} \tag{2.33}$$

That is, to form the Legendre Transform of a function, subtract from the original function the products of each variable to be changed and the derivative of the function with respect to that variable. After that, one can proceed to tidy up by eliminating y in the new function by differentiating. Thus in the case of $y = y(x)$,

$$\Psi = y - px$$

$$\begin{aligned} d\Psi &= dy - p\,dx - x\,dp \\ &= -x\,dp \quad (dy = p\,dx \text{ by definition}) \end{aligned}$$

or

$$\frac{d\Psi}{dp} = -x$$

and in the general case

$$d\Psi = -\sum_i x_i dp_i$$

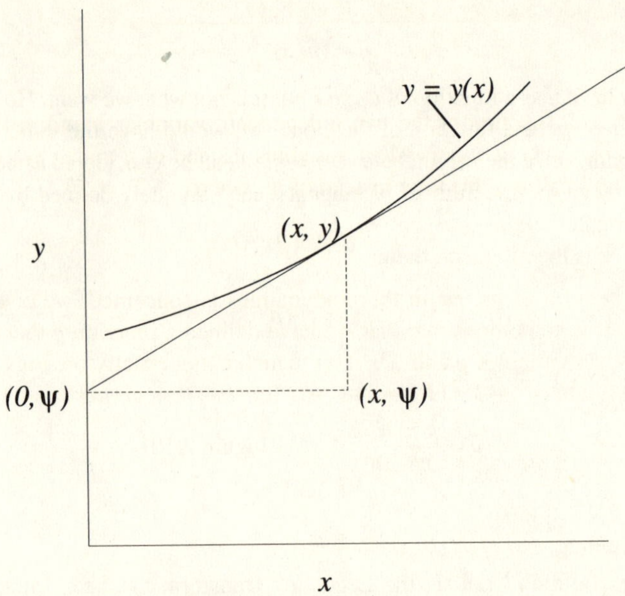

FIG. 2.10. The function $y = y(x)$ and the tangent to the function at an arbitrary point. Ψ is the y-intercept of the tangent, and the slope $p = (y - \Psi)/(x - 0)$.

For example, if

$$y = x^3$$

then

$$p = 3x^2$$

and

$$\begin{aligned}\Psi &= y - x \cdot 3x^2 \\ &= x^3 - 3x^3 \\ &= -2x^3\end{aligned}$$

or since

$$x = (p/3)^{\frac{1}{2}}$$
$$\Psi = -2(p/3)^{\frac{3}{2}}$$

and

$$\begin{aligned}d\Psi/dp &= -(p/3)^{\frac{1}{2}} \\ &= -x\end{aligned}$$

Next, consider another example. If we have

$$x = y^3 - 3z^2$$

we know x as a function of the two independent variables y and z. If we wish to use y and $(\partial x/\partial z)_y$ as independent variables instead of y and z, we define a new function—let's call it g this time instead of Ψ, where

$$g = x - z(\partial x/\partial z)_y$$

Frankly, as far as its use in thermodynamics is concerned, we could stop right there. However, to demonstrate that we have defined a function g that really has as its independent variables y and $(\partial x/\partial z)$, which is not exactly obvious at this stage, we can substitute $(y^3 - 3z^2)$ for x and $-6z$ for $(\partial x/\partial z)$, giving

$$\begin{aligned} g &= y^3 - 3z^2 - z(-6z) \\ &= y^3 + 3z^2 \end{aligned}$$

Then substituting $-(\partial x/\partial z)/6$ for z we have

$$g = y^3 + (\partial x/\partial z)_y^2/12$$

showing that g is indeed a function of y and $(\partial x/\partial z)$.

If we let

$$p = (\partial x/\partial z)_y \qquad (= -6z)$$

we see that

$$\begin{aligned} g &= y^3 + p^2/12 \\ (\partial g/\partial p)_y &= p/6 \\ &= -z \end{aligned}$$

analogous to $\partial\Psi/\partial p = -x$ in the last case, and in conformity with the generalized equation (2.33).

Finally, if we need a function not of y and z or y and $(\partial x/\partial z)$ but of $(\partial x/\partial y)$ and $(\partial x/\partial z)$, we transform both variables, and invent the new function h, such that

$$h = x - y(\partial x/\partial y)_z - z(\partial x/\partial z)_y$$

Again, for thermodynamic purposes, that goes far enough, but as before, we can demonstrate that h is a function of $(\partial x/\partial y)$ and $(\partial x/\partial z)$. Thus, since

$$x = y^3 - 3z^2$$

Let

$$\begin{aligned} p_1 &= (\partial x/\partial y)_z \\ &= 3y^2 \end{aligned}$$

and

$$p_2 = (\partial x/\partial z)_y$$
$$= -6z$$

then

$$h = y^3 - 3z^2 - y(3y^2) - z(-6z)$$
$$= -2y^3 + 3z^2$$
$$= -2(p_1/3)^{\frac{3}{2}} + 3(-p_2/6)^2$$
$$= (-2/\sqrt{27})p_1^{\frac{3}{2}} + p_2^2/12$$

Thus h is a function of p_1 and p_2, and

$$(\partial h/\partial p_1)_{p_2} = -y$$

and

$$(\partial h/\partial p_2)_{p_1} = -z$$

2.3.5. *Jacobian Transformations and Thermodynamic Partial Derivatives*

If we consider the ten most common thermodynamic variables P, V, T, U, S, G, H, A, q and w, there exists a very large number of partial derivatives and relations between their derivatives (see Margenau and Murphy, 1956, p. 15). For example, there are 720 (= 10 × 9 × 8) ways of choosing any 3 different variables from a set of 10; hence there must be 720 partial derivatives of the form $(\partial x/\partial y)_z$ relating these variables. Now we have shown above that any one such partial derivative may generally be related to three other mutually independent derivatives by the following kind of manipulation. Given a function

$$V = V(P, T)$$

then the total differential is

$$dV = \left(\frac{\partial V}{\partial T}\right)_P dT + \left(\frac{\partial V}{\partial P}\right)_T dP \tag{2.6}$$

and the total derivatives of this with respect to T and P, holding a new fourth variable x constant (where x is one of the remaining 7 variables U, S, G, H, A, q or w) are

$$\left.\begin{aligned} \left(\frac{\partial V}{\partial T}\right)_x &= \left(\frac{\partial V}{\partial P}\right)_T \left(\frac{\partial P}{\partial T}\right)_x + \left(\frac{\partial V}{\partial T}\right)_P \\ \left(\frac{\partial V}{\partial P}\right)_x &= \left(\frac{\partial V}{\partial T}\right)_P \left(\frac{\partial T}{\partial P}\right)_x + \left(\frac{\partial V}{\partial P}\right)_T \end{aligned}\right\} \tag{2.19}$$

For an alternative method of obtaining the same relationships, see Denbigh (1966, p. 92). The total number of equations such as (2.19) relating any combination of the 720 derivatives taken four at a time is $\frac{720\times719\times718\times717}{4!} = 11,104,365,420$. To make matters worse, there are many relationships between two or three first order derivatives (such as Maxwell's relations), and many relationships involving second and even higher order derivatives; this makes the total of all possible relations even more staggering. To be sure, we might never need more than a small fraction of these equations and derivatives. However, their total number is so gigantic that there would seem to be no simple way even to tabulate them so we might pick and choose those we do need.

This conundrum was first overcome by Percy Bridgman (1914, 1926), an eminent physicist who is also remembered today by petrologists and geochemists for his many contributions to high-pressure experimental chemistry and physics—most notably the Bridgman unsupported area seal used in many high pressure vessels. Bridgman developed an ingenious method of tabulating the 720 first derivatives, expressing his results in terms of T, P, S and three derivatives that are readily measured, $(\partial V/\partial T)_P$, $(\partial V/\partial P)_T$, and $C_p = (\partial H/\partial T)_P$. Bridgman's approach combined algebra and the manipulation of partial derivatives described earlier in this chapter (for details, see Margenau and Murphy (1956, p. 16)). Some years later, a more general method based on the properties of Jacobian determinants was developed (Shaw, 1935). Further details are given by Margenau and Murphy (1956, p. 17) and Tunell (1960).

Although understanding Jacobians is not particularly difficult, we will save considerable space here by presenting only the conclusions most useful to us, in the form of Table 2.1, which quickly generates all 336 partial derivatives of the form $(\partial x/\partial y)_z$ for the eight variables T, P, V, S, U, H, A and G (omitting q and w). The results presented in Table 2.1 are expressed (after Bridgman) in terms of T, P, S, and the measurable derivatives $(\partial V/\partial T)_P$, $(\partial V/\partial P)_T$, and $C_p = (\partial H/\partial T)_P$ (thermal expansion, compressibility, and heat capacity, respectively). The symbols $(\partial x)_y$ and $(\partial u)_y$ were originally introduced by Bridgman (1914) to represent Jacobians, and are also used by Pitzer and Brewer (1961, p. 667). They immediately suggest the relationship

$$\left(\frac{\partial x}{\partial u}\right)_y = \frac{(\partial x)_y}{(\partial u)_y} \tag{2.34}$$

which is the key to using the Table.

For example, to find $(\partial V/\partial S)_P$ use the second and third rows of the Table to obtain

$$\begin{aligned}\left(\frac{\partial V}{\partial S}\right)_P &= \frac{(\partial V)_P}{(\partial S)_P} \\ &= \frac{T\left(\frac{\partial V}{\partial T}\right)_P}{C_p}\end{aligned} \tag{2.35}$$

This relates an almost unmeasurable quantity to three experimentally accessible variables T, C_p, and $(\partial V/\partial T)_P$.

Tables similar to but more comprehensive than 2.1 are given by Bridgman (1926), Goranson (1930), Shaw (1935), Margenau and Murphy (1956, p. 22) and Tunell

(1960). A small subset of this Table gives Maxwell's relations between partial derivatives of thermodynamic variables (§14.3.1).

Table 2.0 Jacobians giving partial derivatives $(\partial x/\partial y)_z$ for the variables T, P, V, S, U, H, A, G using the relationship $(\partial x/\partial u)_y = (\partial x)_y/(\partial u)_y$. After Pitzer and Brewer (1961).

$$(\partial T)_P = -(\partial P)_T = 1$$

$$(\partial V)_P = -(\partial P)_V = \left(\frac{\partial V}{\partial T}\right)_P$$

$$(\partial S)_P = -(\partial P)_S = \frac{C_p}{T}$$

$$(\partial U)_P = -(\partial P)_U = C_p - P\left(\frac{\partial V}{\partial T}\right)_P$$

$$(\partial H)_P = -(\partial P)_H = C_p$$

$$(\partial G)_P = -(\partial P)_G = -S$$

$$(\partial A)_P = -(\partial P)_A = -\left[S + P\left(\frac{\partial V}{\partial T}\right)_P\right]$$

$$(\partial V)_T = -(\partial T)_V = -\left(\frac{\partial V}{\partial P}\right)_T$$

$$(\partial S)_T = -(\partial T)_S = \left(\frac{\partial V}{\partial T}\right)_P$$

$$(\partial U)_T = -(\partial T)_U = T\left(\frac{\partial V}{\partial T}\right)_P + P\left(\frac{\partial V}{\partial P}\right)_T$$

$$(\partial H)_T = -(\partial T)_H = -V + T\left(\frac{\partial V}{\partial T}\right)_P$$

$$(\partial G)_T = -(\partial T)_G = -V$$

$$(\partial A)_T = -(\partial T)_A = P\left(\frac{\partial V}{\partial P}\right)_T$$

$$(\partial S)_V = -(\partial V)_S = \frac{1}{T}\left[C_p\left(\frac{\partial V}{\partial P}\right)_T + T\left(\frac{\partial V}{\partial T}\right)_P^2\right]$$

$$(\partial U)_V = -(\partial V)_U = C_P\left(\frac{\partial V}{\partial P}\right)_T + T\left(\frac{\partial V}{\partial T}\right)_P^2$$

$$(\partial H)_V = -(\partial V)_H = C_P\left(\frac{\partial V}{\partial P}\right)_T + T\left(\frac{\partial V}{\partial T}\right)_P^2 - V\left(\frac{\partial V}{\partial T}\right)_P$$

$$(\partial G)_V = -(\partial V)_G = -\left[V\left(\frac{\partial V}{\partial T}\right)_P + S\left(\frac{\partial V}{\partial P}\right)_T\right]$$

$$(\partial A)_V = -(\partial V)_A = -S\left(\frac{\partial V}{\partial P}\right)_T$$

$$(\partial U)_S = -(\partial S)_U = \frac{P}{T}\left[C_p\left(\frac{\partial V}{\partial P}\right)_T + T\left(\frac{\partial V}{\partial T}\right)_P^2\right]$$

$$(\partial H)_S = -(\partial S)_H = -\frac{VC_p}{T}$$

$$(\partial G)_S = -(\partial S)_G = -\frac{1}{T}\left[VC_p - ST\left(\frac{\partial V}{\partial T}\right)_P\right]$$

$$(\partial A)_S = -(\partial S)_A = \frac{1}{T}\left\{\left[C_p\left(\frac{\partial V}{\partial P}\right)_T + T\left(\frac{\partial V}{\partial T}\right)_P^2\right] + ST\left(\frac{\partial V}{\partial T}\right)_P\right\}$$

$$(\partial H)_S = -(\partial S)_H = -V\left[C_p - P\left(\frac{\partial V}{\partial T}\right)_P\right] - P\left[C_p\left(\frac{\partial V}{\partial P}\right)_T + T\left(\frac{\partial V}{\partial T}\right)_P^2\right]$$

$$(\partial G)_S = -(\partial S)_G = -V\left[C_p - P\left(\frac{\partial V}{\partial T}\right)_P\right] + S\left[T\left(\frac{\partial V}{\partial T}\right)_P + P\left(\frac{\partial V}{\partial P}\right)_T\right]$$

$$(\partial A)_U = -(\partial U)_A = P\left[(C_p + S)\left(\frac{\partial V}{\partial P}\right)_T + T\left(\frac{\partial V}{\partial T}\right)_P^2\right] + ST\left(\frac{\partial V}{\partial T}\right)_P$$

$$(\partial G)_U = -(\partial U)_G = -V(C_p + S) + TS\left(\frac{\partial V}{\partial T}\right)_P$$

$$(\partial A)_H = -(\partial H)_A = -\left[S + P\left(\frac{\partial V}{\partial T}\right)_P\right]\left[V - T\left(\frac{\partial V}{\partial T}\right)_P\right] + PC_p\left(\frac{\partial V}{\partial P}\right)_T$$

$$(\partial A)_G = -(\partial G)_A = -S\left[V + P\left(\frac{\partial V}{\partial P}\right)_P\right] + PV\left(\frac{\partial V}{\partial T}\right)_P$$

PROBLEMS

1. Show that dV for an ideal gas is an exact differential by integrating equation (2.7) from (P_1, T_1) to (P_2, T_2) by two different paths: (a). $(P_1, T_1) \to (P_1, T_2) \to (P_2, T_2)$ and (b). $(P_1, T_1) \to (P_2, T_1) \to (P_2, T_2)$. The result for both paths should of course be $(V_2 - V_1)$.

2. Find the total differentials of the following functions:

   (a). $u = xyz/(x + y + z)$ (b). $u = \frac{x}{y} + \frac{y}{z} + \frac{z}{x}$

   (c). $u = e^{x+y^2}$

3. Which of the following expressions are homogeneous in x and y and that are homogeneous in x, y and z? What are the degrees of homogeneity?

   (a). $x^2 + y^2 + z^2$ (b). $2xyz + 3yz^2 + z^3$

   (c). $3x^3y^2z + 6x^2y^3 - x^4yz^2$ (d). $e^{(x/y^3)+(x^2/y^3)}$

   (e). $3y^3 + xyz^2 - yz^4$

4. For $u = (x^3/y) + (y^3/x) + xy$, show that $(\partial u/\partial x)_y$ and $(\partial^2 u/\partial x^2)_y$ are homogeneous in the first and zeroth degrees respectively. Verify that

$$x\frac{\partial^2 u}{\partial x^2} + y\frac{\partial^2 u}{\partial x \partial y} = \left(\frac{\partial u}{\partial x}\right)_y$$

This illustrates the fact that differentiation lowers the degree of homogeneity by one.

5. Given that volume is homogeneous first order in the masses of the components, use Euler's Theorem to show that

$$\frac{\partial^2 V}{\partial n_1^2} = -\frac{n_2}{n_1}\frac{\partial^2 V}{\partial n_1 \partial n_2}$$

6. Determine which of the following expressions are exact, and find an integrating factor for the inexact ones.

   (a). $xy\,dx + xy\,dy$ (b). $xy^2\,dx - x^2y\,dy$
   (c). $\frac{x}{y^2}dx - \frac{x}{y^3}dy$

7. Given the function

$$u = y^2 - x^2$$

find the (partial) Legendre Transform $w = w(y, p)$ where $p = (\partial u/\partial x)_y$, and show that $(\partial w/\partial p)_y = -x$.

8. Given the function

$$a = b^3 - 2c^2$$

find the Legendre Transform $f = f(p, q)$ where $p = (\partial a/\partial b)_c$ and $q = (\partial a/\partial c)_b$, and show that $(\partial f/\partial p)_q = -b$ and $(\partial f/\partial q)_p = -c$.

9. Use Table 2.1 to derive an expression for the total differential of the enthalpy in terms of C_p° and $\alpha(= (1/V)(\partial V/\partial T)_P)$. In other words, start with $dH = (\partial H/\partial T)_P dT + (\partial H/\partial P)_T dP$ and find expressions for the two partial derivative terms from Table 2.1. You will find that if in this resulting equation you let $dH = 0$, you get an expression for $(\partial T/\partial P)_H$ identical to that in equation 8.1 (Chapter 8). This derivation is used by Ramberg (1971) in his elegant discussion of the Joule-Thompson effect in a gravitational field.

10. Use Table 2.1 to derive an expression for $(\partial T/\partial P)_S$ (used in §8.2.1, Chapter 8).

3

THERMODYNAMIC TERMS

There are probably, in fact, very few applications of thermodynamic reasoning to natural phenomena where the concept of local equilibrium does not enter the analysis in some way, even though not formally acknowledged.

Thompson (1970)

But no real measurement is reversible, and no real system is completely in equilibrium.

Herzfeld (1962).

3.1. THERMODYNAMICS DEFINED

Thermodynamics is the science that deals with energy differences and transfers between systems, and with systematizing and predicting what transfers will take place. Such fundamental topics naturally find application in all branches of science, and have been of interest since the earliest beginnings of science. In general, since we are dealing with energy transfers between systems, most of what follows has to do with what the entities (equilibrium states) are from which and to which energy is being transferred, and the boundaries or walls through which or by which the transfer is effected. It is in these considerations that we first see the differences between natural systems (reality) and our models of these systems.

3.2. SYSTEM

System refers to any part of the universe we care to choose, whether the contents of a crucible, a cubic centimeter in the middle of a cooling magma, or the solar system. Depending on the nature of the discussion, it must be more or less clearly defined and separated (in fact or in thought) from the rest of the universe, which then becomes known as the system's *surroundings*. At the outset, we will effect an enormous simplification by considering only systems that are unaffected by electrical, magnetic, or gravitational fields, and in which particles are sufficiently large that surface effects can be neglected. Each of these topics can be incorporated into the basic thermodynamic network to be developed, but it is a nuisance to carry them all along from the beginning, and a great deal can be done without considering them at all. More exactly, a great deal can be done if we choose to consider systems where these fields and surfaces play a minor role. Clearly we would not get very far if we tried to understand the solar system without considering gravitational fields. Chemical and geochemical problems however commonly do not need to have these factors included in order to be understood.

In science, when a problem or a phenomenon such as the solar system or the boiling of water is said to be understood, what is usually meant is that we have a model of the phenomenon which is satisfactory at some level, and about which virtually all scientists agree. The model is normally mathematical, and incorporates a number of assumptions or simplifications about the subject or phenomenon being modeled. When the phenomena involve energy levels and transfers, chemical reactions, and so on, the model is invariably thermodynamic to some extent. In constructing such models, we find that there are several kinds of systems that are useful, depending on our purposes. These are as follows.

3.2.1. *Types of System*

- *Isolated Systems* have walls or boundaries that are rigid (thus not permitting transfer of mechanical energy), perfectly insulating (thus preventing the flow of heat) and impermeable to matter. They therefore have a constant energy and mass content, since none can pass in or out. Perfectly insulating walls and the systems they enclose are called adiabatic. Perfectly isolated systems do not occur in nature of course, emphasizing the model nature of these systems.
- *Closed Systems* have walls that allow transfer of energy in or out of the system but are impervious to matter. They therefore have a fixed mass and composition but variable energy levels.
- *Open Systems* have walls that allow transfer of both energy and matter to and from the system. The system may be open to only one chemical species or to several.
- *Simple Systems* are those that contain no interior walls. The simplest of simple systems would be a single phase, having all its properties uniform throughout, but multiphase systems are also simple if there are no impermeable or adiabatic barriers between the phases.
- *Composite Systems* are made up of two or more simple systems, and are thus systems having internal walls.

The term *system* can also of course be applied to a natural situation, rather than a model. We speak of vein systems being open to hydrothermal solutions, or of a crystal becoming a closed system during its cooling history. However, when thermodynamics is applied, the system invariably becomes a *model* system.

3.3. EQUILIBRIUM

We adopt a double-barreled definition of equilibrium:

1. A system at equilibrium has none of its properties changing with time, no matter how long it is observed.
2. A system at equilibrium will return to that state after being disturbed, that is, after having one or more of its parameters slightly changed, then changed back to the original values.

This definition is framed so as to be "operational," i.e., you can apply these criteria to real systems to determine whether they are at equilibrium. And in fact, many real systems do satisfy the definition. For example, a crystal of calcite sitting on a museum shelf obviously has exactly the same properties this year as last year (part 1 of the definition), and if we warm it slightly and then put it back on the shelf, it will gradually resume exactly the same temperature, dimensions, and so on that it had before we warmed it (part 2 of the definition). The same remarks hold for a crystal of aragonite on the same shelf, so that the definition can apparently be satisfied for various forms of calcium carbonate. In other words, many real systems achieve equilibrium. Or do they?

In the following examination of equilibrium states, we will find that although many real systems do satisfy our criteria and can certainly be said to have achieved "equilibrium," a close examination shows that all real systems stop changing before reaching a true minimum energy state, and this is important to an understanding of the thermodynamic model. Barton (1972) and Barton et al., (1963) provide a good discussion of many of the aspects of equilibrium mentioned here.

3.3.1. *Stable and Metastable Equilibrium*

What exactly do we mean by "stable" and "metastable"? The difference is one of energy levels. A system of fixed composition under given conditions is assumed to have one configuration (molecular arrangement; structure) that has a minimum energy level. Configurations that have higher energy levels should spontaneously reorganize themselves so as to lower their energy level (by losing heat, homogenizing, recrystallizing, etc.),[1] but are sometimes prevented from doing so, either by a physical constraint, such as a wall separating one part of the system from another, or by an "energy barrier" of some kind. For example, a mixture of H_2 and O_2 under most conditions contains a great deal more energy than does H_2O, and is metastable with respect to H_2O. Aragonite has a higher energy content per mole than has calcite, and is a metastable form of $CaCO_3$.

The best mechanical analogy is a ball rolling in a series of hills and valleys. In Figure 3.1 we see a ball on a surface having two valleys, one higher than the other. At (a), the ball is in an equilibrium position, which fulfills both parts of our definition—it will stay there forever, and will return there if disturbed, as long as the disturbance is not too great. However, it has not achieved the lowest possible potential energy state, and therefore (a) is a *metastable equilibrium* position. If the ball is pushed past position (b), it will roll down to the lowest available energy state at (d), a *stable equilibrium* state. During the fall, e.g. at position (c), the ball (system) is said to be *unstable*.

In position (b), it is possible to imagine the ball balanced and unmoving, so that the first part of the definition would be fulfilled, and this is sometimes referred to as a third type of equilibrium, admittedly a trivial case, called *unstable equilibrium*. However, it does not survive the second part of the definition, so we are left with only two types of equilibrium, stable and metastable.

[1]Unless of course the system is isolated, in which case it cannot lower its energy level. Systems of this kind are especially interesting and are considered later (Chapter 5).

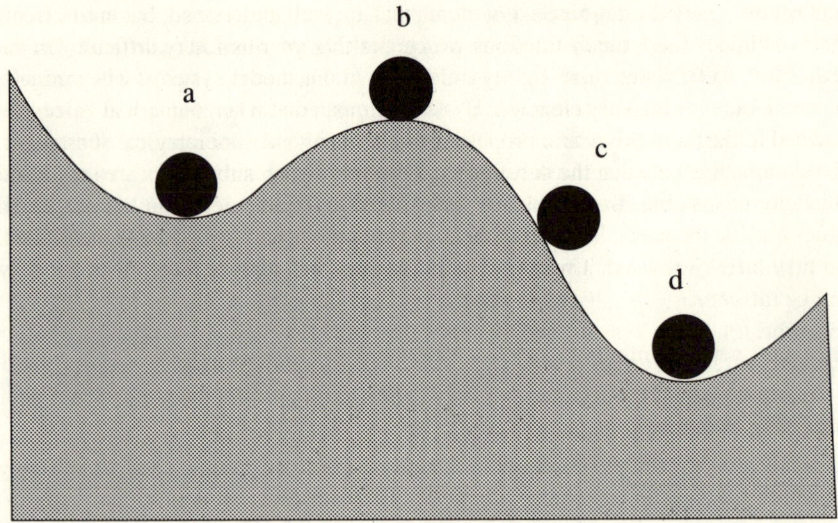

FIG. 3.1. Four positions of a ball on a surface, to illustrate the concept of equilibrium. Position *a*—metastable equilibrium. Position *b*—unstable. Position *c*—unstable. Position *d*—stable equilibrium.

In chemical systems, metastable equilibrium states are quite common because of activation energy barriers which prevent the occurrence of reactions that, if they did occur, would result in a lower energy state. The common examples of hydrogen-oxygen mixtures and polymorphs such as aragonite have already been mentioned. In the mechanical analogy in Figure 3.1, the energy barrier preventing the fall of the ball from (a) to (d) is obviously the energy required to push the ball from (a) to (b).

Another possibility is that the ball could be in position (c), but enclosed in a viscous medium that prevents it from rolling quickly. Depending on the viscosity, we could have any rate of descent, from rapid to not perceivable over millions of years. This state, i.e., one that shows no perceptible change but is or may be actually slowly changing, has also sometimes been described as metastable. Examples would be some types of organic matter, work-hardened metals, glasses, plastics, and so on, that are more accurately thought of as "frozen" unstable states. The second part of the definition is intended to eliminate these states, because the right kind of "disturbance" might briefly "thaw" the system, allowing the ball to roll a little farther down the hill, so that the system does not return to its original state after the disturbance.

Unfortunately, in real systems the distinction between these various states is not always as clear as in Figure 3.1. For example, the ball might be stuck in position (c), but in a rather small valley that is hard to see at the scale of the drawing. The fact that it is released and continues to fall after a disturbance does not necessarily mean that it was not truly metastable. Another problem is that thermodynamics cannot tell us whether or not there are states at even lower energy levels than our "stable" states. These are all problems familiar to the experimentalist. We describe "stable," "metastable", and so on in terms of systems that are well understood, but until a (real) system *is* understood, the distinctions we are making are often very difficult. On the other hand, these distinctions are very clear cut in our model systems. They may be incorrect, but they are very clear cut. By this we mean that when numerical values are assigned to the thermodynamic properties of chemical and mineralogical substances, the relationships between the substances, including which substances are stable and which are metastable, are completely determined and quite clear. But the numerical values may be incorrect, because of the experimental difficulties we have mentioned, and may later be revised. The revised relationships will also be quite clear, but may well be different.

The True Energy Minimum

Finally, we note that for a system to have truly reached the very bottom of its "energy valley," all gradients in temperature, pressure, and composition must be eliminated (a formal proof of this is presented in Chapter 14, but it certainly seems intuitively reasonable). A system can conform completely to our two-part definition of equilibrium, yet not have reduced its energy as much as possible. Thus the museum crystals of calcite and aragonite we spoke of may be at equilibrium as we have defined it, yet have internal gradients in trace elements and isotopes, or various structural defects, or high-energy morphologies (e.g. aragonite needles), and so on. These energy anomalies may be small compared to the energy difference between the two polymorphs, but nevertheless if they are present, each crystal could conceivably reduce its energy content a little further, i.e., get deeper into its "energy valley" by eliminating these features.

This true or idealized energy minimum is also sometimes referred to as the equilibrium state, stable or metastable, and discussions can become confused unless the distinction between the macroscopic, practical kind of equilibrium that we have defined and the more idealized, conceivable kind of equilibrium state is made clear.

3.3.2. *Partial Equilibrium*

Another type, or perhaps sub-type of equilibrium is *partial equilibrium*. That is, systems may appear to be at equilibrium with respect to some processes and not with respect to others. Thus the ball in position (d) in Figure 3.1 is in stable equilibrium as far as its potential energy with respect to the surface under it is concerned, but it may be at the same time out of thermal equilibrium with its surroundings. A system consisting of a layer of oil on top of a layer of water may be in complete thermal equilibrium but not yet at equilibrium with respect to the water vapor pressure above the oil. Adjacent crystals of pyroxene and olivine may have achieved equilibrium with respect to Fe–Mg partitioning, but not with respect to Ni partitioning. All of these and many more cases can be thought of in a practical sense as cases of partial equilibrium. However, a note of caution is required. Not only in each case has equilibrium not been completely achieved, but theoretically at least, equilibrium has not even been partially achieved because as the part of the system that is not at equilibrium changes, the so-called partial equilibrium state also changes in response. For example, in two crystals such as olivine and pyroxene, the equilibrium partitioning of two elements is never theoretically independent of other elements unless they are completely ideal solutions. Thus the gradual equilibration of Ni between the two phases will cause, or should cause, a readjustment of the Fe/Mg ratios, however small. As water diffuses through the oil layer and evaporates into the vapor space above, heat is absorbed, causing a thermal gradient between the system and its surroundings and disturbing the thermal equilibrium. Even in the case of the ball in the valley, as the ball warms up or cools down in reaching thermal equilibrium with its surroundings, it will expand or contract, changing the distance between its center of gravity and the surface it is resting on, and hence its potential energy.

3.3.3. *Local Equilibrium*

There is one more type of equilibrium we need to discuss, *local equilibrium.* In any system that overall is in disequilibrium there may be regions or sub-systems that have reached equilibrium by any test. Thus pyroxene may be in equilibrium with the hornblende that mantles it but not with the magma next to the hornblende; a cubic meter of magma in the center of a pluton may be in equilibrium with the magma next to it, but not with magma close to the margins of the intrusive; and so on. Local equilibrium is thus a kind of regionalized partial equilibrium, and must be a very common situation in geology. It can be argued in fact that because the universe is heterogeneous and in disequilibrium, all natural equilibrium states are in fact only local equilibria. This is not so obvious, or perhaps not very useful, when discussing laboratory systems where great care is taken to control the environment of an experiment. But in natural systems, the system boundaries are always arbitrarily chosen by the investigator (e.g. the cubic meter of magma in the pluton), and the chosen system is always contiguous with other parts of the world with which it is not in equilibrium. Therefore all real systems contain at best only local equilibria, and as we have mentioned above, local or partial equilibrium states have not "really" achieved equilibrium.

3.3.4. *Measurement Sensitivity*

But another aspect of the problem has been raised—that of the sensitivity of our measuring techniques. In each case mentioned above, the changes taking place may well be undetectable. Certainly no electron microprobe operating today would detect any change in Fe–Mg contents of an olivine-pyroxene pair if the Ni contents were to change by a few parts per million. Other cases are easily conceived where we might detect such changes with careful or very modern techniques, but not with the techniques of a few years ago. So are we to say that a given system was in partial equilibrium last year, but now that we have machine X, it is now in disequilibrium?

3.3.5. *"Real" versus "Model" Equilibrium*

It seems clear from this discussion that the question of whether any particular real system had reached equilibrium or not can only be answered in relation to the purposes of the investigator. A given natural system may be in equilibrium for some purposes and not for others. Geologists and chemists often *assume* equilibrium, then apply the thermodynamic model and look at the results. If they make sense, they feel that to that extent the assumption of equilibrium was justified.

Many natural systems achieve equilibrium in a practical sense, i.e., our measurements may not be able to detect any departure from equilibrium. Even if they are slightly out of equilibrium, it doesn't matter much for many purposes. However, probably no one who has looked carefully into mineral compositions and structures would assert that minerals or mineral assemblages had absolutely no gradients in trace elements, isotopes, defects, or that the surface energies were minimized, etc. Experimentalists, who expend prodigious efforts trying to achieve equilibrium in their experiments, would probably be the last ones to assert that any particular experimental charge had actually achieved that state. No real system ever achieves a true energy minimum; there are always imperfections in the real world. This "true energy minimum" is an idealized, or as we call it, a *model equilibrium state*. It is this state, in which the energy of the system is truly minimized, all gradients eliminated, morphologies optimized, and so on, that is used in the thermodynamic model. The distinction between the practical and model equilibrium states is clear conceptually, but may involve only a few joules of energy, i.e., the difference in terms of energy may be quite small. It is the fact that the difference is small that allows thermodynamics to be useful.

So why are we devoting so much discussion to such a small point? The reason will gradually become apparent. This topic is simply the first one in which the distinction between the thermodynamic model and the real world arises. The fact that real systems do achieve "equilibrium" tends to lead to the conclusion that thermodynamics refers to, or even is part of, real systems, and this inevitably leads to confusion with respect to reversible processes, infinitesimals, choice of components, and many other parts of the model. Our point of view is that real systems achieve the kind of practical equilibrium we defined in §3.3, including local and partial equilibria, but that the thermodynamic model uses idealized equilibrium states. If there is not too much difference between them, then the model results are useful in the real world. The definition of "too much difference" depends on the application.

3.4. STATE VARIABLES

Systems at equilibrium have measurable properties. A property of a system is any quantity that has a fixed and invariable value in a system at equilibrium. If the system changes from one equilibrium state to another, the properties therefore have changes that depend only on the two states chosen, and not on the manner in which the system changed from one to the other. This dependence of properties on equilibrium states and not on processes is reflected in the alternative name for them, *state variables*. Recall from the discussion of Euler's theorem in Chapter 2 that *extensive* variables are proportional to the quantity of matter being considered—for example, volume and (total) heat capacity. *Intensive* variables are independent of quantity, and include concentration, viscosity, density, molar heat capacity, and many others. Several important state variables are not measurable in the absolute sense in a particular equilibrium state, though they do have fixed, finite values in these states. However, their changes between equilibrium states are measurable or at least calculable. Reference in the above definition to "equilibrium states" rather than "stable equilibrium states" is deliberate, since as long as metastable equilibrium states are truly unchanging they will have fixed values of the state variables. The thermodynamics of organic compounds, virtually all of which are metastable in an oxidizing environment, is an example of this distinction.

Since state variables have fixed values in equilibrium states and have changes between equilibrium states that do not depend on how the change is carried out, it follows that the differentials of state variables will always be exact differentials, according to our definitions in Chapter 2.

3.5. DEGREES OF FREEDOM

Systems at equilibrium have literally dozens of properties. In addition to those state variables applicable to the system as a whole such as T and V, each phase within the system has a host of properties such as heat capacity, cell size, optic angle, refractive index, and so on, and all can be considered properties of the system according to our definition.

Experience has shown that to describe equilibrium systems, or to distinguish between one equilibrium state and another, it is not necessary to enumerate all the properties of a system. It has been observed that for each equilibrium system, whenever a certain number of properties have been specified, all the others have fixed and unchanging values. For example, for any sample of any gas, whenever the pressure and temperature have been fixed, the volume will be observed to take on a fixed value as well. If the temperature and volume are fixed, the pressure will be observed to have a fixed value also. The gas has a number of other properties too, such as heat capacity, viscosity and so on, and it is possible in principle, if not always in practice, to fix all the properties of the system by fixing any two. (We will describe this notion more formally in a later chapter (§5.4.1) as Duhem's Law.)

Other, more complex systems require more than two variables to be fixed, but all systems have a definite number of properties that must be fixed in order to fix all the others, that is, to define the equilibrium state. This is called the number of independent variables, or the *number of degrees of freedom* possessed by the system. There is a very close analogy, then, between equilibrium systems and the multivariate functions discussed in Chapter 2. There we said that the function $x = x(y, z)$ could be said to have two degrees of freedom since fixing any two of the variables fixes the third. A divariant system then, can be thought of as a multivariate function having dozens of variables, but there exist also a number of equations relating these variables, reducing the independent variables or degrees of freedom to two. We will see what these equations are when we consider the phase rule (§14.6).

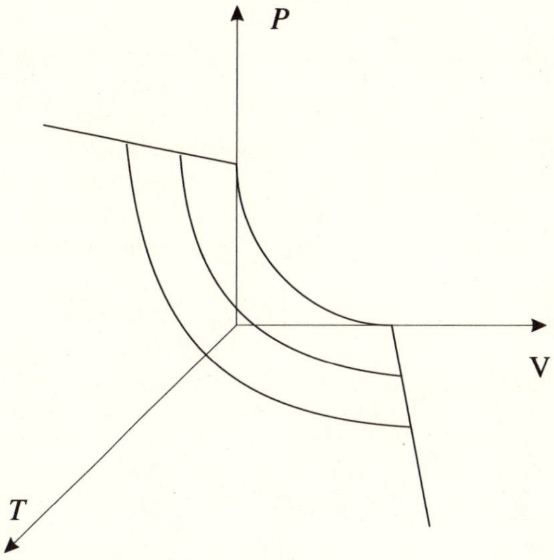

FIG. 3.2. *P-V-T* surface of a simple homogeneous system. Arrows on axes show direction of increases of *P*, *V* and *T*.

3.6. THERMODYNAMIC STATE SPACE

For systems having not more than two degrees of freedom, we can define a *"State Space"* using one or two independent variables and one dependent variable as axes, completely analogous to our "Variable Space" for functions (§2.2.2). Just as functions are represented by lines or surfaces in Variable Space, equilibrium systems are represented by lines or surfaces in State Space.

In Figure 3.2, the stable equilibrium surface of a simple homogeneous system such as a gas is represented schematically in *P-V-T* Space (two independent variables, one dependent). In Figure 3.3 the surface intersects the axes in a pictorially convenient fashion. Thus, for an ideal gas a surface $V = V(T, P) = RT/P$ may be plotted in *P-V-T* state space. This surface represents all equilibrium states attainable by the gas, and the equation $V = RT/P$ is termed an *equation of state*. Usually the term "equation of state" is used exclusively for functions relating the state variables *P*, *V* and *T*. Only equilibrium states are defined by equations of state. Thus the surface $V = V(P, T)$ shown in Figure 3.3 is termed an equilibrium surface. Every stable equilibrium state of the system is represented by a point on this surface, and when the system is not at equilibrium, it cannot be represented in the diagram.

Metastable states can also be represented in state space as additional surfaces sub-parallel to the Stable Equilibrium Surface, since they do not intersect that surface or other metastable surfaces.[2] Figure 3.4 shows a Stable Equilibrium Surface similar

[2]That is, they do not intersect in *P-V-T* space; they *do* intersect in other varieties of thermodynamic state space.

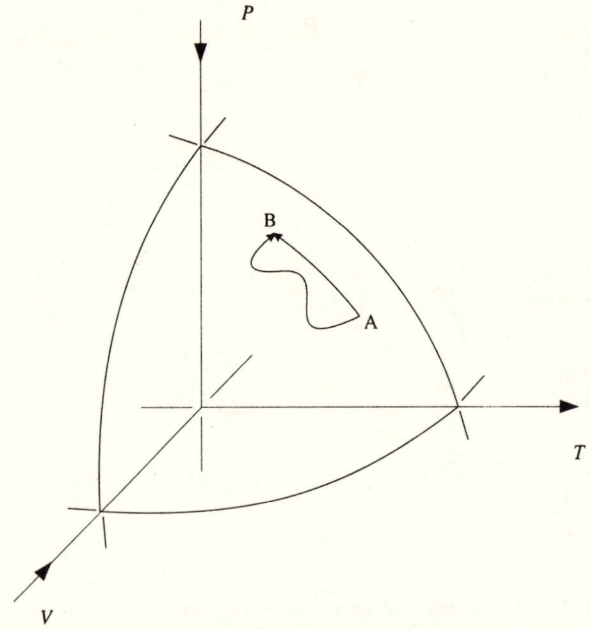

FIG. 3.3. P-V-T surface of a simple homogeneous system, with P and V axes reversed to allow more convenient representation of the surface.

to that of Figure 3.3, but this time it is supposed to represent the mineral graphite. Also shown is the Metastable Equilibrium Surface for diamond.

Clearly, Metastable Equilibrium Surfaces are completely analogous to Stable Equilibrium Surfaces. There is, however, one important difference between the two surfaces. Every metastable system, and corresponding metastable equilibrium surface, has at least one degree of freedom more than the corresponding stable system. To amplify what we mean by this, we will introduce the concept of constraints. In this discussion, we follow the ideas of Reiss (1965) very closely, though not his usage of some terms.

3.6.1. *Constraints and Metastable States*

In mathematics, constraints are functional relationships that must be satisfied in solving a system of equations, as we have seen in Chapter 2. Similarly, if we say for a certain system of fixed composition that $T = 25°\text{C}$, $P = 1$ bar, this is sufficient to fix all the other state variables, and the equations $T = 25°\text{C}$, $P = 1$ bar are analogous to the constraint relationships of mathematics. However, in the case of physical systems, we can speak of constraints as either the values of the variables themselves, or as the physical arrangement that leads to the fixing of a state variable. For example we might use a thermostat to fix the temperature at 25°C and some weights on a piston to fix the pressure. The thermostat and the weights and piston are in a physical sense the

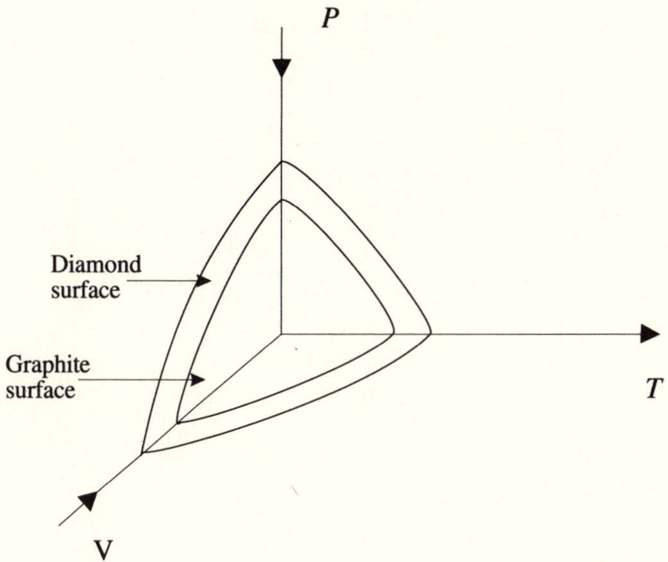

FIG. 3.4. Schematic P-V-T surfaces of graphite and diamond.

constraints on the system that fix all its state variables, and we say that the T and P are *constrained* to certain values. There is *always* a one-to-one relationship between constraints and state variables.

Although, as we have seen, each system has a minimum number of independent variables that must be specified or fixed for the system to be at stable equilibrium, systems frequently have more than this number of variables fixed, and they can then be said to be in metastable equilibrium states. In fact our definition of a metastable state can be restated as *one that has more than the minimum number of constraints necessary to fix the equilibrium state*. To illustrate what we mean by a metastable state, and the wide-ranging nature of the definition, we consider next three examples.

Partitioned Gas Example

Consider the system in Figure 3.5a. We have a given mass of gas with a fixed volume and temperature. As we have seen, this is enough to completely fix all properties of the gas. In Figure 3.5b, we have the same mass of gas, the same total V and the same T but we also have a sliding partition locked in place that produces different gas pressures on each side. To fix the variables of this system we need to specify one more constraint, e,g., T, V and the position of the partition, or T, V and P_1, or T, V and P_2 or T, P_1 and P_2 etc. If we release the constraint (unlock the partition), the partition will spontaneously move to a position where $P_1 = P_2$, and we have the original condition of needing only to fix two constraints. To establish the existence of the third constraint, that is, to push the partition to one side and hold it there, we

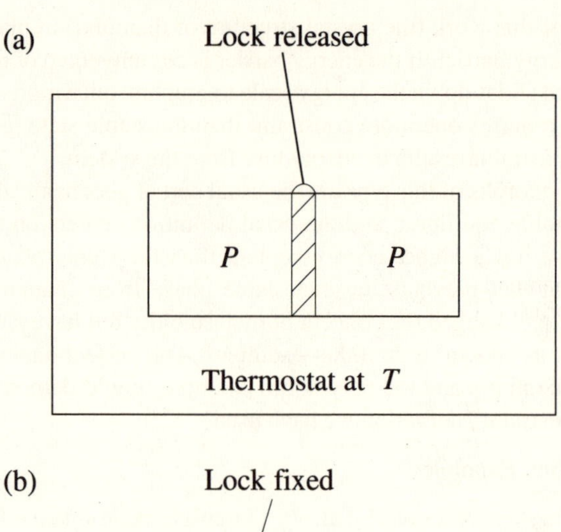

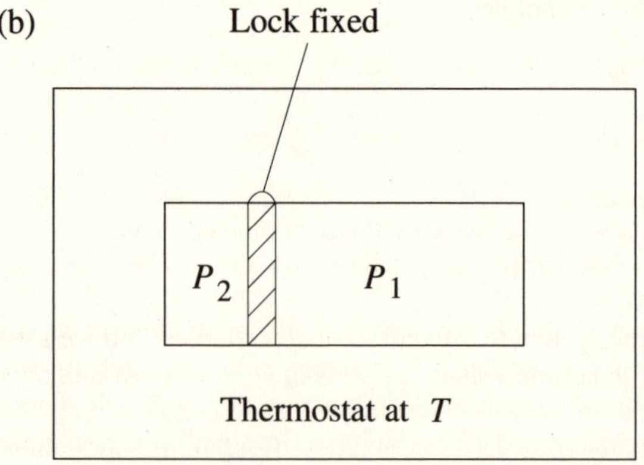

FIG. 3.5. A composite system at fixed T and V. (*a*) Partition is free to move—stable equilibrium. (*b*) Partition locked in place—metastable equilibrium.

must do work on the system, raising its energy content, and when it is released, the energy content of the system is lowered.

Now the system in Figure 3.5b is not the usual sort of example used to illustrate metastable systems, but it is a perfect example for our purpose. All other metastable systems are completely analogous in all respects, with the single exception that we generally have little control over the degree of metastability, or the degree to which a system deviates from stable equilibrium. This lack of control, however, does not change the principles involved.

Polymorph Example

We have been considering the polymorphs calcite and aragonite, but just for a change let's take graphite and diamond. Graphite is the stable form of carbon at the Earth's surface. To fix its properties, in principle we need to specify two variables such as P and T. To produce diamond, we must perform work on the graphite, and we must hold the result of this work (the crystal structure of diamond) in place by means of an activation energy barrier. If the energy barrier is circumvented or released in some manner, diamond slides down the energy scale to graphite once more. The metastable state, diamond, requires one more constraint than the stable state, graphite, and the release of the constraint results in energy loss from the system.

Mineral polymorphs of this type are the usual sort of geochemical examples used to define metastable equilibria, and the usual definition centers on the fact that the metastable phase has a higher energy content than the stable phase. By focusing on the constraint that prevents the metastable phase from changing to the stable phase, we have not changed the concept of metastability but have simply broadened its applicability, as shown in the other examples. The perfect analogy between the partitioned gas example and the polymorph example should demonstrate the utility of the "extra constraint" definition we have used.

Chemical Reaction Example

Consider any chemical reaction A + B = C that might be, for example,

$$\mathrm{C}\,(graphite) + \mathrm{O_2} = \mathrm{CO_2}$$

or, more geologically,

$$\mathrm{CaSO_4}\,(anhydrite) + 2\mathrm{H_2O} = \mathrm{CaSO_4 \cdot 2H_2O}\,(gypsum)$$

At any particular T and P for which A, B and C are not at equilibrium, either A + B is stable or C is stable. If C is stable, consider carefully what A + B means. If C is stable, then obviously the combination of A + B will spontaneously react to form C, unless constrained from doing so. In other words, unconstrained A + B is an *unstable* assemblage. But in applying thermodynamics to A + B and C in order to decide which is stable, we must consider the *equilibrium* properties of A, B, and C. So if C is stable, A and B must be considered separately, not together. The fact that A and B are separated means, in the thermodynamic model, that A + B exists at T and P but is *constrained* from reacting, therefore there is an extra (third) constraint involved, and therefore A + B is a metastable state. The separation of A and B may be effected by various means, such as having a partition between them, or by having A and B in separate containers, but when this separation is ended, or *this constraint is released*, A + B slides down the energy gradient, forming C. In the reverse case, where C spontaneously reacts to form A + B, then A and B are stable together, and no separation is implied.

Some metastable systems of course may have *more* than one extra constraint. Consider A + B = C again, this time where A is H_2 gas, B is O_2 gas, and C is H_2O liquid, where the T is 25°C and the P is 1 bar. If the H_2 and O_2 are in separate parts of a container separated by a partition, they are of course constrained from reacting, and the partition represents a first constraint. If the partition is then removed, the gases mix but they do not react to form water, because there is an activation energy barrier that must be overcome; this represents a second constraint. Finally if a catalyst is introduced, removing this constraint, the gases react to form the stable phase, water.

Another example of metastability, completely different from these, is found in the subject of galvanic cells (Chapter 18).

We therefore define metastable equilibrium states as those having more constraints than the minimum number required to define stable equilibrium for the system. In most cases of interest to Earth scientists, there is one added constraint—an activation energy barrier.

A large part of thermodynamics deals with processes occurring between metastable and stable states, which we consider next.

3.7. PROCESSES

If we now define a *process* as the act of changing a system from one equilibrium state to another, we see that any two equilibrium states of the system may be connected by an infinite number of different processes, because only the initial and final states are fixed; anything at all could happen during the act of changing between them. In proceeding from State A to State B, the system in real life would normally be out of equilibrium while changing between the two states, and thus the process could not be represented in state space. If the equilibrium states are little red lights on the surface, the light at A would be lit until the process began, then it would go out, and some time later, the light at B would come on when the system had settled down in its new conditions.

Much of the following discussion of processes will seem rather futile to anyone looking for a description of what happens in real systems. Reversible, quasistatic and virtual processes just do not occur anywhere in the real world. But we are not trying to describe what happens in real systems. We are describing processes that can be represented by equations in the mathematical model we call thermodynamics.

3.7.1. *Reversible Processes*

Another way of changing the system from state A to state B is also possible. That is, instead of leaving the surface and heading into the realm of disequilibrium, the point representing the system might remain on the surface at all times; or in other words the little red light at A might simply slide over to B, remaining lit. There are still an infinite number of paths it might follow between A and B, whether it remains on the surface or not. All processes leading from A to B that remain on the Stable Equilibrium Surface consist of a continuous succession of stable equilibrium states called *reversible processes*. They are hypothetical limiting cases, because it is in fact impossible for real systems to behave in this way.

We have just defined reversible processes in terms of the stable equilibrium surface, but completely analogous processes are also possible on metastable equilibrium surfaces, and the definition can be extended to include these. In fact, however, most discussions of reversible processes refer to stable equilibrium states and surfaces.

When changes in state variables are integrated (e.g. $\int_{V_1}^{V_2} P\,dV$) the integration, regardless of path, follows the Stable Equilibrium Surface, and hence simulates a reversible process. The net result on the system, however, is the same for irreversible (non-equilibrium) as well as reversible processes, as long as the beginning and end points are the same in each case.

3.7.2. *Irreversible Processes*

Also very important of course are processes leading from a metastable equilibrium state to a stable equilibrium state. These are *irreversible*, because when the (extra) constraint is released, even momentarily, the system is not in balance and proceeds in one direction, towards equilibrium. However, this inexorable progress towards equilibrium can be performed in one jump or can be halted in a succession of (metastable) equilibrium states called a *quasistatic process.*

Quasistatic Irreversible Processes

In many presentations of thermodynamics, quasistatic processes are either not mentioned or are said to be the same as reversible processes. In our usage, however, quasistatic processes are similar to reversible ones in that they are a continuous succession of equilibrium states, but these states are metastable equilibrium states, not stable equilibrium states. A quasistatic process is most easily imagined as resulting when one of the constraints on a metastable equilibrium state is released for an extremely short time, then re-imposed. While the constraint is released, the system changes slightly but irreversibly towards equilibrium, then settles down in its new metastable equilibrium state when the constraint is re-applied. This succession of events is repeated until the final state is reached.

The complete development of this concept, and an understanding of why it is a useful part of the thermodynamic model, must await introduction of the Second Law. However, as an example, consider again the reaction A + B = C in the Chemical Reaction Example in §3.6.1. Under conditions where C is stable, release of the constraint on A + B results, as we said, in the irreversible reaction of A with B to form C. But suppose only a tiny amount of A and B are allowed to react, then they are separated again, or otherwise constrained from reacting, and the reaction products allowed to equilibrate. Then another tiny increment of reaction is allowed, followed by equilibration, and so on.[3] The resulting succession of (metastable) equilibrium states is what we call a quasistatic process. We will see that it is just as fundamental a part of our model as is the reversible reaction, and in the hands of insightful geochemists (and computer programmers) has in recent years become a standard tool of geochemical modeling.

3.7.3. *Virtual Processes*

Following Reiss (1965), we define a *virtual process* as the *inverse* of an irreversible or quasistatic process; in other words it is an irreversible process that goes backwards. Systems will never spontaneously undergo virtual processes, they must be "pushed" by an external energy source. Imposing an additional constraint on a system always involves adding energy to the system, and the process of doing this is the virtual

[3]In Chapter 19 we describe an example where A is K-feldspar and B is water.

reaction. We do not make much explicit use of this kind of reaction, but its existence is implied by many diagrams and statements (see Chapter 5).

3.7.4. *Special Kinds of Processes*

Adiabatic processes occur without any exchange of heat between the system and its surroundings. *Isothermal processes* occur at constant temperature, *isobaric processes* at constant pressure, *isochoric* at constant volume, and *isoplethal* at constant composition. Finally, in a *cyclical* process, initial and final states of the system are identical. Integration in cyclical processes is often symbolized by the integral

$$\oint dx = 0$$

where $\oint$ refers to an integral around any closed path, that is, a process that returns to the initial state. If x is a function of state and thus independent of path, then the value of such cyclical integrals is necessarily zero.

3.8. PHASES, COMPONENTS, SPECIES AND CONSTITUENTS

The matter contained in a system may be grouped into regions having different structures and/or compositions called *phases*. A phase is defined as a homogeneous body of matter, generally having distinct boundaries with adjacent phases and so in principle is mechanically separable from them. The shape, orientation, crystallinity and position of the phase with respect to other phases are irrelevant, so that a single phase may occur in many places in a system. Each mineral in a rock is therefore a single phase, as is a salt solution, or a mixture of gases. Single phase systems are called *homogeneous*, and multiphase systems *heterogeneous*. Some complications in the definition of phases will be taken up in Chapter 14.

It will of course be important in our discussions about thermodynamic systems to be quite specific and precise about their chemical compositions. The term generally used to describe the composition of a system is *component*. The components of a system are defined by the smallest set of chemical formulae required to describe the composition of all the phases in a system. This simple definition requires a surprising amount of explanation and discussion to be completely clear, much of which we postpone to Chapter 14 because it involves concepts not yet introduced. For now, we will simply say that although the *number* of components possessed by a system in a given state is fixed, the *identity* of these components is not. Consider for example the system A–B, in which phases A_2B and AB_2 occur. The compositions of both phases can be described by various amounts of components A and B, or A_2B and B, or A_2 and B_2, or in fact any two compositions on the join A-B, including negative compositions beyond A or B. There are in fact an infinite number of possible choices for any system, but the *number* of components required is fixed. Another thing about components worth mentioning at this stage is that systems can have different numbers of components in different stable equilibrium states. For example pure albite ($NaAlSi_3O_8$) is a one-component system at ordinary conditions, but at high pressures and temperatures it breaks down to jadeite ($NaAlSi_2O_6$) plus quartz (SiO_2), and two components are required to describe the same bulk composition because the number of phases has changed.

Also before proceeding we must explain our usage of the terms *species* and *constituent*. Liquid and gaseous solutions are found to contain, in fact are made up of, molecular and ionic entities called *species* having some existence as identifiable units. Thus air is mostly N_2 and O_2, (which could also be chosen as the components of air), while NaCl in water, two components, has a number of species (Na^+ or $Na(H_2O)_n^+$, Cl^- or $Cl(H_2O)_n^-$, H_2O, H_3O^+, OH^-, etc.) the exact nature of which is still a research topic. Solution components can correspond to species, but need not (we could for example choose N and O or N_{10} and O_{10} as the components of air). Solids and solid solutions present us with a terminology problem, because while in many discussions we refer to various combinations of atoms in a solid, such as MgO in $MgSiO_3$, they are not species in the usual gaseous and aqueous sense, and they are not necessarily components. We cannot properly speak for example of the component MgO in pure $MgSiO_3$, because $MgSiO_3$ has only one component, which is necessarily $MgSiO_3$. We could, however, stretch the point slightly and refer to end-members of solid solutions and to pure minerals as *mineral species*.

We therefore introduce the term *constituent* in order to be able to reserve the use of the term component to its rigorous or "phase rule" sense. A constituent of a phase or of a system is therefore simply any combination of the elements in the system in any stoichiometry. $MgSiO_3$ therefore has only one component but any number of constituents, such as Mg, Mg_2, Mg_{-1}, SiO_2, $(SiO_2)_{-1}$, MgSi, MgSiO, and so on. Constituents are therefore the most general of descriptors of system compositions, including both all species, if such exist, and components, as shown in Figure 3.6.

3.8.1. *Phases, Components and The Model*

Further discussion of these topics is found in Chapters 10 and 14. We leave these other considerations until later because a full discussion is much easier when other thermodynamic concepts such as activity and fugacity have been introduced. One of these considerations will be found to be that although as defined here phases and components appear to be entities that occur in rocks, minerals, solutions, and other parts of the real world, we will show that they are in fact parts of the thermodynamic model, just as much as are equilibrium and reversible processes.

3.9. MATHEMATICS AND PHYSICAL PROPERTIES

We have mentioned several times now that thermodynamics is best viewed as a model, rather than some kind of description of natural processes. In this section we take a closer look at why this must be so. One of the main reasons is that we use physical properties as mathematical variables.

3.9.1. *Using Properties as Variables*

Normally, when constructing a mathematical model to be used in the physical sciences, it is desirable that the functions be single-valued in all independent variables and

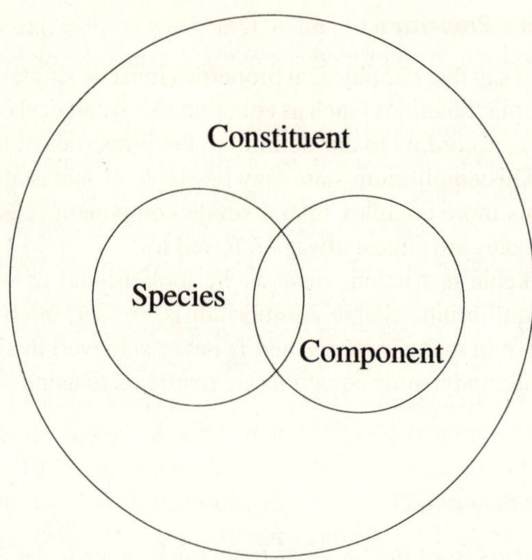

FIG. 3.6. Hierarchy of system compositional terms.

continuous within certain useful ranges of the variables, as well as having continuous higher-order derivatives. In other words, the functions should be "well-behaved". But consider the difference between using pure numbers and physical properties as variables.

The function $x = x(y, z)$ in mathematics will always give the same value or values of x for given values of y and z. You will not get one value of x today and a different one tomorrow. You may get two values of x, but you will get both of them today and both tomorrow. When dealing with physical properties, however, it is conceivable that something quite different might happen. We might have, for example, a function relating pressure (P), temperature (T) and volume (V) of a system

$$V = V(T, P) \tag{3.1}$$

that will always give one value of V for given values of T and P; however, the system itself (that is, if we actually measure V at a particular T and P) might give one value today but a different value tomorrow and a third value the next day. Also, V may be steady and unchanging each time we make the measurement, though steady at different values each time, or it may not be steady but "drifting" to higher or lower values as we try to measure it. This is not a difficult situation to imagine, by the way, and is a common sort of problem for experimentalists. The function (3.1) may be well behaved, but the system is not. Obviously, to be useful, equation (3.1) must be used only in situations where the physical property V is a single-valued function of the physical properties P and T.

3.9.2. *Single-Valued Properties*

What does it mean to say that our physical properties must be single-valued? It means that our thermodynamic functions (such as equation (3.1)) can deal only with systems at equilibrium where, according to our definition, the properties of the system do not change with time. This equilibrium state may be stable or metastable, but generally speaking for systems more complex than a single component (e.g. a pure mineral) stable equilibrium states are almost always referred to.[4]

This is a remarkable restriction, since as we look around us we see rather few systems at stable equilibrium. Stable equilibrium is, in fact, often very difficult or impossible to achieve in many systems, and is never achieved in an absolute sense. Nevertheless, our thermodynamic equations are restricted to using properties in equilibrium states.

3.9.3. *Continuous Properties*

Another problem arises from the fact that function $V = V(T, P)$ represents V as a *continuous* function of T and P. We mentioned in Chapter 2 that many functions are continuous only within certain limits, as illustrated in Figure 2.3. Figure 3.7a shows y as another kind of discontinuous function of x, that is, y only has finite values at certain specific values of x, and not in between. This is all very well for a mathematical function, but if y is the length of a metal bar and x is its temperature, for example, and the points represent our measured values of length at certain temperatures, then the graph is misleading. We know very well that the bar continues to exist and has measurable values at temperatures in between those at which we have chosen to measure it, and we feel perfectly confident in drawing a smooth line through our measured values to represent length at all temperatures (Figure 3.7b), particularly if we have a theory to tell us what the shape of the line should be and if our measurements conform to the theory. To measure the length of a metal bar at a given temperature, we must be sure that the whole bar is at that temperature, i.e., it must be in a state of equilibrium, otherwise the length will be changing as the bar achieves a constant temperature (i.e., as we said above, the properties must be single-valued). Thus every point on the line in Figure 3.7b represents an equilibrium state of our system, and since it is a continuous line, we imply that there is a continuous succession of equilibrium states of our system. The line then represents the lengths we would observe if we could heat the bar to successively higher temperatures (or cool it to successively lower temperatures) while maintaining it at all times at complete equilibrium. However, this is in fact impossible to do. To raise the temperature of the bar, heat must flow into it, which requires a thermal gradient between the outer and inner parts of the bar. As long as the gradients exist, the bar's length will be changing, and it will not be in equilibrium. As soon as the bar re-equilibrates at its new temperature, we may measure its length, but it will be separated from the initial point by some value of ΔT (or Δx in Figure 3.7a).

[4]Except in organic systems, where *metastable* states are often referred to.

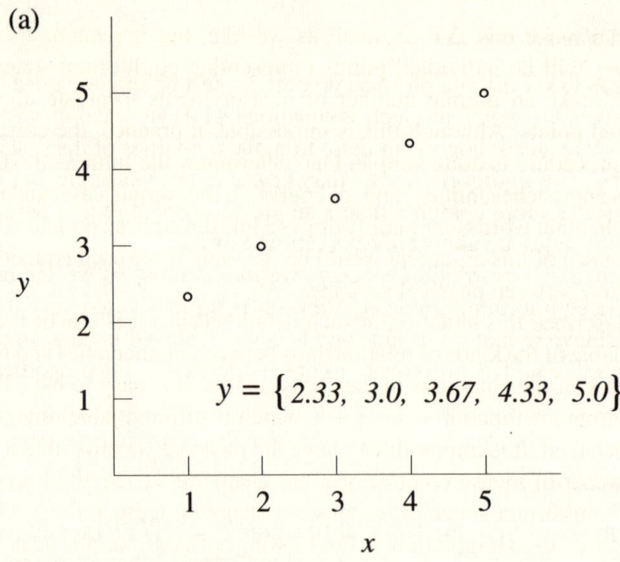

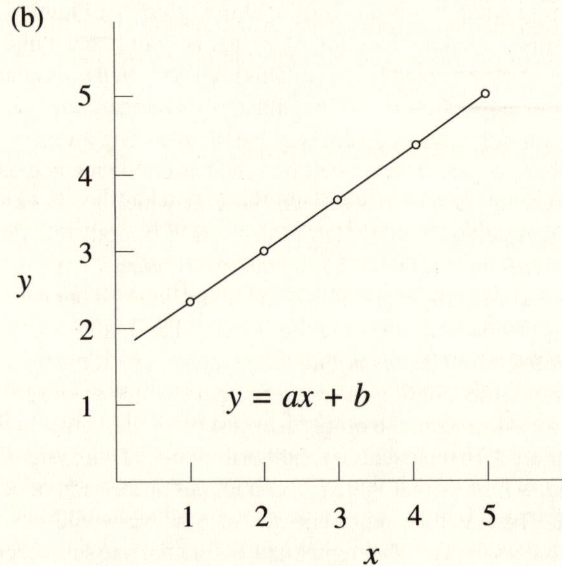

FIG. 3.7. (a) A discontinuous function $y = y(x)$. y only has values at certain fixed values of x. (b) A homogeneous linear function $y = ax + b$ through the points of 3.7a. The existence of the function implies a reversible process.

Now we can make this ΔT as small as we like, but it remains true that the measured values will be individual points representing equilibrium states, and we would have to make an infinite number of measurements to create a continuous line of measured points. Although this is impossible in practice, the corresponding mathematical procedure is quite simple. One determines the limit as the number of determinations approaches infinity, and, of course, in this simple case one obtains the line $y = ax+b$. In other words, one simply draws a line through the points on the graph. Clearly, the essence of this argument would be the same if we used equation (3.1) or any other relation between physical variables.

We have described this situation in such detail because it is one of the simplest examples available of the kinds of relationships between mathematics and reality that make up the bulk of the equations in thermodynamics. It is easy to have the single-valued and continuous function $y = ax + b$, which is differentiable, integrable, and entirely well-behaved. It is impossible to have the physical situation that it describes (an infinite number of measurements, or a bar changing its length in a continuous succession of equilibrium states in response to changing temperature). The mathematical function is an extrapolation, a limit-taking, of possible physical situations, and of course we must do this to be able to apply mathematical techniques (e.g., differentiation, integration) to our measurements. In summary, *continuous mathematical functions using physical parameters as variables* imply *the existence of impossible processes, which we call* reversible processes.

3.9.4. *Thermodynamics as a Mathematical Model*

The continuous succession of equilibrium states represented by the line in Figure 3.7b is one example of a *reversible process*. The explanation of reversible processes takes up considerable space in most texts, and usually it seems to have some connection with entropy changes. However, the importance of reversible processes is much more fundamental than furnishing an explanation for the entropy. It is a direct result of our desire to apply mathematics to physical properties.

The concept of reversible processes often causes difficulties and dissatisfaction in newcomers to the subject, perhaps because of the apparent arbitrariness or physical unreality of such processes. Much of this dissatisfaction can be alleviated by thinking of thermodynamics not as a description of reality or real processes but as a mathematical model, represented by lines and surfaces in thermodynamic state space, and using physical properties as its variables. When real systems achieve equilibrium, which they do only in a practical sense, their properties are interrelated by the model, and therefore changes in real system properties from one equilibrium state to another can also be given by the model. This point of view appears clearly in the following quotation from Pippard (1966, p. 97). He is speaking of a mass of gas that is released from a smaller volume into a larger volume by piercing the wall that separates them.

> It follows from this that when the gas is confined to the smaller volume it has one value of the entropy, when the wall is pierced it has another value, and that it is the act of piercing the wall and not the subsequent expansion that increases the entropy. In the same way when two bodies at different

> temperatures are placed in thermal contact by removal of an adiabatic wall, it is the act of removing the wall and not the subsequent flow of heat that increases the entropy.

Because we have not yet introduced the entropy this quotation may seem out of place, but in fact, as also pointed out by Pippard, the concept applies equally well to any thermodynamic variable. For example, one could say that it is the piercing of the wall and not the subsequent expansion that changes the volume of the gas. By this we mean that the thermodynamic variable volume is fixed by the physical constraints on the system, and once these are fixed, the (model) volume is fixed, whether the gas has achieved this volume or not. In other words, *the properties considered by thermodynamics belong to the model, not to the system being modeled.*

How Can Thermodynamics Be Useful?

The requirements that our functions and properties be single-valued and continuous result in what seem to be rather heavy restrictions in the applicability of thermodynamic functions. If stable equilibrium states are often hard (or theoretically impossible) to achieve, and if the functions imply processes that cannot be carried out, how can thermodynamics be useful? It is useful for two main reasons. First, even if our systems of interest do not reach stable equilibrium, it is of great interest to know what the stable equilibrium state is and what its properties are under given conditions, to know what state the system is "trying to achieve," and how much energy is expended in reaching that state. And second, it is not necessary that processes be carried out reversibly for the functions (that describe the reversible processes) to be useful, because the functions can be integrated to give the changes in properties between equilibrium states regardless of the nature of the real process between these states. This follows from our definitions of equilibrium states (that always have the same properties for a given state), of state variables (that always have exact differentials), and of integrals of exact differentials (that always give the same result regardless of the path followed). For example, in the case of steam engines, thermodynamics was not and is not capable of fully "describing" a single one; however, it is capable of showing what energy output would be available if such engines worked perfectly (thus establishing an efficiency scale), and it can also show what aspects must be changed to improve the efficiency. In mineral systems thermodynamics has found many uses, such as predicting stable mineral assemblages in rocks at various pressures and temperatures. The question of the degree to which individual natural systems have achieved or normally achieve equilibrium is often hotly debated; however, in a general sense it can be confidently stated that a great many natural systems approach stable equilibrium states sufficiently closely that thermodynamic calculations applied to them are very useful. If this were not the case, Earth scientists would have little interest in the subject.

Reversible processes are but one example of a host of concepts of a similarly idealized nature in chemistry and physics—for example, ideal gases and solutions, absolute zero temperature, infinitely dilute solutions, perfect black-body radiation, isolated systems, perfect insulators, and so on. In every case, the adoption of the idealized case simplifies or makes possible the application of mathematics to physical reality. *Mathematics and thermodynamics deal with models of reality, not with reality itself.*

PROBLEMS

1. Look up the definitions of *equilibrium*, *stable*, *metastable*, *reversible* and *quasistatic* in any textbooks on thermodynamics or physical chemistry you have at hand, and reflect on the differences between them and the definitions in this text.
2. Why are the populations of species and components only partly overlapping in Figure 3.6? Give examples.
3. Consider the system $MgO–SiO_2–H_2O$. List two different sets of components, three constituents, three mineral species, and three aqueous species, that might be found in this system.

4

THE FIRST LAW OF THERMODYNAMICS

> ... and now we have the universally accepted doctrine of the equivalence and interconvertibility of mass and energy. Here at last we have the thing that we have been searching for so long: a unique value in any reference frame for the energy of a system and a simple physical significance for the thermodynamically undetermined 'constant of integration' of the energy equations.
>
> Bridgman (1941) p. 92.

> It is important to realize that in physics today, we have no knowledge of what energy *is*.
>
> Feynman, Leighton and Sands (1963, Vol. 1, p. 4-2).

4.1. INTRODUCTION

We have by now amassed sufficient definitions and reviewed enough mathematics to begin discussion of energy levels and transfers. This is a very difficult subject, as evidenced by the fact that more than 50 years of scientific effort were required before the relationships between heat, work, and energy were well understood. A knowledge of the history of development of this understanding by Carnot, Mayer, Rumford, Clausius, Joule, Thompson, and others is very helpful in appreciating the significance of the First Law of thermodynamics.

Part of the difficulty lies in the fact that the concepts of temperature, heat and work are so intuitively familiar to us that we tend to use them without much thought. We have already discussed systems in terms of heat transfer just as if we knew exactly what heat is.

We present here a summary of the present understanding of the relation between heat, work, and energy levels of systems, leading to a definition of the first law of thermodynamics.

4.2. ZEROTH LAW OF THERMODYNAMICS

We are all familiar with the sensation of warmth, that is, that some objects are hotter or colder than others. A large number of instruments called thermometers can be (and have been) devised which will indicate degrees of hotness or coldness of their environments. They have physical properties which vary perceptibly as they become hotter or colder (examples are the volume of a body of fluid at a fixed pressure, the length of a column of mercury in a glass tube, or the voltage produced by a metallic couple); these changes can be assigned arbitrary units of "hotness." The zeroth law says that *two bodies that are in thermal equilibrium with a third body are in equilibrium with each other.* By thermal equilibrium we mean that two bodies are in contact separated by a wall or walls that prevent exchange of mechanical energy or mass, but which still allow the two bodies to interact energetically. Such walls are called diathermal, that is, they allow heat to flow between the bodies. When no further change in the bodies is observed, they are at equilibrium; if one of these is a thermometer, its properties (volume, length, voltage) have been calibrated according to some arbitrary scale, and a reading on this "hotness" or temperature scale may be made. The zeroth law says that if two bodies are placed separately in contact with the thermometer and the same temperature reading results, then the two bodies will be found to be in equilibrium (no changes result) when placed in contact with each other. This innocuous and almost self-evident statement is the basis of the concept of temperature as a useful thermodynamic parameter. This can be shown as follows. Suppose our thermometer is a body of gas at a fixed pressure. As it gets warmer or cooler, its volume changes, and we can conveniently read this volume on a manometer or other gauge attached to the gas. If we suppose the change in volume to be linear between varying states of warmth, then two reference points will suffice to define a temperature scale. To calibrate the thermometer, we first put the gas container in a mixture of ice, water, and ammonium chloride, and when the gas stops changing volume, we mark the manometer level as "zero" temperature. Then we hold the gas container until it reaches thermal equilibrium with a human body, and we mark the manometer level as "96." These two reference states were chosen by D. G. Fahrenheit in 1724 as the basis of his temperature scale. On this scale boiling water gives a reading of 212 and freezing water is 32. The manometer may now be marked off in 96 equal divisions between the 0 and 96 levels established, and other equal divisions may be added above and below. These divisions or "degrees" constitute our temperature scale. Clearly, our degrees are actually a measure of volume in this case, so we don't seem to have made any fundamental advance. But in fact we can follow this calibration procedure using *any* fixed amount of gas and arrive at an equally useful thermometer. Because we can do a similar calibration using many different properties of many substances (such as the length of a metal bar, or the density of a liquid), it appears that the scale of degrees we have established, though it may have arbitrary units, expresses a property of many substances not included in the particular property calibrated. This is recognized by including the degree among the fundamental dimensions of science.

Let us now place our gas thermometer in contact with a second container of another gas having pressure P and total volume V, and we find that the reading on our manometer or temperature scale is θ_1. It has been known since about 1661 that the relation between P and V for a fixed amount of gas at a fixed temperature is $P\mathrm{V} = constant$ (approximately). This is Boyle's Law. Thus by varying the pressure on our gas at θ_1, we will find the volume varying inversely, and a plot of P versus V will describe a hyperbola (Figure 4.1). We could then place our thermometer and second gas container in an oven or refrigerator and we would find that at this different temperature Boyle's Law still holds, but the constant is different. There are in fact a series of *isotherms* on the PV diagram for our gas, one for each temperature reading on our thermometer (Figure 4.1), and there presumably exists a functional relationship between the Boyle's Law constant and θ, hence between PV and θ. Thus

$$P\mathrm{V} = f(\theta)$$

This is called the *equation of state of the gas*, and because θ is a function of two state

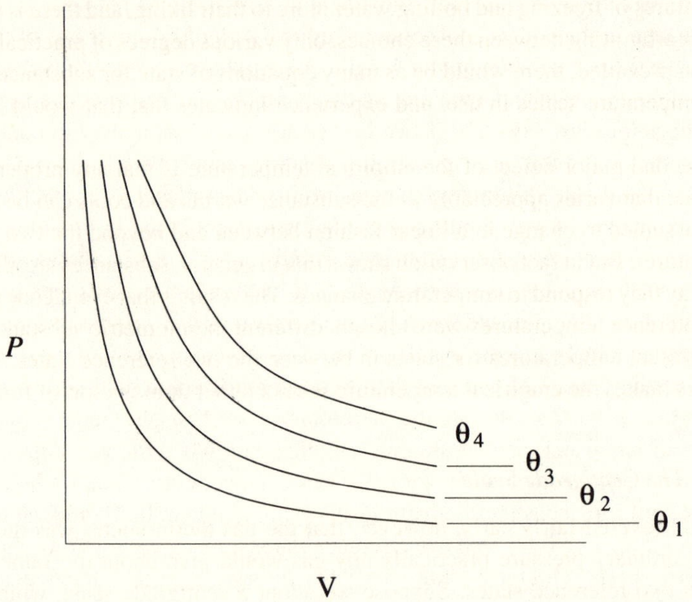

FIG. 4.1. Pressure (P) versus volume (V) for a gas at various fixed values of the empirical temperature.

variables, it is itself a state variable. Since we could carry out this procedure with any gas, or indeed with any substance, then every substance has an equation of state involving P, V, and θ.

It is thus established that temperature is a function of the state of each substance (a state variable) that has the property of taking on the same value for systems in non-adiabatic contact with each other. This is clearly not true of many other state variables including P, V, density, and so on. This underscores the importance of the "diathermal wall," which we said "allowed systems to interact" energetically without allowing mechanical work to be done. That this can be done and that such (heat-conducting) walls exist is a matter of experience.

4.3. TEMPERATURE SCALES

Obviously the empirical temperature scale we have defined has some defects from a practical point of view. The first defect is that although we may feel that Fahrenheit made the only wise choice of reference states in choosing the temperature of a mixture of ice, NH_4Cl, and water and the temperature of the human body, it may happen that someone else will feel that the temperature of the dew at daybreak on the first day of summer and the temperature of the flame of a wax candle would be much better reference states for a temperature scale, and that 666 would be a good number of divisions to have between them. It is even conceivable that others might find the temperatures of freezing and boiling water more to their liking, and there is no logical basis for arbitrating between these choices, only various degrees of practicality. If this situation prevailed, there would be as many equations of state for substances as there were temperature scales in use, and experience indicates that that would be quite a few.

A second major defect of the empirical temperature is that any property of any substance that varies appreciably as the substance warms and cools can be used, and all are assumed to change in a linear fashion between and beyond the two reference temperatures. But in fact observation shows that in general, substances are all different in the way they respond to temperature changes. The result is that even if one particular set of reference temperatures were chosen, different thermometric substances would give different temperatures for states in between the two reference states. This state of affairs makes the empirical temperature seem a rather dubious sort of fundamental property.

4.3.1. *The Centigrade Scale*

It was discovered fairly early, however, that the gas thermometer was quite useful since at ordinary pressure practically any gas would give about the same readings between two reference states. Suppose we adopt a centigrade scale, which strictly speaking simply means that we have 100 divisions between two fixed points. If we have a volume of gas V_0 at the lower reference temperature and a volume V_{100} at the upper reference temperature, then at other temperatures

$$V_\theta = V_0 + \left(\frac{V_{100} - V_0}{100}\right) \cdot \theta$$

where θ is our empirical temperature. Thus

$$\theta = \left(\frac{V_\theta - V_0}{V_{100} - V_0}\right) \cdot 100$$

and we get a value for θ by measuring V_θ when our gas is equilibrated with any particular state at temperature θ. This can also be expressed as

$$\theta = \left(\frac{V_\theta - V_0}{V_0}\right)\left(\frac{V_0}{V_{100} - V_0}\right) \cdot 100$$

The quantity $V_0/(V_{100} - V_0)$ is experimentally determinable, and is found to be 2.73 for most gases when the two reference state temperatures are those of freezing water and boiling water. Then

$$\theta = 273(V_\theta - V_0)/V_0$$

From this it appears that when $\theta = -273$,

$$-273 = 273(V_\theta - V_0)/V_0$$
$$-V_0 = V_\theta - V_0$$

and

$$V_\theta = 0$$

Thus if the linearity of the scale is maintained to very low temperatures, gases equilibrated with an object at −273 degrees on our centigrade gas scale will apparently have zero volume. This must represent an absolute zero on the scale, since we can't imagine negative volumes. In fact we cannot even imagine zero volume for finite quantities of gas (except for ideal gases whose particles occupy zero volume). So it also appears that our absolute zero is not physically attainable.

The suggestion that temperature scales do have an absolute zero which is measurable, although perhaps not physically attainable, would go a long way toward removing the arbitrary nature of our temperature measurements. Fortunately, the truth of these ideas can be established independently of any consideration of individual temperature scales as we have above; Lord Kelvin's demonstration of this will be considered later on.

4.3.2. *Practical and Absolute Temperature Scales*

If freezing water at atmospheric pressure is given a temperature of 0° on our centigrade scale and boiling water 100°, and if we are correct that there is an absolute zero to the degree of coldness attainable at −273 on this scale (−273°C), we could avoid negative numbers by choosing different numbers for the scale, and assigning 0 to absolute zero. This means that water would freeze at 273° and boil at 373°. This scale is called the Kelvin scale, in honor of Lord Kelvin, who first pointed out the existence of the absolute nature of temperatures. We will point out in Chapter 5 that the difference between these scales involves more than just avoiding negative temperatures.

The idea of a temperature scale having two fixed points and an absolute zero exists as an abstract concept, independent of any particular thermometer that might be used in a practical sense. This abstract concept we will call for the moment an "absolute" scale of temperature. As soon as we start to use the scale with real thermometers, we run into problems of calibration, inter-laboratory differences, choice of thermometer, and so on. Therefore in 1927 an international body established the International Temperature Scale, which consisted of temperature values assigned to six reproducible equilibrium states (fixed points) and formulae for calculating temperatures in between the fixed points. The fixed equilibrium states were the ice point, the boiling points of oxygen, water, and sulfur, and the freezing points of silver and gold. The temperatures chosen for these points were the most accurate available, in the sense of conforming most closely to the "absolute" temperature scale, but as techniques improved the values assigned to the fixed points have changed periodically, by international agreement. There is therefore a distinct difference between the abstract "Absolute" Temperature Scale which exists as a kind of absolute truth, and the International Scale, now called the *International Practical Temperature Scale (IPTS)*, which is an ever-evolving approximation of the truth.

Thus thermometers are calibrated according to IPTS, and then give good approximations of the "Absolute" temperature. These temperatures are *expressed* either as °C or as K, depending on whether 0 or 273 (more exactly 273.15) is assigned to the ice point. In 1948, the name of the Centigrade scale was officially changed to the

Celsius scale, retaining the °C symbol, after Anders Celsius, a Swedish astronomer who first used (in 1742) 100 divisions between the freezing and boiling points of water as a temperature scale. The fact that he used 100 for the ice-point and zero for the boiling point was perhaps eccentric, but no more or no less logical or fundamental than the reverse.

The Celsius and Kelvin temperature scales, defined by dividing the temperature interval between freezing and boiling water into 100 divisions (both thus being centigrade scales) had some serious theoretical and practical difficulties. The problem was that even though any two fixed points will serve to define a linear temperature scale, and the freezing and boiling points of water with 100 degrees between them should serve as well as any others, this choice means that the numerical value of the absolute zero temperature in °C, and hence the numerical value of the freezing and boiling points of water themselves in degrees Kelvin were matters of experimental determination. Because of experimental difficulty in precisely reproducing the freezing point of air-saturated water, the temperature of this point was not exactly known, experimental values ranging from about 273.13 to 273.17 K. The boiling point was by definition then exactly 100 degrees higher.

This situation was rectified in 1954 when by international agreement the definition of the temperature scale was changed so that now the triple point of water is fixed by definition at 273.16 on the Kelvin scale. The Kelvin scale now has therefore one "labeled" fixed point instead of two "unlabeled" fixed points with 100 divisions in between. (Alternatively you could think of the new scale as having two fixed points, one at absolute zero and the other at the triple point of water, with 273.16 divisions in between, but since absolute zero is unattainable it is hardly a fixed point in the usual sense.)

Somewhat surprisingly, the triple point of water (the state where ice, water, and water vapor coexist at equilibrium) is quite easily achieved reproducibly in the laboratory, and it is experimentally found to be about 0.01 degrees above the freezing point of water at atmospheric pressure (Figure 4.2). The number 273.16 was therefore chosen so that the freezing point would be 273.15 K, which was the best estimate of that temperature from previous studies, based on the old scale. There was therefore minimum disruption in changing over to the new definition of the Kelvin temperature scale.

The Celsius Scale was defined (1948) as having a temperature of 0.01°C at the triple point of water, so that 0°C is still the temperature of freezing water at one atmosphere pressure. Therefore there are no longer necessarily exactly 100 degrees between the freezing and boiling points of water on either the Celsius or the Kelvin scales, so that neither is now strictly speaking a centigrade scale, and the temperature of boiling water is as much a matter of experiment as any other fixed point. It is of course very close to 100.00°C or 373.15 K, but probably not exactly. The present situation is summarized in Figure 4.3.

In summary, then, the Kelvin temperature scale in use today has one fixed point, the triple point of pure water, assigned a value of 0.01 degrees Celsius and 273.16 degrees Kelvin (more correctly called 273.16 Kelvins). Since there is experimentally 0.01 degree between the triple point and the freezing point of water at atmospheric

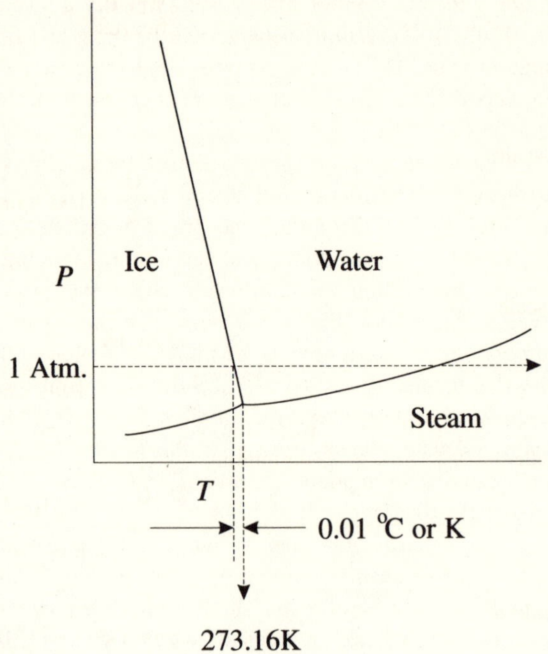

Fig. 4.2. The P - T phase diagram for the system H_20. The temperature of the triple point is defined as 273.16 K.

pressure, the freezing point is 0°C and 273.15 K. Absolute zero is then -273.15°C and 0 K.

Thus the numerical conversion between the two scales is

$$T\ \text{K} = t°\text{C} + 273.15$$

The International Practical Temperature Scale consists of a number of equilibrium states which have been assigned temperatures on the Celsius scale, to facilitate the measurement of temperatures everywhere. Since the number of degrees between the triple point of water and any other equilibrium state, such as the freezing point of molten zinc, is a matter of experimental determination, the temperature values assigned to the chosen fixed points are changed from time to time as techniques improve.

4.3.3. *The Geophysical Laboratory Temperature Scale*

It is of interest to Earth scientists that one of the most prestigious laboratories in the field of experimental petrology and geochemistry, the Geophysical Laboratory of the Carnegie Institution of Washington, has its own practical temperature scale. This was developed in 1914 by Arthur L. Day, based on earlier measurements with a gas thermometer by Day and Sosman, and these same fixed points have been used at the

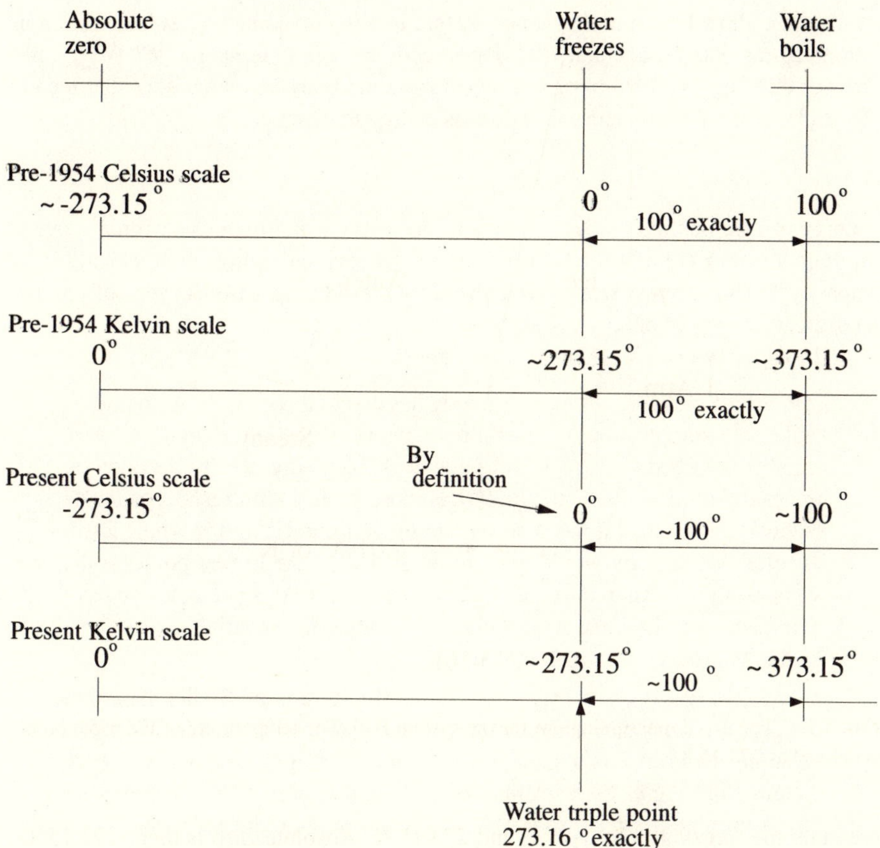

FIG. 4.3. Various temperature scales.

Geophysical Laboratory ever since, in spite of continual changes in the International Practical Temperature Scale (in 1927, 1948, 1969). As a result, experimental results from the Geophysical Lab over the years have a consistent base, and can be easily intercompared. Differences between the Geophysical Laboratory Scale and the IPTS are now all less than 1°C.

4.4. INTERNAL ENERGY

We have attempted to introduce the concept of temperature as a thermodynamic parameter without any necessary reference to the concept of heat, although not necessarily avoiding that term. This was done by saying that diathermal walls exist that allow two bodies to interact energetically while excluding mechanical interaction and mass transfer. Since we limit forms of energy to work and heat (as discussed next), then if energy is transferred and work is eliminated, it follows that diathermal walls must allow heat transfer, whatever that is. Why do some bodies have different temperatures than others, and why does a body change in temperature? We usually answer that heat is added to or subtracted from the body, but what does that mean? To explain this, we had better first discuss energy in general.

4.4.1. *Energy*

Energy can be defined as the capacity to do work, where work is a force moving through a distance, but at first this hardly seems very satisfying. What exactly is "a capacity"? This is most readily visualized by considering various types of energy, and situations where energy transfers occur.

The capacity to do work arises from three sources, giving three types of energy.

1. *Potential Energy*, E_p, is energy a body has due to its position. A brick on a table, for example, could perform work while falling to the floor if it were suitably attached to strings and pulleys. This energy clearly depends on the frame of reference—the brick has more energy with respect to levels below the floor than it has with respect to the floor. Using the whole Earth as reference, a body would have to be at rest on the lowest point on the surface of the earth to have zero potential energy. In this example, we have assumed a gravitational field. Other, or additional, potential energies can be due to electric or magnetic fields.
2. *Kinetic Energy*, E_k, is energy a body has due to its motion. Bodies in motion can perform work during the process of being stopped or slowed down. For example, the brick on the table has zero kinetic energy (again the reference system is important, the brick may have a kinetic energy relative to a moving object, or an extra-terrestrial one) but acquires some during its fall. This kinetic energy does work on the floor by deforming it on impact.
3. *Rest Energy*, E_r, is energy a body has due to its mass. This source of energy was not understood until the introduction of the Special Theory of Relativity by Einstein in 1905. He postulated the essential equivalence of matter and energy, the two being related by

   $$E_r = mc^2$$

   where m is the inertial mass of an object and c the velocity of light. For all objects other than elementary particles, the inertial mass and the ordinary Newtonian rest mass are the same.

Thus the total energy possessed by any macroscopic body or system is

$$E = E_r + E_k + E_p$$

In thermodynamics, we always choose the boundaries of the system to be the reference frame for energy. Thus all systems at equilibrium must have $E_k = E_p = 0$, and any change in energy of the system must be a change in the rest energy, E_r. E_k and E_p above refer of course to the body or system as a whole. Since every macroscopic body is made up of small particles, the rest energy of the body is made up of the total energies of these particles, and each particle itself will have a rest energy, a kinetic energy, and a potential energy. Addition of energy to the system, for example, by heating it or pounding it with a hammer, will result in an increase in the energies of the individual particles, primarily by increasing their velocities. But the overall effect will be to increase the mass of the system, since $E_r = mc^2$, and c is a constant. In other words, not only is there no such thing as an isolated system (since we lack perfectly insulating and rigid walls) but there is no such thing as a closed system, since this was defined as one having a variable energy content but constant mass.

4.4.2. *The Internal Energy*

However, ordinary (i.e., non-nuclear) energy changes result in completely negligible changes in mass, so this is more of a semantic problem than a real one. We will continue to use the concept of a closed system because for almost all our practical and theoretical purposes, the mass is constant. Since we have just one kind of energy for our systems (albeit made up of three kinds on the particle level) we may drop the subscript and simply refer to the energy of the system, or, as is traditional in thermodynamics, the *Internal Energy*. However, in thermodynamics there is no need whatsoever to equate the Internal Energy of a system with its rest energy, E_r. All that is required is that fixed energy levels be attributed to equilibrium states—the numerical value of the energy level is irrelevant—so the Internal Energy is given a different symbol, U (or U, the molar internal energy).

Note that apart from relativity theory, there is no way of computing the total or internal energy of a system, and with relativity theory, one sees that the energy content of macroscopic systems is almost inconceivably large. Although seemingly possible, development of thermodynamics using absolute values of U has never been done since it is not necessary and would add nothing of value to the science, which is quite content to deal only with changes in the energy, as we will see. It is helpful though, to remember that although never given in numerical terms, U in our systems can be thought of as an extraordinarily large quantity, and what is of importance to us are the changes in the last few digits of the number representing it. We repeat however, that this is simply a device to help conceptualize the Internal Energy; U in thermodynamics is in fact not identified with E_r, it is simply an unspecified amount of energy which is fixed in given equilibrium states. The formal relationship is

$$E_r = \mathrm{U} + constant$$

The "constant" is of unspecified size, and is the constant referred to by Bridgman in the quote at the head of this chapter.

Since we do not use absolute values of U or U, we cannot use absolute values of any quantities having U in their equations of definition. This may become a point of some regret if you find yourself puzzling over some unfamiliar standard states later on. Somewhat paradoxically, in spite of being possibly the most fundamental of thermodynamic quantities, Internal Energy or even changes in U are little used in geochemical applications. It is never listed in tables of thermodynamic values, for example, and one rarely needs to calculate ΔU. The reason for this will become apparent as we proceed. It has to do with the fact that we, the users of thermodynamics, have a great predilection for using temperature, pressure, and volume as our principle constraints or measured system parameters. This requires that we use ΔU in slightly modified forms, that is, ΔU modified by what are often relatively small correction factors (such as $P\Delta V$), and these modified forms are given different names and symbols. It is then quite possible to rarely think about ΔU, since it seems only to arise in the development of the First Law. For a better understanding of the subject, however, it is best to realize that in most energy transfers in the real problems that we will be considering, ΔU is by far the largest term involved. Just because we do not usually calculate its value does not mean it is not important (see Chapter 8 for some cases where it *is* calculated).

Since equilibrium states are unchanging in any respect, they evidently have a constant or fixed energy content, and in fact the assertion or supposition that this is the case is at the heart of the First Law of thermodynamics. Before discussing this, let's have a look at how energy may be added to or taken from systems.

4.5. ENERGY TRANSFERS

Systems are capable of different amounts of work at different times, so that evidently energy is capable of entering and leaving systems. We have, of course, implicitly assumed this in our previous discussions and definitions. In classical thermodynamics only two ways of changing energy contents are considered—by adding or subtracting *heat* or *work*.

- *Heat* (q) is the energy that flows across a system boundary in response to a temperature gradient.
- *Work* (w) is the energy that flows across a system boundary in response to a force moving through a distance (such as happens when a system changes volume). [1]

These statements suffice to describe what heat and work are, but do not provide a way of measurement. In fact it has proved quite difficult to provide a rigorous definition of heat in this respect. According to the experts (e.g. Canagaratna, 1969) there is really only one. Heat is that part of any energy transfer that is not accounted for by mechanical work (which has a satisfactory definition: $force \times distance$), and assuming that other forms of energy transfer are negligible. That is,

$$q = \Delta \mathrm{U} - w$$

This turns out to be not very different from the way we actually do think about heat in thermodynamics, as when we subtract a $P\Delta \mathrm{V}$ term from a calorimetry result, and describe the remaining energy term as heat.

[1] Strictly speaking we should introduce q and w for total quantities of heat and work, and q and w for the heat and work per mole of substance affected. As we rarely use these quantities, we follow many other books on the subject in using one symbol for both. Thus in this chapter q and w appear in equations with ΔU, and refer to total heat and work, but in Chapter 8 we calculate values of w in Joules per mole.

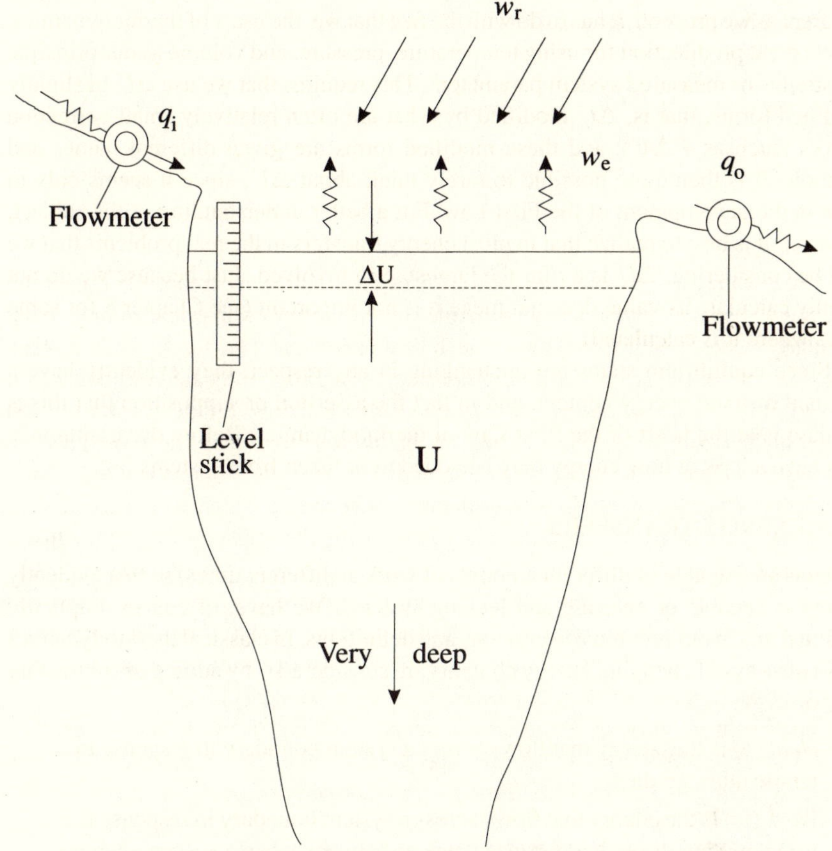

FIG. 4.4. The pond analogy for the First Law.

Heat and work are therefore not separate entities as such, but forms of energy which are transferred in different ways. An enlightening analogy has been offered by Callen (1960): in Figure 4.4 we consider the water in a very deep pond (the amount of water is thus very great but finite and in principle could be exactly measured) to correspond to the internal energy U of a system. Water may be added and subtracted from the pond either in the form of stream water (heat) or precipitation/evaporation (work). Both the inlet and outlet stream water can be monitored by flow gauges, and the precipitation measured by a rain gauge plus knowledge of the surface area of the pond. Evaporation would be trickier to measure, but we may assume either that it is negligible or that we have a suitable measure for it. Now if the volume of stream inlet water over some period of time is q_i and the stream outlet water q_o, the rain w_r and the evaporation w_e, then if there are no other ways of adding or subtracting water, clearly

$$\Delta U = (q_i - q_o) + (w_r - w_e)$$

where ΔU is the change in the amount of water in the pond, which could be monitored by a level indicator as shown. Thus

$$\Delta U = q + w$$

where

$$q = q_i - q_o$$

and

$$w = w_r - w_e$$

Once water has entered the pond it loses its identity as stream- or rain-water. The pond does not contain any identifiable stream-water or rain-water, simply water. Similarly systems do not contain so much heat or work and the terms "heat content" and "heat capacity" are misleading if considered literally. They are hold-overs from the days of the caloric theory, when heat was thought to be a fluid that flowed into cooler bodies from hotter ones. It seems to be intuitive to think of heat as something that systems have, while work is something that systems do, but this is wrong. More than 100 years have passed since Joule demonstrated the equivalence of heat and work, but it seems we often still have difficulty in realizing the implications of this. Just as the water level in the pond can be raised from A to B *either* by stream water alone *or* by rain water alone, Joule showed that a temperature rise in a water bath of so many degrees can be caused *either* by heating (transferring energy due to a temperature difference) *or* by thrashing a paddle wheel about in it (transferring energy by force through distance, i.e., by deformation of the system boundary).

Another implication or assumption in our pond analogy is that water is conserved, i.e., if it disappears somewhere it must reappear somewhere else. The same proposition regarding energy is known as the First Law of thermodynamics.

4.6. THE FIRST LAW OF THERMODYNAMICS

A concise statement of the First Law is *energy is conserved.* This includes systems at equilibrium, which have stable energy contents or levels, but more to the point, it includes energy transfers.

If U is the energy content of a system, and it may gain or lose energy only by the flow of heat (q) or work (w), then clearly, as in the pond analogy, ΔU must be the algebraic sum of q and w. However, we must have some convention as to what direction of energy flow $+q$, $-q$, $+w$ and $-w$ refer to. In the pond analogy we assumed implicitly that addition of water to the pond was positive, whether as stream water or rain water. Thus heat added *to* a system is positive, and work done *on* a system is positive. This convention may be represented as in Figure 4.5a, and is what we call the "scientific" convention—scientists like it because it is internally consistent. It results in the equation previously found,

$$\Delta U = q + w \tag{4.1}$$

Another convention (Figure 4.5b) is to say that heat added to a system is positive, but that work done *by* a system is also positive, or that work done *on* a system is negative.

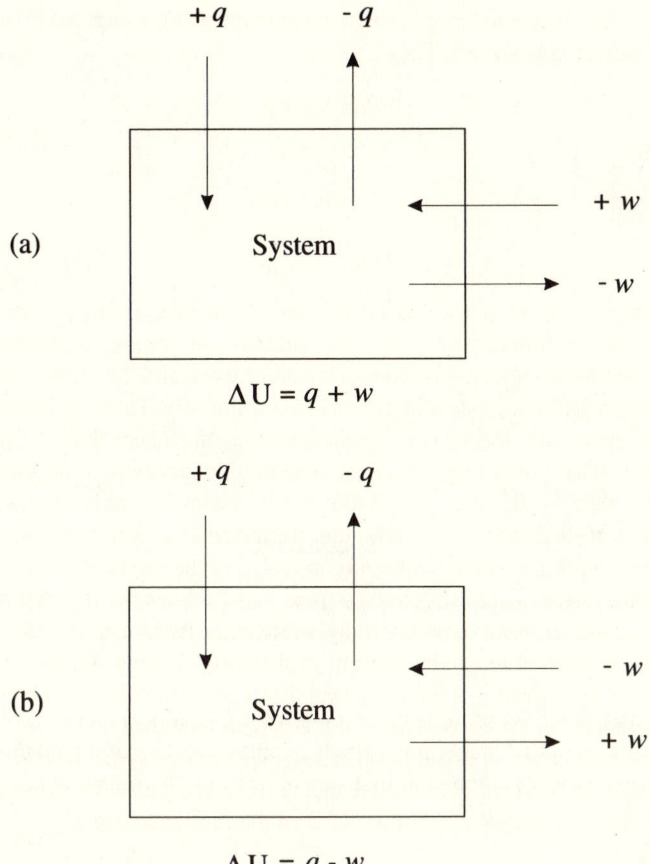

FIG. 4.5. The two commonly used conventions for the sign of q and w, leading to two formulations of the First Law.

This we call the "engineering" convention, because engineers prefer to think in terms of heat engines, and an engine doing work is something positive. This results in the relation

$$\Delta U = q - w$$

Naturally, other conventions are possible (two others, to be exact) but fortunately they are not widely used. In this text we will use the scientific sign convention: any additions of matter and energy to the system are positive in sign and all losses are negative.

Please note that we have not demonstrated the truth of the First Law. It is in fact not capable of "proof." The assertion that energy is always conserved is based on the experience and observations of countless observers over many years, and has become a cornerstone of science, but the fact that it has never been known to fail

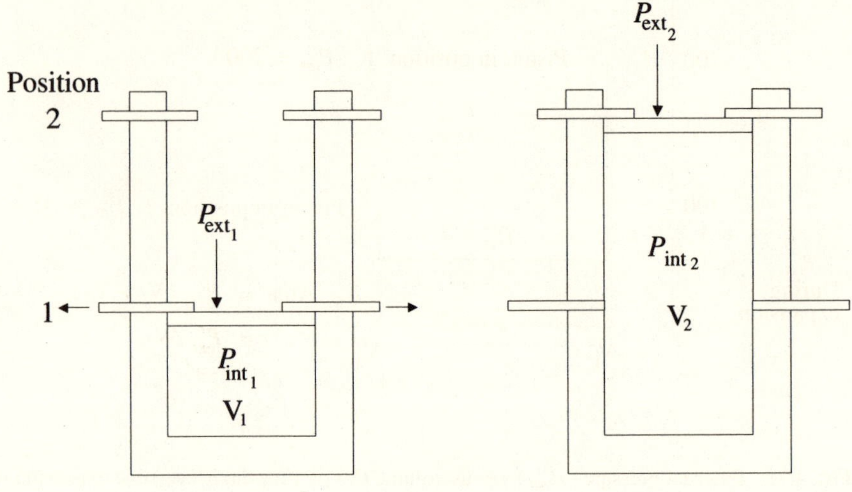

FIG. 4.6. Irreversible expansion of a gas from external pressure P_{ext_1} to P_{ext_2}. During expansion, external pressure is fixed at P' by the weight of the piston.

does not constitute a proof of its correctness. Whether it is in fact a "law" is debated by philosophers, but whatever it is called, an understanding of the relations between energy, heat, and work is the vital first step in learning thermodynamics.

4.6.1. *Work*

We have made the point, demonstrated by Joule long ago, that energy may leave and enter systems either as heat or as work. This is sufficient to demonstrate that q and w are not state variables and hence that dq and dw are not exact differentials. However, it will be best to elaborate on this point somewhat and at the same time increase our understanding of these terms.

There are many configurations for the operation of moving a force through a distance, depending upon whether electric, magnetic or gravity fields, surface tension, and so on are involved. As stated earlier, though, we may consider for now that these sources of work are not present, leaving only the most common sort of work in geological environments, pressure-volume work.

Consider a thermostatted piston-cylinder arrangement as shown in Figure 4.6. The cylinder is fitted with a number of devices which can hold the piston in position at various levels. When the piston is held stationary, the pressure of the gas, P_{int}, is of course exactly balanced by the pressure exerted by the stop devices. Perhaps more exactly, the force exerted by the stops plus the force exerted by the piston mass divided by the area of the piston gives a pressure (P_{ext}) equal and opposite to the gas pressure when the piston is held still. If the stops are removed, then all of a sudden

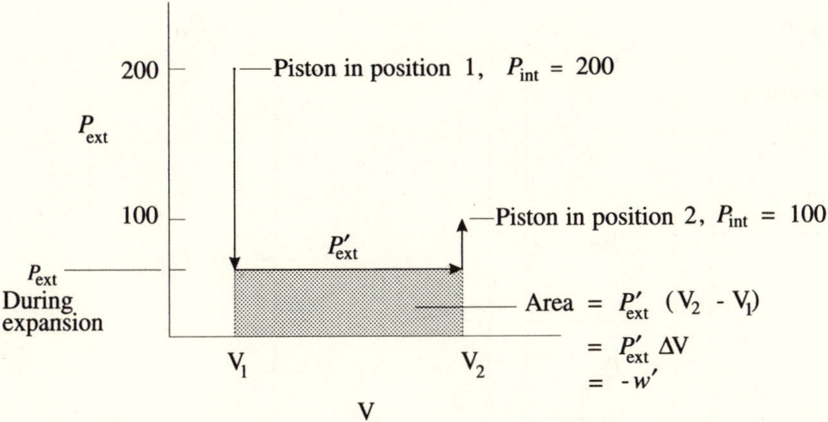

FIG. 4.7. External pressure (P_{ext}) versus volume (V) plot for the irreversible expansion of the gas in Figure 4.6.

P_{ext} is reduced to that produced by the piston mass only, $P_{int} \gg P_{ext}$, and the piston moves up until it encounters more stops—WHAP!—and all of a sudden $P_{int} = P_{ext}$ once more, though at a different (lower) pressure (the experiment has been arranged such that the gas pressure is 100 pressure units at the upper stops, which is position 2, and 200 pressure units at the lower stops, position 1). Real gases tend to cool during expansion, so some heat will flow from the thermostat into the cylinder.

If the piston is well-lubricated and well-constructed, we can ignore friction effects, and the pressure-volume history of the change can be illustrated as in Figure 4.7. The external pressure during expansion is constant, since it is fixed by the mass of the piston. The work done during the expansion is[2]

$$\begin{aligned} w &= force \times distance \\ &= -(P_{ext} \times \mathcal{A}) \cdot \Delta\mathcal{L} \end{aligned}$$

or

$$w = -P_{ext} \cdot \Delta V \tag{4.2}$$

where $\mathcal{A}$ is the area of the piston and $\Delta\mathcal{L}$ the distance it travels, which is seen to be the area under the path of expansion or expansion curve in Figure 4.7. If we repeat the process, but this time we place a weight on the piston (Figure 4.8), exactly the same thing will happen, but as shown in Figure 4.9, more work is done because a greater mass was lifted through the same volume.

Note that the negative sign in equation (4.2) is required to comply with our sign convention—that energy or work added to a system is positive. If a system is

[2]The work done will also include a term ($\frac{1}{2}mv^2$) for the work done in accelerating the piston. If we let the stops be part of the system, this kinetic energy is returned to the system at the upper stops, and can be neglected. [Kivelson and Oppenheim (1966)].

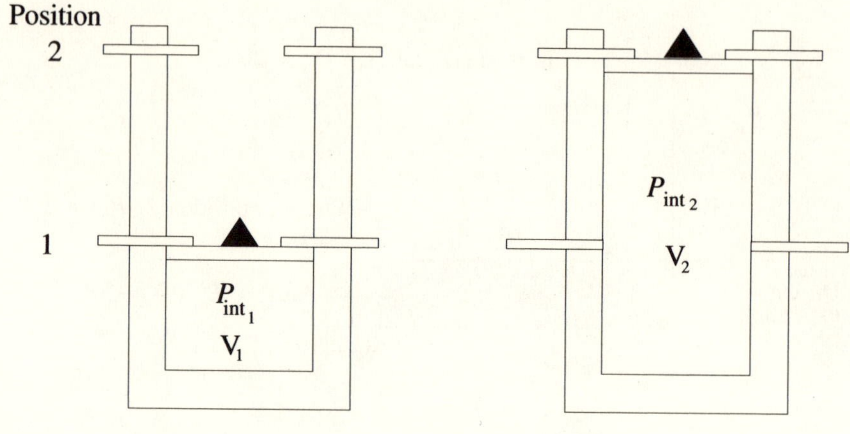

FIG. 4.8. Irreversible expansion of a gas from external pressure P_{ext_1} to P_{ext_2}. During expansion, pressure is fixed at P''.

compressed at constant pressure so that $V_2 < V_1$ and work is done on the system, then the negative sign is required to maintain $w > 0$.

If another weight is added for the next expansion, we may have a total weight that is too great to allow the piston to reach the upper stops (position 2) and it will come to rest (equilibrium) somewhere in between. Then if the second weight is removed, the piston will proceed upward again as before, giving an expansion path as shown in Figure 4.10. Then if we use a lot of weights and remove them one at a time, letting the piston come to rest after each step, we will get a path such as shown in Figure 4.11.

Clearly we are approaching a limit of maximum work obtainable from the expansion of our gas, and clearly too, the maximum will be when we take an infinite number of infinitesimally small incremental steps from V_1 to V_2.

Since we have been letting the piston come to rest or equilibrium after every weight removal, in the limit we will have an infinite number or continuous succession of equilibrium states, giving us another example of a reversible process. In this particular case the name "reversible" is particularly appropriate since at any stage in the expansion the direction of movement can be reversed by changing the external pressure infinitesimally.

In the limit when infinitesimal increments of V are taken, the work of expansion is (see Figure 4.12)

$$w_{rev} = w_{max} = -\int_{V_1}^{V_2} P\,dV \tag{4.3}$$

Here we need make no distinction between P_{ext} and P_{int} because they are never

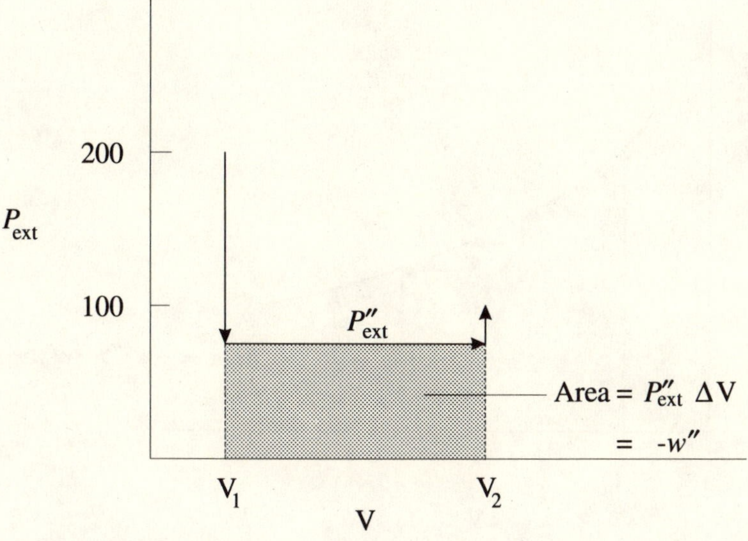

FIG. 4.9. External pressure (P_{ext}) versus volume (V) plot for the process in Figure 4.8.

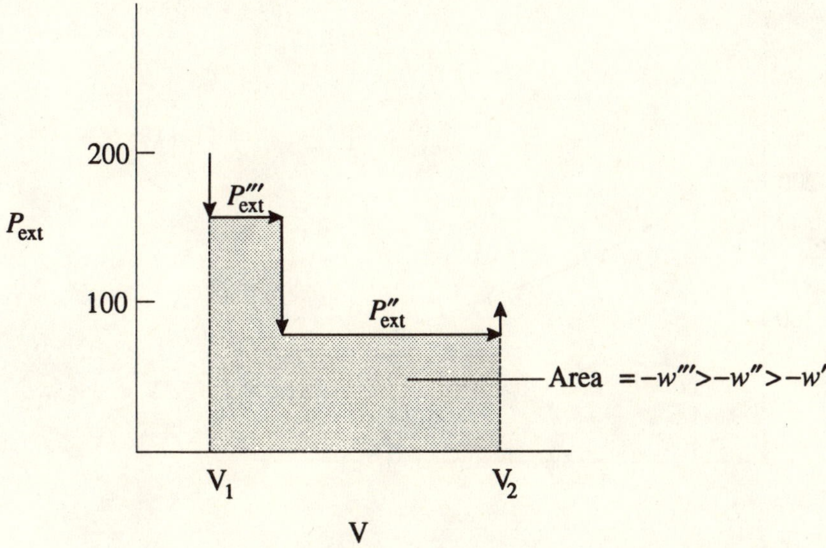

FIG. 4.10. External pressure (P_{ext}) versus volume (V) for a two-stage expansion of gas. After an initial expansion at $P_{ext}^{'''}$, some weight was removed from the piston and the expansion continued at $P_{ext}^{''}$.

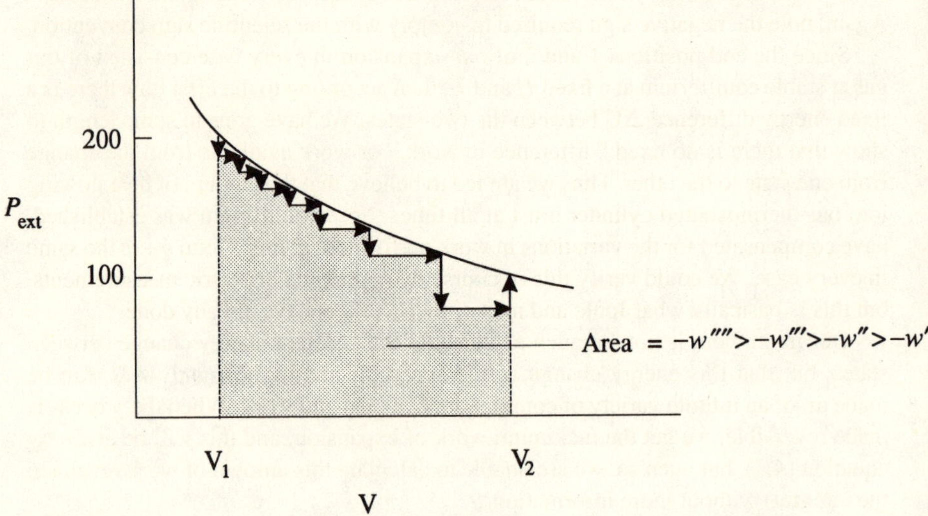

FIG. 4.11. External pressure (P_{ext}) versus volume (V) for a multi-stage expansion of gas. After each constant ($-P_{ext}$) expansion, some weight was removed, allowing a further expansion.

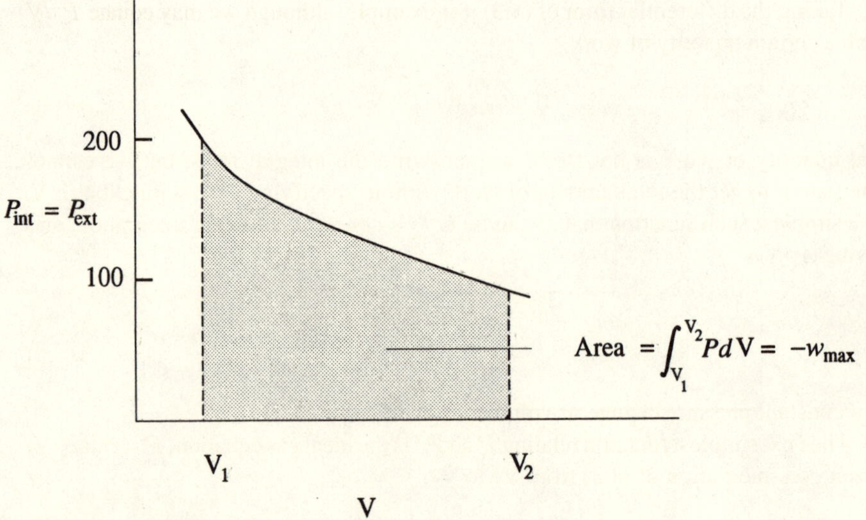

FIG. 4.12. Pressure versus volume for the reversible expansion of a gas. The limiting case where an infinite number of constant P_{ext} steps are taken, giving the maximum area under the curve. During the expansion, internal pressure and external pressure are never more than infinitesimally different, or $P_{int} = P_{ext}$ at all times.

more than infinitesimally different in our continuous succession of equilibrium states. Again, note the negative sign required to comply with the scientific sign convention.

Since the end positions 1 and 2 of our expansion in every case consisted of our gas at stable equilibrium at a fixed P and T, then according to the First Law there is a fixed energy difference ΔU between the two states. We have gone to some length to show that there is no fixed "difference in work," or work available from the change from one state to the other. Thus we are led to believe that the amount of heat flowing into our thermostatted cylinder must at all times, once equilibrium was established, have compensated for the variations in work performed, giving a total $q + w$ the same in every case. We could verify this of course by making calorimetric measurements, but this is basically what Joule and many other workers have already done.

Our intent here is not so much to illustrate the constant energy change between states, but that this energy change, while accomplished by heat and work, can be made up of an infinite variety of combinations of heat and work. When the process is made reversible, we get the maximum work of expansion, and this will be given by equation (4.3), but even so, we are unable to calculate this amount of work (evaluate the integral) without more information.

This is because $P\,d\mathrm{V}$ is not an exact differential.[3] We pointed this out in Chapter 2, where we spoke of differentials in the form $x\,dy$ not being exact. We showed (Figure 2.7) that another way of saying the same thing is to say that the path of integration of $x\,dy$ in x–y space is not fixed, there being no functional relation given between x and y. Exactly the same comments apply to the physical parameters P and V.

Taking the differential form of (4.3), for example, although we may equate $P\,d\mathrm{V}$ with a certain quantity of work,

$$Dw_{rev} = -P\,d\mathrm{V} \tag{4.4}$$

that quantity of work is not fixed; we can write the integral (4.3), but we cannot integrate it to get the total amount of work without specifying P as a function of V. The simplest such function that P can be is $P = constant$, allowing integration, and giving

$$\begin{aligned} w &= -P(V_2 - V_1) \\ &= -P\,\Delta\mathrm{V} \end{aligned} \tag{4.5}$$

in a constant pressure change of volume, as in equation (4.2).

The next simplest function relating P and V is the ideal gas equation, $P = nRT/\mathrm{V}$, giving, on integration of (4.3) from V_1 to V_2,

$$w_{rev} = -nRT\,\ln\frac{\mathrm{V}_2}{\mathrm{V}_1} \tag{4.6}$$

for the work performed during an *isothermal* (note T was held constant during integration), reversible volume change of an ideal gas. This equation would give the

[3]Note, however, that P is an integrating denominator for Dw_{rev}.

work done in our example (Figure 4.12) *provided* that the gas in the cylinder was an ideal gas. Real gases have different functional relationships between P and V, often quite complicated, and this function would have to be known for the particular gas used to evaluate (4.3).

Another thing worth emphasizing is that in any real or non-reversible expansion, as shown in our example, the work obtained is less than the maximum obtainable (from a reversible expansion). Thus in general, rewriting (4.3),

$$w \leq -\int_{V_1}^{V_2} P\,dV \tag{4.7}$$

where the $<$ part of the $\leq$ sign refers to any irreversible change in V. Thus for an *irreversible* expansion from V_1 to V_2, (4.6) becomes

$$w < -nRT \ln \frac{V_2}{V_1} \tag{4.8}$$

This can also be expressed as

$$w \leq w_{max}$$

or

$$w \leq w_{rev}$$

For the opposite case of compressing the gas from 2 back to 1, the inverse series of steps can be employed. Thus, if at position 2 a heavy weight is placed on the piston, it will whap down to the stops at 1 describing a path such as in Figure 4.13. Obviously, much more work has had to be done in compressing the gas than we obtained, even in the reversible case, from expansion. However, by adding a larger number of smaller weights one at a time we can reduce the amount of work required for the compression, gradually approaching the stable equilibrium curve from above, rather than from below as before. In the limit, of course, we find that for a reversible compression the work required is exactly the same as the work available from a reversible expansion.

4.6.2. *Heat*

It might be expected that since

$$\Delta U = q + w$$

and

$$w \leq -\int_{V_1}^{V_2} P\,dV$$

perhaps there is a very similar story for the heat transfers in the gas expansion cases we have been considering. That is, perhaps

$$q \leq -\int_{Z_1}^{Z_2} T\,dZ \tag{4.9}$$

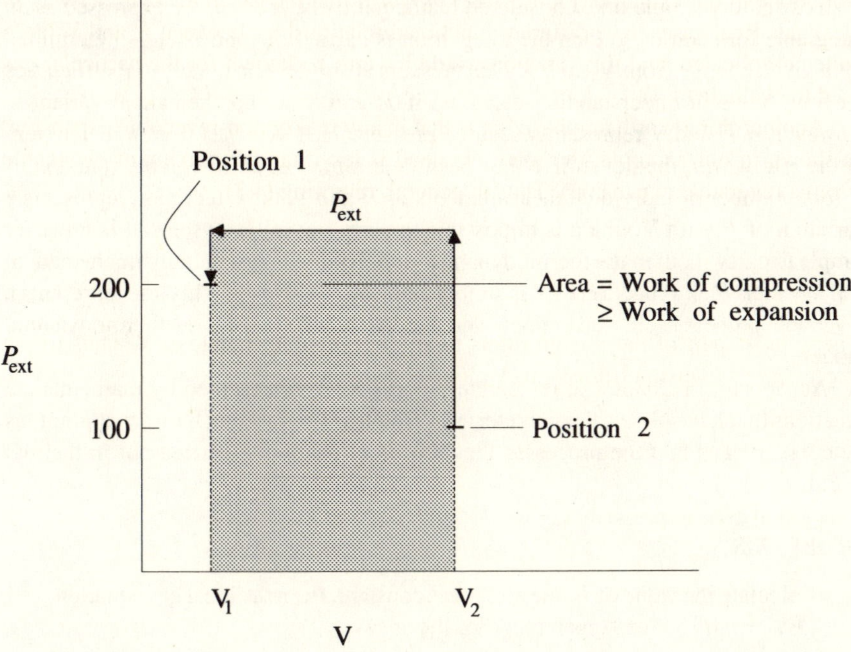

FIG. 4.13. External pressure (P_{ext}) versus volume for the irreversible compression of gas at constant P_{ext}.

where Z is some property of the gas. Then just as P is an integrating denominator for w_{rev}, T would be an integrating denominator for q_{rev}. This is indeed the case (except for a sign change), but we must await the development of the Second Law, which will introduce us to entropy ($-$Z in the above equation). We can then attempt a discussion of the relation between the flow of heat in a calorimeter and the entropy, somewhat analogous to the discussion above on the relation between work and volume changes (§5.2.3).

4.7. THE MODEL AGAIN

In this chapter we have discussed some very practical operations. There is nothing particularly theoretical about gases expanding in cylinders and performing work. It happens countless times every day all over the world. Equations such as (4.2) belong to the real world. However, the result of the limit-taking, when the number of expansions or compressions in a single cycle is increased without limit, is a reversible process that belongs not to the real world but to the thermodynamic model.

The equation

$$w_{rev} = -\int_{V_1}^{V_2} P\,dV \tag{4.3}$$

is an extremely simple one, considered mathematically. If P can be expressed as an integrable function of V, then the integration is carried out and w_{rev} is determined for a given change from V_1 to V_2. This presents absolutely no conceptual difficulties (beyond those in understanding calculus) if P and V are mathematical variables. However, if P and V represent measured pressures and volumes from a real system in the real world, then even if P has been determined as an integrable function of V for a number of individual measurements of P and V, the integration represents a variation of P with V which is impossible to carry out in the system. It is however simple to carry it out in the thermodynamic model, which is essentially mathematical and in which P as a function of V is simply a line in P - V space. This line represents a reversible process, a perfectly simple and understandable facet of the thermodynamic model.

As we said in Chapter 3, reversible processes are represented by mathematical functions that have physical parameters as variables. The fact that they are continuous functions means that the processes they represent cannot be carried out in the real world.

PROBLEMS

1. Calculate the value of R, the molar gas constant, from the ideal gas equation ($PV = nRT$). See Appendix A for the answer.
2. Show that $\mathrm{J\,bar^{-1}}$ is a volume term, and calculate the conversion factor from $\mathrm{J\,bar^{-1}}$ to $\mathrm{cm^3}$. See Appendix A.
3. Verify that $V_0/(V_{100} - V_0)$ is about 2.73 (actually 2.7315) (§4.3.1), using the ideal gas law.

5

THE SECOND LAW OF THERMODYNAMICS

Arguments based on heat engines have little appeal to chemists.

McGlashan (1979), p. 111.

5.1. FALLING BRICKS

We have seen that the first great principle of energy transfers is that energy is conserved. The essence of the second principle or law has to do with energy availability, or with the "directionality" of these transfers (processes). In other words, it is observed that once the constraints on the beginning and ending states are decided upon, processes can only proceed spontaneously in one direction between these states and are never observed to proceed in the other direction unless they are "pushed" with an external energy source.

For example, a brick can fall off a table onto the floor. The potential energy it has on the table is converted to kinetic energy and then to a certain amount of heat and mechanical deformation (work) upon impact. According to the first law, the energy expended on impact will exactly equal the energy the brick had on the table. Bricks have never been observed to spontaneously cool themselves, convert this heat energy into kinetic energy, and fly from the floor to the table. Such events could exactly satisfy the first law, which clearly has nothing to say about why they don't happen—just that if they did happen, energy would be conserved. It would obviously be very useful to have a method of predicting which way a given process would go under given conditions. It would open the way towards systematizing chemical and mineral reactions, for one thing. We could start to predict which minerals would form under given metamorphic conditions, for example, and perhaps even predict their compositions, because all such changes are simply processes that can be considered to go from one equilibrium state to another.

Possibly the greatest single step forward in the history of the development of thermodynamics was the recognition and definition of a parameter, the entropy, which enables such predictions and systematizations to be made.

5.2. THE KELVIN AND CLAUSIUS FORMULATIONS

The traditional way to approach the subject is to state the Second Law as it has been deduced on the basis of years of experience, and then show through use of the Carnot cycle the logical consequences (such as the existence of the entropy and an absolute scale of temperature). Two logically equivalent ways of stating the Second Law are

- The Kelvin formulation:
  It is impossible to construct an engine that, working in cycles, shall produce no effect other than the extraction of heat from a reservoir and the production of work.

For example, you cannot have an engine that will extract heat from the sea and use that energy to power a ship across the ocean, leaving a trail of icebergs in its wake.

- The Clausius formulation:
  It is impossible to construct an engine that, working in cycles, shall produce no effect other than the transfer of heat from a colder to a hotter body.
  For example, you cannot even heat the cabins on the ship by extracting heat from the sea. This slightly simpler formulation eliminates considerations of heat-to-work transfers.

We have regretfully eliminated at this point a discussion of Carnot cycles (a combination of reversible adiabatic and isothermal expansions and compressions of a gas or any other "working substance" arranged in cycles and producing work), which leads to the result that

$$q_1/q_2 = -T_1/T_2$$

for any heat engine operating reversibly, absorbing an amount of heat q_1 at temperature T_1, losing an amount of heat q_2 at a lower temperature T_2, and doing some work in between. Two fundamental conclusions can be deduced from this equation. Neither is immediately obvious, unless you are already steeped in the subject.

5.2.1. *Absolute Temperature*

The first conclusion is that a *Thermodynamic temperature scale* exists which has fixed ratios of temperature between any two equilibrium states. Fixing the temperature of any one equilibrium state then fixes the temperature of all others. This follows from the fact that q_1/q_2 is a fixed number for any two equilibrium states, being independent of the size, shape, or working substance of the (hypothetical) Carnot engine used. These conclusions follow from the derivations we have omitted.

All that need be done is to measure q_1 and q_2 for a Carnot cycle operating between any two equilibrium states having different temperatures, then arbitrarily choose a number for the temperature of one of these states. The temperature of the other state will then be established, and these two temperatures establish a linear scale according to which all other equilibrium states can be measured with a suitable thermometer.

Unfortunately, real Carnot cycles do not exist, so we cannot measure q_1 and q_2, although approximate values could be determined by expending great efforts. Fortunately, we don't have to do this, because the ratio V_1/V_2 or P_1/P_2 of an ideal gas at two equilibrium states 1 and 2 is also equal to T_1/T_2, and hence to q_1/q_2. Ideal gases don't exist either, but the ratio (V_1/V_2 or P_1/P_2) can be measured for real gases at various pressures and the value for $P = 0$ (where it will be equal to that of an ideal gas) determined by extrapolation. This is the way the ratio q_1/q_2 or T_1/T_2 is determined.

When the two equilibrium states are the freezing and boiling points of water, this ratio is about 1.3661, a perfectly fixed, if imperfectly known number. If the temperature of the ice point is called 1000°A (degrees Anderson), the steam point is then 1336.1°A, and this is a valid thermodynamic temperature scale. However not many scientists use °A, perhaps because it is felt that for historical reasons it would be much better to have exactly 100 divisions between the ice and steam points. Thus with

$$T_1/T_2 = 1.3661$$

and

$$T_1 - T_2 = 100$$

it turns out that $T_1 = 373.15$ and $T_2 = 273.15$. Since the triple point of water is 0.01 degrees above the ice point, 273.16 was therefore chosen as its temperature, fixing the Kelvin scale, as discussed in the last chapter.

Clearly, any number of thermodynamic scales exist, but one must be chosen, and the Kelvin scale is that one. Equally clearly, any number of empirical scales exist that are *not* thermodynamic scales, since for them $q_1/q_2 \neq T_1/T_2$. For example, the Celsius scale is not a thermodynamic scale since 100/0 is not even close to 1.3661.

5.2.2. *The Entropy*

The second fundamental conclusion following from

$$q_1/q_2 = -T_1/T_2$$

is that

$$\frac{q_1}{T_1} + \frac{q_2}{T_2} = 0$$

for a single reversible cycle, or

$$\sum_i \frac{q_i}{T_i} = 0 \tag{5.1}$$

for a number (i) of linked reversible cycles.

This is sufficient to suggest the existence of a state variable equal to $(\frac{q}{T})_{rev}$, since this function is conserved in heat engine cycles carried out reversibly. This new state variable is of course the entropy, S, where

$$\Delta \mathrm{S} = \frac{q_{rev}}{T}$$

for a single reversible process represented by the Δ, or more generally,

$$d\mathrm{S} = \frac{Dq_{rev}}{T} \tag{5.2}$$

and

$$\mathrm{S}_B - \mathrm{S}_A = \int_A^B \frac{Dq_{rev}}{T} \tag{5.3}$$

Then, after invoking one of the formulations of the Second Law, it can be shown that for irreversible processes

$$\Delta \mathrm{S} > \int \frac{Dq_{irrev}}{T}$$

and in general

$$\Delta S \geq \int \frac{Dq}{T} \tag{5.4}$$

Equation (5.4) can also be written for summations of steps, which can of course be made as small as we like. Thus (5.4) becomes

$$\Delta S \geq \sum_i \frac{q_i}{T_i} \tag{5.5}$$

For a single step at temperature T, this is essentially the same, except for a sign change, as equation (4.9) in §4.6.2.

This commonly cited result (5.4 or 5.5) is one of the many ways of looking at entropy, and not a very easily understood way in view of the unfamiliar nature of reversible heat engines. It does follow though from (5.4) or (5.5) that if $q = 0$ (i.e. for adiabatic processes)

$$\Delta S_{q=0} \geq 0 \tag{5.6}$$

and here we seem to have something much more useful. According to our derivation, the "equals" sign applies to reversible processes and the "greater than" sign to irreversible or spontaneous processes. Thus, under adiabatic conditions ($q = 0$), S can apparently be used to tell in which direction a given process will be spontaneous ($A \rightarrow B$ or $B \rightarrow A$). $A \rightarrow B$ will be spontaneous and $B \rightarrow A$ impossible if $S_B - S_A > 0$, or $S_B > S_A$, and vice versa. Equation (5.6) is often referred to as the *Entropy Law*. Thus we have defined a state variable that performs the function we were looking for—it gives us the "directionality" of processes. According to the equations we have so far, though, we have no way of measuring it apart from reversible processes, and it is only really useful in adiabatic processes, so we apparently have some way to go before it is a very practical quantity.

5.2.3. *Examples of Entropy Change*

To make expression (5.5) more understandable, let's consider the freezing of water and melting of ice near 0°C. When one gram of water freezes spontaneously (irreversibly) at 272 K, about 333.5 Joules of heat are liberated ($q = -333.5$ J), or according to (5.5),

$$\begin{aligned}\Delta S_{water\rightarrow ice} &= S_{ice} - S_{water} \\ &> \frac{-333.5}{272} = -1.23 \text{ J K}^{-1} \text{ g}^{-1}\end{aligned}$$

or

$$(S_{water} - S_{ice})_{272} < 1.23 \text{ J K}^{-1} \text{ g}^{-1}$$

and when one gram of ice melts spontaneously at 274 K, about 333.5 joules are absorbed ($q = 333.5$ J), or

$$\Delta S_{ice\rightarrow water} > \frac{333.5}{274} = 1.22 \text{ J K}^{-1} \text{ g}^{-1}$$

or

$$(S_{water} - S_{ice})_{274} > 1.22 \text{ J K}^{-1} \text{ g}^{-1}$$

At the equilibrium temperature of 273.15 K no calorimetric measurement is possible, but the equilibrium ΔS has been determined to within 1% by two measurements of irreversible processes.

As another example, consider the problem of determining the change in entropy of a substance X between 300 K and 350 K at one bar. We will suppose that the heat capacity of X is constant at exactly 1 J K^{-1} g^{-1}, so that 50 J are required to heat one gram of X from 300 to 350 K, 25 J to heat it from 300 to 325 K, and so on. It follows that on heating one gram of X from 300 to 350 K in a thermostat (of essentially infinite heat capacity) at 350 K,

$$S_{350} - S_{300} > \frac{50}{350}$$

and on cooling back to 300 K in a similar thermostat at 300 K

$$S_{300} - S_{350} > \frac{-50}{300}$$

from which we conclude that

$$\frac{50}{300} > S_{350} - S_{300} > \frac{50}{350}$$

or

$$0.167 > S_{350} - S_{300} > 0.143 \text{ JK}^{-1}\text{g}^{-1}$$

which means we have determined the ΔS to about one part in six with two measurements. To improve our precision, we could double the number of measurements, and heat and cool in two stages each. That is, because

$$S_{350} - S_{300} = (S_{350} - S_{325}) + (S_{325} - S_{300})$$

then

$$\frac{25}{325} + \frac{25}{300} > S_{350} - S_{300} > \frac{25}{350} + \frac{25}{325}$$

or

$$0.160 > S_{350} - S_{300} > 0.148 \text{ JK}^{-1}\text{g}^{-1}$$

Given enough patience, we could make 50 measurements at one degree intervals, in which case

$$\frac{1}{349} + \frac{1}{348} + \cdots + \frac{1}{301} + \frac{1}{300} > S_{350} - S_{300} > \frac{1}{301} + \frac{1}{302} + \cdots + \frac{1}{349} + \frac{1}{350}$$

or

$$0.1543890 > S_{350} - S_{300} > 0.1539128 \text{ JK}^{-1}\text{g}^{-1}$$

Clearly we are approximating an integral, which in a later chapter we will find is

$$\Delta S = \int \frac{C_p}{T} dT$$

which in this case becomes

$$\begin{aligned}\Delta S &= S_{350} - S_{300} \\ &= 1.0 \times \ln(350/300) \\ &= 0.1541507\ \mathrm{JK^{-1}g^{-1}}\end{aligned}$$

This type of calculation is similar to the one we did for work using the piston-cylinder arrangement, in the sense that we approach the reversible process by taking more and more steps. We reiterate that the physical operation implied by the integral $\int (C_p/T)\,dT$ is a reversible process, which is impossible. This doesn't bother us, however, because the integration involves surfaces in the thermodynamic model, not physical reality. It is helpful though to see, as above, what sequence of physically real measurements could lead to the same result.

It has now been demonstrated how considerations of work and heat changes in cyclical processes combined with a statement of the Second Law lead to the development of a quantity that can distinguish possible from impossible processes under certain conditions. The next step is to combine this with a statement of the First Law to get our first look at a "fundamental equation".

5.3. THE FUNDAMENTAL EQUATION

To begin the development that leads to making S more useful, consider again equations (5.2) and (5.3):

$$d\mathrm{S} = \frac{Dq_{rev}}{T} \tag{5.2}$$

and

$$\mathrm{S}_B - \mathrm{S}_A = \int_A^B \frac{Dq_{rev}}{T} \tag{5.3}$$

or

$$Td\mathrm{S} = Dq_{rev} \tag{5.7}$$

and at constant T,

$$T\,\Delta\mathrm{S} = q_{rev} \tag{5.8}$$

In Chapter 4 we saw that

$$Dw_{rev} = -Pd\mathrm{V} \tag{4.4}$$

or, at constant P,

$$w = -P\Delta\mathrm{V} \tag{4.5}$$

Combining these equations ((5.8) and (4.5)) with the First Law

$$\Delta\mathrm{U} = q + w$$

we have

$$\Delta\mathrm{U}_{T,P} = T\,\Delta\mathrm{S} - P\,\Delta\mathrm{V} \tag{5.9}$$

or from

$$d\mathrm{U} = Dq + Dw$$

we have

$$d\mathrm{U} = T\,d\mathrm{S} - P\,d\mathrm{V} \tag{5.10}$$

or its molar form

$$dU = T\,dS - P\,dV \tag{5.11}$$

This is probably the single most important equation in Thermodynamics, and for this reason is called the *fundamental equation.*[1] It is a truly remarkable equation, the understanding of which is the central theme of this book. This understanding is partly on a theoretical, conceptual level and partly on a more practical level, in the sense that a great deal of data collection and manipulation of equations is aimed directly at allowing the integration of this equation.

At this point we will simply outline some relationships of immediate use, leaving more detailed discussion of this equation for later chapters. From (5.10) and (5.11) we see that

$$(\partial \mathrm{U}/\partial \mathrm{S})_{\mathrm{V}} = T \tag{5.12}$$

and

$$(\partial \mathrm{U}/\partial \mathrm{V})_{\mathrm{S}} = -P \tag{5.13}$$

or

$$(\partial U/\partial S)_V = T \tag{5.14}$$

and

$$(\partial U/\partial V)_S = -P \tag{5.15}$$

And finally, recall that we have shown that for adiabatic ($q = 0$) processes,

$$\Delta \mathrm{S}_{q=0} \geq 0 \tag{5.6}$$

We placed no restriction on whether or not work was done during this process, so we derived this relation for a closed, perfectly insulated system but not necessarily an isolated system. Isolated systems have adiabatically insulating walls as well, so evidently (5.6) applies in isolated systems as well as closed, adiabatic ones.

Thus

$$\Delta \mathrm{S}_{\mathrm{U,V}} \geq 0 \tag{5.16}$$

The subscripts U,V mean that the process that ΔS refers to takes place at constant energy content and constant volume, which means that no heat or mechanical work flows in or out; i.e., the system is isolated. Equation (5.16), which is a product of deduction from the Second Law in this chapter, is used as a postulated beginning point for deductions in Section 5.4.

[1]It is actually one of a number of equally fundamental equations, to be more fully discussed in Chapter 14.

The Meaning of ΔS

We pause here to point out once again the presence of the concept of metastable systems in what we have just written. Equation (5.16) refers to a process taking place adiabatically between an initial state and a final state. For S to be defined in both states, both states must be equilibrium states. If the initial state was a *stable* equilibrium state at a given U and V, no spontaneous change could possibly take place. Therefore, the initial state must be in our terms a *metastable* equilibrium state, having at least one constraint in addition to U and V. When this constraint is released, the system spontaneously changes to the final equilibrium state, which has the same U and V as the initial state, but no added constraint. An example might be some water and NaCl held separately in an isolated system. The fact that they are separated constitutes the added constraint. On releasing this constraint[2] the salt dissolves (resulting in some change in P and T), and the system reaches stable equilibrium at the same U and V.

5.3.1. *The Sign of Entropy*

Although the recognition by Clausius and others that the quantity $\int Dq/T$ was uniquely important was an all-important step in the development of thermodynamics, it is a pity, in retrospect, that it was used to define a state variable in the sense of equation (5.2). That is, if equation (5.2) were

$$dS = -\frac{Dq_{rev}}{T}$$

instead of

$$dS = \frac{Dq_{rev}}{T} \tag{5.2}$$

$$\left(\text{just as} \qquad dV = -\frac{Dw_{rev}}{P}\right)$$

a small blight on the symmetry of thermodynamics would be removed. This will appear in the next section, when we will see that the entropy is just one of a number of state variables that can be used under given conditions to tell which way processes will proceed spontaneously, but they all have the opposite sign to entropy because of equation (5.2). That this is the case is thus a sort of "historical accident."

5.4. THERMODYNAMIC POTENTIALS

Consideration of idealized heat engines (largely omitted), combined with a statement based on experience (the Second Law), allows us to define a parameter, the entropy, which has the useful property of always increasing in adiabatic spontaneous processes. We would like to have similar parameters for other kinds of processes, i.e., to have

[2]Releasing the constraint would be a neat trick if the system was truly isolated. However, all we are really considering is a comparison between the initial and final states of the system, which have the same U and V. What happens in between doesn't matter. In other words, the isolation could be broken, the constraint removed, and isolation restored at the same U and V.

directionality parameters for processes that are not necessarily adiabatic. There are several of these, and there are many ways to derive them, once the "entropy law" (equation (5.6) or (5.16)) has been established. In this section we derive them using the Legendre Transform, introduced in Chapter 2, not only because of the elegance and simplicity of the procedure, but because it helps to show that all the directionality parameters are alike—they are like the various facets of a single cut stone. We also wish to show in what sense they are potential quantities, i.e., why they are called thermodynamic potentials.

5.4.1. *Duhem's Law*

We should, however, introduce two simple concepts to smooth the way. The first of these is the fact that *a system of fixed composition will have its stable equilibrium state determined by fixing a maximum of two state variables*. In a simple mechanical system such as illustrated in Figure 3.1, there is only one way of changing the (potential) energy of the ball (by changing its elevation), so only one variable (elevation) need be fixed to fix the energy state of the system. In thermodynamic systems, there are two ways of changing the energy level (work and heat), so two variables must be fixed. Stable equilibrium is stipulated, because, as we have seen, metastable states have extra constraint variables.

This is referred to as Duhem's law or principle (and incidentally is the origin of the "2" in the phase rule). For example, all of the properties of a system at equilibrium will be fixed at a stated temperature and pressure, or temperature and volume, etc. If the system is capable of changing composition, then of course there are additional compositional variables. The law refers to systems of fixed composition and only the usual thermodynamic variables are included (ignoring electrical and magnetic fields, etc.). The word "maximum" occurs to take care of those cases where, due to the existence of more than one phase, the system is univariant or invariant, as we will discuss in connection with the phase rule.

5.4.2. *Potentials*

The second concept we should be familiar with is that of potentials, or potential quantities. It will be best first to introduce the concept in the sense in which it is used in mechanics, and then see how it can be generalized and used in thermodynamics. Potentials are also used in other areas, such as those involving force-fields (electrostatic, magnetic, etc.), hydrodynamics, and so on, but we will consider only mechanics. In mechanics, a body's potential is identified with its potential energy. A body has a potential or potential energy because a force is acting on it that is capable of causing the body to move. The potential energy is a function of the position of the body, its mass and the force acting on the body.

The force is generally a vector quantity, but if we restrict ourselves to one dimension, the functional relationship between the potential energy of a body E_p, its position r, and the force $\mathcal{F}$ is

$$d\mathrm{E}_p/dr = -\mathcal{F} \tag{5.17}$$

Since the potential E_p is a result of the force $\mathcal{F}$ and will decrease if the body is allowed to move, the terms are given opposite signs. Thus one definition of force could be that it is (the negative of) a space rate of change of energy, and the functional form in equation (5.17) is common to all potential quantities. Any change in the potential will appear as work, either done on the body to increase the potential, or by the body in lowering its potential. Thus

$$\begin{aligned} w &= \int_{r_1}^{r_2} dE_p \\ &= E_{p,r_2} - E_{p,r_1} \end{aligned}$$

where r_1 and r_2 are two positions of the body, and work is considered negative when done by the body.

In the case of a brick of mass m on a table of height h above the floor, the potential E_p is thus

$$dE_p/dh = -mg$$

where g is the gravitational constant, mg the total force on the brick, and h is measured from the brick to the floor. If g is considered constant

$$\int dE_p = -\int mg \cdot dh$$

and

$$E_p = -mgh + constant \tag{5.18}$$

or

$$\Delta E_p = -mg\Delta h$$

If we have a very high table, we may need to consider changes in g as the brick falls.

Simply from the form of equation (5.1) and recalling Chapter 4, one might at this point wonder whether the internal energy, U, is some kind of potential, since

$$(\partial U/\partial S)_V = T \tag{5.12}$$

and

$$(\partial U/\partial V)_S = -P \tag{5.13}$$

so that if S and V could be considered analogous to distances, T and P would be analogous to forces. We will see that this is the case.

Evidently thermodynamics is more complex than mechanics due to the existence of energy transfers in the form of heat as well as work, so that analogies can be dangerous. However, it will be useful to recognize that among the thermodynamic extensive variables there are a number that are clearly energy terms, such as U and its transforms, and others that have been termed "configuration properties," notably S, V, and mole numbers (n), or composition terms generally. Configuration properties are those that depend on the internal arrangement of a system (it will do no harm to leave this definition somewhat vague for the present). That V and n are configuration properties is more or less evident, but that S is also a configuration property will only be clear after the discussion in the next chapter of its basis in the statistics of the particles making up a system. These configuration properties can reasonably be thought of as the more complex analogies of the simple linear configuration property distance (h or r) in our discussion of potential energy.

That the thermodynamic intensive variables P and T (and in fact all the other intensive variables) can be thought of as analogous to mechanical forces is made reasonable by remembering that at equilibrium, no gradients in these properties exist. Therefore, gradients, either within the system or between the system and its surroundings, are an impetus for change, for energy transfer, just as force is an impetus for change in position of the brick. We need not insist too much on the analogy, but it is useful to have this mental classification of thermodynamic variables, and to remember that thermodynamic potentials are energy terms having the form

$$\frac{\partial(\text{energy or potential})}{\partial(\text{configuration})} = \text{intensive variable (thermodynamic force)}$$

Potentials are of particular interest in their extremum values, that is maxima or minima, since this indicates a sort of limit on the system concerned. In the case of a ball rolling (without friction) in a parabolic trough that we considered in Chapter 3 (Figure 5.1),

$$\mathrm{E}_p = \frac{1}{2}\kappa r^2 + constant$$

and

$$\frac{d\mathrm{E}_p}{dr} = \kappa r$$

where κ is a constant that depends on the shape of the parabola as well as the mass of the ball, and r is the horizontal distance of the ball from the apex (the lowest point) of the parabola.

In thermodynamics we are generally interested in stable or "most stable" configurations of systems, corresponding to lowest energy positions. According to the calculus, an extremum (maximum or minimum) is attained when

$$\frac{d\mathrm{E}_p}{dr} = 0$$

or, more simply,

$$d\mathrm{E}_p = 0$$

In the case of the rolling ball,

$$\begin{aligned} d\mathrm{E}_p &= \kappa r \cdot dr \\ &= 0 \text{ when } r = 0 \end{aligned}$$

i.e., when the ball is at the bottom of the trough. We know it is a minimum rather than a maximum because

$$\frac{d^2\mathrm{E}_p}{dr^2} = \kappa > 0$$

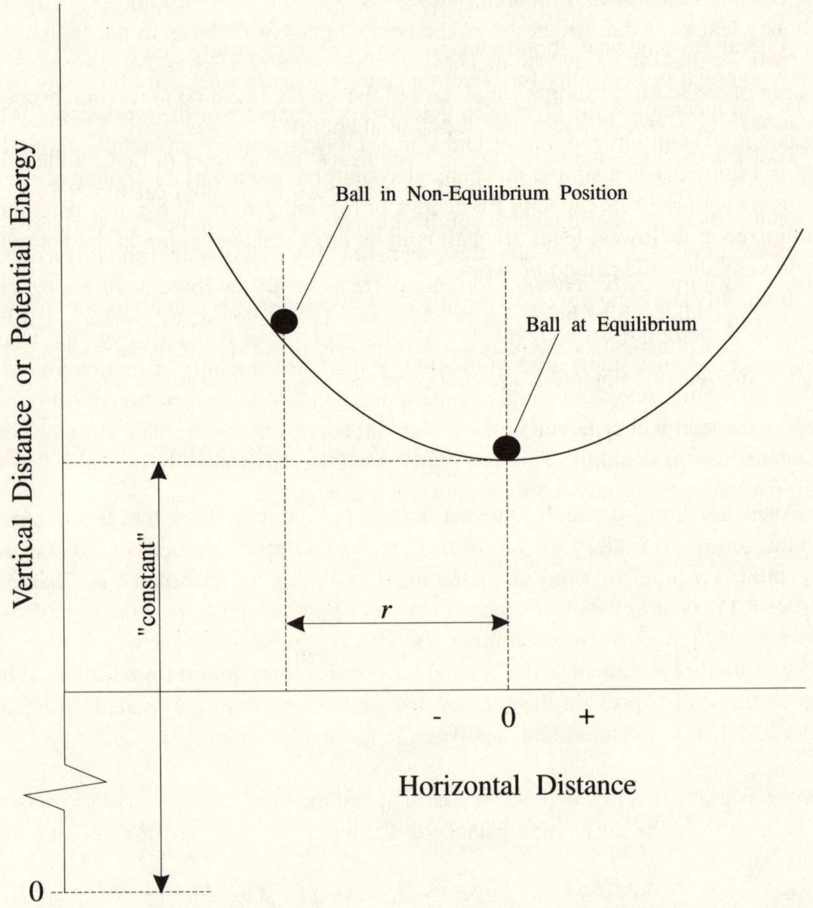

FIG. 5.1. A mechanical analogy for stable (or metastable) equilibrium states (from Figure 3.1), emphasizing the fact that the potential energy of the ball achieves a minimum at the bottom of the valley. This is expressed by the relation $d\mathrm{E}_p = 0$ (E_p is at an extremum at equilibrium), which is closely analogous to thermodynamic relationships such as $d\mathrm{G}_{T,P} = 0$ at equilibrium.

which says that E_p increases for points on the curve away from the bottom of the trough. Note, again, that the expression $d\mathrm{E}_p = 0$ has no necessary connection with infinitesimals. It says simply that at the bottom of the curve, the tangent to the curve is horizontal, and that therefore no matter what the size of r (measured along the tangent), E_p will not change.

Keep this analogy in mind when we discuss equations such as $d\mathrm{G}_{T,P} = 0$, $d\mathrm{U}_{\mathrm{S,V}} = 0$ in the following sections. When we say that G, U, and the other potentials achieve a minimum at equilibrium, it is exactly this "ball in trough" analogy that gives you the best mental picture of what we mean.

5.4.3. *Thermodynamic Potentials*

To generalize somewhat about what we have said concerning potentials, it appears that a potential is a capacity for effecting change, usually a capacity for doing work. In fact, in mechanics, the change in potential is a measure of the work done. When the potential is minimized, the system is in a stable position. For example, the rolling ball in Figure 3.1 is a simple mechanical system having a variable potential energy that is a measure of its capacity for change or for doing work. When this potential is minimized at its lowest level, the ball is in its most stable position at the bottom of the lower valley and can do no work.

It would clearly be a wonderful thing to have analogous potentials for chemical systems. A potential for a chemical system would tell us which of two equilibrium states was the more stable, and would thus be the "directionality" parameter we have been looking for. A system could change spontaneously in the direction of decreasing such a parameter, but never in the other direction, any more than the ball could spontaneously roll uphill. This is exactly what we have already developed in the entropy, but so far it only works for adiabatic processes.

5.4.4. *Entropy as a Thermodynamic Potential*

To begin the development of other directionality parameters, we must describe our entropy function a little more completely. This statement about the entropy can also serve as another statement of the Second Law, and is paraphrased from Callen (1960). It contains some aspects of the entropy that we have not explicitly stated before, and reflects Callen's "postulational" approach to thermodynamics.

> *There exists a homogeneous first order state variable, the entropy, which for isolated systems (those having constant* U *and* V*) achieves a maximum when the system is at stable equilibrium. Entropy and its derivatives are single-valued, continuous and differentiable functions of the other state variables. Entropy is a monotonically increasing function of the energy* U.

For those who skipped Euler's theorem in Chapter 2, the expression "homogeneous first order" in this definition simply means that we are defining S rather than *S* (i.e., total entropy rather than molar entropy). The fact that the entropy achieves a maximum rather than a minimum is the result of a sort of historical accident, as discussed in §5.3.1. It disturbs the "symmetry" of the thermodynamic potentials, since all the others achieve minima, but this creates no serious problems.

We define the new function S[3] and its derivatives as continuous and differentiable to avoid the possibility that even though maximized, it might be ∧-shaped rather than ∩-shaped. This would be mathematically inconvenient.

The mathematical statement that S achieves a maximum for given, U,V, is

$$d\mathrm{S}_{\mathrm{U,V}} = 0$$

[3]We may speak of S either as a parameter or as a function, since all system parameters or state variables have functional relations with all the others.

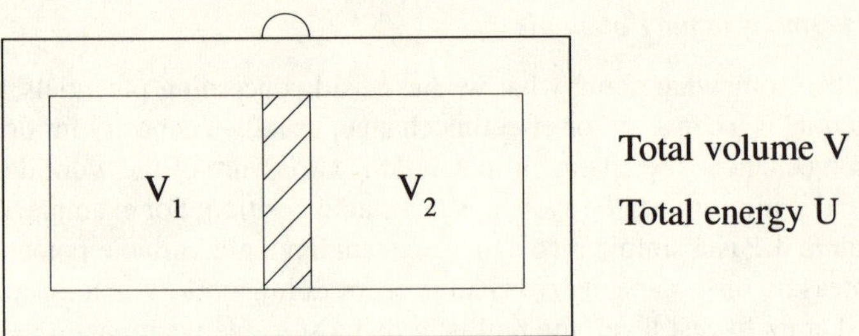

FIG. 5.2. A composite, isolated system. The movable partition is impermeable to matter but conducts heat. Volume $V = V_1 + V_2$

and

$$d^2S_{U,V} < 0 \tag{5.19}$$

We inserted the postulate that S is a monotonically increasing function of U to ensure that $S_{U,V}$ is the only thermodynamic potential exhibiting a maximum; all the others will exhibit minima. To illustrate this and some other things, let us choose a system and look at the stable equilibrium surface.

We must choose a system capable of having states other than the stable equilibrium state, preferably an infinite number of them, and the simplest possible one is probably the gas and piston arrangement previously used in Chapter 3 and repeated in Figure 5.2. The exterior wall is impermeable to energy and rigid, so the system is of constant U and V. The piston is movable and can be locked in any position. It is impermeable to the gas but it conducts heat so that the two sides are at the same temperature. According to our definition of S, the equilibrium position of the piston (that is, when the system has no additional constraints and the piston is free to move) is one of maximum entropy for the system, and any other position has lower entropy. If the two sides have equal amounts of the same gas, the curve illustrating this will be symmetrical.

Then if we consider the same situation but with successively greater energy contents U', U'', and U''', we will have the entropy-volume curves as in Figure 5.3 since S increases with U. In Figure 5.4a, the curves of Figure 5.3 are drawn in three dimensions, and one can now see that the condition that at equilibrium S is a maximum for given U,V implies the condition that U is a minimum for given S, V at a given equilibrium point such as A in Figure 5.4b. (Note that the section here is made at constant total volume, $V = V_1 + V_2$). The curves of Figures 5.3 and 5.4, although mathematically rather simple, are in one respect very curious. For example, at point A in Figure 5.4, we see that $d^2S_{U,V} < 0$, i.e., the curve slopes down on both sides, and we have proposed that our system have these positions by altering the position of the piston. But experience tells us that the piston will move spontaneously in one direction and has to be pushed in the other. Thus the curves, which are just lines in space to the mathematician, have a directionality to the physical scientist.

Thus a system at point A will never spontaneously leave A and go to B (Figure 5.4b); work must be expended to push it there. The system at point B, however, will

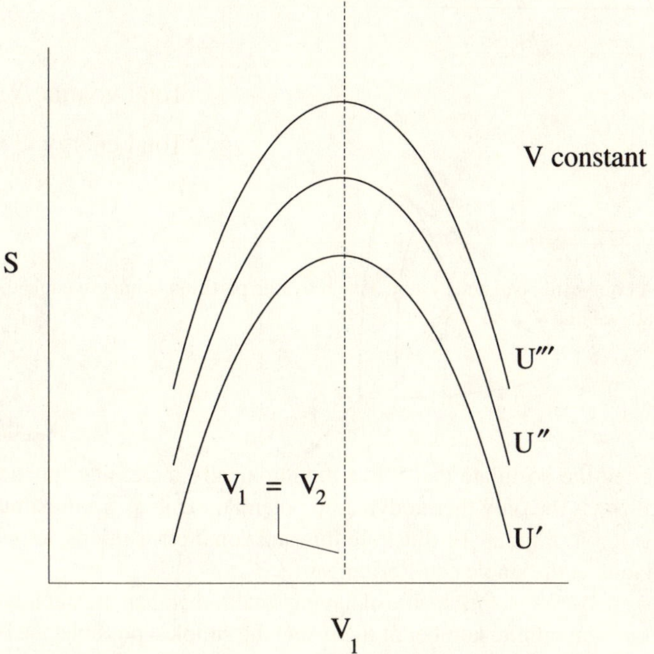

FIG. 5.3. Entropy (S) versus volume (V_1) for the system in Figure 5.2 at three different energy levels, where $U''' > U'' > U'$. Volume V is constant.

travel towards A spontaneously when the piston is unlocked. This difference has led to the identification of processes like A→B, which seem to be implied by statements like $d^2S_{U,V} < 0$, as *virtual processes* and Reiss (1965) has a particularly lucid discussion of this point.

The surfaces in Figure 5.4 must not be confused with the stable equilibrium surface. Only the locus of points at maximum S or minimum U such as A are at stable equilibrium and this locus forms a contour on the stable equilibrium surface as shown in Figure 5.5. All the other points on the surfaces in Figure 5.4 are metastable according to our definitions and would appear at higher U and lower S values having the same total volume V, such as point B in Figure 5.5.

5.4.5. *The Legendre Transform*

In practical applications we rarely have occasion to deal with isolated systems, i.e., those having constant U and V, although in many discussions of entropy, the system plus its "surroundings" are tacitly assumed to be equivalent to isolated systems. We will develop this thought further in a later section.

It turns out, though, to be quite simple to develop additional thermodynamic potentials that give directionality information for systems having other types of constraints, using the Legendre Transform introduced in Chapter 2. This section follows

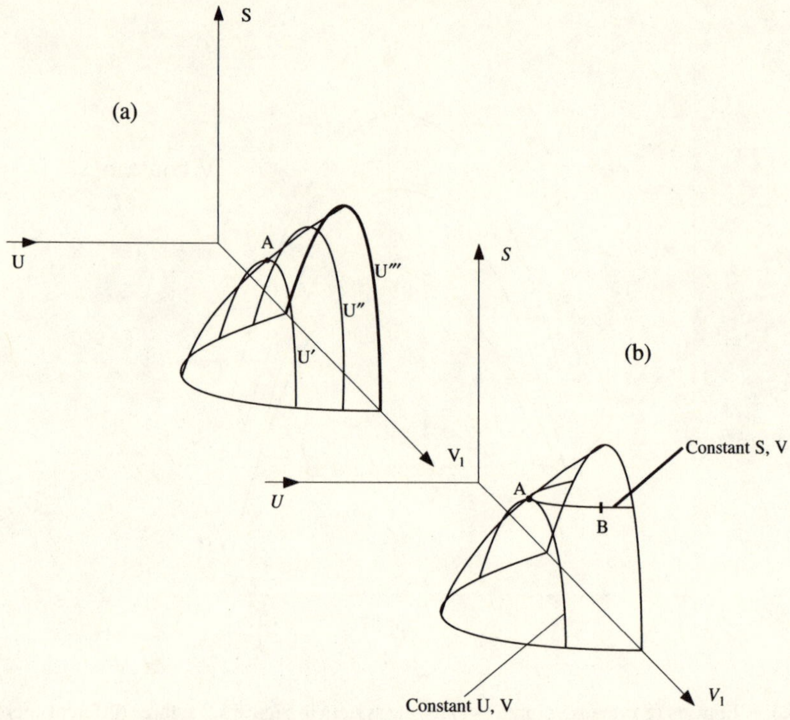

FIG. 5.4. Entropy (S) – Volume (V_1) – Energy (U) plot for the system in Figure 5.2. A three-dimensional version of Figure 5.3 (a) shows constant-U contours. Total volume V is also constant. (b) constant U, V contour and constant S, V contour through the same stable equilibrium point A. Point B represents a metastable equilibrium.

closely the development by Callen (1960).

To start with, we note that in theory we can extract U from the function

$$S = S(U, V)$$

and change it from its status as independent variable to dependent variable, thus

$$U = U(S, V)$$

Incidentally, although the exact nature of the functional relationship between U, S, and V is not needed here, it has been discussed in §5.3 as the "fundamental equation," equations (5.10) and (5.11).

Internal Energy and Volume

We rely on the discussion and diagrams of the preceding section to convince the reader that U is a thermodynamic potential exhibiting a minimum for systems of given S

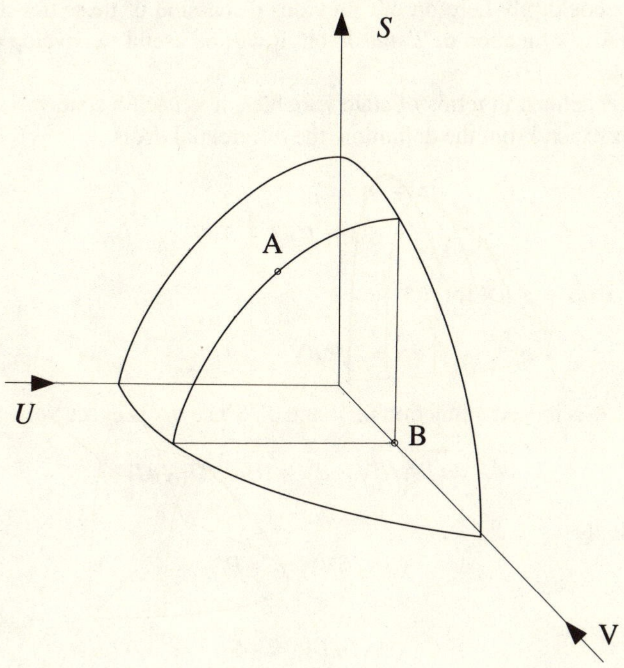

FIG. 5.5. Stable equilibrium surface in entropy (S)–total volume (V)–energy (U) space for a simple homogeneous system. The system in Figure 5.2, having a fixed volume, is represented by the contour through point A. Metastable point B is not on the surface.

and V. Thus

$$dU_{S,V} = 0$$

$$d^2U_{S,V} > 0$$

and $\Delta U_{S,V} < 0$ for spontaneous processes in systems having the same S and V before and after the process. To complete the "symmetry," we might mention that V is equally well a potential for systems of given U and S.

Now, since we know from §5.3 that $(\partial U/\partial S)_V = T$ and $(\partial U/\partial V)_S = -P$, it is clear that Legendre transforms of U(S, V) will lead to functions having more useful variables of constraint (T and P).

Helmholtz Energy

Transforming first the entropy variable, we have

$$\begin{aligned} A &= U - S(\partial U/\partial S)_V \\ &= U - TS \end{aligned}$$

where we use the new symbol A (the Helmholtz work function) to represent the new function. We could rely here on our previous discussion of these transforms to be assured that A is a function of T and V, but it will be useful to develop this here as well.

Since A is defined in terms of state variables, it is itself a state variable, and its differential is exact. From the definition, the differential $d\mathrm{A}$ is

$$\begin{aligned} d\mathrm{A} &= d\mathrm{U} - d(T\mathrm{S}) \\ &= d\mathrm{U} - T\,d\mathrm{S} - \mathrm{S}\,dT \end{aligned}$$

Substituting $T\,d\mathrm{S} - P\,d\mathrm{V}$ for $d\mathrm{U}$,

$$d\mathrm{A} = -P\,d\mathrm{V} - \mathrm{S}\,dT \tag{5.20}$$

showing that A is indeed a function of V and T. Since $d\mathrm{A}$ is exact,

$$d\mathrm{A} = (\partial\mathrm{A}/\partial\mathrm{V})_T d\mathrm{V} + (\partial\mathrm{A}/\partial T)_\mathrm{V} dT$$

So that evidently

$$(\partial\mathrm{A}/\partial\mathrm{V})_T = -P$$

and

$$(\partial\mathrm{A}/\partial T)_\mathrm{V} = -\mathrm{S}$$

It follows from (5.20) that

$$d\mathrm{A}_{T,\mathrm{V}} = 0$$

and thus A exhibits an extremum at equilibrium for systems of given T and V; it is therefore a thermodynamic potential of the kind we have been looking for, i.e., one having more convenient variables.

Since A is a transform of U, it will also exhibit a minimum, rather than a maximum, as will the other transforms to be developed from U.

Thus

$$\Delta\mathrm{A}_{T,\mathrm{V}} < 0$$

for spontaneous processes in systems having the same T and V before and after the process.

Enthalpy

Next, we can transform the other variable in the expression U = U(S, V). Thus we have a new function, the enthalpy H, where

$$\begin{aligned} H &= \mathrm{U} - \mathrm{V}(\partial\mathrm{U}/\partial\mathrm{V})_\mathrm{S} \\ &= \mathrm{U} + P\mathrm{V} \end{aligned}$$

which, if all goes according to plan, should be a thermodynamic potential for systems at given S and P.

To demonstrate this, we write

$$\begin{aligned} dH &= dU + P\,dV + V\,dP \\ &= T\,dS - P\,dV + P\,dV + V\,dP \\ &= T\,dS + V\,dP \end{aligned}$$

also

$$dH = (\partial H/\partial S)_P dS + (\partial H/\partial P)_S dP$$

so that

$$(\partial H/\partial S)_P = T$$

and

$$(\partial H/\partial P)_S = V$$

and

$$dH_{S,P} = 0$$

Also $H_{S,P}$ exhibits a minimum, and $\Delta H_{S,P} < 0$ for spontaneous reactions in systems having the same S and P before and after the reaction. Thus H is a function of S and P and is a thermodynamic potential for systems of constant (S, P). We will see however, that H is much more useful in a completely different context.

Gibbs Energy

Finally, if we transform both S and V, we have the Gibbs free energy function, G, where

$$\begin{aligned} G &= U - S(\partial U/\partial S)_V - V(\partial U/\partial V)_S \\ &= U - TS + PV \end{aligned}$$

and

$$\begin{aligned} dG &= dU - T\,dS - S\,dT + P\,dV + V\,dP \\ &= T\,dS - P\,dV - T\,dS - S\,dT + P\,dV + V\,dP \\ &= -S\,dT + V\,dP \end{aligned} \tag{5.21}$$

and since

$$dG = (\partial G/\partial T)_P dT + (\partial G/\partial P)_T dP$$

then

$$(\partial G/\partial T)_P = -S$$

and

$$(\partial G/\partial P)_T = V$$

and

$$dG_{T,P} = 0$$

where $G_{T,P}$ exhibits a minimum, and G is a thermodynamic potential for systems of given T and P.

Thus $\Delta G_{T,P} < 0$ for spontaneous processes in systems having the same T and P before and after the process.

We now have five (six counting volume) thermodynamic potentials defined in terms of U, S, V, T and P for use in various situations, and there are many others that could be defined. However, we do not even use all the ones we have defined so far, so there is little point in going on, at least until compositional variables are introduced.

To reiterate, systems at stable equilibrium must be described in general in terms of two fixed state variables. If two different equilibrium states having the same values of two state variables exist, then either both are metastable, or one is stable and one is metastable, and the metastable states in fact should be described in terms of three or more constraint variables. Usually, however, the third constraint is an activation energy barrier and is not thought of as a third variable (though in principle it is). For each choice of the two state variables that the two states have in common, there exists a function (another state variable) that is minimized or maximized at stable equilibrium; therefore, by comparing values of this variable one can tell which of the two states is more stable. Note finally that although we have many potentials, the existence of the entropy parameter is the fundamental fact that allows us to define them all. It appears in one way or another in all thermodynamic potentials.

The problem at the moment is that these thermodynamic potentials will have no "feeling of reality" for a reader new to the subject. That is, what is the entropy or free energy, how does one measure these things? Only by actually using these concepts will one become familiar with them. The next section is a first attempt at describing these potentials in more familiar terms.

5.5. POTENTIALS FOR WORK AND HEAT

As indicated earlier, potentials generally are a measure of how much work systems can perform, and this is true for thermodynamic potentials as well, with the exception of entropy. Systems that change at constant U, V can evidently perform no work at all since there is no change in energy and no volume change, but they can increase in entropy. The simplest examples involve gases expanding into a vacuum or into other gases. This type of change shows the effect of entropy in its "pure" form, unalloyed by changes in energy level. With other types of change, one has changes in U or V or both, and hence the capacity for doing work, and in each case the thermodynamic potential shows how much work can be done.

Internal Energy

Recall that

$$q \leq T\Delta S$$

where the equality refers to a reversible process and the inequality to all others (i.e., all real, irreversible processes). Combining this with

$$\Delta U = q + w$$

we have

$$w \geq \Delta\mathrm{U} - T\Delta\mathrm{S} \tag{5.22}$$

For constant entropy processes

$$w \geq \Delta\mathrm{U_S} \tag{5.23}$$

where the equal sign refers to a reversible process. Thus

$$\begin{aligned} w_{rev} &= \Delta\mathrm{U_S} \\ &= (\mathrm{U_B} - \mathrm{U_A})_\mathrm{S} \end{aligned}$$

if A and B are the beginning and final states of the system. Comparing this with the equation developed earlier for mechanical potentials we see that for constant S systems, U is a potential for reversible work. Strictly speaking it is only a potential for work in reversible processes, but it is sometimes just called a potential for work, since a decrease in U is an upper limit to the amount of work available ($-w$) from an isentropic process (equation (5.23)).

This can be made to seem intuitively reasonable rather than a result of manipulation of equations by recalling that the "upper limit" for work transfer is a reversible process, and for this case,

$$q_{rev} = T\Delta\mathrm{S}$$

Therefore, if $d\mathrm{S}$ (hence $\Delta\mathrm{S}$) is zero, q is zero, and all energy change must therefore appear as work (since $\Delta\mathrm{U} = q + w$). In irreversible isentropic processes, even though $\Delta\mathrm{S}$ remains zero, q is not zero but negative (since $q_{irrev} < T\Delta\mathrm{S}$), and will share the energy transfer process with w. Thus $\Delta\mathrm{U} < w$ for real (irreversible) isentropic processes; placing negative signs on both sides of the inequality (to give $-\Delta\mathrm{U} > -w$), we see again that the decrease in U is the maximum work obtainable from any such process (recall that work, heat, and energy outputs are negative in sign). In other words, the actual work obtained from any real isentropic process will always be less than the theoretical limit set by $-\Delta\mathrm{U}_{rev}$. Alternatively we could say that the work input required to make some real isentropic process go will always exceed the amount calculated by the ideal quantity $\Delta\mathrm{U}_{rev}$. In either case the efficiency of the real process is less than the ideal.

We shall see below that the other thermodynamic potentials also serve as upper limits for the work or heat output of real processes in completely analogous fashion. This makes intuitive sense—we might expect any real process with frictional losses and so on to behave less efficiently than a perfectly ideal hypothetical model. This also is one of the limitations of thermodynamics—it can only predict the maximum output (or minimum input) of any process and does so using the idealized model of a perfectly reversible, but ultimately unattainable, process.

Helmholtz Energy

Let us now repeat the above exercise, but this time allow entropy to change at constant T. Then we have again equation (5.22), and if we are gifted mathematically, we immediately recognize the right hand side of (5.22) as the change in A at constant T, since

$$\mathrm{A} = \mathrm{U} - T\mathrm{S} \text{ by definition,}$$

hence

$$d\mathrm{A} = d\mathrm{U} - T\,d\mathrm{S} - \mathrm{S}\,dT$$

and

$$d\mathrm{A}_T = d\mathrm{U} - T\,d\mathrm{S} \text{ at constant } T$$

Integrating between unspecified limits represented by the beginning and final equilibrium states,

$$\Delta\mathrm{A}_T = \Delta\mathrm{U} - T\Delta\mathrm{S} \tag{5.24}$$

and hence from (5.22)

$$w \geq \Delta\mathrm{A}_T$$

for processes at constant temperature. Thus the Helmholtz function, A, is a potential for (reversible) work in isothermal processes. As before then, the decrease in A is an upper limit to the work available from isothermal processes. Since the change in the Helmholtz potential during an isothermal, reversible process equals the total work (whether mechanical, electrical, or whatever) performed, it is often termed the *Helmholtz work function*.

Again, it is perhaps better to be able to see the truth of this by simply looking at equation (5.24) rather than by agreeing with the logic of the manipulations. Thus since $T\Delta\mathrm{S}$ is the maximum heat that can be transferred in a constant T process ($q_{rev} = T\Delta\mathrm{S}$) then $\Delta\mathrm{U} - T\Delta\mathrm{S}$ will be the energy available as work in a reversible isothermal process ($w = \Delta\mathrm{U} - q$); according to our previous discussion, work output is maximized in this case. We note, too, that if $w = 0$, then the inequality $w \geq \Delta\mathrm{A}_T$ requires that

$$\Delta\mathrm{A}_{T,\mathrm{V}} \leq 0 \tag{5.25}$$

where the subscript V is added because if no work is done, then no pressure- volume type work is done, and $\Delta\mathrm{V}$ must be zero in this case. This equation says that A will decrease in any spontaneous reaction at fixed T and V, which does not surprise us since this agrees with the definition arrived at through the Legendre transform. We might emphasize here that we are using the terms "spontaneous" and "irreversible" synonymously so that the $<$ symbol in equation (5.25) refers both to irreversible and spontaneous processes.

Enthalpy

As alluded to above, chemical systems undergoing change (i.e., in which reactions occur) can do various kinds of work. For instance, batteries can do electrical work. While undergoing these reactions, the chemical system invariably has some change in volume, because it is most unlikely that the reaction products would have exactly the same volume as the reactants. This change in volume $\Delta\mathrm{V}$ takes place under some ambient pressure P, so that $P\Delta\mathrm{V}$ work is done during the reaction regardless of whether any other kind of work is done or not—if the reaction is to take place, it cannot be avoided. This "work against the atmosphere" (or against the confining pressure, whatever it is) usually is not *useful*, it simply takes place whether we like it or not, and at atmospheric pressure is often a rather small part of the total energy change. While we can decide to eliminate electrical work or other kinds of mechanical work from our systems, we cannot eliminate this $P\Delta V$ work (unless we consider only constant volume systems, which is not usually very practical). So while $\Delta A_{T,V}$ is a useful measure of *total* work, we need a measure of the work available *other than* this useless $P\Delta V$ work.

Net work other than $P\Delta V$ work can be written

$$\begin{aligned} w_{net} &= (w_{total} - w_{P\Delta V}) \\ &= (w + P\Delta V) \end{aligned}$$

since $w_{P\Delta V} = -P\Delta V$. Recalling inequality (5.22),

$$w \geq \Delta U - T\Delta S \tag{5.22}$$

then

$$w + P\Delta V \geq \Delta U - T\Delta S + P\Delta V \tag{5.26}$$

Now, since

$$\begin{aligned} H &= U + PV \\ dH &= dU + P\,dV + V\,dP \\ dH_P &= dU + P\,dV \\ \Delta H_P &= \Delta U + P\Delta V \end{aligned} \tag{5.27}$$

which looks almost like the right hand side of (5.26). With the additional restriction of constant entropy, $\Delta S = 0$, the right hand side of (5.26) is then equal to $\Delta H_{S,P}$. If all kinds of work other than $P\Delta V$ work are eliminated, $(w + P\Delta V) = 0$, then

$$\Delta H_{S,P} \leq 0 \tag{5.28}$$

Thus enthalpy is a potential for (reversible) work in a constant S, P process, and will always decrease in spontaneous processes in such systems, only $P\Delta V$ work being allowed. Again, the fact that H is a potential for (isobaric, reversible) work is almost self-evident from looking at equation (5.27), since $P\Delta V$ is the maximum pressure-volume work available and ΔU is the rest of the energy change available. In fact, the restriction of constant entropy is not necessary for H to be a work potential. If ΔS is not zero, the amount of energy available for work will be reduced but still maximized in a reversible process.

Gibbs Energy

Finally, since

$$\begin{aligned} G &= U - TS + PV \\ dG &= dU - T\,dS - S\,dT + P\,dV - V\,dP \end{aligned}$$

$$d\text{G}_{T,P} = d\text{U} - T\,d\text{S} + P\,d\text{V}$$

and

$$\Delta\text{G}_{T,P} = \Delta\text{U} - T\Delta\text{S} + P\Delta\text{V}$$

and by reasoning similar to the foregoing,

$$\Delta\text{G}_{T,P} \leq 0 \tag{5.29}$$

Gibbs free energy is a potential for reversible work in constant T-P processes, and always decreases in spontaneous processes. By comparison with (5.26) it is clear that G is a measure of the net work or non-$P\Delta\text{V}$ work. This function therefore contrasts with the Helmholtz work function, which measures total work, including mechanical $P\text{V}$ work. The Gibbs free energy is a particularly useful measure of the electrical or chemical work attainable from a process and is used a great deal with chemical systems where $P\text{V}$ work is often unimportant.

5.5.1. *Heat Potentials*

Since work and heat are equally valid means of transferring energy to and from systems, all these potentials can be construed as heat potentials under some conditions, but only two can be conveniently treated so. Thus, for example, rearranging (5.22) and substituting $q = T\Delta\text{S}$, $w = -P\Delta\text{V}$, gives

$$q \geq \Delta\text{U} + P\Delta\text{V}$$

and

$$q_{\text{V}} \geq \Delta\text{U}_{\text{V}}$$

and U is a potential for reversible heat transfer in constant volume processes.

However, the only heat potential of much interest is the enthalpy, and in fact we could say that the only practical interest in the enthalpy is that it is a potential for heat in constant pressure processes. As before, if the only work is $P\Delta\text{V}$ in a constant pressure process (and remember that at constant P, work is exactly equal to $-P\Delta\text{V}$, not $< -P\Delta\text{V}$), then

$$\Delta\text{H}_P = \Delta\text{U} + P\Delta\text{V}$$

and since

$$\Delta\text{U} = q_P - P\Delta\text{V}$$

therefore

$$\Delta\text{H}_P = q_P$$

Thus it happens that in constant pressure processes, the enthalpy change is exactly equal to q, the total heat flow. Or putting it the other way around, q admits a potential H in constant pressure processes. Please note that because H is a state variable, ΔH is perfectly well-defined between any two equilibrium states. But when the two states are at the same pressure, it becomes equal to the total heat flow during the process from one to the other, and in fact ΔH is in practice rarely used except in these cases (another kind of use, isenthalpic expansions, is discussed in Chapter 8).

We should emphasize that strictly speaking, U, A, H, and G should be spoken of in each case as reversible work potentials, rather than as work potentials, since it is only in the reversible case that the work is equal to the change in the potential, analogous to (5.17). Note though, that H is a potential for heat in isobaric processes, reversible or not.

Finally, we should point out that in fact the products TS and PV are also potential quantities, and are both the products of a "thermodynamic force" term (T, P) and a "configuration" term (S,V). Both have dimensions of energy, and are rather analogous to U in that

$$d(TS)/dS = T$$

and

$$d(PV)/dV = P$$

These potentials also have interesting properties, but further discussion would carry our discussion of thermodynamic potentials beyond what we consider useful for our purposes. The interested reader is referred to Badger (1965) (the source of much of this material) for a thorough discussion.

This ends the development of the thermodynamic potentials as indicators of process direction and as work and heat potentials. Although apart from textbook problems the only "direction parameter" used is the Gibbs free energy, and the only heat potential is the enthalpy, it is interesting and useful in a general sense to see the interrelations and the overall "symmetry" involved in these parameters. We should now have a reasonable grasp of the theoretical meaning of these terms. It only remains to introduce the statistical meaning of the entropy for this to be complete. It is also a great help to know something about how these terms are measured, and we will consider this briefly, too.

5.6. SUMMARY

We have postulated the existence of a function of state, the entropy, which achieves a maximum at constant U and V for a system at equilibrium. Using Legendre transforms we have derived the additional functions of state from the following "fundamental equation":

$$S = S(U, V) \quad \text{or} \quad U = U(S, V)$$

$$dU = T\,dS - P\,dV$$

Partial derivatives:

$$(\partial U/\partial S)_V = T$$

$$(\partial U/\partial V)_S = -P$$

extrema:

$$\Delta S_{U,V} \geq 0 \text{ (maximum)}$$

$$\Delta U_{S,V} \leq 0 \text{ (minimum)}$$

where the equality obtains at equilibrium, and the inequality for spontaneous processes.

The derived functions are:

A = A(T, V) Helmholtz work function
H = H(S, P) Enthalpy
G = G(T, P) Gibbs free energy

With the exception of entropy, each of these functions reaches a minimum at equilibrium, and each decreases during spontaneous processes (at constant values of the subscripted variables). Thus each serves as a thermodynamic potential, or as a measure of the capacity for a system described by the subscripted variables to undergo some change or process. These thermodynamic potentials also provide a measure of the maximum work or heat output available from a process:

$$q \leq T\Delta S$$

$$w_{total} \geq \Delta U_S$$

$$w_{total} \geq \Delta A_T$$

$$w_{net} \geq \Delta G_{T,P} \quad (non - P\Delta V \text{ work})$$

Thus the maximum work obtainable from real (irreversible) processes is always less than the theoretical limit given by the appropriate thermodynamic potential for ideal (reversible) processes. Alternatively, the work or heat input required to make a real process go is always more than the amount given by the thermodynamic potential for the ideal (reversible) process. Don't forget, if the $\geq$ sign in the work equations looks backwards to you, the sign convention we adopted in §4.6 means that $-w$ is the work *obtainable* from a system, so that $w \geq \Delta x$ is the same as $-w \leq -\Delta x$, and is read "$-w$ is less than or equal to the *decrease* in x".[4]

Finally, we observed that the enthalpy change of constant pressure processes exactly equals total heat flow:

$$\Delta H_P = q_P$$

5.7. THE ENTROPY OF THE UNIVERSE

Let us introduce a commonly used alternative way of discussing entropy by considering a pail of water sitting on the floor of your back porch on a winter day. During the night, the temperature goes down to -2°C, but when you go out onto the porch in the morning, you find the water has not frozen. You kick the pail, and the water begins to freeze, and within a short time (a time governed by how fast the heat of crystallization can escape from the pail) the water at -2°C changes to ice at -2°C,

[4]Because the numerical values of both w and the Δ-terms can be either positive or negative, it is easy to become confused by these sign manipulations. If you do, take a numerical example. Thus if ΔA_T is -1000 J (A decreases by 1000 J), then $w_{rev} = -1000$ J, i.e., a maximum of 1000 J is available as work. However, $w_{irrev} > -1000$ J. For example w_{irrev} might be -500 J, where $-500 > -1000$, i.e., $500 < 1000$.

i.e., a spontaneous, irreversible process has occurred. However, it is well known that the entropy of ice is less than the entropy of water under these conditions, so that our spontaneous process has occurred with

$$S_{ice} - S_{water} < 0$$

i.e.,

$$\Delta S < 0$$

which seems to conflict with our previous derivations that insist that entropy increases in spontaneous processes. What has happened?

Within the framework of our discussion, what has happened is that we have ignored our carefully formulated qualification that entropy increases in spontaneous processes *in isolated systems*, or at least those for which $q = 0$. Remember, $\Delta S_{U,V} \geq 0$ requires that U and V be the same before and after the process. Since the pail of water can only freeze if a considerable amount of heat leaves the pail, clearly q is not zero if the water itself is our system. The two states we are comparing (water at $T = -2°C$, $P = 1$ atm, and ice at $T = -2°C$, $P = 1$ atm) have the same T and P, and do not have the same U and V. Therefore, in our terms, instead of examining the entropy change we should examine the Gibbs free energy change involved to see which is the more stable phase. This is because $\Delta G_{T,P} \leq 0$ applies to constant T and P processes. Sure enough, we find that

$$G_{ice} - G_{water} < 0$$

under these conditions, indicating that ice is the stable phase, water the metastable phase, and we released the activation energy constraint on nucleation in the system by kicking the pail. Another way of looking at the situation is to say, in effect, that the entropy law

$$\Delta S > 0$$

applies to all systems, including our pail of water, but that you must take into account the entropy change of the system and its surroundings. The "surroundings" of our pail of water is evidently the rest of the universe, that is, the universe minus the pail of water. Then, the story goes, although the entropy of the contents of the pail decreases, the escape of heat from the pail increases the entropy of the surroundings according to

$$\Delta S \geq q/T$$

so that since heat is added to the rest of the universe at a fixed $T(-2°C)$, the entropy of the "surroundings" must increase. Then it is shown that this increase in entropy of the surroundings is greater than the decrease in entropy within the pail, so that the overall ΔS is positive. This generally leads on to a repetition of Clausius's famous dictum to the effect that although the energy of the universe is constant (First Law) the entropy of the universe is tending towards a maximum (Second Law).

Our view is that it is a mistake to use Clausius's brilliant aphorism as a basis for teaching the concept of entropy, at least as exemplified above, because the "surroundings" of systems is such a vague concept, and its vagueness is transferred to the understanding of entropy. This sort of explanation, following a derivation linked to physically impossible Carnot cycles, has done much to give entropy its well-known shroud of mystery.

It may well be true that the entropy of the universe is increasing (see Chapter 6), but whatever it is doing is quite irrelevant to the study of thermodynamics here on Earth. The difference between the two ways of looking at ΔS presented above essentially involves two different definitions of "the system." In our preferred explanation, the system is the water in the pail, and its entropy decreases spontaneously. In the other view, the system is the universe, by implied hypothesis a closed composite adiabatic system, and the pail a portion of this composite system separated from the rest by diathermal walls. In the overall system, entropy increases. In this view, the choice of system is effectively taken from us—we must choose the universe as our system to preserve the dictum that entropy increases in spontaneous processes.

Needless to repeat, we don't favor this view.

5.8. DIFFERENTIAL INEQUALITIES

We derived equation (5.10) by combining a statement of the First Law

$$d\mathrm{U} = Dq + Dw$$

with

$$Dq_{rev} = T\,d\mathrm{S}$$

and

$$Dw_{rev} = -P\,d\mathrm{V}$$

obtaining

$$d\mathrm{U} = T\,d\mathrm{S} - P\,d\mathrm{V} \tag{5.10}$$

or, dividing by the number of moles in the system,

$$dU = T\,dS - P\,dV \tag{5.11}$$

At first glance, these equations would seem to be restricted to reversible processes, but it is not difficult to see that they are not. The terms U, S, and V in (5.10) are all state variables, and hence the change $d\mathrm{U}$ or dU can be carried out any way at all, and still be equal to $T\,d\mathrm{S} - P\,d\mathrm{V}$. The inequalities introduced by considering an irreversible change, i.e.,

$$q_{irrev} < T\,d\mathrm{S}$$

and

$$-w_{irrev} < P\,d\mathrm{V}$$

must cancel out, so that equations (5.10) and (5.11) are quite general. Consider now equation (5.29)

$$\Delta \mathrm{G}_{T,P} \leq 0 \tag{5.29}$$

or its differential form,

$$d\mathrm{G}_{T,P} \leq 0 \tag{5.30}$$

Because G is defined as U − TS + PV, we can write

$$dG = dU - T\,dS - S\,dT + P\,dV + V\,dP$$

and

$$dG_{T,P} = dU - T\,dS + P\,dV \qquad (5.31)$$

Combining (5.30) and (5.31), we have

$$dU - T\,dS + P\,dV \leq 0$$

or

$$dU \leq T\,dS - P\,dV \qquad (5.32)$$

in which $<$ refers to irreversible processes and = to reversible ones.

We seem to have a contradiction here. First we said that (5.10) could be applied to any change in U, S and V, reversible or irreversible, and now in (5.32), otherwise identical, a $<$ sign has crept in for irreversible processes. What's going on? The difference is that in equation (5.10) we were considering only stable equilibrium states, where specifying two variables is enough to fix all the system properties. If the beginning and ending states of a process are both stable equilibrium states, then equation (5.10) is valid. Such processes are illustrated in Figure 3.3. It doesn't matter whether the system properties stay on the surface during $A \rightarrow B$ (reversible process) or leave the surface at A and return to it at B (irreversible process), as far as equation (5.10) is concerned. An example would be changing a crystal of calcite from 25°C, 1 atm to 50°C, 2 atm.

In this chapter, however, we have been considering irreversible processes of a different sort—those that proceed from a metastable state towards a stable state, like a ball rolling down a hill. In these cases, the $<$ sign comes in, because the potential quantities (thermodynamic potentials) are decreasing during the irreversible change. An example would be aragonite at 25°C, 1 atm changing (irreversibly) to calcite at 25°C, 1 atm.

5.8.1. *Controlled Metastable States*

If you write $\Delta G_{T,P} \leq 0$, the meaning is quite clear: a macroscopic difference between two quantities, $G_{initial}$ and G_{final}. But the differential form $dG_{T,P} \leq 0$ is less clear. It implies that there is a function G of which dG is the differential. If dG refers to an increment of a process leading from a metastable state (having three constraint variables) to a stable state (having two constraint variables), this in turn implies that we have a functional relationship between G and the three constraint variables. This should have the form

$$dG_{irrev} = -S\,dT + V\,dP + [\text{a negative quantity}] \qquad (5.33)$$

or

$$dU_{irrev} = T\,dS - P\,dV + [\text{a negative quantity}] \qquad (5.34)$$

It is very instructive to consider this function explicitly. To do this we need an appropriate system.

Consider a binary alloy of fixed composition in which there are two types of crystallographic sites (geologists may prefer to consider the K-feldspars, which have variable ordering of the Al atoms in tetrahedral and octahedral sites). The two types of atoms may arrange themselves so that each occupies only one type of site (complete order); they may be distributed at random (complete disorder); or there may be a degree of disordering specified by a parameter ϕ.

In any given stable equilibrium state, the degree of disorder of the atoms on the sites will be such that S will be maximized compared to any other degree of disorder at the same U and V, G will be minimized with respect to any other ϕ at the same T and P, H will be minimized with respect to other ϕ values at the same S and P, and so on. The following discussion could be carried out using any of the thermodynamic potentials with their associated state variables, but we have chosen S, U, and V so as to refer directly to equation (5.32).

Experimentally, it might be difficult to measure the properties of the system as a function of the degree of disorder, but one can imagine changing ϕ at an elevated temperature where the order-disorder process proceeds rapidly, then cooling quickly to the desired conditions with ϕ "frozen-in." Of course, this is irrelevant to the thermodynamic model, in which we have no trouble manipulating S, V, ϕ, and other variables. ϕ is simply an additional constraint on the system (that is, additional to our chosen S and V, or U and V). We may therefore speak of the system as achieving equilibrium (no observable changes no matter how long we wait; return to original conditions after slight temporary changes in constraint parameters) with respect to two constraints (minimum U for given S and V—stable equilibrium) or with respect to three constraints (minimum U for constant S, V and ϕ—metastable equilibrium), and as U is now a function of S, V, and ϕ, the function relating them is

$$dU = (\partial U/\partial S)_{V,\phi}dS + (\partial U/\partial V)_{S,\phi}dV + (\partial U/\partial \phi)_{S,V}d\phi \tag{5.35}$$

or

$$dU = T\,dS - P\,dV + (\partial U/\partial \phi)_{S,V}d\phi \tag{5.36}$$

and

$$dU_{S,V} = (\partial U/\partial \phi)_{S,V}d\phi \tag{5.37}$$

This simple system may be represented in thermodynamic state space by a surface (Figure 5.6) on which every point represents a state of the system at stable equilibrium. At every point on the surface the system has a value of ϕ, and these values may be contoured as shown. These ϕ values are the percentage of disorder in the alloy when it has reached stable equilibrium, i.e., minimum U for given S, V. If ϕ is controlled independently of S, U, and V, then the alloy may have either more or less disorder than the stable equilibrium value at a particular S and V, and in either case the energy U must be greater than the stable equilibrium U. Therefore there are surfaces below (on the low S, high U side of the stable equilibrium surface representing the energy of the crystal at various values of S, V, and ϕ. These are metastable equilibrium surfaces in accordance with the definitions in Chapter 3.

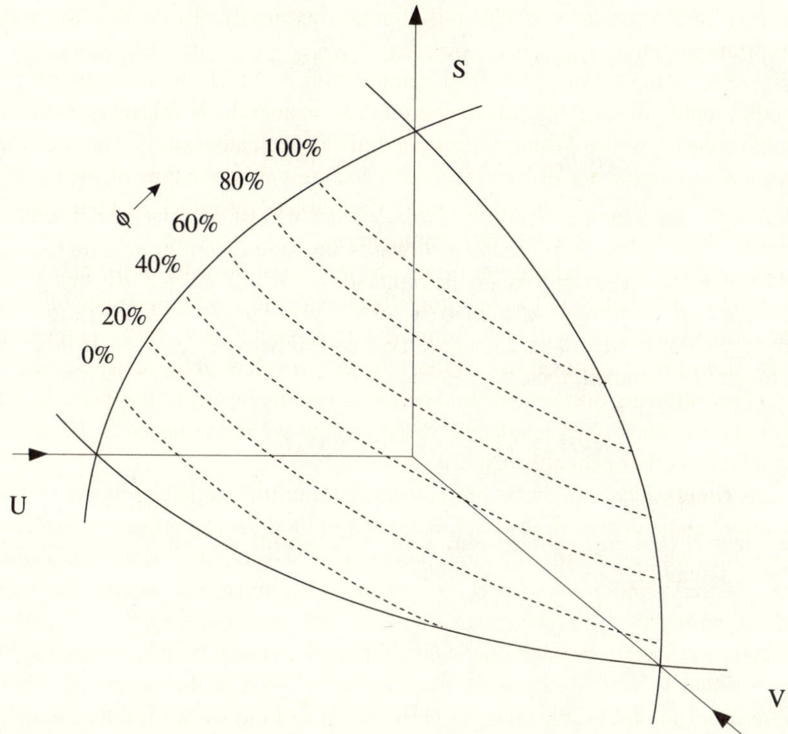

FIG. 5.6. Schematic stable equilibrium surfaces in $S-U-V$ space for a particular alloy sample. ϕ values (contoured) are a measure of atomic disorder in the crystal. The axes are arranged for easy representation of the surface, which continues beyond them.

Now considering a particular metastable alloy having a certain S and V, but a degree of disorder ϕ either greater or less than the equilibrium value at that S, V, a change in ϕ, $d\phi$, towards the stable equilibrium value may be either positive or negative (disorder may increase or decrease), but U must decrease if the process is spontaneous. Therefore $dU_{S,V}$ is negative, as is $(\partial U/\partial\phi)_{S,V}d\phi$ by equation (5.37), and in conformity with equation (5.33).

But degree of disorder is only a particular example of a third constraint that might prevent a system from reaching stable equilibrium, and $d\phi$ is an increment of that extra constraint, an increment of the *progress* of the system towards stable equilibrium. We could generalize this concept of systems approaching stable equilibrium in increments by defining a *progress variable*, ξ, which begins at 0 in the initial state and ends up with a value of 1 (or other convenient number) in the final state. The differential $d\xi$ is therefore an increment of any spontaneous reaction, or more exactly, an increment of any quasistatic irreversible reaction, and the general fundamental equation, which includes processes off the stable equilibrium surface as well as those on the surface, is

$$dU_{irrev} = (\partial U/\partial S)_{V,\xi} dS + (\partial U/\partial V)_{S,\xi} dV + (\partial U/\partial \xi)_{S,V} d\xi$$

or

$$dU_{irrev} = T\,dS - P\,dV + (\partial U/\partial \xi)_{S,V} d\xi \tag{5.38}$$

It follows then that *whatever* the nature of the metastable state from which the spontaneous reaction is proceeding, towards the stable equilibrium surface, there will be (at least) one extra term on the right side of $dU = T\,dS - P\,dV$ that gives the change in U per increment of change (or *progress*) of the system from the metastable state to the stable state, and this extra term will *always* be negative. If this term is not explicitly included, then we can only write $dU < T\,dS - P\,dV$ for the same irreversible reaction.

5.8.2. *The Affinity*

The quantity $(\partial U/\partial \xi)_{S,V}$ turns out to be quite useful and has been given its own name and symbol, the *affinity*, A, where[5].

$$(\partial U/\partial \xi)_{S,V} = -\mathsf{A} \tag{5.39}$$

Thus

$$dU = T\,dS - P\,dV - \mathsf{A}\,d\xi \tag{5.40}$$

When we write the equation in this form it is understood that irreversible processes are being considered; we should perhaps always write

$$dU_{irrev} = T\,dS - P\,dV - \mathsf{A}\,d\xi$$

for clarity, but this is not the usual practice. The affinity has units of cal mol^{-1} or J mol^{-1}, as does U, and (5.40) balances dimensionally because ξ is dimensionless. We can also write

$$d\mathrm{U}_{irrev} = T\,d\mathrm{S} - P\,d\mathrm{V} + (\partial \mathrm{U}/\partial \xi)_{\mathrm{S,V}} d\xi \tag{5.41}$$

and

$$d\mathrm{U} = T\,d\mathrm{S} - P\,d\mathrm{V} - \mathsf{A}\,d\xi \tag{5.42}$$

which also balances if ξ has units of moles, i.e., is extensive.

We will return to this subject in connection with chemical potentials (§14.5), with which the affinity is more commonly connected.

[5]Do not confuse A with the Helmholtz energy, A

PROBLEMS

1. Unknown to most physical scientists, there exists another important meaning for the °C symbol—degrees Crerar. The Crerar Scale has a single, fixed equilibrium state—an aqueous solution 3 molal in NaCl, 0.001 molal in H_2S, at a pH of 4 and saturated with sphalerite (ZnS) at atmospheric pressure and a molality of zinc of exactly 10^{-5}. This state has been found to have a temperature of 100°C on the Celsius scale, and is defined as 1234°C on the absolute Crerar Scale. Find the temperature of freezing water on the Crerar scale. Is the Crerar Scale a Thermodynamic one?
2. If it was decided by an international body that the Crerar Scale should be changed such that it had a value of 100°C at the ice point, how many degrees would it have between the ice and steam points?
3. Calculate $\Delta_r H^\circ$, $\Delta_r G^\circ$ and $\Delta_r U^\circ$ for the reaction $NaAlSiO_4 + 2SiO_2 = NaAlSi_3O_8$ at 25°C, 1 bar (data in Appendix C). Note that $\Delta_r U^\circ$ constitutes the major part of the other two.
4. Look at Appendix C. Why is $\Delta_f G^\circ$ for all compounds negative? Why is S° for everything but aqueous ions positive? Why is $\Delta_f G^\circ$ for all elements zero, except for a few like $S_2(g)$, and why are they not zero?
5. Calculate the entropy of formation from the elements of corundum.
6. Is corundum or gibbsite more stable in the presence of water at 25°C, 1 bar? (To do things like this, you must first write a balanced reaction involving the constituents you are interested in—in this case corundum, gibbsite and water, and then see which way the reaction will go).
7. Which of the three Al hydroxides listed is the most stable in water?

6

STATISTICAL INTERPRETATION OF ENTROPY

> How awkward is the human mind in divining the nature of things, when forsaken by the analogy of what we see and touch directly.
>
> L. Boltzmann, quoted in Bent (1965), p. 176.

6.1. THE CONCEPT OF ENTROPY

As we have seen, thermodynamics is based for the most part on the idea of the conservation of energy (First Law) and the concept of entropy (Second Law). The conservation of energy gives little problem intuitively, but it is quite another story with entropy. Entropy can be considered from the point of view of idealized heat engines operating in cycles, or by deriving some of its inherent properties (Chapter 5). We will see how it is measured and tabulated in Chapter 7. This is all very useful, but doesn't help much in gaining an intuitive grasp of entropy, such as we have for the other thermodynamic parameters.

Just what *is* entropy, anyway? There may not be any definitive short answer to this question. If we had to rely on classical thermodynamics for an answer, we would talk at some length about the *availability* of energy, e.g., the fact that in spite of the tremendous quantity of energy in the ocean, we cannot use any of it to power a ship or to do anything else; the ocean's thermal energy is unavailable unless we provide a reservoir for heat at a lower temperature. This is of course perfectly true, and many useful discussions of the meaning of entropy follow this line of thought, but somehow after all these discussions, the entropy remains somewhat elusive. There is, however, another way to think of entropy that is by far the most useful, and that is from the statistical/probability point of view. This requires that we consider matter from the point of view of the individual particles (atoms, molecules, ions) rather than as macroscopic, homogeneous bodies, and is therefore not a part of classical thermodynamics, but of statistical mechanics.

In this chapter we present the rudiments of this approach, not so that the reader can become proficient in statistical thermodynamics (a considerably more thorough introduction is required for that) but to show how entropy is related to statistical considerations. Statistical mechanics does not exactly *explain* what entropy is, but rather provides a model, quite different from the thermodynamic model, that contains a parameter identical to the entropy of the thermodynamic model in every measurable respect. Equating a parameter in one model with a parameter in another model may not be completely satisfactory as an explanation, but there is no doubt that the statistical point of view is of considerable help in gaining an intuitive grasp of entropy. There is no doubt in our minds that a familiarity with this point of view is essential to a good understanding of thermodynamics. As Nash (1972) says:

> . . . statistical mechanics offers us the immense intellectual satisfaction of rendering transparent what thermodynamics leaves opaque: that is, statis-

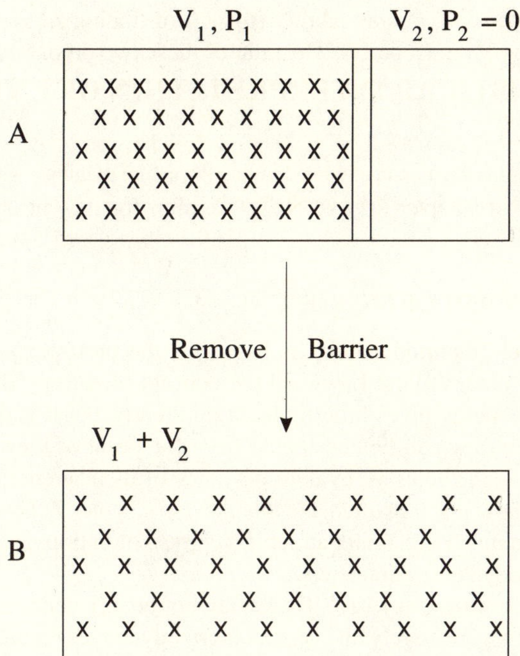

FIG. 6.1. Expansion of a gas into a vacuum.

tical mechanics casts a brilliant light of *interpretation* throughout most of the realm in which classical thermodynamics can afford us little more than phenomenological *description*.

6.2. ENTROPY, PROBABILITY, AND DISORDER

We begin by demonstrating that there is in fact a relationship between entropy and probability. Consider the expansion of an ideal gas containing N_0 molecules from an initial volume and pressure, V_1 and P_1, into an adjacent evacuated chamber of volume V_2 as illustrated in Figure 6.1. The final volume is $V_1 + V_2$. After expansion the probability of finding one particular molecule in the original volume is $V_1/(V_1 + V_2)$. The probability of finding all N_0 molecules in the original volume V_1 at the same time without being constrained to be there is $[V_1/(V_1 + V_2)]^{N_0}$ (since there are N_0 molecules and probabilities are multiplicative). Therefore the probability of the initial state arising spontaneously after expansion is

$$p_{initial} = \left(\frac{V_1}{V_1 + V_2}\right)^{N_0}$$

This is an extremely small probability. For example if $V_1/(V_1 + V_2)$ is 0.5, it is about $1/(2^{10^{23}})$.

After the expansion, the probability (p_{final}) of finding all N_0 molecules in the final volume $V_1 + V_2$ must be 1.0. The ratio of these two probabilities is

$$\frac{p_{final}}{p_{initial}} = \left(\frac{V_1 + V_2}{V_1}\right)^{N_0} \tag{6.1}$$

We will see in Chapter 10 that the expression for the entropy change in this expansion process is

$$\Delta S = nR \ln\left(\frac{V_1 + V_2}{V_1}\right) \tag{10.13}$$

where n is the number of moles of gas. Comparison of (6.1) and (10.13) gives [1]

$$\Delta S = nR \ln\left(\frac{p_{final}}{p_{initial}}\right)^{1/N_0}$$

or

$$\Delta S = k \ln\left(\frac{p_{final}}{p_{initial}}\right) \tag{6.2}$$

where

$$k = nR/N_0$$

The number of molecules divided by the number of moles is Avogadro's number, N_a, so that

$$k = R/N_a$$

Thus there seems to be a relationship between entropy and probability. In the example mentioned above where $V_1/(V_1 + V_2) = 0.5$, (6.2) becomes, for one mole of gas,

$$\begin{aligned}\Delta S &= (R/10^{23}) \ln(2^{10^{23}}) \\ &= R \ln 2\end{aligned}$$

(10^{23} is an abbreviation for Avogadro's number, $6.02 \cdot 10^{23}$.) It is also a fact of common experience that there is a relationship between probability and disorder or degree of randomness. In fact, in looking for a definition of these terms (disorder, randomness), you will invariably wind up defining them in terms of the probability of the occurrence of a given state within a number of possible states. A disordered state of a group of objects is more probable because there are more ways in which a disordered state may be achieved. For example, the probability of receiving cards from all four suits in a bridge hand is much greater than the probability of receiving thirteen spades, simply because there are many different ways that thirteen cards can be selected from a deck of 52 to give at least one card from each suit, but only one way of selecting thirteen spades (the fact that the cards are all spades of course has nothing to do with it, except

[1] Equation 10.13 actually has only V_2 in the numerator, but it is equivalent to what we are here calling $V_1 + V_2$.

to a bridge player; there is only one way to select *any* specific thirteen cards).[2] Because the thirteen spades, or any such given arrangement, is a highly *ordered* arrangement compared to the other, less ordered arrangements, and the probability of its occurrence can be calculated, we have (if the entropy–probability relationship shown above is generally true), a relationship between entropy, probability and disorder in systems. In other words, shuffling an ordered deck of cards increases its entropy, just as mixing two gases increases the entropy of that system. This relationship is the most easily grasped significance of the entropy, but a bit more remains to be said about it.

6.2.1. *Thermodynamic Probability*

We were able to calculate a probability change for the gas in Figure 6.1 because although there was a large number of molecules, each was considered in only two states, V_1 before the change and $V_1 + V_2$ after. Suppose there are more possibilities? We can define the probability of a state in a given system (where the system here is simply a group of objects that have a finite number of possible arrangements, or states) as the number of ways that state can be achieved divided by the total number of possible states. For example two dice, each having six faces, have a total number of 6^2 or 36 possible arrangements. Rolling a 2 can be done in only one way (each die must show a 1), so the probability of getting 2 in a random throw of the dice is $1/36$ or 0.0278. However, there are six ways of throwing a 7 (6 + 1, 5 + 2, 4 + 3, 3 + 4, 2 + 5, 1 + 6) so that the probability of rolling a 7 is 6/36 or 0.167. Probabilities defined in this way (the normal way) thus range from 0 (e.g., the probability of throwing a 1 with two dice) to 1.0 (e.g., the probability of throwing a number between 2 and 12 inclusive).

Of course we are more interested in natural systems such as rocks and minerals than in cards and dice (during working hours, anyway), and these tend to be somewhat more complicated. It is the magnificent achievement of statistical mechanics to show that essentially the same ideas lead to all the usual thermodynamic relationships, when applied to atoms and molecules in their various possible energy levels or quantum states. Thus the probability of a given distribution of molecules among energy levels could be defined as the number of ways of realizing that arrangement divided by the total number of arrangements. In such systems the total number of possible arrangements is an inconveniently large number, so the denominator is eliminated and a "thermodynamic probability" (W) is defined as the number of ways of achieving a given distribution or state of the system.

Inspection of equation (6.2) reveals that for the entropy of each state, there is apparently a relation

$$S = k \ln(probability)$$

[2]"Order" vs. "disorder" is thus seen to be inescapably subjective, unless the rules are clear (Denbigh and Denbigh, 1985). To most people, the concept of a well-ordered room has a fairly clear meaning, and there are limitless ways of disordering the room. But if Mr. X likes to keep his shoes on top of the dresser and the rug on the bed, this constitutes perfect order for him. However even for him, there is only one way of achieving perfect order and limitless ways of disordering the room. In either case, the disordered state of the room is more probable.

so, for our thermodynamic probability W, this becomes

$$S = k \ln W \tag{6.3}$$

This is the equation made famous by Boltzmann in 1896, and inscribed on his tombstone in Vienna. It of course has a more rigorous derivation than the one we have given.

6.2.2. *Calculating Configurational Probabilities*

Consider the following problem: How many different ways can N distinguishable particles be assigned j different energy levels?[3] Let's call these energy levels $\epsilon_1, \epsilon_2, \ldots, \epsilon_j$ and put N_1 molecules into ϵ_1, N_2 into ϵ_2 and so on up to N in ϵ_j. One such arrangement is shown in Figure 6.2 (using 35 molecules and j energy levels). Any one arrangement we have called a configuration, and the probability of a configuration is simply the number of ways it can be achieved. There are $N!$ ways of distributing all N particles among j energy levels.[4] However, rearranging or permuting the particles in any one energy level will not change the overall configuration. Stated differently, the order in which particles were put into a level makes no difference, just the total number it contains. Placing molecule number 32 to the left of number 31 in level ϵ_h of Figure 6.2 makes no difference; more importantly, any two different molecules could be exchanged for numbers 31 and 32 and there would still be two molecules in level ϵ_h. Thus we must divide $N!$ by the number of permutations of particles, $N!$ in each level. The total probability of any one configuration is then the number of ways it can be achieved, or

$$\begin{aligned} W &= \frac{N!}{N_1!N_2!\cdots Nj!} \\ &= \frac{N!}{\Pi_{i=1}^{j} N_i!} \end{aligned} \tag{6.4}$$

This equation is illustrated in Figure 6.3 which shows the 12 different ways in which 4 molecules can be put into 4 different energy levels such that level ϵ_1 contains any two molecules, ϵ_2 none, and ϵ_3 and ϵ_4 contain one each. For this particular configuration equation (6.4) correctly predicts 12 possible arrangements:

$$\begin{aligned} W &= \frac{4!}{2!0!1!1!} \\ &= 12 \end{aligned}$$

[3]For the term "particles" we could substitute "molecules," "atoms," "ions", "subset of the total system," etc. For the term "energy levels" we could substitute "vibrational, rotational, electronic,. . . , energy levels." The assumption that the particles are distinguishable and that there are no restrictions as to how many particles go into any energy level are quantum mechanical approximations; for statistical applications this usually is not a problem. We might for example distinguish the same atoms on different sites in a crystal lattice. For details, see Fowler and Guggenheim (1939) or Mayer and Mayer (1940).

[4]There are N ways of picking the first particle, $N - 1$ ways for the second, $N - 2$ for the third, and a total of $N(N - 1)(N - 2)\cdots(1) = N!$ ways for all N particles.

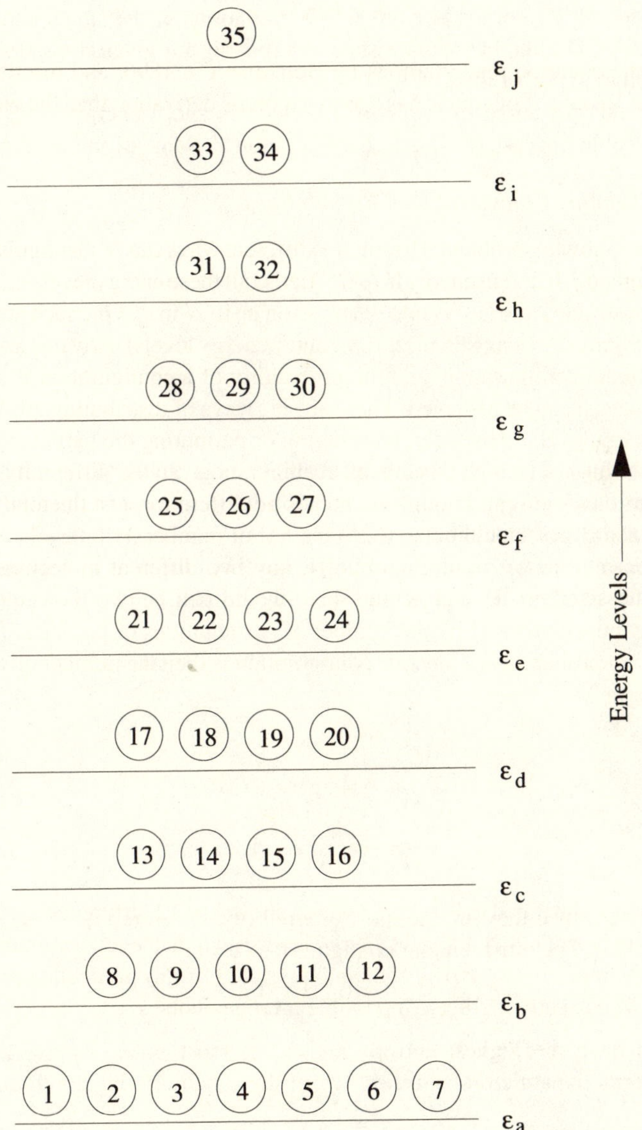

FIG. 6.2. One way of arranging 35 distinguishable molecules (denoted by number) among j different energy levels.

(Note that 0! = 1.) Notice that rearranging the two particles within level ϵ_1 makes no difference. In the jargon of statistical mechanics, each of these 12 arrangements is a microstate; all 12 microstates are simply variations of the single configuration $\epsilon_1^{(2)}, \epsilon_2^{(0)}, \epsilon_3^{(1)}, \epsilon_4^{(1)}$. The number of microstates comprising any given configuration gives its thermodynamic probability.

Now for this particular example of 4 molecules and 4 energy levels there are $4^4 = 256$ possible microstates and 35 configurations.[5] Some of these configurations might be

$$\begin{matrix} \epsilon_1^{(4)} & \epsilon_2^{(0)} & \epsilon_3^{(0)} & \epsilon_4^{(0)} \\ \epsilon_1^{(1)} & \epsilon_2^{(1)} & \epsilon_3^{(1)} & \epsilon_4^{(1)} \\ \epsilon_1^{(2)} & \epsilon_2^{(0)} & \epsilon_3^{(0)} & \epsilon_4^{(2)} \\ \vdots & & & \vdots \end{matrix}$$

and so on. Each would have a specific probability as given by (6.4). Whatever configuration has the largest probability will obviously occur most frequently, and those with low probabilities should occur much less often.

In the present case we can use our knowledge about the relation between entropy and randomness and predict a priori that 4 configurations,

$$\begin{matrix} \epsilon_1^{(4)} & \epsilon_2^{(0)} & \epsilon_3^{(0)} & \epsilon_4^{(0)} \\ \epsilon_1^{(0)} & \epsilon_2^{(4)} & \epsilon_3^{(0)} & \epsilon_4^{(0)} \\ \epsilon_1^{(0)} & \epsilon_2^{(0)} & \epsilon_3^{(4)} & \epsilon_4^{(0)} \\ \epsilon_1^{(0)} & \epsilon_2^{(0)} & \epsilon_3^{(0)} & \epsilon_4^{(4)} \end{matrix}$$

tie for last place since they are the most ordered (lowest entropy). These should all occur least often. The most random configuration is

$$\epsilon_1^{(1)}, \epsilon_2^{(1)}, \epsilon_3^{(1)}, \epsilon_4^{(1)}$$

so it should have the highest entropy and occur most often. Applying equation (6.4), the thermodynamic probabilities, W, of these two limiting configurations are

[5]In general, there are m^N ways of putting N distinguishable objects in m boxes (if there is no limit on the number of objects per box). The joint conditions of distinguishable objects and boxes of limitless capacity are the basis of *Boltzmann statistics*. Alternatively, the number of configurations can be calculated by again placing no restriction on the number of objects in each box, but this time making the objects indistinguishable. The appropriate formula for the number of ways of putting N indistinguishable objects in m boxes is then $(m + N - 1)!/((m - 1)!N!)$, giving 35 in our example ($m = 4$; $N = 4$). Statistical methods based on these two conditions are called *Bose- Einstein statistics*. To calculate the number of ways N indistinguishable objects can be placed in m boxes with no more than one object per box, use *Fermi-Dirac statistics* and the formula $m!/((m - 1)!N!)$. For more details and a particularly clear presentation, see Tinoco et al. (1978, Chap. 11).

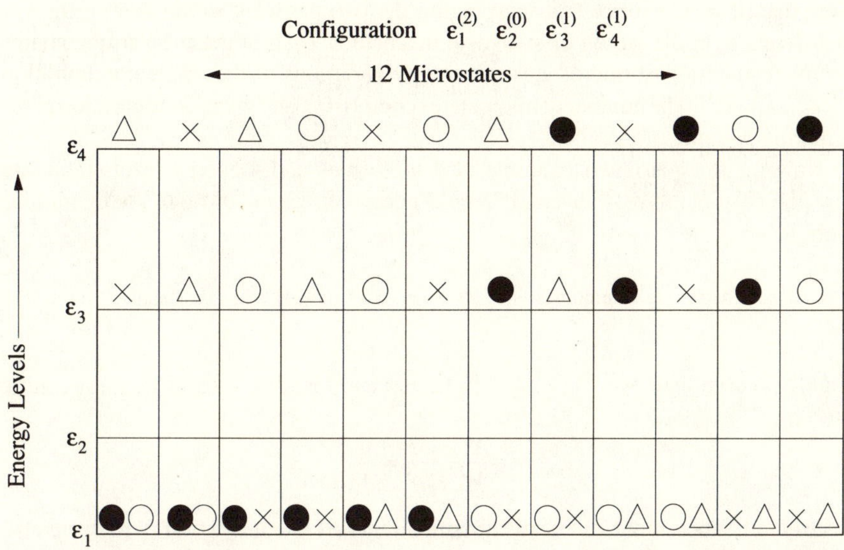

FIG. 6.3. The 12 different ways of achieving that configuration which has 2 molecules in energy level ϵ_1, none in ϵ_2, and 1 each in ϵ_3 and ϵ_4. The system contains a total of 4 molecules designated by symbols ○, •, × and △.

$4!/(4!0!0!0!) = 1$ and $4!/(1!1!1!1!) = 24$, so the most random configuration will occur 24 times more often than the most ordered.

6.2.3. *The Predominant Configuration*

Now comes a very important point in statistical mechanics. For very large N (number of particles) the probability of one configuration will typically predominate so completely over all others that the others can be ignored. This single predominant configuration is considered the equilibrium state. Such a simplification is most helpful since, as we have just seen, one mole of a substance containing roughly 10^{23} particles would produce an absolutely prohibitive number of configurations; instead we have just stated that we need deal only with one predominant configuration. The appearance of a predominant configuration at large values of N can be illustrated simply by tossing a coin enough times.[6] Thus as shown in Figure 6.4, there are $2^4 = 16$ equally probable outcomes of tossing a coin 4 times (since there are 2 possible outcomes each toss). These correspond to the microstates of our previous example; we can also distinguish 5 configurations (see footnote 4 and Figure 6.4). The thermodynamic probability of each configuration can be calculated from equation (6.4) and equals the number of microstates each contains. Clearly configuration III (with 50% heads)

[6]This example is treated in more detail by Nash (1972, Chapter 1). Another quite different and helpful example is given by Callen (1960, Appendix B).

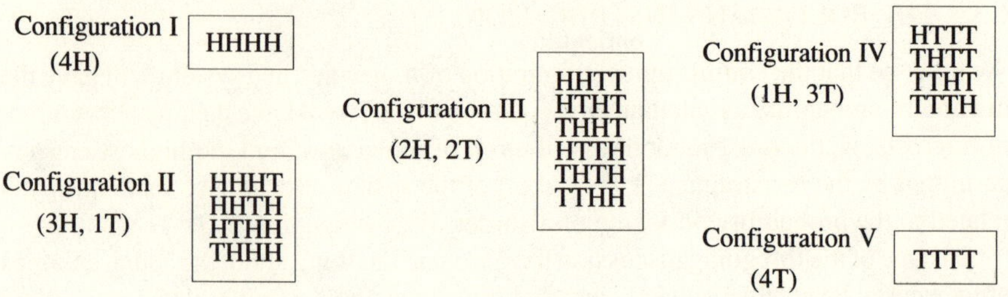

FIG. 6.4. The possible outcomes of tossing a coin 4 times. H = heads, T=tails.

should predominate over II (with 75% heads) and for $N = 4$ tosses the probabilities are

$$\frac{W_{50\% H}}{W_{75\% H}} = \frac{6}{4}$$

Instead of 4 tosses, let us see what happens when we set $N = 6 \cdot 10^{23}$ which is Avogadro's Number. The relative chance of getting 50% heads(H) after $6 \cdot 10^{23}$ tosses is given by (6.4):

$$W_{50\% H} = \frac{N!}{H!T!} = \frac{6 \cdot 10^{23}!}{3 \cdot 10^{23}! \times 3 \cdot 10^{23}!}$$

Similarly, the chance of getting 75% heads would be

$$W_{75\% H} = \frac{6 \cdot 10^{23}!}{4.5 \cdot 10^{23}! \times 1.5 \cdot 10^{23}!}$$

These factorials can be evaluated using Stirling's formula for large N :

$$\ln N! = N(\ln N) - N \tag{6.5}$$

The probability ratio of the two configurations now becomes

$$\ln(W_{50\% H}/W_{75\% H}) = 7.85 \cdot 10^{22}$$

or

$$W_{50\% H}/W_{75\% H} \approx e^{10^{23}}$$

This number is beyond the range of your calculator, and in fact is literally *inconceivably* large, and similar numbers are obtained whatever ratio of heads to tails is used in the denominator. In other words, the 50% heads configuration predominates by such an enormous margin at large values of N that all other configurations are extremely unlikely to occur. Therefore, when working with macroscopic systems containing large numbers of molecules we need only consider one predominant configuration. A rigorous discussion of this point is given by Mayer and Mayer (1940).

6.3. THE BOLTZMANN DISTRIBUTION

We now see that the equilibrium configuration of molecules in a system will have the maximum probability as calculated by equation (6.4). It will have the greatest number of microstates, the most random distribution of molecules, and the highest entropy permitted by the constraints (T, P, V, *etc.*) of the system. Recall also that entropy is related to the probability or degree of disorder, W, by equation (6.3).

In none of the foregoing discussion have we said anything about the actual energies of the energy levels themselves—just that if we know how many molecules there are in each level (as in Figure 6.3) we can calculate the number of possible arrangements (microstates), and the probability and entropy of that configuration. We can take a great step forward at this point by recognizing that quantum mechanics tells us (in principle) the allowed energy levels for these molecules. In fact, many of these levels can be measured directly by spectroscopy. If we know what these energy levels $\epsilon_1, \epsilon_2, \ldots, \epsilon_j$ are, then we should consider a new problem: how will the N particles of the most probable molecular configuration distribute themselves throughout a set of allowed energy levels? This problem was solved, once again, by Ludwig Boltzmann. The result is the Boltzmann distribution—the keystone of statistical mechanics.

The mathematical solution finds that set of energy-level occupation numbers $N_1, N_2, \ldots, N_j$ which maximizes the probability W of a configuration as given in (6.4), subject to the following two constraints: the total number of particles and the total energy (E) must remain constant,

$$\sum_{i=1}^{j} N_i = N \tag{6.6}$$

$$\sum_{i=1}^{j} N_i \epsilon_i = E \tag{6.7}$$

This is a problem in constrained extremals that can be solved using Lagrange's method. The constraining equations (6.6) and (6.7) are multiplied by two arbitrary constraints α and β, added to the logarithm of (6.4) and the desired maximum is given by

$$\frac{\partial}{\partial N_i}\left(\ln W + \alpha \sum N_i + \beta \sum N_i \epsilon_i\right) = 0 \tag{6.8}$$

Omitting the intervening algebra, the solution is

$$N_i/N = g_i e^{-\epsilon_i/kT} \Big/ \sum_i g_i e^{-\epsilon_i/kT} \tag{6.9}$$

This is the most general form of the Boltzmann distribution. It tells us that the fraction of molecules in an energy level ϵ_i increases exponentially with temperature and decreases exponentially with the energy of that level. This is as important for macroscopic systems as the Schrödinger equation is for individual atoms or molecules. The Schrödinger equation, for example, shows that energy levels ($1s$, $2s$, $2p$,...) are possible for a single hydrogen atom; the Boltzmann distribution says that for a large (macroscopic) number of hydrogen atoms, some specific fraction will have the $1s^1$ configuration, another fraction the $1s^0 2s^1$, and so on.

The parameter g that crept in somewhere between equations (6.8) and (6.9) is necessary to correct for the total degeneracy of each energy level. This is the total number of states (particles, molecules, electrons, or whatever) that can have one particular energy ϵ_i. For example, in the Schrödinger model of the hydrogen atom, s, p, and d orbitals are 2-, 6-, and 10-fold degenerate, respectively. An energy level that is g-fold degenerate can have a population g times larger than a single level; therefore, we must multiply equation (6.9) through by the degeneracy, g, of each energy level.

6.3.1. *The Partition Function*

The denominator in equation (6.9) is called the partition function z:

$$z = \sum_{i=1}^{j} g_i e^{-\epsilon_i/kT} \tag{6.10}$$

This is one of the most important parameters in statistical mechanics since it is directly related to the thermodynamic properties of a system. The summation in (6.10) is made over all energy states, so z is a function of the partitioning of all particles among all energies for the equilibrium configuration.

It is important to understand the physical meaning of the partition function since it appears in one form or another in many different applications. The book *Entropy and Energy Levels* by Gasser and Richards (1974, Chapter 3) covers this point clearly and in detail. We can briefly develop some insight here by considering what happens at very low temperatures as all molecules or particles drop to the lowest energy level, or "ground state," $\epsilon_0 = 0$. Let N_0 represent the number of molecules in the ground state, and N_{total} the total number of molecules. The Boltzmann distribution (6.9) can be written

$$N_0/N_{total} = g_i e^{-\epsilon_i/kT}/z$$

or, since in this case $\epsilon_i = \epsilon_0 = 0$

$$z = N_{total} g_0/(N_0)$$

To simplify matters, let us say that there is only one unique ground state with a degeneracy, $g_0 = 1$. Then

$$z = N_{total}/N_0$$

For this limiting case, $z = 1$, since all molecules are in the ground state ($\epsilon_0 = 0$) and $N_{total} = N_0$. At higher temperatures, some of these molecules will jump up to higher energy levels, reducing the number, N_0, in the ground state and raising z accordingly. In fact z can rise to extremely high values at very high temperatures as almost all molecules leave the ground state and a great number of higher energy levels become accessible. So the partition function ranges from 1 to ∞. It measures the distribution

Table 6.1 The Boltzmann Distribution

Level No.	Temperature T			Temperature $2T$			Temperature $3T$		
	ϵ/kT	$e^{-\epsilon/kT}$	N_i for N = 1000	$\epsilon/2kT$	$e^{-\epsilon/2kT}$	N_i for N = 1000	$\epsilon/3kT$	$e^{-\epsilon/3kT}$	N_i for N = 1000
0	0	1.0	632	0.0	1.0	395	0.0	1.0	291
1	1	0.3679	233	0.5	0.6065	240	0.33	0.7165	208
2	2	0.1353	86	1.0	0.3679	145	0.67	0.5134	149
3	3	0.0498	31	1.5	0.2231	88	1.00	0.3679	107
4	4	0.0183	12	2.0	0.1353	53	1.33	0.2636	77
5	5	0.0067	4	2.5	0.0821	32	1.67	0.1889	55
6	6	0.0025	2	3.0	0.0498	20	2.00	0.1353	39
7	7	0.0009	1	3.5	0.0302	12	2.33	0.0970	28
8	8	0.0003	0	4.0	0.0183	7	2.67	0.0695	20
9	9	0.0001	0	4.5	0.0111	4	3.00	0.0498	14
10	10	0.0000	0	5.0	0.0067	3	3.33	0.0357	10
Partition Function		1.58195			2.53111			3.43755	

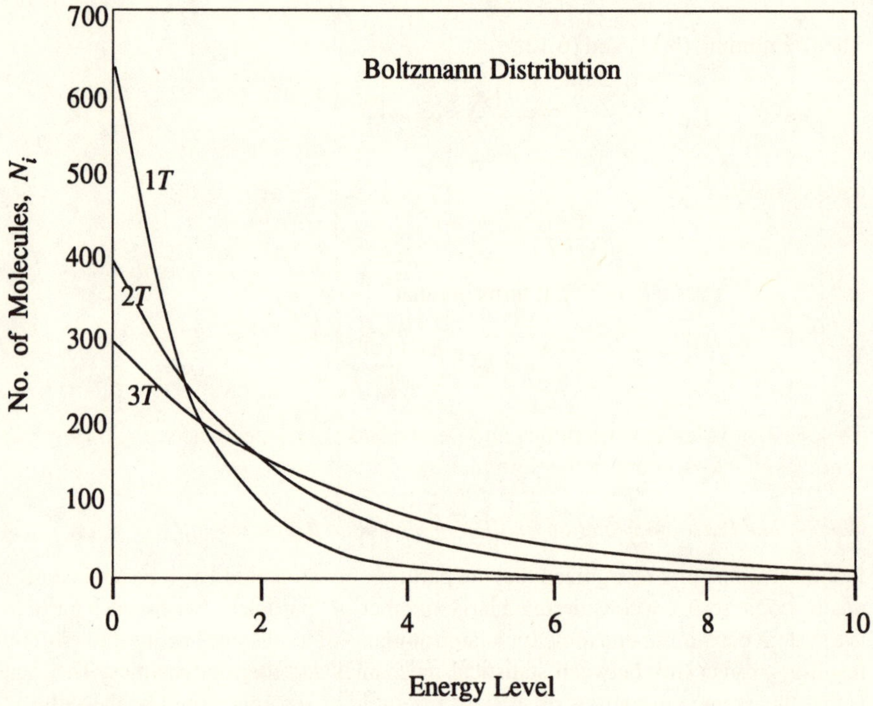

FIG. 6.5. The Boltzmann distribution for three different hypothetical temperatures. Data from Table 6.1.

of molecules among available energy levels, and gives the ratio of the total number of molecules in all levels to those occupying the ground state.

An example of a Boltzmann distribution is shown in Table 6.1 and Figure 6.5 for a set of energy levels with $g = 1$ equally spaced a distance kT apart, for three different temperatures. Note the concentration of particles in the lower levels, gradually spreading to higher levels as T is increased. The meaning of the partition function is particularly easy to see in this example.

Because the partition function is related to the number of particles occupying energy levels above the ground state, it can be used to calculate the *average internal energy*, $\bar{\epsilon}$, of a particle. From equations (6.6) and (6.7) the average energy is

$$\bar{\epsilon} = \frac{\sum N_i \epsilon_i}{\sum N_i} = \sum_i N_i \epsilon_i / N \tag{6.11}$$

Substituting (6.10) into (6.9) gives the more common form of the Boltzmann distribution:

$$N_i/N = g_i e^{-\epsilon_i/kT}/z \tag{6.12}$$

Then combining (6.11) and (6.12)

$$\bar{\epsilon} = \frac{1}{z}\sum_i g_i \epsilon_i e^{-\epsilon_i/kT}$$

and because

$$\frac{d}{d(1/kT)}\left(e^{-\epsilon_i/kT}\right) = -\epsilon_i \cdot e^{-\epsilon_i/kT}$$

and $d(1/(kT)) = -dT/(kT^2)$, it turns out that

$$\bar{\epsilon} = kT^2\left(\frac{\partial \ln z}{\partial T}\right)_V \tag{6.13}$$

The constant volume restriction is imposed because the dependence of ϵ_i on volume is not taken into account in this formulation.

6.3.2. *The Partition Function and Thermodynamic Properties*

J.Willard Gibbs first recognized that the partition function could also be applied on a macroscopic scale by considering a large number of "particles," where each particle is a system containing one mole (or a large amount) of a substance being studied. The result is a direct link between statistical mechanics and thermodynamics. The idea is that the average properties of such an ensemble of systems would be the same as the properties of a single system consisting of molecular particles acting in the same way. Let's therefore consider a large number of systems, each containing one mole of the substance of interest. If the allowed molar energy levels of the systems are $E_1, E_2, \cdots, E_j$, the average energy will be

$$E = \frac{\sum N_i E_i}{\sum N_i} = \sum_i N_i E_i / N$$

Letting

$$Z = g_i e^{-E_i/kT}$$

where [7]

$$Z = z^N$$

then

$$E = kT^2\left(\frac{\partial \ln Z}{\partial T}\right)_V \tag{6.14}$$

[7]This is the relationship between Z and z for an assemblage of identical but distinguishable units. Other relationships between Z and z obtain for other kinds of systems. The evaluation of the macroscopic partition factor Z is the main task in the application of statistical thermodynamics to real systems.

In Chapter 4 we considered that the total energy of a system is given by relativity theory as $E_r = mc^2$ and that the internal energy of thermodynamics is some unspecified smaller quantity such that $E_r = U + constant$. In statistical mechanics the energy of a particle or an energy level is generally taken as the difference in energy between the level considered and a "ground state" of zero energy, quite frequently the state of zero particle vibration, rotation, and translation, i.e., 0 K on the absolute temperature scale. Thus systems have zero energy at 0 K. Thus the E of statistical mechanics is also related to U by $E = U + constant$, but the constant is different. Therefore the E term in (6.14) is actually a ΔE term, and can also be written

$$\Delta E = E - E_0 = kT^2 \left(\frac{\partial \ln Z}{\partial T} \right)_V$$

The *heat capacity at constant volume* is defined as $C_v = (\partial U/\partial T)_V$, or, because $U = E + constant$ (Chapter 4), $C_v = (\partial E/\partial T)_V$. Differentiating (6.14) gives

$$C_v = \frac{k}{T^2} \left(\frac{\partial^2 \ln Z}{\partial (1/T)^2} \right) \tag{6.15}$$

Thus the internal energy and heat capacity are simply related to the change in the partition function with temperature. For certain simple systems such as gases at low temperatures, the partition function can be estimated theoretically. For most systems of geological interest such as minerals and concentrated salt solutions, additional experimental information is required. This might take the form of spectroscopic data on electronic or molecular vibrational frequencies, or direct measurement of some of the non-ideal thermodynamic properties themselves.

6.3.3. *Entropy and the Partition Function*

We should be able to relate S and z because the partition function and Boltzmann distribution were originally derived by combining equation (6.7) for the energy of a system with equation (6.4) for the probability of its equilibrium configuration, W, and we already know that entropy and probability are related by

$$S = k \ln W \tag{6.3}$$

Substituting equation (6.4) for W into (6.3), and applying Stirling's approximation (6.5) for the factorials of large numbers gives

$$S = k(N(\ln N) - N) - k \sum_i (N_i(\ln N_i) - N_i)$$

Substituting version (6.12) of the Boltzmann distribution for N, and (6.10) for the partition function and simplifying, we obtain

$$S = k(N(\ln N) - N) - kN(\ln N - \ln z - 1 - \bar{\epsilon}/kT)$$

or

$$S = N\bar{\epsilon}/T + kN \ln z$$

Letting N be Avogadro's number, the entropy per mole is

$$\begin{aligned} S &= E/T + k \ln z^{N_a} \\ &= E/T + k \ln Z \end{aligned} \tag{6.16}$$

where $Z = z^{N_a}$.

This is the desired link between S and Z and between the molecular world of statistical mechanics and the macroscopic systems of thermodynamics. All other thermodynamic functions can be calculated if we know Z (over a range of temperature and pressure), since E is given by equation (6.14) and S by (6.16). For example, the Helmholtz work function is $A = E - TS$ or

$$\begin{aligned} \Delta A &= A - A_0 \\ &= -kT \ln Z \end{aligned} \tag{6.17}$$

The Gibbs free energy would be simply $\Delta G = \Delta A + P\Delta V$, and so on.

6.4. ENTROPY AND SPONTANEITY

In Chapter 5 we said that Carnot's classic model of an idealized heat engine resulted in the following properties of the entropy function:

$$d\mathrm{S} = Dq_{rev}/T \tag{6.18}$$

$$T\Delta \mathrm{S} \geq q \tag{6.19}$$

$$\Delta \mathrm{S}_{\mathrm{U,V}} \geq 0 \tag{6.20}$$

where the equals signs refer to reversible processes and the inequalities to all real, spontaneous processes. We observed that the increase in entropy for real processes in isolated systems was an all-important criterion of spontaneity. Then this property was used to derive the other useful criteria of spontaneity (which apply to other types of systems): $\Delta G_{T,P} \leq 0$, $\Delta A_{T,V} \leq 0$, $\Delta H_{S,P} \leq 0$, etc. We can now re-examine equations (6.18)–(6.20) in a different light. The following approach was developed by L.K. Nash (1972). We will give an abbreviated, slightly modified version of his treatment.

6.4.1. *Entropy Increase in Isolated Systems*

Let us look at the most important relationship, equation (6.20), first. This time around we know that $\mathrm{S} = k \ln W$. For real chemical systems of geological interest, W is usually far too complicated to estimate theoretically, but the concept it expresses—the degree of disorder—remains invaluable. We have seen that in the equilibrium state, W is maximized since the most random configuration is the most probable. Once again, consider the simple process of mixing two substances contained in an isolated box and initially separated from each other by a barrier. As we have observed, the total entropy of the two subsystems immediately after removing the barrier would be

$$\begin{aligned} S_{total} &= S_1 + S_2 \\ &= k \ln(W_1 \cdot W_2) \end{aligned}$$

where W_1 and W_2 represent the relative disorder (or probability) of each initial subsystem. For the combined systems, mixing or redistribution will occur if there is a possible new equilibrium state with still higher total W_{total}. W_{total} cannot decrease as this would constitute unmixing to a less probable state (for example, two gases separated by a vacuum). At the least, W may remain constant—this would happen, for example, if the two gases were initially identical in every aspect (the equilibrium condition). Thus as the barrier or constraint is lifted,

$$\Delta(W_1 \cdot W_2) \geq 0$$

or

$$\Delta S_{U,V} \geq 0 \tag{6.20}$$

The inequality refers to the approach to equilibrium, and the equality to equilibrium (or real and reversible processes, respectively). The subscripts denoting constant U and V are necessary because the total system is isolated. Therefore equation (6.20) tells us something we already know intuitively—that any process in a completely isolated box will stop or equilibrate at the state of maximum probability. That seems self-evident, but we also realize now that this state represents the maximum permissible disorder or randomness, and the maximum attainable entropy.

6.4.2. *dS versus Dq/T*

Next let us see how the relationship $dS = Dq_{rev}/T$ can be derived statistically. This time we consider the change in W of a system as its energy is increased by reversible input of an infinitesimal amount of heat at constant T, V and total composition (therefore a closed system). Presume that the system adjusts simply by redistributing its molecules among the allowed energy levels. Neither the energy nor the degeneracy of any of these levels is allowed to change.

The probability of the initial configuration is given by equation (6.4), and each individual N_i can be calculated from the Boltzmann distribution. This will then give us an equation relating W to the energies and molecular population of each energy level. Differentiating this equation should produce the desired change in W with the tiny change in total energy and accompanying adjustment of the molecular configuration.

The logarithm of equation (6.4) is

$$\ln W = \ln N! - \sum_i \ln Ni! \tag{6.21}$$

Differentiating and holding the total number of particles N constant yields

$$d \ln W = -\sum_i d \ln N_i! \tag{6.22}$$

Introducing Stirling's approximation (6.5) for the factorials of large numbers, gives for the differential of each $\ln N_i$ on the right hand side,

$$d \ln N_i! = d(N_i \ln N_i - N_i) = \ln N_i \cdot dN_i$$

and therefore,

$$d \ln W = -\sum_i \ln Ni \cdot dNi \tag{6.23}$$

The Boltzmann distribution (6.12) for the ratio of the number of particles in any two states i and 0 is

$$(N_i/N_0) = (g_i/g_0)e^{-(\epsilon_i - \epsilon_0)/kT} \tag{6.24}$$

If we let subscript $_0$ represent the ground state with energy $\epsilon_0 = 0$, then we may rewrite (6.24) as

$$\ln N_i = \ln(N_0 g_i/g_0) - \epsilon_i/kT \tag{6.25}$$

Substituting (6.25) for $\ln N_i$ into equation (6.23) gives

$$d \ln W = -\ln(N_0 g_i/g_0) \sum_i dN_i + \frac{1}{kT} \sum \epsilon_i dN_i \tag{6.26}$$

Note that N_0 is considered fixed by choice of the original configuration for this summation, and the degeneracies g_i and g_0 remain constant. The first term on the right hand side of (6.26) vanishes because the total composition of the system is fixed:

$$\sum_i N_i = N_{total} = constant$$

and therefore

$$\sum_i dN_i = dN_{total} = 0$$

The second summation in equation (6.26) represents the differential of the total energy: $E = \sum_i N_i \epsilon_i$, and $dE = \sum_i \epsilon_i dN_i$. (The ϵ_i are constant since we are not permitting any change in the energies of each energy level, but simply a redistribution of molecules among the original levels.) With these substitutions, equation (6.26) becomes

$$d \ln W = \frac{1}{kT} dE \tag{6.27}$$

This is the desired relationship for the molecular readjustment or change in W caused by a small and reversible change in internal energy at constant temperature, volume, and composition. As would be expected, the relative disorder, W, increases with internal energy.

Now this was a constant volume process, which rules out $P\Delta V$ work. According to the First Law, $dE = Dq + Dw$, the change in internal energy in this case is entirely due to flow of heat, Dq. Recalling that $\mathrm{S} = k \ln W$, we can now rewrite equation (6.27) in terms of entropy and heat, as originally desired:

$$d\mathrm{S} = Dq_{rev}/T \tag{6.18}$$

Thus we have observed that the expected increase in disorder with increasing internal energy expressed by equation (6.27) is directly related to the thermodynamic definition (6.18) of the entropy. It now should make intuitive sense that adding heat to a system will increase its disorder, hence its entropy.

To briefly summarize, we have rederived equation (6.18) by combining the relation $\mathrm{S} = k \ln W = k \ln(N!/\Pi Ni!)$ with the Boltzmann distribution for the energy of the N molecules in all i levels. This gave equation (6.27), which directly relates W (or S) and energy, E. An increase in internal energy, caused by adding heat, necessarily raises both disorder and entropy.

While this provides the conceptual link we wanted between entropy, disorder, and heat, we still have some tidying up to do in our derivation of equation (6.18). As things stand now, this equation only applies to the very restricted kind of process for which it was derived (constant V, T, N, and degeneracy) It remains to be shown that equation (6.18) applies to reversible processes in general.

Imagine two systems A and B separated by a heat-conducting barrier and isolated from the rest of the universe (this might look like the two boxes in Figure 6.1a). Let system A be subjected to exactly the same constraints just used to derive equation (6.18). For system B though, let's lift all restrictions save one—N remains constant, so no chemical reactions are allowed. Heat will flow one way or the other through the barrier until both systems equilibrate. As we observed in rederiving equation (6.20), the two systems will equilibrate at a state of joint maximum probability, $W_\mathrm{A} \cdot W_\mathrm{B}$. If this is a reversible process, the condition for the maximum is

$$dW_\mathrm{A} \cdot dW_\mathrm{B} = 0$$

or

$$d(W_\mathrm{A} W_\mathrm{B}) = 0$$

Substituting $\mathrm{S} = k \ln W$,

$$d\mathrm{S}_\mathrm{A} + d\mathrm{S}_\mathrm{B} = 0 \tag{6.28}$$

Because system A meets all conditions used to rederive equation (6.18), we can substitute $d\mathrm{S}_\mathrm{A} = Dq_{rev}/T$ in equation (6.28). The only heat flow allowed was from A to B or vice versa, so $Dq_\mathrm{A} = -Dq_\mathrm{B}$. Finally, at thermal equilibrium, both systems will have the same temperature. Thus we can now rewrite equation (6.28) as

$$Dq_\mathrm{A}/T + d\mathrm{S}_\mathrm{B} = 0$$

or

$$-Dq_\mathrm{B}/T + d\mathrm{S}_\mathrm{B} = 0$$

or

$$d\mathrm{S}_\mathrm{B} = Dq_\mathrm{B}/T$$

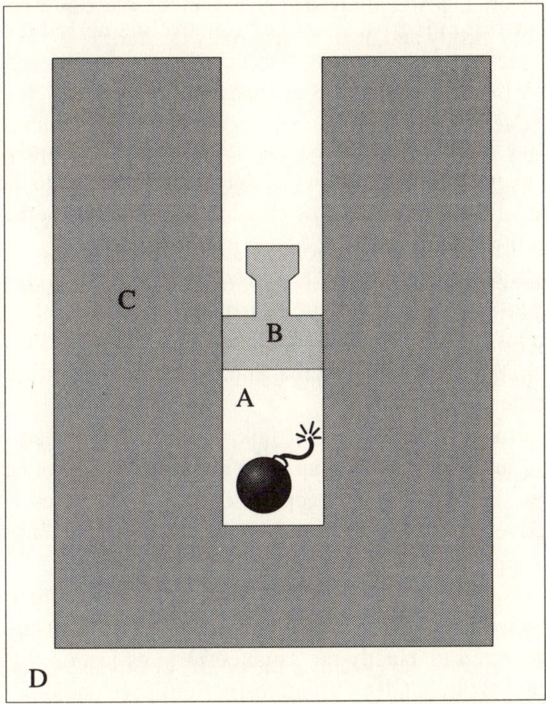

FIG. 6.6. Reaction chamber (A) within a large heat sink (C and B), capped by a sliding piston (B).

Unlike system A, system B is quite general, so the condition (6.18) must also apply to more general cases, which completes the derivation. Thus,

$$dS = Dq_{rev}/T \tag{6.18}$$

applies to *any* closed system, provided no chemical reactions occur, and the process is reversible.

6.4.3. ΔS *versus* q/T

Finally, by generalizing still further and removing all restrictions (including reversibility), we should be able to use the statistical approach to rederive equation (6.19),

$$T\Delta S \geq q \tag{6.19}$$

This time, imagine a chamber in which any kind of physical or chemical process occurs. As shown in Figure 6.6, this could take the form of a deep hole (A) drilled in a large metal block (C) and capped by a heavy metal piston (B). This apparatus might be placed inside a rigid box (D) that is perfectly insulating and impermeable to matter, so that the contents of the box are thermodynamically isolated from the surroundings. The piston is frictionless and free to slide up and down the hole, and it, together with the metal block act as a heat sink or reservoir.

Next, imagine that any kind of process occurs in the reaction chamber. This might be quite irreversible if we want, so let's place an ignited firecracker in the hole and seal everything up. If the firecracker is not too large, it will blow the piston further up (but not out of) the hole. The heat of the explosion will be absorbed by the metal block and piston. At this point we must impose two hypothetical conditions. First, the heat sink must be so enormous, relative to the reaction, that it does not change temperature significantly. Second, the heat of reaction must be transferred reversibly, by infinitesimally small increments, to the piston and block. It doesn't matter that this is impossible, because we are calculating a hypothetical limit. What does matter is that the process in the reaction chamber can be as real (and irreversible) as we wish to make it.

Using the statistical argument once again, we have just seen that for any real process in an isolated system,

$$\Delta W > 0$$

and

$$\Delta \ln W > 0 \tag{6.29}$$

In other words, the state of disorder can only increase in a spontaneous change. We also know that W for the total system is the product of its individual parts:

$$W_{total} = W_{\mathrm{A}} \cdot W_{\mathrm{B}} \cdot W_{\mathrm{C}} \tag{6.30}$$

Again, this is because W measures probability (as well as randomness), and probabilities are multiplicative. For our present three-part system, the mechanical ($P\Delta\mathrm{V}$) work done in raising the piston should have no effect on its molecular configuration. However, the heat absorbed by both piston and block will certainly change their configurations, so we will lump both together as W_{BC}. We fully expect some change of molecular configuration inside chamber A, so the total probability parameter is $W_{total} = W_{\mathrm{A}} \cdot W_{\mathrm{BC}}$. Combining this with relation (6.29), we have for the sum of all processes in the isolated system, $\Delta \ln(W_{\mathrm{A}} \cdot W_{\mathrm{BC}}) > 0$ or

$$\Delta \ln W_{\mathrm{A}} + \Delta \ln W_{\mathrm{BC}} > 0 \tag{6.31}$$

Since $\mathrm{S} = k \ln W$, we can rewrite (6.31) as

$$\Delta \mathrm{S}_{\mathrm{A}} + \Delta \mathrm{S}_{\mathrm{BC}} > 0$$

Because heat is transferred reversibly to B and C, equation (6.19) can be substituted for $\Delta \mathrm{S}_{\mathrm{BC}}$,

$$\Delta \mathrm{S}_{\mathrm{A}} + q_{\mathrm{BC}}/T > 0 \tag{6.32}$$

Finally, the heat lost by reaction chamber A must be exactly that gained by heat sinks B and C, so $q_{\mathrm{A}} = -q_{\mathrm{BC}}$ and (6.32) becomes

$$\Delta \mathrm{S}_{\mathrm{A}} - q_{\mathrm{A}}/T > 0$$

or

$$T\Delta S_A > q_A \tag{6.33}$$

which is the desired relation for the reaction chamber.

Thus we have shown, based on a statistical argument that for any spontaneous, real process, the entropy change or total disorder will always exceed that predicted from heat flux alone. Relationship (6.18), $dS = Dq_{rev}/T$, is therefore a lower limit to this change, applicable only to ideal, reversible processes.

6.4.4. *Summary and Cautionary Note*

Using the statistical mechanical approach, we have been able to rederive equations (6.18)–(6.20) without any mention of steam engines or idealized Carnot cycles. These equations form the basis for much of the rest of thermodynamics, as we have already begun to see in Chapter 5. These few relationships are so useful because they serve as pointers or criteria for the spontaneous direction of any process. Hopefully the statistical approach clarifies much of this, in the sense that we conceive of entropy as a measure of disorder or randomness. The most random permissible state is also the most probable statistically. It is self-evident that spontaneous processes head in the most probable direction; by doing so, they maximize entropy.

However, we should emphasize that when we refer to disorder or probability, we include not only the distribution of gas molecules in a mixture or cards in a deck, but also the distribution of the energy quanta among permissible energy states in the system. If we are shuffling cards or rolling dice this energy distribution does not change, so that entropy change is given by the configuration of the objects themselves. However, in processes in which an energy change takes place, a great deal of the entropy change is accounted for by the changing distribution of energy quanta within the molecules of the system, and the molecules themselves may or may not change their ordering in the same direction. In other words, spontaneous processes, having a positive ΔS, and even in isolated systems, can be associated with an apparent increase in order at the molecular level.

One example will suffice. Consider a kilogram of supercooled water plus vapor at -10°C in an isolated system. When the water spontaneously freezes to ice, the heat released warms the system to 0°C, at which point the process stops at equilibrium. Clearly the system at equilibrium (water plus ice at 0°C) would be said by most observers to be more ordered than the initial state (water at -10°C). We could also use a mass of water to start with such as to give only ice and vapor at equilibrium. This type of situation is common to all supercooled liquids and supersaturated solutions, and in all cases there is no doubt about the increase in entropy, nor in the apparent increase in order. The problem is of course that the entropy change (given by $\int (C_p/T)dT$) is accounted for almost entirely by the change in the energy distributions in the molecules, which have warmed up. In other words, we can be fooled by looking at molecular ordering, while ignoring relatively small temperature (energy) differences.

Thus although there is a relationship between entropy and disorder, it is not always on the macroscopic scale, i.e., it does not always accord with what human observers would call disorder, especially when energy changes are involved. Nevertheless, it is a measure of disorder, and this is usually apparent even on the macroscopic scale, so that it is a useful way to think of entropy.

6.5. THE THIRD LAW

6.5.1. *Background*

According to the Boltzmann relation,

$$S = k \ln W \tag{6.3}$$

the entropy of a perfectly ordered substance must be zero, since there should only be one way of arranging the atoms, ions, or molecules in a perfect crystal ($W = 1$). It seems reasonable to expect that the chance of creating a perfect crystal might improve at temperatures approaching absolute zero where molecular motion ceases. In fact, as $T \to 0\,K$, all molecular and electronic energies should drop to the ground state. If the ground state is completely unique then equation (6.4) predicts $W = N!/\Pi Ni! = N!/N! = 1$ and $S = 0$.

Arguments such as these suggest but do not prove that the entropy of pure crystalline substances must be zero at 0 K. In fact, it can be proved from the Second Law that absolute zero temperature itself is unattainable (although it has been approached very closely); it is also debatable if the ground state of energy is truly unique for complex crystals. However, experimental evidence does support the general idea that the entropies might either vanish or reach a minimum at absolute zero. Even if there were no experimental data whatsoever, it would still be convenient to define a reference point of $S = 0$ at 0 K on the entropy scale, analogous to absolute zero on the temperature scale.

Historically, Nernst originally postulated in 1906 that at $T = 0$ K, the entropy change accompanying any process vanishes.[8] This statement was refined by Planck in 1912, and then by G.N. Lewis in the 1920s, who formulated the Third Law as we know it today: *The entropy of all pure, crystalline, perfectly ordered substances is zero at absolute zero, and the entropy of all other substances is positive.*

Entropy, alone among the thermodynamic functions, is therefore measured on an absolute scale having the reference point zero at 0 K for pure, perfectly crystalline materials. As we shall see, the other functions (G, H, A, U) are measured against an arbitrary standard state and are assigned relative, rather than absolute values. (That is why entropy is designated by an absolute symbol, S°, and all others by relative symbols, e.g., $\Delta_f G^\circ$, in thermochemical tables, as we shall see in Chapter 7). Apart from this, most of thermodynamics as presented here would survive intact if the Third Law had never been discovered.

6.5.2. *Tests of the Third Law*

There are two ultimate experimental tests of the third law. The first might compare the entropies in the limit as $T \to 0\,K$ of perfectly and poorly crystalline forms of the

[8] An interesting first-hand account of the historical development of the Third Law is given by Lewis and Randall, 1st edition, (1923, Chapter 31).

same compound. The entropy of the perfect crystal should approach zero, while that of the imperfect form should have a "frozen in" *residual entropy* attributable to the disorder in its lattice. The Third Law also predicts that the entropies of two different perfectly crystalline forms of the same compound (such as graphite and diamond) should be identically zero at absolute zero, even though one of them is metastable at higher temperature. If the entropy of two such crystals is identically zero, then there should be no entropy change, ΔS_0, when one transforms to the other in the limit as $T \rightarrow 0\,K$.

In fact, it is not possible to make the necessary measurements for these tests at temperatures approaching absolute zero. Instead, heat capacity data at cryogenic temperatures (but still well above 0 K) are extrapolated to 0 K and higher temperature data on phase transformations can be used indirectly as shown in Figures 6.7a and 6.7b. This approach is possible because changes in thermodynamic functions of state such as S depend only on the initial and final states, not on the process used to connect these states.

Referring to Figure 6.7a, we want to know the entropy change for process A $\rightarrow$ D (the difference in entropy between crystalline and non-crystalline glassy glycerol). Since we cannot perform this particular experiment, we use the circuitous route A $\rightarrow$ B $\rightarrow$ C $\rightarrow$ D, which ultimately leads to the same initial and final states. For the process A $\rightarrow$ B we must know or extrapolate the heat capacity C_p of glycerol as a function of temperature from its melting temperature T_m, down to absolute zero. Because $dS = Dq/T$ and $Dq = C_p dT$, we calculate the entropy change for heating glycerol from 0 K to T_m by

$$\Delta_{\mathrm{A}\rightarrow\mathrm{B}} S = \int_0^{T_m} (C_{p_{crystal}}/T)dT$$

Next, crystalline glycerol must be melted at T_m, which involves the heat of fusion $\Delta_m H$ and the entropy change

$$\Delta_{\mathrm{B}\rightarrow\mathrm{C}} S = \Delta_m H/T$$

Finally, liquid glycerol is frozen to glass (hence no heat of fusion), and cooled from T_m to 0 K, and for this,

$$\Delta_{C\rightarrow D} S = \int_0^{T_m} (C_{p_{glass}})/T)dT$$

The entropy change for this whole cycle A $\rightarrow$ B $\rightarrow$ C $\rightarrow$ D must be identical to the direct process A $\rightarrow$ D (since S is a function of state). The experimental result is $\Delta_{\mathrm{A}\rightarrow\mathrm{D}} S = 4.6\ \mathrm{cal\,K^{-1}\,mol^{-1}}$. As expected, the glass has a higher entropy than the crystal because it is more disordered. If the Third Law is correct, and the crystal truly perfect, it will have zero entropy at 0 K, and the glass will have a residual entropy $S_0 = 4.6\ \mathrm{cal\,K^{-1}\,mol^{-1}}$ due to frozen-in defects or randomness.

The second common example, depicted in Figure 6.7b, is less obvious, but really a better test of the Third Law. Here we wish to know if there really is no entropy

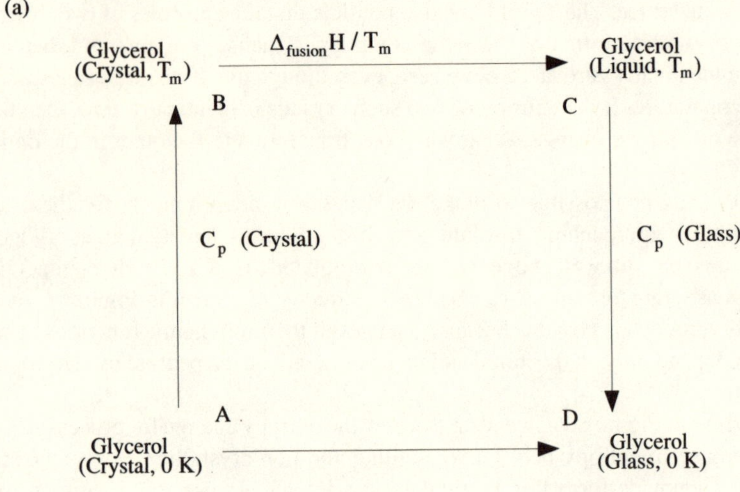

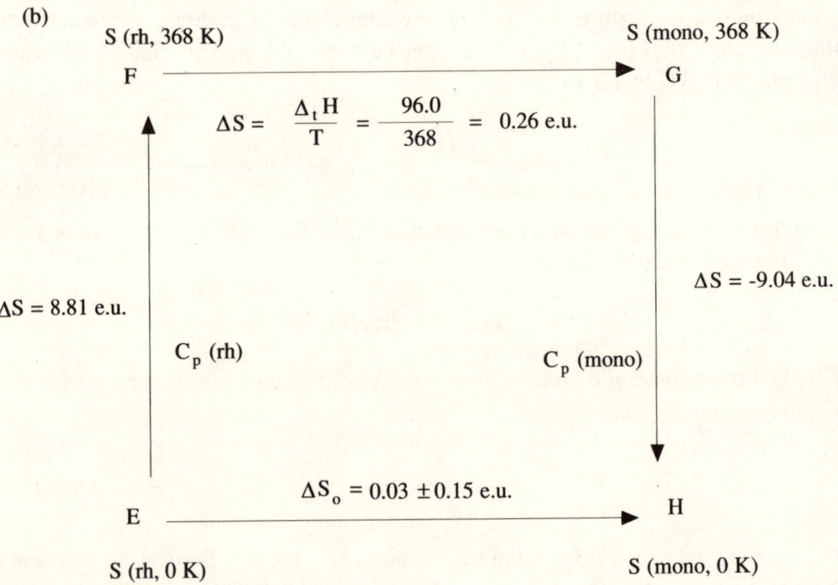

FIG. 6.7. Indirect calculation of the difference in entropy between crystalline and amorphous glycerol (a), and between monoclinic and rhombic sulfur (b) at absolute zero temperature, after Dickerson (1969).

difference between monoclinic and rhombic sulfur at 0 K as predicted. There clearly must be some difference at higher temperatures because the metastable monoclinic form spontaneously changes to the rhombic structure. Once again, we cannot do this experiment at absolute zero, so we follow the hypothetical cycle E $\rightarrow$ F $\rightarrow$ G $\rightarrow$ H : Rhombic sulfur is heated to 368 K, converted to monoclinic sulfur, and then cooled back down to absolute zero. ΔS for this cycle is 8.81 + 0.26 − 9.04 = 0.03 ± 0.15 cal K^{-1} mol^{-1}. It does appear that the entropy change for transformation of rhombic to monoclinic sulfur is zero at 0 K within experimental error.

6.5.3. *Exceptions to the Third Law*

There do exist some minor difficulties in practical application of the Third Law. One interesting problem is statistical mechanical in nature and could never have been predicted on the basis of classical thermodynamics alone. As noted above, the ground state of energy must be unique to give $W = 1$ (and S = $k \ln W = 0$). If it is degenerate in any way (if different but energetically equivalent configurations in geometry, or molecular, electronic, and even nuclear energies are possible), then $W > 1$ and S > 0. In fact different nuclear spins are allowed at $T \rightarrow 0$. As a simple example, frozen H_2 molecules have a residual entropy at 0 K depending on the ratio of two possible nuclear spins. Isotopes of the same element pose another difficulty. Water, for example, typically contains four common isotopes, ^{1}H, ^{2}H, ^{16}O, ^{18}O. Since atoms of the different isotopes are distinguishable and can occupy the same sites in the crystal lattice of ice interchangeably, different arrangements or configurations must be possible in isotopically mixed water (and other compounds) at absolute zero. These different arrangements will give rise to a residual configurational entropy at 0 K. Finally, there are some cases where the molecules of a crystal may exist in more than one possible arrangement at absolute zero. This happens with compounds such as CO, NO, and N_2O where the energy differences $(\epsilon_1 - \epsilon_2)$ between arrangements such as (NNO NNO NNO) and (NNO ONN NNO) are so slight that their relative probability, $e^{-(\epsilon_1 - \epsilon_2)/kT}$ is essentially unity even at low temperatures. Again a random configurational entropy is typically frozen in near 0 K and reflected in the heat capacities measurements from which low-temperature entropies are calculated.

We can easily calculate the maximum value of the residual configurational entropy at absolute zero either in the case of H_2 or nitrous oxide in which there are two possible arrangements per molecule (two nuclear spins for hydrogen, two geometric positions for N_2O). If the system contains Avogadro's number, N_a, of molecules, there are 2 possible states for each, for a total of 2^{N_a} possible arrangements in the crystal. Because S = $k \ln W$, we calculate $S = (R/N_a) \ln 2^{N_a} = R \ln 2 = 1.38$ cal K^{-1} mol^{-1}. This is a small, but not negligible effect.

6.5.4. *Entropies of Ideal Mixing*

The above example is simply a special case of the problem of computing entropies of ideal mixtures, which is very important in dealing with solutions in gases, liquids, and even solids. We can generalize by considering first a mixture of two kinds of molecules A and B. Once again applying equation (6.4), we obtain

$$W = N!/N_A!N_B!$$

for the total number of different ways of arranging these molecules in a mixture. It is assumed here that all N_A molecules of species A are indistinguishable, the same for N_B, and that $N = N_A + N_B$. Applying equation (6.3) for the entropy of this mixture,

$$\begin{aligned} S &= k \ln W \\ &= k(\ln N! - \ln N_A! - \ln N_B!) \end{aligned}$$

Introducing Stirling's approximation (6.5) for factorials of large numbers,

$$S = k(N \ln N - N_A \ln N_A - N_B \ln N_B - N + N_A + N_B) \qquad (6.34)$$

Now we can simplify this by defining a mole fraction as

$$X_A = N_A/N$$

which reduces (6.34) to

$$S = -k(X_A \ln X_A + X_B \ln X_B) \qquad (6.35)$$

Since $k = R/N$, if we set N equal to Avogadro's number in (6.35) then it becomes

$$S = -R(X_A \ln X_A + X_B \ln X_B) \qquad (6.36)$$

Where S is the entropy per mole on mixing two components in the proportion $X_A/X_B = N_A/N_B$. We could rederive this expression in the same way for a mixture of any number n of components, giving the general equation

$$S = -R \sum_{i=1}^{n} X_i \ln X_i \qquad (6.37)$$

This gives the *configurational entropy of mixing* for any number of components. It can be used to calculate residual entropies at absolute zero due to impurities, imperfections, nuclear spin, isotopes, etc., simply by considering the imperfections as one component of a mixture. Equation (6.37) applies equally well to *ideal mixtures* at higher temperatures, as we shall see in Chapter 10.

As an example of the first application, we can use equation (6.37) to calculate the residual entropy of H_2 at absolute zero due to the normal abundance of its two stable isotopes $^1H = 99.9844\%$, $^2D = 0.0156\%$.

$$\begin{aligned} S &= -R[0.999844 \cdot \ln(0.999844) + 0.000156 \cdot \ln(0.000156)] \\ &= 3.03 \times 10^{-3} \text{ cal K}^{-1} \text{ mol}^{-1} \end{aligned}$$

Interestingly, this is a much smaller effect than that arising from nuclear spin multiplicity in solid H_2 calculated above (1.38 cal K^{-1} mol^{-1}).

Finally, we should point out that, while the exceptions to the Third Law noted above may be a headache for scientists who measure calorimetric properties of materials, they pose no practical problems in most chemical applications. Chemical reactions alone do not change nuclear spin, and in many cases do not alter isotope ratios significantly, so that configurational contributions to the entropy of reactants are normally balanced by those of the products in a reaction. In most cases these effects are thermodynamically minor or insignificant.

6.6. THE ENTROPY OF THE UNIVERSE REVISITED

Clausius' statement of the Second Law, "The entropy of the world (universe) tends towards a maximum," has stimulated more discussion than he could have possibly imagined. Thanks in large part to the statistical interpretation, which allows entropies to be calculated from arrangements and probabilities rather than from measured heat capacities, the concept of entropy has found its way into virtually every field of investigation, where it is applied with varying degrees of rigour. For example it has been used to show that Shakespeare's style has roughly half the entropy of Thomas Mann's (e.g., Gasser and Richards, 1974, p. 129). The usefulness, and even the meaning, of many of these applications is somewhat questionable.

In the latter part of the 19th century Clausius' law was also used to predict the "heat death" of the universe. Just as gas molecules left to their own devices in a closed box will reach a state of maximum randomness, it was argued the same thing would ultimately happen to the universe. At some distant time, the entropy of the universe would be maximized, a state of final equilibrium would prevail, and all processes would cease. This rather bleak outlook was shattered in 1929 when Edwin Hubble discovered that the universe is expanding. In a constantly changing universe, total equilibrium is unattainable and entropy must keep changing.

6.6.1. *The Standard Model of the Universe*

Since Hubble's time, and chiefly in the past decade, a theory for the origin and growth of the universe has become so widely accepted that it is now called "the standard model." Very briefly, this model states that the cosmos originated with the "big bang" in which all energy/mass exploded outward from an initial, unthinkably hot and dense fireball. We observe the aftermath of this explosion 10 to 20 $\times 10^9$ years later in the continuing expansion of the universe and the decline of the cosmic background temperature to its present value of 3 K. Depending on the average density of the cosmos (which is not sufficiently well known) the expansion should either continue *ad infinitum* or the whole system will slow down, reverse, and contract ("open" and "closed" universes). In the latter case the universe would collapse to another fireball and then possibly explode again, oscillating between expansion and contraction over great periods of time.

For present purposes let's consider only the first of these two alternatives—the open, perpetually expanding model. Roughly 10^{-2} seconds after the big bang the universe contained a random mixture, mostly of electrons, positrons, neutrinos, antineutrinos, and photons with a tiny, one part per billion admixture of heavier neutrons and protons. At that point, the temperature was 10^{11} K and the density about 4×10^9 g cm^{-3}. The ratio of roughly 10^9 photons per nuclear particle (or nucleon) has continued to the present day. After roughly 3 minutes, the universe cooled to $\sim 10^9\ K$; the positrons had essentially disappeared (leaving a slight excess of electrons in their wake), and neutrons and protons began to combine as simple nuclei (^{1}H, ^{2}H, ^{3}He, ^{4}He). Finally after $\sim 700,000$ years, the temperature fell to 3000 K, cool enough for electrons to associate with nuclei as atoms; at this point stars and galaxies began to form. Living organisms appeared near at least one of these stars after an additional 10^{10} years (very approximate). The universe will continue to expand; in the distant future, the nuclear processes of stars will slowly die out, leaving behind black dwarf and neutron stars, black holes, and innumerable cold planets.

6.6.2. *The Expanding Universe versus the Second Law*

One of the many interesting features about the standard model of the cosmos is that it seems, at first sight, to contradict the Second Law of thermodynamics in every detail. The universe begins in a state of maximum apparent disorder and maximum temperature, and then proceeds systematically toward states of ever-increasing order. Very early on, positrons and an equivalent number of electrons are annihilated; some three minutes later, neutrons and protons combine as nuclei, then after $\sim 700,000$ years, nucleons soak up the few remaining electrons to make atoms. The atoms combine as molecules and then as clumps of matter ranging from dust up to stars, galaxies, and galaxy superclusters. The lighter nuclei combine to make all the heavier ones in stellar nucleosynthesis reactions. The galaxies ultimately collapse as black holes. In the midst of all this, the intricately structured processes of life begin on Earth and probably elsewhere. Finally, the background temperature decreases constantly, asymptotically approaching absolute zero, where the Third Law informs us entropy will be zero or minimized. We cannot get out of this conundrum using the simple analogy of particles expanding in a fixed space (for which entropy does increase) because space itself is expanding: any volume expanding along with a homogeneous region of space contains a constant number of particles and massless particles (or quanta), hence this volume retains constant entropy. This is somewhat analogous to the adiabatic expansion of an ideal gas, for which $\Delta S = 0$.

Living organisms from Protozoans to *Homo sapiens* are the most highly complex and elaborately ordered chemical systems known, at least to us. The internal arrangement of molecules in an organism is necessarily so critical and precise that very few alternatives are allowed. Mutation or random rearrangement of even a single strand of DNA, for example, can have a devastating impact on an entire organism. Since few alternative molecular states are permitted, living things must have extraordinarily low entropies according to (6.3).

If the universe is becoming more ordered in all these ways, how can its entropy "tend toward a maximum"? This is obviously a good riddle, and in recent years it has, in one form or another, attracted the attention of various eminent physicists (see reviews and discussions by Weinberg, 1977; Davies, 1977; Islam, 1979; Dyson, 1979; Frautschi, 1982).

A detailed account of these problems is beyond the scope of this book, despite their fascination. Many aspects of the arguments are accessible only to specialists, but even a superficial reading of the above sources makes it clear that most cosmic events such as nuclear fusion, element formation, and formation, coalescence, and decay of black holes actually generate enormous amounts of entropy, relative to processes familiar to us on Earth. The major factor responsible for this, omitted in simplified accounts such as given above, is that we must take into account not only neutrons, protons, etc., but the enormous number of massless particles generated, such as photons and neutrinos. When this is done, the entropy balance is profoundly changed.

As for the appearance and evolution of life, it is extremely unlikely that this represents a violation of the Second Law because although living matter has a low entropy, the Earth, on which life developed and evolved, is not an isolated system. Although the problem cannot be quantified in the same way, the production of low-entropy living matter is probably no more a violation of the Second Law than the production of low-entropy ice in the bucket in §5.7. As Nash (1971) has pointed out, the existence of highly ordered living organisms is "only a minute upstream eddy in the ongoing torrent of increasing entropy."

6.6.3. *Conclusion*

There is something inherently fascinating about problems relating to the origin and development of life and of the universe, and the entropy problem has long occupied an important role in discussions of them. This has been made possible to a large degree by the enlargement of our understanding of entropy through linking it to the statistical and probability concepts discussed in this chapter.

Discussion of these problems will continue, and presumably our understanding of them will increase, but at the present time it seems that we are justified in using the Second Law at all scales, from microscopic to cosmic.

7

THERMODYNAMIC PROPERTIES OF SIMPLE SYSTEMS

Thermodynamics is an experimental science and not a branch of metaphysics.

McGlashan (1979) p. 111.

7.1. PROBLEM—NO ABSOLUTE ENERGIES

We have now introduced several thermodynamic parameters that are useful in dealing with energy transfers (U, H, G, etc.). We wish now to see how these quantities are measured and where to find values for them. In later chapters we will see how they are used in detail.

However, we have an immediate problem in that we *cannot* measure the energy parameters U, H, G and A, as discussed in Chapter 4. Because we do not know the absolute values of either the total or molar version of these variables, we are forced to deal only with their changes in processes or reactions of interest to us. But we obviously cannot tabulate these changes for every reaction of potential interest; there are too many. We must tabulate some sort of energy term for each pure substance so that the changes in any reaction between them can be calculated. In the example in §5.7 of water at −2°C changing to ice at −2°C, we said that ΔG was negative. How can we know this without carrying out a research program on the thermodynamic properties of ice and supercooled water? We begin by explaining how this is done.

7.2. FORMATION FROM THE ELEMENTS

The problem created by not having absolute energy values is handled very conveniently by determining and tabulating, for every pure compound, the difference between the (absolute) G or H of the compound itself and the sum of the (absolute) G or H values of its constituent elements. In other words, ΔG or ΔH is determined for the reaction in which the compound is formed from its elements (in their stable states). These differences can be determined experimentally in spite of not knowing the absolute values involved. For example, the "Gibbs free energy of formation" (i.e. formation from the elements) of water is

$$\Delta_f G^\circ_{H_2O(l)} = G^\circ_{H_2O(l)} - G^\circ_{H_2} - \frac{1}{2}G^\circ_{O_2}$$

and that for ice is

$$\Delta_f G^\circ_{H_2O(s)} = G^\circ_{H_2O(s)} - G^\circ_{H_2} - \frac{1}{2}G^\circ_{O_2}$$

These are the free energies of reaction ($\Delta_r G$) for the reactions

$$H_2 + \frac{1}{2}O_2 = H_2O(l) \text{ (i.e. water)}$$

and

$$H_2 + \frac{1}{2}O_2 = H_2O(s) \text{ (i.e. ice)}$$

The subscript f signifies formation from the elements and the superscript ° signifies standard conditions of some kind, normally a pressure of one bar and pure substances, but any specified temperature.[1] If both these quantities (that is, the free energies of formation of water and of ice) are known for our conditions of −2°C and one bar, then for the reaction

$$H_2O(l) = H_2O(s), -2°\text{C, 1 atm}$$

we write

$$\Delta_r G = \Delta_r G^\circ = \Delta_f G^\circ_{H_2O(s)} - \Delta_f G^\circ_{H_2O(l)}$$

and we find, as discussed earlier, that this quantity is negative and that therefore ice is more stable than water at −2°C, one bar. It is important to note here that for any balanced reaction, the elements always cancel out. Thus

$$\begin{aligned}\Delta_f G^\circ_{H_2O(s)} - \Delta_f G^\circ_{H_2O(l)} &= G^\circ_{H_2O(s)} - G^\circ_{H_2} - \frac{1}{2}G^\circ_{O_2} \\ &\quad -(G^\circ_{H_2O(l)} - G^\circ_{H_2} - \frac{1}{2}G^\circ_{O_2}) \\ &= G^\circ_{H_2O(s)} - G^\circ_{H_2O(l)} \qquad (7.1)\end{aligned}$$

In other words, the comparison we are making is really between the absolute quantities $G^\circ{}_{H_2O(s)}$ and $G^\circ{}_{H_2O(l)}$, not between arbitrary functions of these quantities. Determining the differences between G or H of compounds and their constituent elements thus allows convenient tabulations to be made which facilitate calculation of ΔG or ΔH for any reaction.

Note too that we write the free energy of oxygen as $G^\circ_{O_2}$ and not G°_O or $\frac{2}{3}G^\circ_{O_3}$. G°_O is not a different way of writing $\frac{1}{2}G^\circ_{O_2}$ but signifies the Gibbs energy of a mole of pure nascent oxygen, and $G^\circ_{O_3}$ is the Gibbs energy of ozone. "Formation from the elements" normally means formation from the elements in their stable states, except at temperatures above the boiling or sublimation temperature of the reference elements. For example, sulfur properties are taken as those of orthorhombic sulfur up to 368.54 K, monoclinic sulfur from 368.54 K up to the melting point at 388.36 K, and liquid sulfur up to the boiling point at 716.9 K. Above the boiling point sulfur exists as a gas containing a mixture of species such as S, S_2, S_4, S_6, S_8, etc., the proportions varying with temperature, but the reference state is chosen as a hypothetical ideal gas composed only of S_2 because it happens that the properties of such a gas are known more accurately than those of the real equilibrium gas phase. Most other elements above their boiling points are treated as ideal monatomic gases.

Finally, we might emphasize that in spite of numerous statements to the contrary, the (absolute) enthalpies and free energies of the elements themselves are not assumed

[1] We will have a great deal more to say about these "standard conditions" in connection with activities, at which point their importance will become clearer. For the moment, the superscript ° basically means that the property is that of a reference substance, i.e. a specially chosen pure substance at some P and T. The P is usually one bar and the T is usually 298.15 K, but both can be any value.

to be zero, or set to zero by convention. The $H(=U+PV)$ and $G(=U-TS+PV)$ of both elements and compounds have finite, large absolute values as discussed earlier. We merely work with the differences between these quantities, and this involves no arbitrary conventions at all. It is however true that this leads to the result that $\Delta_f G°$of the elements themselves in their stable reference states is zero. For example,

$$\begin{aligned}\Delta_f G^\circ_{O_2} &= G^\circ_{O_2}(g) - G^\circ_{O_2}(g)\\ &= 0\end{aligned}$$

Tabulated values of $\Delta_f G°$ and $\Delta_f H°$ for the elements in their reference states at various temperatures therefore appear as a column of zeros—see Table 7.1 for example. This of course does not mean that the absolute values G and H are zero.[2]

Tables 7.1 and 7.2 are copied directly from the thermodynamic compilation of Robie, Hemingway and Fisher (1978), abbreviated as RHF. Many of the other standard thermodynamic data sets discussed later in this chapter are arranged in a similar fashion. We can now begin to examine some of the features of these tables a little more closely. First, we have just observed that $\Delta_f G°$ and $\Delta_f H°$ for the formation of $O_2(g)$ "from the elements" is zero at all temperatures because this is just the difference between the G (or H) of oxygen and the G (or H) of the elements making up oxygen, which is the same thing. We have not yet defined the equilibrium constant K (see Chapter 13), but for completeness we should point out that it is 1.0 and $\log K = 0$ for the reaction for the formation of oxygen from itself, giving us another column of zeros. Note too that the entropy of $O_2(g)$, $S°_T$, given in Table 7.1 is *not* equal to zero at any temperature shown; these are absolute entropies, not entropies of formation from the elements, as discussed in Chapter 6 and again later in this chapter. Table 7.1 is typical of data tables for the elements.

Table 7.2, for halite (NaCl) is more complicated and more typical of natural compounds. As with $O_2(g)$ and all other substances, the entropy $S°_T$ here is the absolute value for the forms of NaCl listed at the top of the Table—pure crystals for temperatures up to 1073.8 K, and liquid NaCl at higher temperatures. Now, however, the free energy, enthalpy, and $\log K_f$ for formation of NaCl from its elements are all non-zero because they refer to the reaction

$$\mathrm{Na} + \frac{1}{2}\mathrm{Cl_2} = \mathrm{NaCl}$$

This appears simple enough, but raises a point of possible confusion concerning the reference state used for the elements. As already noted, this is usually taken to be the most stable form of the elements at each T. For Cl this is simply $Cl_2(g)$ over the entire temperature range shown. However, Na poses a problem because it melts at 370.98 K and boils at 1175 K; hence the most stable reference phase for Na changes from solid to liquid to gas across this temperature range. Similarly, because NaCl melts

[2]Although the "formation from the elements" convention does not in turn require a convention that the free energy (or enthalpy, etc.) of the elements is zero, the convention adopted for the properties of the aqueous ions does. But even in this case, the convention is adopted to simplify things; it is not necessary (see Chapter 17).

at 1073.8 K, the RHF tables list thermodynamic properties for solid NaCl below the melting point and liquid NaCl above it.

The temperatures of all these phase transitions are noted in the footnotes to the Table (for the constituent elements) or in the title (for the compound itself). The phase changes are also indicated by horizontal dashed lines at the appropriate temperatures. The dashed line for the melting point of NaCl goes across the entire table, because all the thermodynamic properties listed refer to solid NaCl at lower T and liquid NaCl at higher T. The dashed lines for phase changes of the elements extend only across the right hand side of the Table, which gives the properties for formation of NaCl from the elements. All other properties in the remaining columns refer to NaCl alone and are hence not affected by changes in the constituent elements. As we shall see, it is necessary to include the enthalpy of melting and vaporization for transitions in the reference states of the elements when calculating $\Delta_f G^\circ$ etc., so these enthalpies are also given in the footnotes. The fact that the reference elements can undergo phase transitions as temperature is changed means that the tabulated values of $\Delta_f G^\circ$ or $\Delta_f H^\circ$ of compounds will have discontinuities at these temperatures, as shown in Figures 7.1 and 7.2. This is an inconvenience in fitting functions to the properties and in interpolation of intermediate values. There are also of course discontinuities at the transition temperatures of the compound itself. It will be pointed out when we discuss phase transitions that two phases of the same composition at equilibrium with each other will have different values of H, but the same values of G (§14.2.4). Thus $\Delta_f H^\circ$ values for compounds as a function of temperature have discontinuities at the phase transition temperatures both of the compounds and of their constituent elements, while $\Delta_f G^\circ$ values have changes in slope at these temperatures, as illustrated in Figures 7.1 and 7.2. $\Delta_f G^\circ$ values also have discontinuities where the properties of one of the constituent elements are not those of the stable form of that element, as discussed above for sulfur vapor (Figure 7.3).

Discontinuities resulting from phase transitions in the reference elements can pose difficulties when interpolating from RHF-type tables, where enthalpies and free energies of formation are usually tabulated for temperature increments of 100°C. The problem can be circumvented to some extent by using the "Gibbs energy function" or Giauque function, $-(G^\circ_T - H^\circ_{298})/T$, which is listed in the fourth column of these tables. The G and H symbols in this function refer to the absolute properties of the compound itself (not to the reference elements). Neither the G nor the H term is known, but the difference can be determined (§7.5.4), and since H°_{298} is a constant and G°_T is continuous (changing slope at phase transitions of the compound), the Gibbs energy function itself must be continuous. It can then be used to calculate free energy changes for reactions given $\Delta_r H^\circ_{298}$, (the standard enthalpy change for the reaction at 298 K), obviating the problems associated with changes in the reference elements because these are not involved. This is discussed in more detail below.

7.3. ENTROPY TABULATION

Entropy differs from enthalpy and free energy in that absolute, rather than relative, values can be tabulated. This is possible because of the Third Law, introduced in

Table 7.1. Thermodynamic properties of oxygen as tabulated by Robie et al. (1978). Note that all properties representing formation from the elements are zero.

Oxygen (Reference State)							Formula Weight 31.999
O_2: Ideal diatomic gas 298.15 to 1800 K.							
					Formation from the Elements		
Temp. K	$\frac{(H^\circ_T - H^\circ_{298})}{T}$ J/mol · K	S°_T J/mol · K	$-\frac{(G^\circ_T - H^\circ_{298})}{T}$ J/mol · K	C°_P J/mol · K	Enthalpy kJ/mol	Gibbs Free Energy kJ/mol	Log K_f
298.15	0.000	205.15	205.15	29.37	0.000	0.000	0.000
Uncertainty		0.04	0.04				
400	7.560	213.87	206.31	30.13	.000	.000	.000
500	12.176	220.70	208.52	31.16	.000	.000	.000
600	15.420	226.47	211.05	32.12	.000	.000	.000
700	17.867	231.48	213.61	32.95	.000	.000	.000
800	19.799	235.93	216.13	33.67	.000	.000	.000
900	21.376	239.94	218.56	34.29	.000	.000	.000
1000	22.694	243.58	220.89	34.82	.000	.000	.000
1100	23.818	246.92	223.10	35.29	.000	.000	.000
1200	24.792	250.01	225.22	35.69	.000	.000	.000
1300	25.644	252.88	227.24	36.05	.000	.000	.000
1400	26.399	255.56	229.16	36.36	.000	.000	.000
1500	27.072	258.08	231.01	36.64	.000	.000	.000
1600	27.678	260.45	232.77	36.89	.000	.000	.000
1700	28.226	262.69	234.46	37.11	.000	.000	.000
1800	28.726	264.82	236.09	37.31	.000	.000	.000

Melting Point	54.35 K	Boiling Point	90.18 K
Enthalpy of Melting	0.444 kJ	Enthalpy of Vaporization	6.816 kJ
$H^\circ_{298}-H^\circ_0$	8.682 kJ	Molar Volume	2478.9200 J/bar
			24789.200 cm^3

Transitions in Reference State Elements

Heat Capacity Equation

$$C^\circ_P = 48.318 - 6.9132 \times 10^{-4}T - 4.2066 \times 10^2\, T^{-0.5} + 4.9923 \times 10^5\, T^{-2}$$

(Equation valid from 298–1800 K)

Table 7.2. Thermodynamic properties of halite (NaCl) as tabulated by Robie et al. (1978).

Halite — Formula Weight 58.443

NaCl: Crystals 298.15 to melting point 1073.8 K. Liquid 1073.8 to fictive boiling point 1791 K.

Temp. K	$\frac{(H^\circ_T - H^\circ_{298})}{T}$ J/mol·K	S°_T J/mol·K	$-\frac{(G^\circ_T - H^\circ_{298})}{T}$ J/mol·K	C°_P J/mol·K	Formation From the Elements: Enthalpy kJ/mol	Formation From the Elements: Gibbs Free Energy kJ/mol	Formation From the Elements: Log K_f
298.15	0.000	72.12	72.12	50.51	−411.260	−384.212	67.313
Uncertainty		0.21	0.21		0.110	0.110	0.019
400	13.100	87.22	74.12	52.34	−413.461	−374.837	48.949
500	21.124	99.09	77.97	54.14	−413.024	−365.229	38.155
600	26.775	109.12	82.34	55.94	−412.352	−355.730	30.969
700	31.070	117.88	86.81	57.73	−411.456	−346.356	25.846
800	34.515	125.71	91.20	59.53	−410.352	−337.136	22.013
900	37.389	132.82	95.43	61.33	−409.064	−328.064	19.040
1000	39.877	139.37	99.49	63.12	−407.593	−319.133	16.670
1073.80	41.702	144.11	102.41	64.45	−380.693	−313.098	15.231
1073.80	67.925	170.33	102.41	68.55	−352.535	−313.098	15.231
1100	67.934	171.89	103.96	68.46	−377.527	−310.967	14.767
1200	67.964	177.83	109.87	68.14	−472.918	−302.951	13.187
1300	67.965	183.27	115.30	67.82	−470.082	−288.905	11.608
1400	67.943	188.29	120.35	67.50	−467.285	−275.089	10.264
1500	67.903	192.93	125.03	67.17	−464.521	−261.445	9.104
1600	67.847	197.26	129.41	66.85	−461.794	−248.002	8.096
1700	67.779	201.30	133.52	66.53	−459.103	−234.720	7.212

Melting Point	1073.8 K	Boiling Point	1791 K
Enthalpy of Melting	28.158 kJ	Enthalpy of Vaporization	164.787 kJ
$H^\circ_{298}-H^\circ_0$	10.611 kJ	Molar Volume	2.7015 J/bar
			27.015 cm^3

Transitions in Reference State Elements

Sodium...... M.P. 370.98, B.P. 1175 K.

Heat Capacity Equations

$$C^\circ_P = 45.151 + 1.7974T \times 10^{-2}\,T$$

(Equation Valid from 298–1073.8 K)

$$C^\circ_P = 72.008 - 3.2228 \times 10^{-3}\,T$$

(Equation Valid from 1073.8–1791 K)

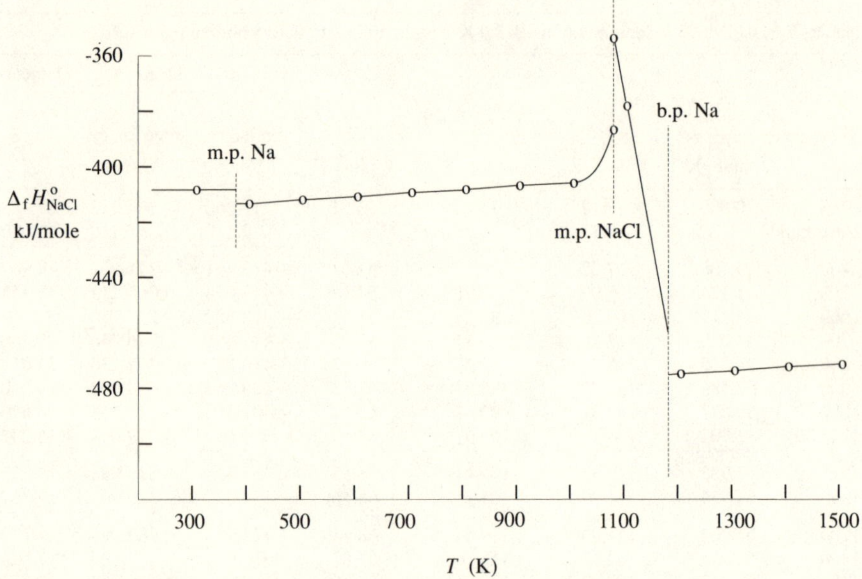

FIG. 7.1. Enthalpy of formation of NaCl from the elements as a function of temperature. Data from Robie, Hemingway and Fisher (1978).

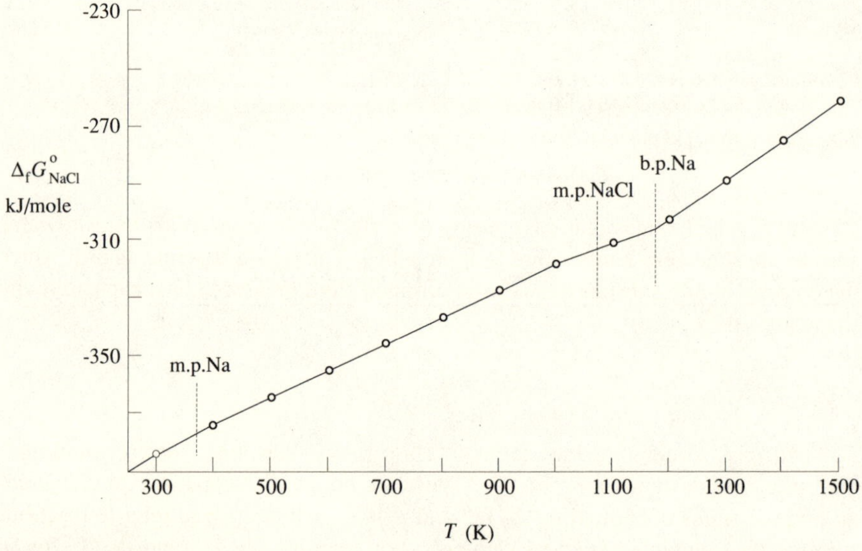

FIG. 7.2. Gibbs free energy of formation of NaCl from the elements as a function of temperature. Data from Robie, Hemingway and Fisher (1978).

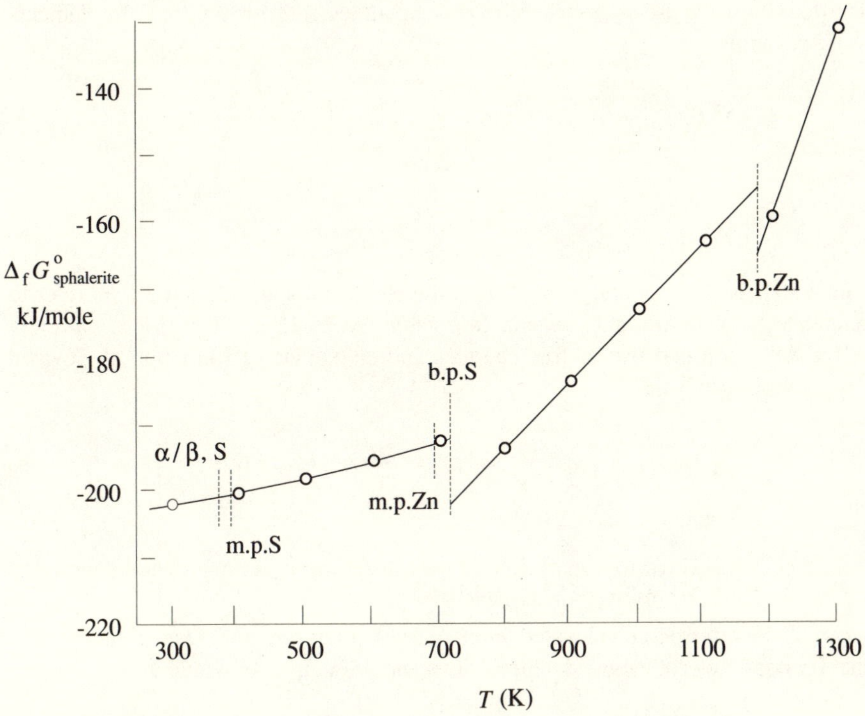

FIG. 7.3. Gibbs free energy of sphalerite from the elements as a function of temperature. Data from Robie, Hemingway and Fisher (1978).

Chapter 6, which states that the entropy of perfect crystalline solids tends towards zero as absolute zero temperature is approached. This sets a baseline against which the entropy of any substance can be determined from measured heat capacities and the relationship

$$S_T = \int_{T=0}^{T} \frac{C_p}{T} dT$$

The entropies so computed are termed "Third Law," "absolute," or "conventional" entropies, and designated S° or S_T° . These are not at all comparable to enthalpies and free energies of formation, $\Delta_f H^\circ$ and $\Delta_f G^\circ$, which instead refer to reactions forming the compound from its elements. For example, $\Delta_f H^\circ$ and $\Delta_f G^\circ$ for an element such as $O_2(g)$ at 25°C are both necessarily zero, while the absolute entropy, S_{298}°, is 205.15 J K^{-1} mol^{-1}. (see Table 7.1).

Entropies of formation from the elements, $\Delta_f S^\circ$, can of course be computed for compounds, and as we shall see below they are used to calculate $\Delta_f G^\circ$ from enthalpy and entropy data, but they are not normally tabulated.

7.4. ΔG AND ΔH AT HIGH T AND P

The definition and use of free energies and enthalpies of formation from the elements as (for example)

$$\left.\begin{aligned} \Delta_f G^\circ_{\mathrm{A_2B},T_r,P_r} &= G^\circ_{\mathrm{A_2B},T_r,P_r} - 2G^\circ_{\mathrm{A},T_r,P_r} - G^\circ_{\mathrm{B},T_r,P_r} \\ &\text{and} \\ \Delta_f H^\circ_{\mathrm{A_2B},T_r,P_r} &= H^\circ_{\mathrm{A_2B},T_r,P_r} - 2H^\circ_{\mathrm{A},T_r,P_r} - H^\circ_{\mathrm{B},T_r,P_r} \end{aligned}\right\} \tag{7.2}$$

where A_2B is a compound formed from the elements A and B, and T_r, P_r refer to reference temperature and pressure (normally 298.15 K and 1 bar) lead naturally to the definition and use of free energies and enthalpies of formation at elevated temperatures as

$$\left.\begin{aligned} \Delta_f G^\circ_{\mathrm{A_2B},T,P_r} &= G^\circ_{\mathrm{A_2B},T,P_r} - 2G^\circ_{\mathrm{A},T,P_r} - G^\circ_{\mathrm{B},T,P_r} \\ &\text{and} \\ \Delta_f H^\circ_{\mathrm{A_2B},T,P_r} &= H^\circ_{\mathrm{A_2B},T,P_r} - 2H^\circ_{\mathrm{A},T,P_r} - H^\circ_{\mathrm{B},T,P_r} \end{aligned}\right\} \tag{7.3}$$

where T, P_r refers to any elevated temperature at 1 bar pressure. One might suppose that for conditions at elevated temperatures and pressures, we would use

$$\left.\begin{aligned} \Delta_f G^\circ_{\mathrm{A_2B},T,P} &= G^\circ_{\mathrm{A_2B},T,P} - 2G^\circ_{\mathrm{A},T,P} - G^\circ_{\mathrm{B},T,P} \\ &\text{and} \\ \Delta_f H^\circ_{\mathrm{A_2B},T,P} &= H^\circ_{\mathrm{A_2B},T,P} - 2H^\circ_{\mathrm{A},T,P} - H^\circ_{\mathrm{B},T,P} \end{aligned}\right\} \tag{7.4}$$

However, while the definitions in equations (7.2) and (7.3) are extensively used, those of equations (7.4) are never used. It is often a difficult task to determine the change of G and H of elements with changing pressure at high temperature, and since the G and H of the elements always cancels out in balanced reactions, there is no real need to do so.

7.4.1. *Standard Free Energies and Enthalpies of Formation at T and P*

Free energies and enthalpies at high temperatures and pressures can thus be defined as

$$\left.\begin{aligned} \Delta_f G^\circ_{\mathrm{A_2B},T,P} &= G^\circ_{\mathrm{A_2B},T,P} - 2G^\circ_{\mathrm{A},T,P_r} - G^\circ_{\mathrm{B},T,P_r} \\ &\text{and} \\ \Delta_f H^\circ_{\mathrm{A_2B},T,P} &= H^\circ_{\mathrm{A_2B},T,P} - 2H^\circ_{\mathrm{A},T,P_r} - H^\circ_{\mathrm{B},T,P_r} \end{aligned}\right\} \tag{7.5}$$

In other words, the free energy and enthalpy of formation of a compound at T and P are calculated from the difference between the absolute G and H values of that compound at T and P and its constituent elements at T and P_r (one bar). The right-hand sides of both these equations are calculated from the known properties of the substance and its elements at 298 K and 1 bar, plus the equations for the change in these properties with T as described below.

These definitions are in common use at the present time, but the fact that the properties of the elements always cancel out in balanced reactions raises the question of why we bother to use the high temperature properties of these elements (equations (7.3) and (7.5)). Would it not be simpler to use the properties of the elements at 298 K, since it would make no difference?

7.4.2. *Apparent Free Energies and Enthalpies of Formation at T and P*

The high temperature properties of the elements have discontinuities as mentioned above that make interpolation very inconvenient in some temperature ranges. Why not define free energies and enthalpies so as to leave the elements at 298 K? Since they cancel out in all balanced reactions, it would make life simpler. Unfortunately there are two conventions for doing this at the present time.

The Benson-Helgeson Convention

The first of these, suggested by Benson (1968) and introduced to the geochemical literature by Helgeson et al. (1978), takes the straightforward approach of defining "apparent" G and H of formation from the elements as the difference between the absolute G or H of a compound at T and P and its constituent elements at T_r, P_r, normally chosen as 298.15 K and 1 bar. Thus for compound A_2B formed from elements A and B,

$$\left.\begin{aligned} \Delta_a G^\circ_{\mathrm{A_2B},T,P} &= G^\circ_{\mathrm{A_2B},T,P} - 2G^\circ_{\mathrm{A},T_r,P_r} - G^\circ_{\mathrm{B},T_r,P_r} \\ &\text{and} \\ \Delta_a H^\circ_{\mathrm{A_2B},T,P} &= H^\circ_{\mathrm{A_2B},T,P} - 2H^\circ_{\mathrm{A},T_r,P_r} - H^\circ_{\mathrm{B},T_r,P_r} \end{aligned}\right\} \tag{7.6}$$

The subscript a rather than subscript f distinguishes these quantities from the traditional $\Delta_f G^\circ$, $\Delta_f H^\circ$. It is naturally important not to use apparent and traditional properties of formation in the same calculation. Equations (7.6) can also be written

$$\Delta_a G^\circ_{\mathrm{A_2B},T,P} = \Delta_f G^\circ_{\mathrm{A_2B},T_r,P_r} + (G^\circ_{\mathrm{A_2B},T,P} - G^\circ_{\mathrm{A_2B},T_r,P_r})$$

and

$$\Delta_a H^\circ_{\mathrm{A_2B},T,P} = \Delta_f H^\circ_{\mathrm{A_2B},T_r,P_r} + (H^\circ_{\mathrm{A_2B},T,P} - H^\circ_{\mathrm{A_2B},T_r,P_r})$$

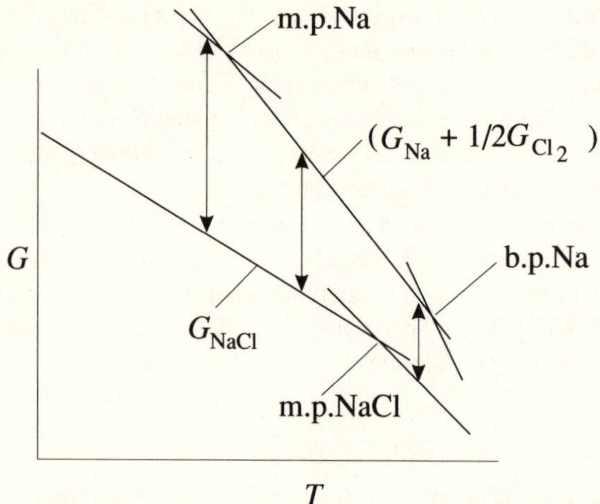

FIG. 7.4. Schematic representation of the absolute G of NaCl and the absolute G of ($Na+\frac{1}{2}Cl_2$) as a function of temperature. The difference between the two curves results in Figure 7.2.

The quantities $\Delta_f G^\circ_{A_2B}$ and $\Delta_a G^\circ_{A_2B}$ change in opposite senses as temperature is raised because, although G of all compounds and elements decreases with increasing T, G of compounds generally decreases more slowly than the total G of the constituent elements, making the difference $\Delta_f G^\circ_{A_2B}$ less negative as T increases (Figure 7.4). $\Delta_a G^\circ_{A_2B}$ on the other hand is increasingly negative with temperature because the elements are fixed at 298 K (Figure 7.5). A plot of $\Delta_a G^\circ_i$ vs. T therefore accurately shows the variation of absolute G_i with T, whereas $\Delta_f G^\circ_i$ vs. T does not.

The Berman-Brown Convention

Although straightforward, the G and H defined by the Benson-Helgeson convention do contain some "excess baggage" in the sense that they contain the entropies of the elements at T_r, P_r, which are contributing nothing to the usefulness of the concept. In other words, equation (7.6) can be rewritten

$$\begin{aligned}\Delta_a G^\circ_{A_2B,T,P} &= G^\circ_{A_2B,T,P} - 2(H^\circ_{A,T_r,P_r} - T_r S^\circ_{A,T_r,P_r}) - (H^\circ_{B,T_r,P_r} - T_r S^\circ_{B,T_r,P_r})\\ &= (G^\circ_{A_2B,T,P} - 2H^\circ_{A,T_r,P_r} - H^\circ_{B,T_r,P_r}) + T_r(2S^\circ_{A,T_r,P_r} + S^\circ_{B,T_r P_r})\\ &= \left\{\Delta_a G^\circ_{A_2B,T,P}(\text{Berman-Brown})\right\} + T_r(2S^\circ_{A,T_r,P_r} + S^\circ_{B,T_r P_r})\end{aligned}$$

Thus

$$\Delta_a G^\circ_{T,P}(\text{Benson-Helgeson}) = \Delta_a G^\circ_{T,P}(\text{Berman-Brown}) + T_r \sum_i \nu_i S^\circ_{i,T_r,P_r}$$

where ν_i is the stoichiometric coefficient of the elements i appearing in the formation from the elements reaction. The Berman-Brown definition can also be written

$$\begin{aligned}\Delta_a G^\circ_{A_2B,T,P}(\text{Berman-Brown}) &= G^\circ_{A_2B,T,P} - 2H^\circ_{A,T_r,P_r} - H^\circ_{B,T_r,P_r}\\ &= \Delta_a H^\circ_{A_2B,T,P} - T \cdot S^\circ_{A_2B,T,P}\end{aligned}$$

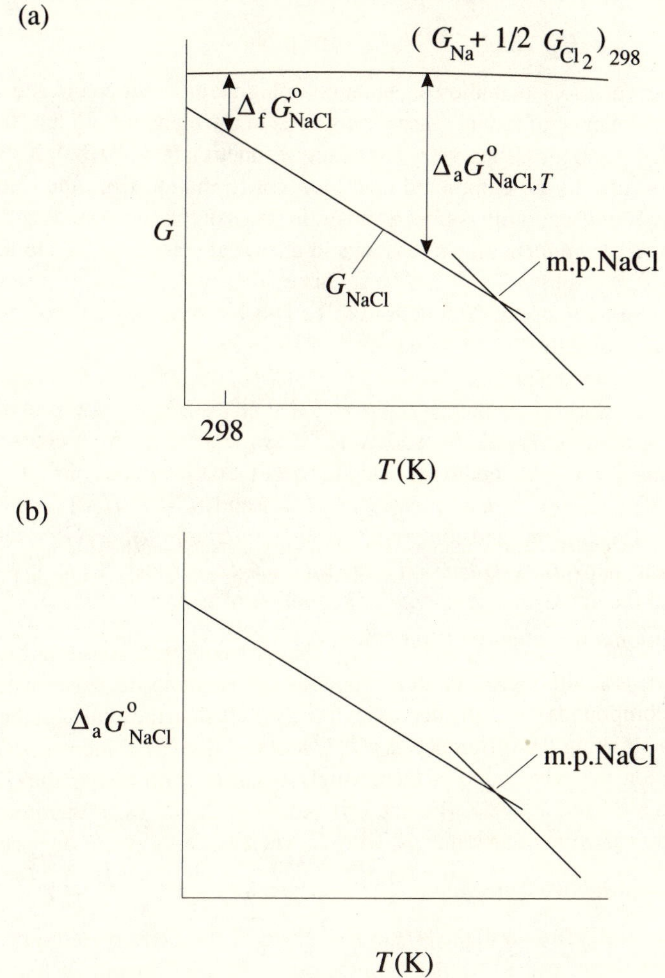

FIG. 7.5. (a) Schematic representation of the absolute G of NaCl as a function of temperature, and the difference between this and the absolute G of (Na + $\frac{1}{2}Cl_2$) which is fixed at its 298.15 K value. This difference we call the apparent free energy of formation. (b). Schematic representation of the difference function $\Delta_a G^o_{NaCl}$ from Figure 7.5a, as a function of temperature.

Both the Benson-Helgeson and Berman-Brown conventions are in common use at the present time. Both are far more sensible than what we have called the standard formation from the elements convention, but there is little to choose between them. Hopefully, one will become dominant in the near future.

The differences between the various definitions we have been discussing is quite often confusing on a first encounter. The best way to understand what is going on is to see how these properties are measured and calculated, which is covered in the following sections.

7.4.3. $(H^\circ_T - H^\circ_{T_r})$ *and* $(G^\circ_T - H^\circ_{T_r})/T$

Two other functions that allow calculation of high temperature ΔH and ΔG without the inconvenience of having phase transitions in the elements are the "heat content" $(H^\circ_T - H^\circ_{T_r})$ and the Gibbs energy or Giauque function $(G^\circ_T - H^\circ_{T_r})/T$. Both of these functions refer to the compound itself and not formation from the elements. They vary slowly and smoothly between transition states of the compounds considered, and calculation of products minus reactants in chemical reactions leads to the quantities $\Delta H^\circ_T - \Delta H^\circ_{T_r}$ and $\Delta G^\circ_T - \Delta H^\circ_{T_r}$, so that evidently $\Delta H^\circ_{T_r}$ for the reaction must be known in order to use these functions at T. This is not usually a problem. T_r in these equations is usually 298.15 K but can be 273.15 K.

Because the quantities H°_T, $H^\circ_{T_r}$, G°_T, and $G^\circ_{T_r}$ refer to the substance of interest (compound, mineral, element), $H^\circ_T - H^\circ_{T_r}$ and $G^\circ_T - H^\circ_{T_r}$ are not formation from the element functions. The $\Delta_r G^\circ$ and $\Delta_r H^\circ$ they give rise to are of course identical to $\Delta_r G^\circ$ and $\Delta_r H^\circ$ obtained from using high temperature values of $\Delta_f G^\circ$ and $\Delta_f H^\circ$ or $\Delta_a G^\circ$ and $\Delta_a H^\circ$, and since$(H^\circ_T - H^\circ_{T_r})$ and $(G^\circ_T - H^\circ_{T_r})/T$ are commonly tabulated for elements and compounds along with traditional $\Delta_f G^\circ$ values, they may be used to convert $\Delta_f G^\circ$,$\Delta_f H^\circ$ to $\Delta_a G^\circ$,$\Delta_a H^\circ$ or vice versa. For example, to calculate the apparent free energy of formation of quartz at T K, $\Delta_a G^{\circ,qtz}_{SiO_2,T}$, given the traditional free energy of formation, $\Delta_f G^{\circ,qtz}_{SiO_2}, T$, we have

$$\begin{aligned}
\Delta_a G^{\circ,qtz}_{SiO_2,T} &= G^{\circ,qtz}_{SiO_2,T} - G^\circ_{Si,T_r} - G^\circ_{O_2,T_r} \\
&= (G^{\circ,qtz}_{SiO_2,T} - G^\circ_{Si,T} - G^\circ_{O_2,T}) \\
&\quad +(G^\circ_{Si,T} - G^\circ_{Si,T_r}) + (G^\circ_{O_2,T} - G^\circ_{O_2,T_r}) \\
&= \Delta_f G^{\circ,qtz}_{SiO_2,T} + (G^\circ_T - G^\circ_{T_r})_{Si} + (G^\circ_T - G^\circ_{T_r})_{O_2} \\
&= \Delta_f G^{\circ,qtz}_{SiO_2,T} + \left(\frac{G^\circ_T - H^\circ_{T_r}}{T}\right)_{Si} \cdot T - \left(\frac{G^\circ_{T_r} - H^\circ_{T_r}}{T_r}\right)_{Si} \cdot T_r \\
&\quad + \left(\frac{G^\circ_T - H^\circ_{T_r}}{T}\right)_{O_2} \cdot T - \left(\frac{G^\circ_{T_r} - H^\circ_{T_r}}{T_r}\right)_{O_2} \cdot T_r
\end{aligned} \tag{7.7}$$

Thus to convert traditional to apparent properties of formation or vice versa, one need only look up the properties $(G^\circ_T - H^\circ_{T_r})/T$ and $(G^\circ_{T_r} - H^\circ_{T_r})/298$ (or $(H^\circ_T - H^\circ_{T_r})$ in the case of ΔH) for the constituent elements.

7.4.4. *Formation from the Oxides*

A convention in common use as an alternative to energies and enthalpies of formation from the elements is energies and enthalpies of formation from the oxides. Thus instead of having

$$\Delta_f G^\circ_{Ab} = G^\circ_{NaAlSi_3O_8} - G^\circ_{Na} - G^\circ_{Al} - 3G^\circ_{Si} - 4G^\circ_{O_2}$$

$$\Delta_f G^\circ_{Jd} = G^\circ_{NaAlSi_2O_6} - G^\circ_{Na} - G^\circ_{Al} - 2G^\circ_{Si} - 3G^\circ_{O_2}$$

and

$$\Delta_f G^\circ_{Ne} = G^\circ_{NaAlSiO_4} - G^\circ_{Na} - G^\circ_{Al} - G^\circ_{Si} - 2G^\circ_{O_2}$$

from which for the reaction

$$2\text{Jd} = \text{Ab} + \text{Ne}$$

we get

$$\begin{aligned}\Delta_r G^\circ &= \Delta_f G^\circ_{\text{Ab}} + \Delta_f G^\circ_{\text{Ne}} - 2\Delta_f G^\circ_{\text{Jd}} \\ &= G^\circ_{\text{NaAlSi}_3\text{O}_8} + G^\circ_{\text{NaAlSiO}_4} - 2G^\circ_{\text{NaAlSi}_2\text{O}_6}\end{aligned}$$

We could also have

$$\Delta_f G^\circ_{\text{Ab}} = G^\circ_{\text{NaAlSi}_3\text{O}_8} - \frac{1}{2}G^\circ_{\text{Na}_2\text{O}} - \frac{1}{2}G^\circ_{\text{Al}_2\text{O}_3} - 3G^\circ_{\text{SiO}_2}$$

$$\Delta_f G^\circ_{\text{Jd}} = G^\circ_{\text{NaAlSi}_2\text{O}_6} - \frac{1}{2}G^\circ_{\text{Na}_2\text{O}} - \frac{1}{2}G^\circ_{\text{Al}_2\text{O}_3} - 2G^\circ_{\text{SiO}_2}$$

$$\Delta_f G^\circ_{\text{Ne}} = G^\circ_{\text{NaAlSiO}_4} - \frac{1}{2}G^\circ_{\text{Na}_2\text{O}} - \frac{1}{2}G^\circ_{\text{Al}_2\text{O}_3} - G^\circ_{\text{SiO}_2}$$

and

$$\begin{aligned}\Delta_r G^\circ &= \Delta_f G^\circ_{\text{Ab}} + \Delta_f G^\circ_{\text{Ne}} - 2\Delta_f G^\circ_{\text{Jd}} \\ &= G^\circ_{\text{NaAlSi}_3\text{O}_8} + G^\circ_{\text{NaAlSiO}_4} - 2G^\circ_{\text{NaAlSi}_2\text{O}_6}\end{aligned}$$

as before.

This convention is somewhat more limited in its application than is formation from the elements. For example, it could not be used for many reactions involving non-oxides such as

$$\text{ZnAl}_2\text{O}_4 + \frac{1}{2}\text{S}_2 = \text{ZnS} + \text{Al}_2\text{O}_3 + \frac{1}{2}\text{O}_2$$

However, for the large range of geologically interesting reactions involving silicates, oxides, carbonates, and sulfates it has one advantage over the formation from the elements convention, and that advantage is that uncertainties associated with formation from the oxides are generally smaller, because the uncertainties associated with forming the oxides from their elements are not included. One must be very careful, of course, not to mix two conventions in a single calculation.

For oxides such as the silicate minerals, the RHF tables list thermodynamic properties twice; first as formation from the elements, and second as formation from the oxides. Table 7.3 is an example for the mineral jadeite showing properties on the right-hand side calculated for formation from the oxides; these are marked with an asterisk as a reminder not to confuse them with formation from the elements.

7.5. PROPERTIES DERIVED FROM CALORIMETRY

7.5.1. *The Measurement of Thermodynamic Properties*

It is very helpful in understanding thermodynamics to know how the thermodynamic data for various materials are actually obtained. In fact, the thermodynamic properties

Table 7.3. Thermodynamic properties of jadeite as tabulated by Robie et al. (1978). The three right-hand columns represent formation from the oxides (not the elements): $\frac{1}{2}Na_2O + 2SiO_2 + \frac{1}{2}Al_2O_3 = NaAl(SiO_3)_2$

Jadeite — Formula Weight 202.140

$NaAl(SiO_3)_2$: Crystals 298.15 to 1300 K.

Temp.	$\left(\frac{H^\circ_T - H^\circ_{298}}{T}\right)$	S°_T	$-\left(\frac{G^\circ_T - H^\circ_{298}}{T}\right)$	C°_P	Formation From the Oxides: Enthalpy	Gibbs Free Energy	Log K_f
K	J/mol · K	J/mol · K	J/mol · K	J/mol · K	kJ/mol	kJ/mol	
298.15	0.000	133.47	133.47	159.95	−162.740*	−159.000*	27.856
Uncertainty		1.25	1.25		2.380	2.400	
400	44.725	184.87	140.14	188.44	−163.140*	−157.660*	20.588
500	75.266	228.87	153.60	205.34	−163.737*	−156.232*	16.321
600	97.977	267.40	169.42	217.10	−164.622*	−154.644*	13.463
700	115.656	301.56	185.90	225.98	−165.873*	−152.891*	11.409
800	129.906	332.22	202.31	233.09	−167.547*	−150.935*	9.855
900	141.711	360.02	218.31	238.99	−169.505*	−148.716*	8.631
1000	151.696	385.47	233.77	244.05	−169.975*	−146.390*	7.647
1100	160.297	408.94	248.64	248.47	—	—	—
1200	167.813	430.74	262.93	252.42	—	—	—
1300	174.461	451.09	276.63	255.99	—	—	—

Melting Point	K	Boiling Point	K
Enthalpy of Melting	kJ	Enthalpy of Vaporization	kJ
$H^\circ_{298}-H^\circ_0$	kJ	Molar Volume	6.0400 J/bar 60.400 cm^3

Transitions in Reference State Oxides

Na_2O . . . . . M.P. 1193 K

SiO_2 . . . . . . Alpha–Beta Transition 844 K

Heat Capacity Equation

$$C^\circ_P = 3.0113 \times 10^2 + 1.0143 \times 10^{-2}\,T - 2.0551 \times 10^3\,T^{-0.5} - 2.2393 \times 10^4\,T^{-2}$$

(Equation valid from 298–1300 K)

of substances can be determined by an enormous variety of experimental approaches. For example, that simple measurement of the vibrational spectra of crystals over a range of temperatures can be used to derive fundamental thermodynamic properties, including those of different isotopes of the same element. The temperature dependence of the electromotive force in redox reactions provides similar information, as we shall see in Chapter 18. Other sources include calorimetry, phase equilibrium and solubility studies, volume and density measurements, electrical conductivity and capacitance, many different spectroscopic techniques, electrochemical techniques, and so on. The list is almost endless, and limited only by the ingenuity of experimentalists.

All these techniques can be seen as various ways of measuring the change in the energy content (ΔU) of a system in one form or another, and since the energy content can be changed by using either heat or work, the experimental methods manipulate q, w, or U. For example, many spectroscopic procedures permit direct measurement of energies, whether vibrational, electronic, nuclear, or kinetic. The other thermodynamic properties such as entropy, heat capacity, enthalpy, and Gibbs free energy are then derived from this measurement as outlined in Chapter 6. Another class of experimental procedures is designed to look at the work term w, usually through the ΔV term in $P\Delta V$. For example, density or molar volume measurements on gases and liquids at different temperatures give rise to fugacities, partial molar volumes, and so on. A third class of techniques is concerned with the measurement of q, or the transfer of energy as heat. These include the many different techniques that are lumped together under the general term "calorimetry."

7.5.2. *Calorimetry*

Of all available methods, calorimetry has been the most important source of thermodynamic data on complex materials, including minerals. For this reason we will focus on the topic of calorimetry in rather general terms, describing how thermodynamic parameters are extracted from calorimeters, and how they are combined to give the tables of data we find in the literature. In this chapter we omit most experimental details for the sake of simplicity. An old, but still useful general reference on the subject is the book edited by McCullough and Scott (1968). The many different kinds of calorimeters are in general designed to measure the heat capacity, C_p, of single materials, or the heat of reaction of mixtures. Knowledge of the heat capacity for a range of temperatures gives the other thermodynamic properties such as S and H directly from the relations $dH = C_pdT$ and $dS = (C_p/T)dT$; Gibbs free energies are then given by $\Delta G = \Delta H - T\Delta S$. It is this kind of direct determination of the fundamental thermodynamic variables that makes calorimetric procedures so useful. Individual kinds of calorimeters are designed for specific temperature intervals ranging from temperatures approaching absolute zero up to magmatic conditions (although no one instrument could span anywhere near this entire range). In general, they are operated at fixed pressure, usually 1 bar.

Unfortunately, calorimeters have not yet been designed to operate over the tremendous ranges of pressures of interest to Earth scientists; this means that they cannot provide direct information on the change in thermodynamic properties with pressure (or on the work term w). Hence these PV-related properties must typically be derived by some other procedure.

The major types of calorimetry to be considered here are:

- Solution calorimetry
    a. HF solution calorimetry
    b. High temperature oxide melt calorimetry
- Drop calorimetry
- Cryogenic calorimetry

Solution Calorimetry

The object in solution calorimetry is to measure the heat liberated when a compound dissolves, then to measure the heat liberated when its component elements or oxides dissolve to give the same final solution composition as did the compound. Thus one has two reactions

$$\text{crystalline compound} + \text{solution}' \rightarrow \text{solution}'' + \text{heat}(\Delta H_1)$$

$$\text{component elements} + \text{solution}' \rightarrow \text{solution}'' + \text{heat}(\Delta H_2)$$

As long as both solution′ and solution″ have the same compositions in both reactions, the reactions may be subtracted giving

$$\text{component elements} \rightarrow \text{crystalline compound} + \Delta H_3$$

where $\Delta H_3 = \Delta H_1 - \Delta H_2$ and is the heat of formation of the compound from its elements.

At ordinary temperatures, the only reasonably convenient solvent capable of dissolving silicate minerals is hydrofluoric acid, HF. At higher temperatures, various oxide melts may also be used, and although the techniques are totally different, the principle is the same.

HF Solution Calorimetry. Actual practice in solution calorimetry is never quite as simple as the two-reaction scheme outlined above, because at least some of the component elements are not suitable for placing directly into a calorimeter. Oxygen, for example, can only be used in the form of oxides, so the heat of formation of the oxides must generally be known by some other method. Alkali metals such as Na and K are highly volatile in the pure form and difficult to handle, and so are used in other forms such as NaCl and KCl. Other elements such as Al and Si could be used in the pure elemental form, but are usually used in other forms such as $AlCl_3.6H_2O$ and SiO_2, so that the heats of formation of these compounds must be obtained in other experiments. The heats of reaction involving gaseous components such as

$$K + \frac{1}{2}Cl_2 = KCl$$

and

$$Si + O_2 = SiO_2$$

are usually determined in a "bomb calorimeter," where the gaseous element is placed in great excess under pressure so that the reaction, which is sometimes explosive or at least highly exothermic, will be sure to go to completion. Once completed satisfactorily, the results are then used in all subsequent HF solution calorimetry measurements.

Then since chloride salts are used, the heats of ionization reactions such as

$$KCl = K^+ + Cl^-$$

must be determined, and similarly with other heat terms such as heats of dilution from one concentration to another, and the heat required to warm various solutions from one temperature to another. Thus to determine $\Delta_f H^\circ_{T_r}$ for a single silicate mineral it is not unusual to have to add algebraically the $\Delta_r H$ of 10 or 12 different reactions. It is therefore imperative that all of them be known with the highest possible precision and accuracy. The art of solution calorimetry is a highly exacting quest for the continual elimination of ever-smaller but ever-present errors, and the construction of ever-improved calorimeters. HF solution calorimetry is described in detail by Robie and Hemingway (1972).

High Temperature Oxide Calorimetry. Calorimetry at temperatures of 600°C to above 900°C using "micro calorimeters" (i.e. apparatus much smaller than conventional HF solution calorimeters) having fused salts such as $2PbO.B_2O_3$ or $3Na_2O.4MoO_3$ as solvents has been developed extensively in recent years (see review in Navrotsky, 1977). These melts dissolve refractory oxides and silicates quite readily, so that determination of the heat of formation of compounds from their oxides can be measured directly. Heats of formation from the elements are not normally attempted by this method. The method is simpler but less precise than HF calorimetry.

Drop Calorimetry

The amount of heat required to raise the temperature of a mole of substance from T_r to T at constant pressure is simply $H_T - H_{T_r}$ (or $H^\circ_T - H^\circ_{T_r}$ for a standard reference substance); again, a difference between two unknown quantities. This quantity is conveniently measured by cooling the substance from T to T_r (usually 298 K, but sometimes 273 K) in a calorimeter and measuring the amount of heat given up by the substance during this process. To do this, a calorimeter is placed directly under a furnace and the sample is dropped from the furnace where it has temperature T, into the calorimeter, where it gives up its heat and achieves temperature T_r, or close to it. Small corrections are then applied to bring the total heat change to $H_T - H_{T_r}$. The procedure is described in detail by Douglas and King (1968).

Values of $H_T - H_{T_r}$ can be combined to give $\Delta_f H^\circ$for substances at high temperatures. Thus for any substance

$$\Delta_f H^\circ_T = \Delta_f H^\circ_{T_r} + \Delta_f (H^\circ_T - H^\circ_{T_r}) \qquad (7.8)$$

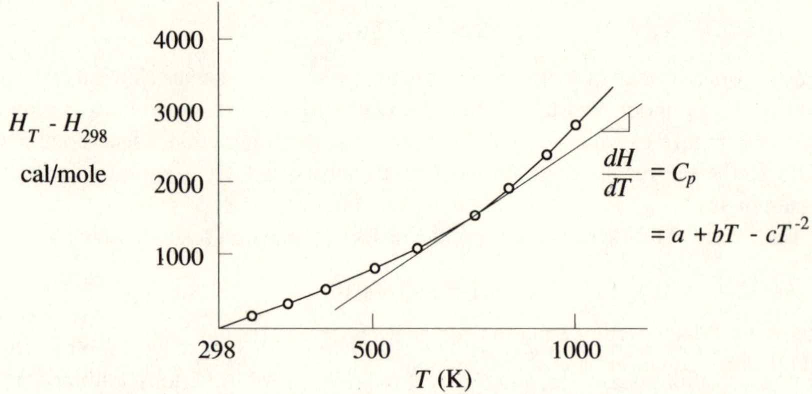

FIG. 7.6. $(H_T^\circ - H_{298}^\circ)$ as a function of temperature. The slope at any point is C_p°.

where Δ_f refers to the reaction in which the substance is formed from its elements. For example,

$$\begin{aligned}\Delta_f(H_T^\circ - H_{T_r}^\circ)_{\mathrm{SiO_2}} &= (H_T^\circ - H_{T_r}^\circ)_{\mathrm{SiO_2}} - (H_T^\circ - H_{T_r}^\circ)_{\mathrm{Si}} - (H_T^\circ - H_{T_r}^\circ)_{\mathrm{O_2}} \\ &= (H_{T,\mathrm{SiO_2}}^\circ - H_{T,\mathrm{Si}}^\circ - H_{T,\mathrm{O_2}}^\circ) - (H_{T_r,\mathrm{SiO_2}}^\circ - H_{T_r,\mathrm{Si}}^\circ - H_{T_r,\mathrm{O_2}}^\circ) \\ &= \Delta_f H_{T,\mathrm{SiO_2}}^\circ - \Delta_f H_{T_r,\mathrm{SiO_2}}^\circ \end{aligned} \tag{7.9}$$

and therefore

$$\Delta_f H_{T,\mathrm{SiO_2}}^\circ = \Delta_f H_{T_r,\mathrm{SiO_2}}^\circ + \Delta_f(H_T^\circ - H_{T_r}^\circ)_{\mathrm{SiO_2}} \tag{7.10}$$

This is the way the tabulated values of $\Delta_f H^\circ$are actually constructed, but of course before this is done, the experimental values of $(H_T - H_{T_r}^\circ)$ for the substance and its elements must be smoothed and interpolated to give values at even temperature intervals. To do this, they are fitted statistically to a function, which is commonly

$$(H_T^\circ - H_{T_r}^\circ) = A + BT + CT^2 + DT^{-1} \tag{7.11}$$

Once the "best fit" values of A, B, C, and D are calculated, $(H_T^\circ - H_{T_r}^\circ)$ may be computed for any desired temperature.

The Heat Capacity. The slope of the plot of $(H_T^\circ - H_{T_r}^\circ)$ versus T is the heat capacity, C_p (Figure 7.6). Thus

$$d(H_T^\circ - H_{T_r}^\circ)/dT = dH_T/dT$$

or simply

$$d(H_T^\circ - H_{T_r}^\circ)/dT = dH/dT$$

since H_{T_r}, although an unknown quantity, is certainly a constant, so that

$$dH_{T_r}/dT = 0$$

Thus

$$\begin{aligned} dH/dT &= C_p \\ &= B + 2CT - DT^{-2} \end{aligned}$$

or

$$C_p = a + bT - cT^{-2} \tag{7.12}$$

where

$$a = B, b = 2C, \text{ and } c = D.$$

This is the Maier-Kelley equation for heat capacities (Maier and Kelley, 1932).[3] Several other equations are also commonly used to fit the heat capacity (see for example Tables 7.1, 7.2 and 7.3). Rather than try to present the rest of the equations in this chapter for all current versions of this equation, we continue to use the Maier-Kelley equation as an example. Commonly used equations derived using other heat capacity expressions are presented in Appendix B.

In machine computations, enthalpies are most conveniently calculated by storing values of a, b, and c for various substances and using the integrated form of the Maier-Kelley equation.

For instance, for the formation of fluorite from its elements we write

$$\mathrm{Ca}(s) + \mathrm{F_2}(g) = \mathrm{CaF_2}(s)$$

$$\Delta_f H^\circ{}_{\mathrm{CaF_2}} = H_{\mathrm{CaF_2}} - H_{\mathrm{Ca}} - H_{\mathrm{F_2}}$$

For each of Ca, F_2, and CaF_2 there is a set of Maier-Kelley coefficients, a, b and c, so that

$$\begin{aligned} C_{p\mathrm{CaF_2}} &= a_{\mathrm{CaF_2}} + b_{\mathrm{CaF_2}}T - c_{\mathrm{CaF_2}}T^{-2} \\ C_{p\mathrm{Ca}} &= a_{\mathrm{Ca}} + b_{\mathrm{Ca}}T - c_{\mathrm{Ca}}T^{-2} \\ C_{p\mathrm{F_2}} &= a_{\mathrm{F_2}} + b_{\mathrm{F_2}}T - c_{\mathrm{F_2}}T^{-2} \end{aligned}$$

and

$$\Delta C_p = \Delta a + \Delta bT - \Delta cT^{-2} \tag{7.13}$$

where

$$\begin{aligned} \Delta C_p &= C_{p\mathrm{CaF_2}} - C_{p\mathrm{Ca}} - C_{p\mathrm{F_2}} \\ \Delta a &= a_{\mathrm{CaF_2}} - a_{\mathrm{Ca}} - a_{\mathrm{F_2}} \\ \Delta b &= b_{\mathrm{CaF_2}} - b_{\mathrm{Ca}} - b_{\mathrm{F_2}} \\ \Delta c &= c_{\mathrm{CaF_2}} - c_{\mathrm{Ca}} - c_{\mathrm{F_2}} \end{aligned}$$

[3]This equation is sometimes written with a $+cT^{-2}$ term instead of the $-cT^{-2}$ term as shown in (7.12). If tabulated values of c are positive, as in Appendix C, equation (7.12) is being used. If they are negative, the $+cT^{-2}$ term is being used.

Integrating

$$d\Delta H = \Delta C_p\, dT$$

we have

$$\int_{T_r}^{T} d\Delta H = \int_{T_r}^{T} \Delta C_p\, dT$$

$$\Delta H_T - \Delta H_{T_r} = \int_{T_r}^{T} (\Delta a + \Delta b T - \Delta c T^{-2}) dT$$

$$= \Delta a(T - T_r) + \frac{\Delta b}{2}(T^2 - T_r^2) + \Delta c\left(\frac{1}{T} - \frac{1}{T_r}\right)$$

If in this equation ΔH_{T_r} is $\Delta_f H^\circ_{T_r}$ and Δa, Δb, and Δc are as above, then ΔH_T is $\Delta_f H^\circ_T$, the standard enthalpy of formation of fluorite at temperature T.

To calculate the *apparent* enthalpy of formation at elevated temperatures, we need the a, b, c values only for the compound of interest and not for its elements. Thus

$$\Delta_a H^\circ_{T,\mathrm{CaF_2}} = H^\circ_{T,\mathrm{CaF_2}} - H^\circ_{T_r,\mathrm{Ca}} - H^\circ_{T_r,\mathrm{F_2}}$$

$$= \Delta_f H^\circ_{T_r,\mathrm{CaF_2}} + a_{\mathrm{CaF_2}}(T - T_r) + \frac{b_{\mathrm{CaF_2}}}{2}(T^2 - T_r^2)$$

$$+ c_{\mathrm{CaF_2}}\left(\frac{1}{T} - \frac{1}{T_r}\right)$$

Effect of Phase Transitions

Equation (7.13) is a perfectly general relationship, but the integration in terms of fit coefficients is not. Fit coefficients such as the a, b, c of Maier-Kelley are only valid for the regions of smoothly changing heat capacity from which they were derived, which means between phase transitions. If between T_r and T the compound of interest goes through a phase transition (polymorphic change, melting, etc.) there is an abrupt change in C_p, and hence in ΔH and ΔS, and the fit coefficients are different above and below the transition temperature. A more complete expression is therefore

$$\int_{T_r}^{T} d\Delta H = \sum\left[\int_{T_{lower}}^{T_{upper}} \Delta C_p\, dT\right] + \sum \Delta_t H \tag{7.14}$$

where $\Delta_t H$ is the enthalpy of transition, and T_{upper} and T_{lower} are the upper and lower temperature limits of the temperature ranges in which T varies smoothly. For example, if there is one transition at temperature T_t, the expression is

$$\int_{T_r}^{T} d\Delta H = \int_{T_r}^{T_t} \Delta C_p\, dT + \Delta_t H + \int_{T_t}^{T} \Delta C_p\, dT$$

Cryogenic Calorimetry

Cryogenic or low-temperature calorimetry is concerned with the measurement of the heat capacity of substances below 298 K, often to within a few degrees of 0 K. This is accomplished by equilibrating the sample at some low T, then introducing a known quantity of heat by means of an electrical heater and observing the resulting change in T, usually a few degrees. Then after correcting for heat losses to the calorimeter, the heat introduced (which is negligibly different from ΔH because the $\Delta(PV)$ term is almost zero) divided by ΔT is plotted against T, from which values of (dH/dT) or C_p are retrieved.

C_p/T is then computed at each T, an extrapolation performed from the lowest measured temperatures to 0 K, and the integral

$$\int_{T=0}^{T_r} \frac{C_p}{T} dT$$

evaluated, usually by integrating an orthogonal polynomial fitted to the C_p/T versus T data.

Then since

$$dS/dT = C_p/T$$

$$\int_{T=0}^{T_r} dS = \int_{T=0}^{T_r} \frac{C_p}{T} dT$$

and

$$S_{T_r} - S_{T=0} = \int_{T=0}^{T_r} \frac{C_p}{T} dT \tag{7.15}$$

The assumption that the entropy of the measured substance at 0 K ($S_{T=0}$) is zero then allows one to give a numerical value for the "absolute" entropy of the substance at 298 K. Alternatively, $S_{T=0}$ can be evaluated or estimated if it is not zero.

Again, if there are phase transitions, a more complete expression would be analogous to (7.14). For one transition at temperature T_t between $T = 0$ and T, this becomes

$$S_T - S_{T=0} = \int_{T=0}^{T_r} \frac{C_p}{T} dT + \Delta_t S + \int_{T_t}^{T} \frac{C_p}{T} dT$$

The variation of C_p/T with T is illustrated in Figure 7.7.

After S_{T_r} has been determined for elements as well as compounds, $\Delta_f S^\circ_{T_r}$, the entropy change for the reaction in which a compound is formed from its elements, is easily calculated if required.

The cryogenic calorimeter which has been in operation at the U.S. Geological Survey for many years has been described by Robie and Hemingway (1972).

7.5.3. *Entropies above 298 K*

For temperatures above 298 K, entropies can be calculated by combining $S^\circ_{T_r}$ and the $(H^\circ_T - H^\circ_{T_r})$ measurements previously described. Since we know $S^\circ_{T_r}$, all we need are values of $S^\circ_T - S^\circ_{T_r}$, which equals $(H^\circ_T - H^\circ_{T_r})/T$.

Thus

$$d(S^\circ_T - S^\circ_{T_r}) = dS^\circ_T \text{ (because } S^\circ_{T_r} \text{ is constant)}$$

$$= d\left(\frac{H^\circ_T - H^\circ_{T_r}}{T}\right)$$

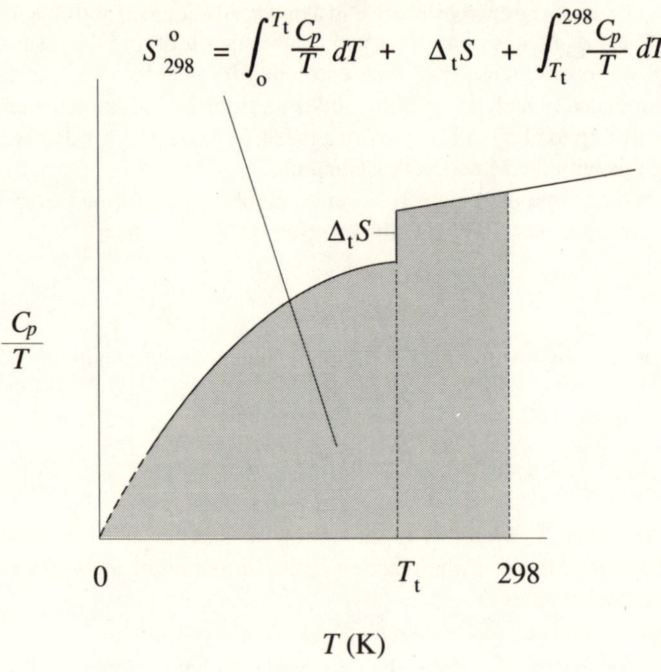

FIG. 7.7. C_p/T as a function of temperature. The area under the curve between 0 and 298 K gives S°_{298}.

so

$$\int_{T_r}^{T} dS^\circ = \int_{T_r}^{T} d\left(\frac{H^\circ_T - H^\circ_{T_r}}{T}\right)$$

The right-hand side is integrated by parts, giving

$$S^\circ_T - S^\circ_{T_r} = \left(\frac{H^\circ_T - H^\circ_{T_r}}{T}\right) + \int_{T_r}^{T} \left(\frac{H^\circ_T - H^\circ_{T_r}}{T^2}\right) dT$$

Since $(H^\circ_T - H^\circ_{T_r})$ and therefore $(H^\circ_T - H^\circ_{T_r})/T^2$ is known as a function of T, the integral can be evaluated, and S°_T values calculated for elevated temperatures. As in the case of ΔH° values, an alternative method usually preferred by Earth scientists using computer programs is to calculate S°_T values, or more likely $\Delta_r S^\circ$ values, at elevated temperatures by means of the Maier-Kelley heat capacity coefficients. In other words, since

$$dS^\circ/dT = C^\circ_p/T$$

then

$$d\Delta_r S^\circ/dT = \Delta_r C^\circ_p/T$$

where $\Delta_r S^\circ$ refers to the entropy change of a balanced chemical reaction. Integrating,

$$\int_{T_r}^{T} d\Delta_r S^\circ = \int_{T_r}^{T} \frac{\Delta_r C_p^\circ}{T} dT$$

Combining this with the Maier-Kelley equation

$$\Delta C_p^\circ = \Delta a + \Delta b T - \Delta c T^{-2} \tag{7.16}$$

we have

$$\Delta_r S_T^\circ - \Delta_r S_{T_r}^\circ = \int_{T_r}^{T} \left(\frac{\Delta_r a}{T} + \Delta_r b - \frac{\Delta_r c}{T^3} \right) dT$$

or

$$\Delta_r S_T^\circ - \Delta_r S_{T_r}^\circ = \Delta_r a \ln\left(\frac{T}{T_r}\right) + \Delta_r b(T - T_r) + \frac{\Delta_r c}{2}\left(\frac{1}{T^2} - \frac{1}{T_r^2}\right) \tag{7.17}$$

In this equation $\Delta_r S_T^\circ$ refers to the entropy change of any balanced chemical reaction at temperature T. If the reaction is the formation of a compound from its elements, $\Delta_r S_T^\circ$ becomes $\Delta_f S_T^\circ$.

The apparent entropy of formation can be calculated from

$$\Delta_a S_T^\circ = \Delta_f S_{T_r}^\circ + \int_{T_r}^{T} \frac{C_p^\circ}{T} dT$$

where C_p°/T refers to the compound only. Thus in the CaF_2 example

$$\Delta_a S_{T,\mathrm{CaF_2}}^\circ = \Delta_f S_{T_r,\mathrm{CaF_2}}^\circ + a_{\mathrm{CaF_2}} \ln\left(\frac{T}{T_r}\right) + b_{\mathrm{CaF_2}}(T - T_r) + \frac{c_{\mathrm{CaF_2}}}{2}\left(\frac{1}{T^2} - \frac{1}{T_r{}^2}\right) \tag{7.18}$$

7.5.4. *Calculation of* $(G_T^\circ - H_{T_r}^\circ)/T$

Values of $(G_T^\circ - H_{T_r}^\circ)/T$ are calculated from tabulated values of $H_T^\circ - H_{T_r}^\circ$ and S_T° from the relation

$$(G_T^\circ - H_{T_r}^\circ)/T = (H_T^\circ - H_{T_r}^\circ)/T - S_T^\circ \tag{7.19}$$

7.5.5. *Calculation of* $\Delta_f G_{T_r}^\circ$, $\Delta_f G_T^\circ$, *and* $\Delta_a G_T^\circ$

Standard free energies of formation from the elements at 298 K are computed from

$$\Delta_f G_{T_r}^\circ = \Delta_f H_{T_r}^\circ - T_r \Delta_f S_{T_r}^\circ$$

and at higher temperatures from

$$\Delta_f G_T^\circ = \Delta_f H_T^\circ - T \Delta_f S_T^\circ$$

or from

$$\Delta_f G^\circ_T = \Delta_f H^\circ_{T_r} + T\Delta_f \left[\frac{G^\circ_T - H^\circ_{T_r}}{T}\right] \tag{7.20}$$

where as before, Δ_f refers to the reaction in which the compound is formed from its elements.

Calculation of the apparent free energy of formation of a compound at elevated temperatures from $\Delta_f G^\circ_T$ and $(G^\circ_T - H^\circ_{T_r})/T$ values was discussed earlier in this chapter. A more direct approach is to write for any compound i

$$\begin{aligned}\Delta_a G^\circ_{T,i} &= \Delta_f G^\circ_{T_r,i} + \int_{T_r}^{T} (\partial G^\circ_i/\partial T)\, dT \\ &= \Delta_f G^\circ_{T_r,i} + \int_{T_r}^{T} -S^\circ_i\, dT \\ &= \Delta_f G^\circ_{T_r,i} + \int_{T_r}^{T} \left[-S^\circ_{T_r,i} - \int_{T_r}^{T} \frac{C^\circ_{p\,i}}{T}\, dt\right] dT \\ &= \Delta_f G^\circ_{T_r,i} - S^\circ_{T_r,i}(T - T_r) - \int_{T_r}^{T} \left[\int_{T_r}^{T} \frac{C^\circ_{p\,i}}{T}\, dt\right] dT\end{aligned} \tag{7.21}$$

The integration in the last term is performed by parts. That is

$$\int u\, dv = uv - \int v\, du$$

where

$$u = \int_{T_r}^{T} \frac{C^\circ_p}{T} dT \quad \text{and} \quad v = T$$

This results in

$$\Delta_a G^\circ_{T,i} = \Delta_f G^\circ_{T_r,i} - S^\circ_{T_r,i}(T - T_r) + \int_{T_r}^{T} C^\circ_{p\,i}\, dT - T\int_{T_r}^{T} \frac{C^\circ_{p\,i}}{T} dT \tag{7.22}$$

which, after substitution of

$$\int_{T_r}^{T} C^\circ_p dT = a(T - T_r) + \frac{b}{2}(T^2 - T_r^2) + c\left(\frac{1}{T} - \frac{1}{T_r}\right) \tag{7.23}$$

and

$$\int_{T_r}^{T} \frac{C^\circ_p}{T} dT = a \ln\left(\frac{T}{T_r}\right) + b(T - T_r) + \frac{c}{2}\left(\frac{1}{T^2} - \frac{1}{T_r^2}\right) \tag{7.24}$$

and collection of terms, results in

$$\begin{aligned}\Delta_a G^\circ_{T,i} &= \Delta_f G^\circ_{T_r,i} - S^\circ_{T_r,i}(T - T_r) \\ &\quad + a_i\left[T - T_r - T \ln\left(\frac{T}{T_r}\right)\right]\end{aligned}$$

$$+\frac{b_i}{2}\left(2TT_r - T^2 - T_r{}^2\right)$$
$$+\frac{c_i\left(T^2 + T_r^2 - 2TT_r\right)}{2TT_r^2} \tag{7.25}$$

This equation could of course also be adapted to the calculation of $\Delta_r G^\circ$by substituting $\Delta_r S^\circ_{T_r}$ for $S^\circ_{T_r,i}$ and $\Delta_r a$, $\Delta_r b$, and $\Delta_r c$ for a_i, b_i, and c_i, where $\Delta_r a$, etc. are the usual product – reactant terms. Thus

$$\begin{aligned}\Delta_r G^\circ_T = {} & \Delta_r G^\circ_{T_r} - \Delta_r S^\circ_{T_r}(T - T_r)\\ & +\Delta_r a\left[T - T_r - T\ln\left(\frac{T}{T_r}\right)\right]\\ & +\Delta_r\frac{b}{2}\left(2TT_r - T^2 - T_r^2\right)\\ & +\frac{\Delta_r c\left(T^2 + T_r^2 - 2TT_r\right)}{2TT_r^2}\end{aligned} \tag{7.26}$$

A special case would be the reaction in which i is formed from its elements, in which case $\Delta_r S^\circ_{T_r}$ becomes $\Delta_f S^\circ_{T_r,i}$, $\Delta_r a$ becomes $\Delta_f a_i$, and so on, and equation (7.26) then gives the traditional free energy of formation as a function of T.

7.6. PROPERTIES OF SOLIDS AT ELEVATED T AND P

Thus far all equations have been derived for conditions of elevated temperature but with the pressure fixed at P_r, normally one bar. For conditions of elevated pressure as well as temperature, an additional term must be added. Equation (7.21) thus becomes, for compound i,

$$\Delta_a G^\circ_{T,P,i} = \Delta_f G^\circ_{T_r,P_r,i} + \int_{T_r}^{T} (\partial G^\circ_i/\partial T)_{P=P_r} dT + \int_{P_r}^{P} (\partial G^\circ_i/\partial P)_{T=T} dP \tag{7.27}$$

for apparent free energies, and *should* become

$$\Delta_f G^\circ_{T,P,i} = \Delta_f G^\circ_{T_r,P_r,i} + \int_{T_r}^{T} (\partial \Delta_f G^\circ_i/\partial T)_{P=P_r} dT + \int_{P_r}^{P} (\partial \Delta_f G^\circ_i/\partial P)_{T=T} dP \tag{7.28}$$

for standard free energies of formation.

However, while the temperature integral can be evaluated using $\Delta_f a_i$, $\Delta_f b_i$, and $\Delta_f c_i$ for the reaction forming i from its elements, resulting in equation (7.25), the pressure integral (which involves both i and its elements) can only be conveniently evaluated for solids. Thus for compounds in which some of the elements are gases at one bar, the evaluation of the pressure integral in (7.28) becomes a decided nuisance, if not impossible, and in fact it is never used. Even for those cases where the reference elements are carried up to T from 298 K, they are left at one bar, and only the compound

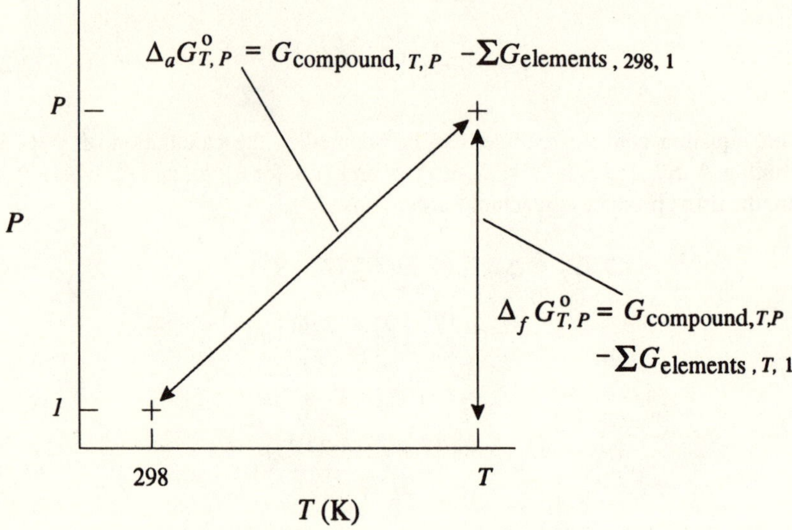

FIG. 7.8. The conceptual difference between $\Delta_a G^\circ$ and $\Delta_f G^\circ$.

of interest itself is considered at elevated pressures. In other words, equation (7.28) becomes

$$\Delta_f G^\circ_{T,P,i} = \Delta_f G^\circ_{T_r,P_r,i} + \int_{T_r}^{T} (\partial \Delta_f G^\circ_i / \partial T)_{P=P_r} dT + \int_{P_r}^{P} (\partial G^\circ_i / \partial P)_{T=T} dP \tag{7.29}$$

In this case $\Delta_f G^\circ_{T,P,i}$ is not exactly what it claims to be. That is, it is not the ΔG of formation of a compound i from its elements, all of which are at temperature T and pressure P. It is the ΔG of the reaction in which i at T, P is formed from its elements which are at T, P_r, and is therefore a sort of apparent free energy of formation. The fact that the pressure integration for the elements is neglected when it becomes difficult points up very well the futility of bothering to include the temperature integral of the elements. The elements might just as well be left back at 298 K, since they all cancel out anyway in all practical applications. This is why apparent free energies and enthalpies as defined earlier are being used increasingly in high T, P applications. As mentioned earlier, carrying the elements up to T from 298 K in $\Delta_f G^\circ$ is not only unnecessary but complicates the fitting and interpolation of high-temperature properties by introducing numerous elemental transition states. The difference between $\Delta_f G^\circ_{T,P}$ and $\Delta_a G^\circ_{T,P}$ is illustrated in Figure 7.8.

7.6.1. *Thermal Expansion and Compressibility*

As shown previously, the derivative of G with respect to P is V, i.e.

$$(\partial G_i / \partial P)_T = V_i$$

so that to calculate the effect of P on G_i we must know how V_i varies as a function of P. Now the variation in volume with temperature and pressure is most commonly measured as the *thermal expansion* at constant P

$$\alpha = \frac{1}{V}\left(\frac{\partial V}{\partial T}\right)_P$$

and the *compressibility* at constant T

$$\beta = -\frac{1}{V}\left(\frac{\partial V}{\partial P}\right)_T$$

where V can be the molar, specific, or total volume, because the volume units cancel. (The $1/V$ term is often replaced by $1/V_0$, where V_0 is the volume in some reference state.) When substance i is a solid phase and thus has relatively small variation of V with both P and T (relative that is to liquids and gases), very little error is introduced by assuming that V_i is a constant at all P, T values. If pressures exceed 10 Kb or so, one may wish to include a compressibility term to allow for the decrease in V_i, but in fact compressibilities and thermal expansions are both poorly known for substances at high P, T. Fortunately, when both P and T are elevated, the errors introduced by the assumption that V is not affected by P or T tend to cancel one another, especially for reactions having solids as both products and reactants. As a result, the assumption of constant V for solids is often adopted for mineral reactions at high P and T, and results in

$$\begin{aligned}\int_{P_r}^{P} (\partial G_i/\partial P)_T dP &= G_{i,P} - G_{i,P_r} \\ &= V_i(P-1) \qquad (7.30)\end{aligned}$$

where V_i is the molar volume of the solid phase in J bar^{-1}, cal bar^{-1}, or equivalent units and P_r = 1 bar. The apparent free energy of a solid phase at elevated P and T is therefore given by equation (7.25) plus this $V_i(P-1)$ term, thus

$$\begin{aligned}\Delta_a G^\circ_{T,P,i} = \Delta_f G^\circ_{T_r,P_r,i} &- S^\circ_{T_r,P_r,i}(T-T_r) \\ &+ a_i\left[T - T_r - T\ln\left(\frac{T}{T_r}\right)\right] \\ &+ \frac{b_i}{2}\left(2TT_r - T^2 - T_r{}^2\right) \\ &+ \frac{c_i\left(T^2 + T_r^2 - 2TT_r\right)}{2TT_r^2} \\ &+ V_i^\circ(P-1) \qquad (7.31)\end{aligned}$$

Similarly, the (so-called) free energy of formation from the elements is

$$\begin{aligned}\Delta_f G^\circ_{T,P,i} = \Delta_f G^\circ_{T_r,P_r,i} &- \Delta_f S^\circ_{T_r,P_r,i}(T-T_r) \\ &+ \Delta_f a_i\left[T - T_r - T\ln\left(\frac{T}{T_r}\right)\right]\end{aligned}$$

$$+\frac{\Delta_f b_i}{2}\left(2TT_r - T^2 - T_r^2\right)$$
$$+\frac{\Delta_f c_i\left(T^2 + T_r^2 - 2TT_r\right)}{2TT_r^2}$$
$$+V_i^\circ(P-1) \tag{7.32}$$

Should it be necessary to calculate apparent enthalpies or entropies of minerals at elevated P and T, the assumption that $(\partial V^\circ/\partial T)_P = 0$ plus the general equations

$$H^\circ_{T,P} = H^\circ_{T_r,P_r} + \int_{T_r}^{T} C_p\,dT + \int_{P_r}^{P}\left[V^\circ - T\left(\frac{\partial V^\circ}{\partial T}\right)_P\right]dP \tag{7.33}$$

and

$$S^\circ_{T,P} = S^\circ_{T_r,P_r} + \int_{T_r}^{T}\frac{C_p}{T}\,dT - \int_{P_r}^{P}\left(\frac{\partial V^\circ}{\partial T}\right)_P dP \tag{7.34}$$

leads to (for apparent enthalpies)

$$\Delta_a H^\circ_{T,P,i} = \Delta_f H^\circ_{T_r,P_r,i} + a_i(T - T_r) + \frac{b_i}{2}(T^2 - T_r^2) + c_i\left(\frac{1}{T} - \frac{1}{T_r}\right) + V_i^\circ(P-1) \tag{7.35}$$

and (for standard enthalpies of formation from the elements)

$$\Delta_f H^\circ_{T,P,i} = \Delta_f H^\circ_{T_r,P_r,i} + \Delta_f a_i(T - T_r) + \frac{\Delta_f b_i}{2}(T^2 - T_r^2) + \Delta_f c_i\left(\frac{1}{T} - \frac{1}{T_r}\right) + V_i^\circ(P-1) \tag{7.36}$$

and (for entropy)

$$S^\circ_{T,P,i} = S^\circ_{T_r,P_r,i} + a_i\ln\left(\frac{T}{T_r}\right) + b_i(T - T_r) + \frac{c_i}{2}\left(\frac{1}{T^2} - \frac{1}{T_r^2}\right) \tag{7.37}$$

or (for standard entropies of formation from the elements)

$$\Delta_f S^\circ_{T,P,i} = \Delta_f S^\circ_{T_r,P_r,i} + \Delta_f a_i\ln\left(\frac{T}{T_r}\right) + \Delta_f b_i(T - T_r) + \frac{\Delta_f c_i}{2}\left(\frac{1}{T^2} - \frac{1}{T_r{}^2}\right) \tag{7.38}$$

All these equations are for solid phases for which V° has been assumed to be constant. For gases and supercritical fluids, this assumption is obviously not reasonable. The variation of the free energies of these gases and fluids is most conveniently handled in a completely different way, that is, by the introduction of a new function, the *fugacity* (see Chapter 11).

7.6.2. *Solid Volumes Not Considered Constant*

The assumption that the volumes of the solid phases are constant is not always acceptable. For example, in reactions involving only solids, both $\Delta_r V$ and $\Delta_r S$ will be small, and because these represent the slope of $\Delta_r G$ versus P and T, small errors in either $\Delta_r V$ or $\Delta_r S$ will mean relatively large errors in $\Delta_r G$ and hence in the T and P of computed phase transitions (Anderson, 1976; Helgeson et al., 1978). There are very few attempts at a systematic treatment of solid volume variations at high T and P, the most useful probably being that of Berman (1988). Berman fit the available data for rock-forming minerals to the expression

$$\frac{V_{P,T}}{V_{P_r,T_r}} = 1 + v_1(P - P_r) + v_2(P - P_r)^2 + v_3(T - T_r) + v_4(T - T_r)^2 \tag{7.39}$$

where v_1–v_4 are fit parameters. The equation has no theoretical basis, and Berman cautions against its use at conditions beyond those used in the derivation of the parameters. Nevertheless, it should lead to improved accuracy of phase equilibrium calculations in the majority of geological situations. To get the change in G from P_r,T to P,T, we need $\int V\,dP$ from P_r to P at temperature T, where V is given by (7.39). Thus

$$\int_{T,P_r}^{T,P} V\,dP = \int_{T,P_r}^{T,P} V_{T_r,P_r}\left[1 + v_1(P - P_r) + v_2(P - P_r)^2 \right.$$
$$\left. + v_3(T - T_r) + v_4(T - T_r)^2\right] dP$$

This is considerably simplified by letting $P_r = 1$. Integrating and collecting terms then results in

$$\int_{T,P_r}^{T,P} V\,dP = V_{T_r,1}\left[\left(1 - v_1 + v_2 + v_3(T - T_r) + v_4(T - T_r)^2\right)(P - 1)\right.$$
$$\left. + \left(\frac{v_1}{2} - v_2\right)(P^2 - 1) + \frac{v_2}{3}(P^3 - 1)\right] \tag{7.40}$$

The right-hand side of equation (7.40) can now be substituted for $V_i^\circ(P - 1)$ in equations like (7.31) and (7.32).

We remind the reader at this point that all the expressions involving a, b, and c in this chapter are derived using the Maier-Kelley expression for the variation of C_p as a function of T. Several other equations are in common use, which will of course change the derived equations. Obviously we cannot derive all the equations for each case; the important thing is to see the dependence of the various functions discussed here on the heat capacity. In Appendix B we present the essential equations for several versions of the heat capacity expression.

7.7. SOURCES OF DATA

For Earth scientists, concerned primarily with the properties of minerals, silicate melts, aqueous solutions and gases at elevated temperatures and pressures, sources of data for thermodynamic calculations presents an unending challenge. This challenge is at two levels—first, do the necessary data exist, and second, if there are several sets of data, which one to choose, i.e., which is "the best"?

In addition to the primary research literature, there is a rather wide range of sources of data in the form of compilations, reviews, and critical comparisons. In an excellent summary of these sources, Nordstrom and Munoz (1985, Appendix F) list 206 sources for all substances except water, and 40 for water alone. Zemaitis et al. (1986, Appendix C) give numerous sources for aqueous systems. Some of the more frequently used data sets are: the Joint Army-Navy-Airforce (JANAF) Thermochemical Tables (Stull and Prophet, 1971, with periodic supplements up to 1982); Robie, Hemingway and Fisher (RHF, 1978); Helgeson, Delaney, Nesbitt and Bird (HDNB, 1978); Berman, Brown and Greenwood (BBG, 1985); Berman (1988); the National Bureau of Standards Tables (Wagman et al. 1982); Helgeson, Kirkham and Flowers (HKF, 1981); and Sillén and Martell (1964, supplement 1971). These data sets differ in many ways. HDNB, RHF, and HKF were prepared by Earth scientists, so they concentrate on naturally occurring compounds (minerals) and solute species, and they are just as concerned with high temperature and pressure conditions as with 25°C, one bar. The others are more comprehensive (Wagman et al. for example contains data for some 26,000 properties of 14,300 substances), but most of these are never found in nature, and their properties are given for the most part only at or close to 25°C.

Another major difference is that HDNB, BBG, and HKF are intended to be used in conjunction with a computer program, rather than as written tables, and this is probably the trend of the future. All modern compilations take great pains to be "self-consistent" in that the values published satisfy the basic thermodynamic relationships among themselves and are consistent with all the experimental data used in deriving them, but again, HDNB, BBG, and HKF being primarily by and for Earth scientists place much more emphasis on high temperature and pressure data in deriving the tabulated values. Thus the data for minerals in HDNB and BBG are consistent with a large body of experimentally-derived phase relations at high temperatures and pressures not considered in the other compilations. A particularly clear discussion of the different approaches, with their advantages and disadvantages, and some of the experimental problems, is given by Hemingway, Haas, and Robinson (1982).

7.8. SIMULTANEOUS EVALUATION OF DATA

The HDNB tables strive for internal consistency by analyzing groups of experimental data for reactions and compounds considered together instead of one at a time. This is a fundamentally different approach from RHF, JANAF and many other thermodynamic sources in which compounds are considered individually. Simultaneous numerical analysis of groups of data for different compounds ensures that derived thermodynamic properties will be internally consistent, simply because they are derived together. This has both advantages and potential disadvantages. A very practical advantage is that phase relations computed from such data sets will be absolutely consistent internally, meaning that thermodynamic relationships in one part of the T-P region considered will be consistent with those in another. A disadvantage is that in practice, the process is so complex that only the originator of the data set can update it. Probably the most difficult part of such work is deciding which one of conflicting data sets to include in the analysis, and it is virtually certain that different investigators will make different choices, and arrive at different internally consistent data sets. There are also various methods of data analysis, weighting and smoothing.

The method used to prepare the HDNB tables is summarized by Helgeson et al. (1978, pp. 34–35). A particularly clear, simplified illustration of this approach is given by Nordstom and Munoz (1985, pp. 417–418). A working knowledge of linear algebra is a distinct advantage in understanding this approach.

7.8.1. *Simultaneous Analysis of All Available Data for Single Compounds*

A second, quite different method for evaluating thermodynamic data makes full use of the fact that all the thermodynamic functions of state (U,G,H,S, A, etc.) are directly related to the heat capacity, C_p. M any of these relationships have been derived already in this chapter. For any one compound, a search of the literature might turn up an enormous variety of measured thermodynamic quantities—enthalpies, entropies, free energies, and so on. Because these seemingly disparate quantities are actually related through C_p, it is possible to analyze them all at once and derive the function $C_p = f(T)$ which provides the best agreement with the entire data set. A statistical procedure using simultaneous multiple regression analysis has been developed for this specific kind of application by Haas and Fisher (1976). The method has been used to compile a thermodynamic data base for minerals characteristic of basalts (Robinson et al., 1982), among other systems.

In their approach, Haas and Fisher added two additional terms to the Maier-Kelly heat capacity equation (7.12) to give

$$C_p = a + 2bT + c/T^2 + fT^2 + g/T^{\frac{1}{2}} \tag{7.41}$$

This is then related to the thermodynamic functions G, H, S, etc., by the equations listed in Appendix B. As we have already observed (with the simpler Maier-Kelly equation above) these functions are all obtained by integrating the heat capacity equation between two temperatures. In practice, considerable care must be taken in fitting regression equations to different kinds of thermodynamic data for single compounds, and the data should be examined quite critically for potentially inaccurate points, because the results can be quite sensitive to inaccurate input data.

7.8.2. *Derivation of Internally Consistent Data Bases Using Linear Programming*

Yet another type of internally consistent thermodynamic data base makes full use of phase equilibrium data as well as all other available thermochemical and physical measurements. The most complete example at present is the Berman, Brown and Greenwood (BBG) data base, described by Berman et al. (1984, 1985, 1986). Like the methods just described, the BBG approach analyzes all data simultaneously and produces an internally consistent set of thermodynamic data. However, the method differs significantly in using the mathematical technique called linear programming to deal with "bracketed" phase equilibrium data.

This particular application of linear programming was introduced and has been succinctly summarized by Gordon (1973, 1977), and we use one of his examples to illustrate the general method here. For details, see the texts by Harbaugh and Bonham-Carter (1970, geologically relevant examples), and Gill et al. (1981, full

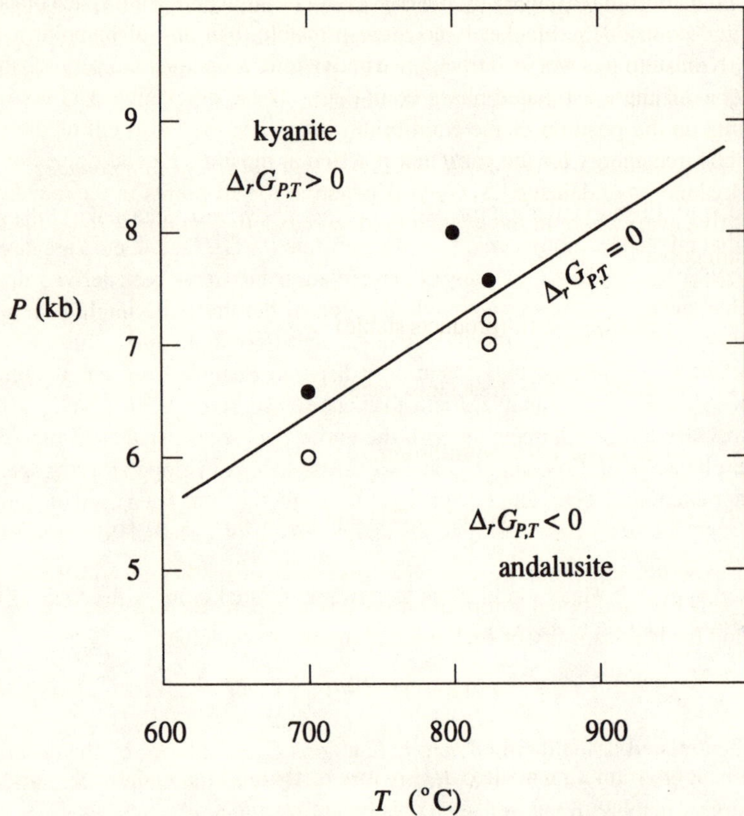

FIG. 7.9. Phase diagram for the reaction kyanite - andalusite showing the experimental data of Richardson et al.(1969). The straight line represents estimated equilibrium conditions for which $\Delta_r G = 0$.

mathematical approach), as well as Berman et al. (1986). The kyanite $\rightleftharpoons$ andalusite reaction is a typical example; the data by Richardson et al. (1969) shown in Figure 7.9 show how the real phase boundary is usually bracketed by experimental points. These experiments are usually carried out by holding material of an appropriate composition at high T and P for a sufficient period of time that complete or at least considerable reaction occurs. The high-pressure vessel is then rapidly quenched and the products are identified. Typically either one mineral or the other is produced, and the experiments are repeated until the brackets are as close together as is feasible experimentally. The same thing happens with multi-phase experiments where more than one mineral or phase occurs on either side of the equilibrium boundary. The experiments should also be reversed, to show that assemblage A can react to form B and vice versa.

Now the position of the equilibrium boundary could be estimated simply by drawing a line through the brackets, and this is how phase diagrams are usually constructed. Unfortunately, the true boundary could lie anywhere within each bracket, and the mid-point of the bracket is no more probable than any other point in the bracket. To make things worse, derived thermodynamic properties are very sensitive to the slopes of these estimated phase boundaries. These bracketing data not only place limits on the position of the equilibrium boundary, they also tell us the sign of the Gibbs free energy for the particular reaction at those P, T conditions. For the reaction kyanite $\rightleftharpoons$ andalusite, $\Delta_r G_{T,P}$ is negative for all points in the andalusite field, positive everywhere in the kyanite field, and exactly equal to zero on the true equilibrium curve:

$$\left.\begin{aligned} &\Delta_r G_{P,T} < 0 \quad \text{(products stable)} \\ &\Delta_r G_{P,T} > 0 \quad \text{(reactants stable)} \\ &\Delta_r G_{P,T} = 0 \quad \text{(equilibrium,} \\ &\qquad\qquad\quad \text{both products and reactants stable)} \end{aligned}\right\} \tag{7.42}$$

For each experimental point we can write one such inequality or constraint. We probably can find additional information on this reaction; for example, the molar volume and the heat capacity of each mineral might be available:

$$V_i = constant \tag{7.43}$$

$$C_{p,i} = a + bT + \cdots \tag{7.44}$$

The linear programming method deals with exactly this kind of data set—one that includes linear inequalities as well as fixed constraints.

Let's see how this works when the free energy of reaction at T and P is written out in full. Consider a reaction involving several different phases, such as

$$a\text{A} + b\text{B} + c\text{C} + \cdots = d\text{D} + e\text{E} + \cdots \tag{7.45}$$

For this, the apparent molar Gibbs free energy change of reaction at P and T is simply

$$\Delta_r G_{P,T} = \sum_{i=1}^{all\ phases} \nu_i \Delta_a G_{i,P,T} \tag{7.46}$$

The parameters ν_i are stoichiometric reaction coefficients. Now if S°_{i,P_r,T_r} and $\Delta_f H^\circ_{i,P_r,T_r}$ are the Third Law entropy and the enthalpy of formation of each pure phase at a reference pressure and temperature P_r and T_r (usually 1 bar and 298.15 K), the free energy of reaction at a general P and T in (7.46) is given by the equation

$$\begin{aligned} \Delta_r G_{P,T} = \sum_{i=1}^{all\ phases} \nu_i \Bigg(&\Delta_f H^\circ_{i,P_r,T_r} - T S^\circ_{i,P_r,T_r} + \int_{T_r}^{T} C_{p_i} dT \\ &- T \int_{T_r}^{T} (C_{p_i}/T) dT + \int_{P_r}^{P} V_i dP + RT \ln a_i \Bigg) \end{aligned} \tag{7.47}$$

This is similar to equation (7.21). It expresses the change in free energy as pressure and temperature are raised from some reference state P_r and T_r to the P and T of interest. The activity term (a_i) is introduced to allow for compositional effects, and is discussed in Chapter 11.

For each data point, we can write one version of equation (7.47) with $\Delta_r G_{P,T}$ less than or greater than zero, depending on whether products or reactants are stable. (This corresponds to the inequalities 7.42). To this we can also add other constraining equations, such as known molar volumes of different phases, and perhaps known heat capacities at T and P (corresponding to equations (7.43) and (7.44)).

These equations can then be analyzed for the range of conditions that satisfies all constraints. This gives the best estimates of the unknown thermodynamic parameters in equation (7.47) for each phase. For example, we might know everything in (7.47) but $\Delta_f H^\circ_{i,P_r,T_r}$ and S°_{i,P_r,T_r} which would then be given by the linear programming procedure. The thermodynamic properties found in this manner are the best estimates consistent with the experimental data. In Terry Gordon's (1977) words, "The resulting values are those that do the least violence to the experimental results."

7.9. APPROXIMATIONS

The equations derived thus far have been based on an explicit formulation of the variation of heat capacity with temperature (that is, the Maier-Kelley formula). It has led to some rather lengthy equations.

There are occasions when heat capacity data are not available, or when the ultimate accuracy is not required. In these cases one may assume that ΔC_p° is not a function of temperature, i.e., that

$$\Delta C_p^\circ = constant$$

or even that

$$\Delta C_p^\circ = 0$$

Assuming that ΔC_p° has a constant value equal to its value at 298 K leads to the expressions

$$\Delta H_T^\circ = \Delta H_{T_r}^\circ + \Delta C^\circ_{p\,T_r}(T - T_r) \tag{7.48}$$

and

$$\Delta S_T^\circ = \Delta S_{T_r}^\circ + \Delta C^\circ_{p\,T_r} \ln(T/T_r) \tag{7.49}$$

which are of course considerably simpler than the corresponding more exact expressions derived earlier, involving the Maier-Kelley coefficients *a*, *b*, and *c*. In fact, however, these equations are little used. It seems that if one goes to the trouble of considering heat capacities at all, one generally goes all the way. Much more popular is the practice of not considering heat capacities at all, i.e., of assuming that $\Delta C_p^\circ = 0$. This of course leads to the supposition that $\Delta H_T^\circ = \Delta H_{T_r}^\circ$ and $\Delta S_T^\circ = \Delta S_{T_r}^\circ$, greatly simplifying equations. For example, the free energy of reaction at high T, normally

$$\Delta_r G_T = \Delta_r H_T - T\Delta_r S_T$$

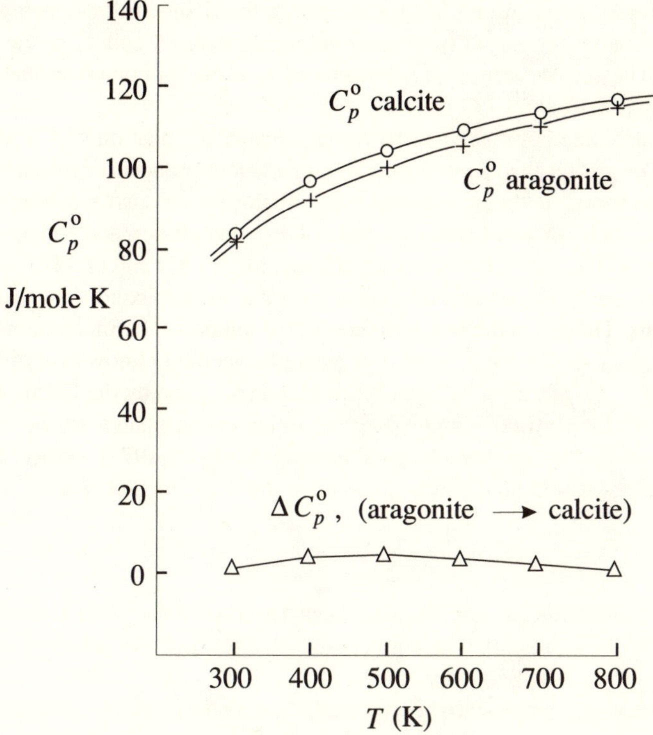

FIG. 7.10. Heat capacities of calcite and aragonite, showing that ΔC_p° varies much less than either C_p°.

becomes

$$\Delta_r G_T = \Delta_r H_{T_r} - T\Delta_r S_{T_r} \tag{7.50}$$

which means that $\Delta_r G$ can be approximated at any T simply by knowing $\Delta_r H$ and $\Delta_r S$ at 298 K. The constant volume assumption for solid phases is not affected by this procedure, so that for high temperatures and pressures (7.50) becomes

$$\Delta_r G_{T,P} = \Delta_r H_{T_r,p_r} - T\Delta_r S_{T_r,p_r} + \Delta_r V(P-1) \tag{7.51}$$

This equation has been widely used for reactions involving only solid phases, because the $\Delta C_p^\circ = 0$ assumption is surprisingly good for many reactions if only solids are involved. As an example the variations of the heat capacities and entropies of calcite and aragonite with temperature are shown in Figures 7.10 and 7.11 (RHF data). Notice that S and C_p for both minerals vary considerably with temperature, but that they vary in parallel. On the same diagrams we have plotted $\Delta_r S$ and $\Delta_r C_p$ for the calcite—aragonite reaction. Neither $\Delta_r S$ nor $\Delta_r C_p$ is exactly constant with temperature, but both are reasonably so, and the value of $\Delta_r C_p$ is also reasonably close to zero.

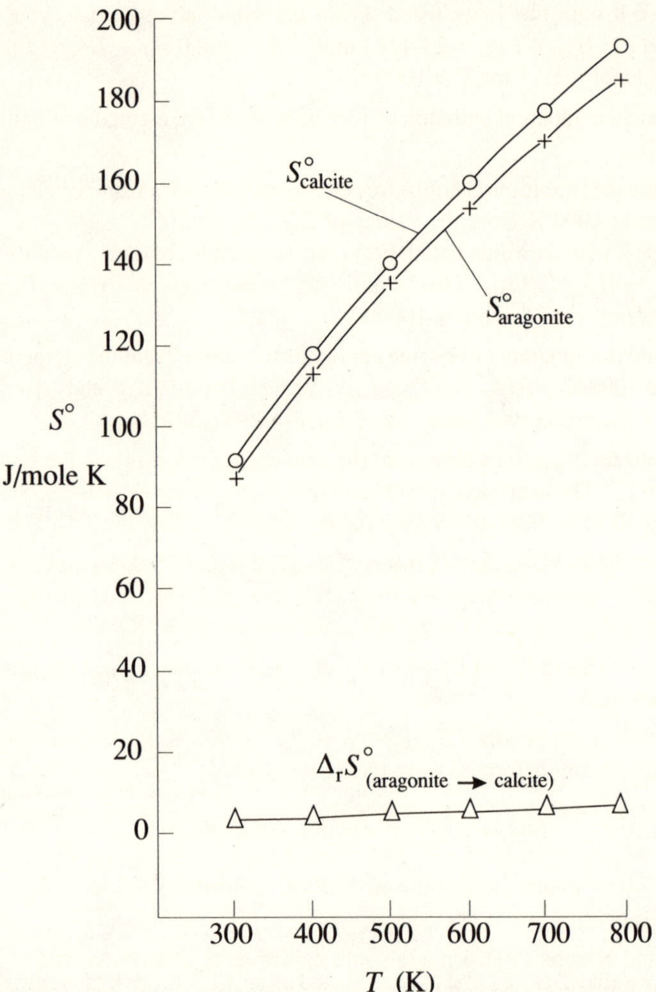

FIG. 7.11. Entropies of calcite and aragonite, showing that ΔS° varies much less than either S°.

This approximation can also fail dismally when liquids, gases or aqueous solutes are involved. As always with approximations the results should be checked to see if they appear reasonable. However, in these cases, other approximations are available, which will be discussed in Chapters 8 and 13.

PROBLEMS

1. Explain to yourself the distinctions between $G, \Delta_r G, \Delta_r G^\circ, \Delta_f G^\circ, \Delta_a G^\circ$.
2. Calculate the (traditional) enthalpy of formation from the elements of halite at 1000 K using values of $(H_T - H_{298})/T$ and $\Delta_f H^\circ_{298}$ in Table 7.2 and

compare it with the value listed. From the same tables of data you will find $[(H_T - H_{298})/T]_{\mathrm{Na}} = 23.428\ \mathrm{J\,mol^{-1}\,K^{-1}}$ and $[(H_T - H_{298})/T]_{\mathrm{Cl_2}} = 25.565\ \mathrm{J\,mol^{-1}\,K^{-1}}$ for $T = 1000$ K.

3. Calculate the apparent enthalpy of formation from the elements of halite at 1000 K.
4. Calculate the (traditional) Gibbs free energy of formation from the elements of halite at 1000 K from the values of $\Delta_f H^\circ_{298}$ and $(G^\circ_T - H^\circ_{298})/T$ and compare it with the value listed. From the same tables of data, you will find $-[(G^\circ_T - H^\circ_{298})/T]_{\mathrm{Na}} = 71.02\ \mathrm{J\,mol^{-1}\,K^{-1}}$ and $-[(G^\circ_T - H^\circ_{298})/T]_{\mathrm{Cl_2}} = 241.19\ \mathrm{J\,mol^{-1}\,K^{-1}}$ for $T = 1000$ K.
5. Calculate the apparent Gibbs free energy of formation from the elements of halite at 1000 K. $-[(G^\circ_T - H^\circ_{298})/T]_{\mathrm{Na}} = 51.30\ \mathrm{J\,mol^{-1}\,K^{-1}}$ and $-[(G^\circ_T - H^\circ_{298})/T]_{\mathrm{Cl_2}} = 233.080\ \mathrm{J\,mol^{-1}\,K^{-1}}$ for $T = 298.15$ K.
6. Calculate $\Delta_f H^\circ_{1000}$ for halite from the heat capacity formula at the bottom of Table 7.2. The heat capacity formula for $\mathrm{Cl_2}(g)$, valid for 298–1800 K, is: $C^\circ_{P\,\mathrm{Cl_2}} = 46.956 - 0.0040158\,T + 9.9274 \times 10^{-7}\,T^2 - 204.95T^{-0.5}$. Sodium melts at 370.98 K. At that T, $[(H_T - H_{298})/T]_{\mathrm{Na}(s)} = 5.808\ \mathrm{J\,mol^{-1}\,K^{-1}}$, and $\Delta_t H^\circ_{371}(fusion) = 2.598\ \mathrm{kJ\,mol^{-1}}$. Below 370.98 K, the heat capacity formula is $C^\circ_{P\,(s)} = 16.82 + .0376T$, and above 370.98 K, it is $C^\circ_{P\,\mathrm{Na}(l)} = 37.482 - 0.019183T + 1.0644 \times 10^{-5}T^2$. You can check your integration in Appendix B.
7. Using data in Appendix C, calculate $\Delta_r H^\circ$ at 900 K for the reaction in which andalusite is formed from the oxides, i.e.

$$\mathrm{SiO_2 + Al_2O_3 = Al_2SiO_5}\ (andalusite)$$

The $\alpha - \beta$ quartz transition at 848 K contributes $\Delta_t H_{\alpha-\beta\ qtz} = 290\ \mathrm{cal\,mol^{-1}}$.

8. Holm and Kleppa (1966) made some calorimetric determinations of the heats of formation of kyanite, andalusite, and sillimanite from their constituent oxides (quartz and corundum) in an oxide melt calorimeter at 968 K. For andalusite, the result for the reaction given above was $\Delta_r H_{968} = -1.99\ \mathrm{kcal\,mol^{-1}}$. Calculate $\Delta_f H^\circ_{298}$ for andalusite, using values of $\Delta_f H^\circ_{298}$ for quartz and corundum and heat capacity coefficients from Appendix C.

8

APPLICATIONS TO SIMPLE SYSTEMS

> There is no way whatever of being sure that any of the ordinary objects of daily life do not have other polymorphic forms into which they may sometime change spontaneously before our eyes. Bridgman (1941) p. 119.

8.1. USEFUL APPLICATIONS

Thus far we have developed just about all the thermodynamic concepts required by Earth scientists with the exception of those needed to deal with solutions. Since all naturally occurring substances are solutions of one kind or another (although some can usefully be treated as pure substances), this is quite an important limitation, and we will proceed to discuss the treatment of solutions in Chapter 10. However, a great deal can be done with the thermodynamics of pure systems, and in this chapter we discuss a couple of applications of the concepts so far developed which are of particular interest to Earth scientists—the thermal effects associated with adiabatic volume changes, and the T-P phase diagrams of pure minerals.

8.2. ADIABATIC VOLUME CHANGES

All systems experience a change in volume in response to changes in pressure. We have discussed this mostly from the point of view of the work accomplished by isobaric volume changes in Chapter 4, but it is even more informative to consider the temperature changes accompanying volume changes. The best way to do this is to consider only cases uncomplicated by heat entering or leaving the system, i.e., adiabatic processes. Such processes, although yet another "hypothetical limiting case," serve as useful end-members in considering actual processes in real systems. The most familiar everyday example is the hand-held bicycle pump, which most cyclists at least know gets quite warm during pumping (compressing air). This process, while not strictly adiabatic (bicycle pumps are not well insulated) is sufficient to show that volume changes can be associated with temperature changes, and it is not difficult to see in this case why—a great deal of energy in the form of work is being added to the gas, and some of it is being used to warm the gas. It seems reasonable to suppose, too, that by reversing the process—suddenly expanding the gas—it would experience a temperature decrease. This much may seem intuitively reasonable, perhaps even obvious. What is not so obvious is the fact, first investigated by Joule and Thompson in 1853, that some substances do not warm but cool during compression, and that in fact all substances have a range of conditions where they warm on expansion and another where they cool on expansion. When the expansions are at constant enthalpy, these two ranges are separated by the Joule-Thompson inversion curve.

From the First Law, $\Delta U = q + w$, we see that adiabatic processes ($q = 0$) are those for which $\Delta U = w$. Thus the final state achieved after such a change will depend entirely on the work done during the change, or conversely, for any two states

Table 8.1 Thermodynamic Data for Water at T and P. Values in Figures 8.1 and 8.4 [from Adcock (1988)]

Point	ΔU° J mol^{-1}	ΔH° J mol^{-1}	V° J bar^{-1} mol^{-1}	S° J K^{-1} mol^{-1}
A (2000b, 700°C)	−235911	−232426	3.601	155.71
B (1027b, 600°C)	−237480	−236372	4.6995	155.71
C (870b, 600°C)	−235910	−234800	5.5506	158.479
D (693.5b, 600°C)	−233728	−232426	7.2397	162.472
E (991.5b, 600°C)	−237155	−236058	4.8571	156.317

connected by an adiabatic path, w is fixed. This leaves quite a variety of possibilities. These fall into two categories, reversible and irreversible. Since there is only one case of reversible adiabatic volume change to consider and an infinite number of irreversible ones, we will start with the reversible case. We will use H_2O at high temperatures and pressures as an example and the P-V-T conditions considered are shown in Figure 8.1 and Table 8.1.

8.2.1. *Reversible Adiabatic or Isentropic Volume Changes*

As explained in Chapter 4, a reversible volume change can be imagined as one in which the pressure in the system is always exactly balanced by a free frictionless piston, which has infinitesimal amounts of weight added to it or subtracted from it (approximated by grains of sand in Figure 8.2a) until the change is accomplished (Figure 8.2a).

It therefore passes through a continuous succession of equilibrium states, a situation which as we have mentioned is very convenient mathematically. While this volume change is taking place, other restrictions may be imagined with no extra effort, such as that the change must take place isothermally or adiabatically. If the reversible change is adiabatic, it is a constant entropy or isentropic process, because $Dq_{rev}/T = dS$, and of course Dq_{rev} is zero. Systems will always cool during an isentropic expansion, because (referring to Table 2.1)

$$\left(\frac{\partial T}{\partial P}\right)_S = \frac{T}{C_p}\left(\frac{\partial V}{\partial T}\right)_P$$

and because T, C_p, and $(dV/dT)_P$ are intrinsically positive, decreasing P at constant S will mean a decrease in T. This is also shown in Figures 8.3 and 8.4. A reversible expansion accomplishes the maximum amount of work, so that for a given isentropic adiabatic expansion w will have its maximum negative value (work done by the system on the environment) and ΔU will also therefore have the maximum negative change possible for a given decrease in P. That means the expanded state will have less energy than the initial state, and less than any other possible adiabatic expansion to the same P or T.

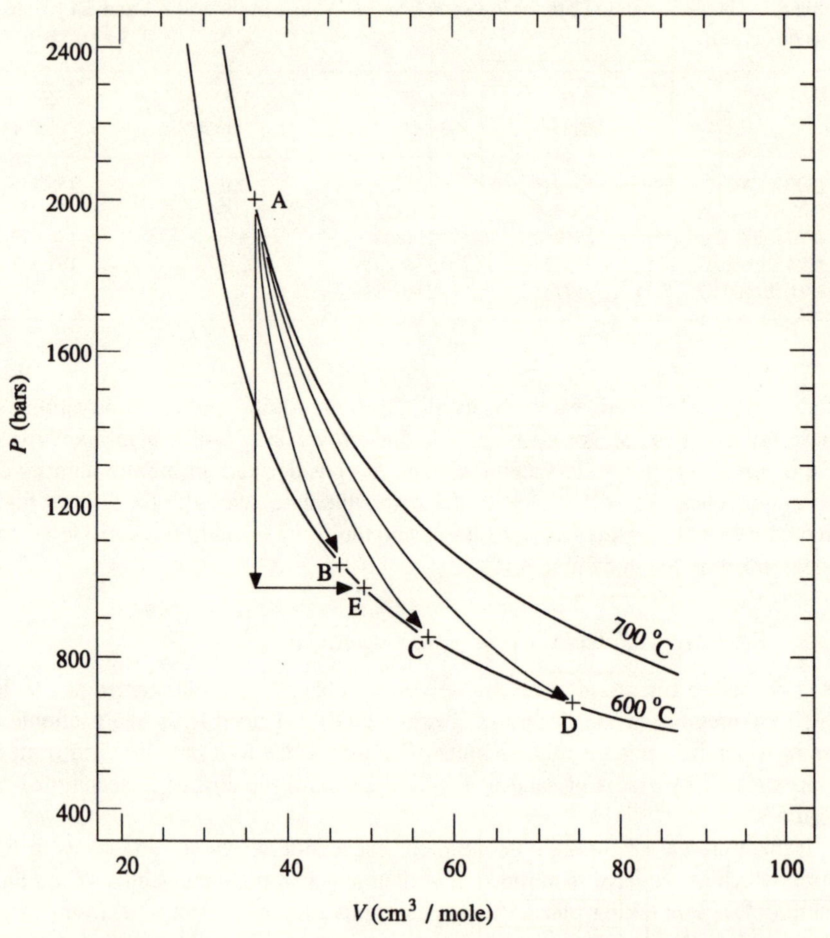

FIG. 8.1. Pressure versus volume for water, showing the 700°C and 600°C isotherms. The effects of various types of adiabatic expansion are shown, starting from point A at 700°C, 2000 bars. A-B: isentropic; A-C: isoenergetic ($\Delta U = 0$); A-D: isenthalpic; A-E: "isobaric". Data in Table 8.1.

To plot the course of this expansion on a P-V diagram we need to have a function for $(dP/dV)_S$ and be able to integrate it. This function is (using Table 2.1)

$$\left(\frac{\partial P}{\partial V}\right)_S = \frac{-C_p/T}{-\frac{1}{T}\left[C_p\left(\frac{\partial V}{\partial P}\right)_T + T\left(\frac{\partial V}{\partial T}\right)_P^2\right]}$$

For an ideal gas this works out to be

$$\left(\frac{\partial P}{\partial V}\right)_S = -\gamma_c \cdot P/V \text{ where } \gamma_c = C_p/C_v$$

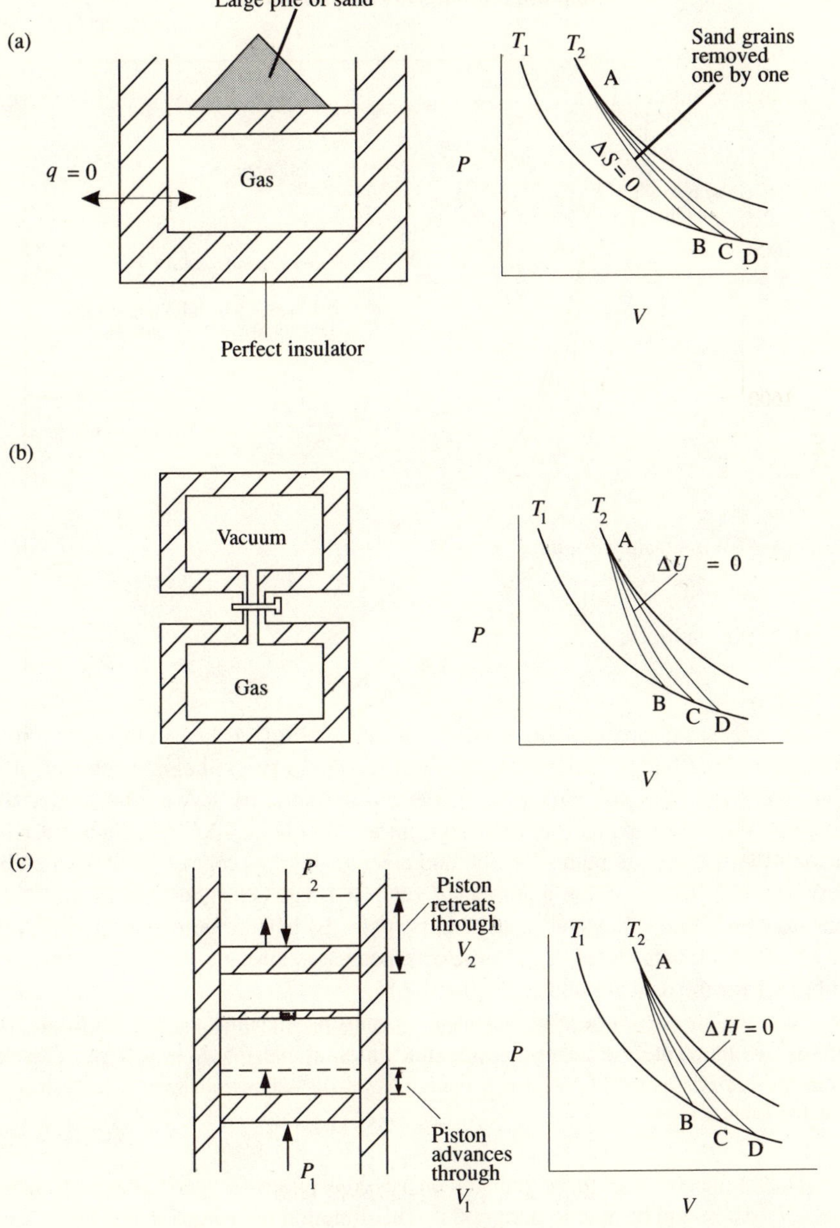

FIG. 8.2. Schematic cartoons illustrating various types of adiabatic volume changes. In each case the points on the PV diagram correspond to those in Figure 8.1. (a) Reversible, Isentropic. (b) Joule Expansion. (c) Joule-Thompson Expansion.

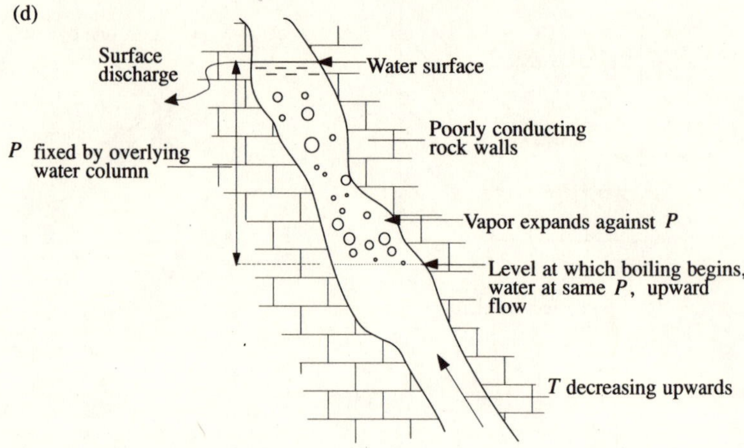

FIG. 8.2. (d) Isenthalpic Boiling.

which has the solution

$$PV^{\gamma_c} = constant$$

if γ_c is constant (Pippard, 1966, p. 62). However, in most real cases the equation of state of natural fluids is either not known or too complex to integrate conveniently. This is certainly the case with water under most conditions, so we have plotted the adiabatic expansion paths for water starting at 2000 bars, 700°C in Figure 8.1 by finding the appropriate points by trial and error, using the program H2O, written by S.W. Adcock, based on the equation of state of Haar, Gallagher and Kell (1984). For example at 600°C, water at 1027 bars (point B) has the same entropy as water at 700°C, 2000 bars (point A), so an isentrope passes through these two points (see Table 8.1 for the data at points in Figure 8.1).

To calculate the work done during the isentropic expansion between these two points, we calculate the change in apparent internal energy, ΔU, between the two points, which is

$$\Delta_a U_{2000b,700^\circ} = -235911 \text{ J mol}^{-1}$$

$$\Delta_a U_{1027b,600^\circ} = -237480 \text{ J mol}^{-1}$$

$$\begin{aligned}\Delta U &= U_\text{B} - U_\text{A} \\ &= \Delta_a U_\text{B} - \Delta_a U_\text{A} \\ &= -1596 \text{ J mol}^{-1} \\ &= w\end{aligned}$$

This is less than half the work that an ideal gas would do under the same conditions.

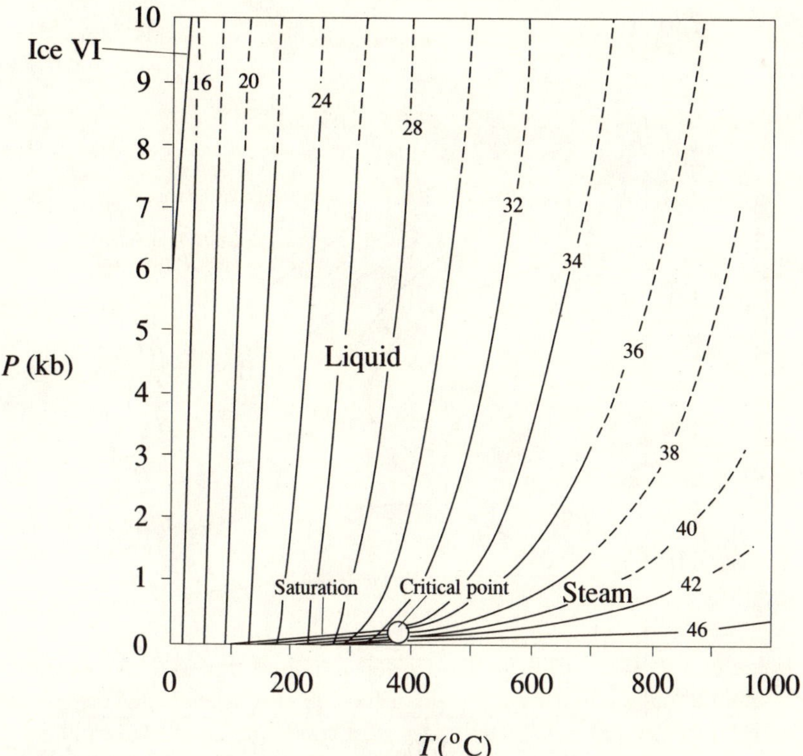

FIG. 8.3. Entropy of water as a function of temperature and pressure. Contours are labelled in cal K^{-1} mol^{-1}. The slope is always positive, showing that isentropic expansion will always cool the fluid. [After Helgeson and Kirkham (1974a)].

The volumes at points A and B are $V_A = 3.60$ J bar^{-1} mol^{-1} and $V_B = 4.70$ J bar^{-1} mol^{-1} (1 cm^3 = 0.10 J bar−1, Appendix A). Approximating the integral $\int P\,dV$ by the area of a trapezoid we get

$$\begin{aligned} w &= -\int P\,dV \\ &\approx -(4.70 - 3.60)(2000 + 1027)/2 \\ &\approx -1665\ \mathrm{J\,mol^{-1}} \end{aligned}$$

which shows, as expected, that the area of the trapezoid is slightly greater than the actual area under the adiabat between A and B. The point is that the work done in this process is in fact shown by the area under the curve, in contrast to the cases considered next.

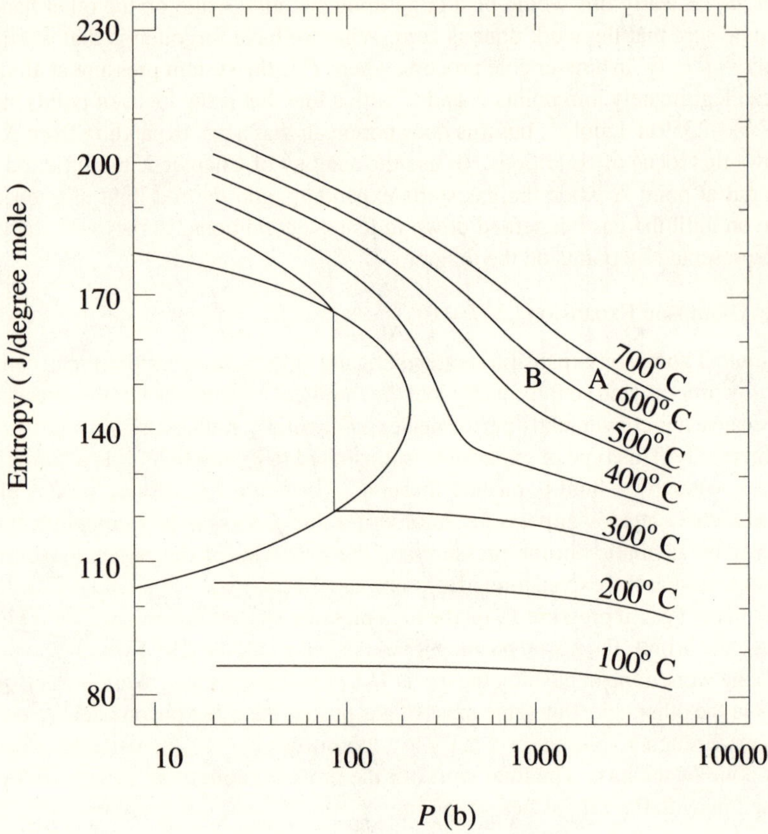

FIG. 8.4. Entropy versus pressure for water. Points A and B correspond to Figure 8.1.

8.2.2. *Irreversible Adiabatic Volume Changes*

There are an infinite number of irreversible adiabatic volume change paths that might be considered, but to keep the book to a reasonable size we will restrict ourselves to three. Each is instructive in its own way.

Joule Expansion ($\Delta U = 0$)

One of the many different kinds of experiments that Joule and others performed in the mid-nineteenth century was to expand gases into an evacuated chamber. This free expansion performs no work, and if adiabatic as well must clearly result in a state having the same internal energy as the initial state. It turns out that water at 600°C, 890 bars (point C in Figure 8.1) has the same apparent internal energy, -235911 J mol^{-1}, (and therefore the same internal energy) as water at 700°C, 2000 bars, and so line AC could be labelled $\Delta U = 0$ (as in Figure 8.2b). This line appears to outline an area on the PV plot that could be evaluated as we did above to give an approximation to

$-\int P\,dV$. Clearly this would be a large finite quantity, while on the other hand we are quite sure that the work done is zero. What we have forgotten is that work only equals $-\int P\,dV$ in a reversible process, where P is the system pressure at all times. We can legitimately join points A and C with a line that is the locus of points having $\Delta_a U = -235911\ \mathrm{J\,mol^{-1}}$, but this does not mean that a gas expanding from A to C follows this locus of conditions. To use the analogy of Chapter 3, the little red light goes out at point A when the gas starts expanding, and the red light at C does not come on until the gas has settled down in its new conditions. In between, the gas is not represented by points on the diagram.

Joule-Thompson Expansion ($\Delta H = 0$)

The Joule-Thompson expansion is interesting not only because it represents a constant enthalpy process, and so falls neatly into the family of expansions we are considering, but because it is conveniently performed experimentally and has proven highly useful commercially. This type of expansion is illustrated in Figure 8.2c. It is a "throttling" process, whereby a fluid is pushed through an orifice (originally a "porous plug") into a chamber at a lower pressure. In order for this process to be isenthalpic, it must not only be adiabatic but the pressures on the two sides of the orifice must remain constant, so that the expansion is from one equilibrium state to another. Consider a given mass of gas at pressure P_1 on the high pressure side of the orifice. As it is pushed through the orifice, the piston pushing it sweeps out a volume (ΔV) which we can call V_1, so the work done in pushing the gas is P_1V_1 (positive because work is done *on* the gas). On the other side, this same mass of gas pushes the other piston back at constant pressure through a volume (or a ΔV) V_2, doing work $-P_2V_2$ (negative because it is work done *by* the gas). The total work for the process is therefore $P_1V_1 - P_2V_2$. We equate this with the ΔU as before, giving

$$\begin{aligned} U_2 - U_1 &= w \\ &= P_1V_1 - P_2V_2 \end{aligned}$$

so

$$U_2 + P_2V_2 = U_1 + P_1V_1$$

or

$$H_2 = H_1$$

$$\Delta H = 0$$

In Figure 8.1, we show that H_2O at 693.5 bars, 600°C (point D) has the same apparent enthalpy of formation (-232426 $\mathrm{J\,mol^{-1}}$) as at 2000 bars, 700°C (point A), so these two points are joined by an isenthalp. The work done in this case is

$$\begin{aligned} U_\mathrm{D} - U_\mathrm{A} &= \Delta_a U_\mathrm{D} - \Delta_a U_\mathrm{A} \\ &= -233728 - (-235911) \\ &= 2183\ \mathrm{J\,mol^{-1}} \end{aligned}$$

meaning that more work is done in pushing the gas on the high pressure side than is gained on the low pressure side. The reader may verify that the same result is obtained by calculating $P_D V_D - P_A V_A$.

Of course there are other ways of achieving a constant enthalpy expansion. All that is required is to do the appropriate amount of work for the given change in U, but the orifice arrangement is the most convenient because it automatically gives the required conditions. Other expansions will in general not be isenthalpic unless carefully designed to be so. The essence of the Joule-Thompson expansion is that it is a *flow* process, pushed from one side. It should not be confused with ordinary piston-cylinder expansions where the system pressure varies during the expansion.

To investigate the variation of temperature during an isenthalpic change of pressure, one is naturally interested in the derivative $(\partial T/\partial P)_H$, called the Joule-Thompson coefficient, μ_{JT}. An easy way to derive an expression for this quantity is to use Table 2.1. From Table 2.1, $(\partial T)_H = -V + T(\partial V/\partial T)_P$, and $(\partial P)_H = C_p$. Then

$$\begin{aligned}\left(\frac{\partial T}{\partial P}\right)_H &= \frac{(\partial T)_H}{(\partial P)_H} \\ &= \frac{-V + T(\partial V/\partial T)_P}{C_p} \\ &= \frac{V(T\alpha - 1)}{C_p}\end{aligned}$$

or,

$$\mu_{JT} = \frac{V(T\alpha - 1)}{C_p} \tag{8.1}$$

where α is the coefficient of thermal expansion, $(dV/dT)_P/V$, and C_p is the isobaric heat capacity. Since both V and C_p are intrinsically positive quantities, the sign of the derivative $(dT/dP)_H$ clearly depends on $(T\alpha - 1)$, which in turn depends on $T\alpha$.

As shown by Waldbaum (1971), all minerals have values of α so small that $T\alpha \ll 1$ for all Earth conditions, so that $(dT/dP)_H$ is always negative, and rocks will always be warmed by isenthalpic expansions (dP negative, dT positive). For most common minerals, the effect is about 20 to 30 degrees per kilobar of pressure change, although if this pressure change is due to a change in depth in the Earth, the effect is less (Ramberg, 1971).

For fluids, which have much larger values of α, $(T\alpha - 1)$ can be positive or negative. Because fluids become less dense and hence show greater thermal expansion at higher temperatures and lower pressures, this is the range of conditions where they exhibit positive values of μ_{JT}. For gases, with the largest α values, this transition takes place at very low temperatures and is of limited interest to Earth scientists. It means, though, that gases invariably have a positive μ_{JT} under ordinary conditions. The most important Earth fluid, water, has a very interesting behaviour in this as in many other respects. In Figure 8.5 we show contours of constant apparent enthalpy of formation of water as a function of P and T. Where the contours have a negative slope, water warms on adiabatic expansion and cools if adiabatically compressed.

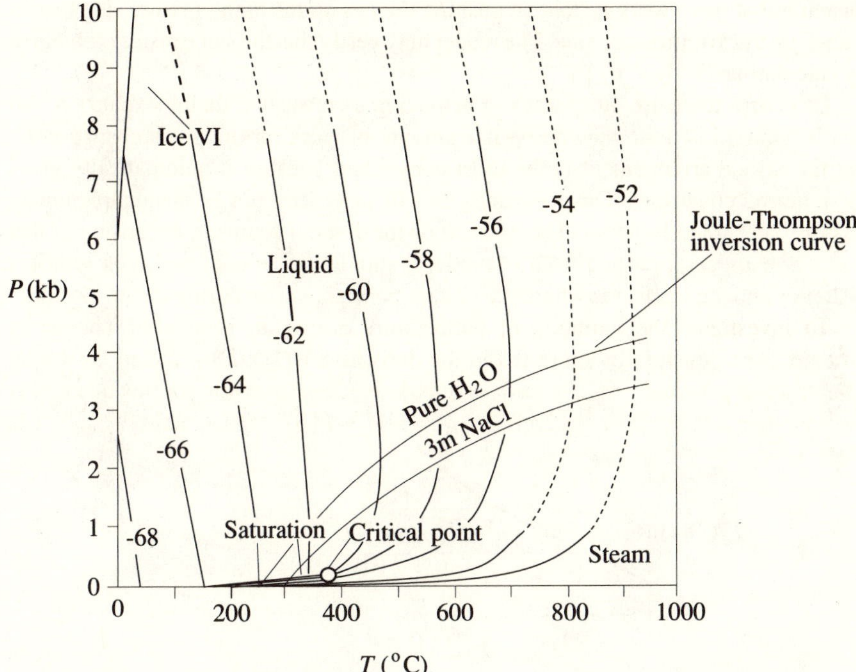

FIG. 8.5. Enthalpy of water as a function of temperature and pressure. Isenthalps are labelled in kcal mol^{-1}. The Joule-Thompson inversion curve is the locus of conditions where the isenthalps change slope from positive to negative. After Helgeson and Kirkham (1974a). The effect of adding NaCl is from Wood and Spera (1984).

Where they have a positive slope, water cools on adiabatic expansion and warms if adiabatically compressed, and the two regions are separated by the Joule-Thompson inversion curve. Much the same information is contained in the enthalpy-pressure diagram (Figure 8.6), where it can be seen that constant enthalpy changes in pressure lead to increases in temperature in one region and decreases in another. The effect of dissolved NaCl on the Joule-Thompson coefficient has been calculated by Wood and Spera (1984), and the effect will be similar for other electrolytes. Because the addition of most electrolytes to water results in a decrease in V and in α, μ_{JT} is smaller, and the net effect is to move the inversion curve to higher temperatures, as shown in Figure 8.5.

As an example of how the Joule-Thompson coefficient might be used, consider some hot spring fluids (approximated by pure water) rising vertically in the crust. When boiling begins, the pressure is 165 bars and the temperature is 350°C. At this point, $V_{H_2O} = 31.35\ cm^3\ mol^{-1} = 0.7493\ cal\ bar^{-1}\ mol^{-1}$, $\alpha = 0.01037\ K^{-1}$, and $C_p^\circ = 43.60\ cal\ K^{-1}\ mol^{-1}$, and

$$\begin{aligned}\mu_{JT} &= 0.7493(623.15 \times 0.01037 - 1)/43.60 \\ &= 0.0939\ K\ bar^{-1}\end{aligned}$$

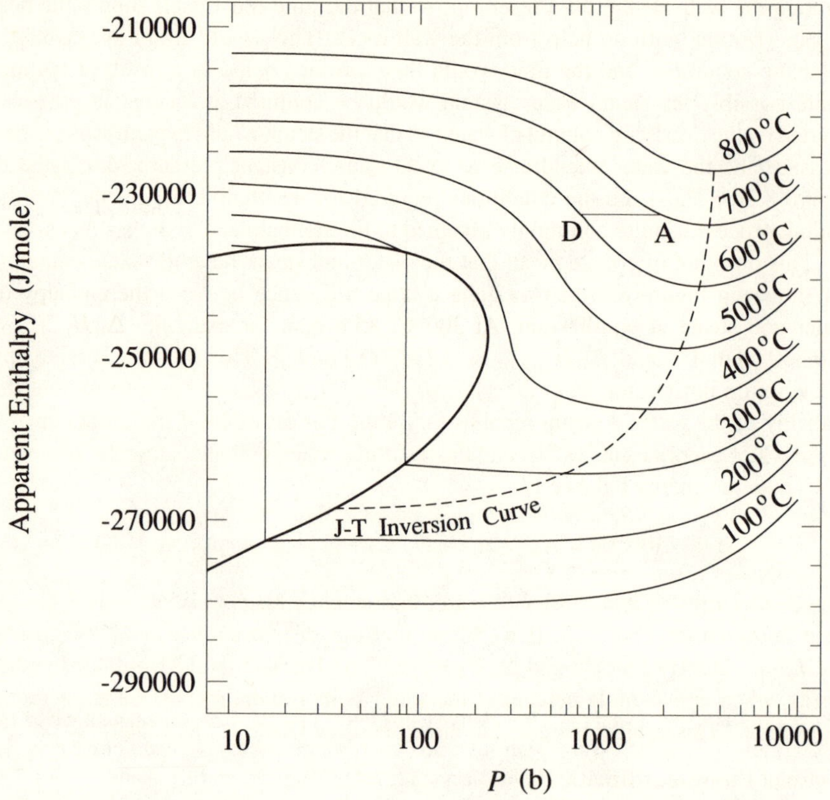

FIG. 8.6. Enthalpy versus pressure for water. Points A and D correspond to Figure 8.1.

meaning that the fluid is cooling at the rate of 0.094°C per bar change in pressure as it rises, simply due to adiabatic decompression. If the water has risen under a hydrostatic head for two kilometres, the pressure change is about 200 bars, giving $200 \times 0.094 \approx 19$ degrees of cooling. Actually, the change is less (about 12°C), because μ_{JT} is not constant but decreases with increasing pressure in this range.

Isenthalpic Boiling. A special case of adiabatic isenthalpic volume change of particular interest to geologists concerned with fluids in the upper levels of the Earth's crust is the case where a phase change, especially boiling, occurs along the cooling path of the fluid. The onset of boiling in a hot aqueous fluid in the crust is the cause of a number of important changes in the properties of the liquid phase because of the partitioning of the aqueous constituents between the two phases. In addition there is a new factor in the heat budget, the heat required to vaporize the liquid to steam, i.e., the heat of vaporization.

Whatever the means of cooling the fluid to the point of phase separation (boiling), the boiling process is generally thought to be fast relative to the conduction of heat through the walls of the fluid reservoir, so that the fluid must itself supply the heat of vaporization, with no help from the wall rocks. This would result in essentially adiabatic conditions and the fluid would therefore be cooled as a result of boiling. Quite possibly the steam-water system would be confined at a constant pressure, exerted by the overlying column of water, so that the steam would expand at a constant pressure and the water would also be at the same constant pressure (illustrated in Figure 8.2d). These are the conditions required for isenthalpic expansion, and the boiling process is quite commonly assumed to be isenthalpic.

This does not of course mean that the water and steam have the same enthalpy. As shown in Figure 8.6 there is quite a large difference between the enthalpy of water and steam at equilibrium. At 300°C, 85.8 bars for example, $\Delta_a H_{water} = -263505\ \mathrm{J\,mol^{-1}}$ and $\Delta_a H_{steam} = -238200\ \mathrm{J\,mol^{-1}}$. This increase in enthalpy of the steam must come from the system itself, which is cooled as a result, but the enthalpy of the water + steam remains the same. For example, if mole fraction X_w of the water vaporizes at 300°C, and the resulting water + steam is thereby cooled to T°C, then the energy balance is

$$\Delta_a H_w(300°\mathrm{C}) = X_w \cdot \Delta_a H_w(T°\mathrm{C}) + (1 - X_w) \cdot \Delta_a H_s(T°\mathrm{C})$$

where subscripts w and s refer to water and steam respectively. To solve this we must know either T or X_w. If we know that the system cooled to 275°C, where $\Delta_a H_w = -265918\ \mathrm{J\,mol^{-1}}$ and $\Delta_a H_s = -237555$ J mol–1 then X_w is 0.92. In other words, when only 8 mole percent of the water boils isenthalpically the temperature drops 25°C. This subject is treated in more detail by Henley et al. (1984).

Constant Pressure Adiabatic Expansion ($\Delta P = 0$)

To round out this discussion, we consider a constant pressure irreversible adiabatic expansion from A to E. This is similar to the expansions considered in Chapter 4 (see Figure 4.8), except that now the piston-cylinder arrangement is perfectly insulated, so that $q = 0$. This expansion is called "constant pressure," although the difference between this expansion and that discussed above as isenthalpic boiling should be carefully noted. In this case the expansion takes place at constant pressure, but this pressure is quite different from the pressure of the initial state. It is strictly speaking not a constant pressure process because there is a large pressure drop between the initial point A and the pressure during the expansion. This pressure was chosen as 991.5 bars so as to have the end point of the expansion on the isotherm at 600°C. Any other pressure could have been chosen with correspondingly different final temperatures. The work done by the expansion is

$$\begin{aligned} w &= -P\Delta V \\ &= -991.5(4.8571 - 3.601) \\ &= -1245.41\ \mathrm{J\,mol^{-1}} \\ &= U_\mathrm{E} - U_\mathrm{A} \\ &= \Delta_a U_\mathrm{E} - \Delta_a U_\mathrm{A} \end{aligned}$$

so

$$\begin{aligned}\Delta_a U_{\mathrm{E}} &= \Delta_a U_{\mathrm{A}} - P\Delta V \\ &= -235911 - 1245.41 \\ &= -237156\ \mathrm{J\ mol^{-1}}\end{aligned}$$

which checks very closely with the actual $\Delta_a U$ of water at this P, T (Table 8.1).

8.3. COMMENTS ON GEOLOGICAL APPLICABILITY

The systematics of adiabatic expansions that we have presented can be seen as simply an exercise in manipulating thermodynamic concepts, but in fact the extent to which fluids circulate at elevated temperatures and pressures in the Earth's crust means that volume changes, both adiabatic and non-adiabatic, are often important in constructing models explaining fluid behavior. Applications of isenthalpic expansion to minerals and rock masses, discussed by Waldbaum (1971), are possible, but to date none have been documented convincingly.

On the other hand, fluids rising through fissures in the upper few kilometers of the crust might well pass by an obstruction, which would have the effect of "throttling" the fluid, hence cooling it whether or not boiling was involved. Of special interest is the fact that there are two different fluid regimes in the crust, an upper regime where fluid pressure is fixed by the mass of the overlying water column, and a lower regime where the fluids are trapped in pores and grain boundaries where they are not free to circulate and where the pressure is fixed by the prevailing stress pattern in the rocks. This stress is thought to be homogeneous at depths where the rocks behave plastically, but may be far from this in the region where the rocks are brittle. The boundaries between these various regimes is a continuous topic of debate, and does not concern our treatment of the thermodynamics involved. The point is that it is unlikely that fluids can pass reversibly between the two regimes; at some point there will be an irreversible release of pressure, with its attendant thermal effects. This change need not be adiabatic, of course, and the question of the heat flow to and from geological processes is one that receives much attention.

8.4. CALCULATION OF SIMPLE PHASE DIAGRAMS

We should recall first of all that we are interested in knowing $\Delta_r G^\circ$ or $\Delta_r G$ for reactions among minerals because if we know either quantity for a mineral reaction over a range of T and P, we can predict which mineral or group of minerals is stable at any particular T and P. This implies that boundaries between mineral stability fields may be calculated in T, P space, and the diagrammatic representations of these are widely used as "phase diagrams." The importance of the Gibbs free energy in these calculations results from our predilection for thinking in terms of the temperature and pressure of various environments on and in the Earth's crust rather than the temperature and volume (in which case the thermodynamic potential of interest would be A), or the entropy and pressure (in which case it would be H), and so on. Some of us take comfort in the fact that should scientists suddenly change their thought framework to one of these other combinations, thermodynamics stands ready for the challenge.

8.5. THE CALCITE–ARAGONITE PHASE DIAGRAM

The simplest type of phase diagram involves a single component that can exist in more than one crystalline form, that is, a mineral having two or more polymorphs. For example, $CaCO_3$ can exist either in the form of calcite or aragonite. Using the equations we have derived, we can calculate either $\Delta_f G^\circ_{T,P}$ or $\Delta_a G^\circ_{T,P}$ for both phases at a number of T, P values, and then at each T, P we can combine them to obtain their difference. In other words, at any T, P we can calculate the free energy of reaction for the reaction

$$CaCO_3(\text{aragonite}) = CaCO_3(\text{calcite})$$

$$\begin{aligned}\Delta_r G &= \Delta_f G^\circ_{calcite} - \Delta_f G^\circ_{aragonite} \\ &= \Delta_a G^\circ_{calcite} - \Delta_a G^\circ_{aragonite} \\ &= G^\circ_{calcite} - G^\circ_{aragonite} \end{aligned} \tag{8.2}$$

Then we can examine these values of $\Delta_r G$. Where $\Delta_r G$ is negative, evidently $G^\circ_{calcite} < G^\circ_{aragonite}$ and calcite is the stable phase, and where $\Delta_r G$ is positive, $G^\circ_{calcite} > G^\circ_{aragonite}$ and aragonite is stable. We could then divide the diagram into two regions, calcite stable and aragonite stable, with a line joining those points for which $\Delta_r G = 0$. We could even contour the diagram with $\Delta_r G$ values. In this sense, a phase diagram is a sort of free-energy map.

Before proceeding with these calculations, we should note two things. First, we can economize on calculations by directly calculating the locus of T, P conditions for $\Delta_r G = 0$, rather than by calculating $\Delta_r G$ for a number of randomly chosen T, P conditions and subsequently noting where $\Delta_r G = 0$ occurs. And second, we can usually deduce the form of the diagram without doing any calculations by looking at the S (or ΔH) and V values for products and reactants. By the form of the diagram, we mean whether the phase boundary ($\Delta_r G = 0$) has a positive or negative slope, and whether the products or reactants for the reaction as written lie above or below the boundary. Thus for a simple polymorphic phase change like calcite–aragonite, there are four possibilities (Figure 8.7). However, LeChatelier's Principle tells us that reactions will generally occur in a direction that tends to absorb the force causing the reaction. Thus high pressures favor smaller molar volumes and high temperatures favor larger entropies and enthalpies.

Since aragonite has the smaller molar volume and calcite has the larger entropy and enthalpy of formation (at 298 K, 1 bar, see Tables 8.2, 8.3), then assuming that these relations hold at other T, P values, Figure 8.7a is the only possibility. The occurrence of aragonite in high pressure, relatively low temperature metamorphic rocks of the blue-schist facies would tend to be consistent with this, but the aragonite found in the shells of some organisms on the sea floor must then be metastable, showing that pelecypods and gastropods are ignorant of the laws of thermodynamics.

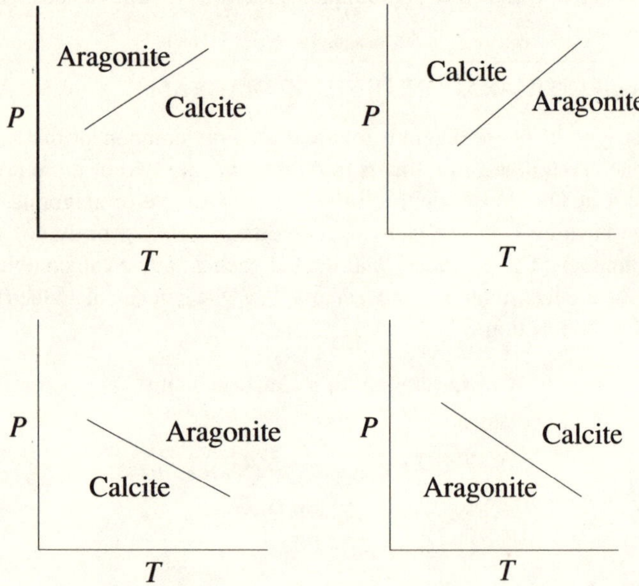

FIG. 8.7. Four possible configurations of the calcite–aragonite phase boundary, but only (a) is in agreement with LeChatelier's Principle.

8.5.1. *The Clapeyron Equation*

Entropy and molar volume can be quantified using a well-known equation describing the slope of any two-phase reaction boundary on a pressure-temperature diagram. From equation (8.2) it follows that

$$dG_{calcite} = dG_{aragonite}$$

on this boundary. From Chapter 5 we know that

$$dG = VdP - SdT \tag{5.21}$$

Table 8.2 Entropies and Heat Capacities for calcite and aragonite. All units are in $\mathrm{J\,K^{-1}\,mol^{-1}}$

T(K)	S°_{calcite}	$S^\circ_{\text{aragonite}}$	ΔS°	$C^\circ_{P\,\text{calcite}}$	$C^\circ_{P\,\text{aragonite}}$	ΔC°_P
298.15	91.71	87.89	3.82	83.87	82.32	1.15
400	118.36	113.75	4.61	97.00	92.67	4.33
500	140.88	135.23	5.65	104.54	99.81	4.73
600	160.43	153.23	6.47	109.87	105.77	4.10
700	177.70	170.68	7.02	114.16	111.18	2.98
800	193.19	185.86	7.33	117.88	116.29	1.59
						avg.=3.15

Table 8.3 Thermodynamic Data for calcite and aragonite

	calcite	aragonite	$\Delta_r G^\circ$ $G^\circ_{cal} - G^\circ_{arag}$ $J\,mol^{-1}$	P $(-\Delta_r G^\circ/\Delta V) + 1$ bars
$S^\circ_{298,1}$				
$cal\,K^{-1}\,mol^{-1}$	22.15	21.56		
V°				
$cm^3\,mol^{-1}$	36.934	34.150		
$cal\,bar^{-1}\,mol^{-1}$	0.88276	0.81622		
$J\,bar^{-1}\,mol^{-1}$	3.6934	3.4150		
$a\ (cal\,K^{-1}\,mol^{-1})$	24.98	20.13		
$b\ (cal\,K^{-2}\,mol^{-1})$	$5.24 \cdot 10^{-3}$	$10.24 \cdot 10^{-3}$		
$c\ (cal\,K^{-1}\,mol^{-1})$	$6.20 \cdot 10^5$	$3.34 \cdot 10^5$		
		HDNB		
$\Delta_f G^\circ_{298,1}$				
$cal\,mol^{-1}$	−270100	− 269875		
$J\,mol^{-1}$			−941	3381
$\Delta_f G^\circ_{400,1}$				
$cal\,mol^{-1}$	−272685	−272391		
$J\,mol^{-1}$	−1140913	−1139685	−1228	4412
$\Delta_f G^\circ_{500,1}$				
$cal\,mol^{-1}$	−275809	−275423		
$J\,mol^{-1}$	−1153984	−1152369	−1615	5802
$\Delta_f G^\circ_{600,1}$				
$cal\,mol^{-1}$	−279437	−278934		
$J\,mol^{-1}$	−1169164	−1167062	−2103	7555
$\Delta_f G^\circ_{700,1}$				
$cal\,mol^{-1}$	−283505	−282869		
$J\,mol^{-1}$	−1186186	−1183526	−2660	9554
$\Delta_f G^\circ_{800,1}$				
$cal\,mol^{-1}$	−287966	−287186		
$J\,mol^{-1}$	−1204581	−1201585	−3266	11731
$\Delta_f G^\circ_{900,1}$				
$cal\,mol^{-1}$	−292780	−291850		
$J\,mol^{-1}$	−1224993	−1221099	−3894	13986
		RHF		
$\Delta_f G^\circ_{298,1}$				
$J\,mol^{-1}$	−1128842	−1127793	−1049	3767
$\Delta_f G^\circ_{400,1}$				
$J\,mol^{-1}$	−1102155	−1100691	−1464	5258
$\Delta_f G^\circ_{500,1}$				
$J\,mol^{-1}$	−1076292	−1074310	−1982	7118
$\Delta_f G^\circ_{600,1}$				
$J\,mol^{-1}$	−1050723	−1048132	−2591	9306
$\Delta_f G^\circ_{700,1}$				
$J\,mol^{-1}$	−1025427	−1022160	−3267	11734

for both minerals, and therefore

$$V_{calcite}dP - S_{calcite}dT = V_{aragonite}dP - S_{aragonite}dT$$

Rearranging this gives

$$dP/dT = (S_{calcite} - S_{aragonite})/(V_{calcite} - V_{aragonite})$$

or for a reaction in general,

$$\begin{aligned} dP/dT &= \Delta_r S/\Delta_r V \\ &= \Delta_r H/T\Delta_r V \end{aligned} \tag{8.3}$$

which is known as the Clapeyron equation. This is an extremely useful relationship, particularly when dealing with multicomponent systems (with many possible reactions between phases), although the ΔS and ΔV terms may each be more complex.

8.5.2. *The Required Equations*

The procedure for calculating a phase boundary is, as just mentioned, to calculate the locus of P, T conditions such that $\Delta_r G = 0$ for the reaction involving the phases on either side of the boundary. To do this we must be able to calculate the apparent or standard free energies of formation of the phases as a function of P and T, as outlined in Chapter 7. This is done for compound i with the equations

$$\Delta_a G^\circ_{i,T,P} = \Delta_f G^\circ_{i,T_r,P_r} + \int_{T_r}^{T} (\partial G^\circ_i/\partial T)_{P=P_r} dT + \int_{P_r}^{P} (\partial G^\circ_i/\partial P)_{T=T} dP \tag{7.27}$$

or

$$\Delta_f G^\circ_{i,T,P} = \Delta_f G^\circ_{i,T_r,P_r} + \int_{T_r}^{T} (\partial \Delta_f G^\circ_i/\partial T)_{P=P_r} dT + \int_{P_r}^{P} (\partial G^\circ_i/\partial P)_{T=T} dP \tag{7.28}$$

For a reaction involving a number of phases or compounds or species as reactants and products, $\Delta_r G$ is simply the sum of the individual $\Delta_r G$ terms ($\Delta_a G$ or $\Delta_f G$) for each compound. (In the following example and those in the next few chapters, each i will be a distinct phase, but in general i may be any *constituent* of the system). Thus equation (7.27) becomes

$$\Delta_r G^\circ_{T,P} = \sum_i \nu_i \Delta_a G^\circ_{i,T,P} \tag{8.4}$$

$$\begin{aligned} &= \sum_i \nu_i \Delta_f G^\circ_{i,T_r,P_r} + \sum_i \nu_i \int_{T_r}^{T} (\partial G^\circ_i/\partial T)_{P=P_r} dT \\ &+ \sum_i \nu_i \int_{P_r}^{P} (\partial G^\circ_i/\partial P)_{T=T} dP \end{aligned} \tag{8.5}$$

where $\sum_i$ signifies summation for all compounds or phases i in the reaction, and ν_i is the stoichiometric coefficient of each compound i, counting products as positive and reactants as negative. For example, if the reaction is

$$SiO_2(s) + 2H_2O = H_4SiO_4(aq)$$

equation (8.4) becomes

$$\Delta_r G^\circ_{T,P} = \Delta_a G^\circ_{T,P,H_4SiO_4} - \Delta_a G^\circ_{T,P,SiO_2} - 2\Delta_a G^\circ_{T,P,H_2O}$$

Since $\sum_i \nu_i \Delta G^\circ_i$ and $\sum_i \nu_i G^\circ_i$ can both be written as $\Delta_r G^\circ_i$, equation (8.5) can also be written

$$\Delta_r G^\circ_{T,P} = \Delta_r G^\circ_{T_r,P_r} + \int_{T_r}^{T} (\partial \Delta_r G^\circ / \partial T)_{P=P_r} dT + \int_{P_r}^{P} (\Delta_r G^\circ / \partial P)_{T=T} dP \qquad \text{cf.(7.27)} \qquad (8.6)$$

This (or equation 8.5) is the general form for the calculation of the free energy of any reaction at T and P, where no changes in phase composition are involved.

Without wanting to be too repetitious, we recall here that in the preceding sentence the "free energy of any reaction" is a short-hand way of saying the difference in Gibbs free energy between products and reactants of the reaction as written. We recall too that except at the equilibrium P and T, one side of the reaction will have a total G that is greater than the other side, i.e., $\Delta_r G$ will be positive or negative. In order to make this comparison, we must consider each set, products and reactants, as separate equilibrium states, one of which is necessarily metastable with respect to the other. The metastable assemblage has one constraint in addition to the designated P and T, the activation energy, which prevents it from sliding down the energy gradient to the stable assemblage.

Inserting the Numbers

Phase boundaries are calculated from equation (8.6), with the additional knowledge that $\Delta_r G = 0$ on the boundary. For the reaction

$$\text{aragonite} \rightleftharpoons \text{calcite}$$

we write

$$\begin{aligned}
\Delta_r G^\circ_{T_r,P_r} &= \Delta_f G^\circ_{Calcite} - \Delta_f G^\circ_{aragonite} \\
&= G_{calcite} - G_{Ca} - G_C - \frac{3}{2} G_{O_2} \\
&\quad -(G_{aragonite} - G_{Ca} - G_C - \frac{3}{2} G_{O_2}) \\
&= G_{calcite} - G_{aragonite} \\
&= -270100 - (-269875) \text{ (HDNB)} \\
&= -225 \text{ cal mol}^{-1} \\
&= -941.4 \text{ J mol}^{-1}
\end{aligned}$$

or

$$\Delta_r G^\circ_{T_r,P_r} = -1128842 - (-1127793) \text{ (RHF)}$$
$$= -1049 \text{ J mol}^{-1}$$

At 298 K, the temperature integral in (8.6) is zero, so we proceed to the pressure integral. In general,

$$\int \left(\frac{d\Delta G}{dP}\right) dP = \int \Delta V \, dP$$

and for solid phases with $\Delta_s V$ constant,

$$\int_{P_r}^{P} \Delta_s V \, dP = \Delta_s V(P - P_r) \qquad (8.7)$$

Now

$$V^\circ_{calcite} = 36.934 \text{ cm}^3 \text{ mol}^{-1}$$
$$= 3.6934 \text{ J bar}^{-1}$$
$$= (3.6934 \text{ J bar}^{-1} \times 0.23901) \text{cal bar}^{-1}$$
$$= 0.88276 \text{ cal bar}^{-1}$$

and

$$V^\circ_{aragonite} = 34.150 \text{ cm}^3 \text{ mol}^{-1}$$
$$= 3.4150 \text{ J bar}^{-1}$$
$$= (3.4150 \times 0.23901) \text{cal bar}^{-1}$$
$$= 0.81622 \text{ cal bar}^{-1}$$

Thus evidently the rate of change of $G_{calcite}$ with P is greater than that of $G_{aragonite}$, and since it starts off at a lower value, there must be some elevated P where they become equal, as illustrated in Figure 8.8.

So

$$\Delta_s V^\circ = V^\circ_{calcite} - V^\circ_{aragonite}$$
$$= 0.2784 \text{ J bar}^{-1} \text{ mol}^{-1}$$
$$= 0.06654 \text{ cal bar}^{-1} \text{ mol}^{-1}$$

and the pressure integral (8.7) is $0.2784(P-1)$ J mol^{-1} or $0.06654(P-1)$ cal mol^{-1}, where P is the pressure required for calcite-aragonite equilibrium at this temperature.

We now have

$$0 = -225 + 0.06654(P - 1) \text{ cal mol}^{-1}$$

from which $P = 3381$ bars (HDNB data, Table 8.3) or

$$0 = -1049 + 0.2784(P - 1)$$

from which $P = 3767$ bars (RHF data, Table 8.3).

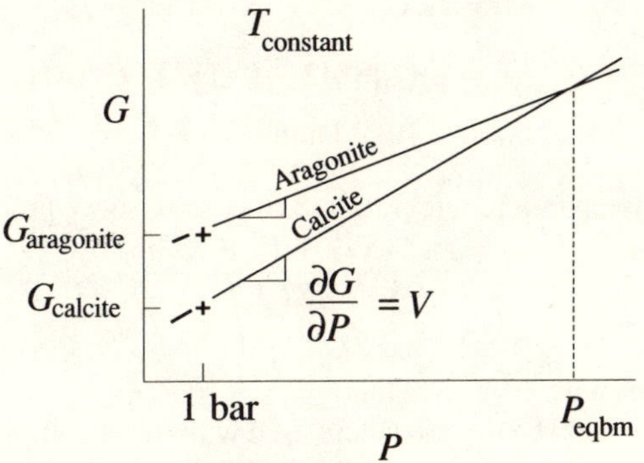

FIG. 8.8. G as a function of P for calcite and aragonite.

This gives us one point on the calcite–aragonite phase boundary. To obtain others at temperatures above 298 K we must include the temperature integral in equation (8.6), which may be easily accomplished by using $S^\circ_{T_r}$ and a, b, c coefficients for each phase as discussed in Chapter 7.

The temperature integral in (8.6) may be written

$$\int_{T_r}^{T} (\partial \Delta_r G^\circ / \partial T)_{P_r} dT = \Delta_r G^\circ_T - \Delta_r G^\circ_{T_r} \text{ at 1 bar}$$

and this may be evaluated by rewriting equation (7.26) as

$$\begin{aligned}\Delta_r G^\circ_T - \Delta_r G^\circ_{T_r} = & -\Delta_r S^\circ_{T_r}(T - T_r) \\ & + \Delta_r a \left[T - T_r - T \ln\left(\frac{T}{T_r}\right)\right] \\ & + \frac{\Delta_r b}{2}(2TT_r - T^2 - T_r^2) \\ & + \frac{\Delta_r c(T^2 + T_r^2 - 2TT_r)}{2TT_r^2}\end{aligned} \tag{8.8}$$

where

$$\begin{aligned}\Delta_r G^\circ_T &= \Delta_a G^\circ_{T,calcite} - \Delta_a G^\circ_{T,aragonite} \\ \Delta_r G^\circ_{T_r} &= \Delta_f G^\circ_{T_r,calcite} - \Delta_f G^\circ_{T_r,aragonite} \\ \Delta_r a &= a_{calcite} - a_{aragonite} \\ \Delta_r b &= b_{calcite} - b_{aragonite} \\ \Delta_r c &= c_{calcite} - c_{aragonite}\end{aligned}$$

The complete expanded form of equation (8.6) for the calculation of phase boundaries when only pure solids are involved is therefore

$$\Delta_r G^\circ_{T,P} = 0$$

$$
\begin{aligned}
&= \Delta_r G^\circ_{T_r,P_r} - \Delta_r S^\circ_{T_r}(T - T_r) \\
&\quad + \Delta_r a \left[T - T_r - T \ln\left(\frac{T}{T_r}\right)\right] \\
&\quad + \frac{\Delta_r b}{2}(2TT_r - T^2 - T_r^2) \\
&\quad + \frac{\Delta_r c(T^2 + T_r^2 - 2TT_r)}{2TT_r^2} \\
&\quad + \Delta_r V^\circ(P - 1)
\end{aligned}
\tag{8.9}
$$

Equation (8.9) would be the procedure used with the HDNB data set, or any other set that gives standard thermodynamic parameters at 298 K, 1 bar, plus Maier-Kelley heat capacity coefficients. The RHF tables list $\Delta_f G^\circ$ at various temperatures above 298 K at one bar, so that these can be used directly to give the first two terms on the right-hand side of equation (8.6). In other words

$$\Delta_r G^\circ_{T,P_r} = \Delta_f G^\circ_{T,P_r,calcite} - \Delta_f G^\circ_{T,P_r,aragonite}$$

which can be obtained from the tables. Then

$$
\begin{aligned}
\Delta_r G^\circ_{T,P} &= 0 \\
&= \Delta_r G^\circ_{T,P_r} + \Delta_r V^\circ(P - P_r)
\end{aligned}
$$

or

$$0 = \Delta_r G^\circ_{T,P_r} + \Delta_r V^\circ(P - P_r) \tag{8.10}$$

from which P may be calculated as before. The results using both procedures for several temperatures are listed in Table 8.3 and shown in Figure 8.9. These results illustrate the difference between the RHF and HDNB data sets. That is, HDNB data will always give a phase boundary that passes through a set of experimentally determined brackets for that boundary, while RHF data may or may not. This is because the HDNB data were calculated using the experimental brackets as constraints, while the RHF data are mostly from other sources such as calorimetric studies. The reader is cautioned not to conclude at this point that because of this the HDNB data set is therefore better for geological or any other purpose. For one thing, there may well be conflicting results from high pressure and temperature experimental studies, and the HDNB data will fit only some of them. And, of course, the nature of the mineral phases used in the phase equilibrium may well have been slightly different,and one cannot always say that those used by the phase equilibrium experimenters are always "better" for geological purposes. There are also by now several other self-consistent sets of data for minerals in the literature, produced using different sources and/or methods, emphasizing the difficulty of knowing the "best " data. Again we remind the reader that the equations used here depend on the choice of the Maier-Kelley formulation for the heat capacity. The equations for other choices are given in Appendix B.

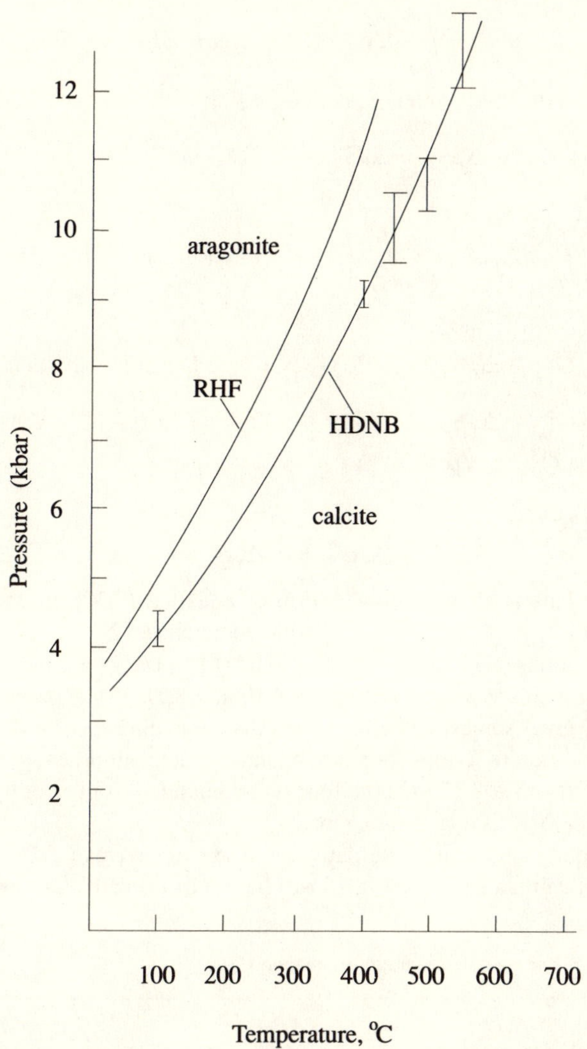

FIG. 8.9. $P - T$ phase diagram for $CaCO_3$, showing calculated phase boundaries and experimental brackets.

8.5.3. *Effect of Approximations on Calculated Phase Boundaries*

To see the effect of the $\Delta C_p^\circ = 0$ and $\Delta C_p^\circ = constant$ approximations on the calculated position of the calcite–aragonite phase boundary, we use equation (7.50) to generate high temperature $\Delta_r G$ values. Thus

$$\left(\frac{\partial \Delta G^\circ}{\partial T}\right) = -\Delta S^\circ$$

so

$$\int_{T_r}^{T} d\Delta G^\circ = -\int_{T_r}^{T} \Delta S^\circ \, dT$$

For the variation of $\Delta_r S^\circ$ with temperature, we have

$$\Delta_r S_T = \Delta_r S^\circ_{T_r} + \Delta_r C^\circ_{p\,T_r} \ln(T/T_r) \tag{7.49}$$

so

$$\int_{T_r}^{T} d\Delta_r G^\circ = -\int_{T_r}^{T} \left[\Delta_r S^\circ_{T_r} + \Delta_r C^\circ_{p\,T_r} \ln(T/T_r)\right] dT$$

$$\begin{aligned}\Delta_r G^\circ_T &= \Delta_r G^\circ_{T_r} - \Delta_r S^\circ_{T_r}(T - T_r) - \Delta_r C^\circ_{p\,T_r} \int_{T_r}^{T} (\ln T - \ln T_r) dT \\ &= \Delta_r G^\circ_{T_r} - \Delta_r S^\circ_{T_r}(T - T_r) - \Delta_r C^\circ_{p\,T_r} \left[T \ln T - (T - T_r) - T_r \ln T_r\right] \\ &\quad + \Delta_r C^\circ_{p\,T_r} \ln T_r (T - T_r)\end{aligned} \tag{8.11}$$

which reduces to

$$\Delta_r G^\circ_T - \Delta_r G^\circ_{T_r} = -\Delta_r S^\circ_{T_r}(T - T_r) \tag{8.12}$$

if $\Delta C^\circ_p = 0$. This is simply another form of equation (7.50), as can be seen by substituting $\Delta_r G^\circ_{T_r} = \Delta H^\circ_{T_r} - T_r \Delta S^\circ_{T_r}$ into equation (8.12).

Use of equations (8.11) (constant ΔC°_p) and (8.12) ($\Delta C^\circ_p = 0$) results in the data in Table 8.4, which are plotted in Figures 8.10 and 8.11. As expected, including a constant ΔC°_p gives somewhat better results than assuming $\Delta C^\circ_p = 0$, but the errors introduced can be quite serious for phase boundary calculations, as shown. For a ΔT of only 400°C (from 298.15 to 700 K) the phase boundary is in error by more than a kilobar whether ΔC°_p is constant or zero.

Another more satisfactory alternative is to use the average ΔC°_p over the ΔT considered, rather than ΔC°_p for 298.15 K. The results using this assumption are also

Table 8.4 $\Delta_r G^\circ_T$ For the calcite–aragonite reaction

T K	$\Delta_r G^\circ_T$ J mol^{-1} (RHF)	$\Delta_r G^\circ_T$ J mol^{-1} (8.11) ΔCp°_{298}	$\Delta_r G^\circ_T$ J mol^{-1} (8.11) ΔCp°_{avg}	$\Delta_r G^\circ_T$ J mol^{-1} (8.12)	P bars (8.11) ΔCp°_{298}	P bars (8.11) ΔCp°_{avg}	P bars (8.12)	P bars Table 8.3
298.15	−1049	−1049	−1049	−1049	3767	3767	3767	3767
400	−1464	−1456	−1488	−1438	5231	5346	5166	5258
500	−1982	−1885	−1999	−1820	6772	7181	6538	7118
600	−2591	−2337	−2573	−2202	8359	9243	7910	9306
700	−3267	−2809	−3200	−2584	10091	11495	9283	11734

$\Delta Cp^\circ_{298} = 1.15$ J mol^{-1}K^{-1}
$\Delta Cp^\circ_{avg} = 3.15$ J mol^{-1}K^{-1}
$\Delta S^\circ_{298} = 3.82$ J mol^{-1}K^{-1}

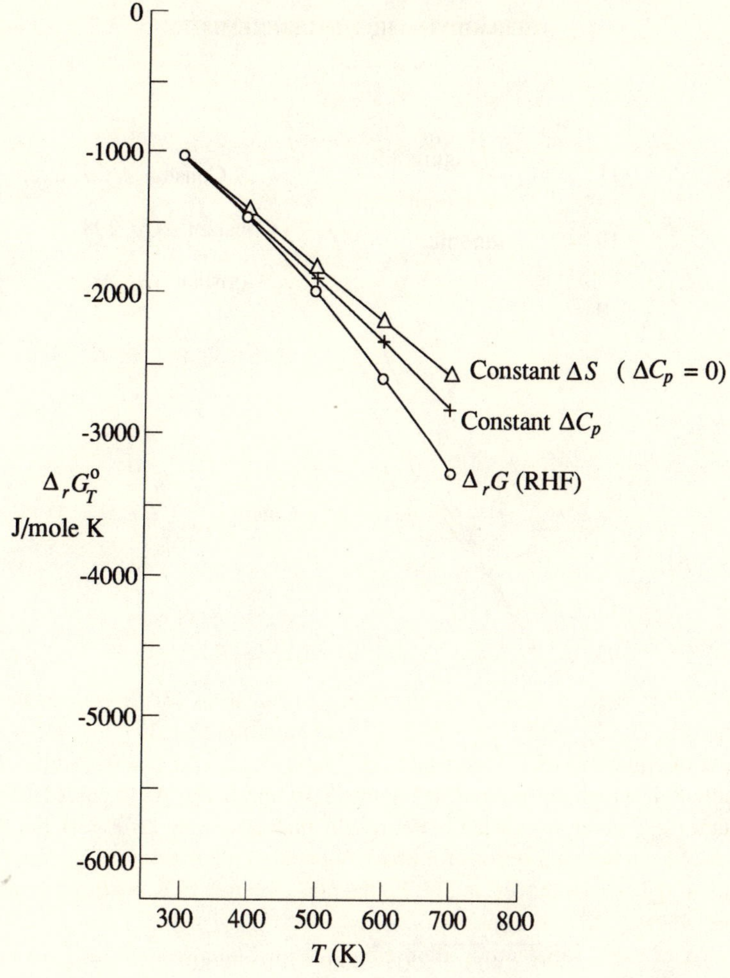

FIG. 8.10. $\Delta_r G$ as a function of temperature, showing effect of approximations. Data in Table 8.3.

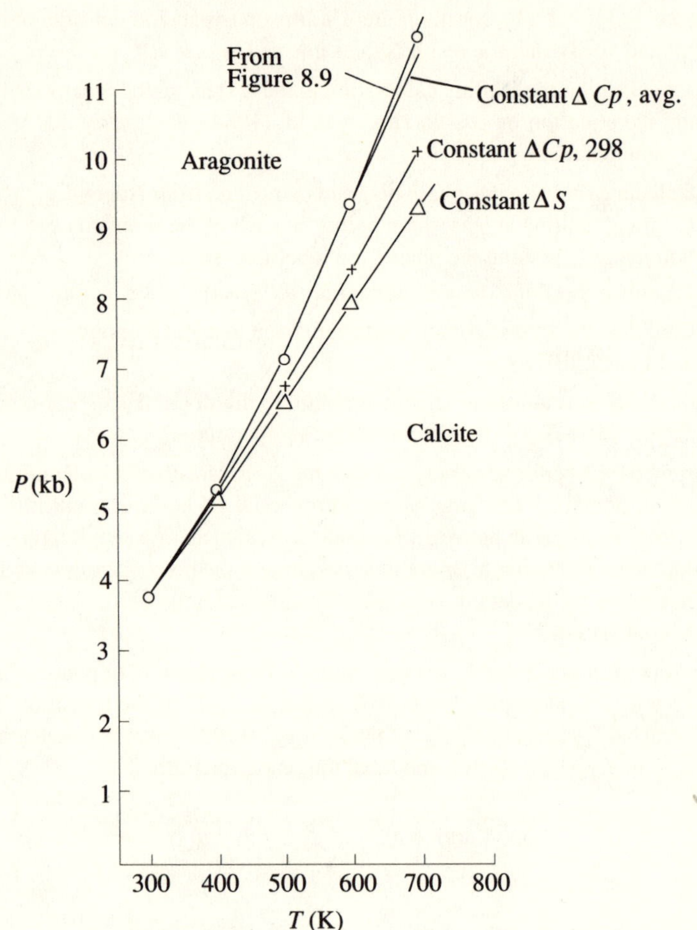

FIG. 8.11. $P - T$ phase diagram for $CaCO_3$, showing the effect of various approximations on the calculated phase boundary.

shown in Figure 8.11, and clearly the errors introduced are quite small. However, if one has enough information to calculate an average heat capacity, it should be possible to also use the Maier-Kelley formulation discussed previously (or one of the several alternatives), and this is preferable.

PROBLEMS

1. Show that the work done by an ideal gas in an adiabatic reversible expansion is $(P_2V_2 - P_1V_1)/(\gamma - 1)$. Verify the statement in the text that an ideal monatomic gas with $\gamma = 1.667$ would do more than twice the work between points A and B (Table 8.1).

2. Calculate $P_D V_D - P_A V_A$ for the Joule-Thompson expansion considered in the text, and verify that the result is the same as $U_D - U_A$.
3. Repeat the Isenthalpic Boiling calculation in the text using the Steam Table data and the equation on p. 4 in Henley et al. (1984), and verify that you get the same result.
4. (a). Calculate the Al_2SiO_5 phase diagram using data from Appendix C, by calculating at least three points on each of the three phase boundaries. Assume the phases are incompressible.

   (b). Calculate the same diagram assuming $\Delta C_p^\circ = 0$.

   (c). Calculate the same diagram using data from Robie, Hemingway and Fisher (1978).
5. What pressure is required to change graphite to diamond at 25°C? Would this pressure increase or decrease at higher temperatures?
6. The origin of red-bed sandstones, in which the grains are coated with minute amounts of hematite, has long been controversial. A key question in the controversy is whether hematite is stable in water at low temperatures. Calculate whether hematite or goethite is stable in the presence of water at 25°C according to the data in Appendix C. At what temperature would the other become stable?
7. If the slope of a phase transition of a mineral from phase α to phase β is -21.0 bar K^{-1} at a temperature of 600 K, the $\Delta_{\alpha\rightarrow\beta}V$ of the transition is +0.150 cal bar^{-1}, and $\Delta_f H^\circ_{600\text{ K}}$ of phase α is -17000 cal mol^{-1}, what is $\Delta_f H^\circ_{600\text{ K}}$ of phase β? Sketch and label the phase diagram.

9

PARTIAL AND APPARENT MOLAR PROPERTIES

Thermodynamics is the kingdom of deltas. Truesdell (1980, p. 1)

9.1. ESSENTIAL TOOLS

To this point we have developed a method of dealing with processes involving for the most part material (phases) of fixed composition. To progress further, we need to develop ways of dealing with the properties of dissolved substances, and with phases of variable composition. In other words, we need to deal with solutions. We have seen that the properties of homogeneous substances (usually, but not necessarily *pure*) are either *total* or *molar* properties (neglecting, as usual, *specific* properties). But obviously substances change drastically when they dissolve. How can we obtain properties for the dissolved substance alone, as opposed to the solution of which it is a part? Properties of dissolved substances are called *apparent molar* and *partial molar* properties, and these are the tools we need to discuss the thermodynamics of solutions in the following chapters.

9.2. PARTIAL MOLAR PROPERTIES

Partial molar properties are defined by partial derivatives (equations 2.22, 9.7), which does not provide a very easy route to understanding them. There is however a highly intuitive way of thinking about partial molar properties. We will use volume as an example because it is readily visualized, but all relations derived can be used equally well for any other state variable.

9.2.1. *Molar Properties as Derivatives*

It will be useful to begin by realizing that we can also use derivatives to define molar properties of pure phases; this mathematical treatment is discussed in Chapter 2. Consider first the volume of a crystalline phase such as NaCl. The volume occupied by the salt, V, varies directly with the quantity of salt, so that a plot of volume versus number of moles of salt is a straight line (Figure 9.1). The slope of this line, $d\mathrm{V}/dn_{\mathrm{NaCl}}$, is obviously equal to the volume of one mole of salt, i.e., the molar volume $V^{\circ}_{\mathrm{NaCl}}$, in this case 27 cm^3 mol^{-1}. Similarly, the volume occupied by one kilogram of water (1001.4 cm^3 at 25°C) added to the volume occupied by the salt changes all the numbers on the Y-axis by 1001.4 cm^3 but does not change the slope or the molar volume of salt (Figure 9.2).[1] However, if we now dissolve the salt in the water we find that the volume of the two mixed together is not equal to the sum of the two separately, because of electrostatic interactions between the charged Na^+ and Cl^- ions and the H_2O molecules. Also, we find that the total volume of the salt solutions

[1] This is the molar volume of salt in a "mechanical mixture" of salt and water, in the sense discussed in Chapter 10

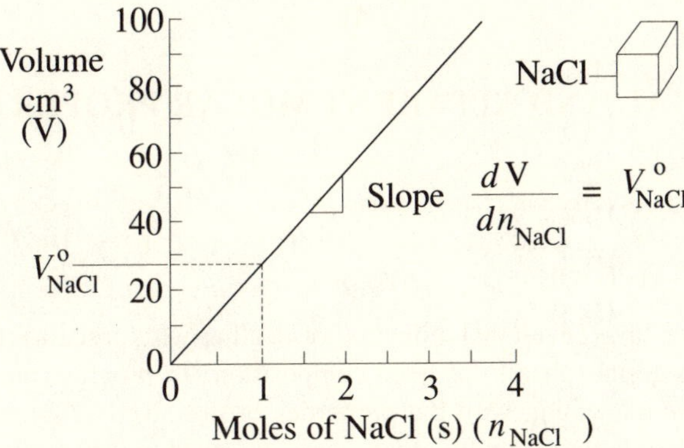

FIG. 9.1. The volume occupied by crystalline NaCl as a function of the number of moles of NaCl.

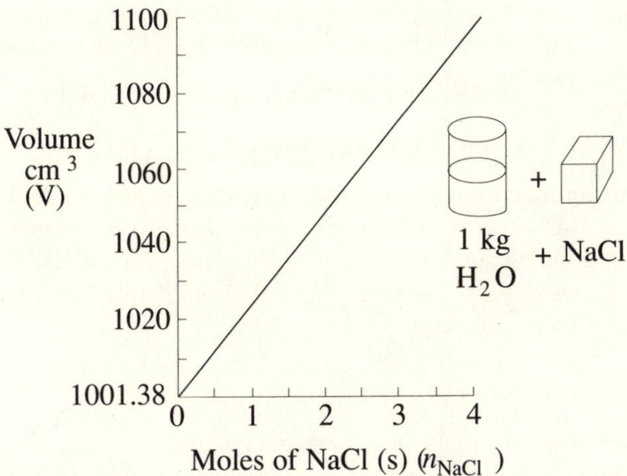

FIG. 9.2. The volume occupied by 1 kg H_2O plus (separately) crystalline NaCl ("mechanical mixture").

is no longer a linear function of the number of moles of salt, but a complex function, which results in a curve as in Figure 9.3. The question still is, what is the volume of a mole of salt in a salt solution of a given composition? Or the volume of a mole of water for that matter? Consider a NaCl solution consisting of 1 mole of NaCl in a kilogram of water (a 1 molal solution). The molar volume is still given by the slope of the V curve at that concentration ($[dV/dn_{NaCl}]_{n_{NaCl}=1}$) because this is the rate of change of V per very small addition of NaCl, expressed in $cm^3\,mol^{-1}$ of NaCl, or in other words, the change in volume of the system per mole of NaCl added when only a small amount of NaCl is actually added. If a mole of salt is actually added, the

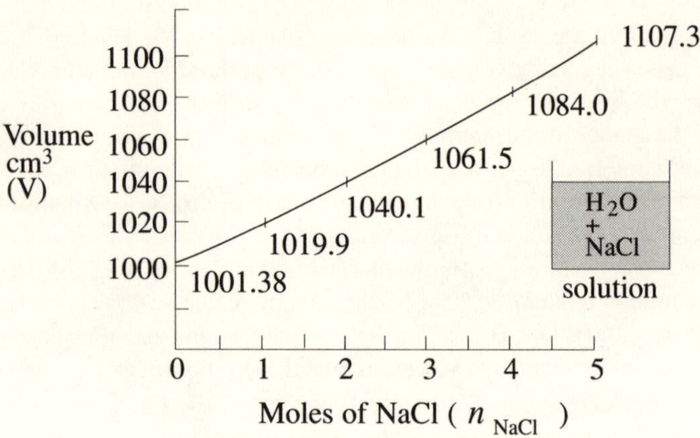

FIG. 9.3. The volume occupied by 1 kg H_2O plus dissolved NaCl.

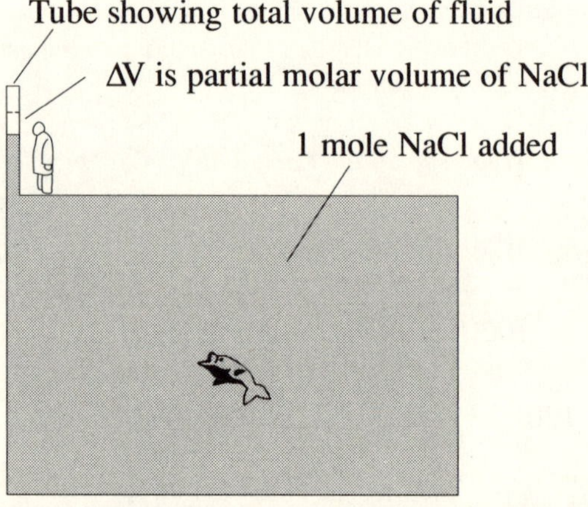

FIG. 9.4. A roomful of 1 molal salt solution. The observer at the tube sees the change in volume caused by adding one mole of salt, which is the partial molar volume of salt in the 1 molal solution.

change in volume is quite different because the concentration of NaCl changes from 1 molal to 2 molal. We wish to have the molar volume of NaCl in a solution of fixed composition.

9.2.2. *Partial Molar Volume*

A better way to look at this is to consider not one kilogram of water, but an extremely large quantity of water—say enough to fill a large room (Figure 9.4). Now let's add enough salt to make the concentration exactly 1 molal, and adjust the volume of the solution so that the room is full, and a little excess solution sticks up into a calibrated tube inserted into the ceiling. By observing changes in the level of solution in the tube, we can accurately record changes in the V of the solution in the room.

Now, when we add a mole of NaCl (58.5 g of NaCl occupying 27 cm^3) to the solution, the change in concentration is very small. In fact, if we can detect *any* change in concentration by the finest analytical techniques available, then our room is too small, and we must find and inundate a larger one. Eventually, we will fill a sufficiently large room with salt solution that on adding 58.5 g of NaCl we are unable to detect any change in concentration—it remains at 1.000 moles NaCl/kg H_2O. But although the concentration remains unchanged, the volume of course does not. The salt added cannot disappear without a trace. The level in the tube in the ceiling changes, and the ΔV seen there is evidently the volume occupied by 1 mole of NaCl in a 1 molal NaCl solution, in this case about 15 $cm^3\ mol^{-1}$ of NaCl. This is, in quite a real sense, the volume occupied by a mole of salt in that salt solution, and has a right to be thought of as a *molar volume* (just as much as 27 $cm^3\ mol^{-1}$ is the molar volume of crystalline salt) rather than as an arbitrary mathematical construct. It is referred to as the *partial* molar volume of NaCl in the salt solution, $\bar{V}_{NaCl}$. The reader will be well advised to think of partial molar properties in this sense, i.e., as *molar* properties of solutes in solutions of particular compositions, rather than in terms of the partial derivative that defines them mathematically (equations 2.22, 9.7).

9.2.3. *Total Volume as the Sum of the Partial Molar Volumes*

Clearly we could also add a mole of water (18.01 g occupying 18.0 cm^3) to our roomful of salt solution and determine its partial molar volume ($\bar{V}_{H_2O}$) as well. This turns out to be about 17 $cm^3\ mol^{-1}$. Since we now know the volumes occupied by both the H_2O and the NaCl, we could calculate the total volume of the solution if we know the masses and hence the number of moles of each component from

$$\mathrm{V} = n_{NaCl}\bar{V}_{NaCl} + n_{H_2O}\bar{V}_{H_2O} \tag{9.1}$$

This will seem like a reasonable conclusion to anyone who recalls our discussion of Euler's Theorem for homogeneous functions in Chapter 2, since V is homogeneous in the first degree in the masses (or mole numbers) of the components NaCl and H_2O. It is, in other words, an extensive state variable.

Equation (9.1) can be seen to be a reasonable conclusion from another point of view as well. This time let's consider not the total volume of the system but the molar volume, which is

$$V = \mathrm{V}/(n_{NaCl} + n_{H_2O})$$

and plot V against the mole fraction of NaCl as in Figure 9.5a. The mole fraction of NaCl in a one molal NaCl solution is $1/(1 + 55.51)$ or 0.018, and drawing a tangent to the curve at this value of X_{NaCl} we obtain the intercepts x and y as shown in Figure 9.5a. Now dividing each term in equation (9.1) by $(n_{NaCl} + n_{H_2O})$ we obtain

$$V = X_{NaCl}\bar{V}_{NaCl} + X_{H_2O}\bar{V}_{H_2O} \tag{9.2}$$

where X_{NaCl} is the mole fraction of NaCl (= $n_{\text{NaCl}}/(n_{\text{NaCl}} + n_{\text{H}_2\text{O}})$). By inspection of Figure 9.5a we see that the molar volume at our tangent point (the one molal solution) is

$$V = x \cdot X_{\text{H}_2\text{O}} + y \cdot X_{\text{NaCl}} \tag{9.3}$$

so that evidently $x = \bar{V}_{\text{H}_2\text{O}}$ and $y = \bar{V}_{\text{NaCl}}$.[2] Thus equations (9.1) and (9.2) express similar relationships, i.e., that we can split the volume of a system into compositionally-weighted values of the partial molar volumes. By implication this is also true for any state variable and any number of components. Thus in general

$$\mathrm{Z} = \sum_i n_i \bar{Z}_i \tag{9.4}$$

and

$$Z = \sum_i X_i \bar{Z}_i \tag{9.5}$$

where Z is any state variable and Z its molar counterpart. Equation (9.4) and the corresponding version of Figure 9.5a will be particularly useful in our discussions involving free energies of solutions.

It will also prove useful to have the equation of the tangent itself. From Figure 9.5b we see that this is (following the general equation of a straight line, $Y = slope \cdot X + intercept$)

$$\begin{aligned} V &= \left(\frac{dV}{dX_{\text{H}_2\text{O}}}\right) X_{\text{H}_2\text{O}} + y \\ &= \left(\frac{dV}{dX_{\text{H}_2\text{O}}}\right) X_{\text{H}_2\text{O}} + \bar{V}_{\text{NaCl}} \end{aligned}$$

or, as an equation for $\bar{V}_{\text{NaCl}}$ we have

$$\begin{aligned} \bar{V}_{\text{NaCl}} &= V - \left(\frac{dV}{dX_{\text{H}_2\text{O}}}\right) X_{\text{H}_2\text{O}} \\ &= V + \left(\frac{dV}{dX_{\text{NaCl}}}\right) X_{\text{H}_2\text{O}} \\ &= V + \left(\frac{dV}{dX_{\text{NaCl}}}\right) (1 - X_{\text{NaCl}}) \end{aligned}$$

[2]If you have difficulty seeing the truth of equation (9.3) by inspection, consider the similar triangles A and B (Figure 9.5b), in which

$$X_{\text{NaCl}}/(x - V) = X_{\text{H}_2\text{O}}/(V - y)$$

so

$$V \cdot X_{\text{NaCl}} - y \cdot X_{\text{NaCl}} = x \cdot X_{\text{H}_2\text{O}} - V \cdot X_{\text{H}_2\text{O}}$$

or

$$V(X_{\text{H}_2\text{O}} + X_{\text{NaCl}}) = x \cdot X_{\text{H}_2\text{O}} + y \cdot X_{\text{NaCl}}$$

and

$$V = x \cdot X_{\text{H}_2\text{O}} + y \cdot X_{\text{NaCl}}$$

because

$$X_{\text{H}_2\text{O}} + X_{\text{NaCl}} = 1$$

(a)

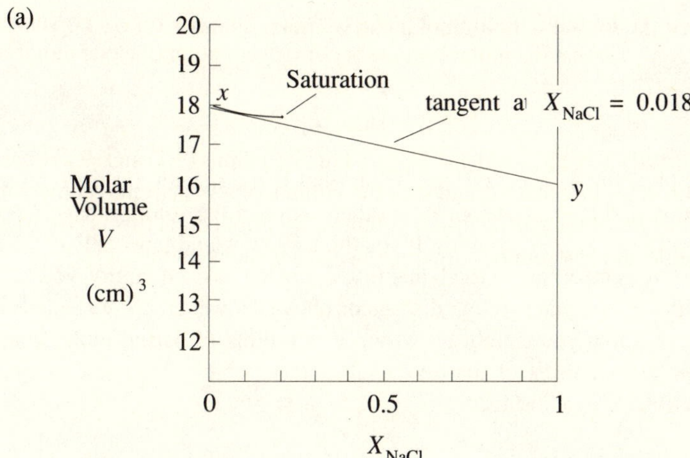

(b)

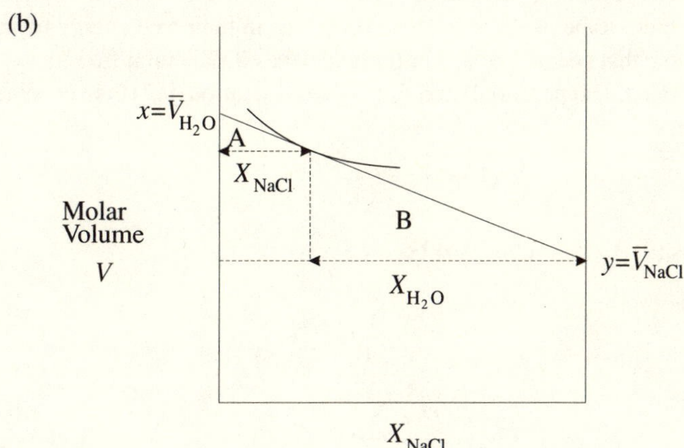

FIG. 9.5. (a) Molar volume of aqueous NaCl solutions as a function of X_{NaCl}. (b) Detail (distorted for visibility) of the tangent to the molar volume curve at 1 molal concentration.

For a generalized binary solution $A - B$, this is

$$\bar{Z}_A = Z + \left(\frac{dZ}{dX_A}\right)(1 - X_A) \tag{9.6}$$

where again Z is any molar state variable and $\bar{Z}$ is its corresponding partial molar property.

9.2.4. *The Formal Definition*

Changing to the generalized state variable Z, we recall from Chapter 2 (equation

(2.22)) that the formal definition of a partial molar quantity is

$$\bar{Z}_i = \left(\frac{\partial Z}{\partial n_i}\right)_{P,T,\hat{n}_i} \tag{9.7}$$

which emphasizes, as we have not so far, that it is a quantity at fixed pressure and temperature, and that all constituents other than i in the solution have fixed values (this is signified by the symbol $\hat{n}_i$). If you think about it, and especially if you note that the derivative is taken of the total quantity, Z, not the molar quantity, you can see that the definition is equivalent to our discussion above. However, it's not exactly obvious, and we repeat that you cannot go wrong if you think of partial molar quantities as simply molar quantities of dissolved substances, which, of course, vary with the concentration of the substance.

9.2.5. *Partial Molar Free Energy*

The equations developed for the partial molar volume and enthalpy can be generalized to all state variables, but given the importance of the Gibbs free energy function, it will be convenient to have some of these equations in their free energy form here.

First we note that because of this importance, the partial molar free energy is given its own name (chemical potential) and symbol (μ). Equation (9.7) then becomes, when Z becomes G,

$$\bar{G}_i = \mu_i = \left(\frac{\partial G}{\partial n_i}\right)_{P,T,\hat{n}_i} \tag{9.8}$$

and equations (9.4), (9.5), and (9.6) become respectively

$$G = \sum_i n_i \mu_i \tag{9.9}$$

$$G = \sum_i X_i \mu_i \tag{9.10}$$

and

$$\bar{G}_A = G + \left(\frac{dG}{dX_A}\right)(1 - X_A) \tag{9.11}$$

These relationships are illustrated in Figure 9.6a, which in principle is similar to Figure 9.5, but there are two important differences. The first is that in 9.6a there are no numbers on the G-axis, because of course no values for G of any substance are known. Nevertheless, we can know the shape of the curve, once the positions of μ_A° and μ_B° are chosen, because the difference between G of the solution and a line joining μ_A° and μ_B° is a measurable quantity. This is the subject of the next chapter.

The other difference is that at both ends, the curve for G of the solution is asymptotic to the vertical axes, which is not the case for the volume curve. This means that at $X_B = 0$, $\mu_B = -\infty$. This is a consequence of the functional relationship between μ_i and X_i to be explored in succeeding chapters.

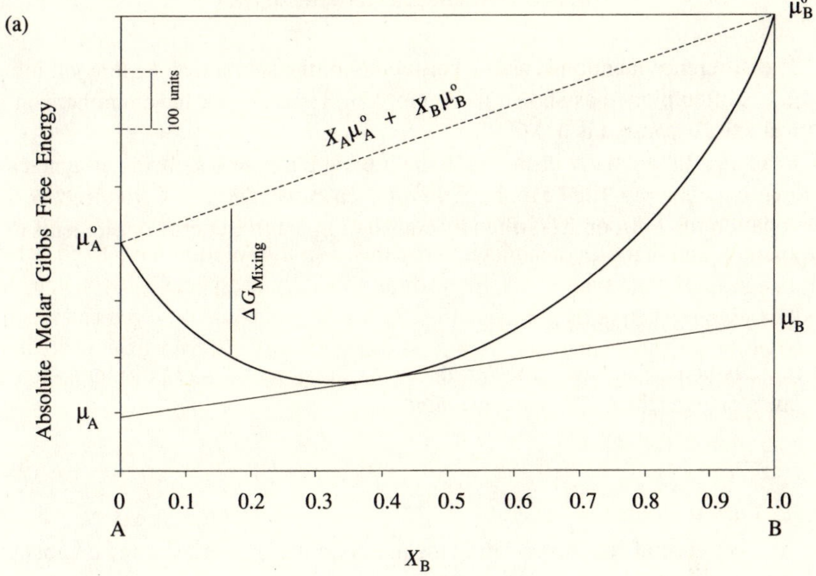

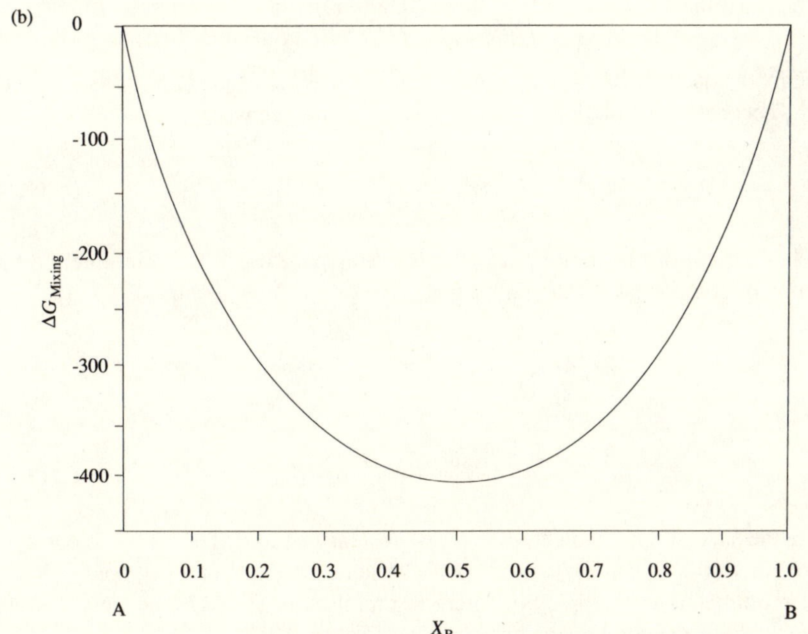

FIG. 9.6. (a) Variation of the molar free energy (G) of an ideal binary solution A–B. Tangent intercepts give the partial molar quantities, as in Figure 9.5. (b) Free energy of mixing, from Figure 9.6(a).

The difference mentioned above between G of the solution and a line joining $\mu_{\rm A}^{\circ}$ and $\mu_{\rm B}^{\circ}$ is often plotted as shown in Figure 9.5b. Here we *can* have numbers on the vertical axis, because it is a ΔG.

Readers will see many instances in the literature of numbers being assigned to μ or to G, e.g. "$\mu_{\rm A} = -1234\ {\rm J\,mol^{-1}}$." In *all* such cases, the μ or G referred to is an abbreviation for a $\Delta\mu$ or ΔG, often by considering the free energy of elements to be zero (which, of course, they are not).

9.2.6. *The Gibbs-Duhem Relation*

Since $\bar{Z}_i$ is homogeneous in the zeroth degree in $n_1, n_2, \ldots$, *etc.*, Euler's theorem (Chapter 2) gives the *Gibbs-Duhem equation*:

$$0 = \sum_i n_i d\bar{Z}_i \tag{9.12}$$

This is a very useful equation relating the partial molar quantities of the components of a system. In a two-component system

$$d\bar{Z}_1 = -(n_2/n_1)d\bar{Z}_2 \tag{9.13}$$

If we fix the quantity of solvent as 1000 g of water, n_1 is fixed at 55.51, and the ratio n_2/n_1 becomes the molality, m_2, or simply m, so we can write

$$\int_{m=0}^{m} d\bar{Z}_1 = -\int_{m=0}^{m} m\,d\bar{Z}_2 \tag{9.14}$$

The lower limit of integration, $m = 0$, represents pure water, or an infinitely dilute solution of component 2. Integration gives

$$\bar{Z}_{1,m} - \bar{Z}_{1,m=0} = -\int_{m=0}^{m} m d\bar{Z}_2$$

or

$$\bar{Z}_1 - \bar{Z}_1^{\circ} = -\int_{m=0}^{m} m\,d\bar{Z}_2 \tag{9.15}$$

where the superscript $^{\circ}$ indicates the pure substance (water). Thus if we know $\bar{Z}_2$ as a function of m, we can evaluate $\bar{Z}_1$ over the same range of compositions.

Unfortunately, the complexities increase rapidly with added components. A useful reference for three-component systems is Schumann (1955).

9.3. APPARENT MOLAR QUANTITIES

Although partial molar quantities are in principle measurable from slopes or intercepts as in Figures 9.3 and 9.5, they are not actually measured in this way. In practice, apparent molar quantities are determined, and the corresponding partial molar quantities are calculated from these. It is standard practice to let component 1 refer to the

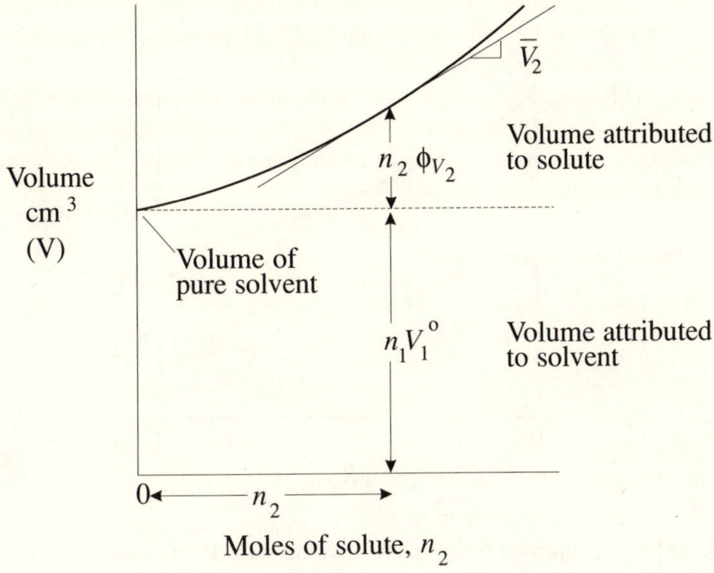

FIG. 9.7. Total volume of a solution as a function of solute concentration showing the difference between partial molar volume and apparent molar volume.

solvent and component 2 to the solute. Then apparent molar quantity ϕ_Z is defined by

$$\phi_Z = (Z - n_1 Z_1^{\circ})/n_2 \tag{9.16}$$

or using volume as an example,

$$\phi_V = (V - n_1 V_1^{\circ})/n_2 \tag{9.17}$$

As illustrated in Figure 9.7, the apparent molar volume is the volume that should be attributed to the solute in a solution if one assumes that the solvent contributes the same volume it has in its pure state. Alternatively,

$$V = n_1 V_1^{\circ} + n_2 \phi_V \tag{9.18}$$

and, dividing by $(n_1 + n_2)$,

$$V = X_1 V_1^{\circ} + X_2 \cdot \phi_V \tag{9.19}$$

and ϕ_V can be seen (Figure 9.8) to be the intercept on the solute axis, which allows calculation of the molar volume of the solution using the lever rule and the molar volume of pure solvent, rather than the partial molar volume of the solvent.

From equations (9.17) or (9.18), we see that the apparent molar volume is known as accurately and as easily as the molar volume or the total volume of a solution whose composition is known, whereas finding the partial molar volume always involves some manipulation of the raw data (such as determining a tangent) and requires a knowledge of a range of compositions, not just a single one.

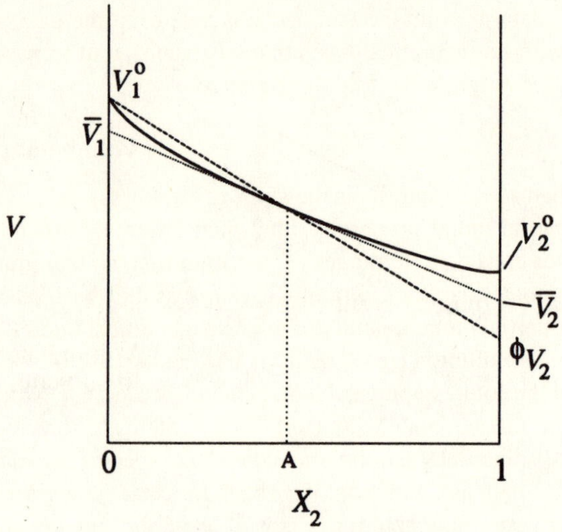

FIG. 9.8. Molar volume of a binary solution versus mole fraction showing the difference between partial molar volume and apparent molar volume for solution at composition A.

9.3.1. *Measurement of Apparent Molar Volume*

If the apparent molar volume is determined for a range of compositions, the partial molar volume may be calculated by differentiation. Suppose measurements of ϕ_V are fitted to the equation

$$\phi_V = a + bm + cm^2 \tag{9.20}$$

where a, b, and c are constants that fit values of ϕ_V to the molality, m. Differentiating equation (9.18) we have

$$\begin{aligned} \left(\frac{\partial V}{\partial n_2}\right)_{T,P,n_1} &= \bar{V}_2 \\ &= n_2 \left(\frac{\partial \phi_V}{\partial n_2}\right)_{n_1} + \phi_V \end{aligned} \tag{9.21}$$

and hence

$$\bar{V}_2 = m \left(\frac{\partial \phi_V}{\partial m}\right)_{n_1} + \phi_V \tag{9.22}$$

which in the case of (9.20) becomes

$$\begin{aligned} \bar{V}_2 &= m(b + 2cm) + a + bm + cm^2 \\ &= a + 2bm + 3cm^2 \end{aligned}$$

which allows calculation of $\bar{V}_2$ at any concentration m. There is absolutely no theoretical rationale for the choice of a second-order polynomial in m (equation (9.20)) to represent the ϕ_V data, beyond the fact that it is able to fit the data fairly closely. Another equation with no theoretical basis that has been found to fit the data well, especially in the dilute range, is the Masson equation

$$\phi_V = \phi_V^\circ + S_V^* m^{\frac{1}{2}} \tag{9.23}$$

where ϕ_V° is ϕ_V when $m = 0$, and S_V^* is the slope of ϕ_V versus $m^{\frac{1}{2}}$.

In fact, many partial molar properties other than $\bar{V}$ are also found to be linear functions of $m^{\frac{1}{2}}$ over considerable ranges of concentration, so that graphical representation of partial molar quantities normally uses $m^{\frac{1}{2}}$ as the abscissa. However, for various reasons, the partial molar quantity of greatest theoretical interest is the partial molar quantity at infinite dilution ($m_2 = 0$), $\bar{Z}_2^\circ$, or $\bar{V}_2^\circ$ in the case of volume. (This is the molar volume of the solute in pure solvent, generally water, where the only solute interaction is with the solvent, none with itself. That is, there are solvent-solute interactions but no solute-solute interactions, since the solute species are so dilute.) Thus the extrapolation of ϕ data as a function of m back to $m = 0$ places great emphasis on measurements in very dilute solutions, as well as on the functional relationship between ϕ and m, and the $m^{\frac{1}{2}}$ relationship is not generally used here, being entirely empirical. An accurate extrapolation of ϕ to $m = 0$ should be based on an adequate knowledge of the theoretical relationship between ϕ and m, and all modern treatments use variations of the Debye-Hückel Theory (§17.4) for this purpose.

9.3.2. *Integral and Differential Heats of Solution*

So far we have considered only the volume as a partial molar quantity. But calculations involving solutes will require knowledge of all the thermodynamic properties of dissolved substances, such as $\bar{H}$, $\bar{S}$, $\bar{C}_p$, and of course $\bar{G}$, as well as the pressure and temperature derivatives of these. These quantities are for the most part derived from calorimetric measurements, that is, of the amount of heat released or absorbed during the dissolution process, whereas $\bar{V}$ is the result of volume or density measurements.

Another difference between $\bar{V}$ and other properties such as $\bar{H}$ and $\bar{G}$ is that absolute values of $\bar{V}$ are obtainable, whereas they are not for $\bar{H}$ and $\bar{G}$, so that we deal always with values of $\Delta\bar{H}$ and $\Delta\bar{G}$. These deltas or differences can mean more than one thing, and it is important to be clear about the nature of the difference in each case.

For instance, if we add a mole of NaCl to a kilogram of pure water and measure the heat absorbed, rather than the volume change, we find that about 917 calories are required to keep the temperature constant at 25°C. That is

$$1000\,\mathrm{g}\;H_2O + 58.5\,\mathrm{g}\;\mathrm{NaCl} = 1\;molal\;solution;\;\Delta H = 917\;\mathrm{cal}$$

If we add another mole of NaCl, we find that this time about 749 *cal* are required, i.e.,

$$1\;molal\;\mathrm{NaCl}\;solution + 58.5\,\mathrm{g}\;\mathrm{NaCl} = 2\;molal\;solution;\;\Delta H = 749\;\mathrm{cal}$$

Also, we expect to find, since H is a state variable, that adding 2 moles of NaCl to a kilogram of water gives the same result as the two above operations combined, i.e.,

1666 cal will required. The complete curve for NaCl additions to water is shown in Figure 9.9. These heats of solution are called "integral heats of solution." The slope of this curve at any particular concentration, $d\Delta H/dn_{\mathrm{NaCl}}$, is called the "differential heat of solution," and is in fact the partial molar heat of solution of NaCl in a solution of that composition. This quantity is represented by $\Delta\bar{H}$, where the Δ refers to the difference in enthalpy between a very large quantity of solution, and that quantity of solution plus one mole of NaCl. Thus for a one molal solution

$$\begin{bmatrix} large\ quantity\ of \\ one\ molal\ \mathrm{NaCl} \\ solution \\ (n_1\ moles\ \mathrm{H_2O}, \\ n_2\ moles\ \mathrm{NaCl}) \end{bmatrix} + 1\ mole\ \mathrm{NaCl} = \begin{bmatrix} large\ quantity\ of \\ one\ molal\ \mathrm{NaCl} \\ solution \\ (n_1\ moles\ \mathrm{H_2O}, \\ n_2 + 1\ moles\ \mathrm{NaCl}) \end{bmatrix}$$

$$\Delta\bar{H} = 794\ \mathrm{cal\,mol^{-1}}$$

whereas at infinite dilution,

$$\begin{bmatrix} large\ quantity\ of \\ pure\ water \\ (n_1\ moles\ \mathrm{H_2O}, \\ 0.0\ moles\ \mathrm{NaCl}) \end{bmatrix} + 1\ mole\ \mathrm{NaCl} = \begin{bmatrix} large\ quantity\ of \\ extremely\ dilute \\ solution \\ (n_1\ moles\ \mathrm{H_2O}, \\ 1\ mole\ \mathrm{NaCl}) \\ 0.00\ molal\ \mathrm{NaCl} \end{bmatrix}$$

$$\Delta\bar{H} = 923\ \mathrm{cal\,mol^{-1}}$$

Partial molar enthalpies of solution also appear as intercepts of tangents on plots of integral heat of solution versus mole fraction, as illustrated in Figure 9.10. A practical consideration in conducting calorimetric heat of solution measurements is that it is frequently much simpler and more accurate to measure heat flows associated with diluting solutions, rather than with making them more concentrated. Dissolving NaCl in a concentrated solution of NaCl in water is a rather slow process, so that the heat is absorbed over a long period of time, and it is difficult to know when the process is complete. Adding water to a concentrated NaCl solution is on the other hand accomplished fairly quickly and the same information can be deduced from these dilution measurements as from "solution" measurements. Of course, what is "dilution" and what is "solution" is to some degree quite arbitrary—a "heat of dilution" on adding water to a NaCl solution could be considered a "heat of solution" of water in the system NaCl–H_2O. Normally, however, water is considered the solvent, and adding water is called dilution. Figure 9.11 shows heats of dilution of $HCl \cdot 3H_2O$, or in other words, integral heats of solution of H_2O in $HCl \cdot 3H_2O$.

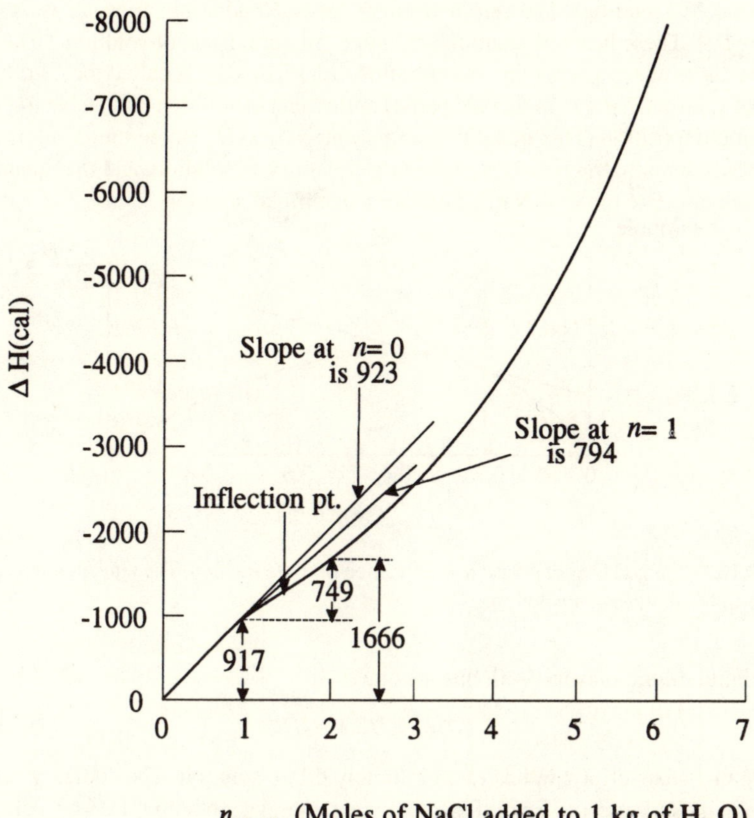

FIG. 9.9. Integral heat of solution of NaCl in water as a function of moles of salt added to 1 kg water. Slopes of tangents are differential heats of solution of salt at that concentration.

9.3.3. *Heats of Solution from Experimental Data*

Apparent Molar Enthalpy

Calorimetric data for solutions are handled in a number of different ways, which can be confusing. In addition to integral and differential heats of solution and the partial molar enthalpy of solution, we also have the apparent partial molar enthalpy, the relative partial molar enthalpy, and the relative apparent partial molar enthalpy. To see how these terms arise, consider the following.

According to equation (9.4) the enthalpy of a two-component solution can be split into partial molar enthalpy terms for the solvent and solute.

$$\mathrm{H} = n_1\bar{H}_1 + n_2\bar{H}_2 \tag{9.24}$$

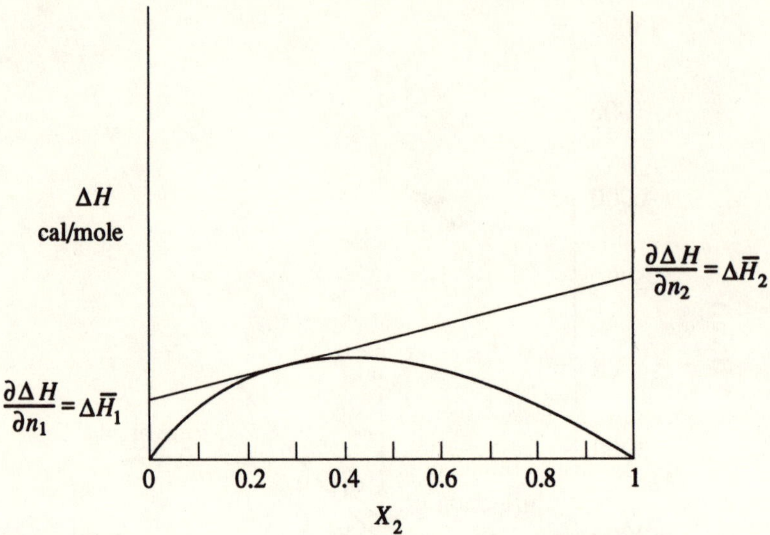

FIG. 9.10. Integral heat of solution as a function of mole fraction. The tangent intercepts are the partial molar heats of solution.

At infinite dilution, as $n_2 \to 0$, this becomes

$$\mathrm{H}^\circ = n_1\bar{H}_1^\circ + n_2\bar{H}_2^\circ \tag{9.25}$$

where the superscript ° indicates an infinitely dilute solution. The enthalpy can also be split into two terms involving the apparent molar enthalpy, as was volume in equation (9.18)

$$\mathrm{H} = n_1\bar{H}_1^\circ + n_2 \cdot \phi_H \tag{9.26}$$

(illustrated in Figure 9.12).[3]

Again considering the case where $n_2 \to 0$ we can write from (9.26)

$$\mathrm{H}^\circ = n_1\bar{H}_1^\circ + n_2 \cdot \phi_H^\circ \tag{9.27}$$

Comparing (9.25) and (9.27) shows that

$$\phi_H^\circ = \bar{H}_2^\circ \tag{9.28}$$

Relative Partial Molar Enthalpy

Thus measuring ϕ_H and extrapolating to infinite dilution would seem to offer a method for determining $\bar{H}_2^\circ$, the most commonly desired quantity. Except for one problem, that is. As emphasized in earlier chapters, absolute values for H and hence $\bar{H}_2^\circ$, ϕ_H, etc. are unknown and unknowable, since the definition of H is

$$\mathrm{H} = \mathrm{U} + P\mathrm{V}$$

[3]Recall that at infinite dilution of component 2, H_1° (in equation 9.26)= H_1° (in Figure 9.12).

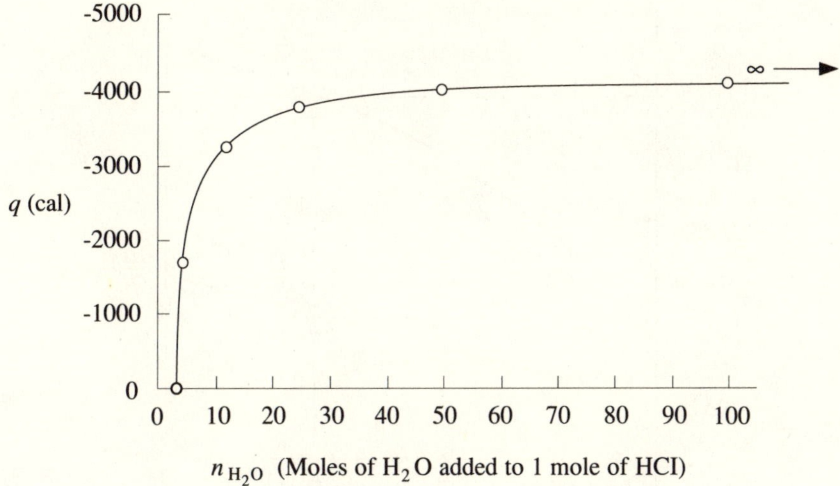

FIG. 9.11. Heats of dilution of $HCl \cdot 3H_2O$.

and U is unknown. We can deal only with differences in U and H, as we have done so far with the integral and differential heats of solution. Therefore in order to be able to use partial molar and apparent molar enthalpies, the infinitely dilute solution is used as a reference point, and subtracting (9.25) from (9.24) we get

$$\mathrm{H} - \mathrm{H}^\circ = n_1(\bar{H}_1 - \bar{H}_1^\circ) + n_2(\bar{H}_2 - \bar{H}_2^\circ) \tag{9.29}$$

Defining

$$\mathrm{L} = \mathrm{H} - \mathrm{H}^\circ$$

$$\bar{L}_1 = \bar{H}_1 - \bar{H}_1^\circ$$

and

$$\bar{L}_2 = \bar{H}_2 - \bar{H}_2^\circ$$

we have

$$\mathrm{L} = n_1\bar{L}_1 + n_2\bar{L}_2 \tag{9.30}$$

where L is the relative enthalpy of the solution, and $\bar{L}_i$ the relative partial molar enthalpy of component i.

Subtracting (9.27) from (9.26) we find

$$\mathrm{H} - \mathrm{H}^\circ = n_2(\phi_H - \phi_H^\circ)$$

and by definition

$$\phi_L = \phi_H - \phi_H^\circ \tag{9.31}$$

Thus

$$\begin{aligned}\mathrm{L} &= n_1\bar{L}_1 + n_2\bar{L}_2 \\ &= n_2\phi_L\end{aligned} \tag{9.32}$$

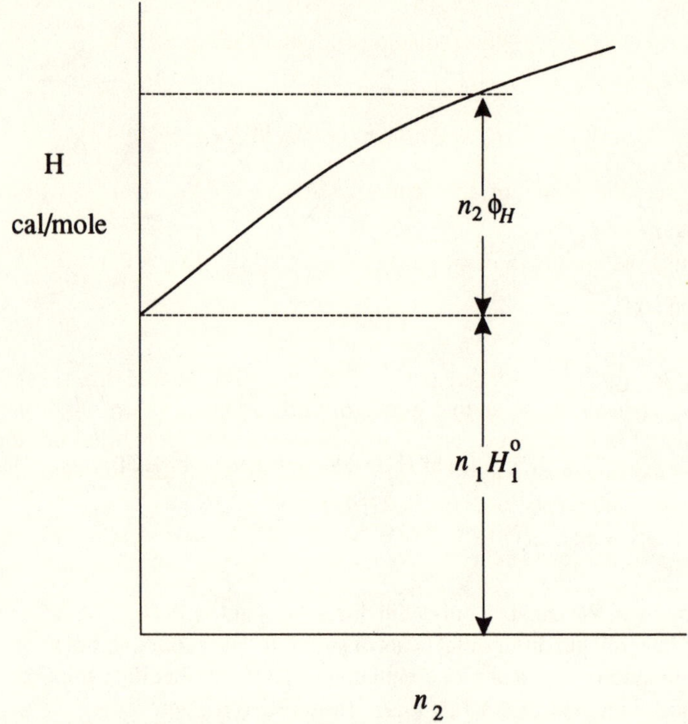

FIG. 9.12. Enthalpy as a function of number of moles of solute. Compare with Figure 9.7. The difference is that volumes are measurable whereas absolute enthalpy is not.

where ϕ_L is called the relative apparent molar heat content or enthalpy. Differentiating equation (9.32) gives

$$\begin{aligned}\left(\frac{\partial \mathrm{L}}{\partial n_2}\right)_{n_1} &= \bar{L}_2 \\ &= n_2 \left(\frac{\partial \phi_L}{\partial n_2}\right)_{n_1} + \phi_L \end{aligned} \tag{9.33}$$

which (finally) gives a theoretically useful quantity ($\bar{L}_2 = \bar{H}_2 - \bar{H}_2^\circ$) in terms of a measurable quantity, ϕ_L. (We recall the important fact that $\bar{H}_2^\circ$ does not refer to pure component 2, but to the value of $\bar{H}_2$ when the quantity of component 2 in solution becomes extremely small.)

$\bar{L}_2$ is one of the more commonly measured calorimetric quantities in solution chemistry, and equation (9.33) is the fundamental basis for these measurements. As mentioned earlier, this is commonly done by measuring heats of dilution rather than of solution. It is related to the temperature derivative of the activity coefficient, as shown in §12.5.1.

PROBLEMS

1. In the next chapter, we show (equation 10.2) that the equation for the curve in Figure 9.6a is

$$G_{solution} = X_A\mu_A^\circ + X_B\mu_B^\circ + RT(X_A \ln X_A + X_B \ln X_B)$$

Show that the slope of this curve, $dG_{solution}/dX_B$, is $\mu_B^\circ - \mu_A^\circ + RT\ln(X_B/X_A)$.

2. Show that the minimum in this curve, at $dG_{solution}/dX_B = 0$, is at

$$X_B = \frac{e^{(\mu_A^\circ - \mu_B^\circ)/RT}}{1 + e^{(\mu_A^\circ - \mu_B^\circ)/RT}}$$

3. To draw Figure 9.6a, μ_A° and μ_B° were arbitrarily assigned values of 100 and 500 cal mol^{-1} respectively, and T was chosen as 298.15 K. Find (a). the slope of the tangent at $X_B = 0.4$. (b). The value of X_B at the minimum. (c). The tangent intercepts (use equation 9.11).

10

IDEAL SOLUTIONS

Let us not grace loose thinking with the word 'model'. H.J. Greenwood (1989).

10.1. SOLUTIONS, MIXTURES, AND MODELS

The chemical constituents of a solution can be varied—added, subtracted and interchanged or substituted for each other—within limits ranging from complete (e.g., gases) to highly restricted (trace components in quartz). Adding or subtracting chemical constituents to or from a phase involves changes in energy, which will be discussed in the following sections. For example, if two components A and B are mixed together, the Gibbs energy of a solution of the two mixed must be less than the sum of the Gibbs energies of the two separately for the spontaneous reaction

$$\mathrm{A} + \mathrm{B} \rightarrow (\mathrm{A}, \mathrm{B})(solution)$$

to take place. That is, if we mix n_A moles of component A and n_B moles of component B, their combined total G is $(n_\mathrm{A}G_\mathrm{A} + n_\mathrm{B}G_\mathrm{B})$ where G_A and G_B are the molar free energies of A and B. If $\mathrm{G}_{(\mathrm{A,B})}$ is the total free energy of the resulting solution, then necessarily

$$\mathrm{G}_{(\mathrm{A,B})} < n_\mathrm{A}G_\mathrm{A} + n_\mathrm{B}G_\mathrm{B}$$

if the solution took place spontaneously.[1] Alternatively, dividing through by $n_\mathrm{A} + n_\mathrm{B}$,

$$G_{(\mathrm{A,B})} < X_\mathrm{A}G_\mathrm{A} + X_\mathrm{B}G_\mathrm{B}$$

where X_A and X_B are the mole fractions.

Thus if A is albite and B is anorthite, then (A,B) is plagioclase, and we say that the plagioclase solid solution is more stable than a "mechanical mixture" of grains of albite and anorthite. On the other hand if A is diopside and B is anorthite, little or no mutual solution takes place because in this case

$$G_{(\mathrm{A,B})} > X_\mathrm{A}G_\mathrm{A} + X_\mathrm{B}G_\mathrm{B}$$

so that no spontaneous solution reaction takes place. The term "mechanical mixture" in this context nicely conveys the idea of quantities of mineral grains mixed together and not reacting, but does not work quite so well if A and B are other things such as water and halite, or water and alcohol. Nevertheless, the term is traditionally used no matter what the nature of the solution constituents, and no harm is done as long as we remember that "mechanical mixture" means that the constituents considered do not react with each other, whatever their physical nature.

[1] If A and B are in their pure reference states, G_A and G_B are G°_A and G°_B, but in general A and B can be anything at all.

It will be helpful to recall our discussion of metastable systems in Chapter 3, where we explained that constituents that are together but do not react or interact in any way, such as those separated by a partition, or those having an activation energy barrier, are examples of metastable systems. Metastable systems have an extra constraint that prevents their constituents from reacting to achieve their lowest potential. In the case of the "mechanical mixture" albite-anorthite it is the activation energy; for water-halite it could be a partition in a composite system. In this chapter, for lack of a better term, we use "mechanical mixture" for this kind of metastable state.

In the following sections we will quantify some of the thermodynamic properties of mechanical mixtures and ideal and non-ideal solutions. As we detail the properties of ideal solutions, it will become clear that they are strictly hypothetical; another thermodynamic concept, like "true equilibrium", which is a limiting state for real systems. Ideal solutions, in other words, are another part of the thermodynamic model, not of reality. It is a useful concept, because real solutions can be compared to the hypothetical ideal solution and any differences described by using correction factors (activity coefficients) in the equations describing ideal behavior. These correction factors can either be estimated theoretically or determined by actually measuring the difference between the predicted (ideal) and actual behavior of real solutions.

10.2. TYPES OF IDEAL SOLUTIONS

10.2.1. *Ideal Gaseous Solutions*

Before proceeding with our numerical treatment, it is helpful to consider the physical picture of an ideal solution. What are the properties of true ideal solutions and why do real solutions not behave this way? The picture differs for gases, liquids, and solids.

Taking the simplest case first, an ideal gas consists of hypothetical, vanishingly small particles that do not interact in any way with each other. They are unaware of the existence of the other particles and there are no forces or energies of attraction or repulsion. An ideal gas must obey the ideal gas law, $P\mathrm{V} = nRT$, where n is the number of moles, T is related to the movement and individual energies of the particles, V is the volume occupied by the particles, and P comes from the only interaction allowed in the system—particles bouncing off the walls or boundaries. A *solution* of two ideal gases will also obey the ideal gas law since the particles of the different constituents remain unaware of all other particles, just as with an ideal single-component gas. Of course, real gases do interact at the atomic scale and can only be expected to approach ideal behavior at very low densities and pressures, or in the limit as $P \to 0$.

10.2.2. *Ideal Liquid Solutions*

Liquids are necessarily more complicated than gases. To start with, they have much greater cohesiveness than gases; for example, a liquid equilibrated with its gaseous vapor develops a meniscus. This boundary has a measurable surface tension caused by the asymmetry of particle interactions in the liquid and minimal interaction of particles in the gas phase above it. Hence a liquid must have significant interaction among its particles—if it did not, it would disperse and become a gas. There is no such thing as an equation of state for the ideal liquid, as there is for the ideal gas. The term "ideal liquid solution" therefore is used in an entirely different sense than the term "ideal gaseous solution." It refers not to the liquid as such but to the mixing properties of the components. In other words, we are concerned simply with changes as the composition of the solution is varied. Recognizing that any liquid must have significant interatomic or intermolecular interactions, an ideal liquid solution will have the following properties:

- The forces of interaction between all molecules, whether of one type or another, must be exactly the same. For example in a liquid solution of constituents A and B, interactions A-A, A-B, and B-B must be identical. This means that all constituents A,B,. . . , must have the same molecular properties (size, charge, polarity, bonding characteristics).
- An ideal liquid solution must obey Raoult's Law and the Lewis Fugacity Rule (just like an ideal gas solution).
- The volume of the solution must equal the sum of the volumes of the pure components before mixing, and no heat is absorbed or given off when the solution is prepared (because such effects are caused by changes in the particle interactions, which we have just ruled out).

Note that ideal gases (and gaseous solutions) have a complete *absence* of intermolecular forces, while ideal liquid solutions have a complete *uniformity* of forces.

10.2.3. *Ideal Solid Solutions*

We shall see shortly that the ideal and non-ideal properties of gaseous, liquid, and solid solutions are all treated in numerically similar fashion. However, the concept of ideal solid solutions is as different from ideal liquid solutions as the liquid is from the gas. This is because a solid is rigid and its component molecules, ions, or atoms are confined to varying degrees to specific structural sites. The degree of structure varies, of course, from glassy to fully crystalline materials, but there is always some ordering of constituents. Whereas ideal gases and gaseous solutions have a complete absence of inter-particle forces and ideal liquid solutions have a complete uniformity, solids must have highly specific interactions between different constituents. We speak of specific sites in crystals, such as tetrahedral silicon–oxygen bonds and octahedral aluminum sites, and the same is true (although to a lesser extent) of glassy solids. The interactions between Si-Si, Si-O, Si-Al, O-O, and Al-O in an aluminosilicate are all quite different. However, within the framework of a perfectly crystalline compound it is frequently possible to substitute one element for another. This substitution and the corresponding solid solution would be ideal *if the two substituting elements or species were completely indistinguishable.* The closest approximation to an ideal solid solution would be the substitution of two isotopes of the same element on the same crystal site. Like ideal gaseous and liquid solutions, an ideal solid solution will also obey Raoult's Law and the Lewis Fugacity Rule (presuming vapor pressures of any constituents could be measured); there would be no heat evolved on mixing the components and the total volume of the solution must simply be the sum of the volumes of the pure constituents before mixing.

10.3. THERMODYNAMICS OF IDEAL SOLUTIONS

Despite the dissimilarities in these hypothetical models of ideal gaseous, liquid, and solid solutions, we see that they share a number of important properties. All ideal solutions, no matter what the phase, have no heat of mixing when prepared from their components, and total volumes must simply be the sum of the individual volumes of the components before mixing.

These relationships, and others, follow directly from the relationship between the partial molar free energy (chemical potential) of a constituent and its mole fraction in an ideal solution

$$\mu_i - \mu_i^\circ = RT \ln X_i \tag{10.1}$$

where μ_i is the chemical potential of some constituent and μ_i° is (in this equation)[2] the chemical potential of the constituent in its pure ($X_i = 1$) form. We will see shortly that (10.1) is *all that is necessary* to define an ideal solution.

From (9.10), the molar free energy of the solution is the weighted sum of the chemical potentials of all components, i.e.

$$G = \sum_i X_i \mu_i$$

Inserting (10.1),

$$\begin{aligned} G_{ideal\ sol'n} &= \sum_i X_i(\mu_i^\circ + RT \ln X_i) \\ &= \sum_i X_i \mu_i^\circ + RT \sum_i X_i \ln X_i \end{aligned} \tag{10.2}$$

At this point we can simplify things by considering a process in which two or more pure components are first mixed together mechanically (brought together in a composite system having a constraint that prevents their mutual interaction), and then allowed to dissolve (the constraint is released) forming a single, ideal solid solution. Before the dissolution, the total free energy of the mechanical mixture has to be the sum of the chemical potentials (or molar free energies) of each of the pure components in the proper proportions:

$$G_{mech\ mix} = \sum_i X_i \mu_i^\circ \tag{10.3}$$

This allows us to break equation (10.2) into two very important parts. One part describes the free energy contribution due solely to the mechanical mixture or sum of free energies of the substances being mixed. This is the first term on the right side

[2]This equation is given a more generalized form in the next chapter.

of (10.2). The remaining term on the right side of (10.2) must describe a change in the total free energy arising from the dissolution process itself, which we could call $\Delta G_{ideal\ dissol'n}$, so that

$$G_{ideal\ sol'n} = G_{mech\ mix} + RT\sum_i X_i \ln X_i \qquad (10.4)$$

and

$$\begin{aligned}\Delta G_{ideal\ dissol'n} &= G_{ideal\ sol'n} - G_{mech\ mix}\\ &= RT\sum_i X_i \ln X_i\end{aligned} \qquad (10.5)$$

"Ideal dissolution" is often called "Ideal mixing."

10.4. ENTROPY OF IDEAL SOLUTIONS

Something new has crept in here, since nothing we have said so far about ideal solutions included a free energy contribution from the solution process itself. What gives rise to this term and what are its physical implications?

We have already observed the answers to these questions in Chapter 6. While there is no volume or enthalpy change when a mechanical mixture dissolves to form an ideal solution, i.e.

$$\begin{aligned}V_{ideal\ sol'n} = V_{mech\ mix} &= \sum_i X_i \bar{V}_i^\circ\\ &= \sum_i X_i V_i^\circ \text{ (if } i \text{ is a pure substance)}\end{aligned} \qquad (10.6)$$

$$\begin{aligned}H_{ideal\ sol'n} = H_{mech\ mix} &= \sum_i X_i \bar{H}_i^\circ\\ &= \sum_i X_i H_i^\circ \text{ (if } i \text{ is a pure substance)}\end{aligned} \qquad (10.7)$$

and

$$\Delta V_{ideal\ dissol'n} = \Delta H_{ideal\ dissol'n} = 0 \qquad (10.8)$$

there *is* a change in entropy. We can figure out exactly what this change in entropy must be by combining (10.5), (10.8) and the relation $\Delta G = \Delta H - T\Delta S$; thus

$$\Delta S_{ideal\ dissol'n} = -R\sum_i X_i \ln X_i \qquad (10.9)$$

and

$$S_{ideal\ sol'n} = S_{mech\ mix} - R\sum_i X_i \ln X_i \qquad (10.10)$$

As in all the preceding equations, the entropy of a mechanical mixture is simply the sum of entropies of the pure, unmixed components:

$$\begin{aligned} S_{mech\ mix} &= \sum_i X_i \bar{S}_i^\circ \\ &= \sum_i X_i S_i^\circ \text{ (if } i \text{ is a pure substance)} \end{aligned} \tag{10.11}$$

so (10.10) becomes

$$S_{ideal\ sol'n} = \sum_i X_i \bar{S}_i^\circ - R \sum_i X_i \ln X_i \tag{10.12}$$

Note that the Gibbs free energy of ideal dissolution (10.5) depends on temperature while the entropy of ideal dissolution (10.9) does not. This has important physical implications which are considered below.

In Chapter 6, equation (10.9) for the entropy change of this dissolution process was derived statistically, starting with

$$\mathrm{S} = k \ln W \tag{6.3}$$

The entropy of ideal dissolution (10.9) can also be easily derived using classical (as opposed to statistical) thermodynamics. This is worth doing here since it provides further insight into the problem. The derivation for ideal gases is very simple, and that for liquids and solids only slightly more complicated. Because we want to look at the effect of volume and pressure changes at constant temperature, we start with the exact differential of S with respect to T and V,

$$d\mathrm{S} = \left(\frac{\partial \mathrm{S}}{\partial T}\right)_{\mathrm{V}} dT + \left(\frac{\partial \mathrm{S}}{\partial \mathrm{V}}\right)_T d\mathrm{V}$$

In an isothermal process, the dT term drops out, but just for the record $(\partial \mathrm{S}/\partial T)_{\mathrm{V}}$ is C_v/T (just as $(\partial S/\partial T)_P$ is C_p/T (§7.3, §7.5.2), and the other derivative $(\partial \mathrm{S}/\partial \mathrm{V})_T$ is equal to $(\partial P/\partial T)_{\mathrm{V}}$. This is one of Maxwell's equations (§14.3.2). Therefore the equation becomes

$$d\mathrm{S} = \left(\frac{\mathrm{C}_v}{T}\right) dT + \left(\frac{\partial P}{\partial T}\right)_{\mathrm{V}} d\mathrm{V}$$

which, for an ideal gas, is

$$d\mathrm{S} = (\mathrm{C}_v/T)dT + (nR/\mathrm{V})d\mathrm{V}$$

For a constant temperature process and because $P\mathrm{V} = nRT$ for ideal gases, this becomes after integration,

$$\begin{aligned} \Delta \mathrm{S} &= nR \ln(\mathrm{V}_2/\mathrm{V}_1) \\ &= nR \ln(P_1/P_2) \end{aligned} \tag{10.13}$$

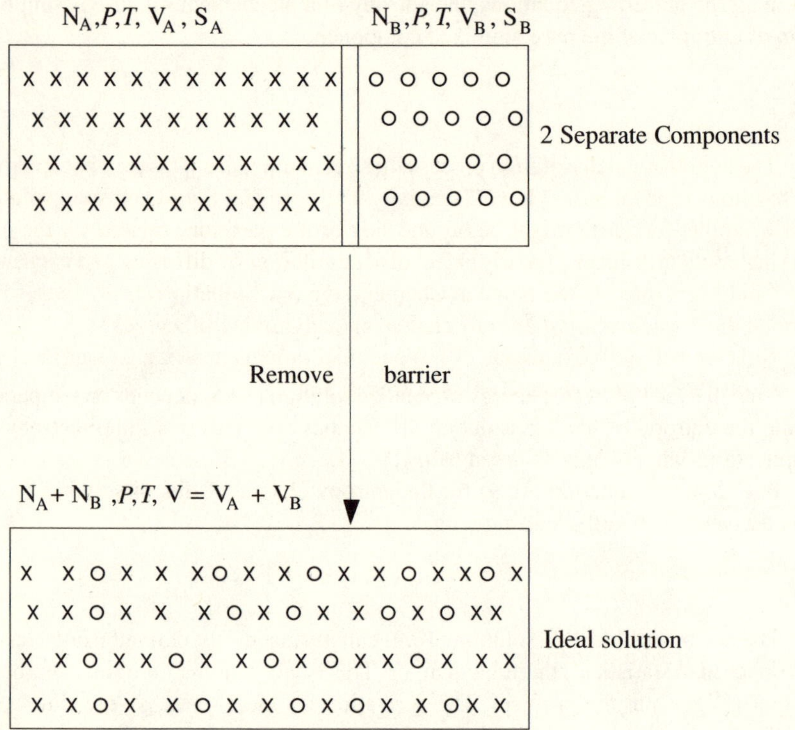

FIG. 10.1. Hypothetical processes in which two ideal gases A and B, originally separated by a partition, mix to form an ideal gaseous solution with a change in molar entropy equaling $-R\sum_i(X_i \ln X_i)$.

Consider the process shown in Figure 10.1 where two ideal gases (originally separated by a partition) mix, forming an ideal solution at fixed total pressure and temperature in an adiabatic container. After mixing, the partial pressure of each gas is given by

$$P_i = X_i P \tag{10.14}$$

where P is total pressure (cf. equation (11.22)). This is Dalton's Law of partial gas pressures (which, as might be expected, only works exactly for ideal gaseous solutions). According to (10.13) and (10.14), the entropy change on expanding both gases from the initial pressure P to their final partial pressures P_i in the solution will be

$$\Delta S = n_A R \ln(P/P_A) + n_B R \ln(P/P_B)$$

or

$$\Delta S = n_A R \ln(1/X_A) + n_B R \ln(1/X_B)$$

Dividing by $(n_A + n_B)$ gives

$$\Delta S_{ideal\ dissol'n} = -RX_A \ln X_A - RX_B \ln X_B$$

which, on generalizing to any number of components, becomes

$$\Delta S_{ideal\ dissol'n} = -R \sum_i X_i \ln X_i \tag{10.9}$$

as previously derived.

The hypothetical dissolution process illustrated in Figure 10.1 need not apply only to solutions of ideal gases. It works equally well for ideal liquid and solid solutions. For example, forsterite might be on one side of the partition, fayalite on the other; mixing could then form a (nearly) ideal olivine solution by diffusing and exchanging Fe^{2+} and Mg^{2+} ions. As we noted in Chapter 6 (§6.5.4), equation (10.9) gives what is termed the *configurational entropy* change of such mixing processes.

Next we will derive equation (10.9) one final time, generalizing to all ideal solutions, whether gaseous, liquid, or solid. Recall that

$$\left(\frac{\partial \bar{G}_i}{\partial T}\right)_{P,n_i} = \left(\frac{\partial \mu_i}{\partial T}\right)$$
$$= -\bar{S}_i \tag{10.15}$$

and, for an ideal solution,

$$\mu_i = \mu_i^\circ + RT \ln X_i \tag{10.1}$$

Hence the partial molar entropy of a component of an ideal solution is

$$\bar{S}_i = \bar{S}_i^\circ - R \ln X_i \tag{10.16}$$

The entropy of the solution is given by the sum

$$\mathrm{S}_{ideal\ sol'n} = \sum_i n_i \bar{S}_i$$
$$= \sum_i n_i \bar{S}_i^\circ - R \sum_i n_i \ln X_i$$

Dividing through by $\sum_i n_i$ we have

$$S_{ideal\ sol'n} = \sum_i X_i \bar{S}_i^\circ - R \sum_i X_i \ln X_i \tag{10.12}$$

as previously derived. Equation (10.9) follows directly from this.

We have just observed that there is an entropy change given by equation (10.9) when two or more components dissolve to make a solution. Because $\Delta G = \Delta H - T\Delta S$, there is a corresponding change in the Gibbs free energy as well. The physical implications make good intuitive sense and are illustrated in Figure 10.2a,b. First, equation (10.3) for a mechanical mixture of two components A and B is a straight line connecting the molar free energies of each of the pure components. The term $\sum_i X_i RT \ln X_i$ in (10.4) for the free energy of the dissolution process itself will always be negative since $X_i < 1$. So, according to (10.5) the free energy of an ideal

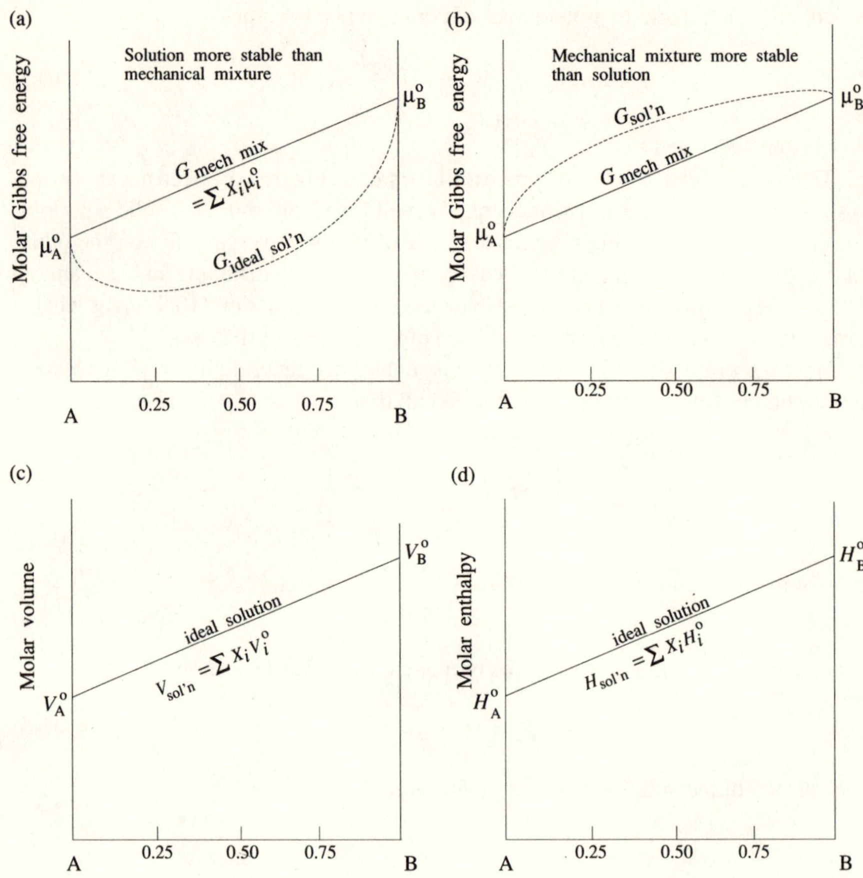

FIG. 10.2. $G - X$, $V - X$, and $H - X$ diagrams for ideal and nonideal solutions.

solution will always be lower than that of a mechanical mixture, or

$$\Delta G_{ideal\ dissol'n} \leq 0$$

always. This makes good sense thermodynamically because the most stable state of a system must always have the lowest free energy. For those cases where a solution is more stable than a mechanical mixture of its components, it will have the lower free energy, as in Figure 10.2a. If instead the mixture is more stable, it will have the lower free energy as in Figure 10.2b. The entropy of the dissolution process itself makes the solution more stable than the mixture—at least for ideal solutions. We shall see below that if the solution is non-ideal, it may have a positive ΔH of dissolution; this often overwhelms the entropy contribution, making the solution less stable than the mixture, (since $\Delta G = \Delta H - T\Delta S$). The entropy of dissolution may be thought of as a sort of driving force pushing towards the stability of solutions. However, its effect may not always be sufficient to produce a true solution if other parameters such as enthalpy work too strongly in the opposite sense.

10.5. ENTHALPY AND VOLUME OF IDEAL SOLUTIONS

We have stated above that the enthalpy and volume change involved in forming an ideal solution from its pure components is zero (equations (10.6) through (10.8)). This makes good intuitive sense and could be given as a definition or criterion of ideal solutions. Both conditions follow from equation (10.1), which is all that is needed to define ideal solutions.

Considering the volume first, differentiate (10.1) with respect to pressure:

$$\begin{aligned}\left(\frac{\partial \mu_i}{\partial P}\right)_T &= \left(\frac{\partial \mu_i^\circ}{\partial P}\right)_T \quad \text{(Recall that } \partial(RT \ln X_i)/\partial P = 0\text{)}\\ &= \bar{V}_i \\ &= \bar{V}_i^\circ = V_i^\circ \end{aligned} \tag{10.17}$$

This shows that the partial molar volume $\bar{V}_i$ of a component of an ideal solution is equal to the partial molar volume of that component in its pure state, $\bar{V}_i^\circ$, which of course is also equal to the molar volume in the pure state, V_i°. It follows that there is no volume change when the pure components dissolve in one another to form the solution, i.e.

$$\begin{aligned}\Delta V_{ideal\ dissol'n} &= \bar{V}_i - V_i^\circ \\ &= 0\end{aligned}$$

It also follows that the volume of an ideal solution is the same as that of a mechanical mixture of the components (a mixture in which no dissolution takes place), and because the volume of such a mixture is

$$V_{mech\ mix} = \sum_i X_i V_i^\circ \tag{10.18}$$

it follows that

$$\begin{aligned}V_{ideal\ sol'n} &= V_{mech\ mix} \\ &= \sum_i X_i \bar{V}_i \\ &= \sum_i X_i V_i^\circ\end{aligned}$$

or

$$\Delta V_{ideal\ dissol'n} = V_{ideal\ sol'n} - V_{mech\ mix} = 0 \tag{10.19}$$

The lack of volume change on ideal dissolution will be referred to in Chapter 11 (§11.9.4) as Amagat's Rule.

For the enthalpy change upon forming an ideal solution, differentiate (10.1) with respect to temperature, after first dividing by T:

$$\begin{aligned}\left(\frac{\partial \mu_i/T}{\partial T}\right)_P &= \left(\frac{\partial \mu_i^\circ/T}{\partial T}\right)_P \\ &= -\frac{\bar{H}_i^\circ}{T^2} + \frac{1}{T}\frac{\partial \bar{H}_i^\circ}{\partial T} - \frac{\partial \bar{S}_i^\circ}{\partial T} \\ &= -\frac{\bar{H}_i^\circ}{T^2}\end{aligned}$$

therefore

$$\frac{\bar{H}_i}{T^2} = \frac{H_i^\circ}{T^2} \tag{10.20}$$

and

$$\bar{H}_i = H_i^\circ$$

This result has many applications to which we will return in subsequent chapters. It parallels the Van't Hoff equation which is used to predict the variation of equilibrium constants with temperature, as we shall see in Chapter 13. Again, this tells us that the partial molar enthalpies of the components in solution are identical to those of the pure components at the same T and P. The enthalpy of the solution is simply the sum of enthalpies of the pure components and there is no enthalpy change on making up the solution from its pure components:

$$H_{ideal\ sol'n} = \sum_i X_i \bar{H}_i = \sum_i X_i H_i^\circ \tag{10.21}$$

$$H_{mech\ mix} = \sum_i X_i H_i^\circ \tag{10.22}$$

$$\begin{aligned}H_{ideal\ dissol'n} &= H_{ideal\ sol'n} - H_{mech\ mix} \\ &= 0\end{aligned} \tag{10.23}$$

These relationships are illustrated schematically in Figures 10.2c,d.

10.6. SUMMARY: IDEAL SOLUTIONS

All ideal solutions obey the same set of rules and equations, whether they be solid, liquid, or gas (an exception is the ideal gas law, obeyed only by ideal gases). The activity coefficient always equals 1.0; the volume and enthalpy changes on dissolution are zero; the total volume and total enthalpy equal the sums of those of the original components mixed together to make the solution; the entropy and Gibbs free energy of dissolution are not zero because the components are more dispersed in the solution than before being mixed together; ideal solutions obey Raoult's Law, Henry's Law, and the Lewis Fugacity Rule. For quick reference, the primary equations describing ideal solution behavior are summarized in Table 10.1. Each equation is given the same number used in the text.

Table 10.1 Summarized Properties of Ideal Solutions

$$a_i = X_i \tag{11.29}$$

$$\mu_i = \mu_i^\circ + RT \ln X_i \tag{10.1}$$

$$G_{ideal\ sol'n} = \sum_i X_i \mu_i^\circ + RT \sum_i X_i \ln X_i \tag{10.2}$$

$$S_{ideal\ sol'n} = \sum_i X_i S_i^\circ - R \sum_i X_i \ln X_i \tag{10.12}$$

Amagat's Rule:

$$V_{ideal\ sol'n} = V_{mech\ mix} = \sum_i X_i \bar{V}_i^\circ or \sum_i X_i V_i^\circ \tag{10.6}$$

$$\Delta V_{ideal\ dissol'n} = V_{ideal\ sol'n} - V_{mech\ mix} = 0 \tag{10.19}$$

$$H_{ideal\ sol'n} = \sum_i X_i \bar{H}_i = \sum_i X_i H_i^\circ \tag{10.21}$$

$$H_{ideal\ dissol'n} = H_{ideal\ sol'n} - H_{mech\ mix} = 0 \tag{10.23}$$

10.7. SPECIES, COMPONENTS, AND CONSTITUENTS

It is now time to amplify our comments in Chapter 3 on the terms used to describe the composition of solutions.

A liquid solution is a mysterious thing. Solids that disappear into liquids, like sugar into tea, obviously must have very different properties in the dissolved state than in the crystalline state, and a great proportion of the effort of physical chemists has always been directed towards identifying the molecular and ionic species in solutions, and the structure of solutions. We now know quite a lot about the dominant species in many solutions, so that, for example, when calcite dissolves into water it is useful to consider that we do not get an aqueous $CaCO_3$ species but Ca^{2+} ions and CO_3^{2-} ions, which hydrolyze (react with water) to varying degrees depending on the solution composition, temperature, and pressure.

To use "classical" Thermodynamics we need absolutely no knowledge of such things. We need only measure macroscopic properties of the bulk solution, such as its bulk composition, and from these we will be able to calculate molar properties of dissolved substances (partial molar properties) just as we do for pure substances.

For example, consider what happens when calcite dissolves in water. When calcite and water are placed together, a spontaneous reaction takes place in which the composition of the calcite is not changed, but the composition of the fluid phase is changed. After a while the reaction stops when the system has reached equilibrium. When calcite dissolves there is no change in its ratio of Ca:C:O, so that we can represent the composition of the dissolved part by $CaCO_3(aq)$ (Figure 10.3). The solution therefore has a composition that can be represented by some proportions of H_2O and $CaCO_3(aq)$, no matter what the nature of the actual species in the solution, and we

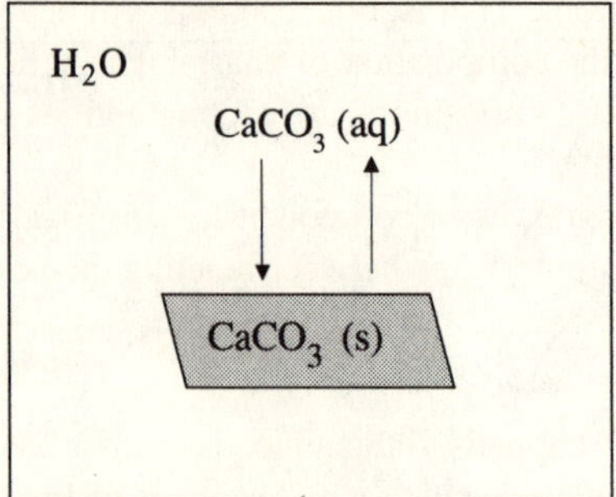

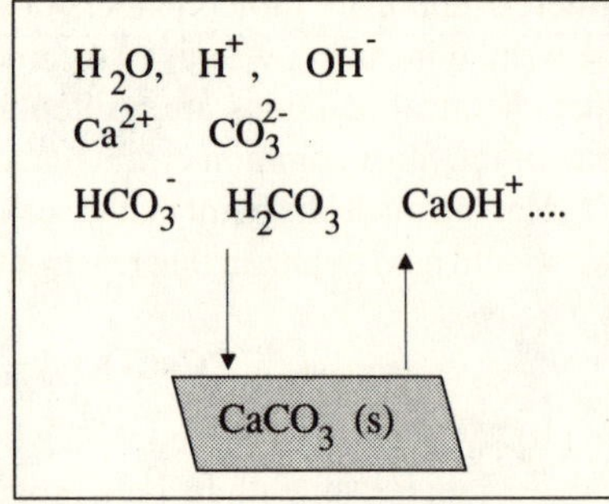

FIG. 10.3. Alternative ways of expressing the equilibrium between calcite and aqueous solution.

can represent the dissolution reaction by

$$CaCO_3(s) = CaCO_3(aq) \tag{10.24}$$

Considering this reaction from the point of view of thermodynamic potentials already developed, we say that the system pure water plus calcite is able to lower its Gibbs free energy at constant pressure and temperature by reacting, that is, some of the calcite dissolves into the water. When the Gibbs energy reaches a minimum, equilibrium is achieved and the reaction stops. During this dissolution process, the $CaCO_3(aq)$ content of the liquid increases and stops increasing at equilibrium. Similarly we know that if the solution was supersaturated with $CaCO_3(aq)$, some calcite would precipitate until the same equilibrium was achieved.[3]

Thus we should consider that $CaCO_3(aq)$, the dissolved calcite, has a certain Gibbs energy per mole that varies with its concentration in the fluid phase. At equilibrium, G for the system is minimized, which can only be true if G of the dissolved $CaCO_3(aq)$ equals the G of calcite. If it is not equal, the system can lower its overall G by either dissolving or precipitating calcite. As mentioned earlier, we need to develop methods for representing and measuring the properties of dissolved substances such as $CaCO_3(aq)$, but we can always represent the dissolved substance by its bulk composition, as in this case, and we don't need to concern ourselves with what "really" happens to the solute. In other words we can deal with the component $CaCO_3(aq)$ in the aqueous phase rather than the various species (Ca^{2+}, CO_3^{2-}, HCO_3^-, $CaOH^+$, etc.) which are known or believed to exist.

Note too, that although H_2O is the other component in this two-component system, and although it makes up the major part of the solution phase, it does not appear in equation (10.24). We do not write

$$x\text{CaCO}_3 + y\text{H}_2\text{O} = (x\text{CaCO}_3 \cdot y\text{H}_2\text{O})(aq)$$

[3]Because of the importance of the $CaCO_3 - H_2O$ system in ocean chemistry, a great deal of experimental work has actually been done on the solution and precipitation of calcite in water. In this discussion we omit reference to a number of experimental problems that are encountered in this work.

where x and y are mole numbers or mole fractions. That is, our reactions will never be written in such a way as to try to represent the composition of phases. The belief that chemical reactions are written using phase compositions as reacting entities is one of the most common errors made by students.

Having made the point that we can write reactions using components, as in (10.24), we should next point out that reactions are also often written between reacting species, e.g.

$$CaCO_3(s) = Ca^{2+}(aq) + CO_3^{2-}(aq) \tag{10.25}$$

In (10.25) we are concerned with what "really" happens to the aqueous or dissolved calcite, rather than with its bulk properties. Actually, what "really" happens includes several other reactions as well, as mentioned earlier, such as

$$CO_3^{2-} + H^+ = HCO_3^- \tag{10.26}$$

and

$$CO_3^{2-} + 2H^+ = H_2CO_3 \tag{10.27}$$

Thus in writing (10.25) we by no means imply that CO_3^{2-} and Ca^{2+} are the only ions in solution, or even that they are the major or dominant ones. They could be quite minor ones. We imply only that they exist, and that when calcite is in equilibrium with aqueous solution, the dissolution of calcite to produce Ca^{2+} and CO_3^{2-} ions will stop. While calcite is dissolving, the CO_3^{2-} ions produced are reacting to produce HCO_3^- and H_2CO_3, and conceivably dozens of other reactions between species major and minor are also taking place, but they all come to equilibrium when calcite equilibrates with the solution. In fact, if one reaction has not equilibrated, then none of them have, in the sense discussed in Chapter 3.

Also, continuing the considerations we had concerning minimizing the Gibbs potential of the solution, evidently we could say that the molar Gibbs energy of the aqueous Ca^{2+} ion plus that of the CO_3^{2-} ion must equal that of calcite at equilibrium. Otherwise calcite would continue to dissolve to produce more ions, or the ions would combine to precipitate calcite.

Combining equations (10.24) and (10.25), we find we can also write

$$CaCO_3(aq) = Ca^{2+}(aq) + CO_3^{2-}(aq) \tag{10.28}$$

We have described $CaCO_3(aq)$ as the total dissolved calcite, with no consideration of how it is split up. But $CaCO_3(aq)$ exists as a species in low concentrations, and equation (10.28) is a perfectly valid expression of the idea that species $CaCO_3(aq)$ must also equilibrate with the ions Ca^{2+} and CO_3^{2-}, and with everything else in the solution. Equation (10.24) can be understood in two ways—either involving $CaCO_3(aq)$ component, or $CaCO_3(aq)$ species. From now on we will distinguish these by using $CaCO_3^\circ(aq)$ for the neutral species, so that

$$CaCO_3(s) = CaCO_3(aq) \tag{10.24}$$

and

$$CaCO_3(s) = CaCO_3^\circ(aq) \tag{10.29}$$

express two different concepts, though in the same equilibrium system. The idea that the $CaCO_3^\circ$ species, in vanishingly small concentration, can have the same molar Gibbs energy as crystalline calcite or as component $CaCO_3(aq)$ may seem a little strange at first, but is perfectly valid. However, its other properties such as volume, heat capacity, and so on are quite different, as are its standard state properties.

Also, we note the fact that this point of view (i.e., regarding $CaCO_3(aq)$ as either a component or a species) is not transferable to solid or crystalline solutions. Thus when we write

$$NaAlSi_3O_8 \text{ (in plagioclase)} = NaAlSi_3O_8 \text{ (in melt)} \tag{10.30}$$

to represent the equilibrium between the albite component of the plagioclase solid solution and the albite component in a silicate melt of granite composition, we may regard $NaAlSi_3O_8$ (in melt) either as a possible species in the melt or as simply one of the melt components, but clearly this choice is not available for $NaAlSi_3O_8$ in the plagioclase. In the crystalline phase we know there is no such thing as a $NaAlSi_3O_8$ species or a $CaAl_2Si_2O_8$ species, only the two components of a crystalline solution.[4] Note again that we do not attempt to represent the composition of the plagioclase or the melt in equation (10.30).

Thus far, we have discussed solutes as either components or species. However, the term "component" has a very definite meaning in thermodynamics, and there are many reactions in which not all the chemical formulae can represent "components" in the true sense. For example, consider coexisting orthopyroxene and olivine, both solid solutions with Mg and Fe end members. An "exchange reaction" can be written

$$MgSiO_3 + Fe_2SiO_4 = FeSiO_3 + Mg_2SiO_4 \tag{10.31}$$

In this case both Fe_2SiO_4 and Mg_2SiO_4 are components of the olivine solid solution, and $FeSiO_3$ and $MgSiO_3$ are components of the enstatite solid solution. But when olivine and pyroxene equilibrate there are not four components, but three, since only three concentrations can be chosen independently, the fourth being fixed by the exchange reaction (10.31). Thus use of the term "component" to characterize the entities in chemical reactions often involves a certain ambiguity, if one adheres to the strict meaning of that term as we intend to do. In (10.31) for example, the terms are all components if olivine and pyroxene were considered separately, but not when they equilibrate. And even then, they represent only one possible choice of components, in the sense that one might choose other compositional variables to represent the system, such as MgO, FeO and SiO_2. Similarly in our albite example, $NaAlSi_3O_8$ is certainly one possible choice as a melt component, but there are others.

Because of this ambiguity, and because of the ambiguity involved when species and components can be represented by the same formula, we will usually refer to chemical reactions between *constituents* of phases. A constituent of a phase or system is any combination of elements in the system in any stoichiometry. Some constituents

[4] Our usage is to restrict the term *species* to liquids and gases, but as we noted in §3.8, *mineral species* is sometimes used for end-members of solid solutions and pure minerals. Other terms are also used.

would be useful choices as components; some represent the dominant species in solution; some represent hypothetical species; some are none of these. All systems therefore contain a very large number of constituents. Our choice of constituents to consider in our chemical reaction equations is governed by our knowledge of the system and our purposes in writing the reactions, and not least by our scientific insight.

PROBLEM

1. Calculate the entropy of ideal mixing (dissolution) involved in combining the three major components of air—79 wt.% N_2, 20 wt.% O_2, and 1 wt.% Ar.

11

FUGACITY AND ACTIVITY

The comprehension of the laws which govern any material system is greatly facilitated by considering the energy and entropy of the system in the various states of which it is capable. Gibbs (1875). From the Dover edition (1961), p. 55.

11.1. THE GIBBS ENERGY OF FLUIDS

In Chapter 7 we saw that the fact that minerals are to a good approximation incompressible means that the effect of pressure on the Gibbs free energy of solid phases is very easily calculated. Thus, in general

$$\int_{P_1}^{P_2} dG = \int_{P_1}^{P_2} V\,dP \tag{11.1}$$

and, because V for solids can often be considered independent of P,

$$G_{P_2} - G_{P_1} = V(P_2 - P_1)$$

This, combined with the more complex integration of dG over a temperature interval at one bar pressure, allowed us to calculate the position of phase boundaries at high pressures and temperatures. The next question is how to evaluate the pressure integral (11.1) when a fluid such as H_2O or CO_2 is involved, either in the pure form, mixed with other fluid components, or reacting with solid phases? Obviously, assuming that the molar volume of a fluid is a constant is not even approximately true, and is unacceptable. A possible way to proceed would be to express V as a function of P in some sort of power series, just as we did for C_p as a function of T (equation 7.12). $V\,dP$ could then be integrated, and we could determine the values of the power series coefficients for each gas or fluid and tabulate them as we do for the Maier-Kelley coefficients.

11.2. DEFINITION OF FUGACITY

Fortunately, thanks to the insight of G.N. Lewis, we can proceed in a simpler and completely different fashion. Lewis in 1901 defined a new function, the fugacity, which can be thought of as a kind of idealized or thermodynamic pressure, which expresses the value of $\int V\,dP$ single-handedly. To see how the inspiration for such a function might have arisen, we consider the form of the volume integral $\int V\,dP$ for an ideal gas. Thus, substituting RT/P for V in (11.1) we have

$$G_{P_2} - G_{P_1} = \int_{P_1}^{P_2} (RT/P)dP \text{ for an ideal gas}$$

$$= \int_{P_1}^{P_2} RT\, d\ln P$$

$$= RT \ln(P_2/P_1) \quad (11.2)$$

If P_1 is 1 bar and this is designated a standard or reference state denoted by a superscript °, then P_2 becomes simply P, and

$$G - G^\circ = RT \ln(P/P^\circ) \quad (11.3)$$

$$= RT \ln P \text{ since } P^\circ = 1 \quad (11.4)$$

Thus for ideal gases $RT \ln P$ all by itself gives the value of $\int_{P=1}^{P} dG$, or in other words of $\int_{P=1}^{P} V\, dP$. Unfortunately, this doesn't work for real gases although it's not a bad approximation at low pressures and high temperatures where real gases approach ideal behavior. However, the *form* of the relationship

$$\int_{P=1}^{P} V\, dP = \int_{P=1}^{P} dG = \int_{P=1}^{P} RT\, d\ln P$$

is sufficient to suggest that we could define a function such that the relationship *would* hold true for real gases. This function is the fugacity, f, where

$$V\, dP = dG = RT d\ln f \quad (11.5)$$

and

$$\int_{P_1}^{P_2} V\, dP = \int_{P_1}^{P_2} dG = RT \ln(f_{P_2}/f_{P_1}) \quad (11.6)$$

Since f appears as a ratio in (11.6), this equation cannot serve as a full definition of f. We must specify how to determine the fugacity at some P and T, then the integral can be used to calculate it at other pressures at that T. To complete the definition, it would be convenient for f to approximate P for gases that behave more or less ideally, that is, at low values of P. We accomplish this by stipulating that

$$\lim_{P\to 0} (f/P) = 1 \quad (11.7)$$

the ratio f/P is called the fugacity coefficient, γ_f.

This means that for an ideal gas, $f = P$, and for gases at low pressures, $f \approx P$. Equations (11.7) plus (11.5) or (11.6) make up the definition of fugacity.

11.3. MEASUREMENT OF FUGACITY

It is obvious from the preceding equations that to measure fugacity, we need to measure the molar volume as a function of P. In other words we have to measure gas densities. Also we should measure these densities down to very low pressures—pressures sufficiently low that we can say $f = P$ to whatever degree of accuracy is required. The greater the accuracy the lower the pressure. But at these low pressures, the volume of a given mass of gas becomes very large and very difficult to measure accurately. Thus, while in principle evaluating $\int V\,dP$ is straight-forward, in practice it is difficult when the lower limit of integration is a very low pressure. To avoid evaluating this integral at very low pressures where $V \to \infty$, it has been found convenient to define the function

$$\begin{aligned}\alpha &= V_{ideal} - V_{real}\\ &= RT/P - V\end{aligned}$$

Thus α is the difference between the molar volume of the gas and the volume it would have if it behaved ideally. Then

$$V = (RT/P) - \alpha$$

and

$$\begin{aligned}dG = V dP &= RT d\ln f\\ [(RT/P) - \alpha]dP &= RT\,d\ln f\\ d\ln f &= d\ln P - (\alpha/RT)dP\end{aligned}$$

Integrating from P_0 to P

$$\int_{P_0}^{P} d\ln f = \int_{P_0}^{P} d\ln P - \frac{1}{RT}\int_{P_0}^{P} \alpha\,dP$$

$$\ln f_P - \ln f_{P_0} = \ln P - \ln P_{P_0} - \frac{1}{RT}\int_{P_0}^{P} \alpha\,dP \qquad (11.8)$$

where P_0 means some unspecified value of P, sufficiently low such that $P = f$ to a good approximation. Thus in (11.8) at this low value of P, the two terms $\ln f_{P_0}$ and $\ln P_{P_0}$ are equal, so that the equation becomes

$$\ln f = \ln P - \frac{1}{RT}\int_{P_0}^{P} \alpha\,dP \qquad (11.9)$$

To integrate (11.9) we still need values of V at low pressures, but evaluation of the integral is now easier, since even though both RT/P and V tend to ∞ as $P \to 0$, their difference does not tend towards zero, but generally to some finite value which becomes constant at low P. Integration, either graphically or by fitting a simple integrable function to α as a function of P, can then be carried out right from $P = 0$ to P, and leads directly to values of f as a function of P at temperature T.

Equation (11.9) is also useful in providing a one-equation definition of fugacity, rather than the two-equation one we gave above. Substituting $(RT/P - V)$ for α and rearranging, we have

$$f = P\exp\left[-\frac{1}{RT}\int_{T,P\to 0}^{T,P}\left(\frac{RT}{P} - V\right)dP\right] \qquad (11.10)$$

This is the equation given by Tunell in his classic paper on fugacity (1931), in which he points out some difficulties with the definition as we have presented it, and shows that equation (11.10) is preferable as a definition of fugacity from a mathematical point of view. Our presentation, which follows the original one by Lewis and Randell (1923), seems preferable from a pedagogical point of view.

The introduction of the α function should not be allowed to obscure the fact that fugacity is obtained by evaluating the function $\int V\,dP$. The α function only simplifies this evaluation.

For a gas in a mixture of gases, the same procedure may be followed, but at each P and T what is required is not the molar volume but the partial molar volume of the gas in the mixture.

11.4. THE COMPRESSIBILITY FACTOR

It has been observed that the compressibility factor Z, which is defined as

$$Z = PV/RT \tag{11.11}$$

is the same for most gases, if the comparison is made when each gas has the same relation to its critical point (principle of *corresponding states*). This was first suggested in the last century by van der Waals. If we define a "reduced temperature"

$$T_r = T/T_c$$

where T is the gas temperature (K) and T_c is its critical temperature (K), and a "reduced pressure"

$$P_r = P/P_c$$

where P is the gas pressure and P_c is its critical pressure, then most gases have the same Z at a given P_r, T_r, within certain ranges.

Now, since

$$\alpha = (1 - Z)RT/P$$

Then from (11.9)

$$\ln(f/P) = \int_{P\to 0}^{P} \frac{Z-1}{P}\,dP \tag{11.12}$$

This shows that f/P, the fugacity coefficient, can be calculated as a function of Z, the compressibility factor, and since Z is known and is the same for most gases in terms of their P_r and T_r, then f/P can be calculated for any gas given its P_c and T_c. Several authors have prepared charts showing f/P as a function of P_r at various T_r. This is a useful device for obtaining fugacities of gases in the lower range of geologically interesting $P-T$ conditions. The charts do not extend far into the metamorphic range of $P-T$ conditions.

11.5. FUGACITY IN EQUILIBRIUM CALCULATIONS

Assuming now that we have access to tables of fugacities or fugacity coefficients for a range of P, T conditions, we now have a means of conveniently evaluating the integral $\int_{P_1}^{P_2} V\,dP$ at any temperature T for a fluid species such as H_2O or CO_2. A large and important class of mineral reactions involves only solid minerals and either water or CO_2, such as

$$KAl_3Si_3O_{10}(OH)_2 = KAlSi_3O_8 + Al_2O_3 + H_2O$$

or

$$CaCO_3 + SiO_2 = CaSiO_3 + CO_2$$

Therefore we need only split the volume integral in our previous equations (Chapter 8) into two parts—one for the solids and one for the fluid. Suppose we have a reaction

$$A(s) = B(s) + C(fluid)$$

where

$$\begin{aligned}\Delta_r V &= V_B + V_C - V_A \\ &= (V_B - V_A) + V_C \\ &= \Delta_s V + V_{fluid}\end{aligned}$$

where $\Delta_s V$ means $\Delta V(solids)$, or the volume difference between the solid constituents only, in this case $V_B - V_A$. Hence

$$\int \Delta_r V\,dP = \int \Delta_s V\,dP + \int V_{fluid} dP \tag{11.13}$$

The pressure integral for the solids is then evaluated using the constant ΔV approximation, and that for the fluid is evaluated using fugacities. Thus

$$\int_{P=1}^{P} \Delta_r V\,dP = \Delta_s V(P-1) + RT\ln(f_P/f_{P=1\,bar})$$

For the muscovite breakdown reaction given above, we can start with equation (8.6)

$$\begin{aligned}\Delta_r G^\circ_{T,P} &= \Delta_r G^\circ_{T_r,P_r} + \int_{T_r}^{T} (\partial\Delta_r G^\circ/\partial T)dT + \int_{P_r}^{P} (\partial\Delta_r G^\circ/\partial P)dP \\ &= \Delta_r G^\circ_{T_r,P_r} - \int_{T_r}^{T} \Delta_r S^\circ dT + \int_{P_r}^{P} \Delta_r V^\circ dP\end{aligned} \tag{11.14}$$

Since we have a, b, and c Maier-Kelley coefficients for water and other fluids as well as for solids, the temperature integral is evaluated as before. Now

$$\begin{aligned}\Delta_r V^\circ &= V^\circ_{\text{Kspar}} + V^\circ_{\text{Cor}} + V^\circ_{\text{H}_2\text{O}} - V^\circ_{\text{Musc}} \\ &= \left(V^\circ_{\text{Kspar}} + V^\circ_{\text{Cor}} - V^\circ_{\text{Musc}}\right) + V^\circ_{\text{H}_2\text{O}} \\ &= \Delta_s V^\circ + V^\circ_{\text{H}_2\text{O}}\end{aligned}$$

and

$$\int_{P_r}^{P} \Delta_r V^\circ dP = \int_{P_r}^{P} \Delta_s V^\circ dP + \int_{P_r}^{P} V^\circ_{H_2O} dP$$
$$= \Delta_s V^\circ (P-1) + RT \ln \left(\frac{f_{H_2O,P}}{f_{H_2O,P=1\,bar}} \right)$$

Therefore all we need to do to calculate phase boundaries for this reaction and any other involving only pure solids and a pure fluid is substitute

$$\Delta_s V^\circ (P-1) + RT \ln \left(\frac{f_{fluid}}{f_{fluid,1\,bar}} \right)$$

for $\Delta_r V(P-1)$ in equation (8.9). This then becomes

$$\begin{aligned}
\Delta_r G^\circ_{T,P} &= 0 \\
&= \Delta_r G^\circ_{T_r,P_r} - \Delta_r S^\circ_{T_r}(T - T_r) \\
&\quad + \Delta_r a \left[T - T_r - T \ln \left(\frac{T}{T_r} \right) \right] \\
&\quad + \frac{\Delta_r b}{2} \left[2TT_r - T^2 - T_r^2 \right] \\
&\quad + \frac{\Delta_r c (T^2 + T_r^2 - 2TT_r)}{2TT_r^2} \\
&\quad + \Delta_s V^\circ (P-1) \\
&\quad + RT \ln \left(\frac{f_{fluid}}{f_{fluid,1\,bar}} \right)
\end{aligned} \tag{11.15}$$

In this formulation, one small computational problem that did not arise in Chapter 8 is that since f_{fluid} is a function of P as well as T, one cannot solve equation (11.15) directly for P having chosen T. One must estimate P, find f_{fluid} for that P, solve for P (call it P') in (11.15), get a new f_{fluid} for P', solve again for P (P''), find f_{fluid} for P'', and so on until there is an insignificant difference between $P^{n'}$ and $P^{(n+1)'}$. This usually requires only four or five iterations, but is still only feasible by machine calculations. A function for determining f_{fluid} as a function of P and T must of course be included in the calculation.

Another point that should be mentioned is that we have been referring throughout to reactions involving pure solids and fluids (e.g., pure water or pure CO_2, but not a mixture of the two), and we have continued to use the $^\circ$ superscript to indicate this. However the $^\circ$ superscript actually has a considerably wider meaning, which we consider in the following chapter. You will then discover that so far we have been implicitly assuming that the standard states for all reactants and products are the pure phases at the T and P of interest.

11.6. ALL CONSTITUENTS HAVE A FUGACITY

The development of the fugacity concept in equations (11.1) to (11.7) is directed toward real gases, using the ideal gas as a model. No mention is made of whether the gas is a pure compound (CO_2, N_2, etc.) or a gaseous solution, and the fugacity could be used for either. In fact of course, fugacities of pure constituents are invariably what is required, as in our examples above. In real applications however, whether in industry or in nature, gases rarely occur in the pure form, so that the determination of fugacities of individual gases in mixtures of gases becomes of importance. In this section we show not only that individual gases in mixtures of gases have fugacities, and how to measure them, but more importantly, that the fugacity concept is not limited to gases, but can be applied to any constituent in any phase.

Integrating (11.5) between P_1 and P_2 we have

$$G_{P_2} - G_{P_1} = RT \ln(f_{P_2}/f_{P_1}) \tag{11.16}$$

which is the fugacity equivalent of (11.2), but whereas (11.2) is restricted to ideal gases, (11.16) holds for any gas. In this equation, as mentioned above, G refers to the molar free energy of either a pure gas or of a mixture of gases. To deal with the case of an individual gas in a mixture of gases we need only have the appropriate free energy term, as evidently the fugacity ratio term will refer to whatever the free energy terms refer to. In other words, if we have a ΔG term for the free energy change of an individual gas in a mixture, then the (f/f) term will also refer to that change for that individual gas. The appropriate molar free energy term for a constituent of a solution is of course the *partial* molar free energy, which is given the special symbol μ. As emphasized in Chapter 9, it is important to realize that μ_i and G_i have essentially the same meaning: both refer to the free energy per mole of some constituent i; it is just that when that constituent is in a solution, its free energy depends not only on T and P but also on its concentration, which necessitates a slightly more complicated definition.

Accepting then that gases in solution can have their free energy changes measured, and accepting too that the change involved need not be restricted to a simple change in pressure but could be a change between any two states (after all, free energy is a state variable), equation (11.16) becomes

$$\mu_i'' - \mu_i' = RT \ln(f_i''/f_i') \tag{11.17}$$

where i is the pure gas solution constituent in question, and $'$ and $''$ are any two stable equilibrium states. If we now accept as well the simple extension of our conclusion (§8.5) that the molar free energy of a constituent ($CaCO_3$) is the same in two phases at equilibrium to the equality of μ_i in all phases at mutual equilibrium (fully discussed in Chapter 14), it follows immediately that if the two states $'$ and $''$ are in equilibrium with each other, $f_i' = f_i''$ because $\mu_i' = \mu_i''$. This provides one method of determining the fugacity of a gas in a mixture of gases; one need only equilibrate the pure gas with the mixture through a membrane permeable only to the pure gas in question as in Figure 11.1a. The fugacity of the pure gas being in principle measurable as we have discussed above, the fugacity of that gas in various mixtures could then be determined. Alternatively, one could perform sufficient density measurements on the gas mixture as a function of P, T, and composition to determine its partial molar volume ($\bar{V}_i$) as a function of P and T, and carry out the calculations described above, but substituting $\bar{V}_i$ for V. In practice, because of the shortage of such data, it is more common in the Earth Sciences to assume that gas mixtures are governed by the Lewis Fugacity Rule, which amounts to assuming that the gases mix together ideally, and which permits calculation of the fugacities of the gaseous constituents of a mixture from its composition without performing difficult density measurements. Alternatively, various algorithms have been suggested to calculate fugacity. Further discussion of this topic is found in Chapter 16.

Fugacities are of practical importance only in gaseous and vapor phases, and hence also for phases equilibrated with them. This is obviously because the measurement methods involve measuring gas densities. Nevertheless, equation (11.17) makes clear that the meaning of the fugacity is not restricted to gases. From (11.17) (combined with (11.7)) we can say that any constituent i that can be assigned a partial molar free energy (μ_i) likewise has a fugacity, and in Chapter 14 we will conclude that this means in fact any constituent in any system.

For liquids and solids the fugacity of the constituents in the condensed phase would evidently be approximately equal to their partial pressures in the vapor phase, since the pressure on the vapor is quite low and in the range where $f \approx P$. Alternatively, any individual constituent of a condensed phase may be imagined as equilibrated with a vapor phase of the pure constituent through a membrane permeable only to that constituent, as in Figure 11.1b. For example, the fugacity of constituent AlSi in albite would equal the partial pressure of AlSi species in the vapor equilibrated with albite. Fugacities of solid phases or of individual constituents in solid solutions are not generally known, but it will be useful to recall that they can be thought of as vapor pressures, either with or without membranes as above.

Nevertheless, while it may often be useful to think of fugacities as partial or vapor pressures because in many systems they are approximately equal, in many other systems they are not, or there may be no vapor pressure at all. Therefore it is best to realize that the fugacity is a parameter in the thermodynamic model, not always an approximation to some real pressure. This idea is taken up again in connection with the oxygen fugacity (§18.5.1).

11.7. FUGACITIES: A SUMMARY

The fugacity is a property of systems and of system constituents that was invented in order to facilitate the evaluation of $\int dG$ for gaseous constituents. However, the fact that its only practical use is for gases does not change the fact that in principle it is a property of *all* system constituents. Each constituent fugacity is therefore a system property or state variable, whether measurable or not. (Our insistence on this point is connected with our development of the concept of activities).

Fugacity is often described as an "escaping tendency," but in most solid and liquid systems it is more intuitively thought of as a vapor pressure. In gaseous systems it can be thought of as (approximately) a partial pressure. Its dimensions are the dimensions of pressure, and it is most often measured in bars or pascals. In spite of its appearing as a ratio in one of its equations of definition (11.6), it is nevertheless an absolute quantity for a given constituent in a given system, as is G, thanks to the other equation

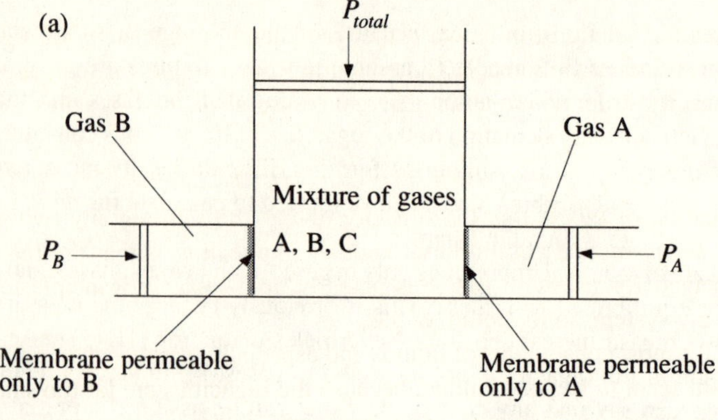

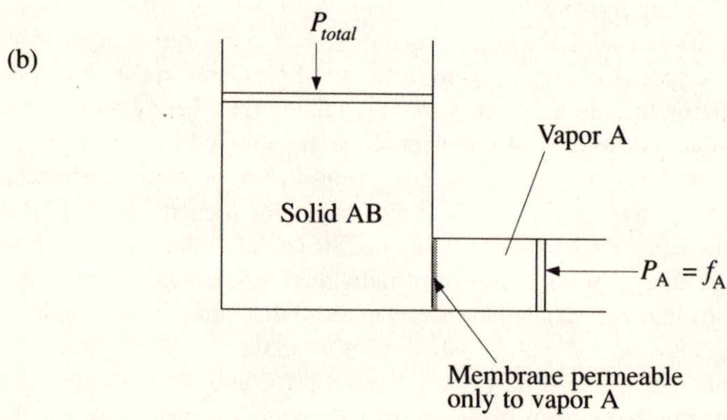

FIG. 11.1. (a) Mixture of gases having total pressure P_t. Membranes are permeable to only one gas, allowing equilibration of the gas across the membrane. (b) Solid AB at pressure P_t. Membrane is permeable only to vapor A. P_A is so low that $P_A = f_A$.

of definition, (11.7). In other words, fugacities do not have standard states in the way that activities do. They are absolute system properties.

11.8. ACTIVITY

Let us summarize our development of the concept of the fugacity, f. Starting with the definition

$$dG = RT\,d\ln f \tag{11.5}$$

and

$$\lim_{P\to 0}(f/P) = 1 \tag{11.7}$$

we found

$$G_{P_2} - G_{P_1} = RT\ln(f_{P_2}/f_{P_1}) \tag{11.16}$$

which expresses the relationship between the free energy and fugacity of a gas at two different pressures at the same T. Generalizing from a single gas to a gas i in a mixture of gases, and from two states at different pressures to any two states $'$ and $''$, this becomes

$$\mu_i'' - \mu_i' = RT \ln(f_i''/f_i') \tag{11.17}$$

One implication of this is that the fugacity of i is the same in any two states or phases that are in mutual equilibrium. Another implication is that the fugacity, although conceived for use with gases, is actually applicable in principle to any constituent of any system, solid, liquid, or gas.

Of course, for most constituents of liquids and especially of solids the usefulness of (11.17) would seem to be very limited, because the fugacities are far too small to measure, and are unknown. However, even in systems where constituent fugacities are unknown, the *ratio* of a constituent fugacity to its fugacity in some other state is quite often a measurable and useful quantity called the *activity*.

Rewriting (11.17) so that state $''$ is any (unsuperscripted) state and state $'$ is a reference state designated by superscript $^\circ$, we have

$$\mu_i - \mu_i^\circ = RT \ln(f_i/f_i^\circ) \tag{11.18}$$

It is important to see that this is a simple generalization of (11.16), and hence a direct result of the definition of fugacity. We now define the activity of constituent i as

$$a_i = f_i/f_i^\circ \tag{11.19}$$

where f_i and f_i° are the fugacities of i in the particular solution or state of interest to us and in some reference state at the same temperature. Thus

$$\mu_i - \mu_i^\circ = RT \ln a_i \tag{11.20}$$

The nature of this reference state will be the subject of considerable discussion in Chapter 12. The fact that activity is defined as a ratio of two quantities, one of which refers to an arbitrary reference state and both of which, in many cases, are unknown, leads to many sources of confusion in using it. A further difficulty is that the same quantity is also defined in a completely different way for dilute solutions, often leaving the erroneous impression that activities come in several varieties, depending on what kind of system one is dealing with. We will take some pains to show that there is only one activity concept, but that it is applied differently to different types of systems, i.e., only the choice of reference state changes. The flexibility of the activity concept is both an indicator of its usefulness and a source of confusion.

11.9. ACTIVITY AND MOLE FRACTION

In this section we will see how the activity, arrived at in such a seemingly abstruse way, turns out to be equal to the mole fraction in ideal solutions, so that

$$\mu_i - \mu_i^\circ = RT \ln X_i \tag{11.21}$$

and how varying the choice of reference state allows convenient application to many types of systems. We do this by examining several laws relating to ideal solutions.

11.9.1. *Dalton's Law*

The simplest imaginable system other than a vacuum is undoubtedly an ideal gas. One mole of ideal gas occupies 22.71 l at 0°C, 1 atm, so that (from the ideal gas law) one mole of ideal gas occupying one liter at 0°C would have a pressure of 22.71 bars (Figure 11.2a). It was an early discovery (Dalton, 1811) that mixtures of gases would exert a pressure equal to the sum of the pressures that each of the constituent gases would have if each alone occupied the same volume. This was established using gases at relatively low pressures where they behave close to ideally, and in fact it is only strictly true for mixtures of ideal gases, which are also then ideal gases. Thus for each constituent gas 1,2,3, etc.

$$P_1\mathrm{V} = n_1RT$$

$$P_2\mathrm{V} = n_2RT$$

$$etc\ldots$$

and for the gas mixture

$$P_{total}\mathrm{V} = \sum_i n_iRT$$

Thus

$$\frac{P_1}{P_{total}} = \frac{n_1}{\sum_i n_i} = X_1$$

$$\frac{P_2}{P_{total}} = \frac{n_2}{\sum_i n_i} = X_2$$

$$etc\ldots$$

or,

$$\left.\begin{array}{lcl} P_1 & = & X_1 \cdot P_{total} \\ P_2 & = & X_2 \cdot P_{total} \\ \vdots & & \vdots \\ etc. & & \end{array}\right\} \tag{11.22}$$

P_1, P_2, etc. are called the *partial pressures* of the solution constituents and equations (11.22) are now normally used as the definition of partial pressure even though in real, non-ideal solutions they give a quantity that is not equivalent to the original meaning, i.e., the pressure a gas would exert if it alone occupied the total volume. If a membrane were available which was permeable to only one of the gases in the system (Figure 11.2b) so that we could measure the pressure in a pure gas that is at equilibrium through the membrane with the same gas in the mixture, we would find that for ideal gases the pressure on the pure gas would be the same as the partial pressure defined above. If a membrane were available for each of the ideal gas solution constituents, the total of the pure gas pressures would equal the pressure on the

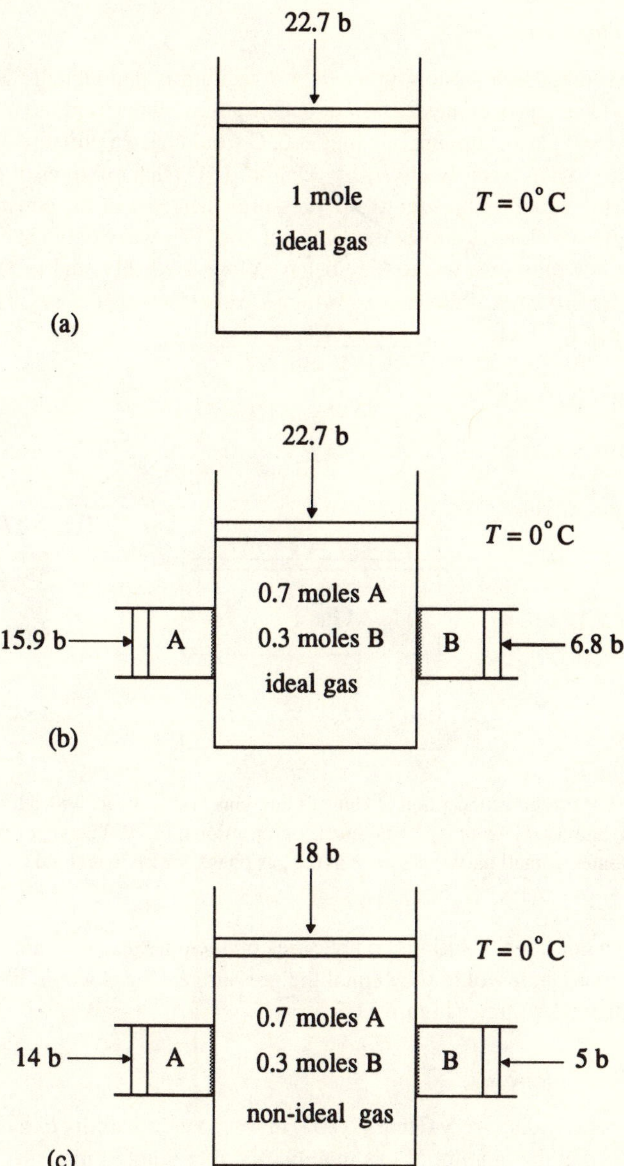

FIG. 11.2. (a) One mole of an ideal gas at 0°C has a pressure of 22.7 bars. (b) An ideal gas solution. 0.7 moles of gas A has a partial pressure of $0.7 \times 22.7 = 15.9$ bars. 0.3 moles of gas B has a partial pressure of $0.3 \times 22.7 = 6.8$ bars. The partial pressure of each gas is the same as the pressure on the pure gas at equilibrium with the solution through a membrane permeable only to that gas. (c) A non-ideal gas solution ($PV \neq nRT$). The partial pressures are defined as $0.7 \times 18 = 12.6$ and $0.3 \times 18 = 5.4$, but these pressures are not observed on the pure gases at equilibrium with the solution through the membranes, and the membrane pressures do not add up to the total pressure.

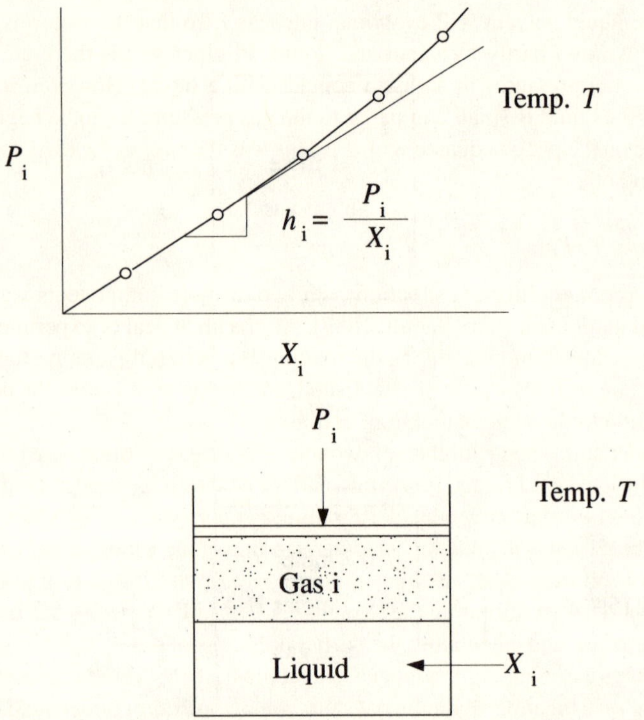

FIG. 11.3. The original conception of Henry's law. Gas i is in contact with a liquid of another composition, and dissolves in it. X_i is directly proportional to P_i. The vapor pressure of the liquid contributes a small partial pressure in the gas phase, which is ignored.

solution. In a non-ideal solution the pressures on the pure gases would not add up to the total pressure, nor would they equal the pressure each gas would have if it alone occupied the total volume (Figure 11.2c).

11.9.2. *Henry's Law*

Another very early discovery (Henry, 1803) involved the solubility of gases in liquids. It was found that the amount of gas that dissolved in a liquid in contact with it was directly proportional to the pressure on the gas (Figure 11.3). Thus

$$P_i = h_i \cdot X_i \tag{11.23}$$

where P_i is the pressure on the gas i, X_i is the mole fraction of i in the liquid, and h_i is a constant (the Henry's Law constant) which varies with temperature and with the nature of the gas i and the solvent. As shown in Figure 11.3 the slope of the solubility curve is a constant at sufficiently low values of P_i and X_i but is usually found to deviate to higher or lower values as P_i and X_i get larger. In this experimental situation, it will be noted that the total pressure is not strictly speaking P_i, because some of the liquid solvent will evaporate into gas i, so that the piston is supported partly by gas i and partly by vaporized liquid. In other words there are always at least two partial pressures in a gas in contact with a liquid. However, if the vapor pressure of the liquid is small compared to the gas pressure, it can be neglected, and the pressure on the piston equated with P_i. This was the case in the early experiments of Henry and others.

11.9.3. *Raoult's Law*

The liquid–vapor equilibrium situation with two or more components was not fully investigated until much later (Raoult, 1887). After a great deal of experimental work, which extended well into the 1900s, the relationship between vapor partial pressures and liquid compositions was well established, and forms what is now the best avenue of introduction to the understanding of activities.

The reader at this stage might well wonder why, if gas solutions are fairly simply treated by the ideal gas law and fugacities, and the next step is to understand liquid and solid solutions, we start now considering liquid–vapor systems. Why not just liquid systems? The reason is that having an understanding of the vapor phase, which at low pressures we can treat as an ideal gas mixture, we can use it as a tool to investigate the liquid phase by equilibrating the two phases. Thus by measuring the composition of the vapor phase and calculating the partial pressures of the components, we are in fact measuring the escaping tendencies (fugacities) of the liquid components. This technique of investigating a complex system, which you don't understand, by equilibrating it with a simpler system, which you do understand, is now a classic technique that has found wide use in chemical and geochemical experimentation.

This work was done by mixing two or more liquid components in known proportions, then equilibrating the liquid with its own vapor and measuring the composition of the vapor. Because the total vapor pressure (also measured) was relatively low (generally well below one atmosphere) the vapor behaved as an ideal gas solution, and the partial pressures of the components could be calculated from their compositions using Dalton's Law. Of the many systems investigated, a very few were found to exhibit a particularly simple relationship between the vapor partial pressures and the liquid composition. In these systems, for all compositions, the partial pressures of the gas constituents were found to be a linear function of their mole fractions in the liquid. That is, in the binary system A–B,

$$P_{\mathrm{A}} = X_{\mathrm{A}}^{liquid} \cdot P_{\mathrm{A}}^{\circ}$$

$$P_{\mathrm{B}} = X_{\mathrm{B}}^{liquid} \cdot P_{\mathrm{B}}^{\circ} \qquad (11.24)$$

where P_i° is the vapor pressure of component i in equilibrium with pure component i. A pressure–mole fraction of liquid diagram can therefore be drawn with straight lines as in Figure 11.4b. The experimental situation for one particular composition is shown in Figure 11.4a. The relationship between liquid and vapor compositions that results from these relations is shown in Figure 11.4c. For example, a liquid having

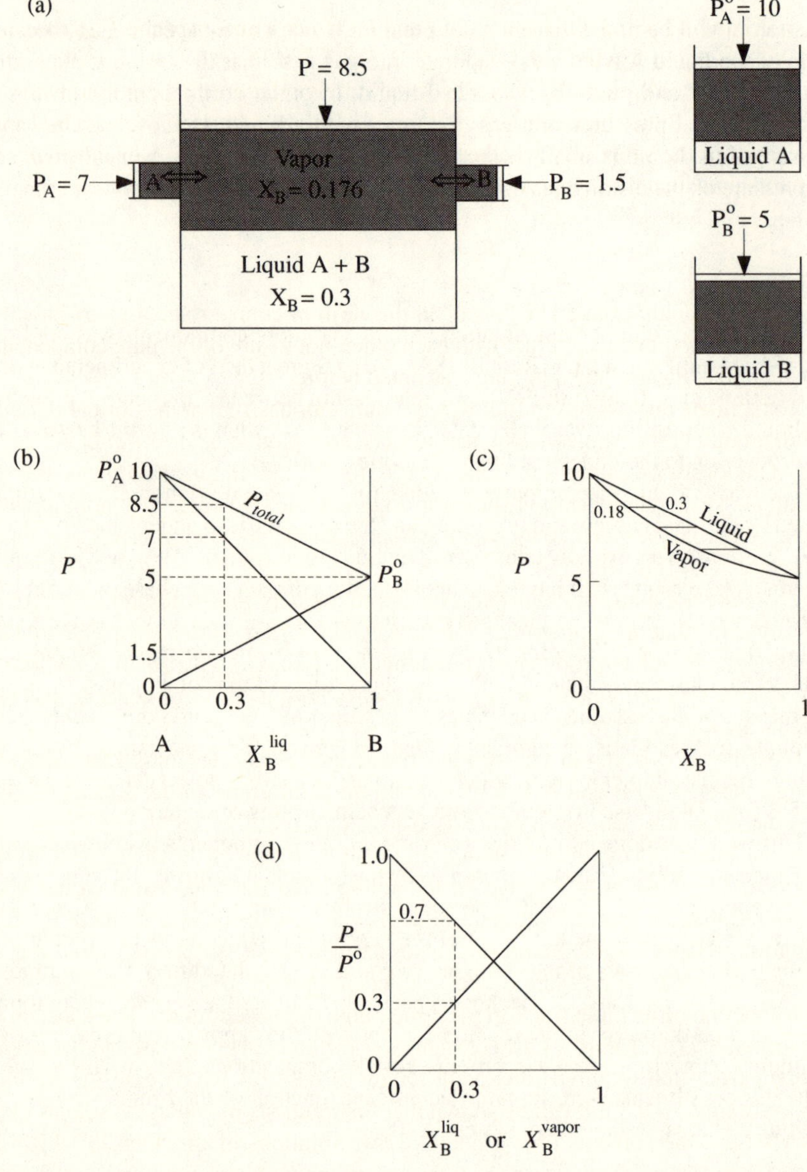

FIG. 11.4. (a) Ideal liquid solution. Pure component A has a vapor pressure of 10 (units are not important—could be millibars). Pure component B has a vapor pressure of 5. The vapor phase above the liquid consists only of the same components, A and B. The pressures on the pure gases through the hypothetical membranes would equal the partial pressures. (b) Observed total pressure and calculated partial pressures in the vapor phase. Pressures observed in Figure 11.4(a) are indicated. (c) Compositions of coexisting liquid and vapor phases. (d) Normalized total pressures and partial pressures. Figure 11.4(b) is converted to this diagram by dividing the partial pressures of each gas by the vapor pressure of the pure gas. The mole fraction axis can now refer to the vapor or to the liquid.

$X_A = 0.7$ and $X_B = 0.3$ will have $P_A = 7$ and $P_B = 1.5$, a total pressure $P_A + P_B = 8.5$, and $X_A^{vapor} = 7/8.5 = 0.82$ and $X_B^{vapor} = 1.5/8.5 = 0.18$.

As mentioned earlier, the only way these simple relationships can hold is for the intermolecular forces between A-A, B-B and A-B to be identical, so that a molecule A behaves in the same way whether it is surrounded mostly by A or mostly by B. Solutions in which this happens are called *ideal solutions* , and the relation

$$P_i = X_i^{liquid} \cdot P_i^{\circ} \tag{11.25}$$

is a form of Raoult's Law. The "ideal" in this term of course refers to way in which the solution components interact with each other, not to any of the pure components. "Ideal" in fact means "ideal mixing", as noted before.

To facilitate discussions to follow, we should emphasize that the normal Raoult's law diagram (11.4b) shows *vapor* partial pressures plotted against *liquid* compositions. Obviously we cannot substitute X_B^{vapor} for X_B^{liquid} in this diagram as they are not equal. Their relationship is shown in Figure 11.4c. However by "normalizing" the vapor partial pressures to the vapor pressures of the pure components, we obtain a diagram (11.4d) in which we *can* use X_B^{liquid} *or* X_B^{vapor}.

There are not many systems that even approximately follow Raoult's Law. Even those systems that do approximately follow Raoult's Law such as benzene–toluene and ethylene bromide–propylene bromide will be found to have small deviations if very accurate measurements are made. Raoult's Law is an ideal concept that real systems are compared to.

Most liquid systems in fact deviate considerably from Raoult's Law behavior because the A-A, B-B and A-B forces are all quite different. In these cases (Figure 11.5) the deviations can be in either a positive or a negative sense, but a region of particular interest is the dilute solution region at each side of the diagram. In dilute solutions, i.e., those at low values of the solute mole fraction, it had earlier been discovered that the (partial) pressure of a gas was proportional to its mole fraction in solution (Henry's Law, Figure 11.3), so that it was not surprising to find that in most cases the partial pressure of the component present in smaller quantity was a linear function of its mole fraction for some range of concentration. The constant of proportionality (the Henry's Law constant) was different in every case, and of course different from that required for Raoult's Law to hold (in which case the Henry's Law constant $h_i = P_i^{\circ}$). It can be shown that for solutions in which the solutes obey Henry's Law, the solvents obey Raoult's Law. Solutions of this kind in which all solutes obey Henry's Law (another ideal concept) are unfortunately also called Ideal Solutions, although they are obviously different from solutions in which solutes and solvents obey Raoult's Law, and we must be explicit in each case as to whether we refer to "Raoultian" or "Henryan" ideality.

11.9.4. *The Lewis Fugacity Rule*

A useful and widely-used variation of Raoult's Law is obtained by expressing it in terms of fugacities instead of pressures. Thus in some homogeneous (one phase)

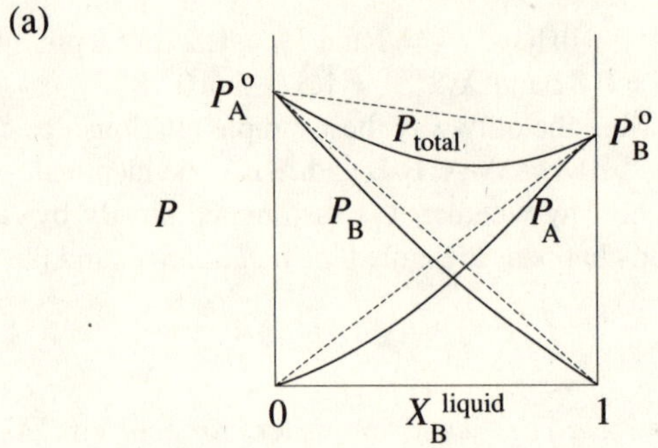

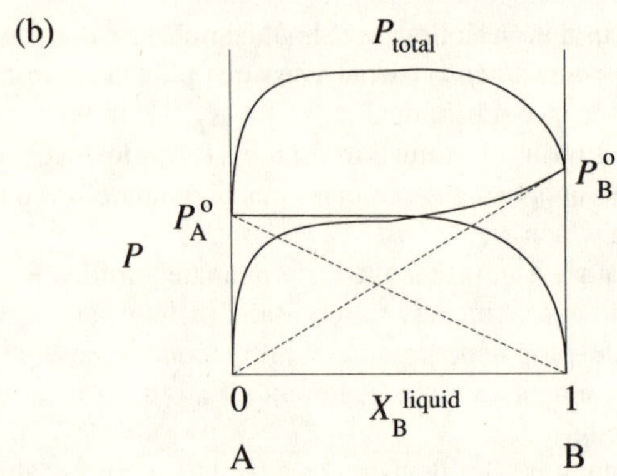

FIG. 11.5. Total vapor pressures and partial pressures in liquid-vapor systems showing (a) negative deviation and (b) positive deviation from Raoult's law.

systems it is found that

$$f_i = X_i \cdot f_i^{pure} \tag{11.26}$$

where f_i and f_i^{pure} are the fugacity of i in the mixture having mole fraction X_i, and the fugacity of pure i respectively. A binary gas system in which both components obey this relationship is shown in Figure 11.6.

As shown by Prausnitz (1969), this relation follows from "Amagat's Rule," which describes the mixing situation where the volume of the solution is the same as the total volume of the individual pure components. In this case the ΔV of mixing is zero, and the partial molar volume $\bar{V}_i$ is equal to the molar volume of pure i, V_i^{pure}, as discussed in Chapter 9. This can be demonstrated by first rewriting equation (11.9) for a pure component i as

$$\ln f_i^{pure} - \ln P = \frac{-1}{RT}\int_{P\to 0}^{P}\left(\frac{RT}{P} - V_i^{pure}\right) dP \tag{11.27}$$

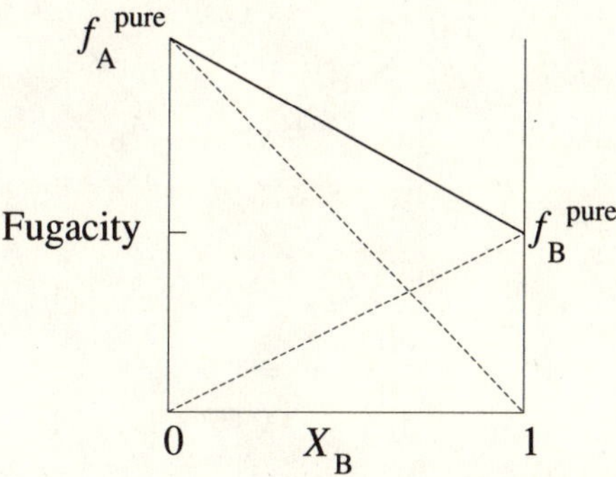

FIG. 11.6. A binary system (solution) that obeys the Lewis Fugacity Rule. The solution can be liquid, solid, gas, or supercritical fluid, although generally this rule is used for the latter two.

and the equivalent equation for component i in a gas solution

$$\ln f_i^{mixture} - \ln(X_i P) = \frac{-1}{RT} \int_{P \to 0}^{P} \left(\frac{RT}{P} - \bar{V}_i \right) dP \tag{11.28}$$

Comparing (11.27) and (11.28) we see that if $\bar{V}_i = V_i^{pure}$, then

$$\frac{f_i^{pure}}{P} = \frac{f_i^{mixture}}{X_i P}$$

and

$$f_i^{mixture} = X_i \cdot f_i^{pure} \tag{11.26}$$

This relationship, the Lewis Fugacity Rule, is a kind of variation of Dalton's Law, and has been widely used to estimate fugacities in gas mixtures (Prausnitz, 1969).

An interesting geological example of this type of solution was demonstrated by Barnes and Ernst (1964) who showed that NaOH in water at 2 Kb and high temperatures follows the Lewis fugacity rule to a rough approximation. Neither NaOH nor H_2O is remotely like an ideal gas under these conditions, but the largely undissociated NaOH dipole and the water dipole have sufficiently little interaction that their volumes change little on mixing. It is likely that many solutes behave approximately in this way at these pressures and temperatures.

Although the Lewis Fugacity Rule is generally used for gas or supercritical solutions, it is particularly interesting to see what results when a condensed phase in equilibrium with such a phase is considered, just as we first considered Dalton's Law and then a condensed phase in equilibrium with a solution obeying Dalton's Law.

11.9.5. *Ideal Mixing and Activity*

Comparison of equations (11.19) and (11.26) shows that for solutions that obey the Lewis Fugacity Rule

$$\begin{aligned} a_i &= f_i / f_i^{\circ} \\ &= X_i \end{aligned} \tag{11.29}$$

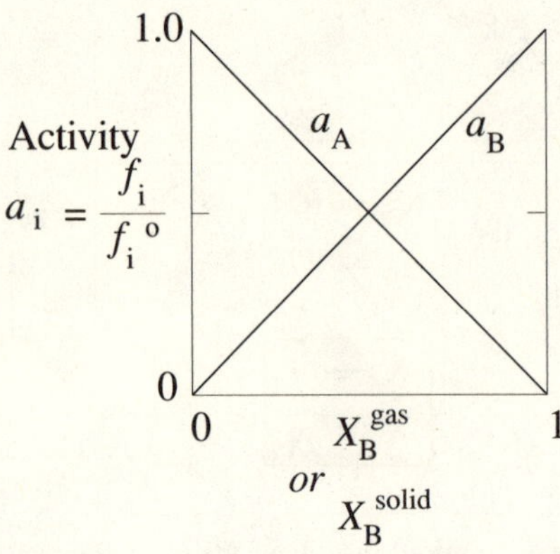

FIG. 11.7. Normalized fugacities in a binary system obeying the Lewis Fugacity Rule. Figure 11.6 is converted to this diagram by dividing the fugacity of each component by the fugacity of the pure component. The mole fraction axis can now be used for the original system or for a phase equilibrated with it.

as long as we let $f_i^\circ = f_i^{pure}$ which is to say that we let pure i at the temperature of the system be the reference state for i, which is the usual case. This is illustrated in Figure 11.7. Combining this with (11.20) we get

$$\mu_i - \mu_i^\circ = RT \ln X_i \tag{11.30}$$

This "normalized" fugacity relationship, where each fugacity is divided by the fugacity of the pure component, can now be used for condensed phases in equilibrium with a gas, just as we saw when considering Raoult's Law (Figure 11.4d), and in fact equation (11.29) is considered another form of Raoult's law. Perhaps surprisingly, a number of crystalline solution systems can be adequately described by equation (11.29); that is, they exhibit ideal mixing. Figure 11.8 shows a way of visualizing this situation.[1]

Similarly, crystalline solutions that show deviations from Raoult's Law (Figure 11.9) will also show adherence to Henry's Law in the dilute solution region where the measured activities are proportional to but not equal to the mole fraction.

To summarize, the laws of Raoult and Henry, originally established for liquid-vapor and liquid-gas systems, give rise to useful generalities for all solutions, gas, liquid, or solid, when restated in terms of fugacities and activities. If fugacities are available, the activity concept is unnecessary. Activity becomes useful in systems where fugacities are unknown, as it can be measured in other ways. Nevertheless,

[1] A crystalline solution in which $a_i = X_i$ can be considered to be following Raoult's Law in its original sense, and not merely the Lewis fugacity rule, because the vapor pressures of solids are so low that the partial pressures of constituents in the vapor phase would undoubtedly equal their fugacities. Where the vapor phase consists entirely of particles having the same composition as the condensed phase (with no complications due to polymerization, dissociation), the fugacity of the condensed phase would equal the vapor pressure itself. This is the situation assumed in Figure 11.8 for pure crystalline A and B.

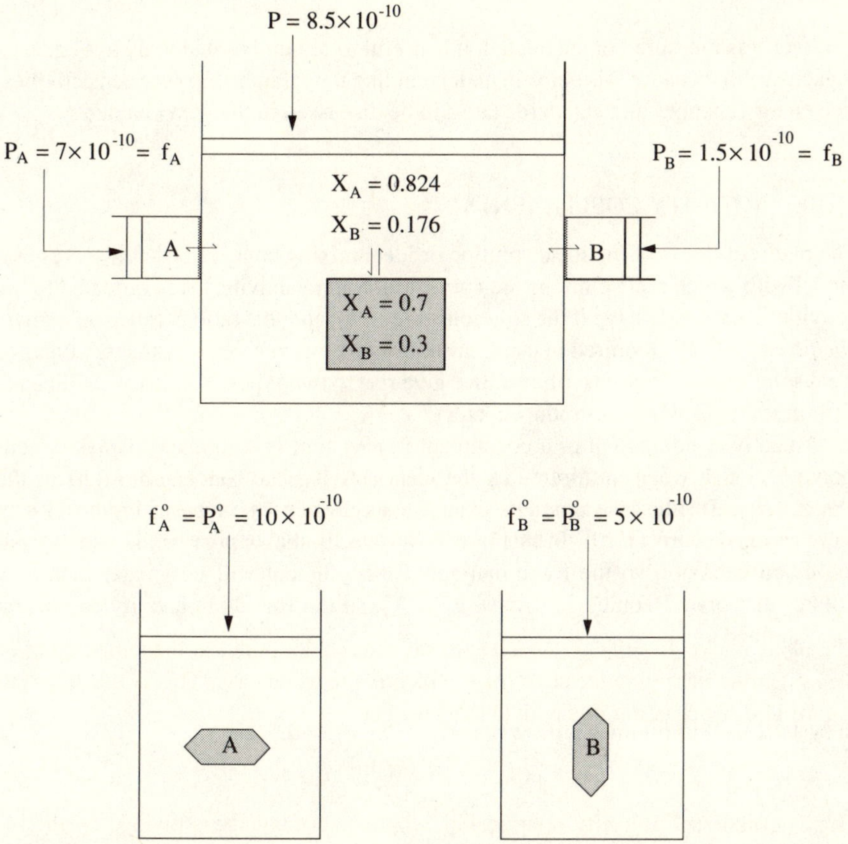

$P = 8.5\times 10^{-10}$

$P_A = 7\times 10^{-10} = f_A$

$P_B = 1.5\times 10^{-10} = f_B$

$X_A = 0.824$

$X_B = 0.176$

A

B

$X_A = 0.7$

$X_B = 0.3$

$f_A^{\circ} = P_A^{\circ} = 10\times 10^{-10}$

$f_B^{\circ} = P_B^{\circ} = 5\times 10^{-10}$

A

B

FIG. 11.8. Ideal crystalline solution in equilibrium with its own vapor. Pure crystal A has a fugacity (assumed equal to its vapor pressure) of $10 \cdot 10^{-10}$ (units are not important, could be bars), and pure crystal B has a fugacity of $5 \cdot 10^{-10}$. The diagram is therefore similar in all respects to Figure 11.4a.

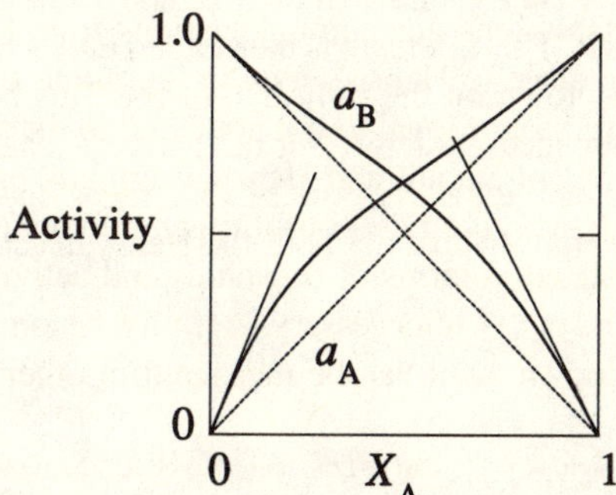

FIG. 11.9. Activities in a non-ideal crystalline binary solution. Activities become proportional to mole fraction in the dilute region, in conformity with Henry's law.

however it is measured or calculated, it is useful to remember that it always equals a fugacity ratio, because this helps in understanding the relationship between activities, free energy changes and standard states, to be discussed in the next chapter.

11.10. ACTIVITY COEFFICIENTS

The real usefulness of the ideal solution or ideal mixing concept is that it serves as a model with which real solutions are compared. Solute activities are compared to the activities they would have if the solution were ideal, and this ratio is called an *activity coefficient*. Care is required in using this number, however, because the two types of ideal solution behavior described above give rise to two types of activity coefficients (still more types will be introduced later).

An activity coefficient of a constituent in a system is a number (always dimensionless) which when multiplied by the ideal activity (the concentration) gives the real activity. To illustrate this let's examine a system of the type in Figure 11.9 in more detail (Figure 11.10). In this binary system, B shows positive deviation from Raoultian behavior, so the Raoultian activity coefficient will be greater than one. Solutes that obey Raoult's Law have $a_i = X_i$, so the Raoultian activity coefficient γ_R is defined as

$$a_i = X_i\gamma_R$$

In system A–B in Figure 11.10, when $X_B = 0.3$, $a_B = 0.5$, so

$$\begin{aligned}\gamma_R &= 0.5/0.3 \\ &= 1.67\end{aligned}$$

Note that the activity of B (0.5) is measured on the scale defined by Raoultian behavior, on the left-hand axis.

On the other hand B shows negative deviation from Henry's Law, and will therefore have a Henryan activity coefficient less than one. A Henryan scale of activities is created by extending the Henry's Law slope defined by dilute solutions of B in A right over to the pure B axis. This intercept is then called unit activity, and this defines a new scale of activities shown on the right-hand axis. This new Henryan activity scale also has $a_i = X_i$ for ideal systems, but the activity scale is different, so that now the real or measured activity at $X_B = 0.3$ is 0.2, measured on the scale defined by Henryan behavior, the right-hand axis. Thus the Henryan activity coefficient, γ_H, defined as

$$a_i = X_i\gamma_H$$

is in this case

$$\begin{aligned}\gamma_H &= 0.2/0.3 \\ &= 0.67\end{aligned}$$

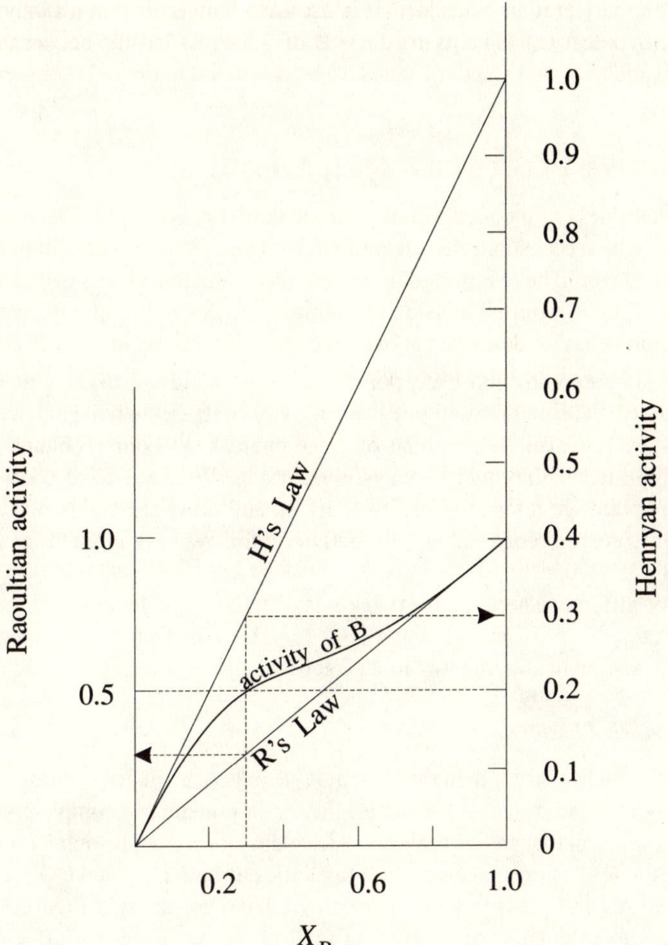

FIG. 11.10. The behavior of component B of Figure 11.9 in more detail, showing the difference between the Raoultian and Henryan activity scales.

If pure B has a fugacity of $5 \cdot 10^{-10}$ bars, as shown in Figure 11.8, this defines f_B° for Raoultian behavior. Thus on the Raoultian scale

$$
\begin{aligned}
a_{\mathrm{B}} &= f_{\mathrm{B}}/f_{\mathrm{B}}^\circ \\
&= f_{\mathrm{B}}/(5 \cdot 10^{-10}) \\
&= 0.5
\end{aligned}
$$

so that at $X_{\mathrm{B}} = 0.3$

$$f_{\mathrm{B}} = 2.5 \cdot 10^{-10} \text{ bars}$$

However on the Henryan scale, which is 2.5 times longer than the Raoultian scale, we have a hypothetical fugacity for pure B of $2.5 \times 5 \cdot 10^{-10}$ bars, so that on the Henryan scale

$$\begin{aligned} a_B &= f_B/f_B^\circ \\ &= f_B/(11.5 \cdot 10^{-10}) \\ &= 0.2 \end{aligned}$$

so that

$$f_B = 2.5 \cdot 10^{-10} \text{ bars}$$

as before.

Obviously we could choose any point on the right-hand axis to call unit activity, to fix our activity scale, and the activities and activity coefficients will vary accordingly. Fugacities are however independent of these choices, and our problems would be greatly simplified if they were always measurable. What we have been doing, in moving the point we refer to as unit activity up and down the right-hand axis, is in fact changing the reference state for the activity term. We have more to say about this in Chapter 12.

A major difference between the two kinds of activity coefficients which should be noted is that $\gamma_R \rightarrow 1$ as $X \rightarrow 1$, but $\gamma_H \rightarrow 1$ as $X \rightarrow 0$. Thus γ_H is generally more useful for constituents having low mole fractions, that is, solutes in dilute solutions.

11.10.1. *A Real Example*

Somewhat more important than the system A–B in geochemistry is the system CO_2–H_2O. At high temperatures and pressures these components are completely miscible, and form a solution that shows a small positive deviation from Raoult's Law. Bowers and Helgeson (1983) present calculated fugacities of both H_2O and CO_2 at a number of temperatures and pressures, from which we have constructed Figure 11.11. The data points are calculated from the data in Bowers and Helgeson (1983), and are shown in Table 11.1. The lines are fitted to the data points by the Margules equations, discussed in Chapter 15. The Henry's Law slope is inconveniently steep in this case, so we do not show its intersection with the pure CO_2 axis. Very commonly this is not required anyway, because in aqueous solution work, the region of interest is the lower left-hand corner, i.e., the dilute-solution region. This topic requires more discussion, which we present in the next chapter.

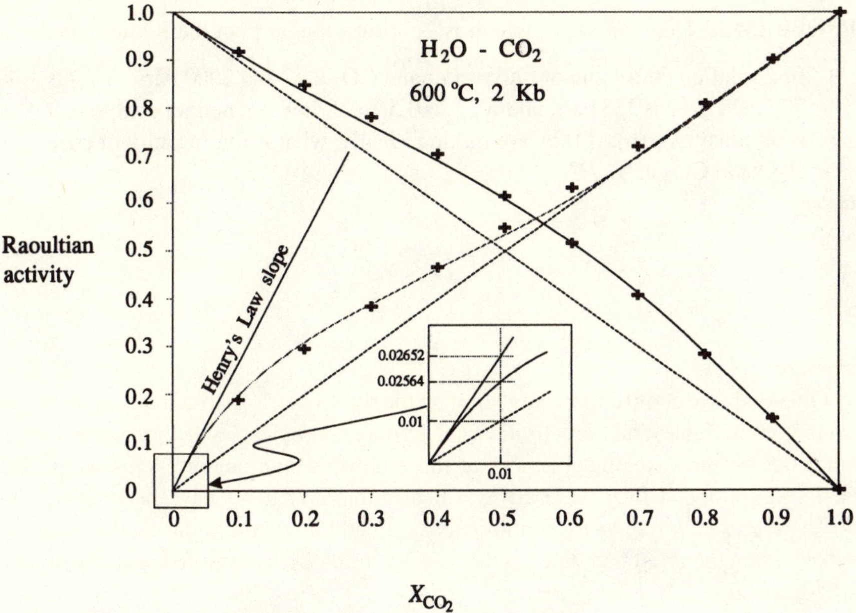

FIG. 11.11. Raoultian activities of H_2O and CO_2 in the binary solution at 600°Cand 2 kb. Data from Bowers and Helgeson (1983). The curved lines are fit to the data with Margules equations, discussed in Chapter 15. The inset refers to a discussion of standard states in Chapter 12.

Table 11.1 Fugacities and activities of H_2O and CO_2 in solution at 600°C, 2kb. Fugacities from Bowers and Helgeson (1983).

X_{CO_2}	P_{CO_2} bars	γ_{f,CO_2}	f_{CO_2} bars	a_{CO_2}	γ_{R,CO_2}	γ_{f,H_2O}	f_{H_2O} bars	a_{H_2O}	γ_{R,H_2O}
0.0	0.0	5.026	0.0	0.0	2.791	0.474	948.0	1.0	1.0
0.1	200	3.371	674.2	0.1872	1.872	0.483	869.4	0.917	1.019
0.2	400	2.664	1065.6	0.2958	1.479	0.503	804.8	0.849	1.061
0.3	600	2.303	1381.8	0.3836	1.279	0.528	739.2	0.780	1.114
0.4	800	2.097	1677.6	0.4657	1.164	0.555	666.0	0.703	1.171
0.5	1000	1.973	1973.0	0.5478	1.096	0.583	583.0	0.616	1.230
0.6	1200	1.896	2275.2	0.6316	1.053	0.612	489.6	0.516	1.291
0.7	1400	1.848	2587.2	0.7183	1.026	0.641	384.6	0.406	1.352
0.8	1600	1.820	2912.0	0.8084	1.011	0.671	268.4	0.283	1.416
0.9	1800	1.805	3249.0	0.9020	1.002	0.702	140.4	0.148	1.481
1.0	2000	1.801	3602.0	1.0000	1.000	0.732	0.0	0.000	1.544

PROBLEM

1. In a solution consisting of only water and CO_2 at T and 2000 bars., f_{H_2O} is 527 bars, f_{CO_2} is 333 bars, and X_{H_2O} is 0.5. (a). Show that neither component is an ideal gas. (b). If they are mixing ideally, what is the fugacity of pure H_2O and CO_2 at T, P?

12

STANDARD STATES

Thermodynamics is not difficult if you can just keep track of what it is you are talking about.

W. F. Giauque, quoted in Bent (1972).

12.1. USEFULNESS OF THE ACTIVITY

At this point we have introduced the activity as a ratio of fugacities (Chapter 11). The fugacity of a constituent, in turn, we saw was a quantity very much like a vapor pressure or partial pressure, which is directly linked to the Gibbs free energy of that constituent, such that a ratio of fugacities leads directly to a difference in free energies. The fugacity was introduced as a means of dealing with gases and gaseous solutions, and it is measured by measuring gas volumes or densities. Nevertheless, there is nothing restricting its use to gaseous constituents, and we suggested that it is very useful to regard the fugacity as a state variable; as a property of any constituent of any system, solid, liquid, or gas, whether equilibrated with a gas or not, and whether measurable or not.

This leads to the easiest approach to understanding activities. The activity of a constituent is the ratio of the fugacity of that constituent to its fugacity in some other state, which we called a reference state. We then showed through consideration of the Lewis Fugacity Rule, which is an extension of Dalton's Law, that for ideal solutions of condensed phases, the activity of a constituent equals its mole fraction, if the reference state is the pure constituent at the same P and T. Deviations from ideal behaviour are then conveniently handled by introducing Henryan and Raoultian activity coefficients.

The utility of these relations would be quite sufficient for retaining the activity in our collection of thermodynamic parameters, but in fact the activity can be applied to a much wider range of conditions, simply by varying the choice of reference state. We now examine the various possible choices of this reference state, and the resulting equations and applications.

In the most general sense, the fugacity and activity concepts satisfy the need to relate system compositions to free energy changes. That a single parameter, the activity, can do this for essentially any system is a tribute to its tremendous versatility.

12.2. ACTIVITIES AND STANDARD STATES

In deriving equation (11.18),

$$\mu_i - \mu_i^\circ = RT\ln(f_i/f_i^\circ) \tag{12.1}$$

we said simply that superscript $^\circ$ designated a reference state, the nature of which would later be examined in some detail. It is now time to do this.

The only mathematical restriction on μ_i° and f_i° in equation (12.1) is that they both refer to the same integration limit, or in physical terms, that they refer to the same equilibrium state. This state has been referred to in various places thus far as a reference state, which it is. We now consider it in more detail, with a more exact definition, and we refer to this more precise concept as a *standard state*. The exact nature of this state is completely a matter of definition, although a few definitions have themselves become "standard" because of their utility. We have used it in discussions of thermodynamic properties such as G_i°, H_i°, etc. to signify that the substance is in its pure state, and we have seen in the two conventions discussed in Chapter 7 that the pressure and temperature of the standard state could be different in different cases. With the introduction of the activity concept, standard states take on added importance because of their use in a wide variety of solutions, temperatures, and pressures, both fixed and variable, and we must now pay more attention to their definition than we have done so far.

Standard states are simply a special sort of reference state for physical properties, made necessary, as we have mentioned several times, by our lack of knowledge of absolute values for the properties U, G, H, and A. Standard states are therefore systems or states of matter under specified conditions. The definition must be sufficiently complete as to determine the thermodynamic parameters of the substance, and therefore must have at least four attributes: 1. temperature 2. pressure 3. composition 4. state of aggregation (solid polymorph, liquid, gas, ideal gas, ideal solution, etc.). Thus "25°C, 1 bar" is *not* a standard state. The question is, what *system* at 25°C, 1 bar?.

Because the goal of the definition is to specify the thermodynamic parameters of the substance, it frequently happens that the standard state chosen is a hypothetical, perhaps physically unrealizable state, because the thermodynamic parameters of such a state are often either well-known or determinable. The importance of these states lies in our knowing their properties, not in being able to actually achieve them. Certain standard states are so commonly used that one need not always elaborate on the definition, i.e., it may be obvious from the context. In other cases it is necessary to be quite specific, and it might even be necessary to specify other factors such as grain size, defect structure, degree of disorder, amount of strain, etc., in order to sufficiently define the system being used as a standard state.

Standard states are either stated or implied in any quantitative discussion or tabulation of free energies, enthalpies, internal energies, or activities, but the following discussion will be based on the use of standard states for activities because of the much wider range of possibilities encountered. Standard states for tabulated free energies, etc. generally do not get any more complicated than the cases already dealt with in Chapter 7.

In the following discussion of standard states we must distinguish between the properties of the standard state (T°, P°, m°, etc.) and those of the state of interest (i.e., the equilibrium state in which the activity of some component or species is of interest to us; T, P, m, etc.) because the two states are often completely different. We will also refer to the state of interest as "the system."

12.3. VARIABLE TEMPERATURE STANDARD STATES

The first thing to note about the standard state as used with activities is that the standard state and the state of interest are virtually always at the same temperature ($T = T^\circ$). Because we are often interested in a series of equilibrium states at different temperatures, we therefore have a corresponding series of standard states, one for each temperature. This can be regarded rather as a single standard state having a variable temperature.

The reason why T must generally equal T° can be seen by considering the integration of

$$dG_i = RT\,d\ln f_i \tag{11.5}$$

or, for solutions,

$$d\mu_i = RT\,d\ln f_i$$

which, if the lower limit of integration is state $^\circ$ and the upper limit is unsuperscripted, gives

$$\int d\mu_i = \int RT\,d\ln f_i \tag{12.2}$$

or

$$\mu_i - \mu_i^\circ = RT\ln(f_i/f_i^\circ) \qquad (11.18), (12.1)$$

During integration, T is held constant, so that f_i and f_i° necessarily refer to states at the same temperature. As temperature is often variable in a set of experimental or theoretical results, this naturally gives rise to a variable temperature standard state. This means that the comparison being made (in this case between μ of i in some state of interest, and μ of i in its standard state) is always between two states having the same T.

But not only are f_i and f_i° necessarily at the same temperature, we must also note that G_i° (or μ_i°) and f_i° are independent of the system pressure. That is to say, they depend on P° but not on P. This can be seen by integrating (12.2) indefinitely rather than between limits as we have done. This gives

$$\mu_i + constant' = RT\ln f_i + constant''$$

where, by comparison with (12.1), we see that both μ_i° and f_i° are constants, and therefore independent of parameters that are allowed to vary during integration, such as the system pressure (P). Once the standard state is chosen, it is a function only of the value of T, held constant during the integration. In physical terms, the value of f_i° and of μ_i°, which are state variables of the standard state, depend on the temperature ($T^\circ = T$), pressure (P°), and composition of the standard state, but they are independent of P, the pressure in other states of interest to us, which may be compared to the standard state.

It might be pointed out too, that there is no logical necessity to be restricted to using equation (12.1) and hence the variable temperature standard state. That is, if for some reason you preferred a fixed temperature standard state, you could modify equation (12.1) to include a term accounting for the variation of μ_i° with temperature. To the best of our knowledge this has never been done, because it adds to the difficulty with no great benefit.

12.3.1. *Standard States Using Fugacities*

We can also see from the same equation why it often proves convenient to choose a standard state for i, which is not only not its most stable state, but one that is extremely hypothetical. If f_i° is set to 1.0, equation (12.1) becomes

$$\mu_i - \mu_i^\circ = RT \ln f_i \tag{12.3}$$

which is a very convenient form of the equation, provided that fugacities of i are available or measurable. The physical significance of setting the denominator to 1.0 is that constituent i is said to be in a state in which the fugacity is 1.0 at all temperatures. The only substance for which this is true is an ideal gas at $P = 1$, so equation (12.3) implies the choice of "ideal gaseous i at one bar and temperature T" as the standard state. This perhaps seems reasonable enough for a gas, but it can be used for any substance including solids and liquids. The only reason it is not universally used is that fugacities (f_i) are not known for many constituents of interest, especially solids[1] and dilute solutes. The idea of using "ideal gas i" as a standard state for solid i or liquid i or a dilute solution of i takes some getting used to, but there is no logical reason not to do it. There are practical reasons of course, as noted.

To put this in other terms, if you have the fugacity of some substance i in some system, then $RT \ln f_i$ is the difference in Gibbs free energy per mole of i in the system at T and i as an ideal gas at T. Whether i could ever come close to existing as an ideal gas is irrelevant. Other examples of hypothetical standard states are discussed below.

12.4. VARIABLE PRESSURE STANDARD STATES

12.4.1. *Standard States Based on Raoult's Law*

To see how standard states having variable pressure as well as variable temperature arise, we first go back to equation (12.1), which for a pure ideal gaseous component 1 having a standard state of "ideal gas at T and 1 bar" (so that $f_i^\circ = 1$ and $f_i = P$) becomes

$$G_1 - G_1^\circ = RT \ln P \tag{12.4}$$

or

$$G_1 - G_1^\circ = RT \ln(RT/V_1) \tag{12.5}$$

where V_1 is the molar volume of component 1. Now

$$V_1 = \mathrm{V}/n_1 \tag{12.6}$$

[1] Interest in the condensation of solids from the solar nebula is rapidly making the fugacity of solids a more relevant topic.

where V is the total volume occupied by n_1 moles of component 1, so

$$G_1 - G_1^\circ = RT\ln(RTn_1/\mathrm{V}) \tag{12.7}$$

If into this volume we now introduce n_2 moles of ideal gaseous component 2, we have seen that this mixture is also an ideal gas, so that now

$$\mathrm{V} = (n_1 + n_2)RT/P \tag{12.8}$$

Substituting for V in equation (12.7) we have

$$\begin{aligned}\mu_1 - G_1^\circ &= RT\ \ln(n_1P/(n_1+n_2)) \\ &= RT\ln(X_1P)\end{aligned} \tag{12.9}$$

or

$$\mu_1 - G_1^* = RT\ln(X_1) \tag{12.10}$$

where

$$G_1^* = G_1^\circ + RT\ln P \tag{12.11}$$

Note that in changing from a pure substance (equation (12.7)) to a solution (equation (12.9)), even though an ideal solution, we change from G_1 to μ_1. We still want to express the Gibbs free energy per mole of component 1, but in solutions this can only be done with μ.

Equation (12.10) can also be used for non-ideal gases as well as for liquid and solid solutions by introducing the Raoultian activity coefficient, thus

$$\mu_1 - G_1^* = RT\ln(X_1\gamma_R) \tag{12.12}$$

or

$$\mu_1 - G_1^* = RT\ln a_1 \tag{12.13}$$

or in general,

$$\mu_i - G_i^* = RT\ln a_i \tag{12.14}$$

or, even more generally,

$$\mu_i - \mu_i^\circ = RT\ln a_i \tag{12.15}$$

which of course becomes

$$\mu_i - \mu_i^\circ = RT\ln X_i \tag{12.16}$$

in the case of the ideal solution.

With the change in form of our equation from (12.1) to (12.14) we should note a corresponding change in the standard state normally used. Because (12.14) can be used for solids and liquids as well as gases, it is clearly more convenient to use as standard state the pure phase ($X_i = 1; \gamma_\mathrm{R} = 1$) at the temperature and pressure of interest. In this case a pure phase will have an activity of one at all temperatures and pressures, which is very convenient in many calculations.

Comparing equations (12.14) and (12.1) we see a distinct similarity, but one important difference. When a_i is used in the form (f_i/f_i°), the standard state free energy is G_i° (or μ_i°) and is independent of P, the pressure on the system. When a_i is converted to X_i , the standard state free energy is G_i^*, which is a function of P (as well as T) as shown by equation (12.11).

Equations (12.10) and (12.14) are examples of expressions that use a *variable pressure standard state*. Although it is theoretically possible to keep P° fixed and to simply add ($RT \ln P$) to the ($\mu - \mu^\circ$) term for each different P considered, in practice one usually considers that the pressure of the standard state (P°) and the pressure on the system or state of interest (P) are the same, so that μ° is a function of the system pressure.

12.4.2. *Standard States Based on Henry's Law*

Because the activities of solutes in dilute solutions can be more closely approximated with Henry's Law than with Raoult's Law, they are traditionally treated separately and use standard states different than those we have so far encountered. There are two variations of usage here, both of the variable pressure type, one required when using mole fractions, and another when using molalities.

Mole Fractions

We derived equation (12.12) from equation (12.10) simply by saying that γ_R could be introduced to take care of deviations from Raoult's Law. Deviations from Henry's Law are similarly accounted for by introducing γ_H instead of γ_R, thus

$$\mu_i - G_i^* = RT \ln(X_i \cdot \gamma_H)$$

or again, more generally,

$$\mu_i - \mu_i^\circ = RT \ln(X_i \cdot \gamma_H) \tag{12.17}$$

Just as we then switched to a standard state having $X_i = 1$ and $\gamma_R = 1$, (the pure phase at T and P), we switch here to one having $X_i = 1$ and $\gamma_H = 1$. But γ_H equals 1 only at infinite dilution of i, so that this standard state is necessarily hypothetical. It is a state, at the T and P of the system of interest, in which pure i has the same properties it has at infinite dilution in the system. On an activity-mole fraction diagram it is represented by the point of intersection of the Henry's Law tangent with the $X_B = 1$ axis. Two possible relationships between the resulting Henryan activities with Raoultian activities are shown in Figures 12.1 and 12.2. This standard state is commonly used in metallurgical and ceramic studies involving solid and liquid solutions.

Molalities

Studies of aqueous solutes traditionally use molalities, which implies use of still another standard state. To develop this subject we will resort to using Euler's theorem, which makes the derivation less than completely intuitive, but basically we are just looking for the relationship between μ and m, the solute molality, in the dilute solution region where Henry's Law is obeyed.

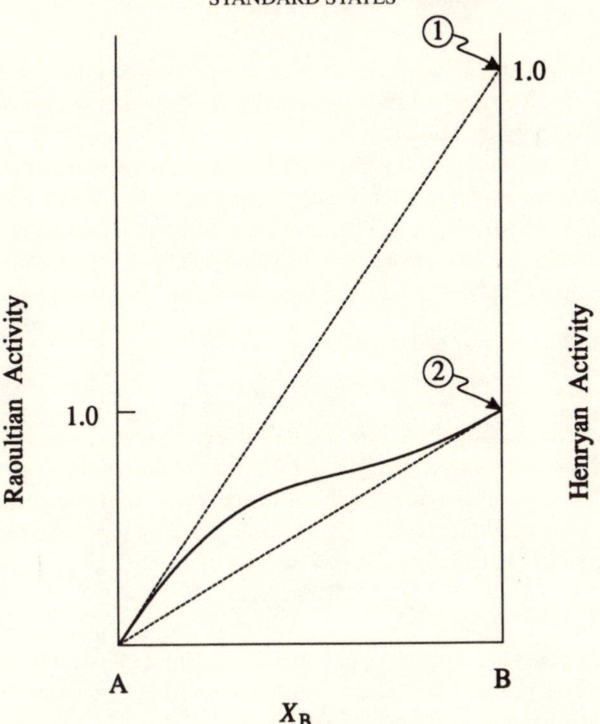

Fig. 12.1. One possible relationship between Henryan and Raoultian activities (essentially the same as Figure 11.10).

Partial differentiation of equation (12.12) with respect to the number of moles of the second component, n_2, gives

$$\left(\frac{\partial \mu_1}{\partial n_2}\right)_{T,P,n_1} = RT\left(\frac{\partial \ln \gamma_R}{\partial n_2}\right)_{T,P,n_1} + RT\left(\frac{\partial \ln X_1}{\partial n_2}\right)_{T,P,n_1}$$

$$= RT\left(\frac{\partial \ln \gamma_R}{\partial n_2}\right)_{T,P,n_1} - RT/(n_1 + n_2)$$

We wish to consider first solutions in which the solute obeys Henry's Law, and it can be shown that in these solutions the solvent must obey Raoult's Law, i.e., $\gamma_R = 1$. If we also assume that n_2 is so much smaller than n_1 that it may be neglected in the $(n_1 + n_2)$ term, then

$$\left(\frac{\partial \mu_1}{\partial n_2}\right)_{T,P,n_1} = -RT/n_1$$

Now since the chemical potential of the solute (μ_2) is homogeneous in the zeroth degree in the masses of the components, and hence in n_1, n_2, we have by Euler's theorem

$$n_1\left(\frac{\partial^2 G}{\partial n_1 \partial n_2}\right) + n_2\left(\frac{\partial^2 G}{\partial n_2^2}\right) = 0$$

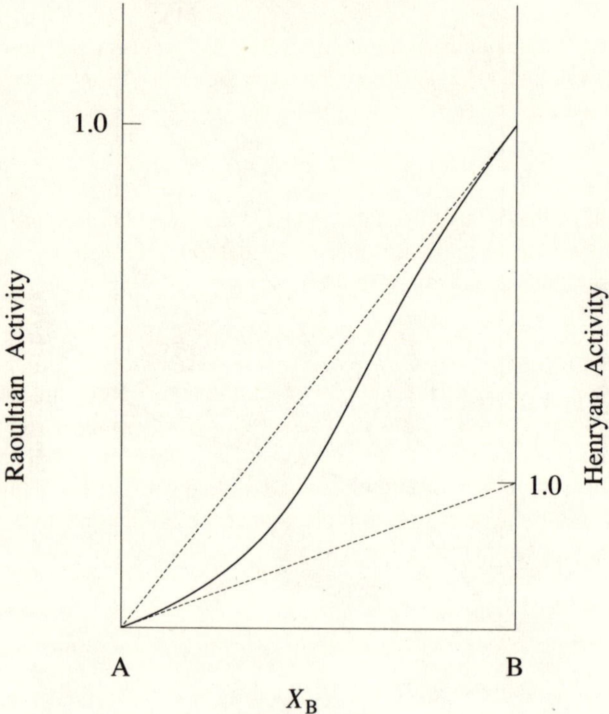

FIG. 12.2. Another possible relationship between Henryan and Raoultian activities.

Thus

$$\left(\frac{\partial^2 G}{\partial n_2^2}\right) = -n_1(-RT/n_1)/n_2$$

or

$$\left(\frac{\partial \mu_2}{\partial n_2}\right)_{T,P,n_1} = RT/n_2$$

If we let

$$n_1 = 1000/M_1$$

where M_1 is the gram formula weight of component 1, n_2 becomes the molality of component 2, which we call simply m. Thus

$$\left(\frac{\partial \mu_2}{\partial m}\right) = RT/m$$

Integrating between two arbitrarily chosen molalities m' and m'', we have

$$\mu_2'' - \mu_2' = RT \ln(m''/m') \tag{12.18}$$

for solutions in which our assumptions are valid. As n_2 gets larger these assumptions are no longer valid, and again a correction term, the Henry's Law activity coefficient, is applied to the concentration terms, giving us

$$\mu_2'' - \mu_2' = RT \ln(\gamma_H m)''/(\gamma_H m)'$$

This equation is analogous to equation (11.17), and we can proceed in exactly the same way as we did in that case, that is, by defining the state represented by the denominator as the standard state, thus

$$\mu_2 - \mu_2^\circ = RT \ln(\gamma_H m)/(\gamma_H m)^\circ \tag{12.19}$$

In this equation, μ° is again a function of both the temperature and the pressure on the state of interest ($T = T^\circ$, $P = P^\circ$) because only m was allowed to vary during the integration. This means that for every change in pressure or temperature of the state of interest, there is a different standard state. (We will not retain the special superscript (*) introduced before to indicate that the standard state is a function of pressure as well as temperature, but will use ° for all standard states, whatever their definition.)

12.4.3. *Activities Independent of Units for a Given Standard State*

Let's now recall equations (12.1), (12.15), (12.17) and (12.19):

$$\mu_i - \mu_i^\circ = RT \ln(f_i/f_i^\circ) \tag{12.1}$$

$$\mu_i - \mu_i^\circ = RT \ln a_i \tag{12.15}$$

$$\mu_i - \mu_i^\circ = RT \ln(X_i \cdot \gamma_H) \tag{12.17}$$

$$\mu_i - \mu_i^\circ = RT \ln(\gamma_H m)/(\gamma_H m)^\circ \tag{12.19}$$

Now recall that μ_i is a state variable with a fixed absolute value for any system at equilibrium. The same is true of course for μ_i°, except that the state is arbitrarily chosen and sometimes does not physically exist. Nevertheless, when the standard state is chosen, μ_i° has a fixed value. It follows that for a given state of interest *and a given standard state*, ($\mu_i - \mu_i^\circ$) is a fixed quantity and therefore

$$\begin{aligned} a_i &= f_i/f_i^\circ \\ &= X_i \cdot \gamma_H \\ &= (\gamma_H m)/(\gamma_H m)^\circ \end{aligned} \tag{12.20}$$

In other words we have not introduced new activity terms, just redefined the old one in units more convenient for use in different situations. The reason that the equivalence noted in (12.20) is not cited more commonly is that usually, on changing from one type of system to another we do not retain the same standard state, but choose a different one. For example we don't usually use the ideal gas as a standard state for an aqueous solute, but if we did, (12.20) would hold. We pursue this idea a little further below, with the help of Figures 12.3 and 12.4.

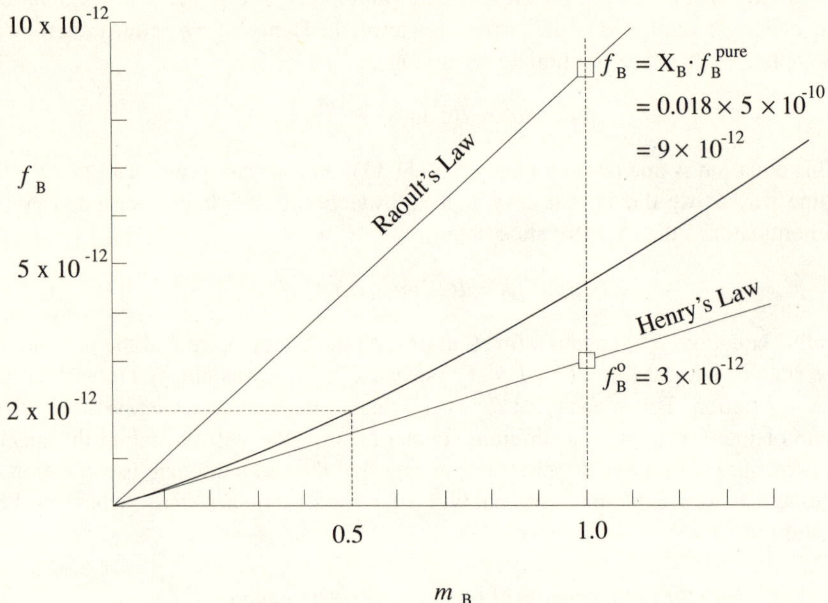

FIG. 12.3. Hypothetical fugacities of solute B dissolved in water, as a function of molality of B.

12.4.4. *The Ideal One Molal Standard State*

We have already seen that in the fugacity definition of activity (equation (12.1)), f_i° can quite commonly take on a wide range of values. It is often defined as 1.0, but is sometimes defined as the fugacity of a pure component (giving that component when pure an activity of 1.0 at all pressures and temperatures). We must now consider possible values for $(\gamma_H m)^\circ$. The most obvious thing to do turns out to be not only the best thing but (almost) the only thing to do. That is, it would seem that setting $(\gamma_H m)^\circ$ equal to 1.0 would simplify things, just as it did in the fugacity definition of activity. This still leaves the question of the individual values of γ_H and m, but we can start by supposing them both to be 1.0. The result is the "hypothetical ideal one molal" standard state ("ideal" with respect to Henry's Law, not Raoult's Law, being understood), and the relation

$$\mu_i - \mu_i^\circ = RT \ \ln(\gamma_H m) \tag{12.21}$$

The physical meaning of this choice can be illustrated by returning first to Figure 11.10. There we have a solute B that has an activity of 0.5 at a mole fraction of 0.3, using our "Raoultian" activity scale based on fugacities. We can look at the same system from a molality point of view by letting component A be water (55.51 moles H_2O/kg), so that n_A = 55.51. At X_B = 0.3, B then has a molality (m_B) of about 23.8, far beyond the concentration at which most solid substances become saturated in water. In other words we cannot simply let Figure 11.10 represent an aqueous

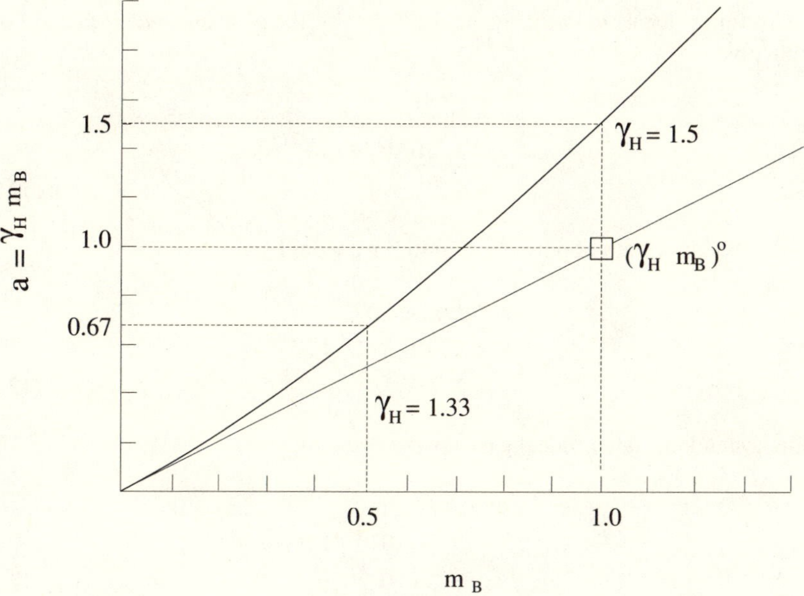

FIG. 12.4. The same system as Figure 12.2, with activities referred to the ideal one molal standard state.

solution. Rather few solids show complete miscibility with water, and those mostly at high pressures and temperatures, so that the mole fraction scale is just not suitable for aqueous solutions.

To discuss the molality scale standard state therefore we move to Figure 12.3, which shows the activity of B at various concentrations in water up to about one molal. We will attempt to show the relationship between the fugacity and molality activity scales by assuming that we not only know the fugacity of pure B ($f_B^{pure} = 5 \cdot 10^{-10}$), as in Figure 11.8, but that we have been able to measure the fugacity of B in the aqueous solutions of B. We have found that the fugacity of B is directly proportional to its molality (it obeys Henry's Law) up to about 0.2 molal, with a slope less than the Raoult's Law slope (the fugacities predicted by Raoult's Law are of course given by $X_B f_B^{pure}$). Extrapolation of the Henry's Law slope shows that if B continued to obey Henry's Law up to one molal, it would have a fugacity of $3 \cdot 10^{-12}$ bars, whereas its measured fugacity at one molal is $4.5 \cdot 10^{-12}$ bars and its fugacity if it obeyed Raoult's Law would be $9 \cdot 10^{-12}$ bars. If we now ask the question "what is the activity of B at a concentration of 0.5 molal?" we can answer in various ways. The measured fugacity of B at 0.5 molal is $2 \cdot 10^{-12}$ bars so the activity of B using the ideal one molal standard state (recalling that we mean Henryan ideality) is

$$\begin{aligned} a_B &= f_B/f_B^\circ \\ &= 2 \cdot 10^{-12}/3 \cdot 10^{-12} \\ &= 0.67 \end{aligned} \tag{12.22}$$

Changing "ideal" to mean "Raoultian" ideality, the ideal one molal standard state would give

$$\begin{aligned} a_{\mathrm{B}} &= f_{\mathrm{B}}/f_{\mathrm{B}}^{\circ} \\ &= 2\cdot 10^{-12}/9\cdot 10^{-12} \\ &= 0.22 \end{aligned} \tag{12.23}$$

Using a standard state of pure crystalline B we would get

$$\begin{aligned} a_{\mathrm{B}} &= f_{\mathrm{B}}/f_{\mathrm{B}}^{\circ} \\ &= 2\cdot 10^{-12}/5\cdot 10^{-10} \\ &= 4\cdot 10^{-3} \end{aligned} \tag{12.24}$$

Using a standard state of ideal gaseous B we would get

$$\begin{aligned} a_{\mathrm{B}} &= f_{\mathrm{B}}/f_{\mathrm{B}}^{\circ} \\ &= 2\cdot 10^{-12}/1.0 \\ &= 2\cdot 10^{-12} \end{aligned} \tag{12.25}$$

Obviously there are many possible choices if one knows the fugacities involved, but of course if we did know the fugacities, we would not use activities and their bothersome standard states at all, we would use the fugacities themselves.

The real situation is of course that in this type of solution no fugacity data are available for the solute, but activity coefficients indicating the deviation from Henry's Law can be either calculated or measured, and they allow us to calculate exactly the same activities and hence $(\mu_i - \mu_i^{\circ})$ values using the ideal one molal standard state. For example at 0.5 and 1.0 molal the activity coefficient (γ_H) is 1.33 and 1.5 respectively, so the activities of B using the ideal one molal standard state ($a_{\mathrm{B}} = \gamma_H m_{\mathrm{B}}$) are 0.67 (compare equation (12.22)) and 1.5 respectively. These values are plotted in Figure 12.4, on a scale that gives the standard state an activity $(\gamma_H \cdot m)^{\circ}$ of 1.0 rather than a fugacity of $3 \cdot 10^{-12}$ bars.

Figure 12.4 corresponds to reality. The lengthy introduction by way of the fictitious Figure 12.3 is simply to emphasize that activities using the ideal one molal standard state are really no different from any other activities. They can be thought of as fugacity ratios, and they are simply another of the wide range of choices available for standard states.

12.4.5. *Mole Fraction—Molality Conversion for Henryan Activity Coefficients*

Because of the fact that molality is not exactly directly proportional to mole fraction except in the limit of infinite dilution, a straight line on an activity-mole fraction diagram (e.g. Figures 12.1, 12.2) will be slightly curved on an activity-molality diagram (e.g. Figures 12.3, 12.4). Therefore a solution that obeys Henry's Law (fugacity or activity exactly proportional to mole fraction) will appear to be slightly non-ideal when plotted on a molal basis. In other words, Henryan activity coefficients based on a molal concentration scale will not exactly equal Henryan coefficients based on a mole fraction scale. We avoided this problem in going from equation (12.18) to (12.19) by simply assuming that the γ_H introduced in connection with mole fractions in (12.17) would be unchanged in the new situation.

The relationship between Henryan coefficients measured on the two concentration scales can be derived as follows. Consider an aqueous solute having a mole fraction X and a molality m. Because the chemical potential of this solute is the same whether we measure concentrations in mole fractions or molalities, we write

$$\begin{aligned}\mu &= \mu_X^\circ + RT\ \ln(\gamma_{H_X} X)\\ &= \mu_m^\circ + RT\ \ln(\gamma_{H_m} m)\end{aligned}$$

where μ_X° is the chemical potential of the solute described above in connection with (12.17), i.e., pure solute having the properties of infinitely dilute solute, μ_m° is the chemical potential of the solute in the ideal one molal solution, and γ_H is now recognized as having two variants denoted by the subscripts X and m. Thus

$$\mu_X^\circ + RT\ln(\gamma_{H_X} X) = \mu_m^\circ + RT\ln(\gamma_{H_m} m) \tag{12.26}$$

and

$$\mu_X^\circ = \mu_m^\circ + RT\ln(\gamma_{H_m} m)/(\gamma_{H_X} X)$$

As $m \to 0$, both γ_{H_X} and $\gamma_{H_m} \to 1$, so $(\gamma_{H_m} m)/(\gamma_{H_X} X) \to (m/X)$. Now the mole fraction of an aqueous solute measured in molal units is by definition

$$X = m/(m + 55.51)$$

where $55.51 = 1000/18.0153$, 18.0153 being the molecular weight of water. So

$$m/X = m + 55.51$$

and

$$\lim_{m \to 0} (m/X) = 55.51$$

Therefore

$$\mu_X^\circ = \mu_m^\circ + RT\ln 55.51$$

Substituting this in (12.26), we have

$$\ln(\gamma_{H_X} X) + \ln 55.51 = \ln(\gamma_{H_m} m)$$

or

$$\ln(\gamma_{H_X}) = \ln\left(\frac{\gamma_{H_m} m}{55.51X}\right)$$

but $m/X = m + 55.51$, so $m/55.51X = 1 + m/55.51$ and

$$\ln(\gamma_{H_X}) = \ln(\gamma_{H_m}) + \ln(1 + m/55.51)$$

or

$$\ln(\gamma_{H_X}) = \ln(\gamma_{H_m}) + \ln(1 + 0.0180153m)$$
$$= \ln(\gamma_{H_m}) + \Gamma_\gamma$$

where the conversion factor is

$$\Gamma_\gamma = \ln(1 + 0.0180153m)$$

For all but large values of m, the difference between γ_{H_X} and γ_{H_m} is quite small, and the conversion is often neglected. Nevertheless activity coefficients based on mole fractions are generally more coherent with the rest of thermodynamics, and are to be preferred in situations where the choice makes any difference.

12.4.6. *Why One Molal? Why Ideal?*

Before leaving this subject there is one final detail to be considered. Having arrived at a molality-based definition of activity (equation (12.20)), we chose a standard state such as to make the denominator disappear, i.e. $(\gamma_H m)^\circ = 1.0$, by defining both γ_H and m as 1.0. Might there be situations where one or other of these could usefully have some other value? This could include cases where the product was still 1.0 as well as cases where the product was some other value. As far as we are aware, no such cases have been presented in the literature, and it is indeed difficult to see how they might arise. The only exception seems to be that occasionally cases involving the determination of the stability constants of aqueous complexes in solutions of constant ionic strength (used to keep activity coefficients constant) are treated by using a standard state of 1 molal in the solute of interest, but with an ionic strength chosen such that activity coefficients cancel out. This could be called a non-ideal one molal standard state.

The main reason that the only really useful choice is to have γ_H equal to 1.0 in the standard state is that values of γ_H in real solutions approach 1.0 in very dilute solutions, so that many properties of ideal solutions can be estimated by measuring them as a function of concentration, and then extrapolating to zero concentration to get their value in an ideal solution, i.e. where $\gamma_H = 1.0$. These are then called "standard state values" of partial molar volume, or partial molar enthalpy, etc. There is no way of deriving properties of solutions with other values of γ_H. The choice of m as 1.0 does not have the same degree of necessity. Any value could be used, adding a constant factor to all attendant values of solute activities and $(\mu - \mu^\circ)$ values. The very dilute or infinitely dilute solutions themselves (m very small or zero) would make rather poor standard states, because although γ_H° values would be 1.0, consideration of equation (12.20) shows that very small or zero values for $(\gamma_H m)^\circ$ would be decidedly inconvenient.

In summary, a value of 1.0 for γ_H° allows us to obtain values for the properties of the standard state, which we mentioned earlier is the essential factor in the choice of standard state. As for m°, any value could be used, but none has any advantage over 1.0. Therefore the hypothetical ideal one molal standard state is in universal use for dilute solution (molality-based) activities.

12.5. EFFECT OF TEMPERATURE AND PRESSURE ON ACTIVITIES

The change in activities as a response to changes in temperature and pressure of the system naturally depends to a large extent on the standard states involved. Note too, that any variation of activity with change in T or P is actually due to variation of the activity coefficient, because these effects are normally calculated for constant composition conditions.

12.5.1. *Temperature*

We have by now defined activity in four different but equivalent ways, useful, generally speaking, for gases, solids, liquids, and solutes respectively. That is,

$$\mu_i - \mu_i^\circ = RT \ln a_i$$

where

$$\begin{aligned} a_i &= f_i / f_i^\circ \\ &= P \cdot \gamma_f / f_i^\circ \end{aligned}$$

or

$$a_i = X_i \cdot \gamma_R$$

or

$$a_i = X_i \cdot \gamma_H$$

or

$$a_i = (m \cdot \gamma_H)/(m \cdot \gamma_H)^\circ$$

Differentiating with respect to temperature, we have

$$\frac{\partial}{\partial T}(R \ln a_i) = \frac{\partial}{\partial T}(R \ln \gamma_i) \tag{12.27}$$

$$= \frac{\partial}{\partial T}\left(\frac{\mu_i}{T} - \frac{\mu_i^\circ}{T}\right)$$

$$= -(\bar{H}_i - \bar{H}_i^\circ)/T^2 \tag{12.28}$$

$$= -\bar{L}_i/T^2 \tag{12.29}$$

where γ_i may be γ_f, γ_R , or γ_H. Thus the temperature derivative of the activity is a simple function of the *relative partial molar enthalpy*, $\bar{L}_i$, which was discussed in Chapter 9 (§9.3.3). The numerical value of $\bar{L}_i$ of course depends on the form of activity being used, because this determines the value of $\bar{H}_i^\circ$. For standard states based on Raoult's Law, that is, where the standard state is the pure substance i at T and P, $\bar{H}_i^\circ$ is the same as H_i°, the molar enthalpy of the substance (an unknown quantity, but as usual it appears in a difference term, so we don't need to know its absolute value). For the ideal gaseous standard state, $\bar{H}_i^\circ$ is evidently the molar enthalpy of an ideal gas. For standard states based on Henry's Law, where $\gamma_H \rightarrow 1$ as X or $m \rightarrow 0$, $\bar{H}_i^\circ$ is the partial molar enthalpy of the solute in the hypothetical pure substance having $\gamma_H = 1$ or the hypothetical ideal one molal solution respectively. Substances in these strange states have partial molar enthalpies (and volumes) equal to that at infinite dilution, hence providing a method of measurement. This can be seen by considering equations (12.27) and (12.28), which show that $\bar{H}_i^\circ$ becomes equal to $\bar{H}_i$ when γ_i is 1.0. Therefore for Henryan standard states where $\gamma_i \to 1$ as X_i or $m \to 0$, $\bar{H}_i^\circ$ must be the partial molar enthalpy of i at infinite dilution, and for Raoultian standard states where $\gamma_i \to 1$ as $X_i \to 1$, $\bar{H}_i^\circ$ must be the partial molar enthalpy (the molar enthalpy) of pure i (confirming what we stated by simple inspection, above). (Note that in the case of multicomponent solutions infinite dilution means infinite dilution of all components, not just of component i). Thus the Henryan standard states, which seem so unattainable, are actually convenient because some of their properties are the same as those of the infinitely dilute solution, and these are obtainable by extrapolation from measurements at finite concentration.

12.5.2. *Pressure*

In considering the effect of pressure on activity, we must recall that the standard state pressure (P°) is not always the same as the system pressure (P), so that the differentiation with respect to pressure is not always completely analogous to differentiation with respect to temperature. First of all, for variable pressure standard states, those that do have $P^\circ = P$, we have

$$\begin{aligned}\frac{\partial}{\partial P}(RT\ln a_i) &= \frac{\partial}{\partial P}(RT\ln\gamma_i)\\ &= \frac{\partial}{\partial P}(\mu_i - \mu_i^\circ)\\ &= \bar{V}_i - \bar{V}_i^\circ\end{aligned}$$

where γ_i may be γ_f, γ_R, or γ_H, and $\bar{V}_i^\circ$ is either the molar volume of pure i or the partial molar volume of i at infinite dilution, depending on the standard state used (see above). Integration of this expression requires a knowledge of the variation of the *relative partial molar volume* $\bar{V}_i - \bar{V}_i^\circ$ with pressure for which there is no general expression. There is also no special symbol for $\bar{V}_i - \bar{V}_i^\circ$ as there is for $\bar{H}_i - \bar{H}_i^\circ$.

However for the fixed pressure standard states,

$$\frac{\partial}{\partial P}(\mu_i^\circ) = 0 \tag{12.30}$$

because changing P does not change the pressure on the standard state (P°). Therefore

$$\frac{\partial}{\partial P}(R\ln a_i) = \bar{V}_i$$

A case of particular interest to us is the activity of solids, for which in many cases the assumption that $\bar{V}_i$ is unaffected by pressure (solid i is incompressible) is reasonable. If $\bar{V}_i$ is a constant, then

$$\int_{P_1}^{P_2} d\ln a_i = \frac{1}{RT}\int_{P_1}^{P_2}\bar{V}_i dP$$

$$\ln(a_i)_{P_2} - \ln(a_i)_{P_1} = \bar{V}_i(P_2 - P_1)/RT$$

Furthermore, if $P_1 = 1$ bar and the standard state of i is pure i at T and one bar, then pure i at P_1 has an activity of 1, and

$$\ln(a_i)_{P_2} = \ln(a_i)_P = \bar{V}_i(P - 1)/RT$$

or for *pure* solids

$$\ln(a_i)_P = V_i(P - 1)/RT \tag{12.31}$$

This permits calculation of the activity of a pure mineral at any pressure, relative to the same mineral at one bar.

For example, consider calcite and aragonite at equilibrium at 3767 bars and 25°C, from our example in Chapter 7. The activity of calcite is, from equation (12.31),

$$\begin{aligned}\ln a_{calcite,3767b} &= 3.6934 \cdot (3767 - 1)/(8.3143 \times 298.15)\\ &= 5.61\end{aligned}$$

so

$$a_{calcite,3767b} = 274$$

and the activity of aragonite is

$$\begin{aligned}\ln a_{aragonite,3767b} &= 3.4150 \cdot (3767 - 1)/(8.3143 \times 298.15)\\ &= 5.19\end{aligned}$$

so

$$a_{aragonite,3767b} = 179$$

Thus $CaCO_3$ in the form of calcite and aragonite, at 3767 bars (according to the data of Robie et al., 1978) and 25°C, although having the same chemical potential has two different activities because there are two different standard states. Recalling that activity is the ratio f_i/f_i°, we see too that the physical meaning of an activity of calcite of 274 is that the fugacity of calcite is increased by a factor of 274 when it is squeezed from 1 bar to 3767 bars. We know this without knowing either f at 3767 bars or f° at one bar for calcite.

12.6. ACTIVITIES AND STANDARD STATES: AN OVERALL VIEW

We have now said everything necessary about activities and standard states, but the overall effect for the newcomer is often one of confusion at this stage. To try to draw the various threads together we consider in Figure 12.5a a hypothetical three-phase equilibrium at temperature T and pressure P. A solid crystalline solution of B in A is in contact with an aqueous solution of A(*aq*) and B(*aq*), which is in turn in contact with a vapor phase containing A(*v*) and B(*v*) in addition to water vapor. We can suppose the dissolution of (A,B)(*s*) to be stoichiometric so that the ratio of A to B is the same in all three phases, but this is irrelevant to our development as we consider only component A. Let's say that for a solid solution composition of $X_A = 0.5$, $X_B = 0.5$, the concentration of A(*aq*) at equilibrium (m_A) is 10^{-2} molal, and the fugacity of A in the vapor (f_A) is 10^{-5} bars. Assuming activity coefficients in the solid and liquid

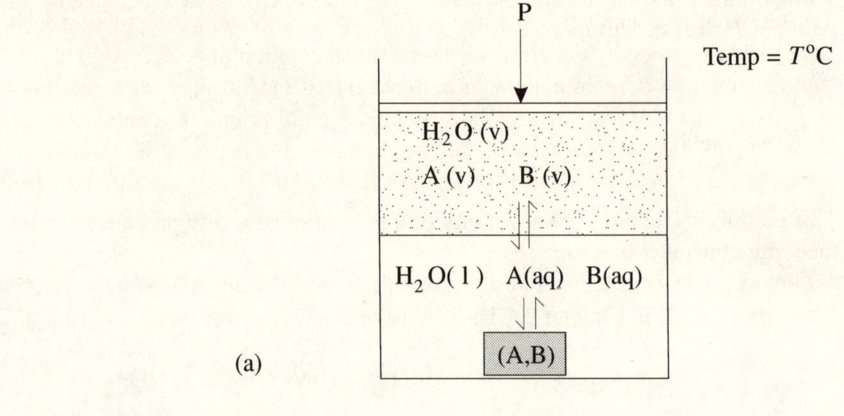

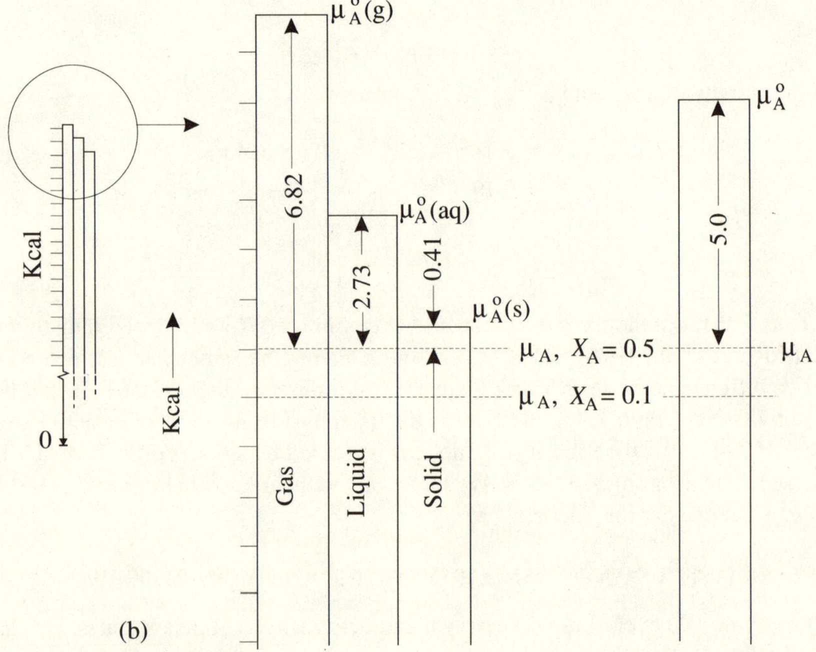

FIG. 12.5. (a) A hypothetical three-phase system at equilibrium at pressure P and temperature T. (b) The top part of the histogram of chemical potentials in kilocalories. The length of the bar for each phase is fixed when the standard state is chosen, and the chemical potential of A in the equilibrium system is represented by a line across the histogram at a level depending on the amount of B in the system. The lengths of the bars shown on the left represent traditional standard states, but any position for the top of the bar could be chosen, such as the one shown on the right, thus defining a new standard state.

phases to be one, the activity of A in the solid solution (using a standard state of pure crystalline A at T and P) is 0.5, the activity of A in the aqueous solution (using a standard state of the hypothetical ideal one molal solution of A at T and P) is 10^{-2}, and the activity of A in the vapor (using a standard state of pure ideal gaseous A at T and one bar) is 10^{-5}. Because the system is at equilibrium, the chemical potential of A (μ_{A}) is the same in each of the three phases, but because the three standard states are different, the standard chemical potential of A (μ_{A}°) is different for the three phases. The difference ($\mu_{\mathrm{A}} - \mu_{\mathrm{A}}^{\circ}$) is calculable from the equations we have just derived. Thus, letting T = 25°C,

$$\begin{aligned}(\mu_{\mathrm{A}} - \mu_{\mathrm{A}}^{\circ})_{solid} &= RT \ln X_{\mathrm{A}} \\ &= 1.987 \times 298.15 \times \ln(0.5) \\ &= -0.41 \text{ kcal mol}^{-1}\end{aligned}$$

$$\begin{aligned}(\mu_{\mathrm{A}} - \mu_{\mathrm{A}}^{\circ})_{aq} &= RT \ln m_{\mathrm{A}} \\ &= 1.987 \times 298.15 \times \ln(10^{-2}) \\ &= -2.73 \text{ kcal mol}^{-1}\end{aligned}$$

$$\begin{aligned}(\mu_{\mathrm{A}} - \mu_{\mathrm{A}}^{\circ})_{gas} &= RT \ln f_{\mathrm{A}} \\ &= 1.987 \times 298.15 \times \ln(10^{-5}) \\ &= -6.82 \text{ kcal mol}^{-1}\end{aligned}$$

It is instructive to consider these differences on a histogram (Figure 12.5b) in which the ordinate is a scale of kcal mol^{-1}, on which we plot the *absolute* chemical potentials of A. These absolute potentials may be very large, so we look at only the tops of the bars in the histogram, and we unfortunately don't know the values of the absolute potentials individually, so we can't put an absolute scale on the ordinate. But we *can* plot the relative positions of the tops of the bars, and the position of the equilibrium chemical potential of A in the system.

If we now consider systems having more and more B in the solid solution (and hence in the other two phases), but always at equilibrium, the histogram bars stay where they are (because we are not changing standard states) but the level of the (absolute) chemical potential of A is lowered, increasing the distance between the top of the histogram bar for each phase and the level of μ_{A}, that is, increasing the (negative) value of ($\mu_{\mathrm{A}} - \mu_{\mathrm{A}}^{\circ}$) as the activity of A is lowered.

This diagram is worth careful thought. It illustrates several things that are useful in understanding activities, chemical potentials, and standard states, such as the absolute nature of chemical potentials and the necessity of using differences, the equality of chemical potentials in each phase, and the arbitrary nature of the standard state. To further illustrate the last point, suppose we choose a new energy level for the standard state more or less at random, such that ($\mu_{\mathrm{A}} - \mu_{\mathrm{A}}^{\circ}$) when X_{A} is 0.5 is −5000 cal mol^{-1}. This implies a value of a_{A} of $10^{-3.67}$, and this in turn defines the physical characteristics of the state we have chosen. If this state is used as a standard state for the vapor, then

$$f_A/f_A^\circ = 10^{-3.67}$$

but

$$f_A = 10^{-5} \text{ bars}$$

so therefore

$$\begin{aligned} f_A^\circ &= 10^{-1.33} \\ &= 0.047 \text{ bars} \end{aligned}$$

so that the standard state is hypothetical ideal gaseous A at T and 0.047 bars.

If it is used for the aqueous phase, then

$$m_A/m_A^\circ = 10^{-3.67}$$

but

$$m_A = 10^{-2} \text{ mol kg}^{-1}$$

so therefore

$$\begin{aligned} m_A^\circ &= 10^{1.67} \\ &= 46.77 \text{ mol kg}^{-1} \end{aligned}$$

so that the standard state is a hypothetical ideal 46.77 molal solution of A in water.

If it is used for the solid phase, then because a mole fraction of A of 0.5 has an activity of $10^{-3.67}$, a mole fraction of 1.0 has an activity of twice this, or $10^{-3.37}$. This implies that the standard state is hypothetical solid A having a fugacity 0.00043 times the normal fugacity of pure crystalline A.

These weird standard states have one very attractive feature, which is that because they all have the same value of μ_A°, the activity of A would always be the same in all three phases at equilibrium. The three standard states could also coexist at equilibrium, if they could exist at all. As mentioned earlier, there is no reason why other concentrations or pressures could not be chosen for the standard states, that is, other than one molal or one bar, as long as ideal behavior is still part of the definition. But these other concentrations or pressures would then appear in all activity calculations and all equilibrium constants, and we would have to give up the convenience of being able to think of gaseous activities as approximate or "thermodynamic" pressures, and of aqueous activities as approximate or "thermodynamic" concentrations. It seems generally more convenient to add a little diversity to standard states, and keep activity expressions simple, as is the present custom.

We hope that this brief glance at other possible standard states, like a science fiction story about other possible worlds, will not only clarify the present way of doing things, but make it more acceptable in the sense that although standard states are a little weird, they could be worse.

12.6.1. *Changing From One Standard State to Another*

Suppose you have the activity of a constituent with respect to a particular standard state, but you need its activity using some other standard state. For example, you might know f_{CO_2} in a fluid, which is equivalent to knowing its a_{CO_2} using an ideal gaseous CO_2 at T, 1 bar standard state, but you want to do speciation calculations so you need a_{CO_2} using the ideal one molal standard state.

There are two ways of making such a change, for say constituent A:

1. Find $\mu_A^{\circ,\,new} - \mu_A^{\circ,\,old}$, and subtract it from your activity expression. Thus,

$$\mu_A - \mu_A^{\circ,\,old} = RT \ln a_A^{old}$$
$$\mu_A^{\circ,\,new} - \mu_A^{\circ,\,old}$$

$$\mu_A - \mu_A^{\circ,\,new} = RT \ln a_A^{new}$$

   An example of this procedure is used in the Problems at the end of the chapter.

2. Calculate a_A directly. To do this you need to know either γ_H or γ_R over a range of compositions. Knowing one usually allows calculation of the other, which is the main requirement in changing standard states.

To illustrate the second method, consider the system $H_2O - CO_2$ at 600°C, 2000 bars, illustrated in Figure 11.11. Let's say you need a_{CO_2} using the ideal one molal standard state for $X_{CO_2} = 0.01$ ($m_{CO_2} = 0.56$). The data in Table 11.1 are insufficient to provide this directly, but they have been fitted with Margules equations (see Chapter 15), and the following data calculated from the fit coefficients. The details are expanded upon in the Problems sections of Chapter 15. The Henry's law slope of the CO_2 activities at $X_{CO_2} = 0.0$ is calculated to be 2.652, and the CO_2 activity according to Raoult's Law at $X_{CO_2} = 0.01$ is 0.02564. If CO_2 obeyed Henry's Law at $X_{CO_2} = 0.01$, its activity would be $0.01 \times 2.652 = 0.02652$, so $\gamma_H = 0.02564/0.02652 = 0.967$, and $a_{CO_2} = 0.967 \times 0.56 = 0.54$ using the ideal one molal standard state, the desired result.

12.7. EFFECT OF SIZE OF THE MOLE ON ACTIVITIES

We will see in discussing the Phase Rule that the choice of components in studying or discussing systems is often not a simple matter. One aspect of this choice which is best presented in the context of activities is the choice of the mole of a solute component, a factor that is important particularly in the case of crystalline solutions.

In the case of gaseous and liquid solutions, the molecular structure of a solute species can often be determined, and it is often best to choose as a component of the solution the species thus determined. For example, nitrogen in the air exists as dimers of N_2, and for most purposes one would use N_2 as one of the components, there being no advantage to using N, N_3, or N_4. In fact for such a solution if we write

$$N_4 = 2N_2 \qquad (12.32)$$

we would undoubtedly be referring to a possible reaction between actual *species* of nitrogen in the gas. The activity of each of these species, if available, would be referred to the "ideal gas at one bar" standard state, and could therefore be represented by its fugacity. The fugacity (and therefore the activity) of species N_2 would undoubtedly be much greater than that of species N_4, and there would be no simple relationship between μ° for N_2 gas and that for N_4 gas. That is, $\mu^\circ_{N_2}$ refers to the Gibbs energy of a mole of pure N_2 gas at T and one bar, and $\mu^\circ_{N_4}$ refers to the Gibbs energy of a mole of pure N_4 gas at T and one bar. The equilibrium constant would be

$$K = (a_{N_2})^2/a_{N_4} \tag{12.33}$$

However, as we tried to make clear in Chapter 10, equations like (12.32) can refer to species *or* to components. If equation (12.32) *did* refer to components N_2 and N_4 rather than to species, the situation is completely different. It is unusual to write a reaction between two different choices of component for the same element or combination of elements, because their relationship is not in question, it is known; but it is just this relationship that we wish to point out now. Obviously a reaction such as equation (12.32) written between two components of the same bulk composition has no physical reality; N_2 and N_4 are two different ways, among an infinite number, of representing nitrogen, or actually of choosing a nitrogen component. We should perhaps not write their relationship as a reaction; there is no need to do so. What is important is that however nitrogen actually occurs, the Gibbs free energy per mole of component N_4 is exactly twice the Gibbs energy per mole of component N_2, because there is twice the mass in N_4 that there is in N_2, and because in this case we are *not* comparing N_2 gas with N_4 gas as before, but two different ways of representing the *same* gas. Similarly, component N_2 and component N_4 have the same standard state—pure nitrogen gas acting ideally at T and one bar. Component N_4 just needs twice as much of it to make up one mole. In fact there is exactly twice the quantity of all extensive parameters, both in the standard state and in any system. Therefore,

$$\mu_{N_4} = 2\,\mu_{N_2} \text{ (true for species, given equilibrium, and components)}$$

$$\mu^\circ_{N_4} = 2\,\mu^\circ_{N_2} \text{ (true for components only)}$$

and

$$(\mu_{N_4} - \mu^\circ_{N_4}) = 2(\mu_{N_2} - \mu^\circ_{N_2}) \text{ (components only)}$$

It follows that

$$RT \ln a_{N_4} = 2RT \ln a_{N_2}$$

and that

$$a_{N_4} = (a_{N_2})^2$$

where N_2 and N_4 are components, and by extension of the argument

$$a_{N_{n\times m}} = (a_{N_m})^n \tag{12.34}$$

This relationship is entirely formal and contains no information about nitrogen itself.

12.7.1. *What Is a Mole of Olivine?*

We derived this relationship for a gas species because it is possible to differentiate between the species N_2 and the component N_2. In crystalline solutions there are no species, but there certainly is a wide choice of components. Consider for example the forsterite–fayalite crystalline solution. Normally we write forsterite as Mg_2SiO_4 and fayalite as Fe_2SiO_4. Is this the best choice of components, or should we choose $MgSi_{0.5}O_2 - FeSi_{0.5}O_2$, or $Mg_4Si_2O_8 - Fe_4Si_2O_8$, or . . . ? [2]

From the discussion of the nitrogen case, we know that

$$a_{Mg_2SiO_4} = (a_{MgSi_{0.5}O_2})^2$$

The critical part of this derivation is the fact that component Mg_2SiO_4 has twice the mass and therefore twice the Gibbs energy of component $MgSi_{0.5}O_2$. This fact is easier to understand in the case of a crystalline solution than in the nitrogen case considered above, because there is no possibility of confusing components with species. Component Mg_2SiO_4 obviously contains twice as many atoms as $MgSi_{0.5}O_2$, but they share exactly the same crystal structure, so the only difference between them is the number of atoms chosen as the mole.

The question now is, what practical difference does this make? Perhaps the easiest way to show this is to consider Raoult's Law, which for a crystalline solution (equation (11.29)) says that

$$a_i = X_i$$

Now you have to realize that the mole fraction X is unaffected by the choice of component problem we have been considering, that is,

$$X_i = X_{2i} = X_{0.5i} \text{ etc.}$$

so that if

$$a_i = (a_{0.5i})^2$$

we have a problem. Obviously, a_i and $a_{0.5i}$ cannot both be equal to X_i. If $a_{0.5i}$ gives a straight line when plotted against X_i, obviously a_i will give a parabola (Figure 12.6). Raoult's Law apparently isn't quite as simple as it appeared.

This problem can only be resolved experimentally. That is, the activity of forsterite and fayalite over a range of compositions can be determined, and this will show which formulation of activity (i.e., a, a^2, etc.) gives a straight line relationship with mole fraction. For this system, this was done by Nafziger and Muan (1967), who found that the activities of components $MgSi_{0.5}O_2$ and $FeSi_{0.5}O_2$ came closest to giving straight lines when plotted against X. The reason for this is simply that individual Mg atoms exchange with individual Fe atoms on the Mg-Fe sites of the crystal, so the

[2]There may be some who are bothered by having "half an atom of silicon" in a component formula. Components are simply the minimum number of algebraic formulae units which can be combined to represent the composition of all parts of a system. (They could thus even be negative quantities or contain negative subscripts). More important in this case, a mole of a component is Avogadro's number of formula units of atoms, so that $MgSi_{0.5}O_2$ does not imply half an atom of silicon, but half of Avogadro's number of silicon atoms.

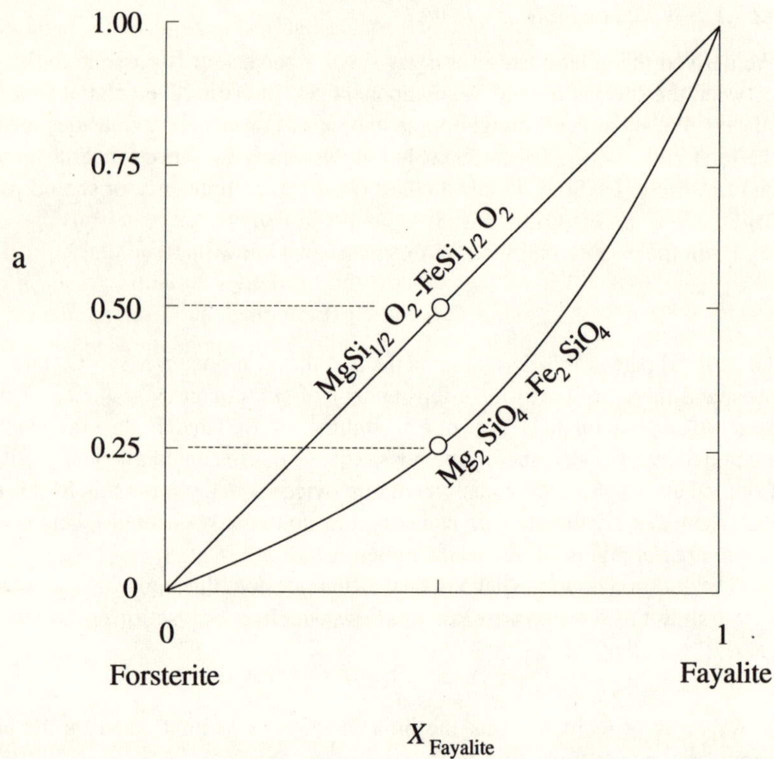

FIG. 12.6. Activities which would be observed in the forsterite-fayalite system if Raoult's Law were followed. If components are $MgSi_{0.5}O_2 - FeSi_{0.5}O_2$, a mole fraction of 0.5 would have an activity of 0.5. If components are $Mg_2SiO_4 - Fe_2SiO_4$, a mole fraction of 0.5 has an activity of $(0.5)^2$. All sets of components follow Raoult's Law, but only one plots as a straight line on an $a - X$ diagram.

components should be written with a single Mg or Fe. If there existed Mg-Mg pairs which exchanged with Fe-Fe pairs in the crystal, we would undoubtedly find that components $Mg_2SiO_4 - Fe_2SiO_4$ would be a better choice, that is, they would come closest to producing straight lines on the $a - X$ plot. It is for exactly the same reason that N_2 is a better choice of component than N_4. However, as such dimer species don't exist in crystals (at least not in any that we will consider), we can assume that the "best" choice of component for exchange reactions in crystals will always be that containing one atom of the exchanging species. Note that this means we would make a different choice depending on which atom of a mineral was being exchanged. For example, if we were interested in substitution of germanium for silicon in forsterite, we would probably consider the components Mg_2SiO_4 and Mg_2GeO_4, while if it were sulfur for oxygen (!) we might use $Mg_{0.5}Si_{0.25}O - Mg_{0.5}Si_{0.25}S$.

The problem is more difficult in other systems. How does one choose components in a complex silicate melt, for example? In a melt there are no stoichiometric restrictions to be observed, but the formal relationship between the activities of various component choices that we have discussed remains true. So if you measure the activity of some component in a melt, and determine the deviations of these activities from Raoult's Law by calculating activity coefficients, the question is, what part of these activity coefficients represents non-ideal behavior, and what part represents a poor choice of components? Generally speaking, extremely large or extremely small activity coefficients mean that the component involved has been badly chosen, which is to say that it does not come very close to representing the "real" situation in the system. In these situations, thermodynamics provides no help whatsoever. It points out the consequences of choices relative to each other, and from there on the investigator is on her own. In other words, the choice of components, as much as the choice of system to investigate, is a part of the "art of doing science," that part which relies on skill and intuition, and can never be taught.

PROBLEMS

1. Consider crystalline A to be in equilibrium with water saturated with A at a pressure of 1234.0 bars and a temperature of 567.0°C. A is very slightly soluble (0.001 moles/Kg H_2O) so that $P_{H_2O} = P_{total}$. The vapor pressure of crystalline A varies with temperature according to the relation

$$\log P(\text{bars}) = (-19,130/T(\text{K})) + 7.65$$

Its molar volume is 22.7 $\text{cm}^3\,\text{mol}^{-1}$ and is essentially constant in the pressure range under consideration. The fugacity coefficient for water at this P and T is 0.541.

   1. What is the activity of crystalline A, using the following standard states:
      (a). pure A(s) at T, P.
      (b). pure A(s) at T and a pressure of 1 bar.
      (c). pure A(s) at T and under its own vapor pressure.
      (d). pure ideal gaseous A at T and one bar.
   2. What is the activity of water, using the following standard states:
      (a). pure water at T and P.
      (b). pure water at T and a pressure of 500 bars (fugacity coefficient is 0.747).
      (c). ideal gaseous water at T and a pressure of 123.456 bars.
      (d). Hypothetical ice at T and a pressure of 1 bar. (Suggest how this could be done, without performing any calculations).
   3. What is the activity of A(aq) ($\gamma_A(aq)$ assumed to be 1.0) using the following standard states:
      (a). an ideal one molal solution of A at T and P.

(b). a 17.2 molal solution of A having an activity coefficient of 0.123, at T and P.

4. (a). Under what conditions is $\mu_{A(s)} = \mu_{A(aq)}$?
   (b). Under what conditions is $\mu^\circ_{A(s)} = \mu^\circ_{A(aq)}$?
5. Calculate the difference in chemical potential between dissolved A at T and P and dissolved A in standard state 3(a).
6. Experiments show that at 9.7 Kb, 1080°C, the system SiO_2-H_2O shows a second critical end point at which quartz and a supercritical fluid of composition 75 wt. % SiO_2, 25 wt. % H_2O coexist at equilibrium. What is the activity of SiO_2 in this fluid, referred to a standard state of quartz at 1080°C and one bar?

2. In §12.3 it says that if you wished to have a constant T standard state, you need to include a term accounting for the change in μ° with T. Is this not backwards? Explain.
3. Equation (15.42) for a $CO_2 - H_2O$ fluid at 600°C, 2000 bars is

$$RT \ln \gamma_{CO_2} = (2W_{G_{H_2O}} - W_{G_{CO_2}})X^2_{H_2O} + 2(W_{G_{CO_2}} - W_{G_{H_2O}})X^3_{H_2O}$$

where $W_{G_{H_2O}} = 1751.294$ and $W_{G_{CO_2}} = 7081.006$. What is the activity of CO_2 in a $CO_2 - H_2O$ fluid having $X_{CO_2} = 0.01$at this T and P, using a standard state of pure CO_2 at the same T and P? What is its fugacity? See Table 11.1 for data.
4. What is the standard state used in the "a_{CO_2}" column in Table 11.1?
5. According to Wellman (1969)[3], the fugacity of NaCl in equilibrium with nepheline ($NaAlSiO_4$) and sodalite ($3NaAlSiO_4 \cdot NaCl$)at 600°C, 1 bar, is $10^{-10.566}$ bars, and the fugacity of pure halite at the same T,P is $10^{-5.780}$ bars. Thus the activity of NaCl at nepheline–sodalite equilibrium at these conditions is $10^{-4.786}$ bars, using a pure halite at T,P standard state. If $\Delta_f G^\circ_{NaCl(aq)}$ is -393133 J mol^{-1} (ideal 1 m standard state), and $\Delta_f G^\circ_{NaCl,\, Halite}$ is -384138 J mol^{-1}, what is the activity of NaCl at nepheline–sodalite equilibrium using the ideal one molal standard state?

[3]Geochim. Cosmochim. Acta, v. 33, pp. 1302–1304.

13

THE EQUILIBRIUM CONSTANT

It is possible to know thermodynamics without understanding it.

Dickerson (1969), p. 387.

13.1. THE MOST USEFUL EQUATION IN THERMODYNAMICS

Consider a chemical reaction, involving any number of reactants and products and any number of phases, which may be written

$$\sum_i \nu_i M_i = 0 \tag{13.1}$$

where M_i represents the chemical formulae of the reaction constituents and ν_i represents the stoichiometric coefficients, negative for reactants and positive for products.[1] An example would be

$$\nu_1 M_1 + \nu_2 M_2 + \nu_3 M_3 = 0 \tag{13.2}$$

where, if M_1 is $SiO_2(s)$, M_2 is H_2O, and M_3 is $H_4SiO_4(aq)$, and $\nu_1 = -1$, $\nu_2 = -2$, and $\nu_3 = 1$, the reaction is

$$SiO_2(s) + 2H_2O = H_4SiO_4(aq) \tag{13.3}$$

Now let's recall (from Chapter 3) what we mean by an equation such as (13.3). If there are no constraints placed on the system containing M_1, M_2, and M_3 other than T and P (or T and V; U and V; S and P; etc.) then M_1, M_2, and M_3 react until they reach an equilibrium state characterized by a minimum in the appropriate energy potential as indicated by expressions like $dG_{T,P} = 0$. A corollary of this equilibrium relationship, to be fully developed in the next chapter, is that the sums of the chemical potentials of the reactants and products must be equal. In the example, this would be

$$\nu_3 \mu_{M_3} = \nu_2 \mu_{M_2} + \nu_1 \mu_{M_1}$$

or

$$\mu_{H_4SiO_4} = \mu_{SiO_2} + 2\mu_{H_2O}$$

or in general terms,

$$\sum_i \nu_i \mu_i = 0 \tag{14.25}$$

No notation is necessary for the phases involved because μ_i must be the same in every phase in the system.

[1] We discuss in the next chapter the fact that the quantities M_i are not restricted to species or components, but may be constituents (defined in Chapters 3 and 10). However in this chapter our examples are confined to minerals and aqueous species.

However, if more than the minimum two constraints apply to the system, then any equilibrium state achieved will be in our terms a metastable state, (14.25) does not apply, and the difference in chemical potential between products and reactants is not zero. In our example, a solution might be supersaturated with H_4SiO_4 but prevented from precipitating quartz by a nucleation constraint, so that $\mu_{H_4SiO_4} - \mu_{SiO_2} - 2\mu_{H_2O} > 0$.

To simplify the notation in the next few equations, let's use a, b, c instead of ν_1,ν_2,ν_3, and A,B,C instead of M_1,M_2,M_3, so that the example reaction is now

$$a\mathrm{A} + b\mathrm{B} = c\mathrm{C}$$

Now because for each constituent

$$\mu_i = \mu_i^\circ + RT \ln a_i \tag{13.4}$$

i.e.

$$\mu_\mathrm{A} = \mu_\mathrm{A}^\circ + RT \ln a_\mathrm{A}$$
$$\mu_\mathrm{B} = \mu_\mathrm{B}^\circ + RT \ln a_\mathrm{B}$$
$$\mu_\mathrm{C} = \mu_\mathrm{C}^\circ + RT \ln a_\mathrm{C}$$

then whatever the number of constraints on the system, products minus reactants can be expressed as

$$\begin{aligned}
\Delta_r\mu &= c\mu_\mathrm{C} - a\mu_\mathrm{A} - b\mu_\mathrm{B} \\
&= c(\mu_\mathrm{C}^\circ + RT \ln a_\mathrm{C}) - a(\mu_\mathrm{A}^\circ + RT \ln a_\mathrm{A}) - b(\mu_\mathrm{B}^\circ + RT \ln a_\mathrm{B}) \\
&= (c\mu_\mathrm{C}^\circ - a\mu_\mathrm{A}^\circ - b\mu_\mathrm{B}^\circ) + RT \ln a_\mathrm{C}^c - RT \ln a_\mathrm{A}^a - RT \ln a_\mathrm{B}^b \\
&= \Delta_r\mu^\circ + RT \ln \left(\frac{a_\mathrm{C}^c}{a_\mathrm{A}^a\, a_\mathrm{B}^b} \right)
\end{aligned}$$

In general notation, this is

$$\Delta_r\mu = \Delta_r\mu^\circ + RT \ln \prod_i a_i^{\nu_i} \tag{13.5}$$

which expresses the difference in chemical potential between any two equilibrium states containing constituents i in terms of a *standard* potential difference and a term involving the activities of the products and reactants in a balanced chemical reaction between constituents i. The $\prod_i a_i^{\nu_i}$ term is given the symbol Q, so that

$$\Delta_r\mu = \Delta_r\mu^\circ + RT \ln Q \tag{13.6}$$

As we have just noted, $\Delta_r\mu$ is not necessarily zero, and is not if the system is in a metastable state, but when the system achieves equilibrium with respect to the minimum two constraints (what we have called *stable* equilibrium), $\Delta_r\mu$ becomes zero, the activities in the Q term take on their stable equilibrium values, and $\prod_i a_i^{\nu_i}$ is called K instead of Q. Thus at stable equilibrium,

$$0 = \Delta_r\mu^\circ + RT \ln K$$

or

$$\Delta_r \mu^\circ = -RT \ln K \tag{13.7}$$

Standard states usually refer to pure substances (except for the aqueous standard states) in which $\mu = G$, so this equation is often written

$$\Delta_r G^\circ = -RT \ln K \tag{13.8}$$

This equation has been called, with some reason, the most useful in chemical thermodynamics, and it certainly merits the most careful attention. Several things about it need comment.

First and most important is the fact that the activity product ratio (K) on the right-hand side is independent of variations in the system composition. It is equal to a difference in standard state free energies, and so is a function only of the variables that affect the standard states involved. Once the physical and compositional aspects of the standard state are chosen, the only variables that affect the standard states are temperature or temperature and pressure, because as we have seen, standard states are always defined as of fixed composition. Therefore K must be independent of the individual values assumed by the activities in the reaction, which is to say it is independent of the system composition. In some cases it is also independent of the system pressure, but this depends on the standard states involved. It is a constant for a given system at a given temperature or temperature and pressure, and is called the *equilibrium constant*. Its numerical value for a given system is not dependent on the system actually achieving equilibrium, or in fact even existing. Its value is fixed when the standard states are chosen, which is not to say that its value is always known. A great deal of experimental effort is expended in determining equilibrium constants, and often through them, of standard state free energies of reaction.

A curious fact about this equation (13.8) is that the two sides of the equation refer to completely different physical situations. It is a numerical equality only. The left-hand side refers to a difference in free energies of a number of different physical states, which may individually be almost anything, but one thing they never are; they are never at mutual equilibrium. The right-hand side, on the other hand, refers to a single system, the constituents of which have achieved mutual equilibrium, or more exactly, to the activity product ratio that would be observed if the system reached equilibrium.

The great usefulness of equation (13.8) lies in the fact that knowledge of a few standard state free energies allows calculation of an indefinite number of equilibrium constants. In case we didn't mention it, the tabulated properties, both apparent and standard, discussed in Chapter 7 are all standard state properties. Furthermore these equilibrium constants are very useful pieces of information about any reaction. If K is very large, it shows that a reaction will tend to go "to completion," i.e., mostly products will be present at equilibrium, and if K is small, the reaction hardly goes at all before enough products are formed to stop it. If you are a chemical engineer designing a process to produce some new chemical, it is obviously of great importance to know to what extent reactions should theoretically proceed. The equilibrium constant of course will never tell you whether reactants will actually react, or at what rate; there may be some reason for reaction kinetics being very slow. It indicates the activity product ratio at equilibrium, not whether equilibrium is easily achievable. Another very useful thing about K is that its variation with temperature and pressure is related to the enthalpy and volume changes of the reaction considered, as will be discussed later in this chapter.

13.2. SPECIAL MEANINGS FOR K

Equilibrium constants are also sometimes equal to system properties of interest, such as vapor pressures, solubilities, phase compositions, and so on. This is because quite often it can be arranged that all activity terms drop out (are equal to 1.0) except the one of interest, which can then be converted to a pressure or composition. In Chapter 14 we will see that wherever this is the case, we are dealing with a "buffered" reaction, and the subject is developed further there.

13.2.1. *K Equal to a Solubility*

Quartz–Water Example

In our quartz example (13.3), the equilibrium constant is

$$K = a_{H_4SiO_4}/(a_{SiO_2}a^2_{H_2O}) \tag{13.9}$$

At this point the expression is perfectly general, valid for any conditions, and K is calculable from equation (13.8) if we know the standard state free energies of the three constituents. If the system we are considering is simply quartz in water at T and P, and if we define our standard states to be pure quartz and pure water at T and P, and ideal one molal H_4SiO_4 at T and P, then both quartz and water have activities of 1.0, and

$$\begin{aligned} K &= a_{H_4SiO_4} \\ &= (\gamma_{H_4SiO_4} \cdot m_{H_4SiO_4})/(\gamma_{H_4SiO_4} \cdot m_{H_4SiO_4})^\circ \\ &= (\gamma_{H_4SiO_4} \cdot m_{H_4SiO_4}) \end{aligned} \tag{13.10}$$

because $(\gamma_{H_4SiO_4} \cdot m_{H_4SiO_4})^\circ$ has been defined as 1.0. Then we need only estimate, calculate, or measure $\gamma_{H_4SiO_4}$ in order to get $m_{H_4SiO_4}$, the solubility of quartz, starting from values of ΔG°. Note that on the left-hand side (of 13.8) we entered the free energy of H_4SiO_4 at a concentration of one molal, and on the right hand side calculated its equilibrium concentration, usually a few ppm. Conversely a measurement of quartz solubility could be used to calculate the free energy of H_4SiO_4 in an ideal one molal solution.

Albite–Nepheline Example

Next consider the reaction

$$0.5\,NaAlSi_3O_8 + 2\,H_2O = 0.5\,NaAlSiO_4 + H_4SiO_4(aq) \tag{13.11}$$

Using the same standard states, that is, the pure minerals and water at T and P, the equilibrium constant expression for this reaction also reduces to

$$K = \gamma_{H_4SiO_4} \cdot m_{H_4SiO_4} \tag{13.12}$$

Let's suppose that a measurement of quartz solubility has been used to obtain the free energy of formation (standard or apparent) of H_4SiO_4 in the ideal one molal standard state. This number can then be used (with $\Delta_f G^\circ$ terms for the minerals) to calculate the equilibrium constant of the albite–nepheline reaction (equation (13.11)), giving the equilibrium silica concentration in a solution that may never have been experimentally determined, or perhaps never existed, and in which quartz is not stable. Thus knowing the solubility of quartz, one could in a similar way calculate the silica concentration in fluids in contact with a variety of mineral assemblages.

This is one demonstration of the power of thermodynamics. But with great power come great possible pitfalls. In this case the possible pitfalls are associated with the extent of our knowledge about the speciation of aqueous SiO_2. We measured the concentration of aqueous silica in equilibrium with quartz, which we *called* H_4SiO_4, and we attached a standard state free energy value to this constituent. Then we wrote the albite–nepheline reaction so as to produce the same constituent ("H_4SiO_4"). Therefore what we calculate from equation (13.12) is not necessarily the silica concentration in water equilibrated with albite and nepheline, but the activity in this solution of whatever species is produced by quartz solubility, which we have called H_4SiO_4. In other words, in the albite–nepheline system, there might be other aqueous species involved, which are not provided for in equation (13.11). We can only calculate the properties of constituents for which we have data. This is self-evident, but can be forgotten.

While we are on the subject, let's look at the same situation in a different way, as another illustration of the use of activities. If we write

$$NaAlSi_3O_8 = NaAlSiO_4 + 2SiO_2$$

we can calculate the activity of silica in the assemblage albite plus nepheline. Thus

$$K = a_{NaAlSiO_4} \cdot a^2_{SiO_2} / a_{NaAlSi_3O_8}$$

Using standard states of the pure minerals albite, nepheline, and quartz at T and P, and with a system consisting of pure albite and nepheline, this reduces to

$$K = a^2_{SiO_2}$$

This will produce a value of a_{SiO_2} of less than one, because quartz is not stable in this system. Suppose a_{SiO_2} works out to be 0.3. What does this mean? How can we have the activity of a mineral in a system where it does not exist? Well, in the first place, we are not actually calculating the activity of quartz. Quartz would have an activity of one in this system, if it were there. 0.3 is the activity of component SiO_2 in the system, and could be thought of as the mole fraction of SiO_2 in a mineral composed of SiO_2 and XO_2, where SiO_2 and XO_2 form an ideal solution. In other words, for quartz to exist in this system, it would have to be diluted with XO_2 to a considerable extent.

Now having the activity of SiO_2 in the albite–nepheline system, we can calculate the H_4SiO_4 activity from equation (13.9). Where previously the silica activity was 1.0 (presence of quartz), now it is 0.3 (presence of albite + nepheline). Thus

$$\begin{aligned} a_{H_4SiO_4} &= K \cdot a_{SiO_2} \cdot a_{H_2O}^2 \\ &= K \cdot a_{SiO_2} \end{aligned}$$

since the activity of water is 1.0. In this case if the SiO_2 activity is 0.3, we calculate that the silica concentration in equilibrium with albite–nepheline is 0.3 times the concentration in equilibrium with quartz. All we have done really is to break up reaction (13.11) into two separate reactions:

$$\begin{aligned} NaAlSi_3O_8 &= NaAlSiO_4 + 2SiO_2 \\ 2SiO_2(s) + 4H_2O &= 2H_4SiO_4(aq) \\ \hline NaAlSi_3O_8 + 4H_2O &= NaAlSiO_4 + 2H_4SiO_4(aq) \end{aligned}$$

but doing it the second way helps to show that it doesn't really matter what we call the silica solute. Whatever it is, we can calculate its activity in the albite–nepheline case. This being the case, we could stop the pretense of knowing what it is ("H_4SiO_4") and just call it "aqueous silica," $SiO_2(aq)$, and write

$$SiO_2(s) = SiO_2(aq) \tag{13.13}$$

instead of equation (13.3). The properties of $SiO_2(aq)$ will evidently differ from those of H_4SiO_4 by exactly twice the corresponding property of water, but the calculations are unaffected by this. This is the approach taken by Helgeson and co-workers (e.g., Walther and Helgeson, 1977), and works fine except in those cases where it is the speciation of aqueous silica that is in question. Thus equation (13.13) is evidently not very useful if you are investigating the state of hydration of the silica complex.

13.2.2. *K Equal to Fugacity of a Volatile Species*

Another common use of the equilibrium constant is to arrange for it to equal the fugacity of a gaseous species in equilibrium with a mineral assemblage. This usually has reference to dehydration, decarbonation, or desulfidation reactions, where all reacting constituents are minerals except one, which is the gas species. For example, consider one of the simplest and most studied reactions

$$Mg(OH)_2(brucite) = MgO(periclase) + H_2O(water)$$

We will use the subscripts b, p, and w to refer to $Mg(OH)_2$, MgO and H_2O respectively. The equilibrium constant for this reaction is

$$K = a_p \cdot a_w / a_b \tag{13.14}$$

If the standard states for the minerals are the pure phases at T and one bar, then at a pressure on the system of one bar the pure minerals will have activities of one at all temperatures, and

$$\begin{aligned} K &= a_w \\ &= f_w / f_w^\circ \end{aligned}$$

If the standard state of the water is ideal gaseous water at T and one bar, then $f_w^\circ = 1$ at all temperatures, and

$$K = f_w \tag{13.15}$$

Thus for this and any similar reaction, the equilibrium constant can always be made equal to the gas fugacity at equilibrium with the (pure) minerals in the reaction. The physical situation is illustrated in Figure 13.1a. The equilibrium constants at various temperatures have been calculated from the data in Robie et al. (1978) (referred to here as RHF) and are shown in Tables 13.1 and 13.3 and plotted in Figure 13.2. We should perhaps emphasize that although K is completely independent of the compositions of the minerals or of the gas phase, it only equals the water fugacity in equilibrium with brucite and periclase when those two minerals are pure.

Note (Table 13.1) that the water fugacity in equilibrium with pure brucite and periclase, while very low (5.47×10^{-7} bars) at 25°C, rises rapidly with increasing temperature to quite high values. At each temperature the water pressure is approximately the same as the fugacity. Since the pressure on the solid phases is only one bar, obviously water having a pressure of greater than one bar cannot coexist with the solids, either stably or metastably. In the situation pictured in Figure 13.1, water at greater than one bar pressure would simply pass through the membrane and exert its pressure on the solids, so that the equilibrium situation implied by the calculations at 300°C and above cannot exist. However, as shown in Figures 13.1b and 13.2, at about 265°C the water fugacity is such that the water pressure is one bar, and the three phases can coexist at the same pressure. With water pressure fixed at one bar, 265°C is evidently the temperature above which brucite would react spontaneously to give periclase and water, and below which the reverse reaction would apply.

13.2.3. *Miscellaneous Special Equilibrium Constants*

Equilibrium constants in a number of other cases assume special meanings and have been given special names. For example ionization constants, stability constants, solubility product constants, hydrolysis constants, and so on are all simply equilibrium constants. In a couple of other fairly common circumstances, approximations to equilibrium constants (made by ignoring activity coefficients) have also been given special names. Thus metamorphic petrologists often use “Distribution Coefficients” (K_D), which are ratios of mole fractions of some constituents in coexisting minerals. These are usually related to equilibrium constants through activity coefficients that are omitted because they are unknown and assumed equal to one. Distribution coefficients can be useful, for example by being uniform throughout an area, hence indicating achievement of chemical equilibrium, and by being compared from one

(a)

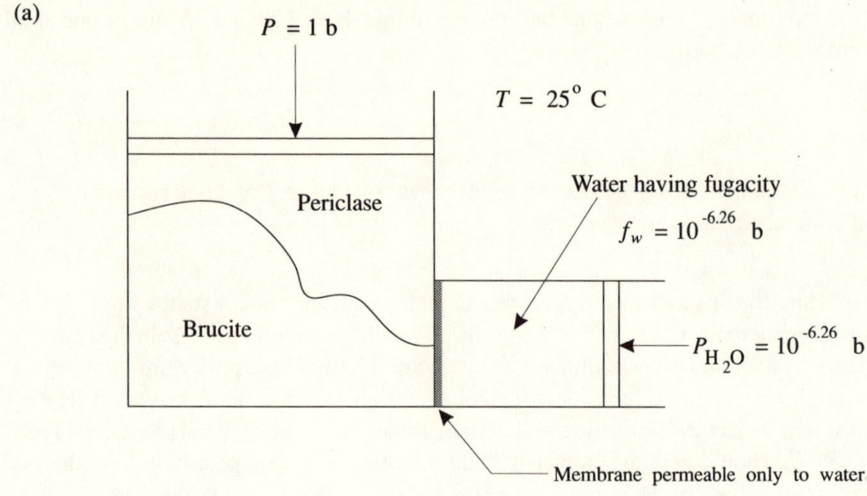

(b)

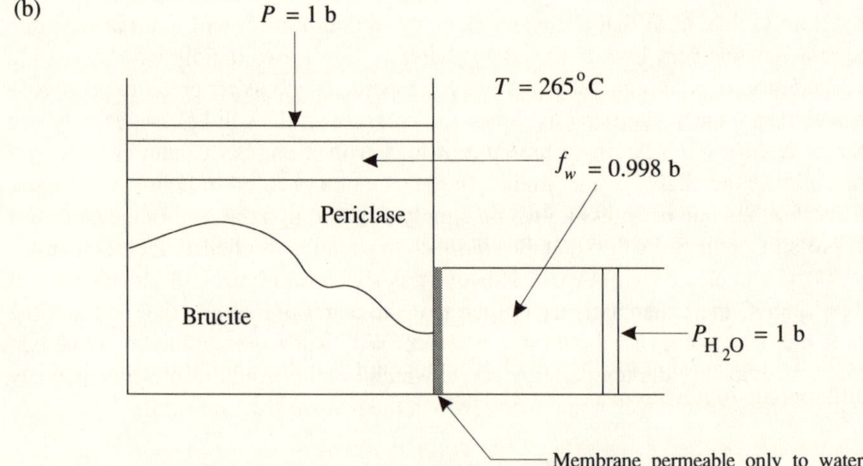

FIG. 13.1. (a) Water having a fugacity (and therefore approximately a pressure) of $10^{-6.26}$ bars at equilibrium with pure brucite and periclase through a membrane permeable only to water. The system is at 25°C. (b) The same system at 265°C. At this temperature the equilibrium water fugacity is one bar, so that water can coexist with the minerals as a phase.

locality to another. They are empirical rather than thermodynamic parameters. Aqueous chemists also avoid dealing with activity coefficients in some cases by keeping the ionic strength of their solutions constant. This supposedly keeps activity coefficients constant so that variations of solution constituents can be investigated by using "stoichiometric equilibrium constants," which are simply equilibrium constants using molalities instead of activities. They are useful at the ionic strength at which they

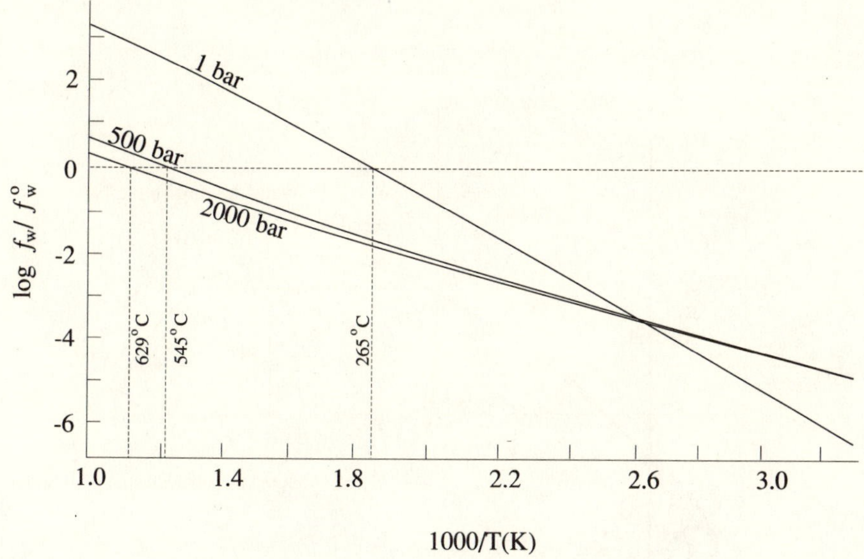

FIG. 13.2. Log of the water fugacity for brucite–periclase equilibrium as a function of reciprocal absolute temperature, at one bar (Table 13.1) and at higher pressures (data for 2000 bars in Table 13.3), using RHF data.

were measured, but would be different at any other ionic strength. They are widely used in the study of complexing reactions.

13.3. CHANGE OF K WITH TEMPERATURE

To find an expression for the effect of temperature on the equilibrium constant, we differentiate $\ln K$ with respect to T. Thus

$$\begin{aligned}\frac{\partial}{\partial T}(R\ln K) &= \frac{\partial}{\partial T}(-\Delta_r G^\circ/T)\\ &= \Delta_r H^\circ/T^2\end{aligned} \tag{13.16}$$

To integrate this expression, we need to know how $\Delta_r H^\circ$ varies with temperature, a subject we investigated in Chapter 7. At that time we had not yet discussed the full implications of the superscript °, but because we have decided that the standard state will always have the same temperature as the system and we need not distinguish between T and T°, the conclusions about effects of temperature reached in Chapter 7 will apply to all standard state properties.

A simple way to visualize the contributions of $\log K_{T_r}$, $\Delta_r H^\circ$ and $\Delta_r C_p^\circ$ to the variation of $\log K$ with T is shown in Figure 13.3. Imagine the firing of a gun, where $\log K_{T_r}$ represents the elevation, $\Delta_r H^\circ$ the inclination, and $\Delta_r C_p^\circ$ controls the curvature of the trajectory.

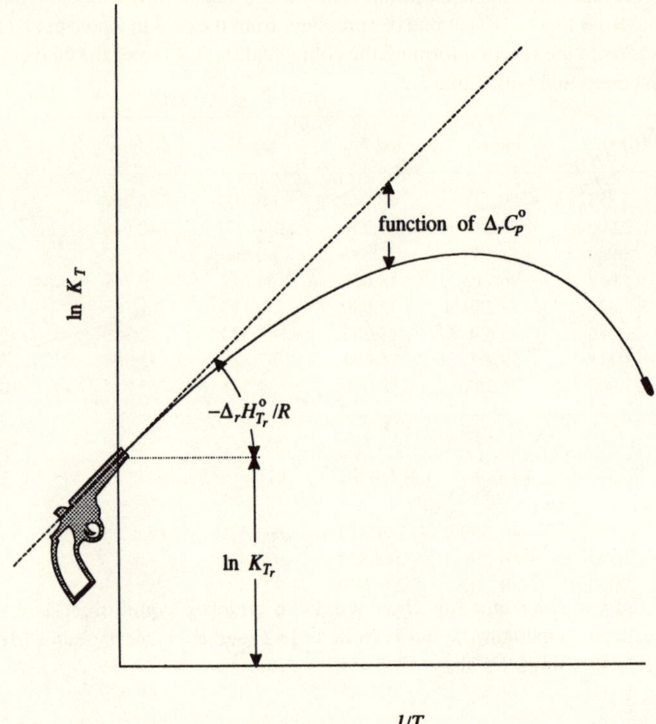

FIG. 13.3. How $\log K_{T_r}$, $\Delta_r H^\circ$ and $\Delta_r C_p^\circ$ contribute to the variation of $\log K$ with temperature.

13.3.1. $\Delta_r C_p^\circ = 0$

To start with the simplest possibility, we could suppose $\Delta_r H^\circ$ to be unaffected by temperature, i.e., to be a constant during integration. This assumption is never true, but is often sufficiently close to the truth (i.e., $\Delta_r H^\circ$ varies little with temperature) that the results are useful, at least for small temperature intervals, and for some purposes. Integrating from T_1 to T_2 with $\Delta_r H^\circ$ constant, we have

$$\ln K_{T_2} - \ln K_{T_1} = -\frac{\Delta_r H^\circ}{R}\left(\frac{1}{T_2} - \frac{1}{T_1}\right) \tag{13.17}$$

Another commonly used expression embodying the same assumption is obtained by substituting $\Delta_r H^\circ - T_r \Delta S^\circ$ for $\Delta_r G^\circ$ in equation (13.8), giving

$$\ln K = -\frac{\Delta_r H^\circ}{RT} + \frac{\Delta_r S^\circ}{R} \tag{13.18}$$

Equations (13.17) and (13.18) are the equations of a straight line having a slope of $-\Delta_r H^\circ / R$ on a plot of $\ln K$ versus $1/T$. On such a plot, if $\Delta_r H^\circ$ is in fact not

Table 13.1 Calculation of the equilibrium constant and standard enthalpy of reaction for the reaction $Mg(OH)_2 = MgO + H_2O$ at one bar pressure, from the data in Robie et al. (1978). The $\log K$ terms refer to the reaction forming the compound (b, p, w) from the elements, as described in connection with Table 7.2

T(K)	$1000/T$	$\log K_p$	$\log K_w$	$\log K_b$	$\log K_r$	$K = f_{H_2O}$
298.15	3.354	99.721	40.044	146.027	−6.262	5.47E−07
400	2.500	72.887	29.236	104.771	−2.648	0.00225
500	2.000	57.180	22.883	80.640	−0.577	0.265
600	1.667	46.712	18.629	64.572	0.769	5.875
700	1.429	39.239	15.580	53.113	1.706	50.816
800	1.250	33.636	13.285	44.533	2.388	244.34
900	1.111	29.280	11.494	37.870	2.904	801.68
1000	1.000	25.757	10.058	32.541	3.274	1879.3
T(K)	$1000/T$	$\Delta_f H^\circ_p$ kJ mol^{-1}	$\Delta_f H^\circ_w$ kJ mol^{-1}	$\Delta_f H^\circ_b$ kJ mol^{-1}	$\Delta_r H^\circ$ kJ mol^{-1}	
298.15	3.354	−601.490	−241.814	−924.540	81.236	
400	2.500	−601.501	−242.836	−924.416	80.079	
500	2.000	−601.302	−243.820	−923.482	78.360	
600	1.667	−601.031	−244.758	−922.180	76.391	
700	1.429	−600.762	−245.634	−920.691	74.295	
800	1.250	−600.528	−245.444	−919.108	73.136	
900	1.111	−600.351	−247.185	−917.475	69.939	

constant, the slope at any temperature is (still) given by

$$\frac{\partial}{\partial(1/T)}(\ln K) = \frac{\partial}{\partial(1/T)}(-\Delta_r G^\circ/RT)$$
$$= -\Delta_r H^\circ/R \qquad (13.19)$$

In Figure 13.2 we see that the change in $\Delta_r H^\circ$ of about 11 kJ (Table 13.1) over the temperature range 298 to 900 K results in a slight curvature in the $\ln K$ versus $10^3/T$ plot, the slope of which at each temperature is equal to $-\Delta_r H^\circ/R$ at that temperature. Nevertheless, over short temperature intervals the plot is fairly linear. Furthermore it is evident that with greater uncertainty in the plotted values of $\ln K$, the nature of the curvature would become uncertain, and one would be justified in approximating the plot with a straight line having a slope giving about the average $\Delta_r H^\circ$ over the temperature interval.

Using the $\Delta_r C^\circ_p = 0$ ($\Delta_r H^\circ$ constant) approximation for aqueous reactions such as ionization constants can lead to very poor results because these reactions commonly have heat capacity terms that vary considerably with temperature. A useful observation is that for a great many aqueous reactions, $\Delta_r C^\circ_p$ can be made much closer to zero, hence giving a much straighter line on a $\log K$ versus $1/T$ plot, by transforming the reaction into the "isocoulombic" form. "Isocoulombic" refers to

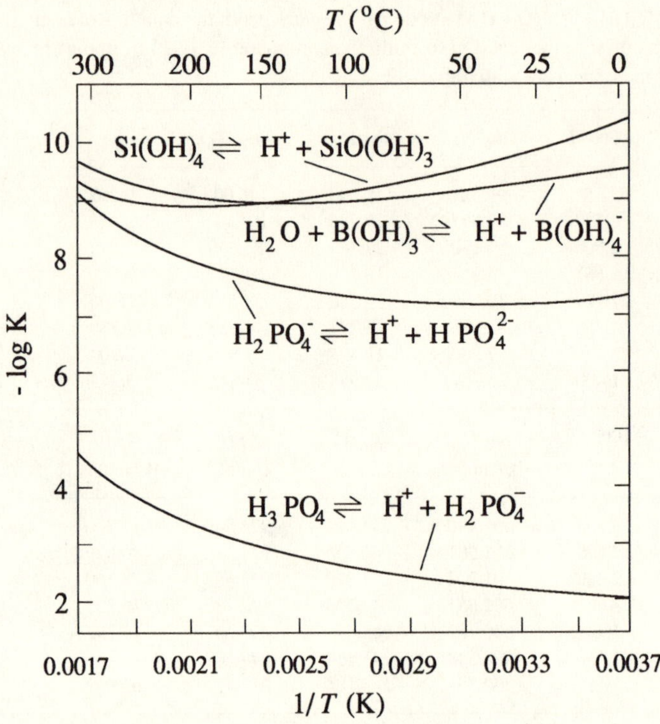

FIG. 13.4. log K for several ionization reactions plotted as a function of the reciprocal of absolute temperature.

an aqueous reaction having the same total charge among the products as among the reactants. For example, consider the reaction

$$H_3PO_4 = H^+ + H_2PO_4^-$$

Log K for this reaction is plotted with some others in Figure 13.4. Note that it has a positive and a negative charge, both on the right hand side. Add to this the reaction

$$OH^- + H^+ = H_2O$$

and we have

$$H_3PO_4 + OH^- = H_2PO_4^- + H_2O$$

which has one negative charge on each side, and therefore a smaller value of $\Delta_r C_p^\circ$. It is now "isocoulombic," and its $\log K$ versus $1/T$ plot is shown in Figure 13.5, along with some others. After extrapolation in this form, the ionization constant of water at higher temperatures must be subtracted to recover the original constant, but this is no problem as the ionization constant of water is reasonably well known. Use of the $\Delta_r C_p^\circ = 0$ approximation for reactions involving ions should always be preceded by this transformation if possible.

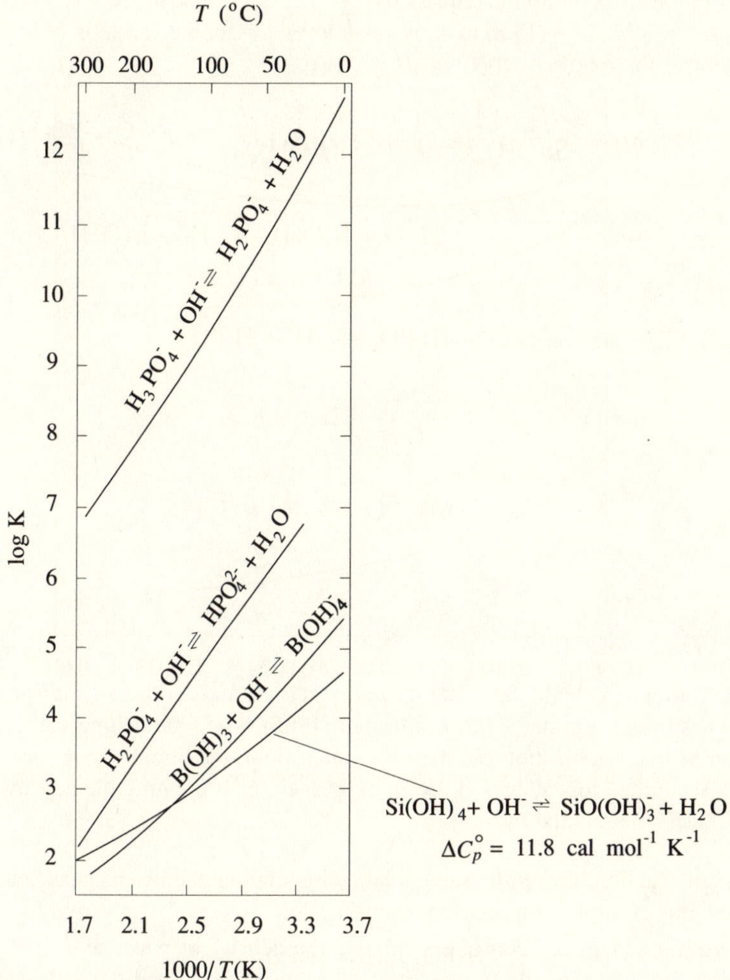

FIG. 13.5. The same reactions as in Figure 13.4 after conversion to the isocoulombic form. Note the much closer approximation to a straight line.

13.3.2. $\Delta_r C_p^\circ$ = *constant*

The $\Delta_r C_p^\circ = 0$ approximation is an obvious choice of options if there are no data whatsoever available for $\Delta_r C_p^\circ$, and can give acceptable results in favorable circumstances. The next possibility is that $\Delta_r C_p^\circ$ is known at 25°C but not at higher temperatures. In this case it is generally better to assume that $\Delta_r C_p^\circ$ is constant as temperature increases. We have already seen the effect of this assumption on the variation of G with T in the calculation of the calcite–aragonite equilibrium in Chapter 8. Because we have an expression for $\Delta_r G^\circ$ as a function of T that includes the

$\Delta_r C_p^\circ = constant$ assumption (equation (8.11)), it is a simple matter to combine this with $\Delta_r G^\circ = -RT \ln K$(13.8) to derive an expression for the variation of $\ln K$ with T. This turns out to be

$$\ln K_T = \ln K_{T_r} - \frac{\Delta_r H^\circ_{T_r}}{R}\left(\frac{1}{T} - \frac{1}{T_r}\right) + \frac{\Delta_r C^\circ_{p_r}}{R}\left(\ln \frac{T}{T_r} + T_r(\frac{1}{T} - \frac{1}{T_r})\right) \quad (13.20)$$

If the various constant terms are combined, we have the expression

$$\ln K = a + b/T + c \ln T \quad (13.21)$$

where a, b, and c are constants with the values

$$a = \ln K_{T_r} + \frac{1}{R}\left(\frac{\Delta_r H^\circ_{T_r}}{T_r} - \Delta_r C^\circ_{p_r} \ln T_r - \Delta_r C^\circ_{p_r}\right)$$

$$b = (T_r \Delta_r C^\circ_{p_r} - \Delta_r H^\circ_{T_r})/R$$

$$c = \Delta_r C^\circ_{p_r}/R$$

13.3.3. *The Density Model*

Another option for calculating K at high temperatures (and in this case high pressures as well) has been suggested by R. E. Mesmer (1985), who found empirically that the variation of many ionization constants in water along the liquid-vapor coexistence curve is accounted for rather well by an expression incorporating the density of the solvent, water, having the form

$$\ln K = p_1 + p_2/T + (p_3 \cdot \ln \rho)/T \quad (13.22)$$

where p_1, p_2, and p_3 are constants and ρ is the density of water at T and P. He suggested, too, that it would work quite well in the supercritical range, because the density term is not restricted in any way to liquid-vapor coexistence.

Because $\ln K$ is proportional to $\Delta_r G^\circ$(13.8) and $\ln \rho$ is specified at specific values of T and P, it follows that an expression relating $\ln K$ and $\ln \rho$ is logically equivalent to one giving $\Delta_r G^\circ$as a function of T and P. In the next chapter we will see that an equation giving $\Delta_r G^\circ$as a function of T and P is called a *fundamental equation*, and that it implicitly contains information on the variation of *all* thermodynamic parameters with T and P. Therefore there are implicit relationships between the parameters p_1, p_2, and p_3 and all other thermodynamic parameters. Anderson et al. (1991) show that for the heat capacity, this relation is

$$\Delta C_p^\circ = \Delta C^\circ_{p_r} \cdot \frac{T(\partial \alpha/\partial T)_P}{T_r(\partial \alpha/\partial T)_{P_r}} \quad (13.23)$$

where α is the coefficient of thermal expansion of H_2O (and V is the molar volume),

$$\alpha = (1/V)(\partial V/\partial T)_P = (\partial \ln V/\partial T)_P = -(\partial \ln \rho/\partial T)_P$$

and $\Delta_r C^\circ_{p_r}$ is the standard heat capacity of reaction in a chosen reference state. The parameters p_1, p_2, p_3 have the values

$$p_1 = \ln K_r + \frac{\Delta H^\circ_r}{RT_r} - \frac{\Delta C^\circ_{p_r} \cdot \alpha_r}{RT_r(\partial\alpha/\partial T)_{P_r}}$$

$$p_2 = -\frac{\Delta H^\circ_r}{R} + \frac{(T_r\alpha_r + \ln\rho_r)\Delta C^\circ_{p_r}}{RT_r(\partial\alpha/\partial T)_{P_r}}$$

and

$$p_3 = \frac{-\Delta C^\circ_{p_r}}{RT_r(\partial\alpha/\partial T)_{P_r}}$$

and the complete expression for $\ln K$ is

$$\ln K = \ln K_r - \frac{\Delta H^\circ_r}{R}\left(\frac{1}{T} - \frac{1}{T_r}\right) + \frac{\Delta C^\circ_{p_r}}{RT_r(\partial\alpha/\partial T)_{P_r}}\left(\frac{1}{T}\ln\frac{\rho_r}{\rho} - \frac{\alpha_r}{T}(T - T_r)\right) \tag{13.24}$$

These and other relationships that follow from the model are summarized in Table 13.2. This empirical formula apparently works well because $(\partial\alpha/\partial T)$ as a function of temperature for water is U-shaped, which gives to the expression for $\Delta_r C^\circ_p$ (if ΔC°_p is negative) an inverted-U shape with a maximum around 100°C, which is the same shape that the C°_p of many aqueous ions have (see Chapter 17). Furthermore the variation of ΔC°_p at higher pressures is also fairly faithfully modeled by the expression for ΔC°_p.

To use the equation, one needs only the values of $\ln K$, ΔH°, and ΔC°_p for the reaction at the reference conditions, (which will frequently be 25°C, 1 bar, but could easily be some other conditions in cases where experimental data at high temperatures or pressures are involved), as well as the density of the solvent, water, at the desired P, T conditions. To obtain estimates of $\Delta_r V^\circ$ and $\Delta_r C^\circ_p$ for the reaction at T, P, values of α and β for the solvent are also required. As an example we show in Figure 13.6 the measured and predicted values of $\log K$ for the ionization of water to 300°C. Several other examples and additional details are given by Anderson et al. (1991). For reactions for which $\log K$, ΔH°, and ΔC°_p at 25°C are available but little else, the Density Model is one of the best ways of obtaining estimates of $\log K$ and other parameters at higher temperatures and pressures.

13.3.4. $\Delta_r C^\circ_p$ *Known*

If the variation of $\Delta_r H^\circ$ with temperature is known, that is, if an expression for $\Delta_r C^\circ_p$ as a function of temperature is available, integration of equation (13.16) results in a more accurate but more cumbersome expression. For example, if the Maier-Kelley expression for $\Delta_r C^\circ_{p_r}$ is used, the result is

$$\ln K_T = \ln K_{T_r} - \frac{\Delta_r H^\circ}{R}\left(\frac{1}{T} - \frac{1}{T_r}\right) + \frac{\Delta_r a}{R}\left(\ln\frac{T}{T_r} + \frac{T_r}{T} - 1\right) + \frac{\Delta_r b}{2R}\left(T + \frac{T_r^2}{T} - 2T_r\right) + \frac{\Delta_r c}{R}\frac{(-T^2 - T_r^2 + 2TT_r)}{2T^2T_r^2} \tag{13.25}$$

The results for other formulations of the heat capacity expression are listed in Appendix B.

Don't confuse $\Delta_r H^\circ$ and $\Delta_r H$. It is important to remember that the enthalpy term given by the slope of log K versus $1/T$ (e.g., Figure 13.2) is a standard state enthalpy of reaction ($\Delta_r H^\circ$), the meaning of which is determined entirely by the standard states of the reaction constituents, and may or may not correspond to an enthalpy that is directly measurable ($\Delta_r H$). In Figure 13.2 the slope is exactly equal at all temperatures to the ΔH° calculable from the tables of Robie et al. (1978) (Table 13.1) because we did not change their standard states in calculating K. Anderson (1970) considers the difference between these terms in more detail.

13.4. CHANGE OF K WITH PRESSURE

We have seen that unlike T and T°, the pressure of the standard state (P°) may or may not be the same as the system pressure (P). Therefore differentiating with respect to system pressure may not be the same as differentiating with respect to standard state pressure. Differentiating ln K with respect to system pressure, we have

$$\frac{\partial}{\partial P}(\ln K) = \frac{-1}{RT}\cdot\frac{\partial}{\partial P}(\Delta_r G^\circ) \tag{13.26}$$

Now, if all standard states involved in the reaction have been defined as being at a constant pressure P, normally 1 bar, these states are unaffected by changes in system pressure P, and

$$\frac{\partial}{\partial P}(\Delta_r G^\circ) = 0 \tag{13.27}$$

and

$$\frac{\partial}{\partial P}(\ln K) = 0 \tag{13.28}$$

Thus in this case the equilibrium constant K is independent of P.

On the other hand, if even one of the reaction constituents has a standard state with a variable pressure, normally $P^\circ = P$, then equation (13.27) is not true, and integration of equation (13.28) requires a knowledge of how $\Delta_r G^\circ$ varies with pressure. We will simplify the following discussion by assuming that reaction constituents having the same physical state (solid, liquid, gas, solute) will have the same kind of standard

Table 13.2 The Density Model Equations

$$p_1 = \ln K_r + \frac{\Delta H_r^\circ}{RT_r} - \frac{\Delta C_{p_r}^\circ \cdot \alpha_r}{RT_r(\partial\alpha/\partial T)_{P_r}}$$

$$p_2 = -\frac{\Delta H_r^\circ}{R} + \frac{(T_r\alpha_r + \ln\rho_r)\Delta C_{p_r}^\circ}{RT_r(\partial\alpha/\partial T)_{P_r}}$$

$$p_3 = \frac{-\Delta C_{p_r}^\circ}{RT_r(\partial\alpha/\partial T)_{P_r}}$$

$$\begin{aligned}\Delta H^\circ &= -R(p_2 + p_3(T\alpha + \ln\rho)) \\ &= \Delta H_r^\circ + \frac{\Delta C_{p_r}^\circ}{T_r(\partial\alpha/\partial T)_{P_r}}\left(T\alpha - T_r\alpha_r + \ln\frac{\rho}{\rho_r}\right)\end{aligned}$$

$$\begin{aligned}\Delta S^\circ &= R(p_1 - p_3\alpha) \\ &= \Delta S_r^\circ + \frac{\Delta C_{p_r}^\circ}{T_r(\partial\alpha/\partial T)_{P_r}}(\alpha - \alpha_r)\end{aligned}$$

$$\begin{aligned}\Delta C_p^\circ &= -RTp_3(\partial\alpha/\partial T)_P \\ &= \Delta C_{p_r}^\circ \cdot \frac{T(\partial\alpha/\partial T)_P}{T_r(\partial\alpha/\partial T)_{P_r}}\end{aligned}$$

$$\begin{aligned}\Delta V^\circ &= -p_3 R\beta \\ &= \frac{\Delta C_{p_r}^\circ \cdot \beta}{T_r(\partial\alpha/\partial T)_{P_r}}\end{aligned}$$

$$\begin{aligned}\Delta G^\circ &= -R(p_1 T + p_2 + p_3 \cdot \ln\rho) \\ &= \Delta G_r^\circ + \Delta H_r^\circ\left(1 - \frac{T}{T_r}\right) + \frac{\Delta C_{p_r}^\circ}{RT_r(\partial\alpha/\partial T)_{P_r}}\left(\alpha_r(T - T_r) + \ln\frac{\rho}{\rho_r}\right)\end{aligned}$$

$$\begin{aligned}\ln K &= p_1 + p_2/T + (p_3 \cdot \ln\rho)/T \\ &= \ln K_r - \frac{\Delta H_r^\circ}{R}\left(\frac{1}{T} - \frac{1}{T_r}\right) + \frac{\Delta C_{p_r}^\circ}{RT_r(\partial\alpha/\partial T)_{P_r}}\left(\frac{1}{T}\ln\frac{\rho_r}{\rho} - \frac{\alpha_r}{T}(T - T_r)\right)\end{aligned}$$

The above equations are for reactions involving aqueous species only. For reactions involving minerals and aqueous species:

$$\begin{aligned}\ln K = \ln K_r &- \frac{\Delta H_r^\circ}{R}\left(\frac{1}{T} - \frac{1}{T_r}\right) + \frac{\Delta a}{R}\left(\ln\frac{T}{T_r} + \frac{T_r}{T} - 1\right) \\ &+ \frac{\Delta b}{2R}\left(T + \frac{T_r^2}{T} - 2T_r\right) + \frac{\Delta c}{R}\left(\frac{-T^2 - T_r^2 + 2TT_r}{2T^2T_r^2}\right) \\ &+ \frac{\Delta C_{p_r}^\circ}{RT_r(\partial\alpha/\partial T)_{P_r}}\left(\frac{1}{T}\ln\frac{\rho_r}{\rho} - \frac{\alpha_r}{T}(T - T_r)\right)\end{aligned}$$

where a, b, c are Maier-Kelley heat capacity coefficients and $\Delta C_{p_r}^\circ$ refers to aqueous species only. In these equations, ρ is the density of pure water

$$\alpha = -(\partial\ln\rho/\partial T)_P$$
$$\beta = (\partial\ln\rho/\partial P)_T$$

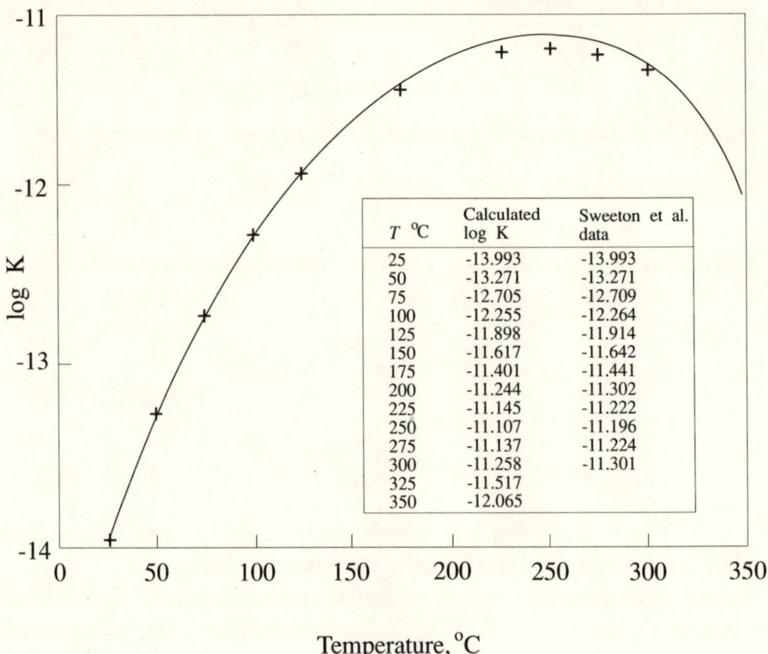

T °C	Calculated log K	Sweeton et al. data
25	-13.993	-13.993
50	-13.271	-13.271
75	-12.705	-12.709
100	-12.255	-12.264
125	-11.898	-11.914
150	-11.617	-11.642
175	-11.401	-11.441
200	-11.244	-11.302
225	-11.145	-11.222
250	-11.107	-11.196
275	-11.137	-11.224
300	-11.258	-11.301
325	-11.517	
350	-12.065	

FIG. 13.6. $\log K$ for the ionization of water as measured by Sweeton et al., and as calculated by the Density Model, using as reference parameters $\log K_{298} = -13.993$; $\Delta H^\circ_{298} = 13340 \text{ cal mol}^{-1}$; $\Delta C^\circ_{p\,298} = -55.3 \text{ cal K}^{-1} \text{ mol}^{-1}$.

state. In practice, this is invariably the case. We first take equation (13.26) a little further by writing

$$-RT \int_{P_1}^{P_2} d\ln K = \int_{P_1}^{P_2} d\Delta_r G^\circ$$

$$-RT(\ln K_{P_2} - \ln K_{P_1}) = \int_{P_1}^{P_2} d\Delta_r G^\circ \tag{13.29}$$

The integration of the standard free energy of reaction in this equation is best done after breaking it up into separate free energy terms for each of the main types of physical states involved in reactions, i.e., condensed phases (solids and liquids), gases and solutes (usually aqueous in our cases). Thus we can write

$$\Delta_r G^\circ = \Delta_r G^\circ_{condensed} + \Delta_r G^\circ_{gases} + \Delta_r G^\circ_{solutes} \tag{13.30}$$

For example, in a reaction such as

$$a\mathrm{A}(s) + b\mathrm{B}(g) + d\mathrm{D}(aq) = e\mathrm{E}(s) + f\mathrm{F}(g) + g\mathrm{G}(aq) \tag{13.31}$$

$$\Delta_r G^\circ_{condensed} = \Delta_r G^\circ(s) = eG^\circ_{\mathrm{E}} - aG^\circ_{\mathrm{A}} \tag{13.32}$$

$$\Delta_r G^\circ_{gases} = \Delta_r G^\circ(g) = fG^\circ_{\mathrm{F}} - bG^\circ_{\mathrm{B}} \tag{13.33}$$

and

$$\Delta_r G^\circ_{solutes} = \Delta_r G^\circ(aq) = gG^\circ_{\mathrm{G}} - dG^\circ_{\mathrm{D}} \tag{13.34}$$

Now we have seen that the change of $\Delta_r G^\circ$ with pressure is in general

$$\frac{\partial}{\partial P}(\Delta_r G^\circ) = \Delta_r V^\circ \tag{13.35}$$

For condensed phases for which the constant volume approximation is appropriate (§10.6.1), this results in

$$\begin{aligned}\int_{P_1}^{P_2} d\Delta_r G^\circ &= \int_{P_1}^{P_2} \Delta_r V^\circ(s) dP \\ &= \int_{P_1}^{P_2} \Delta_s V^\circ dP \\ &= \Delta_s V^\circ (P_2 - P_1)\end{aligned} \tag{13.36}$$

where $\Delta_s V^\circ$ is the volume analogue of G in equation (13.32).

For gases and supercritical fluids, fugacities are normally used, and the standard state is normally chosen as the ideal gas at the system temperature (T) and one bar, i.e., a fixed pressure standard state (P° = 1 bar), so that normally

$$\int_{P_1}^{P_2} d\Delta_r G^\circ_{gases} = 0 \tag{13.37}$$

If however $P^\circ = P$, then f° for each gas is not 1.0 but the fugacity of the pure gas at the system T and P, and f is less than f° because of the presence of other components. In this case it is best to refer to the *activity* of each gas ($= f/f^\circ$), and the change in standard state free energy of the gas components is

$$\int_{P_1}^{P_2} d\Delta_r G^\circ_{gases} = RT \ln Q(g) \tag{13.38}$$

where $Q(g)$ is a term having the form of an equilibrium constant, but involving only the standard state fugacities of the gaseous constituents of a reaction. For example in reaction (13.31)

$$Q(g) = \frac{\left(\frac{(f^\circ_{\mathrm{F}})_{P_2}}{(f^\circ_{\mathrm{F}})_{P_1}}\right)^f}{\left(\frac{(f^\circ_{\mathrm{B}})_{P_2}}{(f^\circ_{\mathrm{B}})_{P_1}}\right)^b} \tag{13.39}$$

At high temperatures f° at P_1 is usually 1.0 (even though the standard state is the real gas and not an ideal gas), so

$$Q(g) = \frac{(f^\circ_{\mathrm{F}})^f_{P_2}}{(f^\circ_{\mathrm{B}})^b_{P_2}} \tag{13.40}$$

Note that when the standard state is fixed at one bar for all gases $Q(g) = 1$, so (13.38) becomes zero as previously noted (13.37).

For solutes no general expression exists to substitute for the volume integral. To integrate partial molar volumes as a function of pressure one must have an analytical expression for the effect of P on V, either derived empirically or from some model (see HKF model, Chapter 17). We are left with the general expression

$$\int_{P_1}^{P_2} d\Delta_r G^\circ_{solutes} = \int_{P_1}^{P_2} \Delta_r V^\circ(aq)dP \tag{13.41}$$

where as before $\Delta_r V^\circ(aq)$ is the volume analogue of $\Delta_r G^\circ(aq)$ in equation (13.34).

Thus in general terms, for the change in K with pressure, integration of (13.29) gives

$$-RT(\ln K_{P_2} - \ln K_{P_1}) = \Delta_s V^\circ(P_2 - P_1) + RT\ln Q(g) + \int_{P_1}^{P_2} \Delta_r V^\circ(aq)dP$$

or

$$\ln K_{P_2} = \ln K_{P_1} - \frac{\Delta_s V^\circ}{RT}(P_2 - P_1) - \ln Q(g) - \frac{1}{RT}\int_{P_1}^{P_2} \Delta_r V^\circ(aq)dP \tag{13.42}$$

Normally of course the expression for the variation of K with P is simpler than this, perhaps because all three states of matter may not be present, but also because it is quite unusual to use a variable pressure standard state for constituents whose fugacities are known or sought, (because this adds complexities rather than simplifying matters), and the $\ln Q(g)$ term is therefore essentially never required. To take a real example, let's consider the brucite–periclase reaction again. We have discussed the variation of the equilibrium constant for the brucite–periclase–water reaction with temperature at one bar, and showed that the equilibrium temperature for the reaction at one bar is about 265°C. Calculation of the equilibrium temperature of dehydration reactions such as this one at higher pressures was discussed briefly in §13.2.2. Here we will discuss the reaction in different terms to demonstrate the relationships between activities, standard states and equilibrium constants.

What is the equilibrium constant for this reaction at say 2000 bars and 25°C?

Case 1:

Consider first the case where the activities of brucite, periclase, and water have their standard states fixed at one bar. According to our discussion above and equation (13.28), the equilibrium constant for the reaction at 2000 bars will be the same as at 1 bar, i.e., $10^{-6.262}$ (Table 13.1). However, the activities of brucite and periclase are not the same as at one bar. They are calculated from equation (12.31) (and shown in Table 13.3):

$$\ln a_{p,2000b} = 1.1248(2000 - 1)/(8.3143 \times 298.15)$$

$$a_{p,2000b} = 2.477$$

and

$$\ln a_{b,2000b} = 2.463(2000 - 1)/(8.3143 \times 298.15)$$

$$a_{b,2000b} = 7.288$$

At 2000 bars, then

$$K = a_p \cdot a_w / a_b$$

$$10^{-6.262} = 2.477 \cdot a_w / 7.288$$

so

$$\begin{aligned} a_w &= f_w/(f_w^\circ = 1) \\ &= 10^{-5.793} \end{aligned}$$

The activity of water we have calculated ($10^{-5.793}$) is in fact the water fugacity, because all standard states are fixed at one bar, which means that f_w° is fixed at 1.0. From this example one can see that because we know the equilibrium constant as a function of T at one bar and it is independent of P, we therefore know it at all P's and T's, and because we can calculate the activities of the solid phases at any P and T, we are able to calculate the fugacity of water required to maintain brucite–periclase equilibrium at any P and T. In the particular case above, squeezing the crystals to 2000 bars raises the water fugacity in equilibrium with the two phases by roughly a factor of three.

Case 2:

Next, consider the case where the mineral standard states are of the variable pressure type, that is, the standard states for brucite and periclase are taken to be the pure phase at the system P and T, while water continues to have a standard state of ideal gaseous water at T and one bar. Because there is essentially no mutual solution between the three phases they are essentially pure when at mutual equilibrium, and the mineral activities are therefore 1.0 at all Ps and Ts. This is only an apparent simplification, because now the equilibrium constant varies with pressure. Its value at 2000 bars, 25°C can be calculated from equation (13.42), thus

$$\begin{aligned} \log K_{2000b} &= \log K_{1b} - \Delta_s V^\circ (P_2 - P_1)/2.303RT \\ &= -6.262 - (1.1248 - 2.463)(2000 - 1)/(2.303 \times 8.3143 \times 298.15) \\ K_{2000b} &= 10^{-5.793} \\ &= f_w \end{aligned}$$

Note that f_w° in the $\ln Q(g)$ term is one, so that term drops out. Now we have a different value for the equilibrium constant because of the different standard states, but this time K equals the water fugacity required for brucite–periclase equilibrium (equation (13.15)).

Case 3:

Finally, consider the case where all three constituents have as their standard states the pure phase at P and T. Now the activities of all three components when pure are one at all Ps and Ts, and $K = 1$ along the univariant equilibrium curve. Activities less than one in any of the phases are attainable by diluting the phase with another substance, but activities greater than one are not possible (except in the case of small grain size or other metastable factors). Naturally, pure water cannot be in equilibrium with pure brucite and periclase except along the univariant curve. At temperatures below the univariant equilibrium, water must be diluted if it is to equilibrate with the pure minerals. The water fugacities found for brucite–periclase equilibrium in the first two cases considered above are converted to this case simply by dividing by the fugacity of pure water at the particular P and T considered. In equation (13.42) this is accomplished by including the ln $Q(g)$ term, which of course requires knowledge of the fugacity of pure water as a function of P and T. For this purpose we have used the values of Haar et al.(1984). Thus at 2000 bars, 25°C,

$$\begin{aligned}\log K_{2000b} &= \log K_{1b} - \Delta_s V^\circ(P_2 - P_1)/2.303RT - \log Q(g) \\ &= -6.262 + 0.469 - \log(0.1287) \\ &= -4.903\end{aligned}$$

$$\begin{aligned}K_{2000b} &= a_w \\ &= 10^{-4.903}\end{aligned}$$

These calculations are summarized for a pressure of 2000 bars in Table 13.3 and illustrated in Figure 13.2.

13.5. FINDING THE UNIVARIANT CURVE USING K

We discussed the determination of the univariant (in this case three-phase) equilibrium curve in Chapter 8 in terms of finding the locus of P and T for $\Delta_r G = 0$. This is always the most general method, and other methods for doing this will be discussed in more detail in Chapter 19, but it is useful here to point out that as long as $\Delta_r G = \Delta_r G^\circ$, which is to say as long as 1. the standard states are the pure phases, and 2. the phases at equilibrium are fairly pure; then $K = 1$ can also be used as criterion for the univariant equilibrium. Thus K can be calculated along isobars (as in Table 13.3) or isotherms and the T or P for $K = 1$ found by interpolation.[2]

When this is done for the calculations based on RHF (1978) that we have been doing (Table 13.3), we find that the univariant curve does not coincide with the best available experimental evidence (which in this case is probably the data of Barnes and Ernst, 1963). The calculated curve lies about 30°C to the low-temperature side of the experimentally determined position. If we perform the same calculations using

[2]In Table 13.3 this has been done by finding the least-squares parameters a, b and c of the second order polynomial $\ln K = a(1/T)^2 + b(1/T) + c$ and finding T for $\ln K = 0$. The slope of $\ln K$ versus $1/T$ $(= -\Delta H^\circ/R)$ then is $2a(1/T) + b$.

Table 13.3 Brucite Dehydration Calculation using RHF data

T(K)	$1000/T$	T°C	logK 1 bar	a_p 2000b	a_b 2000b	f_w 2000b	$\log f_w$ 2000b	f_w° bars	$\log f_w/f_w^\circ$	slope	$\Delta_r H^\circ$ J mol^{-1}	$\Delta_r H^\circ$ cal mol^{-1}
298.15	3.354	25.00	−6.262	2.477	7.288	1.61E−06	−5.793	0.1287	−4.903	−1918	36722	8777
400.00	2.500	126.85	−2.648	1.966	4.395	5.03E−3	−2.299	7.212	−3.157	−2121	40603	9704
500.00	2.000	226.85	−0.577	1.717	3.269	5.04E−01	−0.298	60.78	−2.081	−2240	42876	10248
600.00	1.667	326.85	0.769	1.569	2.683	1.00E+01	1.002	217.1	−1.335	−2319	44391	10610
700.00	1.429	426.85	1.706	1.472	2.330	8.05E+01	1.906	483.6	−0.779	−2375	45473	10868
800.00	1.250	526.85	2.388	1.402	2.096	3.65E+02	2.563	810.8	−0.346	−2418	46284	11062
900.00	1.111	626.85	2.904	1.351	1.931	1.15E+03	3.059	1134.1	0.005	−2451	46915	11213
902.15	1.108	629.00		1.350	1.928				0.000	−2451	46927	11216
1000.00	1.000	726.85	3.274	1.311	1.808	2.59E+03	3.414	1411.5	0.264	−2477	47420	11664

the data of HDNB (1978), we expect the calculated curve to correspond exactly to the experimental data, because the data of Barnes and Ernst were part of the data set used by HDNB to deduce the properties of brucite and periclase. The water data they used (Helgeson and Kirkham 1974a) is not the same as the Haar et al. data used here, but the differences are small. These calculations are summarized for 2000 bars pressure in Table 13.4, and Figure 13.7 shows that the calculated univariant curve does indeed agree with the experimental data. In Figure 13.8 we also show contours of water activity, calculated in the same way as the univariant curve, that is by finding by interpolation at each pressure the temperature at which certain values of (f_w/f_w°) occur. These curves show by how much the univariant curve would be displaced if the water activity were to be changed from 1.0 to lower values, for example by dilution. It is relatively easy to calculate these curves for given activities; what is much more difficult is to know what aqueous solution compositions will actually give these water activities. The determination of the activities of constituents in solutions is the problem of "mixing models," discussed in Chapter 15.

The work of Barnes and Ernst (1963) provides some information on this, because they experimentally determined the position of the brucite–periclase equilibrium not only in pure water but in 5m and 12.5m NaOH solutions. The NaOH does not enter the reaction in any way except to dilute the water, so that we can compare the experimental and theoretical curves, and thus determine the activity of water in 5m and 12.5m NaOH solutions at the P and T values determined by Barnes and Ernst. This comparison is made in Figure 13.8 and Table 13.5, where we see that, as pointed out by Barnes and Ernst, water and NaOH mix in a fairly ideal way under these supercritical conditions, the activity coefficient of water being generally about 0.7 to 0.8. Despite the value of these results, we should point out that if the mixing properties of water–NaOH solutions is the primary interest (it was only one of several goals for Barnes and Ernst), determination of displaced univariant curves is not the most direct way of proceeding. One would normally want to perform experiments on the densities of these solutions, and extract fugacities as described in Chapter 11. This is more easily said than done, however.

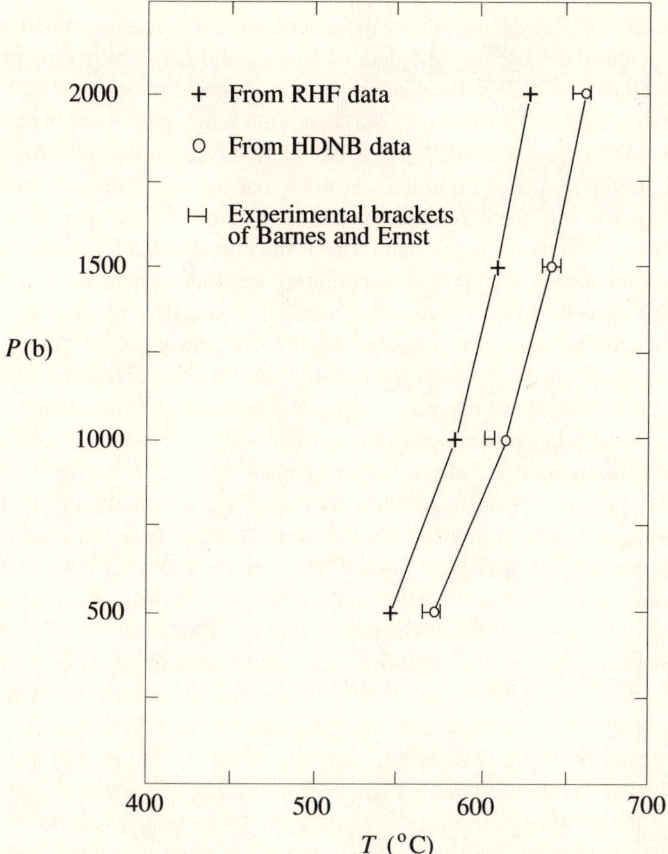

FIG. 13.7. Brucite-periclase-water equilibrium temperatures predicted from RHF and HDNB data at various pressures compared to the experimental brackets of Barnes and Ernst (1963).

Table 13.4 Brucite Dehydration Calculation using HDNB data

T(K)	$1000/T$	T°C	logK 1 bar	a_p 2000b	a_b 2000b	f_w 2000b	$\log f_w$ 2000b	f_w° bars	$\log f_w/f_w^\circ$	slope	$\Delta_r H^\circ$ J mol^{-1}	$\Delta_r H^\circ$ cal mol^{-1}
298.15	3.354	25	−6.262	2.477	7.288	1.61E−06	−5.793	0.1287	−4.903	−1666	31890	7622
373.15	2.680	100	−3.627	2.064	4.889	5.59E−04	−3.253	3.194	−3.757	−1955	37423	8944
473.15	2.113	200	−1.225	1.771	3.496	1.18E−01	−0.930	38.23	−2.512	−2198	42072	10055
573.15	1.745	300	0.302	1.603	2.810	3.51E+00	0.546	163.3	−1.667	−2356	45099	10779
673.15	1.486	400	1.347	1.494	2.410	3.59E+01	1.555	403.4	−1.051	−2467	47266	11287
773.15	1.293	500	2.099	1.419	2.151	1.90E+02	2.280	720.8	−0.578	−2549	48803	11664
873.15	1.145	600	2.659	1.363	1.970	6.59E+02	2.819	1050.6	−0.202	−2613	50019	11955
935.15	1.069	662		1.335	1.884				0.000	−2645	50642	12104
973.15	1.028	700	3.089	1.320	1.838	1.71E+03	3.233	1342.7	0.105	−2663	50985	12186
1073.15	0.932	800	3.426	1.287	1.736	3.60E+03	3.556	1577.9	0.358	−2704	51771	12374

Table 13.5 Calculation of Water Activity in NaOH Solutions

Pressure	Eqbm T°C			f_w° at eqbm temps			Calc'd f_w for b/p eqbm.			$X_w \times f_w^\circ$		$\gamma = f_w/(X_w \times f_w^\circ)$	
bars	H_2O	5m NaOH	12.5m NaOH	H_2O	5m NaOH	12.5m NaOH	H_2O	5m NaOH	12.5m NaOH	5m NaOH	12.5m NaOH	5m NaOH	12.5m NaOH
2000	662	635	595	1240.8	1158.8	1034.7	1240.8	939.7	602.8	1063.0	844.6	0.88	0.71 §
1500	641	613	578	828.9	869.4	785.3	828.9	589.4	373.3	797.5	640.9	0.74	0.58
1000	615	588	560	631.0	615.7	569.5	631.0	460.7	325.2	564.8	464.8	0.82	0.70
500	572	558	538	344.3	367.7	354.2	344.3	287.7	220.2	337.3	289.1	0.85	0.76

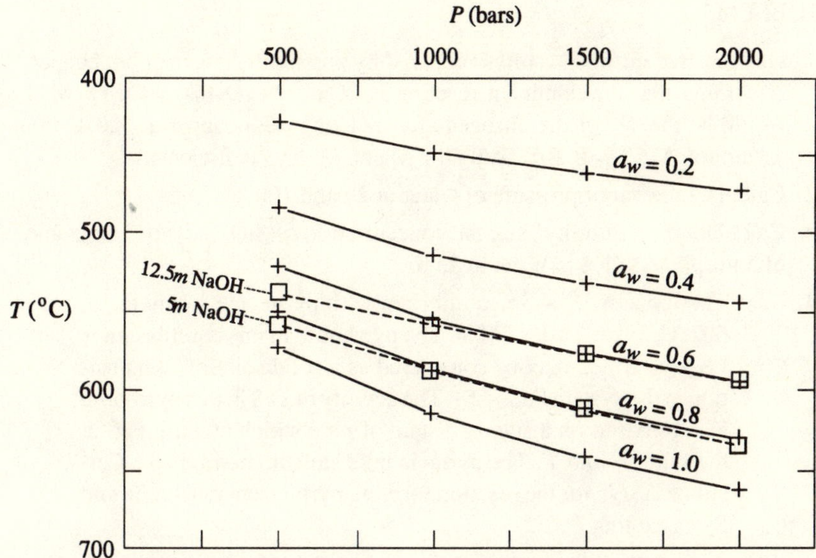

FIG. 13.8. Brucite-periclase equilibrium temperatures as a function of pressure for various water activities. The standard state is pure water at the system P and T. Also shown (dashed lines) are the positions of the brucite-periclase equilibrium in $5m$ and $12.5m$ NaOH solutions from the data of Barnes and Ernst (1963). Superposition of the two sets of curves allows determination of the activity of water in the NaOH solutions.

PROBLEMS

1. Oxygen-free nitrogen is often prepared by passing N_2 gas over hot copper. Assuming that the dominant reaction is $2Cu(s) + \frac{1}{2}O_2(g) = Cu_2O$, what would be the f_{O_2} of the nitrogen after reacting with copper at 600°C (a). assuming $\Delta_r C_p^\circ = 0$. (b). Using the Maier-Kelley coefficients.
2. Calculate the vapor pressure of water at 25 and 100°C.
3. Calculate the solubility (i.e., the concentration of $SiO_2(aq)$) of quartz and of amorphous silica in water at 25°C.
4. (a). The fugacity of S_2 in equilibrium with pyrite and pyrrhotite at 602°C, 1 bar is $10^{-1.95}$ bar. The pyrrhotite in this equilibrium is $Fe_{0.92}S$, which may be considered as a solid solution composition in the system $FeS - S_2$. The activity of FeS in this pyrrhotite is 0.46 based on a standard state of pure stoichiometric FeS at the same P and T. The pyrite is pure stoichiometric FeS_2. Calculate $\Delta_r G^\circ$ for the reaction forming pyrite from pyrrhotite and S_2 gas at this P, T.
   (b). Is the pyrrhotite involved in this free energy term FeS or $Fe_{0.92}S$?
   (c). The fugacity of sulfur in equilibrium with iron and stoichiometric FeS at 602°C, 1 atm. is $10^{-12.5}$ atm. Calculate the standard free energy of formation of pyrite from its elements at 602°C, 1 atm.
5. A wollastonite-bearing contact metamorphic zone is observed adjacent to a granite which has intruded a quartz-bearing limestone horizon. Heat-flow calculations indicate that the maximum temperature achieved at a given distance from the contact is given by

$$T = 760 - 7.66d + 0.0396d^2$$

   where T is the temperature in °C, and d is the distance in feet. Stratigraphic considerations put the pressure at the time of intrusion at 2000 bars. If the contact zone is 50 feet wide, what was the f_{CO_2} in the pore fluid of the limestone? If you assume that the pore fluid was a $H_2O - CO_2$ solution, what do you need to know to calculate its composition?
6. If the solubility of quartz in water at 600°C, 4000 bars is 0.208 m, calculate the silica content of water in equilibrium with albite and nepheline at the same T,P.
7. Calculate the pressure of the plagioclase–omphacitic jadeite–quartz equilibrium at 500 K if the activity of albite in the plagioclase is 0.9 and the activity of jadeite in the omphacite is 0.7.
8. The SiO_2 concentration of seawater is about 7 ppm. If you assume that radiolaria tests composed of amorphous silica are in equilibrium with seawater, calculate the Gibbs free energy of formation from the elements of $H_4SiO_4(aq)$.

14

HETEROGENEOUS AND OPEN SYSTEMS

I have heard the (mischievous) argument that every natural system has as many components as elements in the periodic chart—we only need analyze for them. The corollary is that all natural systems have somewhere in the neighbourhood of 100 degrees of phase rule freedom, discounting isotopes. My reply is that this is nonsense—natural systems don't have components—only our models do, and if we're sensible, we choose model components that have predictive value for the problem at hand. T.M. Gordon (personal communication, 1989)

14.1. INTRODUCTION

Up to this point, with minor exceptions, we have discussed only closed systems, that is, systems having a fixed composition, and have for the most part not bothered to consider whether the system was homogeneous (one phase) or heterogeneous (more than one phase). We must now explicitly consider the implications of having more than one phase, and of the transfer of matter between phases and into and out of the system.

14.2. OPEN SYSTEMS

There are two kinds of open systems that concern us, illustrated in Figure 14.1. In the first kind, the open system is simply a separate phase in a system that is closed overall, illustrated in Figure 14.1a. The phases are free to change composition by exchange of components in response to changes in the conditions (say P and T) of the closed system. The phases in a crystallizing magma are examples of open systems in this sense. In the second kind we distinguish the system from an environment, joined by means of a membrane permeable only to certain components (Figure 14.1b). This enables the system to change composition in response to conditions in the environment, which may be quite different from those in the system. The membrane may be real, as in the case of experimental systems in which hydrogen, having an externally controlled fugacity, diffuses into the system through the platinum walls of the system, or it may be imaginary, as when it is used to help clarify our thinking about metasomatic processes. Of course, if you think carefully about it, the difference between the two cases is not always very clear-cut. For example in the crystallizing magma, quartz and the melt in which it is crystallizing could be considered as the "environment" for the other crystallizing phases, controlling their chemical potential of SiO_2. Nevertheless, the distinction is usually clear enough for our purposes. We will first consider open systems in the first sense in the derivation of the phase rule, and then consider cases of "membrane" or "osmotic" equilibria, which is the kind that the term "open system" has generally come to mean in geochemistry.

(a)

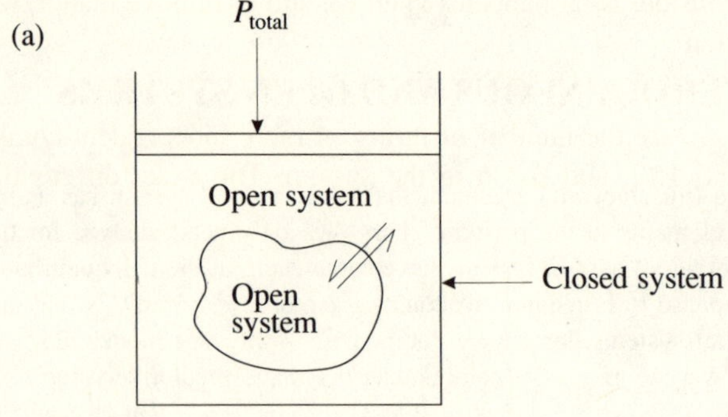

(b)

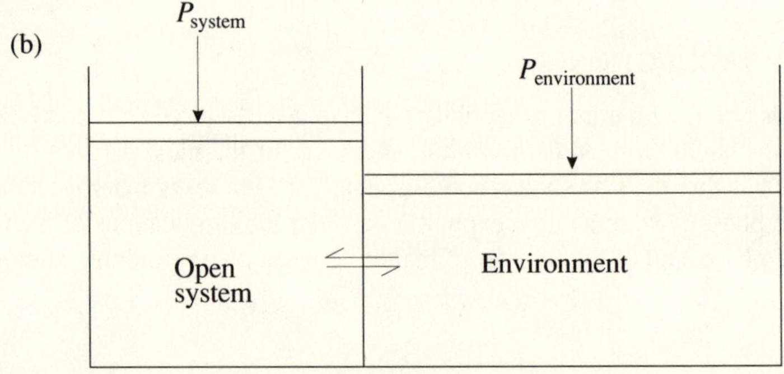

FIG. 14.1. Two types of open systems. (a) A closed heterogeneous system in which each phase is an open system. (b) A closed composite system which is separated into two subsystems by a membrane permeable only to one or some components. One subsystem is called "the system" and the other "the environment."

14.2.1. *Thermodynamic Potentials in Open Systems*

We saw in Chapter 5 that the function U = U(S, V) expressed as a total differential is[1]

$$d\mathrm{U} = \left(\frac{\partial \mathrm{U}}{\partial \mathrm{S}}\right)_{\mathrm{V}} d\mathrm{S} + \left(\frac{\partial \mathrm{U}}{\partial \mathrm{V}}\right)_{\mathrm{S}} d\mathrm{V}$$

and that because $(d\mathrm{U}/d\mathrm{S})_{\mathrm{V}} = T$ and $(d\mathrm{U}/d\mathrm{V})_{\mathrm{S}} = -P$, it may also be written

$$d\mathrm{U} = Td\mathrm{S} - Pd\mathrm{V} \tag{14.1}$$

In this equation we implicitly consider the system to be capable of changes in U, S, and V, and hence can be heated, cooled, or squeezed, but it always has the same

[1] It will be convenient to begin first with the parameters for the total system (i.e., U rather than U, and then explicitly show the changes introduced by using molar properties.

composition. To consider what happens when we add or remove matter, we need to consider the function

$$\mathrm{U} = \mathrm{U}(\mathrm{S}, \mathrm{V}, n_1, n_2, \cdots, n_c)$$

where $n_1, n_2, \ldots, n_c$ are the number of moles of the c independent compositional terms (components) 1, 2, and so on in the system. The exact differential of this function is

$$d\mathrm{U} = \left(\frac{\partial \mathrm{U}}{\partial \mathrm{S}}\right)_{\mathrm{V},n} d\mathrm{S} + \left(\frac{\partial \mathrm{U}}{\partial \mathrm{V}}\right)_{\mathrm{S},n} d\mathrm{V}$$
$$+ \left(\frac{\partial \mathrm{U}}{\partial n_1}\right)_{\mathrm{S,V},\hat{n}_1} dn_1 + \left(\frac{\partial \mathrm{U}}{\partial n_2}\right)_{\mathrm{S,V},\hat{n}_2} dn_2 + \cdots + \left(\frac{\partial \mathrm{U}}{\partial n_c}\right)_{\mathrm{S,V},\hat{n}_c} dn_c$$

or

$$d\mathrm{U} = Td\mathrm{S} - Pd\mathrm{V} + \sum_{i=1}^{c} \left(\frac{\partial \mathrm{U}}{\partial n_i}\right)_{\mathrm{S,V},\hat{n}_i} dn_i \tag{14.2}$$

where n means $n_1, n_2, \ldots, n_c$ (all components), n_i refers to any individual component i, and $\hat{n}_i$ refers to all components except n_i. This equation still represents a plane tangent to a stable equilibrium surface as described previously, except that there are now a number of added dimensions, that is, another dimension for every component.

Similarly, we could write for the functions

$$\mathrm{A} = \mathrm{A}(T, \mathrm{V}, n_1, n_2, \ldots, n_c)$$

$$\mathrm{H} = \mathrm{H}(\mathrm{S}, P, n_1, n_2, \ldots, n_c)$$

and

$$\mathrm{G} = \mathrm{G}(T, P, n_1, n_2, \ldots, n_c)$$

the exact differentials

$$d\mathrm{A} = -\mathrm{S}dT - Pd\mathrm{V} + \sum_{i=1}^{c} \left(\frac{\partial \mathrm{A}}{\partial n_i}\right)_{T,\mathrm{V},\hat{n}_i} dn_i \tag{14.3}$$

$$d\mathrm{H} = Td\mathrm{S} + \mathrm{V}dP + \sum_{i=1}^{c} \left(\frac{\partial \mathrm{H}}{\partial n_i}\right)_{\mathrm{S},P,\hat{n}_i} dn_i \tag{14.4}$$

and

$$d\mathrm{G} = -\mathrm{S}dT + \mathrm{V}dP + \sum_{i=1}^{c} \left(\frac{\partial \mathrm{G}}{\partial n_i}\right)_{T,P,\hat{n}_i} dn_i \tag{14.5}$$

The integrated form of equation (14.2) can be found by considering a single phase that is enlarged in size, its temperature, pressure, and composition remaining unchanged. Under these conditions the values of $\left(\frac{\partial U}{\partial n_i}\right)_{\mathrm{S,V},\hat{n}_i}$ are also unchanged. Integration of (14.2) between any two equilibrium states gives

$$\Delta\mathrm{U} = T\Delta\mathrm{S} - P\Delta\mathrm{V} + \sum_{i=1}^{c} \left(\frac{\partial \mathrm{U}}{\partial n_i}\right)_{\mathrm{S,V},\hat{n}_i} \Delta n_i$$

Let the original values of the internal energy, etc., of the system be U, S, V and n. If the system is enlarged to k times its original size, the final values of these parameters (which are all extensive, or homogeneous degree one in the masses of the components) will be kU, kS, kV, and kn. Thus

$$\begin{aligned}\Delta \mathrm{U} &= k\mathrm{U} - \mathrm{U}\\ &= (k-1)\mathrm{U}\end{aligned}$$

and similarly for ΔS and ΔV. Substituting in the above equation, we have

$$(k-1)\mathrm{U} = T(k-1)\mathrm{S} - P(k-1)\mathrm{V} + \sum_{i=1}^{c}\left(\frac{\partial \mathrm{U}}{\partial n_i}\right)_{\mathrm{S,V},\hat{n}_i}(k-1)n_i$$

and, dividing through by $(k-1)$,

$$\mathrm{U} = T\mathrm{S} - P\mathrm{V} + \sum_{i=1}^{c}\left(\frac{\partial \mathrm{U}}{\partial n_i}\right)_{S,V,\hat{n}_i} n_i \tag{14.6}$$

(this derivation from Denbigh, 1966, p. 93)

Analogous integration of equations (14.3) to (14.5), recalling that dT and dP are zero, gives

$$\mathrm{A} = -P\mathrm{V} + \sum_{i=1}^{c}\left(\frac{\partial \mathrm{A}}{\partial n_i}\right)_{T,\mathrm{V},\hat{n}_i} n_i \tag{14.7}$$

$$\mathrm{H} = T\mathrm{S} + \sum_{i=1}^{c}\left(\frac{\partial \mathrm{H}}{\partial n_i}\right)_{\mathrm{V},P,\hat{n}_i} n_i \tag{14.8}$$

$$\mathrm{G} = \sum_{i=1}^{c}\left(\frac{\partial \mathrm{G}}{\partial n_i}\right)_{T,P,\hat{n}_i} n_i \tag{14.9}$$

A number of interesting relationships are inherent in the equations written thus far; in fact, they contain or imply just about everything there is to be said about (model) heterogeneous equilibria.

First, consider equations (14.6) and (14.9). Because we have defined

$$\mathrm{G} = \mathrm{U} - T\mathrm{S} + P\mathrm{V},$$

and

$$\left(\frac{\partial \mathrm{G}}{\partial n_i}\right)_{T,P,\hat{n}_i} = \mu_i$$

it follows that

$$\left(\frac{\partial \mathrm{U}}{\partial n_i}\right)_{\mathrm{S,V},\hat{n}_i} = \left(\frac{\partial \mathrm{G}}{\partial n_i}\right)_{T,P,\hat{n}_i} = \mu_i$$

and of course similarly for $\left(\frac{\partial \mathrm{A}}{\partial n_i}\right)_{T,\mathrm{V},\hat{n}_i}$ and $\left(\frac{\partial \mathrm{H}}{\partial n_i}\right)_{\mathrm{S},P,\hat{n}_i}$. These are entirely equivalent expressions for the chemical potential, although because of our fondness for the

G function, we tend to emphasize the partial molar property $\left(\frac{\partial \mathrm{G}}{\partial n_i}\right)_{T,P,\hat{n}_i}$. It follows that equations (14.2) to (14.5) can be rewritten

$$d\mathrm{U} = Td\mathrm{S} - Pd\mathrm{V} + \sum_{i=1}^{c} \mu_i dn_i \tag{14.10}$$

$$d\mathrm{A} = -\mathrm{S}dT - Pd\mathrm{V} + \sum_{i=1}^{c} \mu_i dn_i \tag{14.11}$$

$$d\mathrm{H} = Td\mathrm{S} + \mathrm{V}dP + \sum_{i=1}^{c} \mu_i dn_i \tag{14.12}$$

$$d\mathrm{G} = -\mathrm{S}dT + \mathrm{V}dP + \sum_{i=1}^{c} \mu_i dn_i \tag{14.13}$$

and in their integrated forms

$$\mathrm{U} = T\mathrm{S} - P\mathrm{V} + \sum_{i=1}^{c} \mu_i n_i \tag{14.14}$$

$$\mathrm{A} = -P\mathrm{V} + \sum_{i=1}^{c} \mu_i n_i \tag{14.15}$$

$$\mathrm{H} = T\mathrm{S} + \sum_{i=1}^{c} \mu_i n_i \tag{14.16}$$

$$\mathrm{G} = \sum_{i=1}^{c} \mu_i n_i \tag{14.17}$$

Equation (14.17) is also the result of Euler's Theorem applied to G as a function homogeneous in the first degree in the masses of the components.

We next point out that equation (14.17) emphasizes what we said in Chapter 9 about the chemical potential. We said that it should be thought of as the free energy per mole of a dissolved substance, and should not be thought of in terms of derivatives or infinitesimal quantities. Equation (14.17) emphasizes this by showing that the total free energy of a system is simply the sum of the number of moles of each component in the system (n_i) times the free energy per mole of that component (μ_i).

14.2.2. *Conditions of Equilibrium*

Suppose now that we have a system containing a number of phases that may exchange matter in response to changing conditions, but that the system as a whole is of fixed composition. From Chapter 5, we know that if a closed system is constrained to a constant entropy and constant volume, then its energy content will seek a minimum value, and the system will be at equilibrium when $d\mathrm{U}_{\mathrm{S,V}} = 0$. In the present case we still want to consider a system that is closed overall, but within which matter is free to move between phases, i.e., in which the phases are open. Still, because the system is closed overall, the same criterion ($d\mathrm{U}_{\mathrm{S,V}} = 0$) applies. If we denote the various phases in the system by accents, we can consider that during any increment of change of energy $d\mathrm{U}$ the various phases contribute $d\mathrm{U}'$, $d\mathrm{U}''$, etc., so that

$$d\mathrm{U} = d\mathrm{U}' + d\mathrm{U}'' + d\mathrm{U}''' + \cdots \text{ etc.} \tag{14.18}$$

Now if $d\mathrm{U}_{\mathrm{S,V}} = 0$, then from (14.10) and (14.18)

$$\begin{aligned} &T'd\mathrm{S}' - P'd\mathrm{V}' + \mu_1' dn_1' + \mu_2' dn_2' + \cdots + \mu_c' dn_c' \\ &+T''d\mathrm{S}'' - P''d\mathrm{V}'' + \mu_1'' dn_1'' + \mu_2'' dn_2'' + \cdots + \mu_c'' dn_c'' \\ &+T'''d\mathrm{S}''' + \cdots \text{ etc.} \\ &= 0 \end{aligned} \tag{14.19}$$

For the system, S, V, and the quantity of each component are constant, so that

$$\left.\begin{aligned} d\mathrm{S}' + d\mathrm{S}'' + d\mathrm{S}''' + \cdots &= 0 \\ d\mathrm{V}' + d\mathrm{V}'' + d\mathrm{V}''' + \cdots &= 0 \\ dn_1' + dn_1'' + dn_1''' + \cdots &= 0 \\ dn_2' + dn_2'' + dn_2''' + \cdots &= 0 \\ \vdots \qquad\qquad &\quad \vdots \\ dn_c' + dn_c' + dn_c'' + \cdots &= 0 \end{aligned}\right\} \tag{14.20}$$

For this to be true it is necessary and sufficient that

$$\left.\begin{aligned} T' = T'' = T''' &= \cdots \text{ etc.} \\ P' = P'' = P''' &= \cdots \text{ etc.} \\ \mu_1' = \mu_1'' = \mu_1''' &= \cdots \text{ etc.} \\ \mu_2' = \mu_2'' = \mu_2''' &= \cdots \text{ etc.} \\ \vdots \qquad\qquad &\quad \vdots \\ \mu_c' = \mu_c'' = \mu_c''' &= \cdots \text{ etc.} \end{aligned}\right\} \tag{14.21}$$

Equations (14.21) express the conditions for thermal, mechanical, and chemical equilibrium of the system, i.e., that temperature, pressure, and the chemical potential of every component must be constant throughout. We will return to the importance of the equality of chemical potentials at equilibrium shortly.

In this derivation we have tacitly assumed that all components are capable of being independently added or subtracted from every phase, i.e., that all dn_i terms can be positive or negative. Another way of saying this is that all components have been assumed to be in all phases, and that they are free to enter or leave each phase. If this is not the case for some components in some phases, though, nothing is changed. If component 1 does not exist in phase $'$, or more exactly if it cannot be transferred to or from phase $'$, then dn_1' is zero, the $\mu_1' dn_1'$ term is missing from equation (14.19) and the dn_1' term from (14.20), and there is no change in the general conditions of equilibrium, equations (14.21), except that the μ_1' term is missing.

14.2.3. *Equality of Chemical Potentials for All Constituents in All Phases*

It is important to realize that although the above derivation refers exclusively to independent components, the conclusion that μ_i is the same in all parts of the system is true not only for components but indeed for all constituents. This is true because although the $n_1, n_2, \ldots, n_c$ terms in equation (14.19) are independent components, we could rewrite the equation expanding the number of compositional terms to include as many constituents as we like, as long as for each constituent added beyond the number of independent components, we add an additional equation of constraint. These normally take the form of equation (14.25), as we will show below. For example, for a system having components A,B,C, equation (14.19) becomes

$$\sum T dS - \sum P dV + \sum \mu_A dn_A + \sum \mu_B dn_B + \sum \mu_C dn_C = 0$$

where $\sum$ signifies summation over all the phases. If we choose to include constituent AB in the equation, it becomes

$$\sum T dS - \sum P dV + \sum \mu_A dn_A + \sum \mu_B dn_B + \sum \mu_C dn_C + \sum \mu_{AB} dn_{AB} = 0$$

but we must impose an additional constraint

$$\mu_A + \mu_B = \mu_{AB}$$

which keeps the number of possible *independent* variations in composition the same as before, i.e., three. However, the conclusion in equations (14.21) about the equality of μ_i in all phases now includes constituent AB.

Clearly, as there is no restriction on how many constituents may be included in this way, the conclusion that μ_i is the same in all parts of the system holds for all constituents. This is a fairly liberating concept, as it means, eventually, that one is free to write any conceivable balanced reaction and form the equilibrium constant for it, whatever the nature of the system. Whether the reaction is useful or can even be investigated is up to the investigator, not thermodynamics.

14.2.4. *Chemical Potentials Balanced in All Reactions*

It follows from (14.13) and $dG_{T,P} = 0$ that

$$\sum_i \mu_i dn_i = 0 \tag{14.22}$$

for systems at stable equilibrium. We have just seen that these terms in i need not be restricted to components, as long as an additional constraint is imposed for each i beyond the number of components. We now show what this constraint is. Consider now any balanced reaction in this system, which can be represented by

$$\sum_i \nu_i M_i = 0 \tag{14.23}$$

where ν_i are the stoichiometric coefficients, positive for products and negative for reactants, and M_i are the chemical formulae of constituents in the reaction (see §13.1 for an example). This equation is simply a mass balance, as is the fact that for this reaction,

$$\frac{dn_1}{\nu_1} = \frac{dn_2}{\nu_2} = \frac{dn_3}{\nu_3} = \cdots \text{ etc.} \tag{14.24}$$

which expresses the fact that at equilibrium, changes in the mass of one constituent must be reflected by proportional changes in all the others. From (14.24) we can express changes in all constituents in terms of the first one. Thus

$$dn_2 = \left(\frac{\nu_2}{\nu_1}\right) dn_1;\ dn_3 = \left(\frac{\nu_3}{\nu_1}\right) dn_1;\ dn_4 = \left(\frac{\nu_4}{\nu_1}\right) dn_1;\ \cdots;\ dn_c = \left(\frac{\nu_c}{\nu_1}\right) dn_1.$$

Combining this with (14.22) we get

$$\left(\mu_1 + \mu_2\left(\frac{\nu_2}{\nu_1}\right) + \mu_3\left(\frac{\nu_3}{\nu_1}\right) + \cdots + \mu_c\left(\frac{\nu_c}{\nu_1}\right)\right) dn_1 = 0$$

from which

$$(\nu_1\mu_1 + \nu_2\mu_2 + \nu_3\mu_3 + \cdots + \nu_c\mu_c) = 0$$

or

$$\sum_i \nu_i \mu_i = 0 \tag{14.25}$$

which, as we have seen, is applicable to any balanced reaction among any constituents of a system at equilibrium, and expresses the general condition, $dG_{T,P} = 0$, for particular reactions. In other words, the chemical potentials in *all conceivable* balanced reactions among system constituents are balanced (the algebraic sum is zero) at system equilibrium. Only a tiny proportion of these are ever considered important enough to think about or to warrant investigation, but every now and then someone shows the usefulness of a new one, such as when a new complex species is found, or when a new way of writing components is suggested (such as the exchange operators of Thompson, 1982).

14.3. FUNDAMENTAL EQUATIONS

Equations (14.10) to (14.13) have a special significance that needs discussion, but first we note that all of them have at least one extensive property as a differential term. We can obtain an equation having only differentials of intensive properties by differentiating any of the integrated forms of these equations, i.e., (14.14) to (14.17), and subtracting the corresponding differential form, (14.10) to (14.13). The result is the celebrated Gibbs-Duhem relation, or "Gibbs 97," as that is the number it has in Gibbs' original 1875 publication. It is

$$\mathrm{S}dT - \mathrm{V}dP + n_1 d\mu_1 + n_2 d\mu_2 + \cdots + n_c d\mu_c = 0$$

or

$$\mathrm{S}dT - \mathrm{V}dP + \sum_{i=1}^{c} n_i d\mu_i = 0 \tag{14.26}$$

Gibbs 97 (14.26) shares with equations (14.10) to (14.13) the remarkable property that if it is integrable, that is, if sufficient experimental data are available (heat capacities, activity coefficients, partial molar volumes, etc.) that each of the individual terms in any one of the equations can be integrated over the range of conditions of interest, then *all* thermodynamic properties are known for that phase. The distinction between equations that are and that are not fundamental may be seen by comparing an equation between

$$\mathrm{G}, T, P, n_1, n_2, \cdots, n_c$$

which is (14.13), with one between

$$\mathrm{G}, \mathrm{S}, P, n_1, n_2, \cdots, n_c,$$

i.e., with one between

$$\mathrm{G}, \left(\frac{\partial G}{\partial T}\right), P, n_1, n_2, \cdots, n_c \tag{14.27}$$

Obviously, knowing G as a function of T we can calculate $(\partial \mathrm{G}/\partial T)$ or $-\mathrm{S}$, but knowing G as a function of $(\partial \mathrm{G}/\partial T)$ will not permit calculation of T. Changing any of the independent variables in one of the fundamental equations reduces the amount of information it can convey. Equations (14.10) to (14.13) and (14.26) are therefore known as fundamental equations, and (14.13) in particular is the basis for almost all the rest of this book.

The reader will not fail to note that the fundamental equations (14.10) to (14.13) are relations that unite the state variables U,A,H,G with those independent variables that enable them to act as thermodynamic potentials, i.e., $\mathrm{U}(\mathrm{S}, \mathrm{V}, n_1, n_2, \ldots)$; $\mathrm{G}(T, P, n_1, n_2, \ldots)$; and so on. Thus one might say that (14.13) is a fundamental equation because $\mathrm{G}(T, P, n_1, n_2, \ldots)$ is a thermodynamic potential, and that an equation relating the variables in (14.27) is not fundamental because $\mathrm{G}(\mathrm{S}, P, n_1, n_2, \ldots)$ is not a thermodynamic potential.

Before proceeding, we note again that the equations thus far have been entirely in their extensive form. Each can be converted to the intensive form by dividing each term by $(n_1 + n_2 + n_3 + \cdots + n_c)$. S thus becomes S, V becomes V, etc., and the n_i terms become mole fractions, X_i. Equations (14.10) to (14.13) and (14.26) thus become

$$dU = T\,dS - P\,dV + \sum_{i=1}^{c} \mu_i dX_i \tag{14.28}$$

$$dA = -S\,dT - P\,dV + \sum_{i=1}^{c} \mu_i dX_i \tag{14.29}$$

$$dH = T\,dS + V\,dP + \sum_{i=1}^{c} \mu_i dX_i \tag{14.30}$$

$$dG = -S\,dT + V\,dP + \sum_{i=1}^{c} \mu_i dX_i \tag{14.31}$$

$$0 = S\,dT - V\,dP + \sum_{i=1}^{c} X_i d\mu_i \tag{14.32}$$

and (14.17) becomes

$$G = \sum_{i=1}^{c} X_i \mu_i \tag{14.33}$$

All applications to real systems use the intensive form.

14.3.1. *Maxwell's Equations*

For closed systems ($dX_i = 0$), applying the reciprocity relation (§2.2.6) to equations (14.28)–(14.31) results in the set of "Maxwell's equations," which are often useful in manipulating thermodynamic equations. These are

$$\left(\frac{\partial T}{\partial V}\right)_S = -\left(\frac{\partial P}{\partial S}\right)_V$$

$$\left(\frac{\partial S}{\partial V}\right)_T = \left(\frac{\partial P}{\partial T}\right)_V$$

$$\left(\frac{\partial T}{\partial P}\right)_S = \left(\frac{\partial V}{\partial S}\right)_P$$

$$\left(\frac{\partial S}{\partial P}\right)_T = -\left(\frac{\partial V}{\partial T}\right)_P$$

14.4. CONDITIONS FOR MASS TRANSFER

There remain a couple of things to say about the conditions of equilibrium, equations (14.21). One is that just as inequalities (gradients) in T or P are driving forces for the transfer of heat and work, a chemical potential gradient is a driving force for the transfer of matter. Consider for example equation (14.13) in the case of two coexisting phases at a constant T and P, each of which is a homogeneous solution of two components. Thus

$$d\mathrm{G}'_{T,P} = \mu'_1 dn'_1 + \mu'_2 dn'_2 \tag{14.34}$$

$$d\mathrm{G}''_{T,P} = \mu''_1 dn''_1 + \mu''_2 dn''_2 \tag{14.35}$$

If the system is closed, any matter that leaves one phase must enter the other, so that

$$dn'_1 = -dn''_1 \quad \text{and} \quad dn'_2 = -dn''_2$$

Therefore, because $d\mathrm{G} = d\mathrm{G}' + d\mathrm{G}''$,

$$d\mathrm{G}_{T,P} = (\mu'_1 - \mu''_1)dn'_1 + (\mu'_2 - \mu''_2)dn'_2 \tag{14.36}$$

If $\mu'_2 = \mu''_2$,

$$d\mathrm{G}_{T,P} = (\mu'_1 - \mu''_1)dn'_1$$

For the system as a whole, which is closed, the general relation

$$d\mathrm{G}_{T,P} \leq 0$$

still holds, so

$$(\mu'_1 - \mu''_1)dn'_1 \leq 0 \tag{14.37}$$

As before, we see that the condition of equilibrium ($d\mathrm{G} = 0$) is that $\mu'_1 = \mu''_1$. But in addition we see that for $d\mathrm{G}$ to be negative (< 0), which is the criterion for spontaneous change, either

$$(\mu'_1 < \mu''_1) \text{ and } dn'_1 \text{ is positive,}$$

or

$$(\mu'_1 > \mu''_1) \text{ and } dn'_1 \text{ is negative.}$$

In either case, spontaneous change calls for the transfer of mass (in this case component 1) from the phase in which it has a higher potential to the phase in which it has a lower potential. Thus the chemical potential is aptly named, as it tends to cause matter to transfer to regions of lowest potential. As usual, we point out that although there is this tendency, real systems may contain factors that prevent such change from taking place, thus preventing stable equilibrium from being realized.

14.5. THE AFFINITY AND CHEMICAL POTENTIALS

It is appropriate at this point to expand on a concept introduced in Chapter 5—the affinity. We said that irreversible processes begin from some metastable state and end either at stable equilibrium (or "on the stable equilibrium surface") or perhaps in another metastable state closer to stable equilibrium than the first one. As an example we used a binary alloy with varying degrees of disorder ϕ, and we concluded that the expression

$$dU_{irrev} < T\,dS - P\,dV$$

for irreversible reactions can be expanded to

$$dU_{irrev} = T\,dS - P\,dV - \mathsf{A}\,d\xi$$

where, by the normal properties of exact differentials, $\mathsf{A} = -(\partial U/\partial \xi)_{S,V}$, and where ξ, the *progress variable*, indicates the degree of advancement of the system from the initial metastable state towards the final state on some convenient scale. In the alloy example, $d\xi$ was identical to $d\phi$, i.e., increments of reaction were in fact increments of change in the degree of order of the alloy. The reaction must always be, in our terms, a *quasistatic* reaction, that is, an infinite succession of (metastable) equilibrium states along an irreversible path, and to achieve this, the system must be subject to at least one constraint in addition to the two required by Duhem's Law.

In considering open systems, we have introduced the idea of components moving from place to place until they are of uniform potential and balanced throughout the system (14.21; 14.25). But if we consider a metastable state where the chemical potentials are under some constraint and are *not* balanced, then releasing that constraint allows an irreversible redistribution of potentials to occur. This was in fact one of our examples of metastable systems in Chapter 3. (§3.6.1). Now imagining that this irreversible reaction takes place in a series of very small increments ($d\xi$ in moles) gives us another (much more general) way of understanding the affinity.

Comparing (14.10) and (5.42),

$$\mathrm{dU} = T d\mathrm{S} - P d\mathrm{V} + \sum_{i=1}^{c} \mu_i dn_i \tag{14.10}$$

$$\mathrm{dU} = T d\mathrm{S} - P d\mathrm{V} - \mathsf{A} d\xi \tag{5.42}$$

we see that

$$-\mathsf{A} d\xi = \sum_{i=1}^{c} \mu_i dn_i$$

or

$$-\mathsf{A} = \frac{\mu_1 dn_1}{d\xi} + \frac{\mu_2 dn_2}{d\xi} + \cdots \tag{14.38}$$

$$= \sum_{i=1}^{c} \mu_i \frac{dn_i}{d\xi} \tag{14.39}$$

To see what this means, we need a relationship between dn_i and $d\xi$. We have already mentioned reaction increments in this chapter, although we didn't call them that. We said (§14.2.4) that because reactant and product constituents are related by fixed stoichiometries, changes in their masses are related by (14.24),

$$\frac{dn_1}{\nu_1} = \frac{dn_2}{\nu_2} = \frac{dn_3}{\nu_3} = \cdots \text{ etc.} \tag{14.24}$$

But the only natural cause of such changes in these masses is a spontaneous or irreversible reaction such as we are now considering, so it appears natural to identify the increments in (14.24) with increments in the progress variable, $d\xi$. Thus

$$\frac{dn_1}{\nu_1} = \frac{dn_2}{\nu_2} = \frac{dn_3}{\nu_3} = \cdots = d\xi$$

from which it follows that

$$\frac{dn_1}{d\xi} = \nu_1; \quad \frac{dn_2}{d\xi} = \nu_2; \quad \frac{dn_i}{d\xi} = \nu_i \tag{14.40}$$

In other words, increments of ξ are now in moles rather than degree of disorder and the rate of change of each constituent in a reaction is its (dimensionless) stoichiometric coefficient. This may be quite opaque on first reading, but will become almost trivial when understanding dawns. If for example we have a mole of A that wants to change spontaneously into 5 moles of B, the reaction is A = 5B, and $(dn_{\mathrm{A}}/d\xi) = -1$; $(dn_{\mathrm{B}}/d\xi) = 5$. This just says that for every mole of A that disappears, 5 moles of B must appear. This fairly obvious relation now allows us to link ξ with the chemical potentials. From (5.39) we have

$$\mathsf{A} = -\left(\frac{\partial \mathrm{U}}{\partial \xi}\right)_{\mathrm{S,V}}$$

and by a simple transformation of variables (§2.2.7, especially equation 2.18), and summing across all i constituents,

$$\mathsf{A} = -\sum_{i=1}^{c}\left(\frac{\partial \mathrm{U}}{\partial n_i}\cdot\frac{\partial n_i}{\partial \xi}\right)_{\mathrm{S,V}}$$

and because

$$\left(\frac{\partial \mathrm{U}}{\partial n_i}\right)_{\mathrm{S,V}} = \mu_i \quad \text{and} \quad \frac{dn_i}{d\xi} = \nu_i$$

it follows that

$$\mathsf{A} = -\sum_{i=1}^{c}\nu_i\mu_i \tag{14.41}$$

Comparing this with (14.25) we see that $\mathsf{A} = 0$ is another way of expressing the $d\mathrm{U}_{\mathrm{S,V}} = 0$ (or $d\mathrm{G}_{T,P} = 0$) criterion of equilibrium.

For example, for the reaction $O_2 + 2H_2 = 2H_2O$,

$$A = \mu_{O_2} + 2\mu_{H_2} - 2\mu_{H_2O}$$

In a metastable mixture of oxygen and hydrogen, A then expresses the amount by which the chemical potentials could be lowered by spontaneous reaction on release of the metastable constraint, and is in fact the "driving force" of the reaction. The driving force becomes zero when $2\mu_{H_2} + \mu_{O_2} = 2\mu_{H_2O}$.

Another useful relation involving A is obtained by combining (14.41) with (11.20),

$$\mu_i = \mu_i^\circ + RT \ln a_i \tag{11.20}$$

Multiplying by ν_i and summing across all i constituents, we get

$$\sum_{i=1}^{c} \nu_i \mu_i = \sum_{i=1}^{c} \nu_i \mu_i^\circ + \sum_{i=1}^{c} \nu_i RT \ln a_i$$

or

$$-A = \Delta_r G^\circ + RT \ln Q$$

where $Q = \prod_i a_i^{\nu_i}$ as explained in §13.1. It is a term with the form of the equilibrium constant, but in which the activity terms do not have their equilibrium values, but some metastable values. Then by (13.8),

$$\begin{aligned} A &= RT \ln K - RT \ln Q \\ &= RT \ln(K/Q) \end{aligned} \tag{14.42}$$

From what we have said, equations (14.10) through (14.13) can also be written as

$$dU = TdS - PdV - Ad\xi \tag{14.43}$$

$$dA = -SdT - PdV - Ad\xi \tag{14.44}$$

$$dH = TdS + VdP - Ad\xi \tag{14.45}$$

$$dG = -SdT + VdP - Ad\xi \tag{14.46}$$

We have tried to show that the affinity is a convenient representation of how far a system is from stable equilibrium, and increments in the progress variable allows us to consider the system at various stages as it progresses towards stable equilibrium. This "progress" is a part of the model we call a quasistatic reaction—a continuous succession of metastable equilibrium states in an overall irreversible reaction.

Also we might mention again here a point we have made before (Chapter 2) but which can easily be forgotten, and that is that the differential quantities in all our equations, such as (14.43) through (14.46), are not necessarily infinitesimals, and can be of any magnitude. Prigogine and Defay (1954) in their classic treatise emphasize this point explicitly by often using a δ instead of d in their equations. Equation (14.43) for example would become

$$\delta U = T\delta S - P\delta V - A\delta\xi$$

We have not resorted to special notation, but remind the reader at intervals that differentials are defined as of any size.

So far we have considered the progress variable and the affinity in connection with single reactions. It is much more useful to include many simultaneous reactions, in which case the progress variable most conveniently refers to the rate-limiting reaction, i.e., the slowest, of the series of reactions considered (Helgeson, 1979). By calculating the mass changes of all species of interest after each irreversible increment of this ξ, a whole new world of geochemical modeling opens up, which we consider in Chapter 19.

14.6. THE PHASE RULE

The phase rule is a subject at once simple and complex, depending on how you look at it. It receives rather little attention in chemistry texts, because when applied to simple systems in laboratory situations, it is fairly simple. It receives a great deal more attention from geochemists, who find it much more complex to use, because they are interpreting much more complex systems with complex histories millions of years old. Because the requirements of the equilibrium model are so well defined, it is invariably used as a base or starting point in interpreting the petrogenetic history of suites of related rocks, with the effects of thermal, pressure, or chemical gradients superimposed. The phase rule is one of the tools most useful in defining the equilibrium state (textural observation is another), and hence in developing petrogenetic models.

14.6.1. *Phases and Components*

In Chapter 3 we defined a *phase* as a homogeneous body of matter, generally having distinct boundaries with adjacent phases, and generally physically separable from them. Shape and position are irrelevant, so that a phase can occur in one or many places in a system, and can be of just about any size and shape. Liquids and gases often conform quite closely to this definition, and individual minerals are also usually regarded as phases, despite not being really homogeneous. We will discuss some of the difficulties this introduces after explaining the Phase Rule.

The *components* of a system are the smallest number of formulae required to describe all the phases in the system. The word "formula" refers here to any assemblage of elemental symbols, each having any stoichiometry, positive or negative. The problem here is, what are we to regard as the composition of the phase? Do we include trace elements? Again, we postpone discussion, and will discuss the Phase Rule for fairly simple cases where these problems do not appear to arise.

A system having a fixed composition may have a number of equivalent sets of components. For example, the compositions of all phases in the system ABC can be described by the components (A,B,C); (A,ABC,AC); $(A_2B, B, B_{-1}C)$; or any other three points lying in the ABC plane. Any two compositions are insufficient and any four are too many, because one of the four could always be described by a suitable combination of the other three. ABC is therefore a three-component system.

We continue to use the term *constituent* in the sense defined in Chapters 3 and 11. A constituent is any combination of the components of a system, in other words, any formula that can be devised from the elements comprising the system. Systems therefore have an essentially infinite number of constituents, and all components are constituents but not vice versa. We use the term constituent in order to preserve the term component for use in its strict or "Phase Rule" sense, and because the term "species" is not applicable to solids.

14.6.2. *Derivation of the Phase Rule*

A single homogeneous phase such as an aqueous salt (say NaCl) solution has a large number of properties, such as temperature, density, NaCl molality, refractive index, heat capacity, absorption spectra, vapor pressure, conductivity, partial molar entropy of water, partial molar enthalpy of NaCl, ionization constant, osmotic coefficient, ionic strength, and so on. We know however that these properties are not *all* independent of one another. Most chemists know instinctively that a solution of NaCl in water will have all its properties fixed if temperature, pressure, and salt concentration are fixed. In other words, there are apparently three independent variables for this two-component system, or three variables which must be fixed before all variables are fixed. Furthermore, there seems to be no fundamental reason for singling out temperature, pressure, and salt concentration from the dozens of properties available, it's just more convenient; any three would do. In saying this we have made the usual assumption that "properties" means "intensive variables," or that the size of the system is irrelevant. If extensive variables are included, one extra variable is needed to fix all variables. This could be the system volume, or any other extensive parameter.

This intuitive conclusion is fortunately entirely in accord with our fundamental equations. If it is true that a fundamental equation can represent all possible variations of the thermodynamic properties of a homogeneous system, it must have the right number of independent variables. Equation (14.26) (Gibbs 97) contains $c + 2$ intensive variables (c chemical potentials, T and P) and the equation is a relationship between them, so there are evidently $(c + 2 - 1)$ or $c + 1$ independent variables in any homogeneous phase, consistent with our intuition in the water–salt system above. If extensive variables are included, equations (14.10) to (14.13) show that there are $c + 3$ variables in each case, with one linking equation, giving $c + 2$ independent variables. Note that we do not count the coefficients as variables (e.g. T, P, and μ_i in (14.10)) because they are simply derivatives of the other variables and hence are not independent of them. Note too that the same conclusions are reached if we start with the intensive form of the equations, (14.28) to (14.32). We need only recall that the terms $n_1, n_2, n_3, \ldots, n_c$ represent c independent compositional terms, whereas $X_1, X_2, X_3, \ldots, X_c$ represents only $c - 1$ compositional variables, the final mole fraction being fixed if all but one are given. Thus in (14.28) for example there are $c - 1$ independent compositional variables, plus three other intensive variables (U, S, and V), and one linking equation (equation (14.28)), giving $c - 1 + 3 - 1$ or, as before, $c + 1$ independent variables in a single homogeneous phase.

Now consider two phases at equilibrium, say solid NaCl and a saturated salt solution. Again, intuition or experience tells us that we no longer have three independent variables, but two, because for example we cannot choose the composition of the salt solution once T and P are fixed—it is fixed for us by the solubility of NaCl in water. If we then consider the possibility of having a vapor phase in equilibrium with the salt and the solution, we see that we lose another independent variable because we can no longer choose the pressure on the system independently once the temperature is chosen—it is fixed by the vapor pressure of the system. So it would seem that in general we restrict the number of independent variables in a system by increasing the number of phases at equilibrium. This is also predictable from (14.26), because in a multi-phase system while there are still only a total of $c + 2$ variables (T, P and all the μ terms being the same in all phases) there are now a number of relations linking them equal to the number of phases, there being one equation (14.26) for each phase. Therefore if there are p phases, there are $c + 2 - p$ independent variables at equilibrium. These independent variables are also called "degrees of freedom," f, and the result is the Phase Rule,

$$f = c - p + 2 \tag{14.47}$$

We emphasize that the "degrees of freedom" include only intensive variables, and inasmuch as there is a functional relationship, known or unknown, between any intensive variable and all the others, the quantity $c - p + 2$ refers to any combination of the intensive variables of a system. Naturally, in practice, these are normally T, P and concentrations.

Degrees of freedom can also be described as the number of intensive variables that can be changed (within limits) without changing the number of phases in a system. This point of view is perhaps more useful to someone looking at a phase diagram; thus divariant, univariant, and invariant systems correspond to areas, lines, and points in a P-T projection. We prefer however to emphasize the fact that coexisting phases reduce the number of independent variables, and that some systems have all their properties determined. This fact is very useful, as we will elaborate on below, and its explanation in terms of the Phase Rule is a very beautiful example of the interface between mathematics and physical reality.

14.7. BUFFERED SYSTEMS

A great deal of the power and usefulness of the Phase Rule in geochemistry comes from its demonstration of which systems are divariant, and which therefore have all their properties fixed at a given T and P. What do we mean by this? Duhem (Chapter 5) said that any closed system has all properties fixed by fixing only two of them, e.g., at fixed T and P. Thus our homogeneous NaCl solution at fixed T and P has definite fixed values of all its dozens of properties, and this is still true no matter how many components are present; thus sea water also has all its properties fixed at a given T and P. But these systems only have fixed properties by virtue of being closed, which eliminates all changes in bulk composition; this is not what we mean by divariant. A divariant system in our sense at a given T and P will have all its properties fixed including the composition of all phases in the system, regardless of changes in the bulk composition of the system (i.e., changes in the amounts of the components used to define the system, not adding more components). For example, our saturated NaCl solution at T and P has a fixed aqueous phase composition. Changing the number of moles of water or of salt in the system will change the masses of the solid and aqueous phases, but will not change the composition or the properties of either phase. The same cannot be said of seawater. Changing the bulk composition of seawater, which is saturated with one or even with several phases, *does* change the properties of the system. Divariant systems are of course those that have the same number of components and phases. Because the presence of a certain number of phases fixes all the properties of the system at given T and P, including the compositions of all phases, the system has a certain resistance to any process attempting to change these properties, and it will continue to resist change until the process in question succeeds in decreasing the number of phases. The system is therefore said to be "buffered". For example, a rock consisting of gypsum ($CaSO_4 \cdot 2H_2O$) and anhydrite ($CaSO_4$) has two phases and two components, so its properties are fixed at a given T and P. Thus the activity of water is fixed through the relation

$$0.5(CaSO_4 \cdot 2H_2O) = 0.5CaSO_4 + H_2O$$

for which

$$\begin{aligned} K &= (a_{CaSO_4})^{0.5} \cdot a_{H_2O}/(a_{CaSO_4 \cdot 2H_2O})^{0.5} \\ &= a_{H_2O} \quad \text{when } a_{CaSO_4 \cdot 2H_2O} = a_{CaSO_4} = 1 \end{aligned}$$

If the standard states for the solid phases are the pure phases at T and P, their activities are both one, and the equilibrium constant equals the activity of water, i.e., the activity of water is a constant (fixed) as long as (pure) gypsum and anhydrite coexist. Water introduced along cracks in this rock will tend to change the activity of water in the system, but no change will in fact take place (at least in the model) until all the anhydrite is converted to gypsum, at which point the system is no longer buffered and the water activity can assume a new value fixed by the introduced water.

Systems having fewer than two degrees of freedom (that is, one or zero) are also of course buffered in the same way. A pure component at its melting point for example ($c = 1, p = 2, f = 1$) is buffered against changes in temperature or pressure. If heat is added, the temperature does not rise until the solid phase has disappeared. Divariant systems at a given T and P are also in principle buffered against changes in T and P, but unless reactions with a large ΔH or ΔV are involved, which generally means a melting or vaporization reaction, the buffering effect will be small. Thus the principle of buffering should not be confused with the buffering capacity, which depends on the reactions involved and the proportions of the phases. Univariant systems are particularly important in metamorphic petrology where they are known as isograds. Invariant systems are quite rare in nature but at least one is fairly common in the laboratory; the ice–water–water vapor triple point is frequently used in the calibration of thermometers at 273.16 K.

This property (of resisting change) of systems having two (or fewer) degrees of freedom has been known for over 100 years, so one could be forgiven for being surprised at the great change (it's probably exaggerating to call it a revolution) in geochemical thought and practice brought about when Hans Eugster drew attention to its usefulness in experimental and theoretical work. The original application was a demonstration that the assemblages hematite–magnetite–water, quartz–fayalite–magnetite–water, nickel–nickel oxide–water and others could be used in experimental work to control the oxidation state of experimental systems by separating the "buffering system" (one of the above) from the experimental system by a membrane (platinum) permeable to hydrogen. The buffer system, having the same number of components as phases, as well as having an element (Fe, Ni) in two different valence states, fixed the properties of each of its phases, and in particular fixed the oxygen and hydrogen fugacities in the water. The hydrogen diffused through the membrane and controlled the oxidation state of the experimental system, which also included an aqueous phase. This arrangement is illustrated in Figure 14.2. The buffer system thus "imposes" its oxidation state on the experimental system, and will continue to do so for some time even if the experimental system consumes hydrogen or oxygen in its reactions. Since then, innumerable buffering systems and experimental arrangements have been devised, the details of which are not of interest here. What is interesting is the way that the concept of buffered systems has pervaded geochemistry. It is hardly possible today to discuss a geochemical problem without mentioning a buffering reaction of some sort, whereas before Eugster, such reactions were hardly mentioned, and numerous experiments were performed with results of very limited or zero usefulness because the experiments were not buffered, and the experimental conditions were hence not completely defined.

The buffering capacity of systems with $f \leq 2$ explains only part of their significance in geochemistry. Perhaps even more importance lies in the fact that their properties being fixed, they are thus in principle determinable, and a great deal of effort has been expended in determining these properties. One cannot determine the properties of "granite," only of specific granites, but the (equilibrium) properties of coexisting quartz, albite, and K-spar in the system $SiO_2 - NaAlSi_3O_8 - KAlSi_3O_8$ at a given T, P are fixed and determinable. Buffered systems exist not only in the laboratory, but in nature as well, and other natural systems can be seen to have properties intermediate between those of certain buffered systems, so that the buffering concept provides a framework for thinking about natural systems. For example, hematite and magnetite coexist in many iron formations, and rocks containing magnetite but not hematite or wüstite (FeO) are thought of as lying between the magnetite–hematite and magnetite–wüstite buffers in terms of oxidation state. Similarly syenites containing no quartz and no nepheline can be thought of as lying between the quartz and albite–nepheline buffers for SiO_2.

In view of the fact that the majority of experimental applications of the buffering principle have been in the control of specific gaseous or aqueous species, we should perhaps reiterate that in a buffered system *all* properties are fixed, and that completely solid systems are no exception. Thus for example although the assemblage hematite–magnetite buffers the activity of oxygen through the reaction

$$6Fe_2O_3 = 4Fe_3O_4 + O_2$$

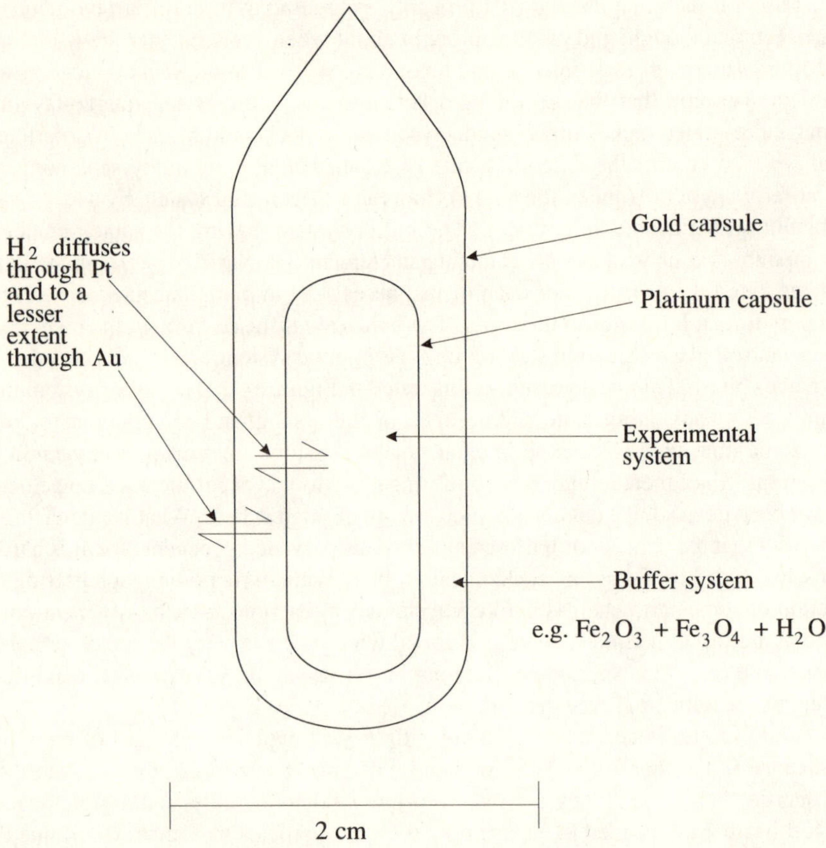

FIG. 14.2. Double-capsule technique designed to hold an experimental system at a fixed, known oxidation state. This arrangement is held in a pressure vessel containing a fluid at high temperature and pressure. The pressure is transmitted to the experimental system by the flexible walls of the gold and platinum tubes, and at the experimental temperature hydrogen, but not other components, is able to diffuse through the metals.

it also buffers the FeO activity through

$$Fe_3O_4 = Fe_2O_3 + FeO$$

the Fe activity through

$$3Fe_3O_4 = 4Fe_2O_3 + Fe$$

the O_3 activity through

$$9Fe_2O_3 = 6Fe_3O_4 + O_3$$

as well as the activity of any others of the infinite number of constituents in the Fe – O system. The system having at least as many phases as components could be thought of as a kind of "absolute" buffer. That is, the properties of all phases are absolutely determined once the remaining variables are chosen, and of course once equilibrium is achieved. The system properties, dozens of them, are state variables, are independent of the system's history, and in principle are determinable, given enough data. We add to this statement once more the cautionary note that in nature a buffering assemblage may not in fact have buffered anything, because of being present in too small a quantity, i.e., the buffering capacity may have been negligible, or because of unfavorable kinetics. Such assemblages are nevertheless always of interest in interpretations of geological history, because they are locations where thermodynamic parameters have a fixed and often known relationship.

Many systems having $f = 3$ or more are also spoken of as buffers. The most common examples are of course the aqueous buffers so commonly used in chemistry, usually to control the pH of a solution. For example, a solution containing sodium acetate and acetic acid will resist changes in pH when an acid or base is added due to the common ion effect. The pH does not change as much as it would if the NaAc and HAc were not there, but it does change, because the system is not an "absolute" buffer in the sense we have been using. Nevertheless, it does resist change and so is buffered. Other systems can be buffered simply due to the large mass involved. Thus the chlorine concentration of the ocean is not controlled by any process or reaction, but pouring salt into the ocean doesn't change it at all; it is buffered by the sheer size of the system.

14.7.1. *Activities in Buffered Assemblages*

It is often useful to be able to calculate the activity of components of mineral assemblages, for example the activity of SiO_2 in the assemblage forsterite–enstatite ($Mg_2SiO_4 - MgSiO_3$), or the activity of Na_2O in the assemblage albite–andalusite–quartz ($NaAlSi_3O_8 - Al_2SiO_5 - SiO_2$). For one thing, the activity of silica in various assemblages has been proposed as a useful way of classifying igneous rocks (Carmichael et al. 1974), but there are as well some geochemical problems where such a calculation is necessary. We discuss here only the calculation of activities of buffered components; the relation between activities, concentrations, and activity coefficients for unbuffered constituents is discussed elsewhere.

For buffered components, the method has already been used above, in the discussion of the activity of water in the assemblage gypsum–anhydrite. In principle, all components, no matter how chosen, are fixed when the variance is zero, which in geological examples is usually when T and P have been chosen for a divariant system. For this type of system, one only has to write the buffering reaction and find the required free energy data to calculate the activity of any component using the equilibrium constant. It is not a question of finding out if the buffering reaction exists; it does. To take a different example, consider the system $SiO_2 - Al_2O_3$ at 25°C, where we will suppose that the only stable compound is andalusite, Al_2SiO_5. With two components, both the assemblages quartz + andalusite and corundum + andalusite are divariant, and hence buffered at 25°C, 1 bar. In Figure 14.3a, this is shown in terms of the $G - X$ section. The free-energy loops shown are entirely hypothetical, as none of these minerals has as yet been shown to deviate from perfect stoichiometry.

To find the activity of Al_2O_3 in the assemblage quartz + andalusite, we need only write

$$Al_2SiO_5 = SiO_2 + Al_2O_3 \tag{14.48}$$

for which the equilibrium constant is

$$K = a_{SiO_2} \cdot a_{Al_2O_3} / a_{Al_2SiO_5}$$

When pure quartz and andalusite having unit activities coexist,

$$K = a_{Al_2O_3}$$

Similarly when corundum and andalusite coexist,

$$K = a_{SiO_2}$$

K is calculated from

$$\Delta_r G^\circ = -RT \ln K \tag{14.49}$$

Using data from RHF,

$$\begin{aligned}\Delta_r G^\circ &= \Delta_f G^\circ_{quartz} + \Delta_f G^\circ_{corundum} - \Delta_f G^\circ_{andalusite} \\ &= -204.646 - 378.082 - (-584.134) \\ &= 1.406 \text{ kcal mol}^{-1}\end{aligned}$$

$$\begin{aligned}\ln K &= -1406/(1.987 \times 298.15) \\ &= -2.37 \\ K &= 0.093 \\ &= a_{Al_2O_3}\end{aligned}$$

Thus the activity of Al_2O_3 in andalusite + quartz (and the activity of SiO_2 in andalusite + corundum) at 25°C is 0.093 (pure corundum standard state). The activity of Al_2O_3 in andalusite + corundum is of course 1.0.

It is instructive to note that the activity of Al_2O_3 is also related to a chemical potential difference by (equation (11.20))

$$\mu^{and+qtz}_{Al_2O_3} - \mu^\circ_{Al_2O_3} = RT \ln a^{and+qtz}_{Al_2O_3} \tag{14.50}$$

and because the standard state substance for Al_2O_3 is corundum,

$$\begin{aligned}\mu^\circ_{Al_2O_3} &= \mu^{corundum}_{Al_2O_3} \\ &= G^{corundum}_{Al_2O_3}\end{aligned}$$

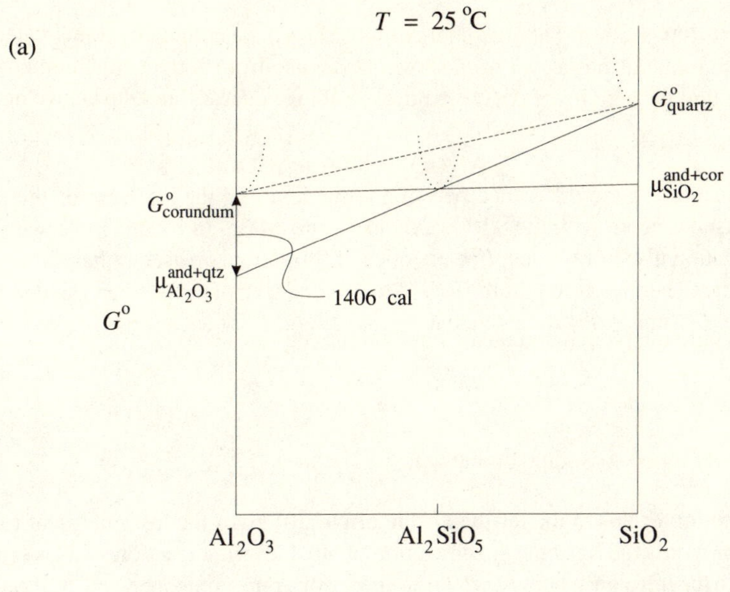

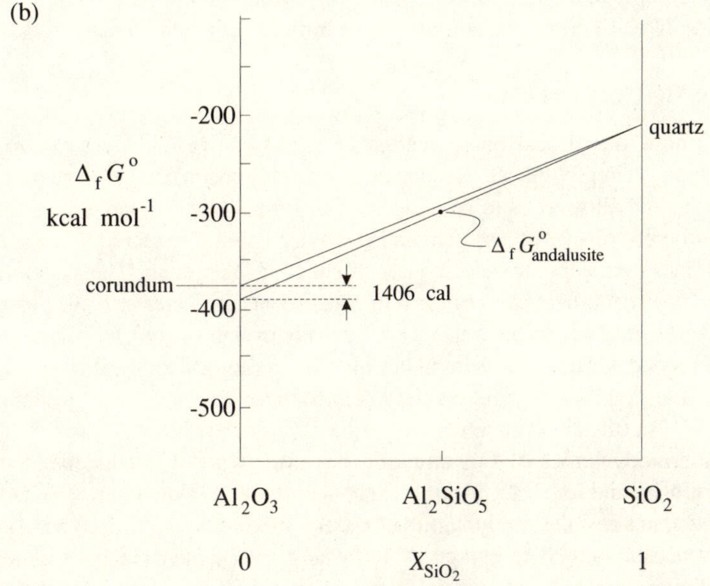

FIG. 14.3. (a) Absolute Gibbs free energy versus mole fraction for the system Al_2O_3-SiO_2 at fixed P and T. The width of the hypothetical free energy curves for each mineral is greatly exaggerated. (b) Gibbs free energies of formation for the same system at 25°C, 1 bar. Note that the free energy of andalusite is 1/2 the tabulated value, because the mole fraction axis requires that the mid-point contain 1/2 a mole of Al_2O_3 and 1/2 a mole of SiO_2. The difference between corundum and the intersection of the extrapolated quartz–andalusite join is greatly exaggerated for visibility.

and therefore

$$\begin{aligned}\mu_{\mathrm{Al_2O_3}}^{and+qtz} - G_{\mathrm{Al_2O_3}}^{corundum} &= RT\ln(0.093)\\ &= -1406\ \mathrm{cal\,mol^{-1}}\end{aligned}$$

This quantity is the difference between corundum and the intercept of the quartz–andalusite tangent with the Al_2O_3 axis in Figure 14.3a. In Figure 14.3b we see that this relation also holds when free energies of formation are used rather than absolute molar free energies as in Figure 14.3a. The equivalence of the left-hand side of (14.50) with $\Delta_r G^\circ$ in (14.49) is made clear by realizing that by equation (14.25)

$$\mu_{\mathrm{Al_2SiO_5}} = \mu_{\mathrm{Al_2O_3}} + \mu_{\mathrm{SiO_2}}$$

so

$$\mu_{\mathrm{Al_2O_3}}^{and+qtz} = \mu_{\mathrm{Al_2SiO_5}} - \mu_{\mathrm{SiO_2}}$$

Substitution of this in the left-hand side of (14.50) gives the free energy of reaction. In other words the free energy of reaction of any buffering reaction such as (14.48) is equal to the difference between the free energy of a pure component and the "buffering tangent intercept" in a diagram like Figure 14.3, which in turn is directly related by an equation like (14.50) to the activity of the buffered component.

14.8. OSMOTIC SYSTEMS

We return now to a discussion of open systems, which we said were of two types. The first type is simply the various phases in a heterogeneous closed system, consideration of which allowed us to develop the full form of the fundamental equations. The second type consists of a system and an environment, connected by a membrane or membranes permeable to selected constituents of the system. The system is thus "open" to its environment because certain constituents can enter or leave the system, and these constituents can have their activities controlled by the environment rather than by the system. This arrangement has obvious geological applications in metasomatic and alteration zones, where a fluid is introduced into a rock (the system) from somewhere else (the environment).

The thermodynamics of this situation has been worked out independently by D.S.Korzhinskii and J.B.Thompson, Jr., and we follow their ideas here. We can begin by showing that a new thermodynamic potential can be derived for this type of system. Consider the osmotic system in Figure 14.4, where we represent the environment by a piston–cylinder arrangement containing pure component B connected to the system, which contains a homogeneous solution of A and B, by a membrane permeable only to B. This system + environment is in fact a composite system as defined in Chapter 3, i.e., a system made up of two sub-systems. In the following discussion we will refer to the part containing both A and B as "the system," and the whole thing as "the composite system" or "the system plus environment." Both the system and the environment have the same temperature, T, but the pressure on the system, P_{total}, and the pressure on the environment, P_{B}, are independent of one another. Obviously

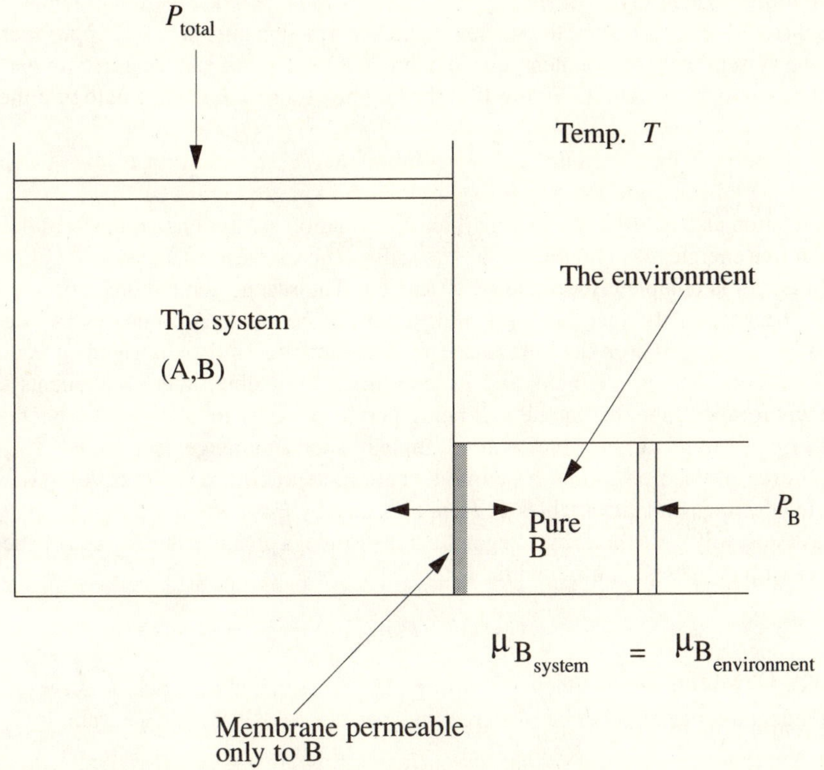

FIG. 14.4. The second type of open system (Figure 14.1b) where the system contains two components A and B, and the environment contains pure component B. The temperature is the same in both but the pressures are independent, so that the chemical potential of B in the system (A,B) is controlled by that in the environment.

there is one extra degree of freedom compared to the system (A,B) considered as a closed system (f = 4 instead of 3), because now two pressure terms instead of one must be fixed to completely determine the system, and we will return to this fact shortly. For now, consider the fundamental equation (14.13) applied to this system, with P_{total} abbreviated to P. Thus

$$d\mathrm{G} = -\mathrm{S}dT + \mathrm{V}dP + \mu_{\mathrm{A}}dn_{\mathrm{A}} + \mu_{\mathrm{B}}dn_{\mathrm{B}}$$

T, P, and n_{A} in the system are fixed, so this reduces to

$$dG = \mu_{\mathrm{B}}dn_{\mathrm{B}}$$

The closed system criterion for equilibrium is $d\mathrm{G}_{T,P} = 0$, but this is obviously inappropriate in this case because μ_{B} is certainly not zero and dn_{B} is not necessarily zero because we have said nothing about P_{B}. There are in fact no conditions pertaining only to the system that will specify the equilibrium state when the environment is attached to it. This is essentially self-evident. A thermodynamic potential appropriate for the system plus environment must evidently contain terms referring to both parts of the composite system. In Figure 14.4 the required four parameters would be either

$$T, P_{total}, P_{\mathrm{B}}, n_{\mathrm{A}}$$

or alternatively

$$T, P_{total}, n_{\mathrm{A}}, \mu_{\mathrm{B}}$$

because μ_{B} is completely defined by T and P_{B}. The second set is more commonly used, because of the fact that μ_{B} could in general be changed by processes other than by changing P_{B} on the pure phase B. The pressure on the environment could be kept constant, for example, and μ_{B} controlled by adding other components to the environment, the membrane still being permeable only to B. The experimental buffering system in Figure 14.4 is an example of such an arrangement.

So the problem is to find a thermodynamic potential for a (composite) system having the independent variables T, P (represented by P_{total} in Figure 14.4), $n_{\mathrm{A}}, \mu_{\mathrm{B}}$. The easiest way to do this is by a Legendre transform of a potential already established for other parameters. Starting with

$$\mathrm{G} = \mathrm{G}(T, P, n_{\mathrm{A}}, n_{\mathrm{B}})$$

we want a new function, which Thompson (1970) has called L, which preserves the potential qualities of G but has different independent variables; i.e., we want

$$\mathrm{L} = \mathrm{L}(T, P, n_{\mathrm{A}}, \mu_{\mathrm{B}})$$

As in Chapter 2 we find the Legendre transform by subtracting from G the product of the old variable and the derivative of the old function with respect to that variable. Thus

$$\mathrm{L} = \mathrm{G} - n_{\mathrm{B}} \left(\frac{\partial \mathrm{G}}{\partial n_{\mathrm{B}}} \right)$$

or

$$\mathrm{L} = \mathrm{G} - n_{\mathrm{B}}\mu_{\mathrm{B}} \tag{14.51}$$

$$= \mathrm{U} - T\mathrm{S} + P\mathrm{V} - n_{\mathrm{B}}\mu_{\mathrm{B}} \tag{14.52}$$

To see more clearly that we indeed now have the function we wanted, we can write, again following the procedure of Chapter 2,

$$d\mathrm{L} = d\mathrm{G} - n_{\mathrm{B}}d\mu_{\mathrm{B}} - \mu_{\mathrm{B}}dn_{\mathrm{B}}$$

Adding to this

$$d\mathrm{G} = -\mathrm{S}dT + \mathrm{V}dP + \mu_{\mathrm{A}}dn_{\mathrm{A}} + \mu_{\mathrm{B}}dn_{\mathrm{B}}$$

we get

$$d\mathrm{L} = -\mathrm{S}dT + \mathrm{V}dP + \mu_{\mathrm{A}}dn_{\mathrm{A}} - n_{\mathrm{B}}d\mu_{\mathrm{B}} \tag{14.53}$$

showing that L has $(T, P, n_{\mathrm{A}}, \mu_{\mathrm{B}})$ as independent variables. L is a function that is minimized at equilibrium for given values of T, P, n_{A} and μ_{B}, in exactly the same way that G is minimized for given values of T, P, n_{A} and n_{B}. It follows too, that because (from (14.17))

$$\mathrm{G} = \mu_{\mathrm{A}} n_{\mathrm{A}} + \mu_{\mathrm{B}} n_{\mathrm{B}}$$

and

$$\mathrm{L} = \mathrm{G} - \mu_{\mathrm{B}} n_{\mathrm{B}} \tag{14.51}$$

that

$$\mathrm{L} = \mu_{\mathrm{A}} n_{\mathrm{A}} \tag{14.54}$$

At constant T, P, and n_{A} , (14.53) becomes

$$d\mathrm{L}_{T,P} = -n_{\mathrm{B}} d\mu_{\mathrm{B}}$$

Since n_{B} is not zero, dL is clearly only zero when $d\mu_{\mathrm{B}}$ is zero, which will of course be when μ_{B} is the same in the system and in the environment.

Generalized forms of (14.51) and (14.54) are

$$d\mathrm{L} = -\mathrm{S}dT + \mathrm{V}dP + \sum_{\mathrm{A}} \mu_{\mathrm{A}} dn_{\mathrm{A}} - \sum_{\mathrm{B}} n_{\mathrm{B}} d\mu_{\mathrm{B}} \tag{14.55}$$

and

$$\mathrm{L} = \sum_{\mathrm{A}} \mu_{\mathrm{A}} n_{\mathrm{A}} \tag{14.56}$$

where $\sum_{\mathrm{A}}$ and $\sum_{\mathrm{B}}$ represent summation over all "system components" and "environment components" respectively. These could now be listed with our other fundamental equations, (14.55) with (14.10) to (14.13), and (14.56) with (14.14) to (14.17).

Similarly, the intensive forms,

$$dL = -S\,dT + V\,dP + \sum_{\mathrm{A}} \mu_{\mathrm{A}} dX_{\mathrm{A}} - \sum_{\mathrm{B}} X_{\mathrm{B}} d\mu_{\mathrm{B}} \tag{14.57}$$

and

$$L = \sum_{\mathrm{A}} \mu_{\mathrm{A}} X_{\mathrm{A}} \tag{14.58}$$

can be listed with the other intensive forms, (14.57) with (14.28) to (14.32) and (14.58) with (14.33).

Having derived the open system potential quantity, L, we must now admit that it is not much used, any more than are the other thermodynamic potentials apart from G. You never see tables of ΔL in the way that you see tables of ΔG or ΔH (you never see tables of ΔU either, but that doesn't mean it's not important). It is not used probably because it is unfamiliar, and the number of real applications may be limited. One application is discussed by Ghiorso (1987) and used by Ghiorso and Carmichael (1987). They used L to calculate the equilibrium composition of melts at given values of T, P and μ_{O_2}; i.e., values of f_{O_2} fixed by various buffer assemblages.

14.9. THE PHASE RULE IN OPEN SYSTEMS

We return now to our observation that the open system in Figure 14.4 has one extra degree of freedom ($f = 4$) compared to the homogeneous fluid phase (A,B) by itself, i.e., if it were not connected to the environment. A number of components could be controlled externally in this way, each with its own membrane, and each would add one degree of freedom to the composite system. That is, each must have its chemical potential specified before the composite system can be fully defined. Therefore if we separate components into two types, those whose potentials are fixed within the system (call them c_s)and those whose potentials are fixed in the environment (c_e) (the nomenclature of these two types of components will be discussed separately below), the phase rule as derived previously (14.47) would be written

$$f = c_s - p + 2,$$

and adding a degree of freedom for each externally controlled component, we have

$$f = (c_s + c_e) - p + 2 \tag{14.59}$$

which can be called an *open system Phase Rule*.

Natural systems, which can exist over a range of temperatures and pressures (such as a granite), must have at least two degrees of freedom (i.e., T and P). Combining the relation $f \geq 2$ with the Phase Rule, (14.47), gives

$$p \leq c \tag{14.60}$$

This says that the maximum number of phases to be expected in any naturally occurring system is given by the number of components in that system. This is called the "Mineralogical Phase Rule," and was pointed out by Goldschmidt in 1911. In open systems,

$$f \geq 2 + c_e$$

so that the Mineralogical Phase Rule becomes

$$p \leq c_s \tag{14.61}$$

where $c_s = c - c_e$. This simply says that the maximum number of phases expected in an open system is the number of components whose potentials are controlled by the system rather than by the environment. As applied to rocks, these are deceptively simple relationships because of the difficulty of distinguishing the two types of components. However, speaking in fairly general terms, the Mineralogical Phase Rule (14.60) shows why most rocks have a relatively small number of phases, relative, that is, to the large number of phases that could be formed from the components present. Similarly, (14.61) shows why metasomatic and altered rocks tend to have fewer phases than their unaltered equivalents.

14.10. MORE ON COMPONENTS AND PHASES

Before proceeding to a discussion of the applications of the Phase Rule in geology, we must unfortunately return to the question of the definition of components and phases.

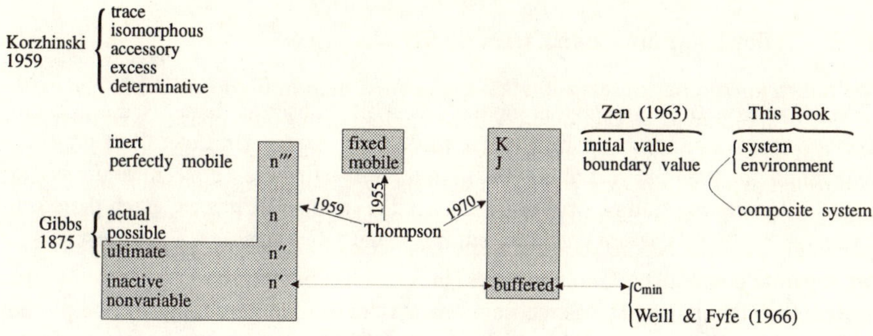

FIG. 14.5. Some types of components suggested by various authors.

14.10.1. *Components*

Figure 14.5 shows a number of the types of components introduced over the years. This profusion of names reflects the difficulty geologists have encountered in applying the Phase Rule to naturally-occurring systems. In one sense, there is only one kind of component—the components we finally choose to include in our genetic models. We subscribe completely to the thought expressed by Terry Gordon at the head of this chapter, i.e., that components are part of the thermodynamic model (as is the Phase Rule), and not part of rocks, and in that sense there is no need to worry about whether trace elements are components or not. If your model needs them, they are; if not, they are not. However, saying this does not completely solve the problem of choosing your components in a specific system. Therefore we include a discussion of the various types of components with the hope of clarifying the choices that must be made. We should also mention that this is one of the topics where the use of linear algebra is most helpful.

To begin with, we note the fairly extensive terminology proposed for those components we have called "system" and "environment." This reflects a dissatisfaction with the early definitions of Korzhinskii and Thompson ("mobile") because it was realized, not least by Korzhinskii and Thompson, that in fact the mobility of components was irrelevant to the distinction actually being made. For example, the activity or chemical potential of SiO_2 could be fixed within the system by the presence of quartz, yet SiO_2 might be quite mobile in the aqueous phase, even escaping into the environment. The real distinction is whether the chemical potential of the component is fixed within the system, and hence by the masses of the components in the system, or fixed in the environment, by whatever means. We apologize for adding yet more names to the list, but we do not suggest that the names we have used are in any way better or more suited to real problems than the others. We used them simply because they were suggested by the arrangement in Figure 14.5, and they should probably not be used outside of the present context. All component types other than those equivalent to our "environment" components are various types of "system" components.

The following definitions are from Korzhinskii (1959).

Trace Components are those components, usually chosen as elements, that occur at very low concentration in the minerals of the system (e.g., V, Cr, Co etc. in a granite), and are not an essential part of any phase in the sense that their concentrations could be varied over wide ranges without changing the number of phases. Each therefore could represent a degree of freedom, but it is usually one that doesn't interest us.

Accessory Components. Some components occur almost entirely in one phase, and at essentially trace levels in all other phases. The commonest examples are P_2O_5 in apatite; TiO_2 in rutile, sphene, or ilmenite; ZrO_2 in zircon; and Fe_2O_3 in hematite. In each case the presence of the component causes a phase to exist, but no degree of freedom is added (changing the amount of Ti in a rock containing rutile, for example, will change the amount of rutile but no other properties of the rock, including the trace levels of Ti in the other phases, assuming equilibrium).

Isomorphous Components are those that substitute freely for one another on a lattice site. If the substitution is truly ideal, i.e., if the mineral properties are unaffected by the substitution, then the two components act as one and can be counted as a single component. No two elements are exactly alike, however, so mineral properties are never completely unaffected by substitution. The question is whether the change is sufficiently small that the two components can be counted as one for the purposes of the Phase Rule. Figure 14.6a illustrates the problem.

Excess Components are those which occur in the "pure" form, as well as in some or all of the other phases in the system. For example SiO_2 occurs in all the silicates present as well as in the form of quartz in many rocks. The difference between an excess component like SiO_2 and an accessory component like TiO_2, which might also occur in the pure form (rutile), is that apart from the rutile, TiO_2 occurs only in trace quantities, while SiO_2 is an essential component (meaning they could not exist without it) of many other minerals. Ignoring TiO_2 and rutile is like removing all rutile from the rock, or pretending not to see it. Ignoring quartz and SiO_2 is like projecting a diagram from the SiO_2 apex of the system; all phases are saturated with quartz, but quartz does not appear in the diagram.

H_2O has also sometimes been treated as an excess component. Although obviously not a phase in the rocks as we see them now (fluid inclusions don't count), many rocks have assemblages that are consistent with the former presence (during metamorphism) of an aqueous phase. Unfortunately, they are also consistent with the possibility that water, although present, did not exist as a phase, because the mineral assemblage could be the same in either case. Most metamorphosed pelitic rocks are of this type (an example of a mineral assemblage that is not the same in these two cases is described below). However, the general observation that the water content of metamorphosed rocks decreases as the grade of metamorphism increases shows that water escapes from rocks during metamorphism. Whether it escapes as a nearly pure water phase, and

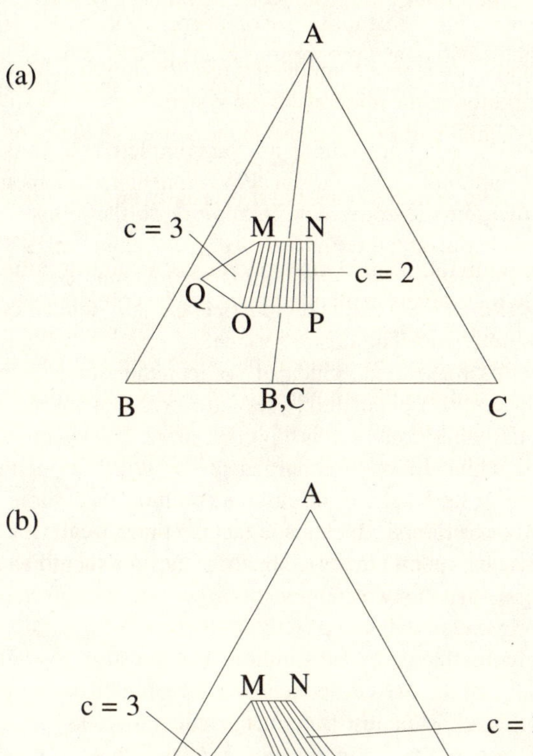

FIG. 14.6. Two coexisting solid solution series M–N and O–P in the system ABC. (a). Components B and C behave identically in both solid solutions, so that any single pair of solid solutions needs only two components, e.g. A and (B,C). If phase Q appears, three components are needed. (b). B and C do not behave identically,and three components are needed in all parts of the system.

is therefore an excess component, or whether it diffuses out along grain boundaries with μ_{H_2O} less than maximum cannot be distinguished by use of the Phase Rule. If it was a phase, it is no longer there; if it was not a phase, its potential was not controlled by the mass of water in the rock but by the nature of the escape channels and the enclosing rocks; i.e., it was environmentally controlled, and the number of mineral phases is the same. Other evidence may of course be brought to bear on the question, such as the demonstration of μ_{H_2O} gradients in certain directions, but generally speaking for those rocks in which water is not a determinative component, the Phase Rule is satisfied either by water as "excess," or water as "environmental."

Determinative Components are those left after subtracting all the components in the above categories. They are thus the set of independent components the abundances of which control the nature and proportions of the phases, other than those eliminated from consideration. The determinative components are those that are used in plotting diagrams of the phase relations.

As shown in Figure 14.5, we have not yet completed the consideration of terms used to describe components. Thompson (1959) considered components in the context of a series of metasomatic zones, and in addition to the terms *actual* and *possible* used by Gibbs, introduced the terms *inactive* (n', changed in 1970 to *buffered* , and called C_{min} by Weill and Fyfe, 1964), *ultimate* (n''), and *nonvariable*. (Gibbs also occasionally used the term "ultimate components," but defined only by the context and slightly different from Thompson's usage).

All these terms for various kinds of components are perhaps useful, and the discussions of Thompson, Weill and Fyfe, and Korzinskii are recommended reading for those particularly interested in this topic. We prefer however to offer the discussion on applications, which follows, unburdened by the details of the definitions and discussions by these authors. Our discussion could only get more difficult to follow if we introduced these terms, which are in fact not much used. Our point of view with respect to the disagreements expressed by these authors should be apparent from our discussion.

14.10.2. *Phases*

Because "component" is an abstract concept, it is fairly easy to accept the idea that components are part of the thermodynamic model. "Phases," on the other hand, are generally regarded as physical objects, such as coexisting water and ice (two phases), or the grains of quartz in a granite (one phase). Actually, however, the word "phase" is used in two senses (not to mention other meanings in other fields). We cannot now change common usage, in which the word is used to mean physical entities as in the examples just given. However, in the thermodynamic model, phases are defined as physically distinct, homogeneous, mechanically separable, and sufficiently large that surface energy is not significant. Each of these criteria presents problems when looking at natural phases.

Minerals show inhomogeneities of many types. Some of these, such as solid and fluid inclusions, alteration products, and so on can be regarded as separate phases, but compositional zoning is very common, as are structural defects and surface features of many types. "Physically separable" is also a troublemaker in some cases. Is a colloidal solution one phase because ordinary filtering will not separate the suspension? And if a method is found to separate a given colloidal suspension, the more general question is how small must particles be, or how large must nuclei get before we have one phase or two phases respectively? A more mineralogical example is exsolution lamellae—how fine must they be to be called one phase? Then there are the mixed-layer clays, where compositional differences can be gradational down to the unit cell level. These questions may be thought to be like asking "when does a stream become a river?," but they are not. There is in principle a phenomenological difference between one phase and two phases, where the "phenomena" are megascopic and can be checked against the Phase Rule. Zen (1962) discusses this problem and suggests a couple of tests (phenomena to look for) in the case of the mixed-layer clays.

There are also cases where matter, even pure matter, exists, yet is not a phase. The best known geological example is perhaps intergranular water, i.e., water that exists along the grain boundaries of a rock. In rocks water is of course not the only substance to accumulate along the boundaries of crystals, and this zone is one of the more interesting relatively unexplored areas of geochemistry. Nevertheless, even if it were pure, water would only be a phase in this position if it exhibited the same properties as would bulk water at the same T and P. The properties of intergranular water have never been measured, but it is certain that they would depend on factors such as the nature of the boundary crystals and the thickness and nature of the water layer. It is also certain that until the layer of water attained some critical thickness, its fugacity would be less than that of bulk water at the same T and P (otherwise it would be a phase) and it is this that most clearly shows its non-phase nature. As usual, we find that nature is very complicated, but the thermodynamic model is quite simple.

We have already mentioned (§3.3.5) some of the reasons that real equilibrium states are only approximations to the true equilibrium state required by the thermodynamic model. These had to do with imperfections in minerals, and by the same reasoning minerals themselves are not part of the model. That is, they are not phases in the thermodynamic sense, despite common usage. However, just as natural systems approach true equilibrium sufficiently closely that thermodynamics can be applied to them, so naturally occurring phases such as minerals approach thermodynamic phases sufficiently closely that thermodynamics can be applied to them; e.g., they usually obey the Phase Rule. Obviously, some minerals approximate a thermodynamic phase more closely than do others; e.g., almost any quartz crystal is more homogeneous than a zoned plagioclase. But on close examination, it will be found that no minerals, and perhaps no solids, satisfy the requirements of the thermodynamic model. Thermodynamic phases are just as much an abstraction as thermodynamic components or thermodynamic equilibrium, or any other part of the thermodynamic model.

14.11. APPLICATIONS OF THE PHASE RULE

Interpretation of the origins of rocks formed by complex processes millions of years ago at high temperatures and pressures, and in which no trace of any fluid phase, if once present, may now remain, is obviously not easy. These ancient systems are considered to have been of both open and closed types, sometimes both at different stages in their development. There are no rules to aid in distinguishing these two kinds of ancient systems, only insight based on geological knowledge and experience, guided by thermodynamic principles. As in many areas of science, two different investigators are very likely to develop two different hypotheses based on the same observations of physical reality. Because a large part of this problem is geological and not thermodynamic, any discussion on our part is necessarily only an introduction, but we can discuss some examples, which at least shed light on the thermodynamics involved.

To begin with, recall the Mineralogical Phase Rules, (14.60) and (14.61), which can also be written

$$p_{max} = c$$

and

$$\begin{aligned} p_{max} &= c_s \\ &= c - c_e \end{aligned}$$

Thus the maximum number of phases to be expected in a natural system is decreased by one for each "environmental" component, compared to the same system having that component as a "system" component. In other words, systems having $f \leq 2$ will lose one phase for every component that is given an arbitrary potential.

14.11.1. $CaSO_4 - H_2O$

This is very well illustrated by $G - X$ diagrams such as Figure 14.7b, for the $CaSO_4 - H_2O$ system shown in Figure 14.7a. The coexistence of the two phases gypsum and anhydrite fixes or buffers μ_{H_2O}, as mentioned earlier. This is a specific, exact value of μ_{H_2O} (which of course must be less than μ for pure water), which is extremely unlikely to be chosen if μ_{H_2O} is chosen at random. In fact if the points representing gypsum and anhydrite in Figure 14.7b are truly points and not areas, the probability of its being picked at random is precisely zero. Therefore if the chemical potential of water is fixed in the environment, with no reference to the requirement for gypsum–anhydrite equilibrium, it will not be fixed at the value required for that equilibrium. Depending on whether its value is greater or less than the value required for gypsum–anhydrite equilibrium, either gypsum or anhydrite respectively will be the stable phase in the system, but the point is that only one phase instead of two will exist in this two-component system at fixed T and P, as a direct result of μ_{H_2O} being fixed externally.

Extensive beds of both gypsum and anhydrite are found in nature, and occurrences of both phases together are not rare. Generally speaking these occurrences are entirely consistent with a surface origin for gypsum, followed by dehydration to anhydrite on deep burial, followed by rehydration to gypsum if erosion brings the deposit back close to the surface. Occurrences of both phases together could represent areas where the dehydration or rehydration process did not reach completion, or where the system became closed. Usually in this case the system is considered as open and gypsum as stable in the wet surface environment and anhydrite as stable in the dryer depth environment, but of course each occurrence should be considered on its own merits.

14.11.2. $Al_2O_3 - SiO_2 - H_2O$

In a ternary system, $G - X$ diagrams become difficult to draw, but can be imagined as having planes tangent to three free energy loops.[2] Instead of attempting this, let's

[2]One of our favorite recollections is of Prof. J.W. Greig lecturing on ternary phase equilibria using beach balls as (inverted) free energy loops and a sheet of Plexiglas as a tangent plane.

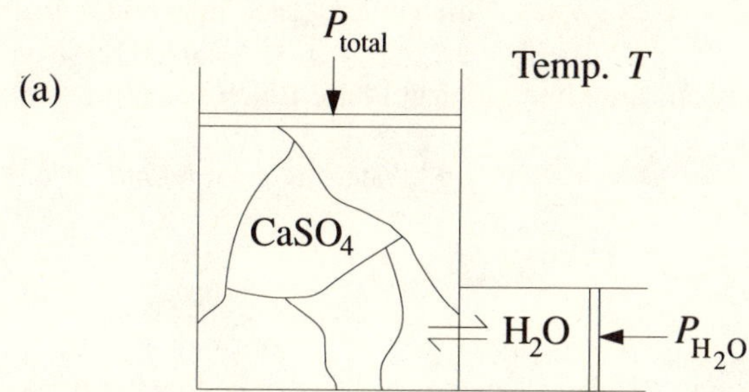

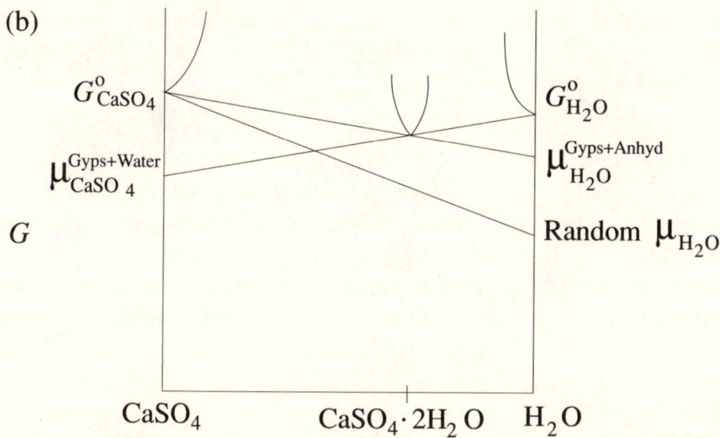

FIG. 14.7. (a) The system $CaSO_4 - H_2O$ with the chemical potential of water controlled externally. (b) $G - X$ diagram for this system, showing that for gypsum and anhydrite to coexist, a specific value of μ_{H_2O} is required, which will not obtain if μ_{H_2O} is selected at random in the environment, or in other words is fixed by some process other than gypsum-anhydrite equilibrium in the environment. Coexisting water and gypsum fix μ_{CaSO_4} at a value too low for anhydrite to exist, and a low value of μ_{H_2O} does not allow gypsum to exist.

consider the system $Al_2O_3 - SiO_2 - H_2O$ in a different way. The possible mineral phases in this system are shown in Figure 14.8. The arrangement of stable joins between these phases depends on the temperature and pressure; we show an arrangement consistent with most of the data of Zen (1961). First let's consider what assemblages one might expect if the system is closed. Given that each assemblage must be capable of stable existence over some range of T and P, no more than three phases should coexist. If water is one of these, then the other two can only be kaolinite–corundum or kaolinite–quartz. All other assemblages lie in the "water-deficient region" (Yoder, 1952), and cannot coexist with a pure water phase. The chemical potential of water is of course buffered in any of these three-phase assemblages.

Suppose now one found a rock having only two of these phases, say kaolinite and quartz. You could not conclude necessarily that water was a phase during metamor-

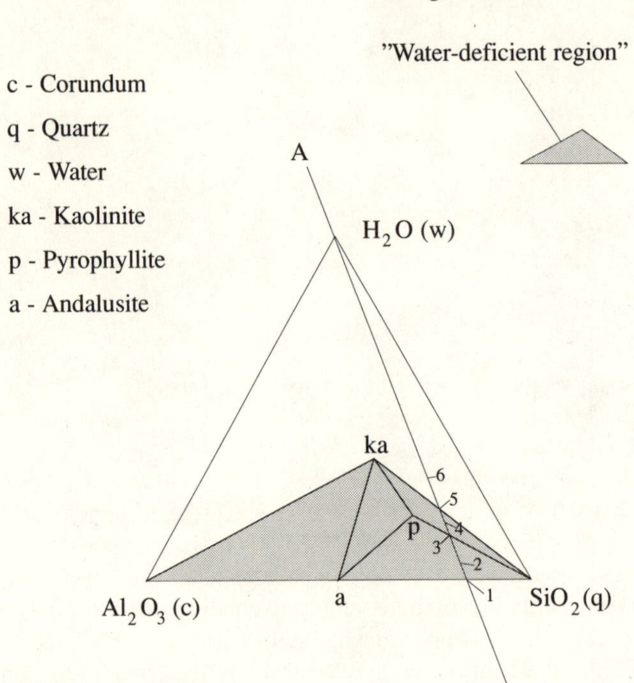

FIG. 14.8. The system $Al_2O_3 - SiO_2 - H_2O$ and some of the phases stable under low grade metamorphic conditions. If water exists as a phase, only corundum+kaolinite or quartz+kaolinite can coexist. Three-mineral assemblages preclude the existence of pure water as a phase. Chemical potentials and mole fractions of water along the section A-B are shown in Figure 14.10.

phism, because you do not know the bulk composition at that time. If it lay within the H_2O–kaolinite–quartz triangle, water would have been a phase, but perhaps the composition lay (as it does now) on the kaolinite–quartz join. In this case the system requires only two components, not three, for a complete description of the composition of all phases, so that two phases would be the maximum expected. The assertion that a water phase existed would have to be based on other evidence. Similarly, other two-phase assemblages can be considered as two component systems, but a different set of components is required in each case, and if these assemblages are found in geologically close proximity, one has to decide whether one is dealing with a number of two-component systems or with one three-component system. "Geologically close proximity" is a purposely vague term, because no criteria other than mutual contact are available for distinguishing (what was formerly) a single system from two separate systems. For example, Figure 14.9 represents what one might see in a

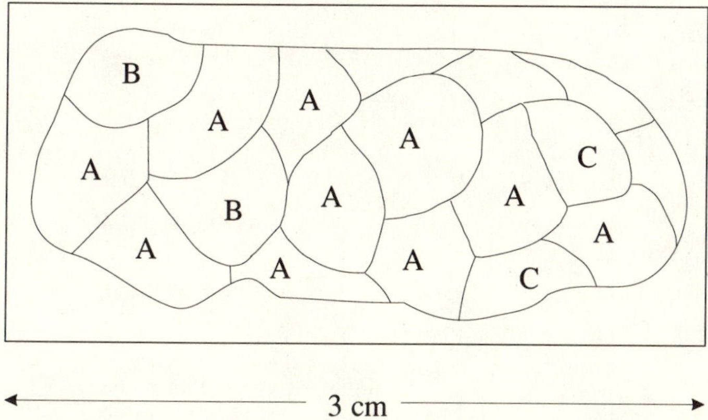

FIG. 14.9. Thin section of a rock containing three minerals, A,B and C. B and C both have contacts with A, but B and C are never in contact. Is this one system or two systems?

thin section consisting of three minerals, A,B, and C. Mineral B occurs only at one side and C on the other side. Both coexist with A but B and C never coexist. Are these two systems AB and AC, or one system ABC? There is no way of knowing, given only this evidence, but geologists usually put together evidence from many sources. For example the spatial distribution of the assemblages and their relationship to other features such as sources of heat, compositional gradients, etc. would be expected to fit into some overall hypothesis of origin. In other words, the existence of a number of different two-component systems in close proximity, all of which belonged to the same three-component system, would call for some explanation.

One explanation could be that the system was not closed, but open to water. If the chemical potential of water was set by the environment, only two phases would normally coexist in the three-component system. Consider the section A-B in Figure 14.8, along which we have plotted both X_{H_2O}, the mole fraction of water, and μ_{H_2O} , its chemical potential, in Figure 14.10. Each three-phase assemblage has a fixed μ_{H_2O} but a range of X_{H_2O}, and each two-phase assemblage has the reverse, a fixed X_{H_2O} but a range of μ_{H_2O} values over which it can exist. This illustrates the fact that if μ_{H_2O} is the controlled variable, only two phases can coexist, whereas if X_{H_2O} is the deciding factor, three phases will coexist. If the mineral assemblages formed in an environment having a gradient in μ_{H_2O} caused by some regional factor, one would expect to find a series of two phase assemblages arranged in a sequence of increasing or decreasing μ_{H_2O}. Interpreting these assemblages as a series of closed two-phase assemblages would not violate the Phase Rule, but one would be expected to come up with an explanation for such a remarkable occurrence in terms of closed systems.

14.11.3. *Pyrophyllite Example*

The reason for presenting this hypothetical analysis of the Al_2O_3–SiO_2–H_2O system is that just such a study was made by Zen (1961) of seven pyrophyllite deposits in

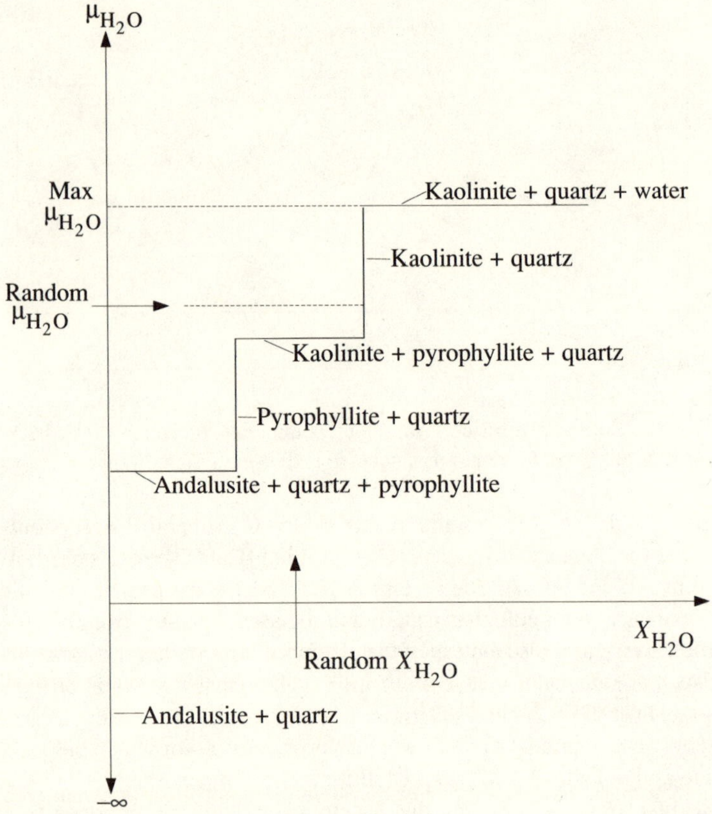

FIG. 14.10. Chemical potential and mole fraction of water for the numbered assemblages shown in Figure 14.8. Random choice of μ_{H_2O} results in a two-mineral assemblage and fixes X_{H_2O}; random choice of mole fraction results in a three-mineral assemblage and fixes μ_{H_2O}.

North Carolina. Many of the minerals found in these deposits do not belong to this system, but each mineral not belonging to the system requires the presence of one additional component (e.g., muscovite–K_2O; paragonite–Na_2O; clinozoisite–CaO; chloritoid and/or chlorite–FeO and MgO; hematite–Fe_2O_3; fluorite–F; ilmenite–Ti; pyrite–S and so on), so each is an accessory phase as defined above and can be ignored. The phases in the Al_2O_3–SiO_2–H_2O system found in the deposits occurred for the most part in assemblages of two and three phases; rarely (twice) four phases were seen in a single thin section, but no more than three phases were ever seen in contact. The common and widespread occurrence of three-phase assemblages is inconsistent with their formation in an open system with μ_{H_2O} externally controlled, and the two-phase assemblages showed no consistent spatial relationship that might indicate control by μ_{H_2O}. These observations helped Zen to conclude that the previous hypothesis of origin, hydrothermal alteration of volcanic slates, was incorrect, and to

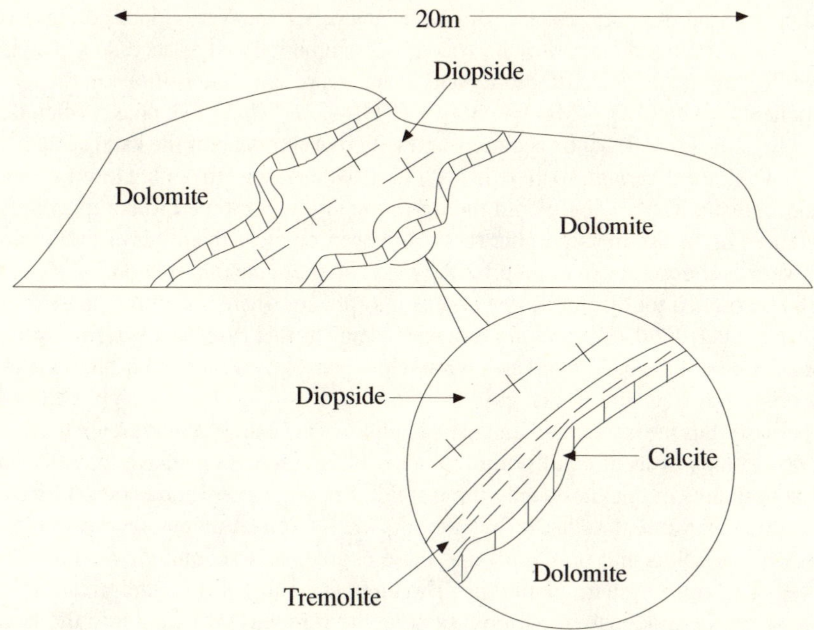

FIG. 14.11. Sketch of a road-cut exposing a deformed layer of diopside in dolomite. The diopside has zones of tremolite and calcite at its contact. Quartz occurs interstitially. Five minerals, five components. No problem?

propose a hypothesis based on the closed system metamorphism of weathered slates. Note that the meaning of the "water deficient region" takes on a slightly different meaning if open system conditions are considered. The fact that assemblages in this region cannot coexist with a pure water phase is not changed, but there could theoretically be an indefinite amount of water in the system, as long as its potential was controlled externally (or more exactly, as long as μ_{H_2O} did not have its maximum value). However in the absence of another volatile component, which would permit the existence of a fluid phase in which the water could be dissolved (but which would give a total of four components instead of three), water can only exist in the hydrous minerals and along grain boundaries where its properties are poorly known, but where its fugacity is less than that of pure water as a phase. Thus in most such cases the term "water deficient region" retains its original meaning in open as well as closed systems.

14.11.4. *Marble Example*

Another example is taken from the study of Cermignani and Anderson (1973) of a zoned silicated dolomitic marble. This rock occurs in several outcrops, each showing ridges of coarse diopside separated from the host dolomite by a zone of tremolite and a zone of calcite (Figure 14.11). Minor calcite and quartz occurs interstitially in the diopside and tremolite zones. All phases are very pure. Very minor amounts of other phases (phlogopite, pyrite) also occur but can be ignored as accessory phases. Dolomite, calcite, diopside, tremolite and quartz represent compositions in the five-component system $CaO - MgO - SiO_2 - H_2O - CO_2$. These five phases often occur in the same (zoned) hand specimen, and with five components the hand specimen system would be divariant, so there is no "Phase Rule reason" to doubt closed system metamorphism. The zoning would then presumably be attributed to compositional differences in the pre-metamorphic rock. However, no more than four of these minerals were ever observed in close proximity, e.g., in the same thin section. This could also be accounted for by closed system metamorphism, where one of the phases was a (supercritical) fluid, and now no longer present. In this case "the system" would be smaller than the zoned hand specimen. However (again) these four phases were only observed once, and in that case the diopside was not fresh but highly altered, and perhaps thus the remnant of an unstable phase. This could be interpreted to mean that in domains of local equilibrium, which would be quite extensive laterally but only centimetres in thickness, only three solid phases are ever found coexisting stably. If this is the case, the Phase Rule permits the interpretation that there were four phases (three solids and one fluid) and three degrees of freedom: T, P, and μ_{H_2O} or μ_{CO_2}, i.e., open system conditions. The authors preferred this interpretation for various non-thermodynamic reasons, such as the difficulty of explaining the symmetrical compositional zoning by closed system metamorphism, the likelihood that the competent diopside would fracture providing channels for water migrating from adjacent rocks, and also accounting for the symmetrical zoning, and the occurrence of tremolite–calcite veins not discussed here.

Our point here is not at all whether Cermignani and Anderson (1973) or Zen (1961) were right or wrong in their interpretations. Our point is that the Phase Rule is a useful tool in petrogenetic interpretations, but that it is only a tool. The crucial parts of any interpretation are made by the investigator, not the Phase Rule or any other tool. Obviously, in the cases cited, a crucial decision is choosing the system in which local equilibrium prevails. In metamorphic rocks it often appears to be smaller than the hand specimen in at least one dimension. In hydrothermally altered zones it can be very large. The difference seems to be related to the existence or non-existence of a fluid phase during the rock-forming or altering process, but considerations like this are irrelevant to the thermodynamic model.

PROBLEMS

1. In this question we use absolute free energies. Their relative magnitudes are more or less reasonable, but of course in real life you never get to deal with quantities like these. The exercise is useful, however, because you should get used to the idea that free energies and chemical potentials are finite, absolute quantities, even if unknown.
   Mineral α-ABC has the following hypothetical absolute properties, in units mol^{-1}.

°C	bars	G°_{ABC}	G°_{A}	G°_{B}	G°_{C}
25	1	470	300	150	30
25	P	1550	1000	500	100
T	1	180	90	30	10
T	P	1260	850	400	90

(a) Calculate the Gibbs free energy of formation ($\Delta_f G^\circ$) of α-ABC at T and P from the elements (A, B and C) according to the "traditional" convention, and according to the convention adopted by Helgeson (Benson - Helgeson convention).

(b) Define a third convention and calculate $\Delta_f G^\circ$ using it.

(c) Show that α-ABC is stable with respect to its elements at both 25°C,1 bar and T, P.

(d) What is unrealistic about the data?

(e) Calculate the equilibrium constant for this reaction at T, P. Use $R = 0.1$ units $\mathrm{K}^{-1}\ \mathrm{mol}^{-1}$.

(f) If the molar volume of α-ABC at 25°C, 1 bar is 0.5 units $\mathrm{bar}^{-1}\ \mathrm{mol}^{-1}$, calculate P. If T is 500°C, calculate S°_{ABC} (assumed constant).

(g) A polymorph, β-ABC, has a smaller molar volume and a higher entropy. What would you expect the P-T phase diagram for ABC to look like, if at 25°C, 1 bar
   - (i) $G_{\alpha-\mathrm{ABC}} < G_{\beta-\mathrm{ABC}}$
   - (ii) $G_{\alpha-\mathrm{ABC}} > G_{\beta-\mathrm{ABC}}$

(h)
   - (i) If α-ABC is equilibrated with element A at T, P, what is the chemical potential of A in ABC?
   - (ii) If α-ABC is equilibrated with both elements A and B at T, P, what is the chemical potential of each in ABC?
   - (iii) What happens if you try to equilibrate ABC with A, B and C?
   - (iv) What is the chemical potential of B in ABC which is equilibrated with A and C at T, P?

(i) Calculate the activity of B in the assemblage (ABC + A + C) at T, P using two different standard states for B. Use $T = 500$°C; $R = 0.1$ units $\mathrm{K}^{-1}\ \mathrm{mol}^{-1}$.

15

SOLID SOLUTIONS

> It is questionable, however, that any one physical model will prove appropriate for the great variety of crystalline solutions in nature. — E. Jewett Green (1970)

15.1. TYPES OF SOLUTIONS

Real solutions of practical interest to Earth scientists do not behave ideally, although some do come fairly close. The problem of course lies in the stringent and unrealistic physical models we have prescribed for ideal solutions. The molecules of a gas do interact with each other, molecular forces within mixed component liquids really are non-uniform, and the different ions substituting for each other in solids are never exactly alike. So why bother defining an ideal solution in the first place if real systems do not behave that way? In fact, the ideal solution is a very useful artifice. It is something simple against which the behaviour of real solutions can be measured and compared.

Our most fundamental definition of an ideal solution was[1]

$$a_i = X_i \text{ (activity coefficient always } = 1.0) \tag{11.29}$$

With this as our reference, we can define a non-ideal solution as one for which the activity coefficient of each component i differs from unity

$$a_i = X_i\gamma_i \qquad (\gamma_i \neq 1.0, \text{ in general}) \tag{15.1}$$

The activity coefficient is the single quantity that expresses all deviations from non-ideality for each component of a solution. As we shall see, parameters other than the activity coefficient itself are frequently used to describe non-ideal behavior, but these could, if we wished, be related back to (15.1). Note that we say $\gamma_i \neq 1.0$ *in general;* there are times when γ_i = 1.0 for specific conditions (one set of T, P, X_i, etc.) even in highly non-ideal systems. This is just coincidental and certainly does not mean that the system is ideal at that particular point—the activity coefficient would have to be unity under all possible conditions for that to be the case.

In this chapter we will describe some of the non-ideal properties of solids, and the set of Margules equations commonly used for solid solutions. However, although solid solutions have been the main area of application of the Margules equations in the Earth Sciences to date, they work equally well for non-ionic liquid and gaseous solutions. Aqueous electrolyte solutions are sufficiently different that we will give them a later chapter of their own (Chapter 17).

[1] The use of activity always implies a standard state. $a_i = X_i$ implies that the standard state of i is pure, stable i at the T and P of interest.

We will examine two types of real solid solutions in this section—representing both slight and extreme non-ideal behavior. Recall that

$$\Delta G = \Delta H - T\Delta S$$

Now for an ideal solution, $\Delta H_{ideal\ dissol'n} = 0$ and

$$\begin{aligned} \Delta G_{ideal\ dissol'n} &= -T\Delta S_{ideal\ dissol'n} \\ &= RT\sum_i X_i \ln X_i \end{aligned} \tag{10.5}$$

The relationship between G, H, and S of ideal solutions is shown in Figure 15.1. This uses the binary (or two-component) olivine system as an example. This system is relatively ideal, so the three ideal curves of Figure 15.1a should approximate reality in this case.

15.1.1. *Regular Solutions*

Next imagine that the two substituting ions (Mg^{2+} and Fe^{2+}) have significantly different bonding energies, but that their charges and radii remain similar. If they still can disperse fairly randomly through the crystal lattice, then their $\Delta S_{dissol'n}$ will remain close to ideal. A good knowledge of the crystal structure and the nature of the cations is required to allow educated guesses as to whether this might be true. Also, the volume change, $\Delta V_{dissol'n}$, for different Mg/Fe ratios should remain relatively ideal, since the radii of the cations are about the same (once again, in real life there could be a slight change because the different bonding energies might distort the lattice). However, in contrast with entropy and volume, if the bonding energies of the two ions are different, the enthalpy and free energy of dissolution are certainly going to be non-ideal.

This new, non-ideal model is only one step removed from an ideal solution. All restrictions remain the same except that the intermolecular forces are no longer uniform (hence G and H will be non-ideal, but S should remain ideal, and V nearly so). The special conditions we have just described define what is called a *regular solution.*

Regular Solutions

$$\left.\begin{aligned} \Delta S_{reg\ dissol'n} &= \Delta S_{ideal\ dissol'n} = -R\sum_i X_i \ln X_i \\ \Delta V_{reg\ dissol'n} &\approx 0 \\ \Delta H_{reg\ dissol'n} &\neq 0 \\ \Delta G_{reg\ dissol'n} &\neq 0 \\ &= \Delta H_{reg\ dissol'n} - T\Delta S_{ideal\ dissol'n} \end{aligned}\right\} \tag{15.2}$$

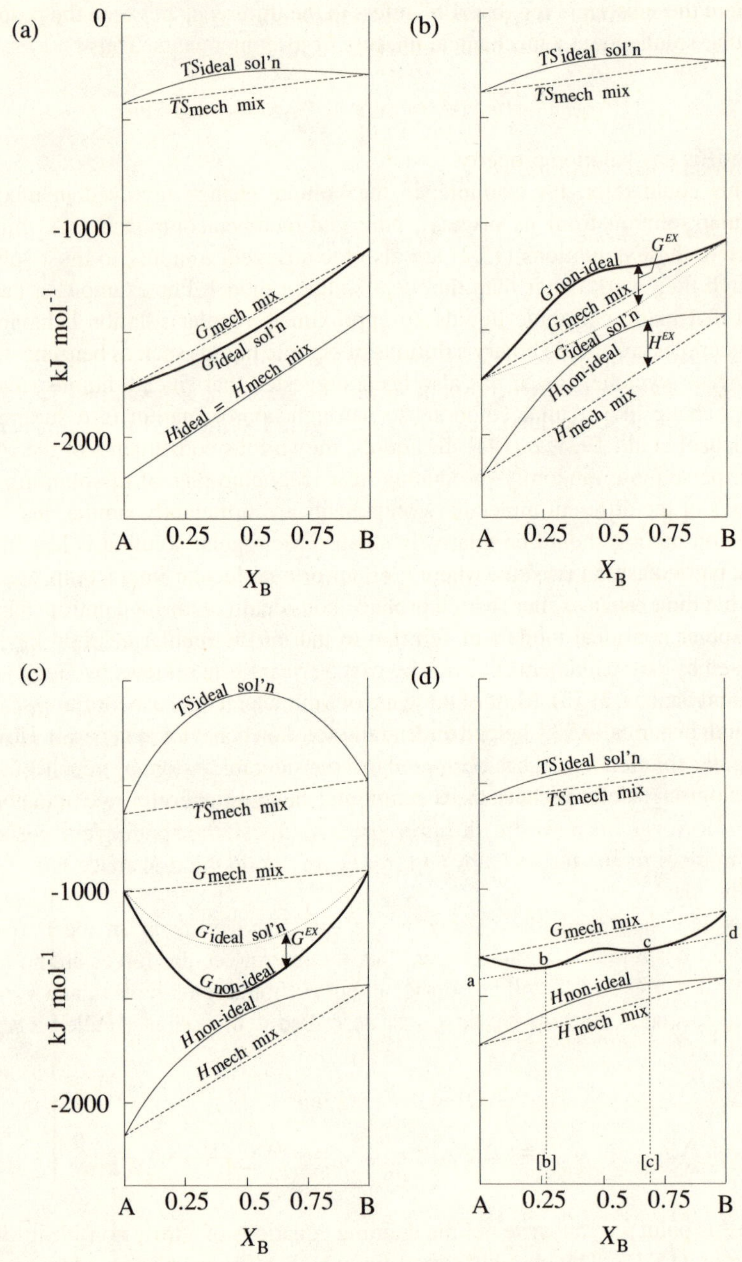

FIG. 15.1. Energy–composition diagrams for two-component mixtures or solutions. (a) Ideal solution at 1000 K between the two components A and B. Molar enthalpy, Gibbs free energy, and TS are solid lines, and properties of mechanical mixtures are dashed lines. (b) Similar to (a), except enthalpy is non-ideal, producing non-ideal free energy curve. (c) Similar to (b) but at a higher T. (d) Similar to (b) but $H^{EX}(= H_{non\ ideal} - H_{mech\ mix})$ is smaller.

Here the subscript "reg dissol'n" refers to the difference between the property of a regular solution and a mechanical mixture of its components. Thus

$$Z_{reg\ dissol'n} = Z_{reg\ sol'n} - Z_{mechanical\ mixture}$$

where Z is any solution property.

This could refer, for example, to the volume change involved in mixing up a regular solution from its original, pure end-member components. As might be expected, these conditions (15.2) are also too stringent to apply to most solutions, although they work better than the ideal solution model. For example, it has been found that many non-ionic liquids do approximate regular solution behavior. The best examples are simple binary solutions of organic liquids such as benzene–toluene or benzene–cyclohexane. It has also been suggested that silicate liquids (magmas) might behave like regular solutions to a rough approximation (see summary by Carmichael et al., 1974, p. 194). In liquids, the various constituents of the solution can disperse more randomly (producing near-ideal entropies of dissolution), and if the sizes of the different molecular components are sufficiently similar, the volume change on mixing should be relatively small. True regular solution is less likely in solids, particularly in crystals, where non-uniform molecular interactions can create preferred ion exchange sites, which produce non-ideal volume and entropy changes. Many other non-ideal models in addition to that of the regular solution have been proposed by physical chemists over the past 50 years (see reviews by Guggenheim, 1952, and Saxena, 1973). Most of these are of limited practical use for Earth scientists (although historically they helped understand solution behavior in general). However, the regular solution model has been used to a considerable extent by geochemists and mineralogists, chiefly because of its simplicity, and we shall refer to it again below.

Non-ideal Solutions

$$\left.\begin{aligned}
a_i &= X_i\gamma_i \qquad (\gamma_i \neq 1,\ \text{in general})\\
\Delta S_{non-ideal\ dissol'n} &= -R\textstyle\sum_i X_i \ln a_i\\
\Delta V_{non-ideal\ dissol'n} &\neq 0\\
\Delta H_{non-ideal\ dissol'n} &\neq 0\\
\Delta G_{non-ideal\ dissol'n} &= \Delta H_{non-ideal\ dissol'n} + RT\textstyle\sum_i X_i \ln a_i
\end{aligned}\right\} \qquad (15.3)$$

At this point we can write out the defining equations of a truly non-ideal solution (equations (15.3)). This would be one for which *all* properties in addition to the enthalpy differ from the ideal values. Here again, the subscript "non-ideal dissol'n" refers to the difference between a non-ideal solution and a mechanical mixture of its pure components:

$$Z_{non-ideal\ dissol'n} = Z_{real\ sol'n} - Z_{mechanical\ mixture}$$

where again Z is any solution property.

Returning to Figure 15.1, let us see how a non-ideal regular solution of two components behaves at different temperatures and compositions. In Figure 15.1b, we have indicated that the solution enthalpy is more positive than the ideal model in Figure 15.1a, and entropy remains the same. Since $\Delta G = \Delta H - T\Delta S$, the free energy increases with ΔH. Here, ΔH is so large that it overwhelms the $T\Delta S$ term. The heavy line representing $G_{non-ideal}$ is now more positive than $G_{mechanical\ mixture}$ for all compositions. This means that a mechanical mixture is now more stable than a solution, and in this case the two end-member minerals would coexist as completely immiscible phases. This is extreme non-ideal behavior—the complete opposite of the case in Figure 15.1a. Positive enthalpy of mixing (H^{EX}) indicates a repulsion between the two components. Figure 15.1c is another extreme example, this time for a similar system at very high temperature. This is drawn for the same degree of non-ideal enthalpy and for ideal entropy at higher T. But now the entropy term in $\Delta G = \Delta H - T\Delta S$ overwhelms the enthalpy, and the picture is reversed. The free energy of the solution is now lower than that of the mechanical mixture and the solution has become stable for all compositions from pure A to pure B. This time entropy is the predominant influence because the temperature has increased sufficiently. Intuitively this seems reasonable because particles move faster at higher temperatures and tend to be more randomly ordered (have higher entropy). The entropy of ideal dissolution as given by equation (10.9) is independent of T, reaching a maximum of 5.763 J $mol^{-1}K^{-1}$ at a mole fraction of 0.5. The entropy contribution to the total free energy of the system is given by $-T\Delta S$. As indicated in Figure 15.2, this changes by almost an order of magnitude from 298 to 2000 K. That is why solubilities and the stabilities of solutions tend to increase at higher temperatures.

Figures 15.1b and c illustrate the two opposite cases where either enthalpy or entropy dominates the free energy function. The most difficult situation is that shown in Figure 15.1d, where entropy and enthalpy approximately balance each other. This produces a peculiar double-humped free energy curve as illustrated. Here the total free energy of the system is always lower than that of a mechanical mixture of the two pure end-members. However, this time there cannot be a complete solid solution from pure A to pure B as in the previous case. This is because we can draw a tangent (*abcd* on the figure) that touches the free energy curve at two points (*b* and *c*). This means that two phases can coexist (having compositions $X_{\rm B}^{b}$ and $X_{\rm B}^{c}$) in which the chemical potentials of each component are the same. That is, the chemical potential of A in both phases is given by the intercept a, and the chemical potential of B in both phases is given by the intercept at *d*.

Figure 15.3a is an enlargement of this free energy curve. It is sufficiently important (and complicated) that much of the rest of this chapter will be spent discussing it. Look first at the tangent *abcd* to the free energy curve in Figure 15.3a. In Chapter 9 (§9.2.3) we showed that a partial molar quantity of a constituent in a solution is given by the Y-intercepts of the slope (dZ/dX_i) of property Z versus composition X_i at the composition of interest. Thus we could find the partial molar free energy of components A and B in binary solutions or mixtures by drawing a tangent to the free energy curves in Figures 15.1 or Figure 15.3. In Figure 15.3a, we can draw a single

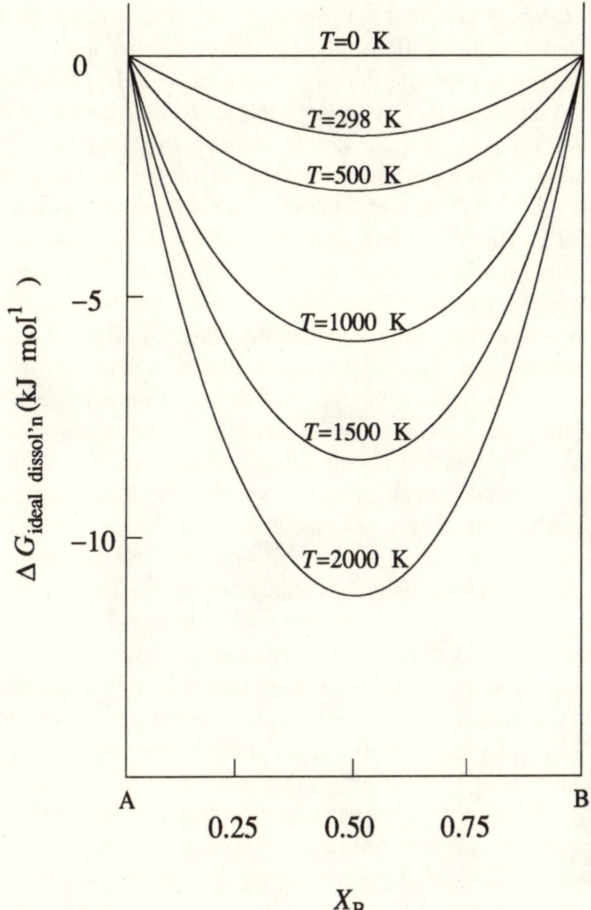

FIG. 15.2. Gibbs free energy of dissolution for ideal binary solutions from 0 to 2000 K. $\Delta G_{ideal\ dissol'n}$ and $G_{ideal\ sol'n}$ decrease markedly at higher T, making solutions more stable.

tangent that touches the free energy curve at the two points *b* and *c*. The chemical potentials μ_B^b and μ_B^c are read directly at point *d*; those for μ_A^b and μ_A^c are read at point a. The superscripts *b* and *c* refer to solutions with intermediate compositions X_B^b (for point *b*) and X_B^c (for point *c*) as shown on the horizontal axis. For all compositions between *b* and *c*, a mechanical mixture of these two solutions has the lower free energy and is therefore more stable (line *ab*). Outside these compositions, a single solution is more stable. For example, between points a and *b*, the total free energy curve lies below line *gb* (which represents the free energy of a mechanical mixture of pure A and the solution at *b*).

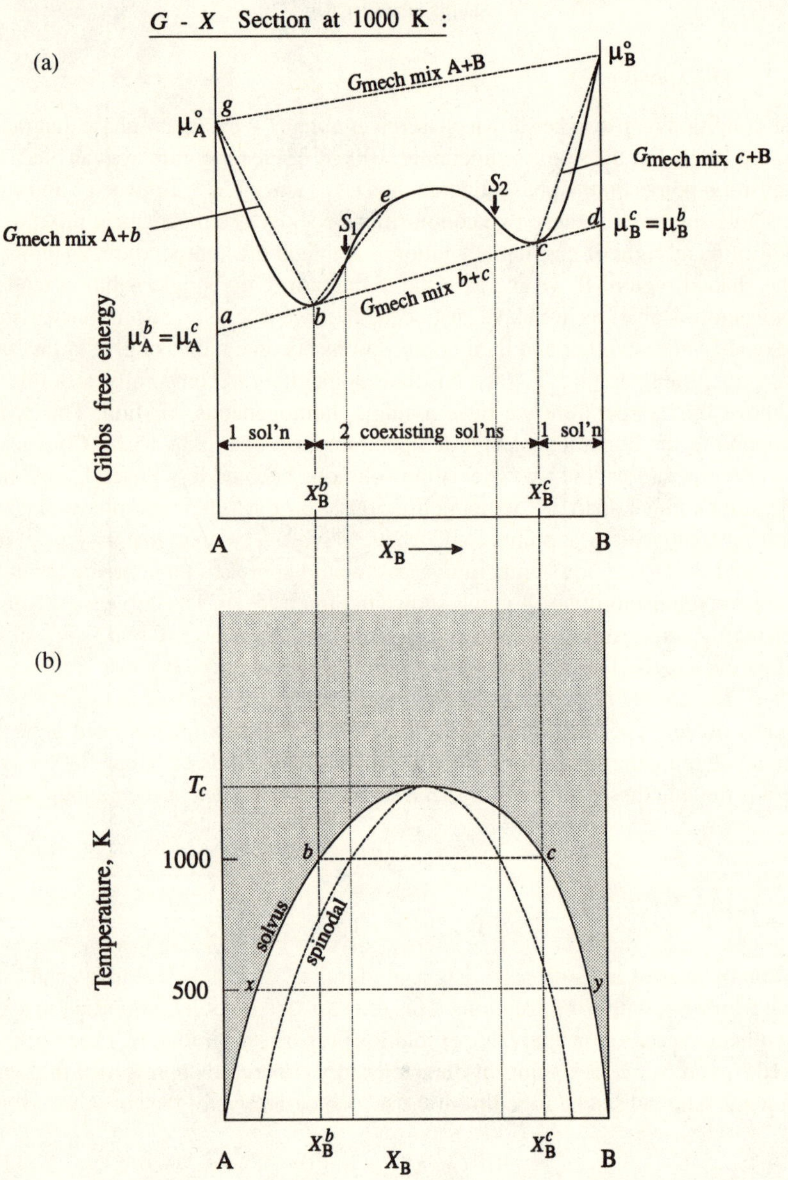

FIG. 15.3. (a) Enlargement of the free energy curve in Figure 15.1d, illustrating range of compositions for which a mechanical mixture is more stable than a solution, and vice versa. Between points b and c, a mechanical mixture of intermediate compositions X_B^b and X_B^c is more stable than a solution. Between a and b and between c and d, a solution is more stable. Intermediate compositions X_B^b and X_B^c are shown on the composition axis. S_1 and S_2 are spinodal points. (b) $T - X$ phase diagram showing intermediate compositions X_B^b and X_B^c at 1000 K from above figure. Shaded area represents region where a single solution is stable. Two solutions exist inside the solvus curve (unshaded region) with compositions given at each T by horizontal tie lines (e.g., at 500 K, two immiscible solutions coexist at compositions x and y. Solvus is solid line, spinodal is dashed line. T_c is critical temperature or consolute point.

15.1.2. *The Solvus*

These conditions can also be shown on a conventional $T - X$ binary phase diagram as in Figure 15.3b. For the single temperature, 1000 K of the top figure, we can place two composition points on the phase diagram at b (X_B^b) and c (X_B^c). Between b and c two immiscible solutions of these two compositions coexist as a mechanical mixture. To the left of b and right of c a single solution is stable at this temperature, as indicated by the shaded region. If we increase the temperature, the single solution tends to become more stable, as we have just seen; the two coexisting solutions dissolve more readily in each other and their compositions become increasingly similar, until at the critical temperature T_c (or consolute point) they become fully miscible. At and above this temperature we have a single, homogeneous solution. This would correspond to the hypothetical high-temperature case of Figure 15.1c. Conversely, at lower temperatures, the two coexisting solutions become less miscible, and their compositions move apart. At temperature 500 K on our schematic phase diagram, coexisting compositions at points x and y have moved so far apart that they are almost pure A and B. At very low temperatures, we would approach the extreme condition of complete immiscibility of A and B shown in Figure 15.1b. The curve traced out by the changing compositions of the two coexisting solutions at different temperatures is called the solvus. This is also often called the binodal curve because it marks the points of the double tangents to the free energy curve. It is possible to have phase diagrams of this type for liquids as well as solids. Much work has been done, for example, defining the solvus for coexisting silicate and sulfide-rich magmatic liquids since it is thought these can produce metal sulfide deposits such as the Sudbury nickel ores.

15.1.3. *The Spinodal Curve*

The double-hump shape of the free energy curve in Figure 15.3a has one other implication of interest to Earth scientists, metallurgists, materials scientists, and other people who work with solid solutions. This is a more subtle point, important in growing synthetic crystals or in analyzing exsolution properties in minerals. The two points marked S_1 and S_2 are flex points of the free energy curve. Flex points (like the center point in the letter S) can be found from the second derivative of the curve, i.e., where

$$(\partial^2 G/\partial X^2)_{T,P} = 0 \tag{15.4}$$

Consider the line bS_1e drawn on Figure 15.3a. Between b and S_1 the free energy curve lies below this line. If a solution with a composition somewhere between these two points were cooled to 1000 K (as on the diagram) or lower, the mechanical mixture b+c might not exsolve as expected. This is because, for very narrow composition ranges between b and S_1, $\bar{G}_{soln} < \bar{G}_{mech\ mix}$. The solution might persist as one metastable phase to lower temperatures despite the fact that a mechanical mixture is really more stable. To nucleate two immiscible phases, as it is supposed to thermodynamically, the original solution would have to diffuse ions and change composition slightly. A slight change in composition would actually raise the total free energy up towards line bS_1.

It would take a major change in composition, such as somehow nucleating another phase with composition over towards point *c*, before phases would spontaneously exsolve. In general, for compositions between the point *b* and S_1 (and similarly S_2 and *c* on the other side), exsolution will be sluggish and might not occur at all. In contrast, for all compositions between S_1 and S_2 any possible mechanical mixture has a lower free energy than the bulk solution (line S_1e for example). Here a slight change in composition produced by diffusion and nucleation drops the system to a lower free energy and a mechanical mixture should spontaneously exsolve. The solution is unstable relative to this mixture even for the slightest change in composition. This all presumes, of course, that the solution is cooled sufficiently slowly for nuclei to form and that diffusion of ions in the solid is rapid enough that the unstable structure is not frozen in. These are really kinetic, rather than purely thermodynamic questions. The compositions S_1 and S_2 are called *spinodal points* or *spinodes*. The trace of their composition with changing temperature is called the *spinodal curve* and roughly parallels that of the solvus as shown in Figure 15.3b. Inside the spinodal curve, exsolution should occur in a cooling solution (provided there are no other kinetic problems). Between the spinodal and solvus curves, metastable single solutions are likely to persist in place of exsolved mechanical mixtures.

There is a possible point of minor confusion here which we should clarify. The spinodes can be found from the second derivative (15.4) of the free energy curve. However, the solvus points a and b do not lie exactly on the minima of this curve, and cannot be found by setting the first derivatives equal to zero. This is apparent from the geometry of the tangent $abcd$ in Figure 15.3a. For most practical applications the differences involved here are likely to be small. A plot of $\bar{G}^{EX}$ (instead of $\bar{G}$) will have the required value of zero for the same first derivatives at the binodal points (since G^{EX} is zero for both pure components, and the slope of the line joining the two end-members is zero).

15.1.4. *Further Thoughts on Non-ideal Enthalpies, Entropies, and Chemical Potentials*

Figure 15.3a illustrates quite nicely one of the essential conditions for chemical equilibrium derived in Chapter 14. The chemical potentials of any one component or constituent must be exactly the same in all coexisting phases at equilibrium, no matter what the composition. Note that in this figure two phases of composition X_{B}^{b} and X_{B}^{c} coexist between points *b* and *c*. The chemical potential of component A is identical in both of these phases, and so is that of component B. This is apparent from the slope of the tangent $abcd$ and its intersections a and d with the vertical composition axes where $\mu_{\mathrm{B}}^{b} = \mu_{\mathrm{B}}^{c}$ and $\mu_{\mathrm{A}}^{b} = \mu_{\mathrm{A}}^{c}$. We have said above that the ideal entropy of dissolution or of mixing any number of components to make a solution is given by $-R\sum_i X_i \ln X_i$ (equation 10.9). This is independent of both pressure and temperature. The increasing importance of entropy at higher temperatures and its effect on making solutions more stable comes from the $-T\Delta S$ in the expression for ΔG. All non-ideal entropies must be less than $-R\sum X_i \ln X_i$ because this represents complete randomness. In olivines, for example, it has been found that there is a slight distortion of the ideal lattice so that Mg ions prefer one site and Fe ions another to a very slight degree (Runciman et al., 1973). This means that olivines must have dissolution entropies somewhat lower than the ideal quantity, and slightly non-ideal free energies as have been calculated by Nafziger (1973). Since entropy drives solutions towards increased stability, any crystal where there is a high preference for different, specific ions on different sites should have a tendency toward reduced solid solution.

Non-ideal enthalpies of dissolution can be either negative or positive, and there is no fixed allowable maximum as with entropy. Negative enthalpies of dissolution tell us that the solution gives off heat as it forms and that there must be an attractive force between the substituting species, or an energy preference for specific ions on specific lattice sites. This tends to make solutions more stable, according to $\Delta G = \Delta H - T\Delta S$. When the enthalpies are unusually negative an intermediate compound might form. Positive enthalpies of dissolution work exactly the opposite way—the solutions are endothermic, the substituting constituents repel each other, and a mechanical mixture may be more stable than the solution (as we saw in Figure 15.1b).

15.2. ACTIVITIES OF COMPONENTS IN SOLID SOLUTIONS

We should insert a reminder here to be careful when calculating activities of components in crystalline solutions that you use consistent chemical formulas. Recall from Chapter 12 (§12.7) that

$$a_i = (X_i\gamma_i)^n \tag{15.5}$$

for component i, where n is the number of exchangeable sites and γ_i is the activity coefficient. As we have seen, the value of n depends on the formula used for the solution itself. For example, if we write the formula of olivine as $(Fe, Mg)_2SiO_4$ then the activity of the end-member component Fe_2SiO_4 in olivine is calculated with $n = 2$ since the formula contains 2 exchangeable sites. If we write the formula instead as $(Fe, Mg)Si_{0.5}O_2$, then $n = 1$ for the activity of $FeSi_{0.5}O_2$ in olivine. Think of n as representing the number of times the exchanging constituent occurs in the formula written for the solution. One must also be careful when selecting the standard state thermodynamic properties to be used in such calculations that they agree with the formula for the mineral. For example, the free energy at 298 K given by Robie et al.(1978) for fayalite written as Fe_2SiO_4 is -1379.375 kJ mol^{-1}. In calculations where the formula is halved, and written with one exchangeable site as $FeSi_{0.5}O_2$, the free energy is also halved and equals -689.687 kJ mol^{-1}. The same applies to all other thermodynamic properties—their tabulated values depend on the formula chosen to represent one mole of the substance.

15.2.1. *Activities with Multi-site Mixing*

With olivine the substitution of Mg^{2+} and Fe^{2+} occurs on the same crystal site (to a good approximation). What happens if there is more than one type of site on which mixing can occur? This is very common with silicates such as micas, hornblendes, garnets, and so on. With olivine of the composition and formula $(Fe_{0.75}Mg_{0.25})_2SiO_4$, and presuming ideal mixing (completely random arrangement of Mg and Fe), the configurational entropy is given by (15.3) and (15.5), i.e.,

$$\begin{aligned} S_{dissol'n} &= -RX_{\mathrm{Fe}} \ln X_{\mathrm{Fe}}^2 - RX_{\mathrm{Mg}} \ln X_{\mathrm{Mg}}^2 \\ &= -8.314 \times 0.75 \times 2 \times \ln(0.75) - 8.314 \times 0.25 \times 2 \times \ln(0.25) \\ &= 9.350\ \mathrm{J\,K^{-1}\,mol^{-1}} \end{aligned}$$

Here the activity of $FeSi_{0.5}O_2$ would be 0.75 according to (15.5). The activity of the fayalite component written as Fe_2SiO_4 is $(0.75)^2$.

Next let's take a mineral such as garnet where substitution can occur on two different groups of sites instead of one. The general formula for garnet can be written $X_3Y_2Si_3O_{12}$ (remember that this choice for the formula now fixes our choice of n in equation (15.5)). The X sites are cubic (8-coordinated) and we will assume they accept Mg^{2+}, Fe^{2+}, and Ca^{2+} ions. The Y sites are octahedral (6-fold), and let's assume they accept only Al^{3+} and Fe^{3+}. The Si site is tetrahedral (4-fold), and it can also hold Al^{3+} ions. The total configurational entropy for this mineral must be the sum of the entropies on all three of these sites, or

$$\begin{aligned} \Delta S_{dissol'n} = &-3R(X_{\mathrm{Mg^{2+}}} \ln X_{\mathrm{Mg^{2+}}} + X_{\mathrm{Fe^{2+}}} \ln X_{\mathrm{Fe^{2+}}} + X_{\mathrm{Ca^{2+}}} \ln X_{\mathrm{Ca^{2+}}}) \\ &-2R(X_{\mathrm{Al^{3+}}} \ln X_{\mathrm{Al^{3+}}} + X_{\mathrm{Fe^{3+}}} \ln X_{\mathrm{Fe^{3+}}}) \\ &-3R(X_{\mathrm{Si^{4+}}} \ln X_{\mathrm{Si^{4+}}} + X_{\mathrm{Al^{4+}}} \ln X_{\mathrm{Al^{4+}}}) \end{aligned}$$

Here n equals 3, 2, and 3 on the cubic (X), octahedral (Y) and tetrahedral (Si) sites, respectively.

One can see from this that a general equation for the configurational entropy or entropy of dissolution involved in forming a solution is

$$\Delta S_{dissol'n} = -R \sum_{j=1}^{k} n_j \sum_i X_{i,j} \ln X_{i,j} \tag{15.6}$$

Here k is the total number of non-equivalent sites (3 for the X, Y, and Si sites in this example); n_j is the number of times the constituent in question appears in the formula for the jth site in the solution (when considering single ions such as Mg^{2+}, Fe^{3+}, and Si^{4+}, etc., as in this example, $n = 3$ for the X sites, 2 for the Y's, and 3 for the Si sites); $X_{i,j}$ is the mole fraction of the ith constituent (or ion in this example) on the jth site (X, Y, or Si site). One thing to watch for in applying this equation is to make sure that a charge balance is maintained. In other words, the substitutions should be independent of one another. See Ulbrich and Waldbaum (1976) for a detailed discussion of this topic.

For garnet, if we wish to calculate the activity of some molecular species or end-member component such as pyrope ($Mg_3Al_2Si_3O_{12}$), the situation is one step more complicated than it was with olivine. The partial molar Gibbs free energy of pyrope in the garnet solution (or its chemical potential μ_{py}^{gt}), like total entropy, must sum the dissolution energies from each of the exchanging X, Y, and Si sites, or

$$\begin{aligned}\mu_{py}^{gt} &= \mu_{py}^{\circ} + 3RT\ln X_{\mathrm{Mg}} + 2RT\ln X_{\mathrm{Al}} + 3RT\ln X_{\mathrm{Si}} \\ &= \mu_{py}^{\circ} + RT\ln(X_{\mathrm{Mg}}^3 X_{\mathrm{Al}}^2 X_{\mathrm{Si}}^3)\end{aligned}$$

Thus

$$\begin{aligned}a_{gt}^{py} &= X_{\mathrm{Mg}}^3 X_{\mathrm{Al}}^2 X_{\mathrm{Si}}^3 \\ &= X_{py}^{gt}\end{aligned}$$

In other words, $n = 1$ in (15.5) for the activity of the component pyrope in garnet, because this species appears once in the formula as written for garnet. To express pyrope activity in terms of the individual ions we need to use the product above and the values $n = 3$, 2, and 3 for the X, Y, and Si sites; as before, this is because the ions occur 3 times in the X sites, 2 times in the Y, and 3 times in the Si sites for the formula $X_3Y_2Si_3O_{12}$ chosen to represent garnet. Nordstrom and Munoz (1985, p. 156) give an example calculation. Keep in mind, too, that so far we are assuming ideal solutions, or ideal mixing of components.

15.2.2. *Substitution of the Same Cation on Several Non-equivalent Sites*

In the olivine example above, two different cations exchange on one equivalent site. With garnet we had three different sites (X, Y, and Si) each of which could exchange different ions; here, Al^{3+} could substitute on either the octahedral Y sites or the tetrahedral Si sites. To round out the possibilities, we should discuss how to handle such situations, where one species (ion, molecule, or whatever) substitutes on two or more different sites in a crystal—something which is very common in minerals.

This time we will use the mineral orthopyroxene, $(Fe, Mg)SiO_3$, as an example. In nature, this forms a nearly pure binary solution of the two end-member minerals enstatite ($MgSiO_3$) and ferrosilite ($FeSiO_3$). The crystal structure contains two different octahedral sites, M1 and M2, over which the Mg^{2+} and Fe^{2+} ions are distributed. Using exactly the same reasoning as with the garnet example, we can write the formula for orthopyroxene as $(Fe, Mg)_2Si_2O_6$. We might also think of this in terms of the two structural sites, $M1M2Si_2O_6$, where both sites contain some of the total Mg^{2+} and Fe^{2+}. Then the activity of the ferrosilite end-member, written as $Fe_2Si_2O_6$ would be

$$\begin{aligned}a_{ferrosilite}^{opx} &= a_{\mathrm{Fe_2Si_2O_6}}^{opx} \\ &= (X_{\mathrm{Fe(M1)}})(X_{\mathrm{Fe(M2)}}) \\ &= (X_{ferrosilite}^{opx})^2\end{aligned}$$

If you prefer to write the formulas for orthopyroxene and ferrosilite with the more conventional stoichiometries $(Fe, Mg)SiO_3$ and $FeSiO_3$, then the activities become

$$\begin{aligned}a_{ferrosilite}^{opx} &= a_{\mathrm{FeSiO_3}}^{opx} \\ &= (X_{\mathrm{Fe(M1)}})^{0.5}(X_{\mathrm{Fe(M2)}})^{0.5} \\ &= X_{ferrosilite}^{opx}\end{aligned}$$

This is because the structural formula for this stoichiometry would have to be written as $M1_{0.5}M2_{0.5}SiO_3$, making $n = 1/2$ in equation (15.5). In these examples, as always, $a_i = X_i\gamma_i$ and the γ_i's have been omitted for simplicity. In these cases of multi-site mixing, γ_i can differ for each specific site.

Using the above reasoning, we can write a general equation for the activity of a cation i (or of its pure end-member) distributed over j different crystal sites:

$$a_{cation\ i}^{in\ mineral} = \prod_j (a_{i,j}^{n_i}) \tag{15.7}$$

where n_i is the number of times cation i appears in the formula for the jth site.

Let's continue one step further with this example and calculate the total entropy of dissolution

$$\Delta S_{dissol'n} = S_{real\ sol'n} - S_{mech\ mix}$$

This is the entropy change involved in mixing the two pure end-member components $(Fe_{M1}Fe_{M2})Si_2O_6$ and $(Mg_{M1}Mg_{M2})Si_2O_6$ to make a pyroxene solution of some intermediate composition. We will use equations (15.3) for non-ideal entropy and (15.7) for activities of the individual ions on each site, thus

$$\begin{aligned}\Delta S_{dissol'n}^{opx} &= -R\sum_i X_i \ln a_i \\ &= -R\left[\frac{X_{\text{Fe(M1)}} + X_{\text{Fe(M2)}}}{2}\right]\ln(a_{\text{Fe(M1)}}\cdot a_{\text{Fe(M2)}}) \\ &\quad -R\left[\frac{X_{\text{Mg(M1)}} + X_{\text{Mg(M2)}}}{2}\right]\ln(a_{\text{Mg(M1)}}\cdot a_{\text{Mg(M2)}})\end{aligned} \tag{15.8}$$

The free energy of dissolution is calculated from

$$\Delta G_{dissol'n} = -T\Delta S_{dissol'n} \tag{15.9}$$

We have actually come full circle here, and equation (15.8) is really the same as (15.6) for multi-site solutions, with one important adjustment: in (15.8) we have divided the concentrations by 2. This is because we have added the mole fractions for two sites together and the concentration used in (15.6) or (15.3) must be that in the total crystal. For example, if the two sites here contained nothing but Fe^{2+}, we would get the impossible mole fraction of 2 for total Fe in the crystal, without this step. You could also think of the two concentration terms in the square brackets of (15.8) as the average mole fractions of Fe^{2+} and Mg^{2+} relative to the other cations in the total crystal.

What we have just done suggests a new and more general form of equation (15.6) for configurational entropy or entropy of dissolution in multi-site, multi-component crystalline solutions, which is

$$S_{dissol'n} = -R\sum_i \sum_{j=1}^{k_i} (n_j/k_i)X_{i,j} \ln a_{i,j} \tag{15.10}$$

Here i refers to cation i; j is the jth crystal site; k_i is the maximum number of sites on which ion i can substitute; n_j is the number of times ion i appears in the formula for site j, and $a_{i,j}$ and $X_{i,j}$ are the activity and mole fraction of ion i in site j. This formula allows for ions that substitute on different numbers of sites. Our previous warning about maintaining a charge balance still applies.

In carrying out calculations for multi-site mixing such as this, you can either use equation (15.10) directly, or work through the problem stepwise and ion by ion as with (15.8); the two methods are equivalent. Further details are given by Powell (1977). It is possible using analytic techniques such as X-ray diffraction, Mossbauer spectroscopy and infra-red, Raman and UV-visible spectroscopy to analyze for the concentration of different ions on specific crystal sites. Thus the activities, configurational entropies, and other such properties of solid solutions can often be calculated. This provides detailed information on the stability and structure of crystalline solutions.

15.3. EXCESS THERMODYNAMIC FUNCTIONS

We have now seen that for real non-ideal solutions all the thermodynamic properties such as G, S, H, V and the internal energy U can differ significantly from the ideal values. This deviation from ideality can be conveniently expressed as a difference from the ideal quantities. The differences are called *excess thermodynamic functions*:

$$V^{EX} = V_{real} - V_{ideal} = V_{real} - \sum_i X_i V_i^\circ \tag{15.11}$$

$$H^{EX} = H_{real} - H_{ideal} = H_{real} - \sum_i X_i H_i^\circ \tag{15.12}$$

$$U^{EX} = U_{real} - U_{ideal} = U_{real} - \sum_i X_i U_i^\circ \tag{15.13}$$

$$S^{EX} = S_{real} - S_{ideal} = S_{real} - \left(\sum_i X_i S_i^\circ - R\sum_i X_i \ln X_i\right) \tag{15.14}$$

$$G^{EX} = G_{real} - G_{ideal} = G_{real} - \left(\sum_i X_i G_i^\circ - RT\sum_i X_i \ln X_i\right) \tag{15.15}$$

The excess free energy and excess enthalpy are shown on the energy-composition diagrams of Figures 15.1b and c. If there is some way of measuring the real properties of a solution, then the excess properties can be calculated very easily. At first sight, this may appear to be an unnecessary new set of equations or a new notation for old variables. However, excess properties actually simplify notation enormously. This is because we can treat all excess functions entirely like the usual thermodynamic functions, using exactly the same equations and relations. We do not have to write out the right-hand sides of equations such as (15.14) and (15.15) but can simply use the excess functions alone. For example, we can rewrite the fundamental equation (5.11) as

$$dU^{EX} = TdS^{EX} - PdV^{EX}$$

and similarly,

$$\begin{aligned} G^{EX} &= U^{EX} + PV^{EX} - TS^{EX} \\ &= H^{EX} - TS^{EX} \end{aligned}$$

All other thermodynamic relations can be written with excess quantities in this manner, simply substituting the excess property for the usual variable. As with all thermodynamic variables, this permits us to completely determine all excess properties simply by measuring a necessary minimum number of properties. Therefore if excess volumes, free energies, and enthalpies are known for a solution at different temperatures and compositions, then all other properties can be calculated for a range of T, P, and composition. Similarly, if we have measured excess free energies over a range of temperatures, it is not necessary to measure the excess entropy; it is already known from

$$(\partial G_i^{EX}/\partial T)_{P,X_i} = -S_i^{EX}$$

15.3.1. *Relationship Between Excess Properties and the Activity Coefficient*

Excess properties are just another way of representing the activity coefficient and are used because they tend to simplify notation. We now derive the relationships between the activity coefficient and excess free energy, enthalpy, entropy, and volume.

Taking the free energy first, consider an ion i in any non-ideal solution. Starting with

$$a_i = X_i \gamma_i \tag{15.1}$$

we can write (15.1) as

$$\bar{G}_{real,i} = \mu_{real,i} = \mu_i^\circ + RT \ln X_i \gamma_i \tag{15.16}$$

for real solutions, and as

$$\bar{G}_{ideal,i} = \mu_{ideal,i} = \mu_i^\circ + RT \ln X_i \tag{15.17}$$

for ideal solutions. Now, defining

$$\bar{G}_i^{EX} = \bar{G}_{real,i} - \bar{G}_{ideal,i}$$

analogous to our quantity

$$G_i^{EX} = G_{real,i} - G_{ideal,i} \tag{15.15}$$

the relationship between γ_i and G_i^{EX} is then

$$\bar{G}_i^{EX} = RT \ln \gamma_i \tag{15.18}$$

The excess partial molar entropy, enthalpy, and volume can now be obtained from familiar partial derivatives of free energy as follows:

$$\begin{aligned} \bar{S}_i^{EX} &= -(\partial \bar{G}_i^{EX}/\partial T)_{P,X} \\ &= -R \ln \gamma_i - RT(\partial \ln \gamma_i/\partial T)_{P,X} \end{aligned} \tag{15.19}$$

$$\bar{H}_i^{EX} = -T^2 \left(\frac{\partial(\bar{G}_i^{EX}/T)}{\partial T} \right)_{P,X}$$
$$= -RT^2(\partial \ln \gamma_i/\partial T)_{P,X} \tag{15.20}$$

$$\bar{V}_i^{EX} = (\partial \bar{G}_i^{EX}/\partial P)_{T,X}$$
$$= -RT(\partial \ln \gamma_i/\partial P)_{T,X} \tag{15.21}$$

So the remaining partial molar excess quantities can all be expressed as simple functions of the activity coefficient.

We could write all non-ideal properties of solutions in terms of activity coefficients. For electrolyte (salt) solutions we usually do use activity coefficients, but for solid and gaseous solutions Earth scientists have traditionally used other approaches based on excess properties. This is not to say the other approaches are fundamentally different or better. They are simply different ways of representing the same physical properties. However, we have just seen that using excess functions instead of activity coefficients can simplify notation, and this is always an advantage. The next section discusses the method most commonly used to express the activity coefficients and excess functions of solid solutions.

15.4. THE MARGULES EQUATIONS

There is a practical problem in using either activity coefficients or excess functions to describe non-ideal solid solutions—they are both dependent on composition. For example, the activity coefficient of a solution is a function of T, P, and composition, as is the excess free energy. To properly describe the behavior of a solution in terms of these variables, we would have to specify the exact values of γ_i or G_i^{EX} for each composition at every T and P. With the advent of computers this is no longer the problem it once was. The variation of γ_i with T, P and X_i can be described by an equation that is fitted to available data by statistical regression analysis or some other procedure. The trick now is to find the most compact equation with the proper mathematical form for the data. If the data have the shape of a parabola, then we should use a parabolic equation and not something like an exponential function. The best equation will give a closer fit to the data, require fewer coefficients, and can often be extrapolated with less uncertainty than some other, less appropriate function. As might be imagined, there have been a great many suggestions over the years for equations representing the non-ideal properties of solid solutions. Many of these approaches have been summarized by Guggenheim (1952), Prigogine and Defay (1954), Saxena (1973), and Grover (1977). A good short summary is in Brown (1977). In most of these examples the regression parameters are not dependent on composition (although they may depend on T and P). For example, we might use the polynomial

$$\ln \gamma_i = a + bX_i + cX_i^2 + \cdots$$

where the coefficients a, b, c, etc., depend on T and P alone. We could then fit this equation to activity coefficient data for a range of compositions at each T and P.

The set of equations most frequently used in the Earth sciences to represent non-ideal solid solutions evolved over the past century from a study first published in 1885 by Margules. While Margules' original work was on the partial vapor pressures over binary solutions, his general method is now used widely in the materials sciences and in metallurgy and became popular with mineralogists through the work of James B. Thompson, Jr. and his colleagues since the 1960s. Thompson (1967) has provided a detailed review and derivation of what are called the Margules equations, and a useful summary has been written by Grover (1977). Here we will try to show somewhat unconventionally how the Margules equations can be developed from a good guess at an appropriate, compact equation for regular solutions. With one assumption, we will then expand this into a more general equation for truly non-ideal solutions. Next we will show that the general equation has the same form as the virial equation of state, which works well for gas mixtures, suggesting that this is why the Margules equations also work with solid solutions.

15.4.1. *Margules Equations and Slightly Non-ideal, Symmetric Solutions*

Let's begin with the simplest kind of non-ideal solution, one which is practically ideal. As a model we might use the regular solution defined by equations (15.2), with ideal entropy, near-ideal volume, and slightly non-ideal enthalpy and free energy. We need the simplest equation that adequately describes the excess free energy or activity coefficient of such solutions, an equation with the same geometric shape as the properties themselves. G^{EX} for a regular solution will have something close to a parabolic shape—this can be seen from Figures 15.1b and 15.1c. That means we need some kind of $y = a\,x^2$ function. Since excess properties are zero for pure components, this function should approach zero as the mole fraction, X_A or X_B of either of the two components, approaches 1.0. Finally, with a regular solution of two similar components such as benzene and toluene, we might expect the solution to be most non-ideal when the components are mixed in equal proportions (because that's when non-uniform interactions between the two species are maximized). That means the equation should have a maximum or minimum at the 1:1 composition. The simplest equation which satisfies all these conditions is

$$G^{EX} = W_G X_{\mathrm{A}} X_{\mathrm{B}} \tag{15.22}$$

This is one of the Margules equations for a symmetrical, regular (or slightly non-ideal) solution. It gives the total excess free energy per mole as a parabolic function of concentration of the two components, and is therefore often called a *two-suffix Margules equation*. You may also see this referred to as a *one-parameter Margules equation* because of the single coefficient W_G. It is symmetrical about the 1:1 composition of the two components. As an example, the schematic solvus we have drawn in Figure 15.3b is symmetrical about $X_{\mathrm{A}} = X_{\mathrm{B}} = 0.5$, each side being a mirror image of the other. If we were to draw the excess free energy curve for this example, it would have a shape similar to the solvus, and so would all the other excess properties.

The parameter W_G has units of energy and is independent of composition, but varies with T and P. It can be thought of as the energy necessary to interchange a mole of A with one of B in the mixture, without changing composition. Similar to the enthalpy of dissolution described above, if W_G is positive, then molecules A and B repel each other (or prefer energetically to be with molecules of the same type); this can lead to immiscibility and exsolution. If W_G is negative, there is an energy preference for A and B to associate in the solution, and this can produce intermediate compounds. The solution, which fits a two-suffix Margules equation, can be regular as defined by equations (15.2) (with ideal entropy), but this is not absolutely necessary; it just has to have symmetric behavior. This has caused some minor confusion in the literature because many authors say that any solution conforming to (15.22) is "regular," by which they probably mean "slightly non-ideal". Thompson (1967), in his paper marking the beginning of Earth science interest in this subject, got around this problem by distinguishing *strictly regular solutions* from ordinary *regular solutions*—the latter need only fit equation (15.22) while the former also have to satisfy equations (15.2).

By analogy with (15.22) we can now write the other excess functions for this symmetric slightly non-ideal solution as

$$G^{EX} = W_G X_A X_B \tag{15.22}$$

$$V^{EX} = W_V X_A X_B \tag{15.23}$$

$$H^{EX} = W_H X_A X_B \tag{15.24}$$

$$S^{EX} = W_S X_A X_B \tag{15.25}$$

$$U^{EX} = W_U X_A X_B \tag{15.26}$$

We have already seen that the excess properties relate to each other in exactly the same way as do the usual thermodynamic functions. The same thing now applies to the Margules parameters, since they are simply another way of writing excess properties. Thus, as we showed with the excess functions, we can now write relations such as the fundamental equation (5.11) to interrelate Margules parameters:

$$dW_U = T\,dW_S - P\,dW_V$$

And all the other thermodynamic relations also apply, such as

$$(\partial W_G/\partial T)_{P,X} = -W_S$$

and

$$\begin{aligned} W_G &= W_U + PW_V - TW_S \\ &= W_H - TW_S \end{aligned}$$

15.4.2. *Activity Coefficients in Symmetrical Solutions*

All these W parameters should also be independent of composition. This gives them a built-in advantage over the activity coefficient, which, as we noted above, is necessarily dependent on composition as well as on T and P. Remember that there is nothing mystical in all this. We have not suddenly gotten rid of bothersome variables such as T, P, and composition. What we are doing is analogous to fitting a regression equation such as

$$G = a + bX + cX^2 + \cdots + dT + eT^2 + \cdots + fP + gP^2 + \cdots$$

to free energy (or other) data. We are then using the regression parameters $a, b, c, \ldots$, which are independent of X, T, and P if the equation fit our data adequately.

Next we can finally see how the activity coefficient relates to the Margules equations for this case. Recall from Chapter 9 that the partial molar quantity of one component in a binary solution can be obtained graphically from the tangent (as with the chemical potentials $\mu_{\rm B}$ and $\mu_{\rm A}$ in the coexisting solutions of Figure 15.3a). From equation (9.6), the partial molar free energy or chemical potential of component A in a solution of A and B is given by

$$\mu_{\rm A} = G + (1 - X_{\rm A})(dG/dX_{\rm A})_{P,T} \tag{15.27}$$

where G is the free energy per mole of the solution. We can obtain G from (15.15) and (15.22):

$$\begin{aligned} G_{real} &= G_{ideal} + G^{EX} \\ &= X_{\rm A}G^\circ_{\rm A} + X_{\rm B}G^\circ_{\rm B} + RT(X_{\rm A}\ln X_{\rm A} + X_{\rm B}\ln X_{\rm B}) \\ &\quad + W_G X_{\rm A} X_{\rm B} \end{aligned} \tag{15.28}$$

Applying (15.27) to (15.28), inserting $X_{\rm B} = (1 - X_{\rm A})$ and remembering that chemical potentials are the same as molar (for pure substances) or partial molar free energies one obtains

$$\mu_{\rm A} = \mu^\circ_{\rm A} + RT\ln X_{\rm A} + W_G X^2_{\rm B} \tag{15.29}$$

Now, from (15.16) we know that for non-ideal solutions,

$$\mu_i = \mu^\circ_i + RT\ln X_i\gamma_i \tag{15.16}$$

and from (15.18) we have the relationship between γ_i and G^{EX}_i,

$$\bar{G}^{EX}_i = RT\ln\gamma_i \tag{15.18}$$

Combining the above three equations, we can finally write the relationship between the activity coefficient and the Margules parameter W_G :

$$\left.\begin{aligned} RT\ln\gamma_{\rm A} &= W_G X^2_{\rm B} = \bar{G}^{EX}_{\rm A} \\ RT\ln\gamma_{\rm B} &= W_G X^2_{\rm A} = \bar{G}^{EX}_{\rm B} \end{aligned}\right\} \tag{15.30}$$

This has the same parabolic form as equations (15.22) to (15.26), with the change that only one component appears in the Margules term, rather than both. The excess free energy of one component is directly related to the squared concentration of the other. We will see with the analogous equation for asymmetrical solutions below that (15.30) has the form of a truncated virial equation, which is known to work well for gas mixtures. Notice also that here we have the (partial molar) excess property of one component in a binary solution, whereas equations (15.22) to (15.26) were for the excess property of the total solution. Equation (15.30) is a very simple and convenient way of getting the activity coefficient, but few solutions can be expected to behave so neatly. In fact, the two-suffix equations do work fairly well with nearly ideal liquid solutions such as benzene-toluene. Unfortunately though, most real solutions (and particularly most solid solutions) are asymmetric. Something more complicated than a simple parabola is necessary to fit this kind of behavior. One possibility would be to extend (15.30) to a polynomial such as

$$RT \ln \gamma_i = aX^2 + bX^3 + cX^4 + \cdots$$

where the series is carried out to the power that best fits the data. This approach is called the *Redlich-Kister expansion* and works quite well for many non-ideal systems (for further details, see Prausnitz et al. 1986).

15.4.3. *Margules Equations and Non-ideal, Asymmetric Solutions*

Because most real solutions do not follow the parabolic, symmetrical form of the two-suffix Margules equations, it is necessary to add more flexibility to the equations by adding more coefficients and more terms. One way to do this is to simply define a second W_G term of the type in (15.22), then combine the two W_G terms in a linear equation in X. This can be imagined as the "mixing" of two solutions, each symmetrical but having different W_G's:

$$G^{EX}_{sol'n\ 1} = W_{G_1} X_1 X_2 \tag{15.31}$$

$$G^{EX}_{sol'n\ 2} = W_{G_2} X_1 X_2 \tag{15.32}$$

Now mix X_2 moles of the first solution with $X_1 (= 1 - X_2)$ moles of the second, stir well, and make a new solution, which is described by the sum of (15.31) and (15.32):

$$G^{EX}_{sol'n} = X_1 (W_{G_2} X_1 X_2) + X_2 (W_{G_1} X_1 X_2) \tag{15.33}$$

What we have now is the (weighted) sum of two parabolic equations to describe a non-ideal, asymmetric solution. Because of the second parameter, this is frequently called a *two-parameter Margules equation*. It seems reasonable that this might have an appropriate form because (15.33) still goes to zero as either X_1 or X_2 approaches 1.0 (the excess free energy must be zero for pure substances); also with two parameters and two parabolas we should be able to fit a distorted parabolic shape, and that's what we might expect the excess free energy to look like for many real systems. In fact, two-parameter equations of this type do fit experimental data on real systems quite closely, and even work well with minerals that are sufficiently non-ideal that they have miscibility gaps (or a solvus region). We might also expect these two-parameter Margules equations to have the shape of non-ideal solutions because they are similar in form to virial equations of state. To show how and why this is so, we first have to discuss virial equations.

15.4.4. *The Virial Equation*

The virial equation of state was originally developed to describe the $P-V-T$ behavior of non-ideal gases. Like the Margules equations, it appeared in the literature around the turn of the century in forms such as that given by Kammerlingh-Onnes in 1901:

$$PV = RT + B(T)P + C(T)P^2 + D(T)P^3 + \cdots \tag{15.34}$$

where $B(T)$, $C(T)$ and $D(T)$ are the second, third, and fourth virial coefficients, etc. The "first" virial coefficient is the gas constant R.

At first sight this looks like nothing more than a polynomial expansion of the ideal gas law. However, it turns out to have real physical significance, and the form of the equation follows directly from statistical mechanics. The details can be found in most textbooks on statistical mechanics (see, for example, Mayer and Mayer, 1940; Hill, 1960, Chapter 15). We will outline the underlying theory very briefly here because virial equations (or similar approaches) appear several times in this book—see for example the discussion of the Pitzer equations for the non-ideal properties of salt solutions in Chapter 17 and Chapter 16 on gas mixtures.

Remember that ideal gases should have no inter-particle interactions. This gives the "first" virial coefficient, R, and the ideal gas law. Unfortunately, the atoms or molecules of real gases do interact. The energy E of one mole of a real gas is the sum of the kinetic and intermolecular potential energies of all its molecules, $E = \mathrm{K} + \mathrm{U}$.

In theory, if we knew the functional form of each kind of molecular interaction (such as coulombic interactions $\propto r^{-1}$, dipole interactions $\propto r^{-6}$, interaction between neutral particles $\propto r^{-6}$, "overlap" repulsion at very close distances $\propto r^{-9}$ to r^{-12}, and so on, where r is a distance), then we could sum these all up for all particles in the system. We would then plug this into the Boltzmann equation (6.9) or the partition function (6.10), and then, as we observed in Chapter 6, we would know all thermodynamic properties of our gas. This sounds too good to be true, and unfortunately it is. We do not yet know all the functions for all the potentials or energies between real molecules and atoms. Even if we did, we could not solve the unthinkable number of equations necessary to express all interactions between all Avogadro's number of particles in a mole of solution—this would be worse than the notorious n-body interaction problem because we have more than one kind of interaction. In recent years there has been considerable progress with this kind of problem using computer simulation methods based on molecular dynamics (putting N particles in a hypothetical box and simulating their behavior over time given certain molecular interactions). However, for present purposes we need to find some approximate mathematical solution—one equation that, at least in theory, describes all non-ideal interactions. To do this we start by writing the partition function (6.10) of the Boltzmann equation as

$$Z = \sum e^{-(\mathrm{K}+\mathrm{U})/kT} = \sum e^{-\mathrm{K}/kT} \sum e^{-\mathrm{U}/kT} \tag{15.35}$$

The kinetic term here can be evaluated from temperature and particle mass, and it is the second summation that causes the problems. Because we are working with an enormous number of particles, we can write the molecular potentials U as continuous, integrable functions, rather than worrying about discrete, quantized energy levels:

$$Z' = \int \cdots \int \exp\left(-\mathrm{U}(x_1, y_1, z_1, \ldots, z_N)/kT\right) dx_1, dy_1, dz_1, \ldots, dz_N \quad (15.36)$$

Z' is called the configuration integral. The integration is made over the x, y, z coordinates for each of the N molecules, so there are $3N$ separate integrations in (15.36). Quite obviously, if N is even a fraction of Avogadro's number we need to simplify this somehow. One method is to factor the total energy U into sets of terms, each representing the interaction energy of a certain number of molecules or "clusters." These clusters can be written for the interaction of particles in pairs, in groups of three, four, and so on. The configuration integral can then be approximated by the sum of these separate terms:

$$\exp(-\mathrm{U}/kT) = 1 + \sum_{pairs} \mathrm{U}_{i,j} + \sum_{triples} (\mathrm{U}_{ij} + \mathrm{U}_{jk} + \mathrm{U}_{ij}\mathrm{U}_{jk}\mathrm{U}_{ik})$$
$$+ \sum_{quadruples} (\mathrm{U}_{ij}\mathrm{U}_{kl} + \cdots) + \cdots \quad (15.37)$$

Now the remarkable thing about virial equations is that the coefficients B, C, D, etc. in (15.34) are related to the clusters or sums. For example, the second virial coefficient is related to the term summing pairwise interactions in (15.37) by

$$B(T) = \frac{1}{2}\left(V - \sum_{i}^{N}\sum_{j}^{N} -\mathrm{U}_{ij}/kT\right) \quad (15.38)$$

where V is the volume of the system. This relationship is used to estimate the interaction potentials U_{ij} of very dilute, nearly ideal gases. If (15.37) is truncated after the pairwise term, it gives the van der Waals equation for non-ideal gases mentioned in the next chapter (equation 16.6). In a virial equation such as (15.34), the first term describes ideal behavior ($PV = RT$ for a gas); the second term (B) describes non-ideal interactions of pairs of molecules; the next term (C) applies to interactions of molecules considered three at a time, and so on. *The general form of a virial equation is the (P-V-T-X) equation of state for an ideal substance followed by an ascending polynomial in one of the variables.*

Unfortunately, there are so many different kinds of molecular interactions, and many of them poorly known, that (15.37) cannot be used to estimate the properties of real gases in general. However, this does mean that virial equations have the appropriate form to describe or fit real gases. The combined virial equations for two or more gases in a mixture often describe mixtures or solutions of gases quite well. It does not require too much of a leap of faith to suggest that virial equations might have the proper form to fit liquid and solid solutions as well. We shall see in Chapter 17 that the Pitzer equations for electrolyte solutions have the virial form and can be used to describe extremely complex aqueous salt solutions. It turns out that the Margules equations (both symmetric and asymmetric) also have the form of a virial equation and this is probably why they work so well at fitting solid solutions.

To show that the Margules equations have the form of a virial equation, rearrange (15.33) for the total excess free energy of a real, asymmetric solution, remembering that $X_1 = 1 - X_2$:

$$G^{EX}_{real} = 0 + W_{G_2}X_2 + (W_{G_1} - 2W_{G_2})X_2^2 + (W_{G_2} - W_{G_1})X_2^3 \tag{15.39}$$

Because excess free energy is zero for ideal solutions, this has exactly the required form—the first term on the right hand side (zero) is the value of G^{EX} for an ideal solution, and the remaining terms are an ascending polynomial in the variable X_2. This is a virial equation carried to the fourth coefficient. For symmetric solutions, $W_{G_2} = W_{G_1}$ and (15.39) reduces to

$$G^{EX}_{real} = 0 + WX_2 + (-W)X_2^2 \tag{15.40}$$

This is a virial equation carried only to the third coefficient (which is why it doesn't fit more general, asymmetric solutions). It is interesting that Margules wrote equations such as these before virial equations first saw the light of day, and certainly long before the virial coefficients were interpreted in terms of molecular interactions. Margules' initial equation (which you can find reproduced by Grover, 1977, p. 74) turned out to be a very good, educated guess for an expression giving partial vapor pressures over binary solutions; this obeyed Raoult's and Henry's laws at low concentrations of either component.

15.4.5. *Activity Coefficients in Asymmetric Systems*

We can find the activity coefficient of one component in a solution described by (15.33) using the same reasoning as above for solutions that fit the single-parameter equation (15.22). First, as with (15.28), we need to write out the expression for the total free energy per mole of such a solution:

$$\begin{aligned} G_{real} &= G_{ideal} + G^{EX} \\ &= X_1G_1^\circ + X_2G_2^\circ + RT(X_1 \ln X_1 + X_2 \ln X_2) \\ &\quad + X_1(W_{G_2}X_1X_2) + X_2(W_{G_1}X_1X_2) \end{aligned} \tag{15.41}$$

Next we find the chemical potential of either component as before by applying equation (15.27) to (15.41), then using relationships (15.16) and (15.18), rearranging, and remembering that $X_1 = 1 - X_2$:

$$\left.\begin{aligned} RT \ln \gamma_1 = \bar{G}_1^{EX} &= (2W_{G_2} - W_{G_1})X_2^2 + 2(W_{G_1} - W_{G_2})X_2^3 \\ RT \ln \gamma_2 = \bar{G}_2^{EX} &= (2W_{G_1} - W_{G_2})X_1^2 + 2(W_{G_2} - W_{G_1})X_1^3 \end{aligned}\right\} \tag{15.42}$$

This is convenient because, like equations (15.30) it gives the activity coefficient (or excess free energy) of a non-ideal solution for different concentrations in terms of the two Margules coefficients W_{G_2} and W_{G_1}, which are constants for any T and P (they are independent of concentration but vary with T and P). If the fit is satisfactory, this obviates any need to tabulate activity coefficients (and related properties) as a function of concentration. Note that (15.42) also has the form of a virial equation.

If excess free energies have been fit with equation (15.41) to derive W_{G_1} and W_{G_2}, then these are the W_{G_1} and W_{G_2} values which appear in both equations in (15.42). However, it is also possible to use (15.42) to fit independently calculated activity coefficients or partial excess free energies, in which case W_{G_1} and W_{G_2} will have different values in the two versions of (15.42). This was the procedure followed in calculating the curves in Figure 11.11.

Equation (15.42) also shows the interesting fact that when $X_1 = 1$ ($X_2 = 0$), $RT \ln \gamma_2 = W_{G_2}$. The same is true in the symmetrical case, equation (15.30). But

$$\begin{aligned} RT \ln(\gamma_2^{X_2=0}) &= RT \ln(\gamma_2^{X_2=0} \cdot (X_2 = 1)) \\ &= \mu_2 - \mu_2^* \end{aligned}$$

where μ_2^* refers to the standard state discussed in connection with equation (12.17), and represented by the point of intersection of the Henry's law slope at $X_2 = 0$ with the $X_2 = 1$ axis (point 1 in Figure 12.1, if we let component B be component 2). At this same composition ($X_2 = 1$), a standard state of pure component 2 at T, P, would result in $\mu_2 - \mu_2^\circ = RT \ln X_2 = 0$, where μ_2° refers to this new standard state (point 2 in Figure 12.1) Subtracting $\mu_2 - \mu_2^\circ = 0$ from $\mu_2 - \mu_2^* = W_{G_2}$ gives $\mu_2^\circ - \mu_2^* = W_{G_2}$, and similarly for W_{G_1}. Therefore the W_G coefficients are related to the difference in free energy between these two standard states. Similar relations hold for the other parameters W_H, W_V and so on.

15.4.6. *Margules Equations for Ternary and Higher Order Systems*

In the examples so far, we have used binary (two-component) solutions exclusively. Exactly the same reasoning may be used to derive equations for ternary, quaternary, and even more complex solutions. The equations become more unwieldy with each added component, and the properties they represent cannot be illustrated in two-dimensional graphs. However, the equations can be used in computer programs to generate properties of multicomponent systems for given conditions, and they should continue to fit real solutions adequately for the same reasons discussed above. The main problem with these higher order systems is not in using the Margules equations, but that a lot of experimental data are required if the fit is to be useful, and rather few solid solutions have been sufficiently investigated in this respect.

Rather than writing out all equations for all properties, we will save space by deriving only the free energy equations for ternary systems. The same approach can then be used to derive equations for other properties or for solutions containing more than three components. First, for a *ternary symmetric solution*, we can follow the analogous procedure used to derive (15.33) and imagine three regular binary solutions with a total of three components (1+2; 1+3; 2+3) being mixed to form a single ternary regular solution (1+2+3). The total excess free energy is the sum of the two-suffix equations (15.31) for each of these three solutions:

$$G^{EX}_{sol'n} = W_{G_{12}} X_1 X_2 + W_{G_{23}} X_2 X_3 + W_{G_{13}} X_1 X_3 \tag{15.43}$$

Similarly, we can derive the analogous equation for a *ternary, asymmetric solution* by summing expressions (15.33) for a mixture of three asymmetric binary solutions of the same three components:

$$G^{EX}_{sol'n} = \left(W_{G_{23}} X_2^2 X_3 + W_{G_{32}} X_3^2 X_2\right) + \left(W_{G_{13}} X_1^2 X_3 + W_{G_{31}} X_3^2 X_1\right) + \left(W_{G_{12}} X_1^2 X_2 + W_{G_{21}} X_2^2 X_1\right) \tag{15.44}$$

The brackets here show the original three asymmetric binary solutions we have "mixed." Equation (15.44) is also used with a seventh term, a constant, on the right-hand side.

Notice that (15.44) reduces to (15.43) by setting $W_{G_{ij}} = W_{G_{ji}}$ and that symmetric models therefore have half the number of W coefficients. A slightly more complex version of this equation, where the three terms are weighted in proportion to molar composition is proposed by Grover (1977, p. 81); this is similar to the weighting technique used to derive the binary, asymmetric equation (15.33) above. Notice that the number of Margules W parameters increases three times relative to a binary system for both the symmetric and asymmetric equations. For a quaternary system they would increase four times, and so on. This means that we are going to need a great many data points in multi-component systems so as not to over-fit the data. For example, you would not want to fit 7 data points in a quaternary system with an 8-parameter Margules equation analogous to (15.44). In practice, this imposes an upper limit on the number of components one might wish to consider if data for a system is limited.

Fortunately, it is not often necessary to consider solid solutions of more than three or four components. Many minerals have fewer than four major components. Minor components do not have a very significant effect on excess properties of the overall solution, as you can see from equations (15.43) and (15.44) for ternary systems above. For more detail, see Guggenheim (1952), Prigogine and Defay (1954), Saxena (1973), Thompson (1967), and Grover (1977). An algorithm for computer calculation of Margules equations based on higher degree virial equations and applicable to multicomponent systems is given by Berman and Brown (1984).

15.5. APPLYING THE MARGULES EQUATIONS

The Margules equations such as those in Table 15.1 can be fitted by standard least-squares regression analysis to data for real solutions. For example, if data for the total free energy of a binary asymmetric solution is available over a range of compositions at different T and P, you could fit equation (15.41) for G_{real} (or the equation for G^{EX} in Table 15.1) and obtain W_{G_1} and W_{G_2} as regression parameters. The same could be done with the equations for excess enthalpy, entropy, and so on. This permits construction of phase diagrams and determination of thermodynamic properties based on the best statistical analysis of experimental data. However, this kind of complete thermodynamic analysis is possible only with fairly extensive data sets.

15.5.1. *Estimating Margules Parameters: Symmetrical Solvi*

In Figure 15.3, two phases of composition b and c coexist within the solvus at 1000 K. Recall that the chemical potential of any one component must be the same in all coexisting phases at equilibrium. Therefore,

$$\begin{aligned} \mu'_{\mathrm{A}} &= \mu''_{\mathrm{A}} \\ \mu'_{\mathrm{B}} &= \mu''_{\mathrm{B}} \text{ at } 1000\ \mathrm{K} \end{aligned} \tag{15.45}$$

The superscripts $'$ and $''$ denote the two phases, and A and B are the two components. Combining (15.45) and (15.29) gives

$$RT \ln X'_{\mathrm{A}} + W_G (X'_{\mathrm{B}})^2 = RT \ln X''_{\mathrm{A}} + W_G (X''_{\mathrm{B}})^2$$

Rearranging and using $X_{\mathrm{B}} = 1 - X_{\mathrm{A}}$, we can solve for W_G:

$$W_G = \left(RT \ln(X'_{\mathrm{A}}/X''_{\mathrm{A}})\right) / \left((1 - X''_{\mathrm{A}})^2 - (1 - X'_{\mathrm{A}})^2\right) \tag{15.46}$$

A completely analogous equation applies to component B.

From (15.46) we are able to estimate W_G from the known compositions of two coexisting phases within a solvus at each temperature. Note that W_G will be different for every temperature. Plugging W_G into (15.28) then gives the $G_{real} - G_{ideal}$ of the solution at each temperature. The chemical potential of either component at T $(\mu - \mu^\circ)$ comes from (15.29). The excess free energies or activity coefficients of either component in each mineral is given by (15.30). If we solve for W_G at all temperatures along the solvus for which there are data, and write W_G as a function of T, then we could obtain W_S from

$$(\partial W_G/\partial T)_P = -W_S \tag{15.47}$$

and W_H from

$$W_G = W_H - TW_S \tag{15.48}$$

Recalling that $H = U + PV$ and that $W_H = W_U + PW_V$, we can also write the approximation $W_U \approx W_H$ because the PV term for solids is generally negligible at low pressure (1 bar) relative to U and H. Note that this requires that our solvus, hence W_G and W_H were all measured initially at a sufficiently low P, such as 1 bar. We now have estimates for W_G, W_S, W_H, and W_U, based on the variation of solvus composition with temperature. If we also need to know the properties of the system as a function of pressure, several options are available. With luck, there might be data for the solvus at different pressures. We could then find W_G as a function of P, and by analogy with (15.47), calculate W_V :

$$(\partial W_G/\partial P)_T = W_V \tag{15.49}$$

Alternatively, we could estimate $V^{EX} (= V_{real} - V_{ideal})$ from X-ray data on the single-mineral solid solution just above the solvus (more on this below). Knowing

V^{EX} as a function of P, we can then get W_V by fitting (15.23) to our data. Finally, if neither solvus nor volume versus P data are available, you could make the relatively safe assumption (used above to estimate W_U from W_H) that the PV (or PW_V) energy term for crystals is negligible at 1 bar or low pressures, and ignore it altogether. If so, you should test this assumption with a trial calculation, using data from a tabulation such as Robie et al. (1978).

One way or another, we now have estimates for W_G, W_S, W_H, W_U, and W_V. If our regression equations fit the data nicely, and if the data are good, then our initial assumption about the T and P dependence of these parameters should hold fairly well: W_G varies with T and P, W_H with P, and the others are independent of T and P. This means we should be able to write an equation of state for the behavior of this binary mineral system over a range of T and P using the general equation $G = U + PV - TS$ with coefficients on the right-hand side that do not change with T and P. Given

$$W_G = W_U - TW_S + PW_V$$

we can write

$$G^{EX} = U^{EX} - TS^{EX} + PV^{EX} \tag{15.50}$$

using the relations between the excess functions and the W coefficients in Table 15.1. This single equation can be used to generate phase diagrams and to predict the thermodynamic behavior of our mineral system. We should add the warning that if thermodynamic properties such as activities are derived from solvi in this manner, there is also a possibility of considerable error. This is because we have implicitly assumed in deriving W_S and W_H from W_G and the solvus, that the entropy and enthalpy functions fit the Margules equations without any real experimental verification. In the absence of real data on these other functions, the procedure outlined above is a best guess. However, if enthalpy, entropy, volume or other data are available, then they should certainly be used in preference to derive the W_H, W_S, W_V Margules parameters directly.

15.5.2. *Estimating Margules Parameters: Asymmetrical Solvi*

The above procedure required that we have a symmetrical solvus to work with. We have already observed that the feldspars, as most minerals, show asymmetrical behavior and we need to extend the same methods to this more general case. This follows exactly the same reasoning as the above section, but the equations turn out to be much less convenient to use.

Starting with an asymmetrical solvus, we wish to calculate the two Margules parameters W_{G_1} and W_{G_2} in (15.33). Once again, the chemical potentials of each component must be the same in all coexisting, equilibrated phases (equations 15.45). This time, the excess free energy is given by (15.41) rather than (15.28). Combining (15.45) and (15.41), and remembering that

$$\mu_i = \mu_i^\circ + RT \ln X_i + G^{EX} \tag{15.51}$$

gives the following two equations in the two unknowns W_{G_1} and W_{G_2} (changing from components A and B to components 1 and 2):

$$RT \ln X_1' + (2W_{G_2} - W_{G_1})(X_2')^2 + 2(W_{G_1} - W_{G_2})(X_2')^3$$
$$= RT \ln X_1'' + (2W_{G_2} - W_{G_1})(X_2'')^2 + 2(W_{G_1} - W_{G_2})(X_2'')^3 \qquad (15.52)$$

$$RT \ln X_2' + (2W_{G_1} - W_{G_2})(X_1')^2 + 2(W_{G_2} - W_{G_1})(X_1')^3$$
$$= RT \ln X_2'' + (2W_{G_1} - W_{G_2})(X_1'')^2 + 2(W_{G_2} - W_{G_1})(X_1'')^3 \qquad (15.53)$$

Given the compositions of the two coexisting phases, the above two equations can be solved for W_{G_1} and W_{G_2}. You could simply use one of many generalized linear/non-linear equation solvers available for computers, or use rearranged versions of (15.52) and (15.53), which give W_{G_1} and W_{G_2} directly in terms of composition (see Thompson, 1967, *p*. 355; and Eugster et al., 1972, p. 164, for example). From this point on, the procedure could be identical to that described for symmetrical solvi, except that the asymmetrical equations (15.41) (total free energy), (15.42) (activity coefficients), and (15.51) (chemical potentials) are required. Note also that we now have two W parameters for each thermodynamic variable and they should be treated separately. Thus (15.47) for an asymmetric solution becomes

$$(\partial W_{G_1}/\partial T)_P = -W_{S_1}$$

and

$$(\partial W_{G_2}/\partial T)_P = -W_{S_2}$$

This applies to all steps such as (15.48), (15.49) and so on, which are used to calculate one Margules parameter from another. Finally, a general equation of state such as (15.50) could be derived to describe the mineral system, using the same assumptions and with the same warnings about accuracy.

15.5.2. *Calculating Solvi, Spinodal Curves, and Consolute Points*

Once a general equation of state involving W_G versus P and T such as (15.50) is available, it can be used to calculate a statistically smoothed solvus. This is the solvus that provides the best least-squares fit to the experimental data. With caution, it can also be used to extrapolate beyond the T and P range of the experimental data (this is safer if the excess properties such as W_H and W_V are derived directly from real data and not calculated from W_G or otherwise estimated). First, an equation for the total free energy of the system as a function of T, P, and concentration is derived using (15.50) for G^{EX} and the relation

$$G_{real} = G_{ideal} + G^{EX} \qquad (15.15)$$

The solvus or binodal curve is obtained from the first derivative of the excess free energy,

$$(\partial G^{EX}/\partial X_1)_{P,T} = 0 \qquad (15.54)$$

Referring to Figure 15.3b, you can see this is because the two immiscible solutions (labeled b and c in the figure) are caused by two depressions in the free energy curve. If the solvus is only to be calculated at the same fixed P as the data points, then the derivative in Table 15.1 suffices. If you need the solvus over a range of T and P, then it will be necessary to use a full equation such as (15.50).

As already discussed above, the *spinodal points* at T and P are calculated from the second derivative to the total free energy curve (15.39):

$$(\partial^2 G/\partial X_1^2) = 0 \tag{15.4}$$

The derivative equals zero because the spinodes are flex points in the free energy curve (see points S_1 and S_2 on Figure 15.3a, for example). Again, if the spinodal curve is only required for the P at which the experimental data were collected, then the derivative in Table 15.1 is sufficient; otherwise an equation based on (15.50) and (15.15) is necessary for G. Finally, the *consolute point* may either be calculated simply by filling in the entire solvus or by setting the second and third derivatives of the total free energy equal to zero and solving simultaneously (see Thompson, 1967, p. 356; Grover, 1977, p. 78; and Thompson and Waldbaum, 1969, Appendix B). As an example, we have reproduced the solvus and spinodal curves for the two-mica system muscovite ($KAl_3Si_3O_{10}(OH)_2$) and paragonite ($NaAl_3Si_3O_{10}(OH)_2$) from a study by Eugster et al. (1972), which uses practically the same method outlined above—see Figure 15.4. These two micas are partially miscible at temperatures below about 800°C. In this study, the two coexisting micas were synthesized experimentally and then analyzed by electron microprobe. The volumes of both minerals were measured for a range of intermediate compositions. These two data sets were fitted to regression equations from which the Margules parameters for G, U, S, H, and V could be derived. The theoretical approach differed only slightly from that outlined above: the minerals were synthesized at sufficiently high pressures (2 kbar) that W_U could not be approximated by W_H (1 bar) but was calculated from $W_U = W_H - PW_V$. From this limited data set an equation similar to (15.50) was calculated for muscovite–paragonite solid solutions as a function of P, T and composition:

$$\begin{aligned} G^{EX} &= (3082.1 + 0.0822P + 0.1698T)X_{pg}X_{ms}^2 \\ &\quad +(4163.9 + 0.1259P + 0.3954T)X_{ms}X_{pg}^2 \end{aligned}$$

This equation permits cautious extrapolation to conditions beyond the T, P range of the data base. It also gives the best least-squares fit to the experimental data, as illustrated in Figure 15.4. Without this theoretical analysis it would be dangerous at best to "eyeball fit" the solvus, particularly near the consolute point where there are no data, and the spinodal could not be calculated at all. Extrapolating the solvus and spinodal to other P, T conditions would be practically unthinkable. This illustrates the great power of thermodynamics at extracting maximum information from physical observation. In this case, a complete equation of state for a two-component mineral solution was extracted from the compositions of six pairs of coexisting minerals and from limited information on their volumes.

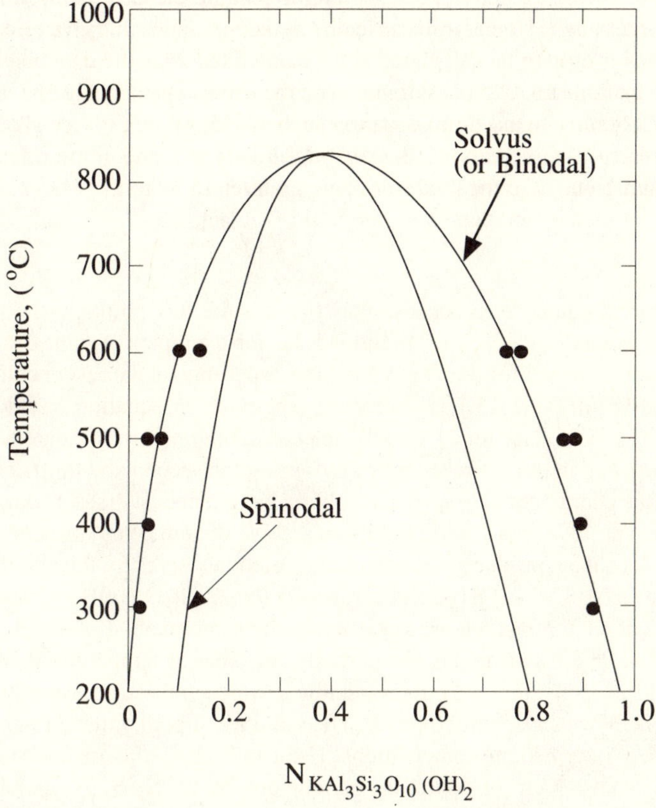

FIG. 15.4. Calculated solvus and spinodal curve for muscovite-paragonite at 2.07 kbar. Experimental data are the solid dots. After Eugster et al. (1972).

15.5.4. *Conclusion*

The use of Margules equations in dealing with solid solutions is largely a matter of empirical curve-fitting, and is best learned through practice. It is not of much help in actually understanding solid solutions, nor in dealing with the really complex solid solutions common in geology, such as the amphiboles. It is best at systematizing relatively simple binary solutions, fluid and solid, and when data are sufficient, ternary solutions. Beyond this, the notation gets very cumbersome, and the data required become very great. Nevertheless, within its range of usefulness, the Margules approach is very useful indeed, as witnessed by the large literature on the subject. Many different ways of expressing the relationships and different notations have been used; in fact the equations are chameleon-like in their variety of appearances, which can be confusing. There are also a number of similar approaches by other authors which we have not tried to cover. They all share one feature, however, and that is that in the absence of sufficient data the thermodynamic parameters they give can be quite erroneous, and there can be considerable differences in the results given by the various models.

Better than any textbook or summary such as we have presented is the series of four papers by Thompson and Waldbaum on the synthesis of data for the alkali feldspars (Thompson and Waldbaum, 1968, 1969; Waldbaum and Thompson, 1968, 1969). This elegant blend of critical thermodynamic thinking with the Margules approach should be studied by anyone interested in this subject.

Table 15.1 Margules Equations

Symmetric, slightly non-ideal binary solutions ("Regular")

$$G^{EX} = W_G X_1 X_2$$

$$V^{EX} = W_V X_1 X_2$$

$$H^{EX} = W_H X_1 X_2$$

$$S^{EX} = W_S X_1 X_2$$

$$U^{EX} = W_U X_1 X_2$$

$$\begin{aligned} G_{real} &= G_{ideal} + G^{EX} \\ &= X_1 G_1^\circ + X_2 G_2^\circ + RT(X_1 \ln X_1 + X_2 \ln X_2) + W_G X_1 X_2 \end{aligned}$$

$$\mu_1 = \mu_1^\circ + RT \ln X_1 + W_G X_2^2$$

$$RT \ln \gamma_1 = \bar{G}_1^{EX} = W_G X_2^2$$

Asymmetric binary solutions ("Sub-Regular")

$$G^{EX} = (W_{G_1} X_2 + W_{G_2} X_1) X_1 X_2$$

$$V^{EX} = (W_{V_1} X_2 + W_{V_2} X_1) X_1 X_2$$

$$H^{EX} = (W_{H_1} X_2 + W_{H_2} X_1) X_1 X_2$$

$$S^{EX} = (W_{S_1} X_2 + W_{S_2} X_1) X_1 X_2$$

$$U^{EX} = (W_{U_1} X_2 + W_{U_2} X_1) X_1 X_2$$

$$\begin{aligned} G_{real} &= G_{ideal} + G^{EX} \\ &= X_1 G_1^\circ + X_2 G_2^\circ + RT(X_1 \ln X_1 + X_2 \ln X_2) \\ &\quad + (W_{G_1} X_2 + W_{G_2} X_1) X_1 X_2 \end{aligned}$$

$$RT \ln \gamma_1 = \bar{G}_1^{EX} = (2W_{G_2} - W_{G_1}) X_2^2 + 2(W_{G_1} - W_{G_2}) X_2^3$$

$$RT \ln \gamma_2 = \bar{G}_2^{EX} = (2W_{G_1} - W_{G_2}) X_1^2 + 2(W_{G_2} - W_{G_1}) X_1^3$$

$$\mu_1 = \mu_1^\circ + RT \ln X_1 + RT \ln \gamma_1$$

$$\left(\frac{\partial G}{\partial X_2}\right)_{P,T} = (\mu_2^\circ - \mu_1^\circ) + RT \ln \frac{X_2}{X_1} + W_{G_1} X_2 (3X_1 - 1) - W_{G_2} X_1 (3X_2 - 1)$$

$$\left(\frac{\partial^2 G}{\partial X_2^2}\right)_{P,T} = \frac{RT}{X_1 X_2} + 2W_{G_1}(1 - 3X_2) + 2W_{G_2}(1 - 3X_1)$$

(= 0 for spinodes)

Ternary solutions

Symmetric

$$G^{EX} = W_{G_{12}} X_1 X_2 + W_{G_{23}} X_2 X_3 + W_{G_{13}} X_1 X_3$$

$$RT \ln \gamma_1 = X_2^2 W_{12} + X_3^2 W_{13} + X_2 X_3 (W_{12} + W_{13} - W_{23})$$

$$RT \ln \gamma_2 = X_1^2 W_{12} + X_3^2 W_{23} + X_1 X_3 (W_{12} + W_{23} - W_{13})$$

Table 15.1 *Continued*

Asymmetric

$$G^{EX}_{sol'n} = \left(W_{G_{23}}X_2^2X_3 + W_{G_{32}}X_3^2X_2\right) + \left(W_{G_{13}}X_1^2X_3 + W_{G_{31}}X_3^2X_1\right) + \left(W_{G_{12}}X_1^2X_2 + W_{G_{21}}X_2^2X_1\right)$$

Quaternary solutions

Symmetric

$$RT\ln\gamma_1 = X_2^2W_{G_{12}} + X_3^2W_{G_{13}} + X_4^2W_{G_{14}} + X_2X_3(W_{G_{12}} + W_{G_{13}} - W_{G_{23}}) + X_2X_4(W_{G_{12}} + W_{G_{14}} - W_{G_{24}}) + X_3X_4(W_{G_{13}} + W_{G_{14}} - W_{G_{34}})$$

$$G^{EX} = W_{G_{12}}X_1X_2 + W_{G_{13}}X_1X_3 + W_{G_{14}}X_1X_4 + W_{G_{23}}X_2X_3 + W_{G_{24}}X_2X_4 + W_{G_{34}}X_3X_4$$

Asymmetric

$$G^{EX} = \sum_{i=1}^{4}\sum_{j=1}^{4} W_{G_{ij}}(X_i^2X_j)\delta_{ij} \quad \begin{pmatrix} \delta_{ij} = 0 \;\; if \;\; i = j \\ \delta_{ij} = 1 \;\; if \;\; i \neq j \end{pmatrix}$$

... and so on for higher order systems.

PROBLEMS

1. Waldbaum and Thompson (1969) found the following equation for the Ab–Or solid solutions:

$$G_{mixing} = G_{ideal\ mixing} + (6327 + 0.093P - 4.632T)X_{\mathrm{Ab}}X_{\mathrm{Or}}^2 + (7672 + 0.112P - 3.857T)X_{\mathrm{Or}}X_{\mathrm{Ab}}^2$$

where P is in bars and T in Kelvins. Calculate and plot G_{mixing} as a function of X from pure Ab to pure Or at $T = 200, 300, 400, 500, 648, 800$°C, 1 bar. Do the same for the activities of the Ab and Or components. Get the binodal points by drawing tangents to the curve minima. Compare these with calculated $dG/dX_{\mathrm{Ab}} = 0$ points. Construct the $T-X$ solvus. Compare your curves with those in Waldbaum and Thompson (1969). Best done in a spreadsheet program using lots of compositional points, to get smooth curves.

2. Calculate $G^{EX}_{\mathrm{CO_2}}$ for the 11 data points in Table 11.1. This is best done with a computer spreadsheet, but can be done by hand.

3. Fit these G^{EX} values to equation (15.42). This is best done with a multiple linear regression program, using $X^2_{\mathrm{CO_2}}$ and $X^3_{\mathrm{CO_2}}$ as the independent variables. Comparing the fit coefficients with (15.42), calculate $W_{G_{\mathrm{CO_2}}}$ and $W_{G_{\mathrm{H_2O}}}$ for CO_2. Answers in Problem 3 at the end of Chapter 12.

4. Show that $W_{G_{\mathrm{CO_2}}}$ is the same as $\mu^\circ - \mu^*$ as discussed in the text.

16

GASEOUS SOLUTIONS

> Theories, as we all know are almost by definition mathematical, and surely come only after a body of experimental and observational knowledge has been accumulated and digested. There is no way to create a 'theory of nature' without first understanding nature in terms of observation and experiment. Nature is about magnets, currents, falling bodies, steam engines, spectral lines, and the like, not about groups, bifurcations, and invariances. Felix Bloch as a young man once told Heisenberg that space is a field of linear operators. 'Nonsense', replied Heisenberg, 'space is blue and birds fly through it.'
>
> Kac (1982)

16.1. METHODS OF DESCRIBING GASES

The procedures described in Chapter 15 are well suited to solid and liquid solutions and could also be applied to gases, but in fact, other approaches are generally used. The main reason for this is partly historical; much work was done early in the history of physical chemistry on the behavior of gases, and these methods have continued to evolve to the present day.

We have also just seen that the Margules equations become very unwieldy with multi-component systems. Because true gases are completely miscible, natural gases often contain many different components, so the Margules approach is not very suitable. Unfortunately, the most successful alternative methods described in this section are also quite unwieldy; however, they do not become much more complicated for multi-component gases than they are for the pure gases themselves, and this is a definite advantage. We have seen that with real, non-ideal gases, all the thermodynamic properties are described if we know the T, P, and the fugacity coefficient. For gaseous solutions, the fugacity coefficient for each component generally depends on the concentrations and types of other gaseous species in the same mixture. All gases, whether pure or multi-component, should approach ideality at higher T and lower P; conversely, non-ideality is most pronounced in dense, low-temperature gases where intermolecular forces are strongest. The challenge here is to find an equation of state that can adequately cover this range of conditions for gases of many different constituents.

In the following discussion we first briefly outline some of the equations of state used to describe pure gases. We will introduce these from the molecular point of view since this helps understand the physical basis (and limitations) of each model. Each of these equations of state can then be applied to mixtures of gases using a set of rules which we describe at the end of this section.

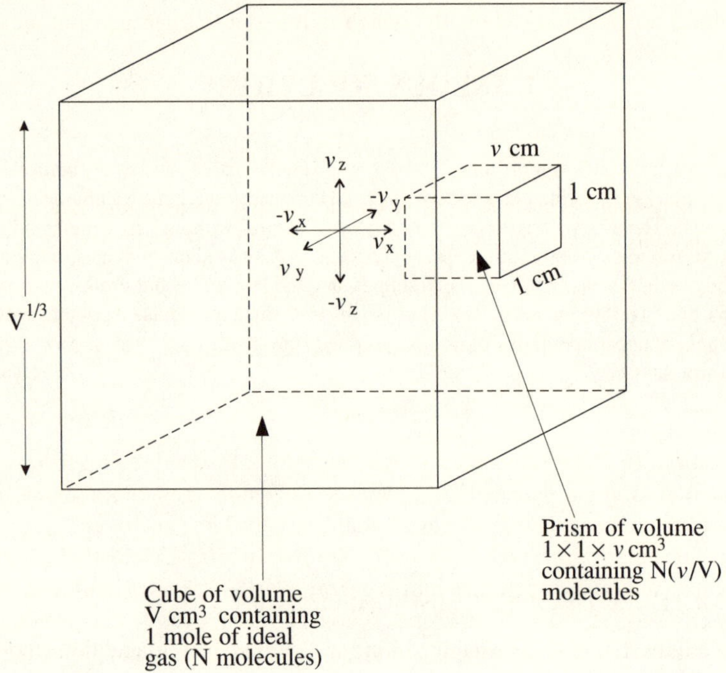

FIG. 16.1. Cubic box containing one mole of ideal gas and coordinate system showing movement v_x, $-v_x$, v_y, $-v_y$, v_z, $-v_z$ of molecules each second.

16.2. SOME EQUATIONS OF STATE FOR PURE GASES

16.2.1. *Ideal Gas*

The ideal gas law may be derived from a simple picture of molecular motion shown in Figure 16.1. Consider a cubic box of volume $V(\text{cm}^3)$ containing one mole, or Avogadro's number N, of ideal gas molecules. The only interactions allowed are collisions with the container walls. The molecules move in all directions, but let's simplify things by assuming that at any one time each face has 1/6 of the molecules heading towards it with an average velocity v (cm sec^{-1}). A 1 cm^2 face will therefore be struck by 1/6 of the molecules in the adjacent prism of volume $1 \times 1 \times v(\text{cm}^3)$ every second, each molecule having the momentum Mv/N (where M is molecular weight). There should be $N(v/V)$ molecules in this prism, of which $Nv/(6V)$ are heading towards this one face. Since the gas is ideal, we can presume that the molecules bounce perfectly off the wall, returning with exactly the same velocity in the opposite direction, for a total change in momentum of $2Mv/N$ per molecule. Now what we think of as pressure is the outward force per unit area on the walls of this box exerted by these impacting molecules, or the total change in momentum per second per cm^2 for the outer face of our prism. (The rate of change of momentum $d(Mv)/dt$ equals mass times acceleration ($M\,dv/dt$) which is force, and force per unit area equals pressure.) Hence

$$P = (Nv/6V)(2Mv/N)$$
$$= (2/3V)(Mv^2/2) \tag{16.1}$$

This is interesting because $Mv^2/2$ is kinetic energy. Temperature is a measure of molecular motion or kinetic energy, ($Mv^2/2 \propto T$) and if we write this proportionality as

$$Mv^2/2 = (3R/2)T \tag{16.2}$$

where R is a constant, then (16.1) becomes the ideal gas law,

$$PV = RT \tag{16.3}$$

Equation (16.2) has some interesting implications. First, it requires that the kinetic energy of an ideal gas be exactly the same at a given temperature, regardless of the nature of the gas. For example, all gases should have the same kinetic energy regardless of their molecular weights when P is sufficiently low or T sufficiently high that they approach ideality. The magnitude of this energy is actually quite significant: At 25°C it is (3/2)(8.3147)(298) = 3.720 kJ mol^{-1} and it is 15.88 kJ mol^{-1} at 1000°C. This is the translational (lateral) energy of a molecule, and for monatomic gases like Ar and He it is the total kinetic energy. Polyatomic gases also have additional rotational and vibrational energies. The velocities of the molecules themselves may be calculated from (16.1) and these too are considerable: The average speed of H_2 molecules at 25°C is 1770 m sec^{-1} which is about as fast as a rifle bullet.

If we differentiate (16.2) with respect to T, we obtain

$$C_v = (\partial U/\partial T)_V$$
$$= (3/2)R \tag{16.4}$$

This is at fixed volume because we are considering a specific container of gas.

The relationship between C_v and C_p can be seen by using Table 2.1 to derive an expression for $(\partial U/\partial T)_V$, which is C_v. This turns out to be

$$C_v = C_p + T\left(\frac{\partial V}{\partial T}\right)_P^2\left(\frac{\partial P}{\partial V}\right)_T$$

which for an ideal gas is simply

$$C_p - C_v = R \tag{16.5}$$

In other words, the heat capacities C_v and C_p are constants for ideal gases. This takes us quickly into the worlds of statistical mechanics and the kinetic theory of gases, but also raises several points of direct thermodynamic interest. First, (16.4) and (16.5) apply only to monatomic ideal gases; these can move independently (or have three degrees of translational freedom) along each of the three orthogonal axes, so each translation contributes roughly $R/2$ to C_v. Polyatomic molecules may also rotate and vibrate, and to an approximation each such motion should also add $R/2$ to the heat capacities. This is a possible way to estimate heat capacities of gases. In fact, the approximation works quite well for monatomic gases, all of which have C_v close to the predicted value of 12.5 J mol^{-1} deg^{-1}; however, it does not apply well to polyatomic gases except at very high temperatures, and is particularly bad with rotational and vibrational motions.

Equations (16.2) and (16.4) can also be used to help understand the physical significance of heat capacity. Temperature may be thought of as a measure of molecular motion or of the kinetic energy of molecules. Energy is stored by molecules as rotational, vibrational, and translational movements. For ideal gases, the increase in this motion (or temperature) with the addition of energy is always the same and given by (16.4). For real substances with specific intermolecular interactions, this will vary from one substance to another in general, and will often vary with T and P (which influence molecular interactions). Substances with higher heat capacities will not show as strong an increase in molecular motion (or T) as those with low heat capacities for the same heat input (U). Crystalline materials should have the highest heat capacities and monatomic inert gases the lowest, because with the former, more heat energy has to be added to cause the same degree of particle motion (due to the tightly bound nature of the lattice itself).

16.2.2. *The van der Waals Equation*

The ideal gas law does not work well for real gases at even moderate pressures. Two of its main problems were recognized by van der Waals in the last century, and appropriate corrections were incorporated into his famous equation of state of 1873:

$$P = \underbrace{\frac{RT}{V-b}}_{repulsion} - \underbrace{\frac{a}{V^2}}_{attraction} \tag{16.6}$$

The quantity $V - b$ corrects for the volume occupied by molecules of finite volume (molecules of an ideal gas are supposed to be vanishingly small). The parameter b is called the excluded volume per mole, since the volume available for movement of any one molecule is V minus the volume b of all other molecules in the same space. This crowding raises the pressure relative to an ideal gas according to $P = RT/(V - b)$; b may be thought of as a correction for crowding or repulsion of squeezed molecules. The second term in (16.6) corrects for intermolecular attractions. As illustrated in Figure 16.2, the forces between all gas molecules somewhere in the center of a container should be the same in all directions, on the average. However, the average forces on molecules at the outer wall of the container are directed inward; this must reduce the pressure exerted by these outermost molecules on the wall. The concentration of the outermost zone of molecules is proportional to the gas density, or to $1/V$, as is the concentration of molecules in the next inward zone (presuming concentrations in both zones are the same—a potential source of error). The inward attraction is proportional to the number of molecules

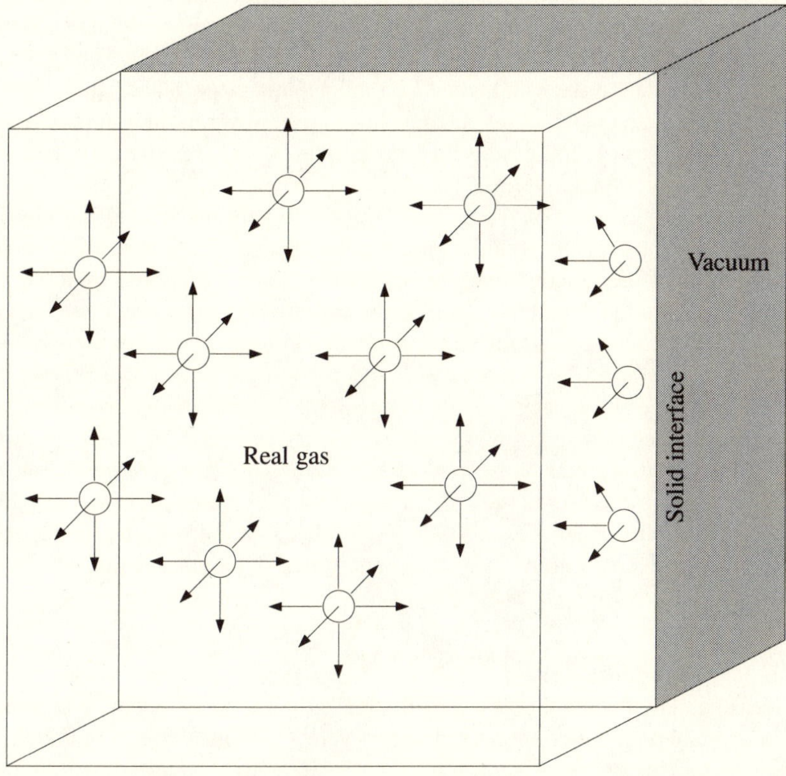

FIG. 16.2. Intermolecular forces within a real gas separated from a vacuum by a solid face. Within the center of the gas, average forces on each molecule are uniform. At the face, forces are all inward, reducing effective gas pressure outward.

in both zones or to $1/V^2$. Introducing the proportionality constant a, this inward pull should reduce the outward force or pressure by $-a/V^2$, and this is the second term in (16.6). Thus the van der Waals equation added a repulsive correction (first term and parameter b) and an attractive correction (second term and parameter a) to the equation of state for an ideal gas. The van der Waals equation fits the P–V–T behavior of real gases better than the ideal gas law, but still fails at moderate to high pressures where intermolecular forces are stronger. We would expect it to work a little better for two reasons: First, it is based on a simplistic, but not unreasonable physical picture of a real gas and should therefore have an appropriate mathematical shape. Second, it contains two adjustable parameters a and b which can be fitted to data by regression analysis or other procedures. This gives it flexibility and "personalizes" it for different gases. You can judge for yourself how well the van der Waals equation works for a relatively simple, non-polar gas such as CO_2 in Figure 16.3.

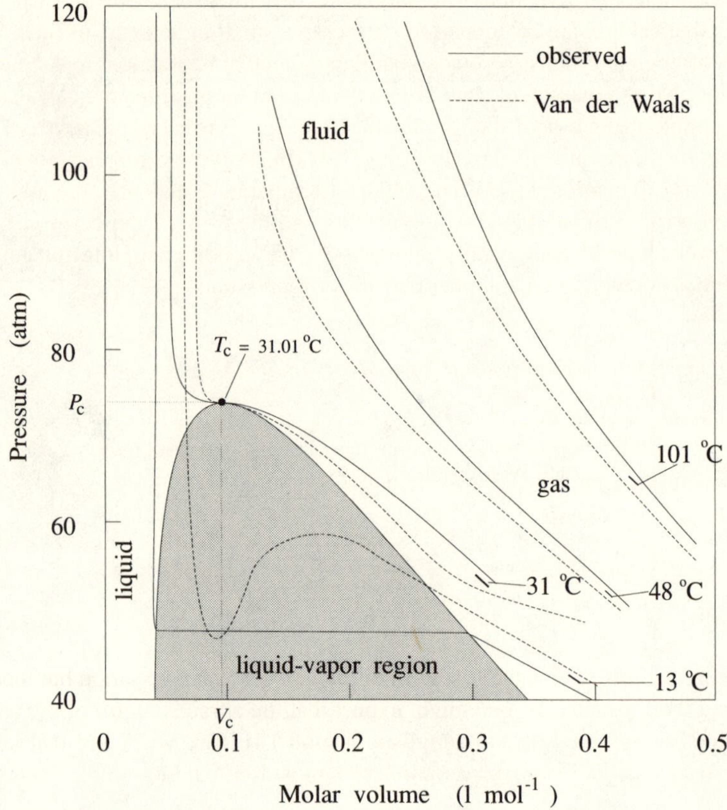

FIG. 16.3. Observed behavior of CO_2 near the critical point and behavior predicted by the van der Waals equation of state. After Barrow (1966).

There have been hundreds of modifications of the van der Waals equation over the intervening century. Most of these follow the same approach, correcting for repulsive and attractive forces. The Redlich-Kwong equation considered below is one of the more successful modifications, and now it too has been modified a great many times. Historically, this is one of the two main directions taken in the search for increasingly better gas equations of state.

The second direction has been statistical-mechanical. This usually follows the kind of reasoning described above for *virial equations of state* (equation 15.34). The interactions of all particles are calculated by summing interactions of pairs, triples, quadruples, and so on. In fact, what results is another regression equation with an appropriate shape and adjustable parameters with possible physical significance. Such equations are fit to real data; at the present time it is not feasible to calculate the parameters theoretically and then predict the behavior of a specific gas (except for highly idealized systems). With the advent of very fast computers it has been possible to develop molecular dynamics simulations in which the behavior of all particles in a hypothetical box are calculated for successive small increments of time. These methods offer great promise for understanding molecular interactions and structures, but do not give an equation of state of practical use in thermodynamic calculations.

The simplest equation of state applicable to real gases is based on the *principle of corresponding states* already described in §11.4. This uses the compressibility factor, $Z = PV/RT$ (equation 11.11), and reduced temperature and pressure. With this method you can estimate fugacity coefficients and the P–V–T properties of gases in the lower range of geologically interesting P–T conditions. Unfortunately, the method does not work well at higher densities (or pressures).

16.2.3. *The Redlich-Kwong Equation and its Modifications*

The equations of state most commonly used at the present time to represent gas behavior at moderate to high pressures are based on a simple modification of the van der Waals equation published by Redlich and Kwong in 1949:

$$P = \underbrace{\frac{RT}{(V-b)}}_{repulsion} - \underbrace{\frac{a}{T^{0.5}V(V+b)}}_{attraction} \tag{16.7}$$

As you can see, this retains the first term of the van der Waals equation but modifies the second term. Intuitively, one might expect that the attractive term of the van der Waals equation is based on too simplistic a model (Figure 16.2) and that a more complicated function is probably required. To quote Redlich himself (1976, p. 48),

> There is no really good theoretical justification for the two changes in the denominator of the a-term but the agreement with observed data is considerably improved . . .

In other words, the form of equation (16.7) is more like that of real gases than the van der Waals equation. By adjusting the van der Waals equation in this way, any physical significance the a and b parameters might originally have had becomes even more remote (but does not completely disappear—see Vera and Prausnitz, 1972). We will refer to the attractive and repulsive terms of the Redlich-Kwong (RK) and modified Redlich-Kwong (MRK) equations for convenience and to be consistent with the literature. However, it is really more appropriate to think of (16.7) as an empirical regression equation with a good shape for fitting the P–V–T data of gases.

In the original RK equation, the parameters a and b were constants, independent of P and T. Since the RK equation first appeared in print, it too has undergone many modifications in an attempt to further improve its fit. Many of these modified Redlich-Kwong equations are summarized by Holloway (1977) and Kerrick and Jacobs (1981), and involve writing the a and b parameters as a function of T, P, or both T and P. The MRK equation of Kerrick and Jacobs fits most of the data for H_2O and CO_2 at elevated P and T to better than $\pm 1\%$ and is chosen for discussion here.

The MRK equation of Kerrick and Jacobs keeps b constant but writes a as a function of both P and T :

$$P = \underbrace{\frac{RT(1+y+y^2-y^3)}{V(1-y)^3}}_{repulsion} - \underbrace{\frac{a(P,T)}{T^{0.5}V(V+b)}}_{attraction} \tag{16.8}$$

where $y = b/4V$ (b is constant) and $a(P,T) = c(T) + d(T)/V + e(T)/V^2$ and $z(T) = z_1 + z_2T + z_3T^3$ (where z is c, d, or e).

Here you can see that the repulsion term holds b constant but is more complicated (which gives it better flexibility). The a parameter in the attraction term is written as a simple polynomial function of $1/V$ (hence of P) and of T. This is a MRK equation with 10 adjustable parameters, so naturally it fits real data considerably better than the original RK equation, which has only two parameters.

As a final point, notice that the van der Waals equation (16.6) and its modifications such as the RK and MRK equations are written with T and V as independent variables. This is inconvenient for Earth scientists since we usually have some idea of P and T, but rarely V. Calculating P given V and T is trivial with all these equations; however, calculating V given P and T becomes a (small) exercise in numerical analysis and requires a computer. For equations such as (16.7) or (16.8) you might do this by successive iteration. As an example, rewriting (16.7) with V on the left-hand side gives

$$V = (RT/P) + b - \frac{a(V-b)}{PT^{0.5}V(V+b)}$$

This can be solved for V at given P and T by plugging T, P, and an initial guess at V in the right-hand side. Using the new, improved V for the second guess, repeat until successive calculated V's stop changing significantly.

16.2.4. *Estimating the van der Waals and Redlich-Kwong Parameters from Critical Conditions*

The a and b parameters in the RK and van der Waals class of equations are best estimated from a regression fit to an overall P–T–V data set for a gas, as described above. However, there is a much quicker way of estimating the same parameters based on the critical behavior of the gas. If these critical data are known with sufficient accuracy and precision, the estimates should be very good, and this approach is used quite frequently when the utmost accuracy is not required. Recall that the critical properties of a gas can also be used to estimate fugacities using the principle of corresponding states discussed in Chapter 11.

You can see from Figure 16.3 that the critical point at the top of the two-phase region occurs at a flex point in the critical isotherm of 31°C. In addition, the tangent to the top of the two-phase region should be horizontal. Thus the first and second derivatives of the critical isotherm must both be zero at the critical point. The van der

Waals equation itself should also work at the critical point, so we can write the three following equations:

$$P_c = \frac{RT_c}{V_c - b} - \frac{a}{V_c^2}$$

$$(\partial P/\partial V)_{T_c} = 0 = \frac{-RT_c}{(V_c - b)^2} + \frac{2a}{V_c^3}$$

$$(\partial^2 P/\partial V^2)_{T_c} = 0 = \frac{2RT_c}{(V_c - b)^3} - \frac{6a}{V_c^4}$$

Solving these three equations for the critical constants gives

$$T_c = 8a/27bR$$

$$V_c = 3b$$

$$P_c = a/27b^2$$

and solving for the van der Waals parameters, we find

$$a = 3P_cV_c^2$$

and

$$b = V_c/3$$

where

$$V_c = 3RT_c/8P_c$$

We can apply exactly the same reasoning to the RK equation (16.7) and obtain

$$a_{RK} = 0.4275R^2T_c^{2.5}/P_c$$

and

$$b_{RK} = 0.0866RT_c/P_c$$

Thus the a and b constants in both the van der Waals and the RK equations can be calculated quite easily given sufficiently good data on the critical temperature and pressure of a fluid. Alternatively, you can calculate the critical conditions for a gas directly from the a and b parameters for either equation of state.

16.3. APPLYING EQUATIONS OF STATE TO GASEOUS MIXTURES

With ideal gases Dalton's and Amagat's laws apply, as described in Chapter 11, and the fugacity of each component exactly equals its partial pressure. Thus we can write

$$V_i = X_iV = \bar{V}_i \text{ (Amagat's Rule)}$$

and

$$f_i = P_i = X_iP_{total} \text{ (Dalton's Law)} \tag{11.22}$$

These approximations can only be used for real gases at very low pressures, as we have already observed.

Moving one step closer to reality, the Lewis Fugacity Rule (described in Chapter 11) is frequently used to approximate the behavior of real gas mixtures

$$f_i = X_i \cdot f_i^{pure} \text{ (Lewis Fugacity Rule)} \tag{11.26}$$

This means that if you know the fugacity of one of the pure components of a gas solution at the same total P and T as the solution itself, then you can calculate its fugacity in the solution. This is analogous to Dalton's Law, but uses fugacities rather than pressures. It is a fair approximation for components at low concentrations and pressures, but presumes that fugacities are independent of the types of gases in the mixture. This is a real problem—for example, we cannot expect the fugacity of H_2O to be the same in H_2O-CH_4 and H_2O–HCl gaseous solutions (the molecular interactions should differ quite significantly). For more accurate estimates something more rigorous is going to be necessary.

This more detailed approach typically uses one of the equations of state for pure gases mentioned above. Using the van der Waals equation (16.6) as an example, the a and b parameters must be known for each of the pure components. If no experimental data are available for the mixture of these components, then we might be able to devise some means of estimating new values of a and b that could apply to the mixture. Many so-called mixing rules have been devised to do this (see, for example, Prausnitz et al. 1986), but the most popular remain those originally suggested by van der Waals. Since the repulsive b parameter is supposed to be a measure of the volume of the molecules, a simple averaging over all m different gas species might work here:

$$b_{mix}(X_1, X_2, \ldots) = \sum_{j=1}^{m} X_j b_j \tag{16.9}$$

The a (molecular attraction) parameter for the mixture is obtained by averaging over all possible kinds of molecular pairs:

$$a_{mix}(X_1, X_2, \ldots) = \sum_{j=1}^{m} \sum_{k=1}^{m} X_j X_k a_{jk} \tag{16.10}$$

where

$$a_{jk} = (a_j a_k)^{\frac{1}{2}} \tag{16.11}$$

In the last two equations, a_{jk} is supposed to describe the interaction between one molecule of type j and another of type k. For $j = k$, a_{jj} is simply the van der Waals a parameter for the pure gas j. The problem here is how to find a term $a_{jk}(j \neq k)$ for the interaction between unlike molecules when there are no data on the mixture itself. Berthelot suggested the geometric mean assumption of equation (16.11) in the last century on strictly empirical grounds, and it was used by van der Waals in his own work on gas mixtures. For some cases this assumption can be theoretically justified (Prausnitz, 1969, p. 45), but why it works is still a bit of a mystery. As can be imagined, there has been much discussion and controversy over this point (see Prausnitz, 1969, p. 156, for details). We recommend that you use it as a last resort when you only have data on the pure gas components for a mixture. If there are some data for the gas mixture, then it is safer to use (16.9) and (16.10) to define a_{mix} and b_{mix} and to derive the a_{jk} parameters by regression, ignoring (16.11) entirely.

The same mixing rules can be applied to the other equations of state mentioned above. For the RK and MRK equations there is no change whatsoever since the a and b parameters can be handled exactly as with the van der Waals equation. Thus equations (16.9) to (16.11) can be used to calculate the a and b parameters for a gas mixture described by a RK or MRK equation.

The mixing rules for a virial equation of state are similar but slightly more complicated, because the equation itself has a different basis than the various outgrowths of the van der Waals equation. Writing a virial equation in terms of the compressibility factor Z, we have

$$\begin{aligned} Z &= PV/RT \\ &= 1 + B/V + C/V^2 + D/V^3 + \cdots \end{aligned} \tag{16.12}$$

Given a mixture of m different gaseous components $i, j, k, \ldots$ one equation (16.12) can be written for each of the pure gases. We can write an equation for the mixture by combining the $B, C, D, \ldots$ parameters for each gas according to the following rules:

$$B_{mix} = \sum_{i=1}^{m} \sum_{j=1}^{m} X_i X_j B_{ij} \tag{16.13}$$

$$C_{mix} = \sum_{i=1}^{m} \sum_{j=1}^{m} \sum_{k=1}^{m} X_i X_j X_k C_{ijk} \tag{16.14}$$

The sums here are made over all the m different species in the gas mixture. The fourth and higher virial coefficients continue in an analogous fashion.

Notice that these parameters for mixtures work the same way as with pure gases: the B_{mix} coefficient takes into account interactions of all pairs of molecules, the C_{mix} coefficient all triples, and so on. This is absolutely rigorous, and no assumptions are introduced in (16.13) and (16.14) that degrade the statistical-mechanical basis of the virial equation (see Hill, 1960; Prausnitz, 1969, p. 99). As with the RK and MRK equations, the same difficulty arises in evaluating the parameters representing interaction between different gas species. For example, expanding (16.14) for a mixture of two gases i and j gives

$$C_{mix} = X_i^3 C_{iii} + 3X_i^2 X_j C_{iij} + 3X_i X_j^2 C_{ijj} + X_j^3 C_{jjj} \tag{16.15}$$

Here the coefficients C_{iii} and C_{jjj} are those for the pure gases, but the coefficient C_{iij} refers to interactions between two molecules of i and one of j, and C_{ijj} refers to a similar i–j–j interaction. You might attempt to calculate these coefficients from data for pure gases by an assumption such as (16.11) but this is questionable in the first place and would also upset the theoretical basis of your virial equation. Again, it is far preferable to derive these interaction parameters from regression analysis of real data for two- and three-gas systems if the data are available.

A potential major advantage of the virial equation over RK and related equations is that it might be possible to describe complex gas mixtures of many components without having data for the interactions of all species in the mixture. Virial equations such as (16.12) are always ascending polynomials that are truncated at some point. We will see in Chapter 17 that the Pitzer virial equations used to describe multicomponent aqueous salt solutions can be cut off at the third (C) term, yet still work remarkably well with solutions containing many more than three components. In other words, it is usually not necessary to consider anything more complicated than three-particle interactions with complex salt solutions. Since the non-ideal interactions of gases should be simpler, we might expect the same truncation to work with gases. For example, suppose data are available for binary and possibly ternary mixtures of 10 different gases. If a virial equation to the fourth coefficient C adequately describes the mixture of all 10 gases, then the B_{mix} and C_{mix} parameters could be derived from data on binary and ternary combinations of these gases. Data would not be required on the P–V–T–X properties of the 10 component system itself (which is good because that would be a formidable undertaking). This should be tested for geologically interesting gases. Unfortunately, obtaining P–V–T data simply for pure gases at elevated T and P of geological interest has proven formidable enough, and there are almost no data on binary gas mixtures for similar conditions at the present time. However, virial equations could be tested on lower-P systems such as volcanic and geothermal gases and the atmosphere itself, where they should work quite well.

16.3.1. *Calculating Fugacities from Gas Equations of State*

A complete P–V–T–X equation of state contains sufficient information to describe many of the thermodynamic properties of a system. For a gas, the thermodynamic property of greatest interest is usually its activity or fugacity; this is required for all calculations in which the gas is involved in chemical reactions. We have already mentioned some of the simpler methods of estimating gas fugacities for ideal or nearly ideal gases (Dalton's law, the Lewis fugacity rule, and the principle of corresponding states), and these are summarized in Table 16.1. For gases at higher densities it is necessary to use a more rigorous equation of state such as a virial, van der Waals, RK, or MRK equation. As we have pointed out above, these are written with T and V as independent variables (rather than T and P as usually preferred by Earth scientists). We have chosen to write the thermodynamic functions used in this book in terms of T and P wherever possible. However, most of these functions can be rederived using T and V as independent variables; methods for interchanging variables in this manner are outlined in Chapter 2, and more information can be obtained from Beattie and Stockmayer (1942), Beattie (1955), Prausnitz (1969), and Prausnitz et al. (1986). As derived by Prausnitz (1969, p. 41), the fugacity coefficient for a pure species is given by

$$RT \ln \gamma_i^{pure} = \int_V^\infty \left(\frac{P}{n_i} - \frac{RT}{V} \right) dV - RT \ln Z + RT(Z-1) \tag{16.16}$$

For a species of a gas mixture, the fugacity coefficient is

$$RT \ln \gamma_i^{mix} = \int_V^\infty \left[\left(\frac{\partial P}{\partial n_i} \right)_{T,V,n_j} - \frac{RT}{V} \right] dV - RT \ln Z \tag{16.17}$$

As usual, Z is the compressibility factor $Z = PV/RT$. To solve for the fugacity, substitute P into (16.17) or (16.16) using the above equations of state for gases with T and V as independent variables. The integration gives an equation for the fugacity coefficient for a species of a gas mixture. It is essential that the mixing rules (such as 16.9 to 16.11) used in evaluating this integral be exactly the same as those used in defining the original equation of state for the mixture (this might seem obvious, but has caused some problems—see Flowers (1979) for a discussion). The results are given in Table 16.1, which includes expressions for fugacity based on the van der Waals (16.6), virial (16.12), RK (16.7), and MRK (16.8) equations of state.

The fugacity equations in Table 16.1 are arranged in order of increasing complexity. The RK and MRK expressions for fugacity are particularly difficult to work with. This is unfortunate, because it means that estimating the fugacity of a gas component in a mixture is a small research project in itself at the present time. The procedure here requires that you first fit your choice of MRK equation to data for pure gases, deriving expressions for the a and b parameters. With the Kerrick and Jacobs MRK (16.8) as an example, you then calculate the parameters a, b, c, d, and e for the mixture using the mixing rules (16.9) to (16.11). If sufficient data exist for the gas mixture you should derive these parameters directly from a regression fit to the data since the mixing rules themselves are questionable. Whichever route you choose, the MRK equation for the mixture would be (16.8) with a_m, b_m, c_m, d_m, and e_m for the mixture substituted for a, b, c, d, and e for the pure gases. You must then solve (16.8) for V_m (molar volume of the mixture) at the T–P–X conditions of interest; this has to be done by a numerical method such as successive iteration on V_m as described above. All parameters are then plugged into the last equation in Table 16.1 to calculate the fugacity of one gas species in this mixture at this particular T, P, and concentration.

The MRK equations for fugacity are unwieldy because of the required integration (16.17) of the original equation. There is clearly room for improvement here. Hopefully future research will provide more tractable equations of state for gas mixtures. For example, it would be helpful to start with an equation which gives a simpler integral. One promising approach might be to use a virial equation, which gives the relatively simple expression for fugacity in Table 16.1. This has several further advantages as already mentioned above—it is based on a reasonable statistical-mechanical model, and it should be applicable to complex mixtures using mixing parameters derived from simpler systems. At the present time, alternatives to MRK equations (such as virial equations) have not received sufficient attention in geological applications.

Aside from their inconvenience, overly complex equations have a potential built-in source of error. To paraphrase Prausnitz (1969, p. 46), more complicated equations with more adjustable parameters may work better for pure gases, but simpler equations frequently work better for gas mixtures. This is because mixing rules (such as 16.9 to 16.11) are required to predict the properties of mixtures from those of the pure gases, and the rules themselves are uncertain. The more parameters you have, the more mixing rules you will require. Also, in extrapolating multi-parameter equations beyond the region of experimental data you must be sure that the equation continues to behave smoothly (or has the proper shape). There is always a possible problem in regression analysis that equations with many parameters might behave erratically outside the field of real data.

With these warnings in mind, let's now look at several examples using different equations of state. The first example, from Prausnitz (1969, p. 45) considers the fugacity of hydrogen at 50°C and 300 bars total pressure in a mixture of 20 mole % H_2, 50% CH_4, and 30% ethane. Referring to Table 16.1, the ideal gas law and Dalton's law of partial pressures give a hydrogen fugacity of 60 bars, the Lewis fugacity rule gives 71 bars, and the van der Waals equation 113 bars. The van der Waals estimate is preferred, of course, and the error of 60% to 90% involved in the other two estimates shows how careful you should be with these approximations. A second example from Prausnitz (1969, p. 172) compares the measured and estimated solubilities of liquid decane in nitrogen gas at 50°C as calculated from decane fugacity and the Henry's Law coefficient (Figure 16.4). The virial equation exactly fits experimental data while the ideal gas, and Lewis fugacity rule approximations fail badly.

Calculations for CO_2–H_2O mixtures using the MRK equation of Kerrick and Jacobs are illustrated in Figure 16.5a. The effect of adding methane to this system is illustrated in Figure 16.5b. These diagrams were calculated as described above from the fit of the MRK equation (16.8) to data for the pure gas systems, and checked against existing data for the binary systems H_2O–CO_2 and H_2O–CH_4. Unfortunately, there are no data for the ternary system at high P and T, and data for the binaries is limited, so that all three diagrams contain unverified extrapolations.

We should now have sufficient insight to be able to explain the predicted behavior of these gas mixtures qualitatively. Looking at the two CO_2–H_2O diagrams, for example, you can see that both gases display positive deviations from ideality. This is expected because the two molecules are quite dissimilar, and in fact the calculation at 400°C shows a miscibility gap at 20 and 30 kbar (see following discussion). Notice also that for each diagram non-ideality increases at higher pressures and lower temperatures. This is also expected since intermolecular interactions become more intense at higher P (and density) and lower T. The CO_2 molecule is linear and non-polar while H_2O is highly polar and interacts through hydrogen bonding; this explains in part why H_2O displays a greater departure from ideality than CO_2 on all diagrams. Finally, you can see that both H_2O and CO_2 approach ideal Raoult's Law behavior as they become the predominant component in the solution.

Figure 16.5c shows the calculated effect of CH_4 on the predicted activities at 400°C and 25 kbar. Notice that the effect on water at these conditions is essentially zero (at higher T's water activity increases slightly at higher CH_4 concentrations).

Table 16.1 Gas Fugacities for Various Equations of State

Ideal Gas

$$f_i = P_i$$

where

$$P_i = X_i P_{total} \qquad \text{Dalton's Law}$$

Ideal mixture of real gases

$$f_i = X_i f_i^{pure} \qquad \text{Lewis Fugacity Rule}$$

Principle of Corresponding States, pure gas

$$\ln f/P = \int_{P \to 0}^{P} \frac{(Z-1)}{P} dP$$

where $Z = PV/RT$ (compressibility factor). Z is similar for most gases at low to moderate P–T conditions when P and T are expressed as the reduced variables P_r, T_r (see Chapter 11). This can be extended to gas mixtures using the pseudocritical approximation described by Pitzer and Brewer (1961, Appendix 2).

Virial Equation

$$\begin{aligned}\ln \gamma_i &= \ln f_i/X_i P \\ &= \frac{2}{V}\sum_{j=1}^{m} X_j B_{ij} + \frac{3}{2}\cdot\frac{1}{V^2}\sum_{j=1}^{m}\sum_{k=1}^{m} X_j X_k C_{ijk} - \ln Z_{mix}\end{aligned}$$

where the B_{ij} and C_{ijk} parameters are given by the mixing rules (16.13) and (16.14). V is the molar volume and Z_{mix} the compressibility factor of the mixture. The number of different species is m, and γ_i is the fugacity coefficient.

van der Waals Equation

$$\begin{aligned}\ln \gamma i &= \ln f/X_i P \\ &= \ln \frac{V}{V - b_{mix}} + \frac{b_i}{V - b_{mix}} - \frac{2\sqrt{a_i}\sum_{i=1}^{m} X_j\sqrt{a_j}}{VRT} - \ln Z_{mix}\end{aligned}$$

where the a and b parameters for the mixture and for each component are defined by the mixing rules (16.9), (16.10), and (16.11). Z_{mix} is the compressibility factor and V the molar volume of the mixture. The number of different species is m, and γ_i is the fugacity coefficient.

Redlich-Kwong Equation (original, unmodified)

$$\begin{aligned}\ln \gamma_k &= \ln f_k/X_k P \\ &= \ln \frac{V}{V-b} + \frac{b_k}{V-b} - \frac{2\sum_{i=1}^{m} X_i a_{ik}}{RT^{3/2}b}\ln\frac{V+b}{V} \\ &\quad + \frac{ab_k}{RT^{3/2}b^2}\left(\ln\frac{V+b}{V} - \frac{b}{V+b}\right) - \ln\frac{PV}{RT}\end{aligned}$$

where a and b parameters are defined by the mixing rules (16.9) and (16.10) but not by (16.11). Other parameters as above.

Table 16.1 (*continued*)

Modified Redlich-Kwong Equation of Kerrick and Jacobs (a) Pure Gas

$$\begin{aligned}\ln\gamma &= \ln f/P \\ &= \frac{8y-9y^2+3y^3}{(1-y)^3} - \ln Z - \frac{c}{RT^{3/2}(V+b)} \\ &\quad - \frac{d}{RT^{3/2}V(V+b)} - \frac{e}{RT^{3/2}V^2(V+b)} \\ &\quad + \left(\frac{c}{RT^{3/2}b}\cdot\ln\frac{V}{V+b}\right) - \frac{d}{RT^{3/2}bV} \\ &\quad + \left(\frac{d}{RT^{3/2}b^2}\cdot\ln\frac{V+b}{V}\right) - \frac{e}{RT^{3/2}2b^2} \\ &\quad + \frac{e}{RT^{3/2}b^2V} - \left(\frac{e}{RT^{3/2}b^3}\cdot\ln\frac{V+b}{V}\right)\end{aligned}$$

Parameters y, b, c, d, and e are defined in (16.8). Other variables as above.

(b) Gas Mixture

$$\begin{aligned}\ln\gamma_i &= \ln f_i/X_iP \\ &= \frac{4y_m-3y_m^2}{(1-y_m)^2} + \left(\frac{b_i}{b_m}\cdot\frac{4y_m-2y_m^2}{(1-y_m)^3}\right) \\ &\quad - \frac{b_ic_m}{RT^{3/2}b_m(V_m+b_m)} + \left[\frac{b_ic_m}{RT^{3/2}b_m^2}\cdot\ln\left(\frac{V_m+b_m}{V_m}\right)\right] \\ &\quad - \left[\frac{P_c}{RT^{3/2}b_m}\cdot\ln\left(\frac{V_m+b_m}{V_m}\right)\right] - \frac{P_d}{RT^{3/2}b_mV_m} \\ &\quad + \left[\frac{P_d}{RT^{3/2}b_m^2}\cdot\ln\left(\frac{V_m+b_m}{V_m}\right)\right] \\ &\quad + \frac{b_id_m}{RT^{3/2}b_m(V_m+b_m)} + \frac{2b_id_m}{RT^{3/2}b_m^2(V_m+b_m)} \\ &\quad - \left[\frac{2b_id_m}{RT^{3/2}b_m^3}\cdot\ln\left(\frac{V_m+b_m}{V_m}\right)\right] - \frac{P_e}{RT^{3/2}2b_mV_m^2} \\ &\quad + \frac{P_e}{RT^{3/2}b_m^2V_m} - \left[\frac{P_e}{RT^{3/2}b_m^3}\cdot\ln\left(\frac{V_m+b_m}{V_m}\right)\right] \\ &\quad + \frac{b_ie_m}{RT^{3/2}2b_mV_m^2(V_m+b_m)} - \frac{3b_ie_m}{RT^{3/2}2b_m^2V_m(V_m+b_m)} \\ &\quad + \left[\frac{3b_ie_m}{RT^{3/2}b_m^4}\cdot\ln\left(\frac{V_m+b_m}{V_m}\right)\right] \\ &\quad - \frac{3b_ie_m}{RT^{3/2}b_m^3(V_m+b_m)} - \ln Z\end{aligned}$$

y_m and b_m for the mixing are derived from y_i and b_i for each pure gas species from (16.9). The remaining parameters c_m, d_m, and e_m are evaluated from c_i, d_i, and e_i for each pure gas species and the mixing rules (16.10) and (16.11). Other variables as above.

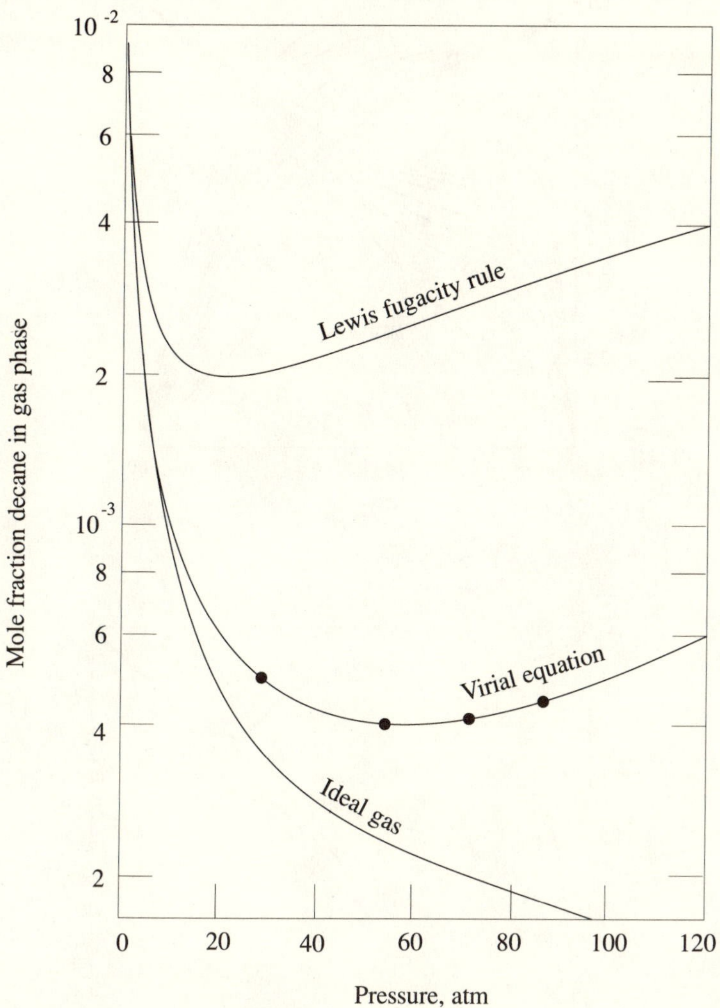

FIG. 16.4. Solubility of liquid decane in nitrogen gas at 50°C, calculated from Henry's Law coefficient and decane fugacities based on ideal gas and Lewis Fugacity Rule approximations and the virial equation. Experimental data are solid dots. After Prausnitz (1969).

However, CH_4 significantly reduces the activity of CO_2 in these mixtures. The CO_2 and CH_4 molecules are similar and non-polar, and both are quite different from H_2O. Jacobs and Kerrick (1981) attribute the predicted activities to simple dilution of CO_2 by CH_4 (lowering its effective activity); the H_2O molecule presumably interacts in somewhat similar fashion with both CO_2 and CH_4 so that a change in the CO_2/CH_4 ratio should have little effect on water activity. The addition of CH_4 to CO_2–H_2O fluids has geological implications evident in this figure. First, the predicted CO_2–H_2O

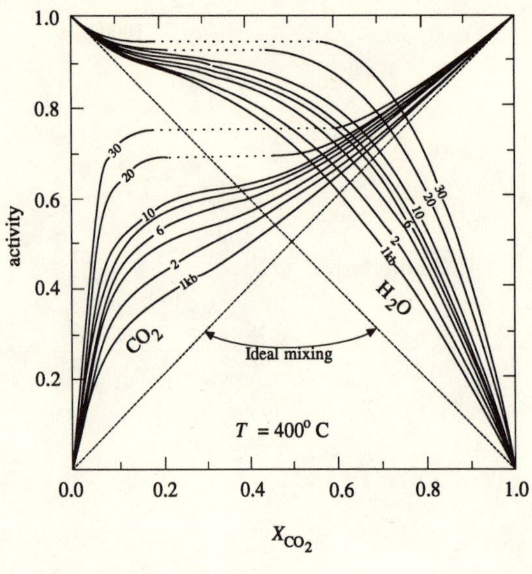

(b)

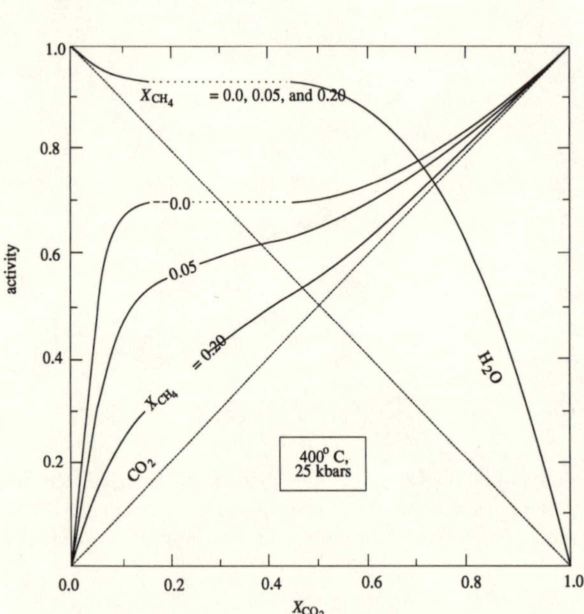

FIG. 16.5. Activities in CO_2–H_2O gas mixtures predicted by a MRK equation, after Kerrick and Jacobs (1981). (a) 400°C and 1–30 kbar. Dotted lines indicate predicted miscibility gap (coexistence of H_2O-rich liquid and CO_2-rich vapor). (b) Predicted H_2O and CO_2 activities in H_2O–CO_2–CH_4 mixtures at 400°C, 25 kbar. Calculated for CH_4 mole fractions of 0.0, 0.05, and 0.20. Dotted curves imply a miscibility gap of H_2O-rich liquid and CO_2-rich vapor. After Jacobs and Kerrick (1981).

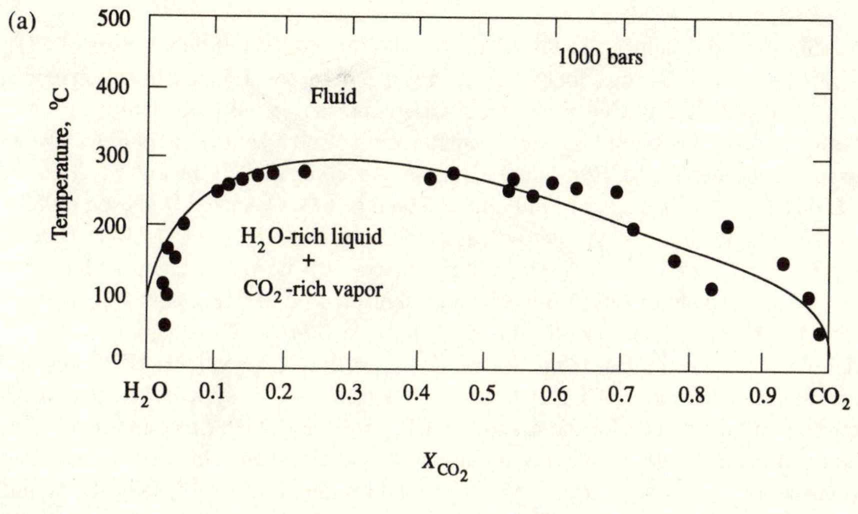

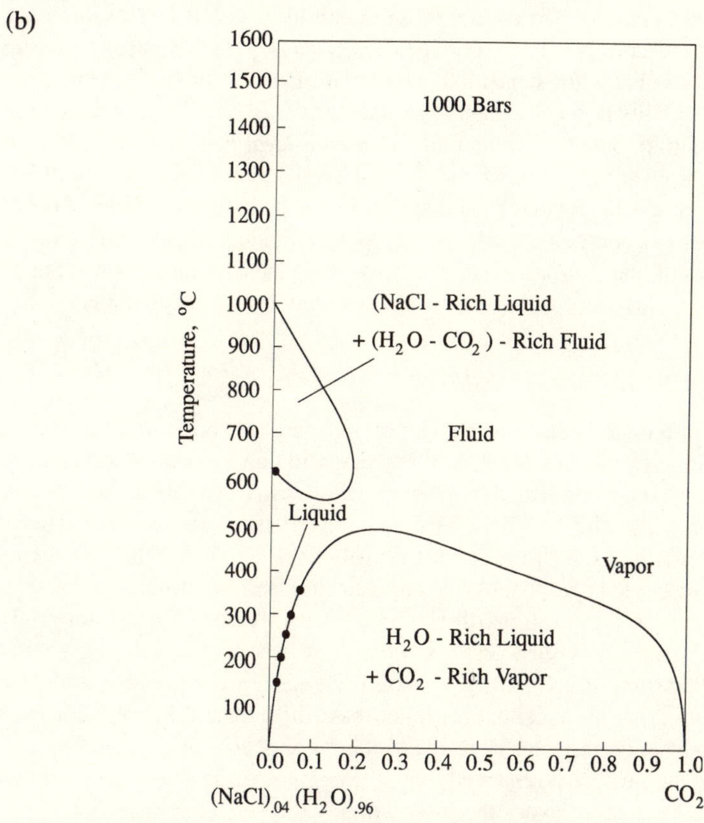

FIG. 16.6. (a) CO_2–H_2O solvus at 1 kbar. Solid line is fit of a MRK equation to experimental data (solid dots). After Bowers and Helgeson (1983). (b) Effect of 12 wt.% NaCl on the CO_2–H_2O solvus at 1 kbar. After Bowers and Helgeson (1983).

miscibility gap disappears at this T and P with the addition of only a trace of CH_4. The lowering of CO_2 fugacity by CH_4 means that decarbonation reactions can persist to significantly higher CO_2 pressures if CH_4 is present in the fluid. This means that fluid compositions should be well characterized before reactions involving CO_2 are used to estimate T and P of mineralization. The comparable effect of NaCl on the CO_2–H_2O solvus is displayed in Figure 16.6a,b from Bowers and Helgeson (1983).

Here Figure 16.6a shows the solvus predicted above for pure CO_2–H_2O systems occurring as low as 1 kbar at temperatures below 300°C, in agreement with experimental data. To obtain this fit Bowers and Helgeson used a MRK equation that differs from that of Kerrick and Jacobs. It is essentially the original RK equation (16.7) with the a parameter written as a function of T (but not P). To extend this MRK equation to systems containing NaCl, they fit pseudobinary H_2O–CO_2 systems at different salt concentrations, solved for the a and b MRK parameters each time, and then wrote an expression for the variation of a and b with NaCl content. This trick is necessary because the RK class of equations is intended for gases (or low-density fluids) and cannot be applied to liquids, particularly those as non-ideal as NaCl–H_2O solutions. As expected, the effect of NaCl on CO_2–H_2O solutions is quite dramatic—compare Figure 16.6a and 16.6b. For the same total pressure of 1 kbar, 12 wt. % NaCl raises the two-phase immiscibility region by roughly another 200°C. This makes sense because a NaCl–H_2O solution must be considerably more non-ideal than H_2O alone due to strong, ionic solvent–solute interactions; the NaCl–H_2O rich liquid phase will become even more unlike the relatively simple CO_2 phase, and immiscibility will occur over a wider range of conditions. Again, the geological implications are considerable, because activities of both CO_2 and H_2O are changed by the addition of NaCl; this in turn affects the conditions for which any reactions involving either component occur.

16.4. SOME CONCLUDING COMMENTS

We have seen in this chapter that the behavior of both gaseous and solid solutions can be described with equations of state fitted by regression analysis to observed data. The more rigorous (and complicated) of these equations can be extrapolated with care beyond the field of experimental data. The equations of state most commonly used have appropriate mathematical forms to describe observed P–T–V-X behavior. This is because they are based on reasonable, if simplistic, models of molecular interaction.

At the present time, the equations of state used for gas mixtures are more unwieldy than those used for solids. This is a paradox because the behavior of gases is generally much simpler than that of solids (e.g., miscibility gaps are not possible with true gaseous mixtures). Also, the equations for solids make full use of the interrelationships between thermodynamic variables while the equations most commonly used for gases do not. For example, we have observed that all Margules parameters are related by the same equations as the primary thermodynamic variables (such as $W_G = W_U + PW_V - TW_S$). This is not practical with the van der Waals, RK, MRK, and related equations.

It seems that the equations most often used to describe solid solutions have developed or evolved further than those used to describe gases. Why should this be so?

The difference probably stems from the fact that an equation of state can be written for an ideal gas, but that nothing like this is possible with a liquid or solid (we even remarked above that defining an ideal, one-component liquid or solid is not really feasible or useful). Starting with $PV = RT$, other more realistic equations of state for gases evolved by simple adjustments, each giving improved descriptions of the P–V–T behavior of gases. Unfortunately, these modified equations (from the van der Waals equation on to MRKs) have to be integrated (as in 16.16 and 16.17) to give fugacities and activities, and this can produce very complex expressions. With solids and liquids there was nothing like the ideal gas law to begin with, and development of equations of state for these phases followed a different (and, in some ways, opposite) route. Here we start by defining *deviation* from ideality through excess functions, and then go on to derive the P–V-T equation of state.

There will always be room for improvement of any general equation of state, but at the present time this would be most useful for gases. It might be possible to start, as with liquids and solids, by defining excess functions of suitable form (such as a virial equation), and develop these into more practical, and less complex, equations of state for gases and gas mixtures. This is not a straightforward problem by any means. Recall that Margules' original equations were actually designed to describe vapor pressures over liquid solutions. The development of the Margules equations roughly followed the path we have just outlined, and they should apply quite well to simple gas solutions. Yet something other than the Margules kind of approach is really required for gas mixtures because the number of adjustable parameters and the equations themselves become too cumbersome with multi-component systems. The virial approach looks promising because of its proven effectiveness with simple gases and with very complicated aqueous electrolyte solutions, but here there is a problem selecting mixing rules for molecular interactions of unlike species. We close this chapter leaving these unresolved problems as a challenge for the future.

17

AQUEOUS ELECTROLYTE SOLUTIONS

Of all known liquids, water is probably the most studied and least understood . . .

Franks (1972, p. 18)

17.1. INTRODUCTION

In dealing with the thermodynamic properties of ions we have one difficulty in addition to those encountered in dealing with compounds and elements. For compounds and elements we found that although we could measure absolute values for some properties, others such as enthalpy and the other energy terms contained an undetermined constant. We got around this by using the concept of "formation from the elements." It would of course be very convenient to also have thermodynamic properties of individual ions, but because positively and negatively charged ions cannot be separated from each other to any significant extent, their individual properties cannot be measured. To get around this, we need an additional convention, while retaining the formation from the elements convention. In addition we have certain problems in dealing with the activities and activity coefficients of electrolytes and individual ions.

In the following section we discuss the problems of activities of ionic species. We follow the presentation of Klotz (1964), and include the HCl example used by Pitzer and Brewer (1961), and an expanded consideration of the choice of solute components. Following that we discuss the conventions used to obtain numerical values for the state variables of individual ions.

17.2. ELECTROLYTE AND SINGLE ION ACTIVITIES

17.2.1. *Algebraic Example*

We begin by demonstrating that the basic approach is not arbitrarily chosen by chemists with a view to confusing students, nor is it dictated by the electrically charged nature of ions. It is dictated by the *algebraic* consequence of the fact that when neutral solute molecules dissociate into charged particles, *the number of solute particles is increased*. For example, when one mole of the undissociated solute AB(aq), which can be treated using Henry's Law, Raoult's Law, and the rest of the equations developed in previous chapters, becomes instead one mole of A(aq) plus one mole of B(aq), certain consequences develop that have nothing to do with whether A(aq) and B(aq) are electrically charged or not.

To illustrate this we consider first the simultaneous equations

$$\left.\begin{aligned} m/p &= k_1 \\ n^2/m &= k_2 \end{aligned}\right\} \tag{17.1}$$

where k_1 and k_2 are known constants and m, n, and p are three variables. With three variables and two relations between them, the situation is univariant, and we can solve for two variables given the value of the third one. It follows too that

$$p = n^2/(k_1 k_2)$$

or simply

$$p \propto n^2$$

Also, when m is small,

$$(m + n)^2 \approx n^2$$

so that

$$p \propto (m + n)^2 \text{ when } m \text{ is small}$$

In Table 17.1 we have calculated values of m and n for various values of p, and you will note that the constant of proportionality between p and n^2 (which is $1/k_1 k_2$) is truly constant, while the "constant" of proportionality between p and $(m+n)^2$ is not truly constant but approaches the value $1/k_1 k_2$ as m gets smaller (this approximation gets much better as k_2 gets larger). These relations are shown in graphical form in Figure 17.1.

If you can follow this simple scheme, then you understand the basis of one of the more confusing conventions in thermodynamics. As indicated in headings of Table 17.1, the particular form of equations (17.1) was chosen to model the situation shown in Figure 17.2. That is, p represents the fugacity of a solute component AB, m represents the molality of the undissociated solute species AB°(aq), and n represents the molality of both species A(aq) and B(aq) (necessarily equal molalities) formed by the breakdown of AB(aq). $(m + n)$ then represents the molality of solute component AB as it would be measured by chemical analysis of the solution. The constants k_1 and k_2 represent respectively the Henry's Law constant (actually its reciprocal) and the ionization constant of AB. It follows from this argument that for a solute component such as AB which is highly dissociated into A and B, $f_{\mathrm{AB}} \propto m_{\mathrm{AB}}^2$, and the point of beginning with this purely algebraic example is to show that this relationship has an algebraic and not a chemical nor a conventional basis.

All we have done, really, is to retain the Henry's Law relation

$$f_{\mathrm{AB}} \propto m_{\mathrm{AB}^\circ}$$

for *species* AB°, and to show that in cases where m_{AB° is a very small fraction of the total AB in solution,

$$f_{\mathrm{AB}} \propto m_{\mathrm{AB}}^2$$

where m_{AB} here is the molality of *component* AB.

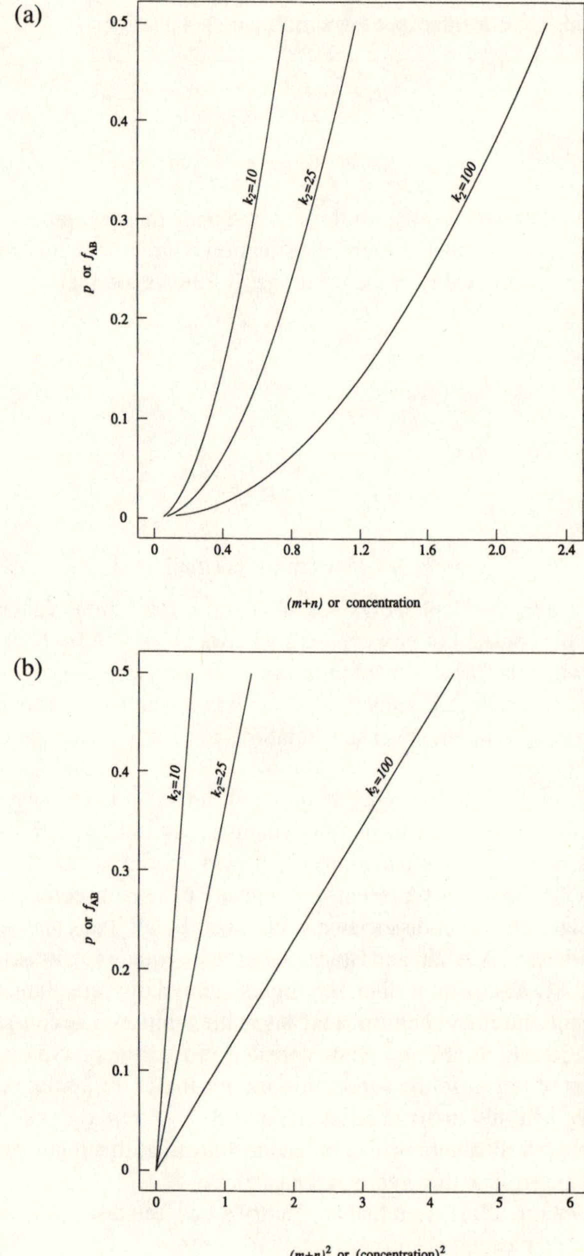

FIG. 17.1. (a) p versus $(m + n)$ for various values of k_2 where m, n, p, and k_2 are related by the equations in the text. p and $(m + n)$ can be thought of as representing fugacity and concentration of a binary electrolyte in solution. (b) p versus $(m + n)^2$ for various values of k_2 where m, n, p, and k_2 are related by the equations in the text. Data in Table 17.1.

Table 17.1 The Algebraic Relationship between Fugacity and (Concentration)^2

$$k_1 = m/p = m_{AB}/f_{AB} = 0.1$$
$$k_2 = n^2/m = m_A m_B/m_{AB} = 10$$

thermodynamic variable:						
f_{AB}	m_{AB}	m_A	m_B	$m_A + m_{AB}$	$1/k_1k_2$	
algebraic variable:						
p	m	n	n	$m+n$	p/n^2	$p/(n+m)^2$
100	10.00	10.00	10.00	20.00	1.00	0.25
50	5.00	7.07	7.07	12.07	1.00	0.25
10	1.00	3.16	3.16	4.16	1.00	0.58
1	0.10	1.00	1.00	1.10	1.00	0.83
0.9	0.09	0.95	0.95	1.04	1.00	0.83
0.8	0.08	0.89	0.89	0.97	1.00	0.84
0.7	0.07	0.84	0.84	0.91	1.00	0.85
0.6	0.06	0.77	0.77	0.83	1.00	0.86
0.5	0.05	0.71	0.71	0.76	1.00	0.87
0.4	0.04	0.63	0.63	0.67	1.00	0.88
0.3	0.03	0.55	0.55	0.58	1.00	0.90
0.2	0.02	0.45	0.45	0.47	1.00	0.92
0.1	0.01	0.32	0.32	0.33	1.00	0.94
0.09	0.01	0.30	0.30	0.31	1.00	0.94
0.08	0.01	0.28	0.28	0.29	1.00	0.95
0.07	0.01	0.26	0.26	0.27	1.00	0.95
0.06	0.01	0.24	0.24	0.25	1.00	0.95
0.05	0.01	0.22	0.22	0.23	1.00	0.96
0.04	0.00	0.20	0.20	0.20	1.00	0.96
0.03	0.00	0.17	0.17	0.18	1.00	0.97
0.02	0.00	0.14	0.14	0.14	1.00	0.97
0.01	0.00	0.10	0.10	0.10	1.00	0.98
0.001	0.00	0.03	0.03	0.03	1.00	0.99

To show that the example is in fact representative of real situations, we take the example of HCl in water used by Pitzer and Brewer (1961). Pure HCl is a gas that on dissolving in water dissociates to a very large degree into the charged particles (called) H^+ and Cl^-. The solubility of HCl in water is so great that even with very sensitive techniques, the concentration of HCl reaches about 4 molal before the vapor pressure of HCl above the solution becomes large enough to be measurable. In Table 17.2 we show some values for the vapor pressure (essentially equal to the fugacity) of HCl for solutions from 4 to 10 molal in HCl.

These solutions are a little too concentrated to illustrate the dilute solution properties we are discussing, but they serve as a basis for calculating the fugacity of HCl in more dilute solutions. This is done using values of $\gamma_\pm$ (to be defined shortly; see §17.2.4), which are measured electrochemically.[1]

The resultant p_{HCl} or f_{HCl} values are shown in Table 17.2, and shown plotted against m_{HCl} and m^2_{HCl} in Figure 17.3. With a dissociation constant of approximately

[1] Values of $k\gamma_\pm$ were derived from measured p_{HCl} values of 4 to 7 molal HCl by calculating $p^{1/2}_{HCl}/m_{HCl}$. These were divided by $\gamma_\pm$ at each concentration to obtain a value of k. The average value of k in this range is 0.000704, and this value was used to calculate f_{HCl} at each concentration as $(0.000704 \cdot \gamma_\pm \cdot m_{HCl})^2$.

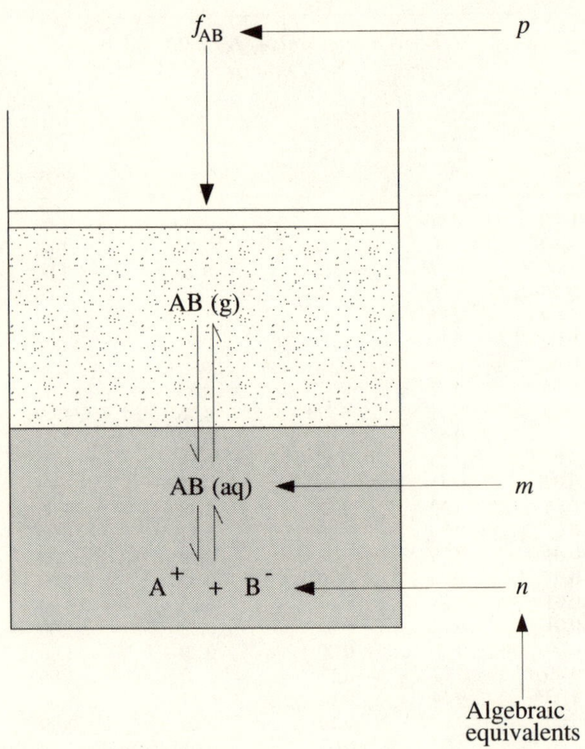

FIG. 17.2. A gas AB having fugacity f_{AB} equilibrated with a solution in which AB is largely dissociated into A and B. The algebraic variables m, n, and p correspond to the thermodynamic variables as shown.

10^7 at 25°C, HCl is clearly a good candidate for the $(m + n) = n$ approximation.

To reiterate, if HCl dissolved in water as a single undissociated particle, then the normal Henry's Law formulation would apply, i.e.,

$$f_{HCl} = h_{HCl} m_{HCl}$$

where h_{HCl} is the Henry's Law constant. But because HCl breaks into two particles, f_{HCl} is proportional not to m_{HCl} but to m^2_{HCl}. The reason for this is essentially algebraic.

17.2.2. *Activities of Electrolyte Components*

For non-electrolytes, we saw that the next step was to define a standard state such that f° was the fugacity of the solute in an ideal one molal solution. Another way of saying this is that f° is the (Henry's Law) constant of proportionality for $f_{AB} \propto m_{AB}$

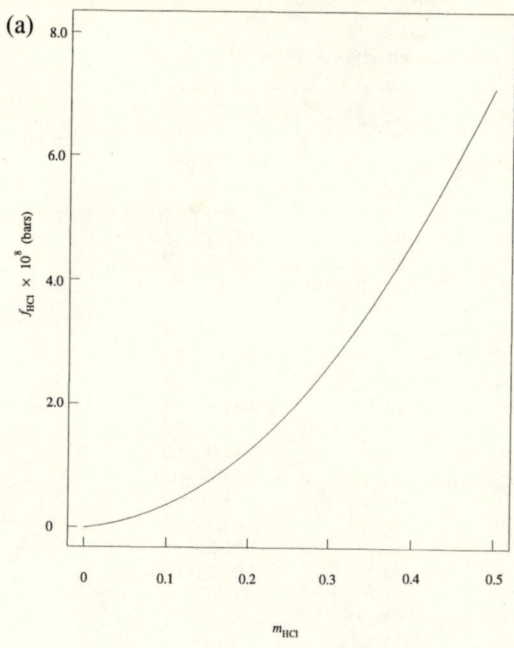

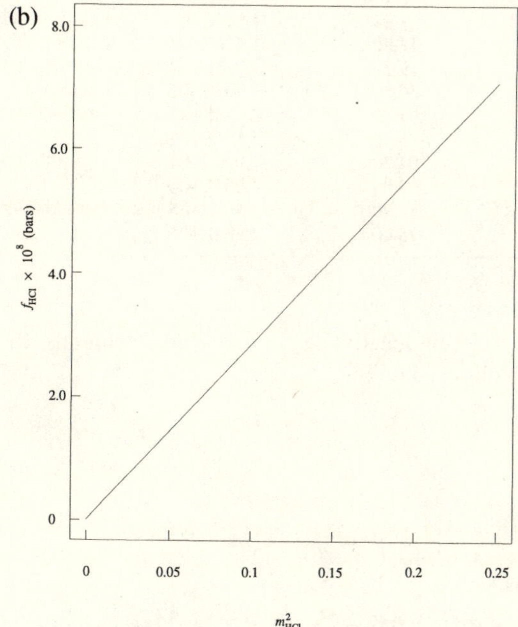

FIG. 17.3. (a) Fugacity of HCl (calculated) versus molality of HCl at 25°C. (b) Fugacity of HCl (calculated) versus molality of HCl^2 at 25°C. Data in Table 17.2.

Table 17.2 Activities in Aqueous Hydrochloric Acid Solutions at 25°C.

m_{HCl}	$\gamma_{\pm,\text{HCl}}$	$a_{\pm,\text{HCl}}$	a_{HCl}	f_{HCl} measured	f_{HCl} calculated
equivalent algebraic variable: $m + n$					p
0.0005	0.9750	0.000487	2.377E−07		1.178E−13
0.0010	0.9656	0.000966	9.324E−07		4.623E−13
0.0020	0.9521	0.001904	3.626E−06		1.789E−12
0.0050	0.9285	0.00464	2.155E−06		1.069E−11
0.01	0.9048	0.00905	8.187E−05		4.059E−11
0.02	0.8755	0.0175	3.066E−04		1.520E−10
0.05	0.8304	0.0415	1.724E−03		8.548E−10
0.10	0.7964	0.0796	6.343E−03		3.145E−09
0.20	0.7667	0.1533	2.351E−02		1.166E−08
0.30	0.7560	0.2268	5.144E−02		2.550E−08
0.40	0.7550	0.3020	9.120E−02		4.522E−08
0.50	0.7571	0.3785	1.433E−01		7.105E−08
0.60	0.7630	0.4578	2.096E−01		1.039E−07
0.70	0.7720	0.540	2.920E−01		1.448E−07
0.80	0.7830	0.626	3.924E−01		1.945E−07
0.90	0.7950	0.716	5.119E−01		2.538E−07
1.00	0.8090	0.809	6.545E−01		3.245E−07
1.50	0.8962	1.344	1.807E+00		8.960E−07
2.00	1.009	2.018	4.072E+00		2.019E−06
3.00	1.316	3.948	1.559E+01		7.728E−06
4.00	1.762	7.048	4.967E+01	2.426E−05	2.463E−05
5.00	2.38	11.90	1.416E+02	7.066E−05	7.021E−05
6.00	3.22	19.32	3.733E+02	1.867E−04	1.851E−04
7.00	4.37	30.59	9.357E+02	4.640E−04	4.640E−04
8.00	5.90	47.20	2.228E+03	1.125E−03	1.105E−03
9.00	7.94	71.46	5.107E+03	2.573E−03	2.532E−03
10.00	10.44	104.40	1.090E+04	5.600E−03	5.404E−03
12.00	17.25	207.00	4.285E+04		2.125E−02
14.00	27.30	382.20	1.461E+05		7.243E−02
16.00	42.40	678.40	4.602E+05		2.282E−01

when the solution is acting ideally, i.e., in very dilute solutions. In other words, for non-electrolytes in dilute solutions,

$$f_{\text{AB}} = f^{\circ}_{\text{AB}} m_{\text{AB}}$$

and

$$a_{\text{AB}} = f_{\text{AB}} / f^{\circ}_{\text{AB}} = m_{\text{AB}}$$

or for *any* solution,

$$a_{\text{AB}} = f_{\text{AB}} / f^{\circ}_{\text{AB}} = \gamma_{\text{AB}} m_{\text{AB}}$$

Using this same line of thought for electrolytes, we choose a standard state such that f° is the constant of proportionality for

$$f_{\text{AB}} \propto m^2_{\text{AB}}$$

Thus

$$f_{AB} = f^{\circ}_{AB} m^2_{AB}$$

and

$$a_{AB} = f_{AB}/f^{\circ}_{AB} = \gamma_{AB} m^2_{AB}$$

In the case of HCl, this becomes

$$\begin{aligned} a_{HCl} &= f_{HCl}/f^{\circ}_{HCl} \\ &= \gamma_{HCl} m^2_{HCl} \end{aligned}$$

or, in very dilute solution where $\gamma_{HCl} \to 1.0$,

$$a_{HCl} = m^2_{HCl} \tag{17.2}$$

17.2.3. *Choice of Solute Component*

Before going on to define single ion activities and activity coefficients, let's pause to reflect on the similarity between the case considered here (a completely dissociated electrolyte in water), and the olivine solid solution case considered in Chapter 12 (§12.7). The physical systems are completely different, but the thermodynamic problem is almost identical, the only significant difference being that in the olivine case the concentrations were measured by mole fractions and ideality consisted in conforming to Raoult's Law, while here concentrations are measured in molality and ideality is represented by Henry's Law. Apart from that, the problem in both cases consists in choosing a solute component that is appropriate to the situation.

In Chapter 12 we showed the purely formal relation between component activities

$$a_i = (a_{0.5i})^2$$

and showed that in the case of $Mg_2SiO_4 - Fe_2SiO_4$ solid solutions, component Mg_2SiO_4 plots as a parabolic curve on the $a - X$ plot, while component $MgSi_{0.5}O_2$ comes close to plotting as a straight line. In the HCl – H_2O case we find that component HCl is strongly curved on a plot of a versus m, but gives a straight line when plotted as a versus m^2. We are entitled to ask why, in the case of HCl, did we not behave consistently and plot component $H_{0.5}Cl_{0.5}$ versus m instead of changing the concentration axis from m to m^2? Obviously $a_{H_{0.5}Cl_{0.5}}$ versus m_{HCl} would result in a straight line too, because $a_{H_{0.5}Cl_{0.5}} = a^{\frac{1}{2}}_{HCl}$.

We can answer this in two ways. First, we *will* introduce and use component $H_{0.5}Cl_{0.5}$ (and analogous components for other compounds), but it is given a special symbol ($a_{\pm}$) for reasons that will become apparent. The other way to answer the question is to point out that although we need no knowledge of the molecular nature of a system to be able to apply thermodynamics to it, we would be silly to ignore such knowledge when it is available. Our rationale for using m^2 rather than m is based on overwhelming experimental evidence that many binary electrolytes split completely into two particles on dissolving in water. This provides the *reason* why a_{HCl} versus m^2_{HCl} works better than a_{HCl} versus m_{HCl}, just as the observation that

Mg and Fe exchange as single atoms in olivine provides the reason why $a_{\mathrm{MgSi_{0.5}O_2}}$ versus X works better than $a_{\mathrm{Mg_2SiO_4}}$ versus X. Thermodynamics doesn't care which component we use, but points out that if we are pleased by a_{HCl} versus m^2_{HCl}, we can get the same results from $a_{\mathrm{H_{0.5}Cl_{0.5}}}$ versus m_{HCl}.

In general then, we see again that the choice of components is an important part of any thermodynamically based analysis or theory. Certain choices will "work better" or be more appropriate than others, and there will always be a reason for this. Although the reason *may* be that the component corresponds closely to the actual chemical species (N_2 certainly works better than N_4 in gas mixture equations), this is not necessarily the case as we have seen in both the olivine and $HCl - H_2O$ examples.

17.2.4. *Relationship of Solute Activity to Single Ion Activities*

A plot of $a^{\frac{1}{2}}_{\mathrm{HCl}}$ (or $a_{\mathrm{H_{0.5}Cl_{0.5}}}$, and called $a_{\pm,\mathrm{HCl}}$ in the electrolyte literature) versus m_{HCl} is shown in Figure 17.4 (data in Table 17.2). Appropriately, the activity coefficients measurable as the ratio $a_\pm/m$ on this diagram are called $\gamma_{\pm,\mathrm{HCl}}$. The reason for using the notation $a_{\pm,\mathrm{HCl}}$ for $a_{\mathrm{H_{0.5}Cl_{0.5}}}$ can be seen by continuing the molecularly-based reasoning we were pursuing before we stopped to consider the similarity to the olivine case. Going back to the relation [2]

$$a_{\mathrm{HCl}} = m^2_{\mathrm{HCl}} \tag{17.3}$$

which holds for infinitely dilute solutions, we can next write

$$a_{\mathrm{HCl}} = m_{\mathrm{H^+}} m_{\mathrm{Cl^-}} \tag{17.4}$$

for the same conditions, because each mole of HCl breaks down completely to H^+ and Cl^-. At this point we introduce the concept of activity coefficients for individual ions, and define them such that they approach 1.0 in infinitely dilute solutions, just as we did with the activity coefficients of undissociated solutes. They are a useful mental concept in spite of not being measurable. At infinite dilution, then,

$$\gamma_{\mathrm{H^+}} = \gamma_{\mathrm{Cl^-}} = 1.0$$

and under other conditions they take on values such that

$$a_{\mathrm{HCl}} = (m_{\mathrm{H^+}}\gamma_{\mathrm{H^+}})(m_{\mathrm{Cl^-}}\gamma_{\mathrm{Cl^-}}) \tag{17.5}$$

remains true. Therefore

$$a_{\mathrm{HCl}} = a_{\mathrm{H^+}} a_{\mathrm{Cl^-}} \tag{17.6}$$

or

$$a_{\mathrm{H^+}} a_{\mathrm{Cl^-}} / a_{\mathrm{HCl}} = 1.0 \tag{17.7}$$

[2]It is helpful in understanding the relations in this section to keep firmly in mind that HCl refers to *component* HCl. *Species* HCl would be HCl° (§12.7). m_{HCl} therefore refers to the total or analytical concentration of HCl, sometimes called m_{total}.

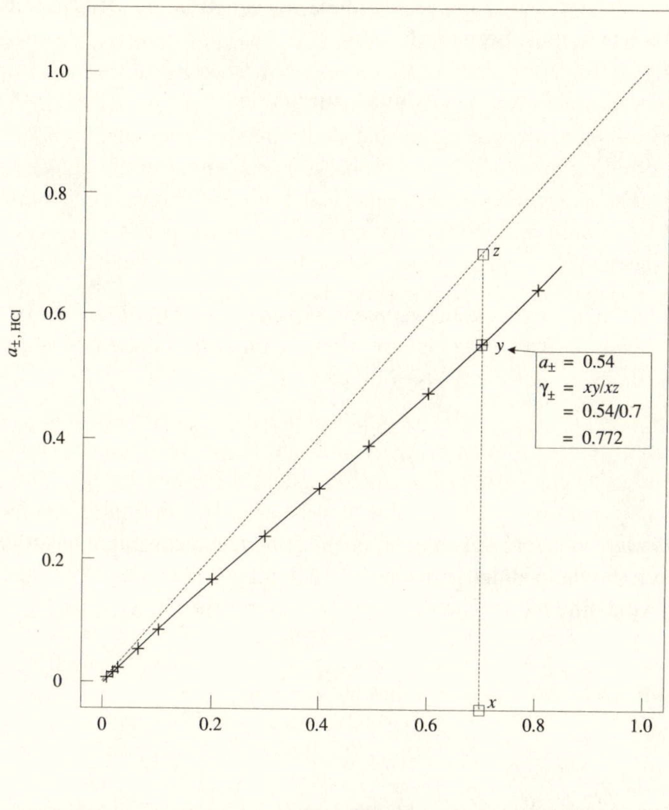

FIG. 17.4. Stoichiometric mean ionic activity of HCl ($a_{\pm,\mathrm{HCl}}$) versus molality of HCl. Crosses are data points from Table 17.2; dotted line is Henry's Law, having an activity of 1.0 at 1.0 molal HCl. The stoichiometric mean ionic activity coefficient at $m_{\mathrm{HCl}} = 0.7$ is $xy/xz = 0.772$.

This defines the equilibrium constant for the reaction between *component* HCl and the ions H^+ and Cl^-. It follows then that for the reaction

$$\underbrace{\mathrm{HCl}}_{component} = \mathrm{H^+} + \mathrm{Cl^-};\ \Delta_r G^\circ = 0 \tag{17.8}$$

As usual, it is best to see the truth of a relationship by understanding it rather than by seeing no fault with its derivation. In this case this can be accomplished by realizing that in the ideal one molal standard state to which $\Delta_r G^\circ$ refers, the solute component HCl consists in solution entirely as H^+ and Cl^-, therefore the G of component HCl has no choice but to be identical to $G_{\mathrm{H^+}} + G_{\mathrm{Cl^-}}$, from which it follows that $\Delta_r G^\circ = 0$.

In the solution chemistry literature it is usual to refer to the solvent in a binary system as component 1, and to the solute (as normally written) as component 2. Therefore in this case a_{HCl} could also be referred to as a_2.

We can now see why $a_{\mathrm{H}_{0.5}\mathrm{Cl}_{0.5}}$ is called $a_{\pm,\mathrm{HCl}}$. It comes from the concept of individual ion activities, because if

$$a_{\mathrm{HCl}} = a_{\mathrm{H}^+} a_{\mathrm{Cl}^-} \tag{17.6}$$

then

$$\begin{aligned} a_{\mathrm{HCl}}^{\frac{1}{2}} &= (a_{\mathrm{H}^+} a_{\mathrm{Cl}^-})^{\frac{1}{2}} \\ &= a_{\pm,\mathrm{HCl}} \end{aligned} \tag{17.9}$$

which means that $a_{\pm}$ refers to the *geometric mean* of the activities of the two particles (+ *and* −) resulting from dissociation, which in turn gives rise to its other name, the *stoichiometric mean ionic activity.* Similarly,

$$\begin{aligned} \gamma_{\mathrm{HCl}}^{\frac{1}{2}} &= (\gamma_{\mathrm{H}^+} \gamma_{\mathrm{Cl}^-})^{\frac{1}{2}} \\ &= \gamma_{\pm,\mathrm{HCl}} \end{aligned} \tag{17.10}$$

the *stoichiometric mean ionic activity coefficient*. The measurable quantity $\gamma_{\pm,\mathrm{HCl}}$ is considered to be the geometric mean of two unmeasurable quantities, γ_{H^+} and γ_{Cl^-}. Similarly, we define the geometric mean of the ion concentrations as

$$m_{\pm} = (m_{\mathrm{H}^+} m_{\mathrm{Cl}^-})^{\frac{1}{2}}$$

so that

$$a\pm = m_{\pm}\gamma_{\pm}$$

The great convenience of being able to think in terms of individual ionic properties means that we always look at ionic solutions this way, rather than as a simple changing of solute component.

17.2.5. *Unsymmetrical Electrolytes*

The discussion so far has focused on HCl, but all symmetrical electrolytes will have the same relationships between m, a, and γ. By symmetrical electrolytes we mean those in which both ions have the same charge, so that on dissolution equal numbers of positive and negative ions result. Other examples are NaCl, $MgSO_4$, HNO_3, and so on. However some complications develop for unsymmetrical electrolytes, i.e., those giving unequal numbers of positive and negative ions, such as Na_2SO_4, $AlCl_3$, and so on. To begin with f, a_2 and $a_{\pm}$, the definitions parallel those for the symmetrical case. Following our algebraic argument, it is found theoretically that for Na_2SO_4, $f \propto m^3$, while for $AlCl_3$, $f \propto m^4$, giving rise to the expressions

$$a_2 = (a_{\mathrm{Na}^+})^2 (a_{\mathrm{SO}_4^{2-}}) \text{ for } \mathrm{Na_2SO_4}$$

and

$$a_2 = (a_{\mathrm{Al}^{3+}})(a_{\mathrm{Cl}^-})^3 \text{ for } \mathrm{AlCl_3}$$

which lead for the same reasons as before, to

$$a_{\pm,\mathrm{Na_2SO_4}} = \left((a_{\mathrm{Na^+}})^2(a_{\mathrm{SO_4^{2-}}})\right)^{\frac{1}{3}}$$

and

$$a_{\pm,\mathrm{AlCl_3}} = \left((a_{\mathrm{Al^{3+}}})(a_{\mathrm{Cl^-}})^3\right)^{\frac{1}{4}}$$

which gives

$$a_{\mathrm{Na_2SO_4}} = a^3_{\pm,\mathrm{Na_2SO_4}}$$

and

$$a_{\mathrm{AlCl_3}} = a^4_{\pm,\mathrm{AlCl_3}}$$

Klotz (1964) points out that these definitions of activity for unsymmetrical salts imply new and rather strange standard states for these electrolytes. If we insist on having, for example,

$$a_{\mathrm{Na_2SO_4}} = (a_{\mathrm{Na^+}})^2(a_{\mathrm{SO_4^{2-}}})$$

it follows that in the binary system ($H_2O - Na_2SO_4$) at very dilute concentrations of Na_2SO_4,

$$\begin{aligned} a_{\mathrm{Na_2SO_4}} &= (2m_{\mathrm{Na_2SO_4}})^2(m_{\mathrm{Na_2SO_4}}) \\ &= 4m^3_{\mathrm{Na_2SO_4}} \end{aligned}$$

This means that we cannot say, as before (§17.2.2), that

$$\lim_{m\to 1}(f/m^3) = f^\circ$$

because this leads to

$$a_{\mathrm{Na_2SO_4}} = m^3_{\mathrm{Na_2SO_4}}$$

unless we let $\gamma_{\mathrm{Na_2SO_4}} \to \frac{1}{4}$ as $m^3 \to 0$. In order to preserve the convenience of having $\gamma \to 1$ as $m \to 0$ for all electrolytes, the standard state is chosen such that its fugacity, f°, is 1/4 of $h_{\mathrm{Na_2SO_4}}$, where $h_{\mathrm{Na_2SO_4}}$ is the Henry's Law constant for Na_2SO_4 (see Figure 17.5). This results in $\gamma_{\pm,\mathrm{Na_2SO_4}}$ being defined as

$$\gamma_{\pm,\mathrm{Na_2SO_4}} = \frac{a_\pm}{4^{\frac{1}{3}}m} \tag{17.11}$$

and to retain the expression

$$\gamma_\pm = \frac{a_\pm}{m_\pm}$$

we define the *stoichiometric mean ionic molality* as

$$\begin{aligned} m_\pm &= \left((m_{\mathrm{Na^+}})^2(m_{\mathrm{SO_4^{2-}}})\right)^{\frac{1}{3}} \\ &= \left((2m_{\mathrm{Na_2SO_4}})^2(m_{\mathrm{Na_2SO_4}})\right)^{\frac{1}{3}} \\ &= 4^{\frac{1}{3}}m_{\mathrm{Na_2SO_4}} \end{aligned}$$

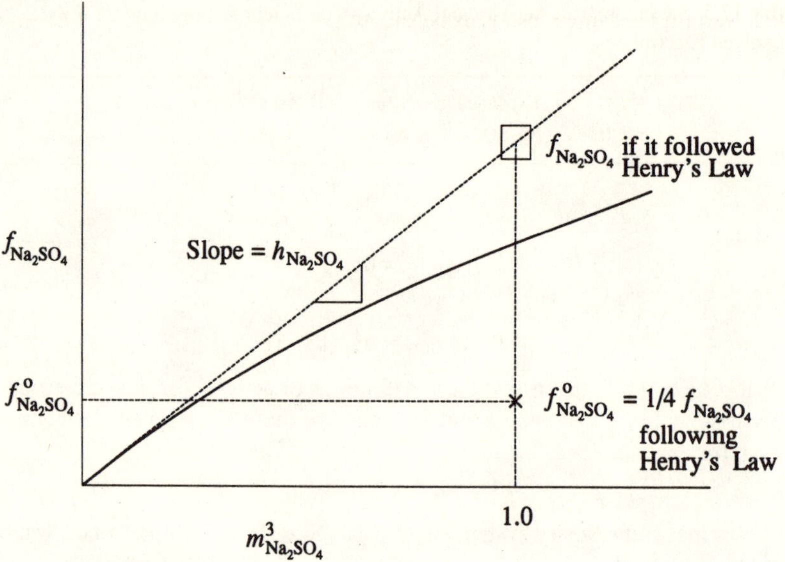

FIG. 17.5. Schematic representation of the fugacity of an unsymmetrical salt (Na_2SO_4) versus $m^3_{Na_2SO_4}$, illustrating the fact that the standard state chosen is not given by the Henry's Law slope at 1.0 m as in Figure 17.3, but 1/4 of this.

This is all quite confusing on the first run-through, but is quite logical. See Klotz (1964) for additional discussion. Table 17.3 contains a summary of these relationships.

Example: From Robinson and Stokes (1968, p. 478) we find that in a 2.0m solution, $\gamma_{\pm,CaCl_2}$ at 25°C is 0.792. Therefore

$$a_{Ca^{2+}} = 2 \times 0.792 = 1.584$$

$$a_{Cl^-} = 2 \times 2 \times 0.792 = 3.168$$

$$a_{CaCl_2} = 1.584 \times 3.168^2 = 15.897$$

$$\text{or} \quad a_{CaCl_2} = 4 \times 0.792^3 \times 2^3 = 15.897$$

$$a_{\pm,CaCl_2} = 15.897^{\frac{1}{3}} = 2.514$$

The importance of activity coefficients is evident in Figure 17.6, which shows measured $\gamma_{\pm}$ versus molal concentration for selected salts in 350°C hydrothermal solutions. Activity coefficient corrections (that is, correction to the concentration to get the activity) of one to two orders of magnitude such as these are not at all uncommon in aqueous systems.

17.2.6. *General Relationships*

These relationships can be looked at more generally by beginning with the equations for the chemical potential of electrolyte solutes, based on the three different concentration scales:

Table 17.3 Stoichiometric Activity and Activity Coefficient Expressions for Single Dissolved Electrolytes

Expanded from Klotz (1964, Table 21.1)

		NaCl	Na_2SO_4	$AlCl_3$	$MgSO_4$	$A_{\nu_+}B_{\nu_-}$
f_2	$\propto$	m^2	m^3	m^4	m^2	$m^{\nu_++\nu_-}$
a_2	=	a_+a_-	$a_+^2a_-$	$a_+a_-^3$	a_+a_-	$a_+^{\nu_+}a_-^{\nu_-}$
	=	$\gamma_\pm^2 m^2$	$4\gamma_\pm^3 m^3$	$27\gamma_\pm^4 m^4$	$\gamma_\pm^2 m^2$	$(\nu_+^{\nu_+}\nu_-^{\nu_-})\gamma_\pm^{(\nu_++\nu_-)}m^{(\nu_++\nu_-)}$
$a_\pm$	=	$(a_+a_-)^{\frac{1}{2}}$	$(a_+^2a_-)^{\frac{1}{3}}$	$(a_+a_-^3)^{\frac{1}{4}}$	$(a_+a_-)^{\frac{1}{2}}$	$(a_+^{\nu_+}a_-^{\nu_-})^{1/(\nu_++\nu_-)}$
a_+	=	$\gamma_\pm m$	$2\gamma_\pm m$	$\gamma_\pm m$	$\gamma_\pm m$	$\gamma_\pm \nu_+ m$
a_-	=	$\gamma_\pm m$	$\gamma_\pm m$	$3\gamma_\pm m$	$\gamma_\pm m$	$\gamma_\pm \nu_- m$
$m_\pm$	=	m	$4^{\frac{1}{3}}m$	$27^{\frac{1}{4}}m$	m	$m(\nu_+^{\nu_+}\nu_-^{\nu_-})^{1/(\nu_++\nu_-)}$
$\gamma_\pm$	=	$(\gamma_+\gamma_-)^{\frac{1}{2}}$	$(\gamma_+^2\gamma_-)^{\frac{1}{3}}$	$(\gamma_+\gamma_-^3)^{\frac{1}{4}}$	$(\gamma_+\gamma_-)^{\frac{1}{2}}$	$(\gamma_+^{\nu_+}\gamma_-^{\nu_-})^{1/(\nu_++\nu_-)}$
	=	$a_\pm/m_\pm$	$a_\pm/m_\pm$	$a_\pm/m_\pm$	$a_\pm/m_\pm$	$a_\pm/m_\pm$

$$\left.\begin{aligned}\mu_2 &= \mu_2^\circ + RT\ln\frac{f}{f^\circ}\\ &= \mu_2^\circ(X) + RT\ln\frac{\gamma_{H_X}X_2}{\gamma_{H_X}^\circ X_2^\circ}\\ &= \mu_2^\circ(M) + RT\ln\frac{\gamma_{H_c}M_2}{\gamma_{H_c}^\circ M_2^\circ}\\ &= \mu_2^\circ(m) + RT\ln\frac{\gamma_{H_m}m_2}{\gamma_{H_m}^\circ m_2^\circ}\end{aligned}\right\}\qquad(17.12)$$

Conversion of activity coefficients from one concentration scale to another is accomplished with the formulae in Table 17.4. These equations are derived by Monk (1961, p. 32) and Robinson and Stokes (1968, p. 32).

Because the total free energy of the solute must equal the sum of its parts (cations + anions),

$$\left.\begin{aligned}\mu_2 &= \nu_+\mu_+ + \nu_-\mu_-\\ \mu_2^\circ &= \nu_+\mu_+^\circ + \nu_-\mu_-^\circ\end{aligned}\right\}\qquad(17.13)$$

Expanding the top equation in (17.13),

$$\mu_2 = \nu_+(\mu_+^\circ + RT\ln a_+) + \nu_-(\mu_-^\circ + RT\ln a_-)$$

Combining this with

$$\mu_2 = \mu_2^\circ + RT\ln a_2$$

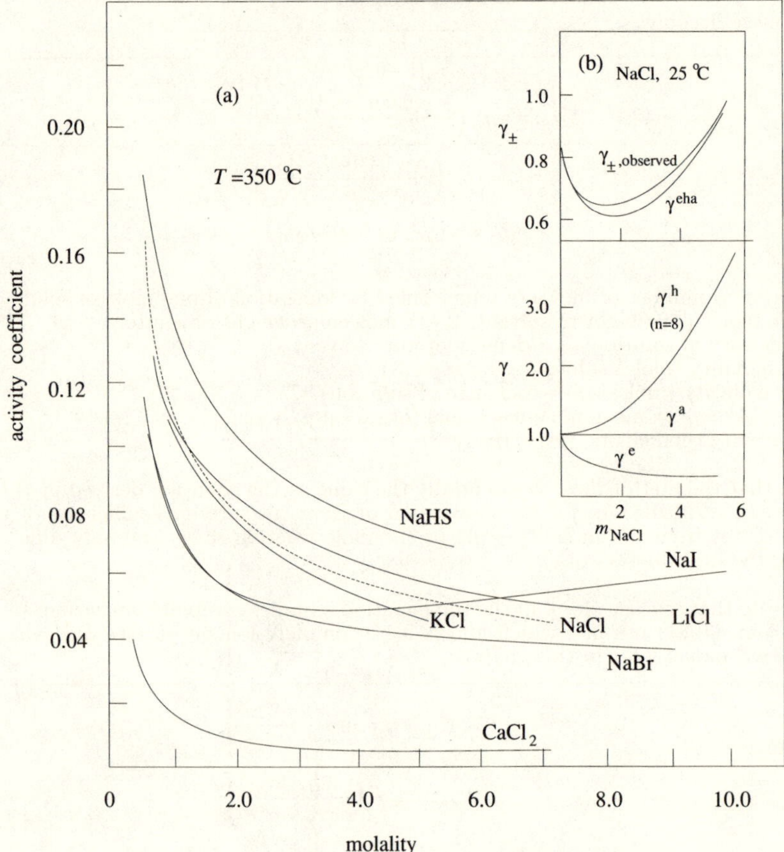

FIG. 17.6. (a) Mean molal stoichiometric ion activity coefficients for selected salts at 350°C versus total salt concentration. Note that coefficients of the 1:1 salts are similar but that the coefficient for $CaCl_2$ is much smaller. Activity coefficients for all salts at this temperature are roughly independent of concentration above $2m$. (b) Insert shows contributions of electrostatic, hydration, and association corrections to the measured activity coefficients of NaCl at 25°C. Symbols are defined in §17.6.1. Data from Wood et al. (1984).

gives

$$a_2 = a_+^{\nu_+} a_-^{\nu_-} \tag{17.14}$$

Defining

$$\left.\begin{aligned} a_+ &= \gamma_+ m_+ \\ a_- &= \gamma_- m_- \end{aligned}\right\} \tag{17.15}$$

Table 17.4 Activity Coefficient Interrelationships

$$\gamma_{H_X} = \gamma_\pm(1 + 0.001\nu W_A m)$$

$$\gamma_{H_X} = \gamma_{H_c}\left(\frac{\rho - 0.001M(\nu W_A - W_2)}{\rho_0}\right)$$

$$\gamma_\pm = \gamma_{H_c}\left(\frac{\rho - 0.001M \cdot W_2)}{\rho_0}\right)$$

$$\gamma_{H_c} = \gamma_\pm \frac{\rho}{\rho_0}(1 + 0.001mW_2)$$

Where ν = number of moles of ions formed by ionization of one mole of solute.
W_A = molecular weight of solvent; W_2 = molecular weight of solute.
ρ = density of solution; ρ_0= density of pure solvent.
m = molality (moles solute per kg solvent).
M = molarity (moles solute per litre of solution).
$\gamma_{H_X}, \gamma_{H_m}, \gamma_{H_c}$ = mean mole-fractional (or rational), molal and molar stoichiometric ion activity coefficients, respectively.

Note that the first of these is essentially the same as the equation derived in §12.4.5 ($\Gamma_\gamma = \ln(1 + 0.0180153m)$) for the conversion of Henryan activity coefficients of non-electrolytes from the molality scale to the mole fraction scale, the only difference being the introduction of ν.

Note too that our use of νm in the top equation assumes complete ionization. Helgeson et al. (1981) use m^*, which allows for incomplete ionization (see discussion of the HKF model later in this chapter).

where

$$\left.\begin{aligned} m_+ &= \nu_+ m \\ m_- &= \nu_- m \end{aligned}\right\} \tag{17.16}$$

and

$$\gamma_\pm = (\gamma_+^{\nu_+} \cdot \gamma_-^{\nu_-})^{\frac{1}{\nu}} \tag{17.17}$$

where $\nu = \nu_+ + \nu_-$. The rest of the equations in Table 17.3 follow directly.

These relations can also be expressed generally using the notation of Helgeson et al. (1981). The terms m and $(\nu_+ + \nu_-)$ in the table above are equivalent to m_k and ν_k respectively in the following equations.

$$\gamma_{\pm,k} = \left(\prod_j \gamma_j^{\nu_{j,k}}\right)^{1/\nu_k}$$

$$\begin{aligned} a_k &= \prod_j \left(\gamma_j \nu_{j,k} m_k\right)^{\nu_{j,k}} \\ &= \gamma_{\pm,k}^{\nu_k} m_k^{\nu_k} \prod_j \nu_{j,k}^{\nu_{j,k}} \end{aligned}$$

$$\begin{aligned} a_{\pm,k} &= a_k^{1/\nu_k} \\ &= \gamma_{\pm,k} m_k \left(\prod_j \nu_{j,k}^{\nu_{j,k}} \right)^{1/\nu_k} \end{aligned}$$

where j represents the ions of component k, ν_j represents the stoichiometry of ion j, and

$$\nu_k = \sum_j \nu_{j,k}$$

For example, if k is $CaCl_2$,

$$\nu_{Ca^{2+}} = 1, \quad \nu_{Cl^-} = 2, \quad \nu_k = 3$$

and

$$\begin{aligned} a_{CaCl_2} &= (\gamma_{Ca^{2+}} m_{CaCl_2})^1 (\gamma_{Cl^-} 2 m_{CaCl_2})^2 \\ &= 4(\gamma_{Ca^{2+}} \gamma_{Cl^-}^2) m_{CaCl_2}^3 \end{aligned}$$

For mixtures of electrolytes, with and without common ions, the notation becomes more complex. Helgeson et al. (1981) give the most comprehensive set of equations available.

17.2.7. *Stoichiometric versus Ionic Properties*

The word "stoichiometric" is included in the names given $a_\pm$, $\gamma_\pm$, and $m_\pm$ to indicate that they are derived according to the formulae given, which, as we have seen, are based on the supposition that the electrolyte dissociates *completely* into ions. If the electrolyte dissociates only partly into ions, these quantities may still be derived, but they won't represent what they have been described as representing, i.e., the geometric mean of the properties of the free ions.

Therefore other classes of activities and activity coefficients are defined ("mean ionic activities" and "mean ionic activity coefficients," without the "stoichiometric"), which are related to the stoichiometric quantities by the degree of dissociation, α, where

$$\alpha = m_j / m$$

Thus

$$\gamma_j = \alpha \cdot \bar{\gamma}_j$$

where $\bar{\gamma}_j$ is the mean ionic activity coefficient of the jth ion, and for component k,

$$\gamma_j \cdot m = \bar{\gamma}_j \cdot m_j$$

$$\bar{\gamma}_{\pm,k} = \left(\prod_j \bar{\gamma}_j^{\nu_{j,k}} \right)^{\frac{1}{\nu_k}}$$

and similarly for the other properties.

We have introduced the factor α to account for the degree of dissociation, but as pointed out quite eloquently by Pitzer and Brewer (1961) in their revision of the text by Lewis and Randall, the value of α varies with the method used to measure it; a situation not uncommon in studies of complex phenomena controlled by molecular interactions. Therefore it is to a large extent up to the investigator whether to use stoichiometric or ionic properties. Stoichiometric properties may be used whether or not ionic association is important, and have the advantage of not requiring estimates of α.

17.3. NUMERICAL VALUES FOR SINGLE-ION PROPERTIES

Having sorted out some definitions of single-ion activities and activity coefficients, we should now take a look at how numerical values are assigned to the thermodynamic properties of single ions. In essence, the procedure followed is a slight variation of the "formation from the elements" procedure discussed in Chapter 7.

It will be convenient to continue with HCl in water as our example, so that the question becomes how do we arrive at numerical values for all the thermodynamic properties of the ions H^+ and Cl^-? The answer begins with the equation for the formation of H^+ and Cl^- from the elements,

$$\frac{1}{2}H_2(g) + \frac{1}{2}Cl_2(g) = H^+ + Cl^- \tag{17.18}$$

The standard free energy change for this reaction is found by combining data from three other reactions. First, the association of hydrogen and chlorine to give HCl gas has the properties

$$\frac{1}{2}H_2(g) + \frac{1}{2}Cl_2(g) = HCl(g)$$

$$\log K = 16.70;\ \Delta_r G^\circ = -95299\ \text{J}$$

(data from Wagman et al., 1982) and the partitioning of HCl between liquid and vapor phases gives

$$HCl(g) = HCl(aq)$$

$$\log K = 6.29;\ \Delta_r G^\circ = -35929\ \text{J}$$

Combining these gives

$$\frac{1}{2}H_2(g) + \frac{1}{2}Cl_2(g) = HCl(aq)$$

$$\log K = 22.99;\ \Delta_r G^\circ = -131228\ \text{J}$$

We have already shown that for the reaction

$$HCl(aq) = H^+ + Cl^-$$

$$\log K = 1;\ \Delta_r G^\circ = 0$$

so it follows that for the reaction we want,

$$\frac{1}{2}\mathrm{H_2}(g) + \frac{1}{2}\mathrm{Cl_2}(g) = \mathrm{H^+} + \mathrm{Cl^-}$$

$$\log K = 22.99;\ \Delta_r G^\circ = -131228\ \mathrm{J}$$

At this point we can go no further without making some arbitrary decision, i.e., without formulating a convention for ionic properties. The decision is basically to define the free energy of formation of the chloride ion as

$$\begin{aligned}\Delta_f \bar{G}^\circ_{\mathrm{Cl^-}} &= \bar{G}^\circ_{\mathrm{Cl^-}} + \bar{G}^\circ_{\mathrm{H^+}} - \frac{1}{2}G^\circ_{\mathrm{H_2}} - \frac{1}{2}G^\circ_{\mathrm{Cl_2}} \\ &= -131228\ \mathrm{J}\end{aligned} \tag{17.19}$$

Once this decision is made for one ion, such as $\mathrm{Cl^-}$, the logjam is broken and values for all other ions can be derived. For example, we know that

$$\mathrm{Na}(s) + \frac{1}{2}\mathrm{Cl_2}(g) = \mathrm{Na^+} + \mathrm{Cl^-}$$

$$\Delta_r G^\circ = -393133\ \mathrm{J}$$

so that

$$\bar{G}^\circ_{\mathrm{Na^+}} + \bar{G}^\circ_{\mathrm{Cl^-}} - G^\circ_{\mathrm{Na}} - \frac{1}{2}G^\circ_{\mathrm{Cl_2}} = -393133\ \mathrm{J}$$

Combining this with

$$\bar{G}^\circ_{\mathrm{Cl^-}} - \frac{1}{2}G^\circ_{\mathrm{Cl_2}} = -131228 + \frac{1}{2}G^\circ_{\mathrm{H_2}} - \bar{G}^\circ_{\mathrm{H^+}}$$

we get

$$\begin{aligned}\bar{G}^\circ_{\mathrm{Na^+}} - G^\circ_{\mathrm{Na}} &= -393133 - (\bar{G}^\circ_{\mathrm{Cl^-}} - \frac{1}{2}G^\circ_{\mathrm{Cl_2}}) \\ &= -393133 - (-131228 + \frac{1}{2}G^\circ_{\mathrm{H_2}} - \bar{G}^\circ_{\mathrm{H^+}})\end{aligned}$$

so that

$$\bar{G}^\circ_{\mathrm{Na^+}} - G^\circ_{\mathrm{Na}} + \frac{1}{2}G^\circ_{\mathrm{H_2}} - \bar{G}^\circ_{\mathrm{H^+}} = -261905\ \mathrm{J} \tag{17.20}$$

and this quantity is called $\Delta_f \bar{G}^\circ_{\mathrm{Na^+}}$.

So in general, just as the free energy of formation of a compound is the absolute free energy of the compound *minus* the sum of the free energies of the constituent elements, so the free energy of formation of an ion is the absolute free energy of the ion *minus* the free energy of the corresponding element, *plus* the quantity $Z_j(\frac{1}{2}G^\circ_{\mathrm{H_2}} - \bar{G}^\circ_{\mathrm{H^+}})$, where Z_j is the ionic charge. This is generalized formally as

$$\Delta_f \bar{\Xi}_{j,conventional} = \bar{\Xi}_{j,absolute} - \frac{1}{\nu_i}\Xi_i + Z_j(\frac{1}{2}\Xi_{\mathrm{H_2}} - \bar{\Xi}_{\mathrm{H^+}}) \tag{17.21}$$

where Ξ represents G, H, A, or U, and i represents the pure element, having stoichiometry ν_i, corresponding to ion j. Just as with the "formation from the elements" convention for compounds, the "excess baggage" (in this case $-\frac{1}{\nu_i}\Xi_i + Z_j(\frac{1}{2}\Xi_{H_2} - \bar{\Xi}_{H^+})$) always cancels out in balanced equations.

The same rules apply for all other thermodynamic properties of ions. For instance,

$$\Delta_f \bar{V}^\circ_{Ca^{2+}} = \bar{V}^\circ_{Ca^{2+}} - V^\circ_{Ca} + V^\circ_{H_2} - 2\bar{V}^\circ_{H^+}$$

$$\Delta_f \bar{C}^\circ_{p\,Al^{3+}} = \bar{C}^\circ_{p\,Al^{3+}} - C^\circ_{p\,Al} + \frac{3}{2}C^\circ_{p\,H_2} - 3\bar{C}^\circ_{p\,H^+}$$

$$\Delta_f \bar{S}^\circ_{O^{2-}} = \bar{S}^\circ_{O^{2-}} - \frac{1}{2}S^\circ_{O_2} - S^\circ_{H_2} + 2\bar{S}^\circ_{H^+}$$

Of course, for $\bar{V}$, $\bar{S}$ and $\bar{C}^\circ_p$ for which absolute values for the elements are available, "formation from the element" properties are not usually tabulated or used. Thus for instance

$$\Delta_f \bar{S}^\circ_{Na^+} = \bar{S}^\circ_{Na^+} - S^\circ_{Na} + \frac{1}{2}S^\circ_{H_2} - \bar{S}^\circ_{H^+}$$
$$= 73.132\ \mathrm{J\,K^{-1}\,mol^{-1}}$$

But S°_{Na} is known to be 51.210 J K^{-1} mol^{-1} and $S^\circ_{H_2}$ is 130.684 J K^{-1} mol^{-1}, so

$$\begin{aligned}\bar{S}^\circ_{Na^+,conventional} &= \bar{S}^\circ_{Na^+,absolute} - \bar{S}^\circ_{H^+}\\ &= 73.132 + 51.210 - 130.684/2\\ &= 59.0\ \mathrm{J\,K^{-1}\,mol^{-1}}\end{aligned}$$

Therefore just as for compounds, the properties of the elements are not involved in the definition of the entropies, volumes, and heat capacities of the ions. The formal statement is

$$\bar{\Xi}_{j,conventional} = \bar{\Xi}_{j,absolute} - Z_j\bar{\Xi}_{H^+} \tag{17.22}$$

where $\bar{\Xi}$ represents $\bar{S}$, $\bar{V}$, or $\bar{C}_p$.

17.3.1. *Possible Additional Conventions*

It is possible to effect some simplification in the equations defining the thermodynamic properties of the ions by introducing additional conventions (a convention can be defined somewhat facetiously as a convenient assumption that we know is not true). If, for example, we decide that the absolute free energies and enthalpies of all pure elements are to be set at zero, then the defining equation for free energies and enthalpies (equation 17.21) becomes the same as that for $\bar{S}$, $\bar{V}$, and $\bar{C}_p$ (equation 17.22). If in addition we define all properties of the hydrogen ion as zero, then the conventional ionic properties become the same as the corresponding absolute properties, and we could have stopped at equation (17.19).

Another possibility is to assume that

$$(\frac{1}{2}G^\circ_{H_2} - \bar{G}^\circ_{H^+} - \bar{G}^\circ_e) = 0 \tag{17.23}$$

where G_e° is the free energy of the electron. This is universally adopted in electrochemistry, so as to derive values for half-cell potentials. If (17.23) is adopted, but not necessarily the assumption of zero free energy for the elements, (17.19) becomes

$$\Delta_f \bar{G}^\circ_{\mathrm{Cl}^-} = \bar{G}_{\mathrm{Cl}^-} - \frac{1}{2}\bar{G}_{\mathrm{Cl}_2} - \bar{G}_e$$

which is the free energy change for the reaction

$$\frac{1}{2}\mathrm{Cl}_2 + e = \mathrm{Cl}^-$$

which is the simplest and most straightforward representation of the formation of the chloride ion from the element. Straightforward, that is, if you overlook the fact that the thermodynamic properties of electrons are unknown, and that it cannot serve as a basis for measurement. Nevertheless, (17.23) is a convenient and often-used convention.

We prefer to retain the more complete definitions given here (17.21; 17.22), with no assumptions regarding the properties of the elements and ions. The simplification introduced by these assumptions is minimal, and the possibility for confusion is increased. In addition, it is important to realize that thermodynamics is in no way dependent on the assumption that the energies or enthalpies of the elements are zero, which is obviously untrue, and is one of the many factors lending a veil of uncertainty over thermodynamic proceedings. Nor is it dependent on the truth of equation (17.23), which will be discussed in more detail in the next chapter.

17.4. THE DEBYE-HÜCKEL THEORY

In 1923 Peter Debye and Erich Hückel published two remarkable papers that described an a priori method of calculating the activity coefficient of electrolytic solutes in dilute solution. Without doubt this was one of the major breakthroughs in electrolyte solution theory.

The problem addressed by Debye and Hückel (1923 a,b) was to describe an ionic solution using the primitive model described above (solvent is an ideal structureless dielectric fluid; solute ions are spherically uniform particles with charges located at their centers; the primary interactions are considered to be long-range Coulombic forces). In a non-ionic solution, thermal motions cause all particles to be randomly distributed. However, in ionic solutions, long-range Coulombic forces cause each ion to be surrounded by a fluctuating group of ions of opposite charge, forming an "ionic atmosphere." Without the thermal motion that is always present, the ions in an electrolyte solution would actually assume the ordered structure of an ionic crystal. The model Debye and Hückel used included the competing effects of thermal motion (causing disorder) and Coulombic interaction (promoting order).

The complete derivation of the D-H theory is readily available in many standard references (e.g., Harned and Owen, 1958; Monk, 1961; Robinson and Stokes, 1968; as well as in an English translation of the original papers).[3] It is helpful in using the D-H

[3]"The Collected Papers of P.J. Debye" (1954). Interscience Publishers Inc., New York.

theory to understand the physical model and assumptions used and the corresponding limitations, but it is too lengthy for inclusion here. The final result is

$$\log \gamma_{H_X} = \frac{-A_M |Z_+ Z_-| \sqrt{I_M}}{1 + B_M å \sqrt{I_M}} \tag{17.24}$$

where

$$A_M = \left(\frac{2\pi N_a}{1000} \right)^{\frac{1}{2}} \cdot \frac{e^3}{2.3036 (k\epsilon_0 T)^{\frac{3}{2}}} \tag{17.25}$$

and

$$B_M = \left(\frac{8\pi N_a e^2}{1000 k \epsilon_0 T} \right)^{\frac{1}{2}} \tag{17.26}$$

N_a in the above is Avogadro's number. The quantity I_M in (17.24) is termed the "molar ionic strength" and is given by the sum:

$$I_M = \frac{1}{2} \sum_i M_i Z_i^2 \tag{17.27}$$

Notice that in this form, a mole-fractional (or rational) activity coefficient is calculated from molar concentrations. To be absolutely rigorous this is how the equation would be applied. As we have observed above, the molal concentration scale is by far the most practical for use with aqueous solutions over ranges of T and P. Equation (17.24) can be rewritten in molal units using the conversions discussed above.

The Debye-Hückel Equation for mean molal stoichiometric ion activity coefficients is:

$$\begin{aligned} \log \gamma_{H_m} &= \log \gamma_\pm \\ &= \frac{-|Z_+ Z_-| A \sqrt{I}}{1 + å B \sqrt{I}} \end{aligned} \tag{17.28}$$

where

$$\begin{aligned} A &= \frac{1.8248 \times 10^6 \rho_0^{\frac{1}{2}}}{(\epsilon_0 T)^{\frac{3}{2}}} \\ &= 0.5095 \text{ at } 25°\text{C} \end{aligned} \tag{17.29}$$

and

$$\begin{aligned} B &= \frac{50.292 \rho_0^{\frac{1}{2}}}{(T\epsilon_0)^{\frac{1}{2}}} \\ &= 0.3284 \text{ at } 25°\text{C} \end{aligned} \tag{17.30}$$

The parameters ρ_0 and ϵ_0 refer to the density and dielectric constant of pure water at the T and P of interest. Z_+ and Z_- are the valences of the cation and anion constituents of the salt, å is the D-H distance of closest approach in units of angstroms. The product åB usually approximates unity. The D-H A and B parameters, calculated over a wide range of P and T are tabulated by Helgeson and Kirkham (1974, pp. 1202 and 1256). I is the molal ionic strength, defined by the following sum over all anions and cations:

$$I = \frac{1}{2}\sum_i m_i Z_i^2 \tag{17.31}$$

The D-H equation (17.28), as we have seen, takes into account only long-range electrostatic interactions between ions, and thus cannot be expected to work in solutions above a certain limiting concentration. It may easily be calculated that only 2 or 3 solvent molecules separate individual ions in a 1 molar solution. In practice the D-H equation works reasonably well to concentrations of approximately 10^{-2} to $10^{-1}m$ but no higher.

At very low concentrations the second term in the denominator of (17.28) becomes insignificant and equation (17.28) reduces to

$$\log \gamma_{\pm} = -|Z_+Z_-|A\sqrt{I} \tag{17.32}$$

This is called the Debye-Hückel limiting law and has the advantage of being simpler and not including the adjustable å parameter; however, because of this, it cannot be used at concentrations above approximately 10^{-3} m. It is often used to calculate the theoretically best slope at the zero concentration axis when extrapolating experimental data as a function of ionic strength back to zero concentration.

The å parameter is adjustable and is usually determined by regression-fitting the D-H equation (17.28) to experimental data for individual salts. These have been tabulated for different compounds by Kielland (1937), and Butler (1964, pp. 434-5). These compilations give å for both anion and cation; where these values are not the same, the mean of the å values for cation and anion provides the best fit with observed activities (Butler, 1964, p. 436), and we recommend this approach, although others have also been used.

A well-known problem with using an adjustable å parameter for individual salts or ions is that if this is done for the individual solutes of a multicomponent solution, $dG_{solution}$ will not be an exact differential, because the cross-differentiation criterion applied to the activity terms of the total differential (§2.2.6) will not be satisfied. To satisfy this criterion, the å parameter, or the åB product, in the denominator of the D-H equation should be the same for all solute components of a mixed electrolyte solution. This is one reason why several modifications and extensions of the D-H equation use a constant (often 1.0) for the åB term (Guggenheim, 1935).

17.5. MEASUREMENT OF ACTIVITY COEFFICIENTS

Many different techniques can be used to measure activity coefficients of electrolytic solutes, many of which are electrochemical. Most of these methods are reviewed in standard references (e.g., Harned and Owen, 1958; Monk, 1961; Robinson and Stokes, 1968, Ch. 8). Solutions of geological interest can span considerable temperature ranges, and it has often proven simplest experimentally to measure the activity of the

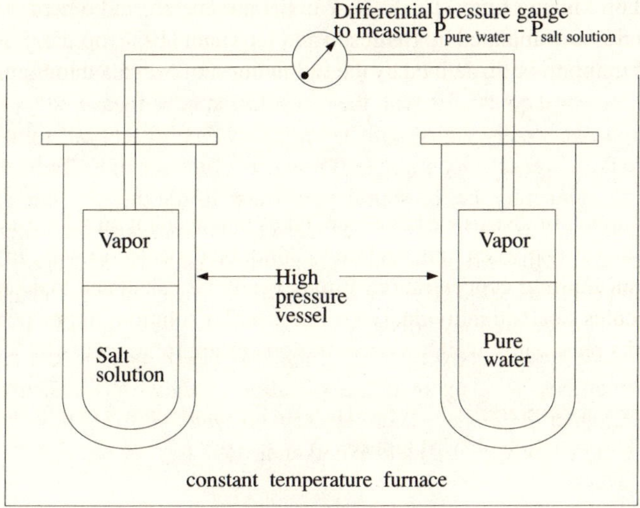

a. Differential Pressure Method

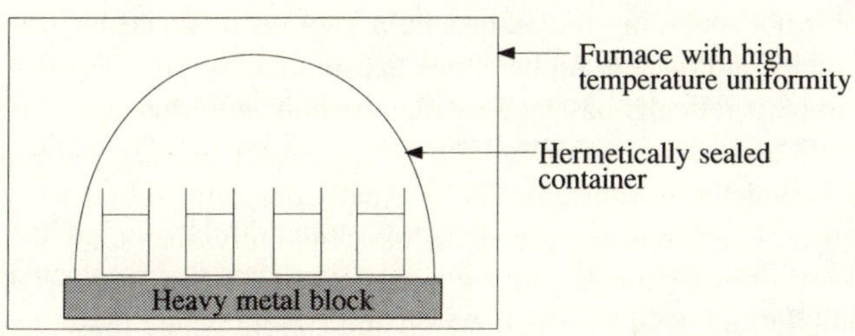

b. Isopiestic Method

FIG. 17.7. Two experimental techniques for measuring stoichiometric ion activity coefficients in hydrothermal solutions. (a) Determination by measuring difference in vapor pressures between solutions of known total salt concentration and pure water. (b) Isopiestic measurement in which sample solutions containing known weights of salts are equilibrated with a standard solution for which activity coefficients have been independently measured. In a sealed system water activity is everywhere the same at equilibrium; measured salt concentrations in each sample container give the desired activity coefficients from the known activity of water in the standard salt solution.

solvent (water) and then use that to calculate solute activities. Most of the available data on activity coefficients of salt components in hydrothermal solutions have been obtained in this general manner.

In all such methods, the solvent activity is measured over a range of solute concentrations. For example, solvent activity for a vapor-saturated aqueous salt solution can be easily determined from the decrease in vapor pressure of the salt solution relative to that of pure water, using the apparatus illustrated in Figure 17.7a. This technique has been used by Liu and Lindsay (1972), Wood et al. (1984), and others to measure activity coefficients of many alkali halides and other simple salts to temperatures of 350°C. Another approach termed the *isopiestic* method is illustrated in Figure 17.7b. In this example several containers with the salt solution to be measured (NaCl in the illustration) are hermetically sealed with a container of a different salt solution (KCl here) for which the water activity is already known as a function of KCl concentration. The fugacity of water must be the same everywhere inside the hermetic system at equilibrium; water will actually enter or leave each container until the water activity in each is identical. At the end of the experiment, the (changed) KCl concentration is measured in that container; this gives the water activity in each of the other containers for the (changed) concentrations of NaCl.

The theoretical method used in relating the water activity to solute activity in each of these methods is an interesting application of the Gibbs-Duhem equation introduced in Chapter 9 (§9.2.6). There are several ways of doing this; for further details see Liu and Lindsay (1972) and Wood et al. (1984).

17.5.1. *Activity Coefficients of Neutral Molecules*

The determination of the activity coefficients of species that exist dominantly as neutral molecules, such as $SiO_2(aq)$, $H_2S(aq)$ and $CO_2(aq)$, is much simpler. In these cases it is usually possible to establish a two-phase equilibrium between the substance in its pure state (solid or gaseous) and the substance in its aqueous or dissolved state. This leads to a simple and rigorous determination of the activity coefficient in solutions of varying composition.

For example, consider H_2S gas in equilibrium with $H_2S(aq)$. The first ionization constant of H_2S is about 10^{-7}, so that species other than molecular H_2S can be neglected in this connection. The reaction of interest is

$$H_2S(g) = H_2S(aq)$$

for which the equilibrium constant is

$$\begin{aligned} K &= a_{H_2S(aq)}/a_{H_2S(g)} \\ &= m_{H_2S}\gamma_{H_2S}/f_{H_2S} \end{aligned}$$

f_{H_2S} can either be taken as equal to the partial pressure of H_2S or calculated from this quantity, and m_{H_2S} is measurable as the total sulfur content of the solution. Then if K is known, γ_{H_2S} can be obtained in any solution simply by measuring f_{H_2S} and m_{H_2S}. For solids such as SiO_2 which dissolve to neutral species, the situation is even simpler, as the activity of the solid can be taken as 1.0. Normally, K is obtained by performing the experiment in pure water, and assuming that in this case, $\gamma_i = 1$. Therefore, activity coefficients obtained in this way are, actually values of

$$\gamma_i(in \text{ solution of interest, i.e., a salt solution})/\gamma_i(in \text{ pure water}).$$

If the "solution of interest" is a solution of salt B in water, having concentration m_B, there is a simple relationship between γ_i and m_B,

$$\log \gamma_i = k_S m_B$$

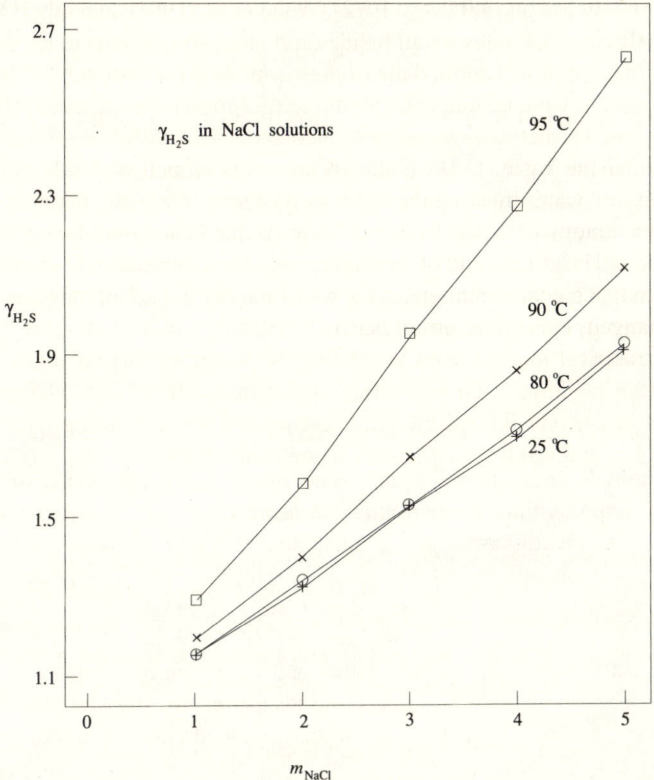

FIG. 17.8. Activity coefficient of $H_2S(aq)$ at various temperatures as a function of NaCl concentration. Data from Barrett et al. (1988).

where k_S is called the Setchénow coefficient. The m_B term has been replaced in some treatments by the "true ionic strength" (§17.7.1).

An extensive review of neutral solutes in aqueous salt solutions is given by Randell and Failey (1927a,b,c. See also Long and McDevit (1952) and Oelkers and Helgeson (1991)). A recent application of this method is the work of Barrett et al. (1988), who obtained activity coefficients for H_2S in NaCl solutions up to $5m$ and 95°C (shown in Figure 17.8), and calculated coefficients for some other gases. As expected, activity coefficients of this type are much closer to 1.0 than those for charged particles in the same situation.

Note that we have only spoken of neutral species of the type that can be obtained as the *dominant* species in a solution; activity coefficients for the neutral species of weak electrolytes and other neutral species in a matrix of charged particles constitute a more difficult problem. Their activity coefficients are usually assumed to be 1.0, or are taken as equal to those of some other neutral species such as H_2S or CO_2 under the same conditions. The activity coefficients of neutral species in electrolyte solutions at elevated temperatures and pressures is a subject of continued study (Oelkers and Helgeson, 1991).

17.6. ACTIVITY COEFFICIENTS AT HIGHER CONCENTRATIONS

Activities predicted by the D-H equation decrease monotonically with solute concentration. Measured activity coefficients typically decrease at first, but then increase at higher concentrations. This indicates that the simple Coulombic "primitive model" used by the D-H theory must be inadequate in more concentrated solutions. There have been many theoretical attempts to model the additional interactions that occur at high concentrations—see summaries by Friedman (1962), Pytkowicz (1979), and Helgeson et al. (1981).

17.6.1. *Factoring Activity Coefficients into Long- and Short-Range Contributions*

Of all the complex interactions which come into play at higher concentrations, two of the most important are ion hydration and ion association. It is easily shown that both the long-range coulombic effect (calculated by the D-H equation) and ion association decrease ion activity, while hydration, which effectively reduces the amount of solvent, increases it. Unfortunately, there is little agreement on hydration numbers for various ions, but using reasonable literature values, Wood et al. (1984) show that mean ionic activity coefficients having the right shape can consistently be calculated at high temperatures and at high concentrations using only these three factors. Using the notation γ_i^e for the electrostatic or long-range ion-ion interaction correction (D-H effect) for ion i, γ_i^a for the correction for ion association, and γ_i^h for the hydration correction, the combined effect is $\gamma^{eha} = \gamma^e \gamma^h \gamma^a$. The relative contributions of these factors at 25°C for NaCl are shown in Figure 17.6. Breaking the activity coefficient into these three factors is thus a useful mental concept, despite the complexity of the actual situation.

17.7. ESTIMATING ACTIVITIES OF INDIVIDUAL IONS

We pointed out above that activity coefficients of individual cations or anions cannot be measured because of the requirement that solutions remain electrically neutral. Instead, the activity coefficient of the total solute (NaCl, $CaCl_2$, etc.) is measured, and that is used to calculate mean activity coefficients, $\gamma_\pm$. These are a kind of average of the activity coefficients for the individual cation and anion and are not the true values for the individual ions. This inability to estimate activity coefficients of individual ions is a problem because it is specific ion activities (rather than activities of total salts) that are most frequently required in thermodynamic calculations.

One possibility is to use the D-H equation (17.28) or any of its variations for a single ion. In this case, the D-H equation becomes:

$$\log \gamma_i = \frac{-A Z_i^2 \sqrt{I}}{1 + \mathring{a} B \sqrt{I}} \tag{17.33}$$

where γ_i, Z_i, and å are properties of the specific ion, but I is the ionic strength (17.31), which includes *all* ionic species. This approach cannot be expected to apply rigorously because, as we have seen, the D-H theory was developed for an ion and its atmosphere of oppositely charged species. Of course, all solutions become more ideal at lower concentrations, so there is some small justification for using this approach with very dilute solutions.

17.7.1. *Activities of Minor Components in Concentrated Solutions*

One of the most difficult problems in working with natural aqueous solutions is assigning activities to trace or minor components that occur in concentrated salt solutions. An example might be calculating the activity coefficient of ppm ($\sim 10^{-5}m$) concentrations of metal ions in hydrothermal solutions containing 1 to 5 molal concentrations of salts such as NaCl, KCl, and $CaCl_2$.

The approach most frequently used by geochemists over the past several decades for calculating activities of minor components in concentrated salt solutions was suggested by Helgeson (1969). This was an outgrowth of earlier work by other eminent chemists such as Scatchard and Harned, summarized by Pitzer and Brewer (1961, pp. 326, 578 and Appendix 4). The idea is to define a deviation function $\dot{B}$ ("B-dot") as the difference between observed and predicted activity coefficients for an electrolyte such as NaCl. This was redefined by Helgeson as

$$\begin{aligned}\dot{B}(I_a) &= \frac{\log \gamma^{obs}_{\pm,\text{NaCl}} + (A|Z_+Z_-|I_a^{\frac{1}{2}}/(1 + B\text{å}_{\text{NaCl}}I_a^{\frac{1}{2}})}{I_a} \\ &= (\log \gamma^{obs}_{\pm,\text{NaCl}} - \log \gamma^{\text{D-H}}_{\pm,\text{NaCl}})/I_a \end{aligned} \tag{17.34}$$

The second term in the numerator is the D-H equation (17.28) and the first term is the experimentally observed activity coefficient for NaCl at the concentration m_{NaCl} of interest. The ionic strength I_a includes a correction for ion association (formation of NaCl°) and is called "true ionic strength" by Helgeson. Various methods of calculating I_a are described by Brimhall and Crerar (1987, p. 280).

$\dot{B}$ has been derived by fitting (17.34) to activity coefficient data for NaCl and is tabulated as a function of T by Helgeson (1969). Helgeson et al. (1981) provided revised values called b_γ for NaCl and several other salts, and for NaCl as a function of P as well as T as part of a more comprehensive study. The activity coefficient of a minor or trace species (say several ppm $FeCl^+$) in a concentrated NaCl solution is then given by rearranging (17.34):

$$\log \gamma_{trace\ i} = \frac{-AZ_i^2\sqrt{I_a}}{1 + \text{å}B\sqrt{I_a}} + \dot{B}(I_a)\cdot I_a \tag{17.35}$$

So (17.35) gives an estimate of the stoichiometric activity coefficient of a trace component, γ^{eha}, which includes the three major non-ideality corrections in a concentrated NaCl solution. In (17.35), I_a is the association-corrected ionic strength including all components of the solution, the parameters Z_i and å refer to the trace component (not NaCl), and $\dot{B}$ is the value for the dominant salt in the solution, NaCl. Because (17.35) gives the stoichiometric activity coefficient, the activity of the trace component would be calculated from its total analytical concentration $m_{i,total}$ in the solution:

$$a_{trace\ i} = \gamma^{eha}_{trace\ i} \cdot m_{total\ i} \tag{17.36}$$

More recently, Helgeson et al. (1981) have revised equation (17.35) so that the $\dot{B}$ parameter explicitly includes a hydration parameter derived from the Born equation and a second parameter for remaining short-range interactions. Some difficulties in using this approach were discussed by Barrett and Anderson (1988). This approach will undoubtedly be further pursued in the next few years.

17.8. CONCENTRATED, MIXED-SALT SOLUTIONS

17.8.1. *A Statistical Mechanical Approach to a Thermodynamic Problem*

Much of the theory underlying the following equations was developed for gases and extended to electrolyte solutions largely by Joseph Mayer (Mayer and Mayer, 1940; McMillan and Mayer, 1945; Mayer, 1950). For a comprehensive summary, see Friedman, (1962).

Some of the necessary background has already been discussed in Chapter 15 (§15.4.4), and we will only outline the major points here. Recall that in theory we could plug expressions for all known interaction energies into the Boltzmann equation (6.9) and we should then know all thermodynamic properties of our system. In reality this is unthinkable because we would generate more than an Avogadro's number of equations for one mole of solution, creating something worse than the famous n-body interaction problem.

Mayer discovered an ingenious way out of this conundrum. Recall from Chapter 15 that the thermodynamic properties of a system of N molecules at volume V and temperature T can be evaluated from the configuration integral Z'

$$Z'(N, \mathrm{V}, T) = \int \cdots \int \exp\left(-\mathrm{U}(x_1, y_1, z_1, \ldots, z_N)/kT\right) dx_1, dy_1, dz_1, \ldots, dz_N \tag{15.36}$$

where U is the total molecular interaction energy and the integration is carried out over the volume V for the coordinates of each molecule. This is an impossible integral, because it is made over the x, y, z coordinates of each molecule, and we could be dealing with an Avogadro's number of molecules!

Fortunately, this can be simplified by factoring the total energy U into terms, each of which represents the interaction energy of a certain number of molecules or "clusters." We can consider interactions of molecules or clusters in groups of two, three, four, etc., and rewrite the configuration integral as the sum of these terms.

$$\exp(-\mathrm{U}/kT) = 1 + \sum_{pairs} \mathrm{U}_{i,j} + \sum_{triples} (\mathrm{U}_{ij} + \mathrm{U}_{jk} + \mathrm{U}_{ij}\mathrm{U}_{jk}\mathrm{U}_{ik}) + \sum_{quadruples} (\mathrm{U}_{ij}\mathrm{U}_{kl} + \cdots) + \cdots \tag{15.37}$$

Now in Chapter 15 we also introduced virial equations of state, which were originally developed to describe the $P - V - T$ properties of non-ideal gases around the turn of the century. An example is the equation

$$\frac{PV}{RT} = 1 + \frac{B_2}{V} + \frac{B_3}{V^2} + \cdots \tag{17.37}$$

It was discovered four decades later that these equations fit real data so well because they had the form of a cluster expansion such as (15.37) above. The first term in a virial equation always represents ideal behavior ($PV/RT = 1$); in the second term, B_2 represents the non-ideal contribution from pairwise interactions of molecules; B_3 gives the interactions of triples, and so on. The virial coefficients can be calculated from known interaction potentials, or, alternatively, can be used to estimate these potentials from observed $P - V - T$ behavior. We observed in Chapter 16 that virial equations fit the non-ideal behavior of gaseous solutions very well, and in Chapter 15 we saw that the Margules equations used to describe non-ideal behavior within solid solutions also have the form of virial equations. It should not be too surprising that expressions with this general form also work well with aqueous electrolyte solutions.

Aqueous solutions can be modeled by writing a virial equation such as (17.37) in which osmotic pressure replaces pressure. Friedman (1962) describes applications of cluster expansion theory, which include long-range Coulombic potentials as well as short-range square-well potentials that operate when unlike ions approach within the diameter of a water molecule. These models are mathematically quite cumbersome and are not easily used for routine calculations. They do predict the non-ideal behavior of simple electrolytes such as NaCl quite admirably at moderate concentrations; however, they use the square-well potential as an adjustable parameter and so retain some of the properties of the D-H equation with an added adjustable term. For this reason these are not truly a priori models.

17.8.2. *The Pitzer Equations*

In the 1970s, Kenneth Pitzer and his associates developed a theoretical model for electrolyte solutions combined the D-H equation with additional terms in the form of a virial equation. This has proven to be extraordinarily successful at fitting the behavior of both single- and mixed-salt solutions to high concentrations. Recent summaries of this model are provided by Pitzer (1979, 1987), Harvie and Weare (1980), and Weare (1987), and much of the following discussion is adapted from these articles, particularly those by Harvie and Weare (1980) and Pitzer (1987).

The Pitzer model adds a virial expansion to a simplified version of the D-H equation and begins by describing the total excess free energy of an electrolyte solution as

$$\mathrm{G}^{EX}_{total}/RT = n_w \left(f(I) + \sum_i \sum_j \lambda_{ij}(I) m_i m_j + \sum_i \sum_j \sum_k \mu_{ijk} m_i m_j m_k \right) \tag{17.38}$$

Here n_w is kilograms of water and m_i is molality of species i. The term $f(I)$ is a version of the D-H equation dependent only on ionic strength (with no adjustable å parameter). The quantities $\lambda_{ij}(I)$ and μ_{ijk} are second and third virial coefficients added to account for short-range interactions at higher concentrations. $\lambda_{ij}(I)$ applies to interactions of pairs of ions i and j, and μ_{ijk} to interactions of ions i, j, k three at a time (μ_{ijk} has been found to be independent of ionic strength and equals zero if i, j, and k are all anions or cations). Pitzer (1987) and Harvie and Weare (1980) note that higher virial coefficients are required only for extremely concentrated solutions, so the series (17.38) is stopped at the third coefficient.

Recall that

$$\left(\frac{\partial \mathrm{G}^{EX}_{total}/n_w RT}{\partial m_i}\right)_{P,T,n_w} = \mu_i^{EX}/RT = \ln \gamma_i \tag{17.39}$$

Equation (17.38) can therefore be rewritten in terms of activity coefficients for specific ions and the osmotic coefficient of the solvent by taking derivatives with respect to the number of moles m_i of each ionic constituent per Kg water (see Pitzer, 1987). These derivatives contain several functions (B, C, Φ and Ψ) defined below, and are as follows.

Osmotic Coefficient

$$\begin{aligned}(\phi - 1) = \frac{2}{\sum_i m_i} \Bigg(& \frac{-A^{\Phi} I^{\frac{3}{2}}}{1 + bI^{\frac{1}{2}}} \\ & + \sum_c \sum_a m_c m_a (B^{\Phi}_{ca} + ZC_{ca}) \\ & + \sum_{c<c'} \sum m_c m_{c'} (\Phi^{\Phi}_{cc'} + \sum_a m_a \Psi_{cc'a}) \\ & + \sum_{a<a'} \sum m_a m_{a'} (\Phi^{\Phi}_{aa'} + \sum_c m_c \Psi_{aa'c}) \Bigg)\end{aligned} \tag{17.40}$$

Activity Coefficients for Cations:

$$\begin{aligned}\ln \gamma_M = z_M^2 F & + \sum_a m_a (2B_{Ma} + ZC_{Ma}) \\ & + \sum_c m_c \left(2\Phi_{Mc} + \sum_a m_a \Psi_{Mca}\right) \\ & + \sum_{a<a'} \sum m_a m_{a'} \Psi_{Maa'} \\ & + |z_M| \sum_c \sum_a m_c m_a C_{ca}\end{aligned} \tag{17.41}$$

Activity Coefficients for Anions

$$\begin{aligned}\ln \gamma_X = z_X^2 F + \sum_c m_c(2B_{cX} + ZC_{cX}) \\ + \sum_a m_a \left(2\Phi_{Xa} + \sum_c m_c \Psi_{Xac}\right) \\ + \sum_{c<c'}\sum m_c m_{c'} \Psi_{cc'X} \\ + |z_X| \sum_c \sum_a m_c m_a C_{ca}\end{aligned} \tag{17.42}$$

In these equations, m_c is the molality of cation c, which has charge z_c (and analogous quantities m_a, a, and z_a apply to anions). The original Pitzer model takes no explicit account of ion association, so all molalities m_i used in the above expressions refer to the total analytic concentration of species i. Subscripts M, c, and c' refer to cations, and X, a, and a' to anions; M and X to the ion being considered, c and a to ions being indexed or summed, and c' and a' to ions other than c and a. Thus, the summation $\sum_c$ is taken over all cation species, but the double summation

$$\sum_{c<c'}\sum$$

is made over all distinguishable pairs of *different cations* c and c' (such as Na^+ and Ca^{2+}, or Na^+ and K^+, etc.). This unusual notation is simply meant to eliminate the inevitable pairing an ion with itself if the usual double summation is used. This will become clear if you work through the example given further on.

The quantities B and C are adjustable parameters derived from measured activity coefficients on single salt + water solutions. The parameters Φ and Ψ come from data on aqueous systems containing two salts. Referring to the original virial equation (17.38), B and Φ are combinations of the second virial coefficients λ_{ij}, and C and Ψ are combinations of the third coefficients μ_{ijk}. These parameters are defined explicitly below.

The last term, first line, in equation (17.40) is a modified D-H equation without the å parameter; Note that the D-H parameter A^Φ used in the Pitzer equations is numerically slightly different from that in the usual D-H equation:

$$A^\Phi = \frac{1}{3}\left(\frac{2\pi N_a \rho_0}{1000}\right)^{\frac{1}{2}} \frac{e^3}{(k\epsilon_0 T)^{\frac{3}{2}}} \tag{17.43}$$

where N_a is Avogadro's constant, e is the electronic charge, k is the Boltzmann constant, and ρ_0 and ϵ_0 are the density and dielectric constant of pure water.

In equations (17.41) and (17.42) for the activity coefficients, F is a function summing the D-H equation and additional terms:

$$F = -A^\Phi \left(\frac{I^{\frac{1}{2}}}{1 + bI^{\frac{1}{2}}} + \frac{2}{b}\ln(1 + bI^{\frac{1}{2}})\right)$$

$$+\sum_c\sum_a m_c m_a B'_{ca} + \sum_{c<c'}\sum m_c m_{c'} \Phi'_{cc'}$$

$$+\sum_{a<a'}\sum m_a m_{a'} \Phi'_{aa'} \tag{17.44}$$

At 25°C, $A^{\Phi} = 0.392$ and b (= åB in D-H) is taken as 1.2. All other parameters in (17.43) and (17.44) are as previously defined.

All coefficients B_{MX} in the Pitzer equations (17.40) to (17.42) and (17.44) vary with ionic strength I. For electrolytes of valence 1-1 and 1-2 (e.g., NaCl and Na_2SO_4) they are written in terms of two regression parameters specific to the electrolyte, $\beta^{(0)}$ and $\beta^{(1)}$, a parameter α which depends on the type of electrolyte (for 1–1, 1–2 and 2–1 salts, $\alpha = 2.0$), and the ionic strength:

$$B^{\Phi}_{MX} = \beta^{(0)}_{MX} + \beta^{(1)}_{MX} e^{-\alpha I^{\frac{1}{2}}} \tag{17.45}$$

$$B_{MX} = \beta^{(0)}_{MX} + \beta^{(1)}_{MX} g(\alpha I^{\frac{1}{2}}) \tag{17.46}$$

$$B'_{MX} = \beta^{(1)}_{MX} g'(\alpha I^{\frac{1}{2}})/I \tag{17.47}$$

The functions g and g' in the above expressions are:

$$g(x) = 2\left(1 - (1+x)e^{-x}\right)/x^2 \tag{17.48}$$

$$g'(x) = -2\left(1 - (1 + x + \frac{1}{2}x^2)e^{-x}\right)/x^2 \tag{17.49}$$

where $x = \alpha I^{\frac{1}{2}}$. The parameters $\beta^{(0)}_{MX}$ and $\beta^{(1)}_{MX}$ are derived from regression fits to osmotic coefficient data for aqueous solutions of single salts (see Wood et al., 1984, p. 676 for an example).

With electrolytes of higher valence such as 2-2 (e.g., $CaSO_4$), the B^{Φ}_{MX} expressions are augmented with an additional term (which is required to account for increased ion association)

$$B^{\Phi}_{MX} = \beta^{(0)}_{MX} + \beta^{(1)}_{MX} e^{-\alpha_1 I^{\frac{1}{2}}} + \beta^{(2)}_{MX} e^{-\alpha_2 I^{\frac{1}{2}}} \tag{17.50}$$

$$B_{MX} = \beta^{(0)}_{MX} + \beta^{(1)}_{MX} g(\alpha_1 I^{\frac{1}{2}}) + \beta^{(2)}_{MX} g(\alpha_2 I^{\frac{1}{2}}) \tag{17.51}$$

$$B'_{MX} = \beta^{(1)}_{MX} g'(\alpha_1 I^{\frac{1}{2}})/I + \beta^{(2)}_{MX} g'(\alpha_2 I^{\frac{1}{2}})/I \tag{17.52}$$

For these higher-valence electrolytes $\alpha_1 = 1.4$ and $\alpha_2 = 12.0$.

The function Z in equations (17.40)–(17.42) is defined as

$$Z = \sum_i m_i |z_i| \tag{17.53}$$

The parameters C_{MX} in equations (17.40) to (17.42) are related to tabulated parameters derived from data on aqueous single salt systems

$$C_{MX} = C^{\Phi}_{MX}/2\sqrt{|z_M z_X|} \tag{17.54}$$

The parameters Φ^{ϕ}_{ij} and Ψ_{ijk} are derived from two-salt data and correct for interactions between ions of the same sign. The Ψ_{ijk} parameters are assumed independent of ionic strength and apply when i and j are different anions and k is a cation, or when i and j are different cations and k is an anion. The second virial coefficient terms Φ^{ϕ}_{ij} are defined by

$$\Phi^{\phi}_{ij} = \theta_{ij} + {}^E\theta_{ij}(I) + I \cdot {}^E\theta'_{ij}(I) \tag{17.55}$$

$$\Phi_{ij} = \theta_{ij} + {}^E\theta_{ij}(I) \tag{17.56}$$

$$\Phi'_{ij} = {}^E\theta'_{ij}(I) \tag{17.57}$$

The terms ${}^E\theta_{ij}(I)$ and ${}^E\theta'_{ij}(I)$ are functions solely of the ionic strength, and account for mixing for unsymmetrical electrolytes when the ions i and j have differing amounts of charge of the same polarity (+ or −). These parameters are zero when i and j have the same charge. Harvie and Weare observed that including these terms significantly improved agreement with real data for complex systems such as Na-Ca-Cl-SO_4-H_2O. Equations used to calculate these terms are given by Pitzer (1975, 1987) and summarized by Harvie and Weare (1980). Pitzer (1987, Appendix A) shows these terms graphically as a function of I, which is useful for checking your calculations.

17.8.3. *Summary of the Pitzer Model*

We have presented the Pitzer equations describing activity and osmotic coefficients of multi-component salt solutions, but probably no brief summary will suffice to get you started if you actually want to use these equations to model real systems. The most useful summaries are probably Pitzer (in Pytkowicz (1979) and in Carmichael and Eugster (1987)) and Weare (in Carmichael and Eugster (1987)).

The truly remarkable thing about the Pitzer equations (17.40) to (17.42) is that the above parameters derived from one- and two-salt systems can be used with extraordinary success to predict behavior in systems containing many more ionic components. Use of the Pitzer equations to predict activities in very complex salt solutions requires data for the single-salt parameters $\beta^{(0)}_{MX}, \beta^{(1)}_{MX}, \beta^{(2)}_{MX}$, and C^{ϕ}_{MX} for all possible combinations of cations M and X in the multi-component system. For example, if the system you wish to model contains Na^+, K^+, Cl^-, and Br^- you would require activity coefficient determinations (which give the above single-salt parameters) for each of the salts NaCl, KCl, NaBr, and KBr dissolved one at a time in water. The two-salt parameters would have to be derived from measurements on systems containing all combinations of these salts taken two at a time. From these relatively simple measurements, it is then possible to predict the thermodynamic behavior of the Na–K–Cl–Br system from dilute solutions to quite high concentrations. This approach has been used for systems containing 10 or more components.

17.8.4. *Example Applications to Concentrated Brines*

One of the first applications of the Pitzer equations to highly complex natural systems were made by Weare, Harvie, and their associates on marine evaporites (Harvie and

Weare, 1980; Eugster et al., 1980; Harvie et al., 1982, 1984; Brantley et al., 1984; Weare, 1987). These calculations, coupled with the free energy minimization method for computing chemical equilibria (described in Chapter 19) successfully predicted the mineral assemblages formed during the evaporation of seawater to almost complete desiccation. It is difficult to conceive of a more complicated test of the Pitzer model. Remember that these calculations represent a prediction from experimental data on systems containing no more than two different salts at a time.

To illustrate the use of the equations, we will take the simpler problem of the solubility of anhydrite in a concentrated NaCl solution. In natural solutions, such as evaporitic brines, you can analyze the total Ca content and the total sulfate, along with everything else, but to compare the ion activity product ($a_{\mathrm{Ca^{2+}}}a_{\mathrm{SO_4^{2-}}}$) to the solubility product in order to determine if the solution is over-saturated or under-saturated, you need the activity coefficients of Ca^{2+} and SO_4^{2-}. This is where Pitzer equations come in. Let's say we want to know if anhydrite is over- or undersaturated in a solution $3m$ in NaCl and $0.01m$ in $CaSO_4$. For the activity coefficient of Ca^{2+} in this solution, equation (17.41) becomes

$$\begin{aligned}\ln\gamma_{\mathrm{Ca^{2+}}} = {} & 4F + m_{\mathrm{Cl}}(2B_{\mathrm{CaCl}} + ZC_{\mathrm{CaCl}}) + m_{\mathrm{SO_4}}(2B_{\mathrm{CaSO_4}} + ZC_{\mathrm{CaSO_4}}) \\ & + m_{\mathrm{Na}}(2\Phi_{\mathrm{CaNa}} + m_{\mathrm{Cl}}\Psi_{\mathrm{CaNaCl}} + m_{\mathrm{SO_4}}\Psi_{\mathrm{CaNaSO_4}}) + m_{\mathrm{Cl}}m_{\mathrm{SO_4}}\Psi_{\mathrm{CaClSO_4}} \\ & + 2(m_{\mathrm{Na}}m_{\mathrm{Cl}}C_{\mathrm{NaCl}} + m_{\mathrm{Na}}m_{\mathrm{SO_4}}C_{\mathrm{NaSO_4}} + m_{\mathrm{Ca}}m_{\mathrm{Cl}}C_{\mathrm{CaCl}} + m_{\mathrm{Ca}}m_{\mathrm{SO_4}}C_{\mathrm{CaSO_4}})\end{aligned}$$

The function F is defined in equation (17.44), and on expanding is

$$\begin{aligned}F = {} & -A^{\Phi}\left(\frac{I^{\frac{1}{2}}}{1 + bI^{\frac{1}{2}}} + \frac{2}{b}\ln(1 + bI^{\frac{1}{2}})\right) \\ & + m_{\mathrm{Na}}m_{\mathrm{Cl}}B'_{\mathrm{NaCl}} + m_{\mathrm{Na}}m_{\mathrm{SO_4}}B'_{\mathrm{NaSO_4}} + m_{\mathrm{Ca}}m_{\mathrm{Cl}}B'_{\mathrm{CaCl}} + m_{\mathrm{Ca}}m_{\mathrm{SO_4}}B'_{\mathrm{CaSO_4}} \\ & + m_{\mathrm{Na}}m_{\mathrm{Ca}}\Phi'_{\mathrm{NaCa}} + m_{\mathrm{Cl}}m_{\mathrm{SO_4}}\Phi'_{\mathrm{ClSO_4}}\end{aligned}$$

The equation giving $\ln\gamma_{\mathrm{SO_4^{2-}}}$ (17.42) looks quite similar. Filling in the values of the parameters from Table 17.5 and solving, we find $\gamma_{\mathrm{Ca^{2+}}} = 0.3499$, $\gamma_{\mathrm{SO_4^{2-}}} = 0.03190$, giving an Ion Activity Product (IAP) of $0.01 \times 0.3499 \times 0.01 \times 0.0319$ or 1.11×10^{-6}. The solubility product K_{sp} is 4.96×10^{-5}, so the solution is undersaturated with anhydrite. Further details of the calculation are given in the problems at the end of the chapter. Monnin (1990) has done the calculation for the same solution with a slightly different result, $((\gamma_{\mathrm{Ca^{2+}}}\gamma_{\mathrm{SO_4^{2+}}})^{\frac{1}{2}} = 0.1130)$, because he included provision for a $CaSO_4^\circ$ ion pair after the work of Møller (1988). Many such variations and new applications are now appearing, because of the evident accuracy and usefulness of this approach.

17.8.5. *Application to Other Thermodynamic Properties*

The same methods used to describe activity coefficients here can also be applied to other thermodynamic properties such as excess volumes, enthalpies, entropies, heat capacities, and so on by manipulating the defining equation (17.38) appropriately (see Pitzer, 1987). Experimental data useful in deriving the ion-interaction parameters of

Table 17.5 Parameters for the Pitzer Equations (Pitzer, 1987)

Single Electrolyte Solution Parameter Values

Cation	**Anion**	$\beta_{ca}^{(0)}$	$\beta_{ca}^{(1)}$	$\beta_{ca}^{(2)}$	C_{ca}^{ϕ}
Na	Cl	.0765	.2644	—	.00127
Na	SO_4	.01958	1.113	—	.00497
Ca	Cl	.3159	1.614	—	−0.00034
Ca	SO_4	.20	3.1973	−54.24	—

Common-Ion Two Electrolyte Parameter Values

c	c'	$\theta_{cc'}$	$\Psi_{cc'\mathrm{Cl}}$	$\Psi_{cc'\mathrm{SO_4}}$	$\Psi_{cc'\mathrm{HSO_4}}$	$\Psi_{cc'\mathrm{OH}}$	$\Psi_{cc'\mathrm{HCO_3}}$	$\Psi_{cc'\mathrm{CO_3}}$
Na	Ca	.07	−.007	−.055	—	—	—	—
a	a'	$\theta_{aa'}$	$\Psi_{aa'\mathrm{Na}}$	$\Psi_{aa'\mathrm{K}}$	$\Psi_{aa'\mathrm{Ca}}$	$\Psi_{aa'\mathrm{Mg}}$	$\Psi_{aa'\mathrm{MgOH}}$	$\Psi_{aa'\mathrm{H}}$
Cl	SO_4	0.03	0.00	−0.005	−0.002	−0.008	—	—

the Pitzer equations can come from such sources as freezing point depression, boiling point elevation, vapor pressure, isopiestic concentrations, E.M.F., enthalpy of dilution, enthalpy of mixing, enthalpy of solution, heat capacity, volumetric data, and solubility. With such data and the general Pitzer model it then becomes possible to quantitatively describe most physical and thermodynamic properties of complex, multi-component systems based on data collected in simpler one- and two-solute aqueous solutions. This represents a truly major advance in our ability to model highly complicated aqueous electrolyte solutions.

For example, excess volumes can be derived from the derivative

$$\left(\frac{\partial \mathrm{G}^{EX}/n_w}{\partial P}\right)_{T,n_w} = V^{EX} \tag{17.58}$$

This approach has been used quite successfully, for example, by Christophe Monnin (1989) to model densities of concentrated, multi-component natural brines based on volumetric data from one- and two-salt systems. Monnin's predictions agree very well with observations made on natural brines in settings such as the Red Sea and Persian Gulf sabkha regions.

17.9. THE HKF MODEL FOR AQUEOUS ELECTROLYTES

The other major contribution to the systematization of our knowledge of aqueous electrolyte solutions at elevated temperatures and pressures takes a completely different approach. This was presented in a series of four papers by H.C. Helgeson and co-workers between 1974 and 1981, with fairly extensive modifications added by Tanger and Helgeson (1988). We present here an outline of this model, with some explanation and comparison with the Pitzer model. We refer to it as the HKF or revised

HKF model, after the three authors of Part IV of the series of papers just mentioned, Helgeson, Kirkham, and Flowers.

17.9.1. *Overall Structure of the HKF Model*

The HKF model is semi-empirical, in the sense that it uses a number of empirical parameters within a framework suggested by fundamental physics and thermodynamics. The variation of the Gibbs free energy of individual ions with T, P, and composition can be represented by writing the total differential of the (partial molar) free energy of the jth ion, giving

$$d\bar{G}_j = -\bar{S}_j dT + \bar{V}_j dP + (\partial\mu_j/\partial n_j)dn_j$$

where j is an ion in an aqueous solution of any composition. Changes in $\bar{G}_j$ due to changes in T, P or n_j are found by integrating this equation,

$$\int_{T_r,P_r,n_0}^{T,P,n} d\bar{G}_j = -\int_{T_r}^{T} \bar{S}_{j,P_r,n_0} dT + \int_{P_r}^{P} \bar{V}_{j,T,n_0} dP + \int_{n_0}^{n} d\mu_{j,T,P} \qquad (17.59)$$

This is illustrated in Figure 17.9. The first two integrals on the right-hand side take place at concentration n_0. If we equate this with the standard state (ideal one molal solution), then $\bar{S}_j$ and $\bar{V}_j$ become $\bar{S}_j^\circ$ and $\bar{V}_j^\circ$ respectively. The third integral takes care of departures from standard state conditions (change of composition) at T and P. Integration of $\int_{T_r}^{T} \bar{S}_j^\circ \, dT$ requires knowledge of $\bar{C}_{p\,j}^\circ$, and integration of $\int_{n_0}^{n} d\mu_{j,T,P}$ $(= \mu_j - \mu_j^\circ = RT \ln a_j)$ requires knowledge of γ_j, so that the minimum information needed to know how the Gibbs energy of an ion j varies with T, P, and composition is how $\bar{C}_{p\,j}^\circ$ and $\bar{V}_j^\circ$ vary with T and P respectively, and how γ_j varies with composition.

17.9.2. *The Born Functions*

Born (1920) showed that free energy change associated with removing an ion j of radius r_j and charge Z_je from a vacuum and placing it in a solvent of dielectric constant ϵ is

$$\Delta\bar{G}_{s,j}^\circ = \frac{N_a(Z_je)^2}{2r_j}\left(\frac{1}{\epsilon} - 1\right)$$

where N_a is Avogadro's number. For an enlightening discussion of this and related topics, see Bockris and Reddy (1970). This simple equation, based on coulombic forces only, predicts enthalpies of hydration that are in reasonable agreement with experiment. That is, reasonable in the sense that they are of the right order of magnitude, but they are otherwise rather inaccurate. This is not surprising, because the hydration process must be much more complicated than provided for in the Born model. However the fact that it succeeds as well as it does suggests that it contains a large part of the truth, and might serve as the basis of a more satisfactory model. In fact it serves as the basis of the HKF model, which uses the Born function to provide the free energy of solvation (or hydration in aqueous solvent) in a strict sense (not including the energy effects of disrupting the water structure in the neighbourhood

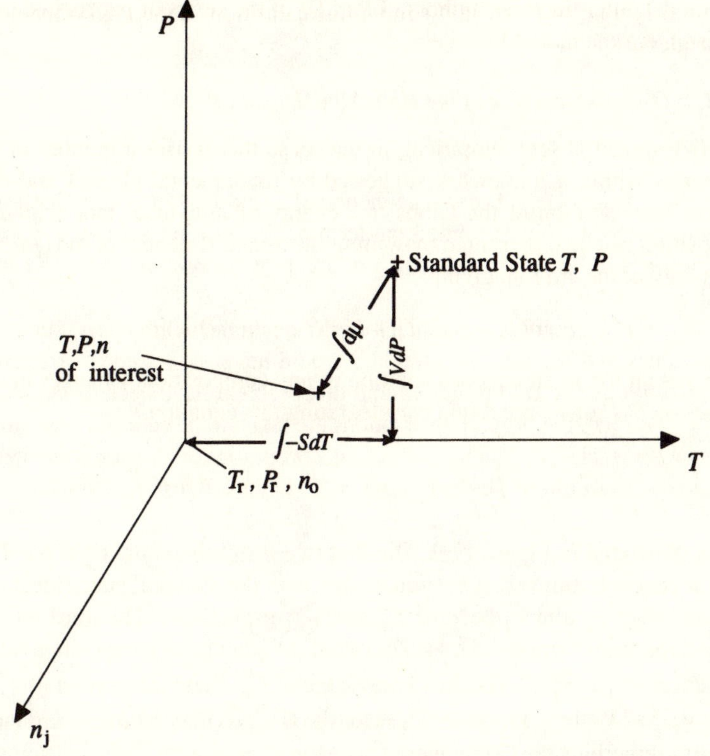

FIG. 17.9. Representation of the integration of $d\bar{G}_j$ from reference conditions to conditions of interest, where j is an ion in an aqueous solution of any composition. The path from T_r, P_r, n_0 to T, P, n is represented by equation (17.59).

of the hydrated ion). In other words, the Born function is taken as providing $\Delta\bar{G}$ for the process

$$\left\{\begin{array}{c} j \text{ in water with disrupted local} \\ \text{structure but before} \\ \text{formation of inner sphere} \\ \text{of strongly attached water molecules} \end{array}\right\} \rightarrow \left\{\begin{array}{c} j \text{ in water with inner} \\ \text{hydration sphere formed,} \\ \text{plus local disrupted} \\ \text{structure} \end{array}\right\}$$

A number of theoretical difficulties in equating the Born function with this process are believed to be accommodated in the $r_{e,j}$ parameter, which in the Born model is the ion radius, but in the HKF model is an adjustable parameter called the *effective* ionic radius. The $r_{e,j}$ parameters were originally related to crystallographic ionic radii

($r_{x,j}$) and ionic charge Z_j in a simple linear fashion in the HKF model, and were independent of T and P

$$r_{e,j} = r_{x,j} + 0.94|Z_j| \text{ for cations}$$

$$r_{e,j} = r_{x,j} \text{ for anions}$$

In the revised model, $r_{e,j}$ became a function of T and P (Tanger and Helgeson, 1988), and is discussed below. An empirical term having several adjustable parameters to account for all other effects of adding an ion to water is then added to complete the model.

The Born function for individual ions is calculable theoretically, but as properties of individual ions are not measurable, it could never be tested experimentally. The two conventions required to get around this problem have been discussed earlier in this chapter. First, the properties of anions are taken to be equivalent to those of their corresponding acids, so that for instance the properties of Cl^- are equivalent to those of HCl(aq). Therefore we define a "conventional" electrostatic Born parameter as

$$\begin{aligned} \omega_j &= \frac{N_a(Z_j e)^2}{2r_{e,j}} - \left(\frac{N_a(Z_{H^+}e)^2}{2r_{e,H^+}}\right) \cdot Z_j \\ &= \frac{N_a(Z_j e)^2}{2r_{e,j}} - 0.5387 Z_j \end{aligned} \tag{17.60}$$

where 0.5387 is the value of $\frac{N_a(Z_j e)^2}{2r_{e,j}}$ for the H^+ ion (Helgeson and Kirkham, 1976). The Born function then becomes

$$\Delta\bar{G}^\circ_{s,j} = \omega_j \left(\frac{1}{\epsilon} - 1\right) \tag{17.61}$$

which is the difference between the $\Delta\bar{G}$ of solvation of ion j (j in vacuum $\rightarrow$ j in water)and the $\Delta\bar{G}$ of solvation of H^+ (H^+ in vacuum $\rightarrow$ H^+ in water). As described above, in the HKF model, this is equated with the $\Delta\bar{G}$ of formation of the inner hydration sphere only. This is sometimes called the *conventional* $\Delta\bar{G}$ of solvation, although since it is the only one we can deal with in a practical way, it is more often just called the $\Delta\bar{G}$ of solvation.

The second convention required is that the properties of individual ions are related to the properties of electrolytes by the additivity rule

$$\bar{\Xi}^\circ_k = \sum_j \nu_{j,k} \bar{\Xi}^\circ_j \tag{17.62}$$

where $\bar{\Xi}$ is any partial molar property and $\nu_{j,k}$ represents the stoichiometry of the jth ion in the kth electrolyte. An example would be $\bar{V}^\circ_{CaCl_2} = \bar{V}^\circ_{Ca^{2+}} + 2\bar{V}^\circ_{Cl^-}$. Thus for example, because the properties ($\bar{\Xi}$) of the Cl^- ion are identified with those of HCl they are "known," and $\bar{\Xi}$ for the Ca^{2+} ion is then obtained from

$$\bar{\Xi}_{Ca^{2+}} = \bar{\Xi}_{CaCl_2} - 2\bar{\Xi}_{Cl^-}$$

In summary, because properties are additive, and the properties of anions are known by convention, then all ionic properties are knowable, in the conventional sense.

It follows from the additivity convention that we can write an analogous Born function for aqueous electrolytes, k, as

$$\Delta\bar{G}^{\circ}_{s,k} = \omega_k \left(\frac{1}{\epsilon} - 1\right)$$

where

$$\omega_k = \sum_j \nu_{j,k}\omega_j$$

In other words, ω_k for electrolytes is simply the sum of its ω_j terms, e.g.

$$\omega_{\mathrm{CaCl_2}} = \omega_{\mathrm{Ca^{2+}}} + 2\omega_{\mathrm{Cl^-}}$$

(Note that in this summation, the properties of the hydrogen ion cancel out). It follows too that the fit parameters in the model to be described are also additive, so that the procedure is to develop the fit parameters for data on electrolytes, then to split them into additive parameters for individual ions. In the following equations, if there is no j or k subscript, the equation applies to both, as long as the appropriate parameters are used.

17.9.3. *The $r_{e,j}$ Term*

The (conventional) entropy and volume changes of solvation according to the Born model are then derived by differentiation, which in the original HKF model was a simple procedure because $r_{e,j}$ was independent of T, P, which meant that ω_j was also independent of T, P. Thus in the original model,

$$\begin{aligned}\Delta\bar{S}^{\circ}_s &= -\left(\frac{\partial\Delta\bar{G}^{\circ}_s}{\partial T}\right)\\ &= -\omega\frac{\partial}{\partial T}\left(\frac{1}{\epsilon} - 1\right)\\ &= \omega Y \qquad (17.63)\end{aligned}$$

and

$$\begin{aligned}\Delta\bar{V}^{\circ}_s &= \left(\frac{\partial\Delta\bar{G}^{\circ}_s}{\partial P}\right)\\ &= \omega\frac{\partial}{\partial P}\left(\frac{1}{\epsilon} - 1\right)\\ &= -\omega Q \qquad (17.64)\end{aligned}$$

and

$$\Delta\bar{C}^{\circ}_{P\,s} = T\left(\frac{\partial\Delta\bar{S}^{\circ}_s}{\partial T}\right)$$

$$
\begin{aligned}
&= \omega T \frac{\partial Y}{\partial T} \\
&= \omega T X \qquad (17.65)
\end{aligned}
$$

where Y, Q, and X are Born coefficients, i.e., temperature and pressure derivatives of the dielectric constant of water, derived and calculated by Helgeson and Kirkham (1974a). However, in the revised model, $r_{e,j}$ becomes a rather complex function of T and P, giving rather more complex expressions for the solvation terms, containing partial derivatives of ω_j. It is helpful in seeing the overall structure of this model to look first at the simple forms of the solvation contributions (above), and then to regard the more complex forms as unfortunate complications.

In the revised model,

$$
\left.
\begin{aligned}
r_{e,j} &= r_{x,j} + |Z_j|(0.94 + g) \ \text{for cations} \\
r_{e,j} &= r_{x,j} + |Z_j|g \ \text{for anions}
\end{aligned}
\right\} \qquad (17.66)
$$

where, as before, Z_j is the ionic charge and $r_{x,j}$ is the crystallographic ionic radius. The added term g is a complex function of T and P obtained by regression of volume and heat capacity data for aqueous NaCl. This results of course in more complex expressions for the solvation terms, which are (Tanger and Helgeson 1988, Appendix G)

$$
\begin{aligned}
\Delta \bar{S}^\circ_s &= -\left(\frac{\partial \Delta \bar{G}^\circ_s}{\partial T}\right)_P \\
&= \omega Y - \left(\frac{1}{\epsilon} - 1\right)\left(\frac{\partial \omega}{\partial T}\right)_P \qquad (17.67)
\end{aligned}
$$

$$
\begin{aligned}
\Delta \bar{V}^\circ_s &= \left(\frac{\partial \Delta \bar{G}^\circ_s}{\partial P}\right)_T \\
&= -\omega Q + \left(\frac{1}{\epsilon} - 1\right)\left(\frac{\partial \omega}{\partial P}\right)_T \qquad (17.68)
\end{aligned}
$$

and

$$
\begin{aligned}
\Delta \bar{C}^\circ_{p\,s} &= T\left(\frac{\partial \Delta \bar{S}^\circ_s}{\partial T}\right) \\
&= \omega T X + 2TY\left(\frac{\partial \omega}{\partial T}\right)_P - T\left(\frac{1}{\epsilon} - 1\right)\left(\frac{\partial^2 \omega}{\partial T^2}\right)_P \qquad (17.69)
\end{aligned}
$$

Note that in regions where $g = 0$ (essentially where $T < 150°\text{C}$ or $P > 2000$ bars), the T and P derivatives of ω in these expressions become zero (because $r_{e,j}$ is independent of T, P when $g = 0$), and the solvation terms resume their original fairly simple forms.

The $r_{e,j}$ term in the HKF model essentially takes the place of the å term in the D-H model, and because it is different for each ion, the overall model suffers from the problem mentioned in §17.4, i.e., that $dG_{solution}$ is not an exact differential. However, the authors consider that the error introduced is acceptable in view of other sources of uncertainty.

17.9.4. *The Empirical Part*

After defining the Born function as described above, comparison of experimental values of $\Delta\bar{V}^\circ$ and $\Delta\bar{C}_p^\circ$ with calculated values of $\Delta\bar{V}_s^\circ$ and $\Delta\bar{C}_{p\,s}^\circ$ showed that the discrepancies could be fitted with functions of the form

$$\Delta\bar{V}_n^\circ = a_1 + a_2 f(P) + a_3 f_1(T) + a_4 f(P) f_1(T) \tag{17.70}$$

and

$$\Delta\bar{C}_{p\,n}^\circ = c_1 + c_2 f_2(T) \tag{17.71}$$

where subscript n stands for non-solvation, and

$$f_1(T) = 1/(T - \Theta)$$

$$f_2(T) = 1/(T - \Theta)^2$$

$$f(P) = 1/(\Psi + P)$$

In the original HKF model, Θ was a fit parameter for each ion having values usually ranging from about 200 to 260 K. Recent studies of supercooled water reviewed by Angell (1982, 1983; references in Tanger and Helgeson, 1988) however show that 228 ± 3 K is a singular temperature at which several properties approach $\pm\infty$, and in the revised model Θ takes on the fixed value of 228 K. The Ψ parameter is also fixed at 2600 bars.

17.9.5. *The Model for $\Delta\bar{V}^\circ$ and $\Delta\bar{C}_p^\circ$*

Combining the Born and empirical parts of the model gives

$$\begin{aligned}\bar{V}^\circ &= \Delta\bar{V}_n^\circ + \Delta\bar{V}_s^\circ \\ &= \underbrace{a_1 + \frac{a_2}{\Psi + P} + \frac{a_3}{T - \Theta} + \frac{a_4}{(\Psi + P)(T - \Theta)}}_{non-solvation\ part} \\ &\quad \underbrace{- \omega Q + \left(\frac{1}{\epsilon} - 1\right)\left(\frac{\partial \omega}{\partial P}\right)_T}_{solvation\ part}\end{aligned} \tag{17.72}$$

for the (conventional) standard partial molar volume of ion j or electrolyte k as a function of T and P, and

$$\bar{C}_p^\circ = \Delta\bar{C}_{p\,n}^\circ + \Delta\bar{C}_{p\,s}^\circ$$

$$= \underbrace{c_1 + \frac{c_2}{(T-\Theta)^2}}_{non-solvation\ part}$$

$$\underbrace{+\omega TX + 2TY\left(\frac{\partial\omega}{\partial T}\right)_P - T\left(\frac{1}{\epsilon}-1\right)\left(\frac{\partial^2\omega}{\partial T^2}\right)_P}_{solvation\ part} \tag{17.73}$$

for the (conventional) standard partial molar heat capacity of ion j or electrolyte k as a function of T only.

No extra effort need be expended to determine the effect of pressure on $\Delta\bar{C}_p^\circ$ because this information is included in the expression for partial molar volume already obtained. That is, because

$$\left(\frac{\partial\bar{S}}{\partial P}\right) = -\left(\frac{\partial\bar{V}}{\partial T}\right)$$

it follows that

$$\left(\frac{\partial\bar{C}_p^\circ}{\partial P}\right) = -T\left(\frac{\partial^2\bar{V}^\circ}{\partial T^2}\right)$$

This gives

$$\bar{C}^\circ_{p\ P} - \bar{C}^\circ_{p\ P_r} = \int_{P_r}^{P} -T\left(\frac{\partial^2\bar{V}^\circ}{\partial T^2}\right)dP$$

or

$$\bar{C}^\circ_{p\ P,T} = \bar{C}^\circ_{p\ P_r,T} + \int_{P_r}^{P} -T\left(\frac{\partial^2\bar{V}^\circ}{\partial T^2}\right)dP$$

which on integration turns out to be

$$\bar{C}^\circ_{p\ P,T} = \bar{C}^\circ_p$$

$$= \underbrace{c_1 + \frac{c_2}{(T-\Theta)^2} - \left(\frac{2T}{(T-\Theta)^3}\right)\left[a_3(P-P_r) + a_4\ln\left(\frac{\Psi+P}{\Psi+P_r}\right)\right]}_{non-solvation\ part}$$

$$\underbrace{+\omega TX + 2TY\left(\frac{\partial\omega}{\partial T}\right)_P - T\left(\frac{1}{\epsilon}-1\right)\left(\frac{\partial^2\omega}{\partial T^2}\right)_P}_{solvation\ part} \tag{17.74}$$

17.9.6. *Expressions for $\Delta\bar{S}^\circ$, $\Delta\bar{H}^\circ$ and $\Delta\bar{G}^\circ$*

Having expressions for the temperature and pressure effects on $\bar{C}_p^\circ$ and $\bar{V}^\circ$, straightforward, if somewhat lengthy, integration gives expressions for $\Delta\bar{S}^\circ$, $\Delta\bar{H}^\circ$, and $\Delta\bar{G}^\circ$, which can refer either to an ion j or an electrolyte k, depending on the fit parameters used in the expression. Thus

$$\bar{S}^\circ_{P,T} - \bar{S}^\circ_{P_r,T_r} = \int_{T_r}^{T}\frac{\bar{C}_p^\circ}{T}dT - \int_{P_r}^{P}\left(\left(\frac{\partial\bar{V}^\circ}{\partial T}\right)_P\right)_T dP \qquad (cf.7.34)$$

$$
\begin{aligned}
&= c_1 \ln\left(\frac{T}{T_r}\right) \\
&\quad -\frac{c_2}{\Theta}\left[\left(\frac{1}{T-\Theta}\right)-\left(\frac{1}{T_r-\Theta}\right)+\frac{1}{\Theta}\cdot\ln\left(\frac{T_r(T-\Theta)}{T(T_r-\Theta)}\right)\right] \\
&\quad +\left(\frac{1}{T-\Theta}\right)^2\left[a_3(P-P_r)+a_4\ \ln\left(\frac{\Psi+P}{\Psi+P_r}\right)\right] \\
&\quad +\omega Y-\left(\frac{1}{\epsilon}-1\right)\left(\frac{\partial\omega}{\partial T}\right)_P-\omega_{P_r,T_r}Y_{P_r,T_r} \qquad (17.75)
\end{aligned}
$$

$$
\begin{aligned}
\bar{H}^\circ_{P,T}-\bar{H}^\circ_{P_r,T_r} &= \int_{T_r}^{T}\bar{C}^\circ_p dT+\int_{P_r}^{P}\left[\bar{V}^\circ T\left(\frac{\partial\bar{V}^\circ}{\partial T}\right)_P\right]_T dP \qquad (cf.7.33) \\
&= c_1(T-T_r) \\
&\quad -c_2\left[\left(\frac{1}{T-\Theta}\right)-\left(\frac{1}{T_r-\Theta}\right)\right]+a_1(P-P_r)+a_2\ln\left(\frac{\Psi+P}{\Psi+P_r}\right) \\
&\quad +\left(\frac{2T-\Theta}{(T-\Theta)^2}\right)\left[a_3(P-P_r)+a_4\ \ln\left(\frac{\Psi+P}{\Psi+P_r}\right)\right] \\
&\quad +\omega\left(\frac{1}{\epsilon}-1\right)+\omega TY-T\left(\frac{1}{\epsilon}-1\right)\left(\frac{\partial\omega}{\partial T}\right)_P \\
&\quad -\omega_{P_r,T_r}\left(\frac{1}{\epsilon_{P_r,T_r}}-1\right)-\omega_{P_r,T_r}T_rY_{P_r,T_r} \qquad (17.76)
\end{aligned}
$$

$$
\begin{aligned}
\bar{G}^\circ_{P,T}-\bar{G}^\circ_{P_r,T_r} &= -\bar{S}^\circ_{P_r,T_r}(T-T_r)+\int_{T_r}^{T}\bar{C}^\circ_p dT-\int_{T_r}^{T}\frac{\bar{C}^\circ_p}{T}dT+\int_{P_r}^{P}\bar{V}^\circ dP \qquad (cf.7.22) \\
&= -\bar{S}^\circ_{P_r,T_r}(T-T_r)-c_1\left[T\ln\left(\frac{T}{T_r}\right)-T+T_r\right] \\
&\quad +a_1(P-P_r)+a_2\ \ln\left(\frac{\Psi+P}{\Psi+P_r}\right) \\
&\quad -c_2\left(\left[\left(\frac{1}{T-\Theta}\right)-\left(\frac{1}{T_r-\Theta}\right)\right]\left(\frac{\Theta-T}{\Theta}\right)-\frac{T}{\Theta^2}\ln\left(\frac{T_r(T-\Theta)}{T(T_r-\Theta)}\right)\right) \\
&\quad +\left(\frac{1}{T-\Theta}\right)\left[a_3(P-P_r)+a_4\ \ln\left(\frac{\Psi+P}{\Psi+P_r}\right)\right]+\omega\left(\frac{1}{\epsilon}-1\right) \\
&\quad -\omega_{P_r,T_r}\left(\frac{1}{\epsilon_{P_r,T_r}}-1\right)+\omega_{P_r,T_r}Y_{P_r,T_r}(T-T_r) \qquad (17.77)
\end{aligned}
$$

At this point we have shown how the HKF model develops expressions for the standard state parameters $\bar{V}^\circ$ and $\bar{C}^\circ_p$ and hence $\bar{S}^\circ$, $\bar{H}^\circ$, and $\bar{G}^\circ$ at high temperatures and pressures. The standard state universally used is the ideal one molal solution, which means that these parameters refer to the properties of ions or electrolytes in infinitely dilute solutions. You might suppose that therefore they would not be of much use to geochemists interested in natural solutions, which are often quite concentrated, but you would be wrong. The standard state properties allow the calculation of the equilibrium constant for reactions involving ions at high T, P, and thus permit the general nature of many important processes to be understood, even in cases where activity coefficients are unknown. Of course for quantitative calculation of ionic concentrations and mass transfers in such cases, activity coefficients are also required.

17.9.7. *Contributions of the Solvation and Non-solvation Terms*

A striking feature of the partial molar volumes and heat capacities of aqueous electrolytes is their inverted-U shape as a function of temperature. Experimental data that cover a sufficiently large range of temperature invariably exhibit a maximum, generally somewhere between 50 and 100°C. This is illustrated in Figure 17.10, which shows the HKF representation of the partial molar volume and heat capacity of NaCl. There is no universally accepted reason for this maximum, although the existence of singular temperatures for water at -45°C (228 K, Angell, 1982,1983) and 374°C (the critical temperature) makes it seem entirely reasonable that thermodynamic parameters of solutes in water should approach $\pm\infty$ at these limits, and therefore reasonable that they should exhibit extrema (or inflection points) between these temperatures.[4]

The revised HKF model is constructed such that the non-solvation contribution to $\bar{V}^\circ$ and $\bar{C}_p^\circ$ dominates at low temperatures and becomes $-\infty$ at 228 K, and the solvation contribution dominates at high temperatures. The contributions of the solvation and non-solvation parts of the partial molar volume of Na^+ are compared in Figure 17.11, and in Figure 17.12 the solvation and non-solvation contributions to the partial molar heat capacity of HCl are shown as a function of temperature. This illustrates quite nicely how the two contributions combine to produce a maximum, and it can easily be imagined how the shape of the combined curve is controlled by the fit parameters of the two contributions. Of course, the two contributions do not always cross in such a pedagogically convenient way. In Figure 17.13 we show the two contributions to the partial molar volume of Cl^- as a function of temperature; the same features are present, but the relative contributions of the two parts of the model are quite different.

17.9.8. *Illustration of the Equivalence of the Properties of Aqueous* HCl *and* Cl^-

The availability of the HKF model makes simple the demonstration of a feature of single ion properties that we have explained previously. By convention the properties of aqueous anions are taken to be those of the corresponding strong acid, unless there are complicating factors. This means, for example, that the properties of the chloride ion are those of aqueous HCl, and the properties of the sulfate ion are those of aqueous H_2SO_4. The conventions involved in this were discussed in §17.6.1.

[4]Don't confuse the properties of *component* NaCl in Figures 17.10, 17.14 with the properties of *species* $NaCl^\circ$ in program SUPCRT92 (Johnson, et al., 1992). Properties of component NaCl are given in the original HKF publications.

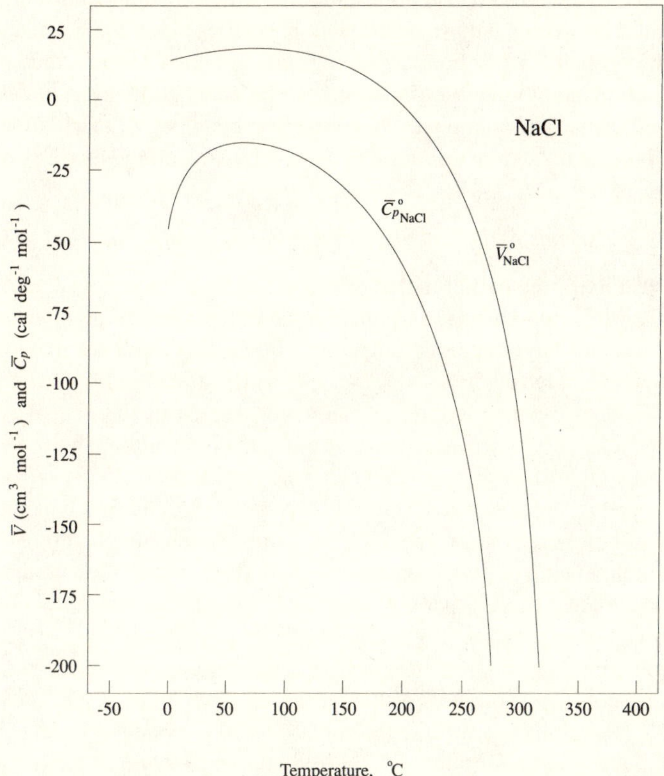

FIG. 17.10. The standard partial molar volume and standard partial molar heat capacity of aqueous NaCl as represented by the HKF model, showing the characteristic inverted-U shape and steep negative slopes at high and low temperatures.

In Figure 17.14 we show the HKF representation of the partial molar heat capacities of Na^+, Cl^-, and NaCl, together with the experimental data. The data shown for the chloride ion are in fact the same data of Tremaine, Sway and Barbero (1986) for HCl shown in Figure 17.12, because of course no data for the chloride ion itself exist, or will ever exist. Some of the data for NaCl are also shown, and it becomes clear then how "data" for Na^+ are derived. The curve for the partial molar heat capacity of Na^+ is derived from the two lower curves from equation (17.62), which in this case is

$$\begin{aligned}\bar{C}^o_{p\,\mathrm{NaCl}} &= \bar{C}^o_{p\,\mathrm{Na^+}} + \bar{C}^o_{p\,\mathrm{Cl^-}} \\ &= \bar{C}^o_{p\,\mathrm{Na^+}} + \bar{C}^o_{p\,\mathrm{HCl}}\end{aligned}$$

17.10. COMPARISON OF PITZER AND HKF MODELS

These two models present a fascinating contrast in their approach to a complex problem. Pitzer stands back, as it were, from the details of ionic interactions, and builds up an empirical model of complex solutions from data on the simpler binary systems of which it is composed. No data as to individual ionic processes are required

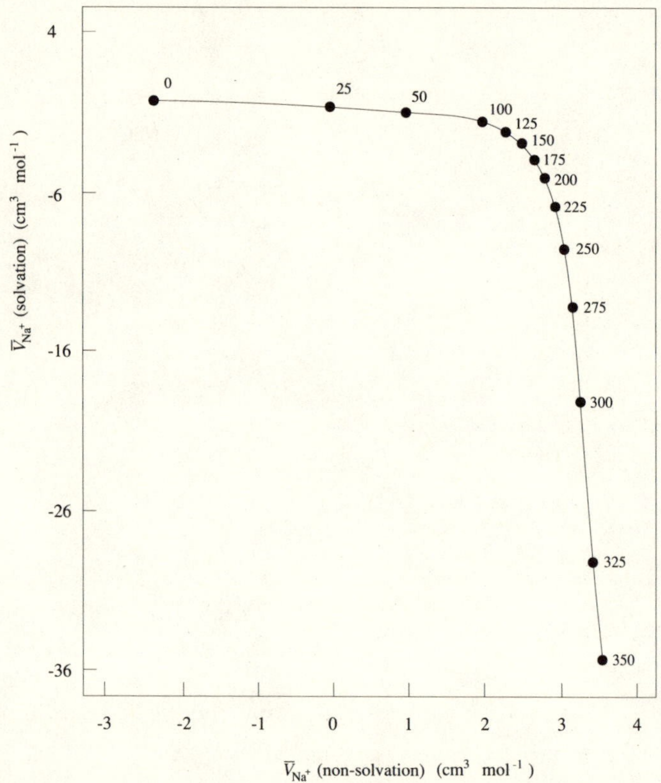

FIG. 17.11. Solvation versus non-solvation contribution to the standard partial molar volume of aqueous Na^+ at saturation pressure, according to the HKF model. Labels are T°C.

or derived from the model; they are all buried in the magnitudes of the parameters. This empirical model works very well at predicting the macroscopic properties of complex solutions, apparently because the form of the equations he has chosen to use suits the problem very well. Helgeson, on the other hand, prefers to think about ionic processes in detail, breaking them down into component parts that can be modeled and then assembled into a large framework. There are empirical parameters in these component parts, to be sure, but the component parts are all interrelated by standard thermodynamic equations.

In a broad way, the approaches are complimentary, because the Pitzer equations have no provisions for manipulating or adjusting the standard state parameters of solution components, but concentrate on the effects of changes in concentration. HKF on the other hand is primarily a model interrelating standard state parameters. The HKF model does incorporate an activity coefficient model (in Part IV; Helgeson et al, 1984), but it remains to be seen how useful it is. There are advantages and disadvantages to both approaches. The magnificent achievements of Harvie and Weare

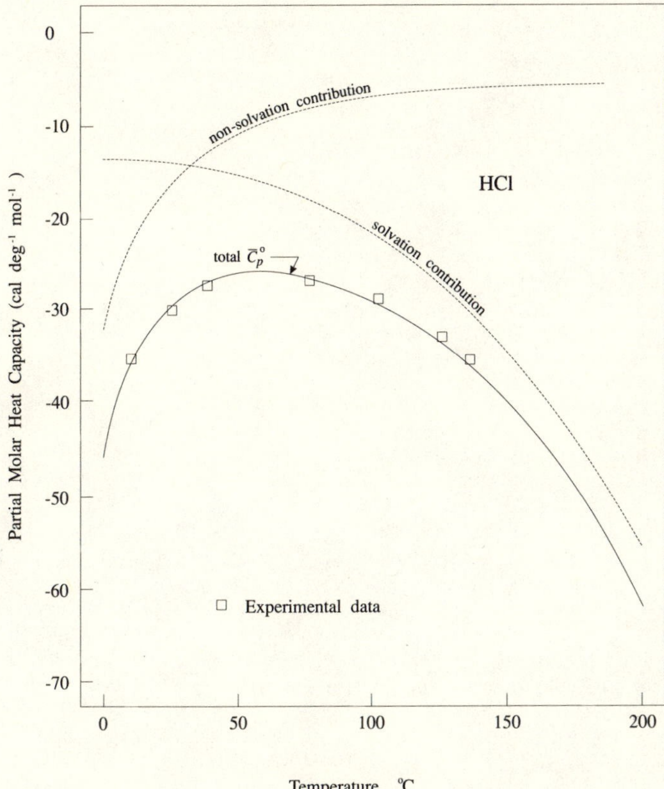

FIG. 17.12. The standard partial molar heat capacity of aqueous HCl as a function of temperature. Squares are the experimental data of Tremaine et al. (1986), and the solvation and non-solvation contributions, which add to the line fitting the data, are from the HKF model. Note how the shapes of the two contributions combine to give the inverted-U shape of the measured heat capacity.

and others in modeling evaporitic brines with the Pitzer model is at the moment well beyond the range of the HKF model, but in providing standard thermodynamic parameters for individual ions to high temperatures and pressures the HKF model is reaction-oriented, and hence fits the way most geochemists think. It has therefore been more widely used at present. Another difference is that the Pitzer model is best used within the range of conditions covered by the experimental systems used, while the HKF model is better suited to extrapolation.

It will be of great interest to follow the progress of these and other models in advancing our understanding of geochemical phenomena in the years to come.

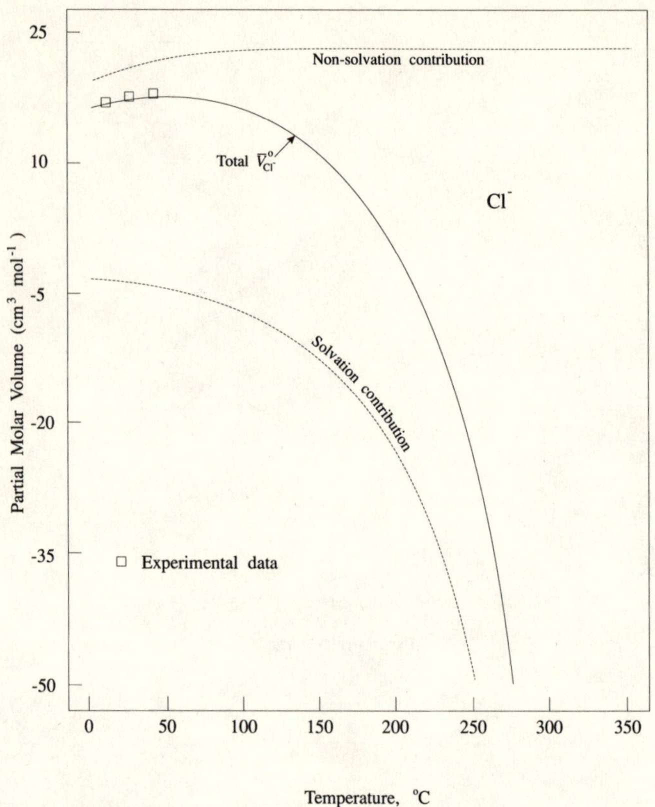

FIG. 17.13. The standard partial molar volume of aqueous HCl as a function of temperature. Squares are experimental data, and the solvation and non-solvation contributions are from the HKF model for Cl^-.

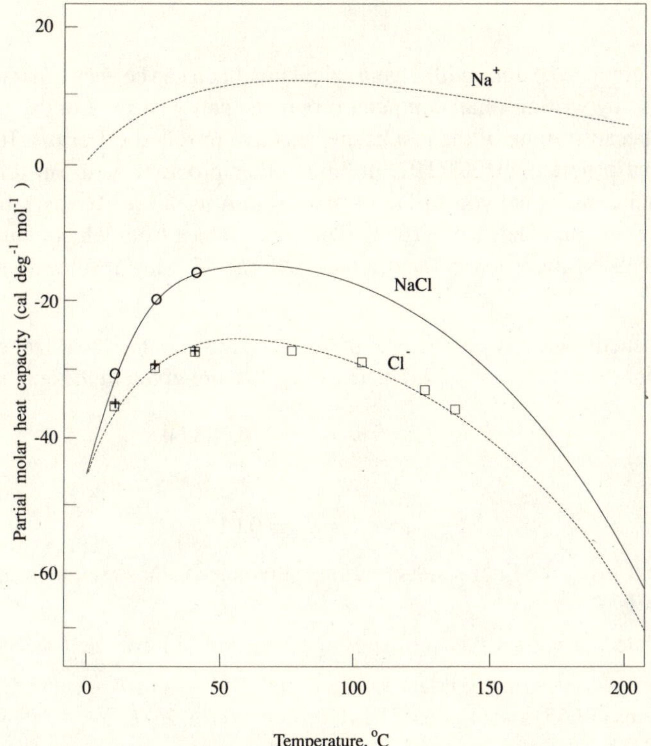

FIG. 17.14. The standard partial molar heat capacities of aqueous NaCl, Na^+ and Cl^- as a function of temperature. The experimental data for Cl^- are in fact those for aqueous HCl of Tremaine et al. (1986) from Figure 17.12, illustrating the fact that the heat capacity of H^+ is zero by convention (see discussion in text). The heat capacity of aqueous NaCl is a measured quantity, and that of aqueous Na^+ is obtained from the difference of the NaCl and HCl (or Cl^-) data.

PROBLEMS

These problems can be done with a hand calculator, but it can be very frustrating. They are best done by writing small computer programs, partly because of the complexity and partly because some of the results are sensitive to round-off errors. If you have access to the program SUPCRT92, or some other program, you can do the HKF problems quite easily, but you will not learn much. A useful reference is the book by Bowers, Jackson and Helgeson (1984). This gives tables from which you can solve these problems easily, but with the original HKF model. Most results are not greatly different.

1. Calculate the activity coefficients of Ca^{2+} and SO_4^{2-} using the Pitzer equations for the example in §17.8.4. The only data not given in the text are

$$^{E}\theta_{\mathrm{Na,Ca}} = {}^{E}\theta_{\mathrm{Cl,SO_4}} = -0.08384$$

   and

$$^{E}\theta'_{\mathrm{Na,Ca}} = {}^{E}\theta'_{\mathrm{Cl,SO_4}} = 0.01355$$

   The text gives the final results; various parameter values are given in Appendix F.
2. Calculate the values of $\omega_{\mathrm{Ca^{2+}}}$ and $\omega_{\mathrm{SO_4^{2-}}}$ at 25°C, 1 bar, and at 300°C, $P = P_{sat}$. You will need data for Ca^{2+} and SO_4^{2-} from Appendix C, and equations (17.65) and (17.60). The value of g is 0.0 at 25°C and -0.004651 at 300°C. The quantity $N_a e^2/2$ is often called η, and has the value 1.66027×10^5Å cal mol^{-1}.
3. Calculate $\Delta_a G°$for anhydrite, Ca^{2+}, and SO_4^{2-} at 25 and 300°C, using equation (17.77) for the ions and (7.25) for anhydrite. The Born parameter Y_{P_r,T_r} is -5.81×10^{-5} at 25°C, 1 bar, and the dielectric constant ϵ is 19.99 at 300°C and 78.46 at 25°C.
4. Calculate K_{sp} for anhydrite at 25°C (verifying the value given in the text), and at 300°C.
5. Calculate the dissociation constant for NaCl(aq) at 25°C, 1 bar. Calculate this constant at 500°C, 1 and 2 kbar, obtaining the necessary values of ϵ and Y from Tanger and Helgeson, (1988). A graph of K_{NaCl} versus T is shown by Sverjensky (1987).

18

REDOX SYSTEMS

The question therefore arises as to what defines P_{O_2} of a complex mineral assemblage and when and where changes of P_{O_2} are likely to occur during metamorphism.

Eugster (1959), p. 423.

18.1. OXIDATION STATE AS A PRIMARY INTENSIVE VARIABLE

Up until this point we have dealt with familiar intensive variables such as temperature, pressure, density, and molar thermodynamic properties (molar entropies, free energies, and so on). There exists another, equally important intensive variable that we have used implicitly, but have not yet discussed in sufficient detail—the oxidation state of a system. This involves concepts and applications so useful to Earth scientists that we devote a complete chapter to this single variable.

Except for nuclear processes, most chemical behavior is determined by electron distributions and interactions. From this point of view, the oxidation state of an atom is among the most fundamental of all its properties. Most elements can exist in multiple valences with each state usually displaying quite different behavior from the others. As an example, consider the element sulfur. This has the electron configuration [Ne]$3s^2 3p^4$ and has valences ranging between the two extremes −2 and +6 (which complete the inert gas configurations [Ar] and [Ne], respectively). The chemical differences among sulfur species spanning this range are enormous, from H_2S through the polysulfides S_n^{2-}, to elemental S and intermediate oxides such as SO_3^{2-} (sulfite) and $S_2O_3^{2-}$ (thiosulfate), up to SO_4^{2-}. Each of these species occurs in different geochemical environments and participates in reactions uniquely determined by its valence state. Other obvious examples are carbon, the transition metals, and the lanthanides and actinides. Only the inert gases, halogens, alkali metals, and alkaline earths are relatively immune to these effects in nature, where they usually retain the same valence state; however, even these elements produce compounds of unusual valence well known to inorganic chemists.

18.2. OXIDATION STATES OF NATURAL SYSTEMS

At a given T and P, the oxidation state of a closed system at equilibrium is, in common with all other properties, determined by the bulk composition, but it is a property that is remarkably useful in characterizing systems, so it is singled out for special treatment. Two methods of measuring and reporting the oxidation state of natural systems are commonly used: the oxidation potential (Eh or pe) and the oxygen fugacity (f_{O_2}). Oxidation potentials are generally easier to use at Earth surface conditions (near 25°C, 1 atm.) mainly because here it is convenient to use an electrochemical method for measurement, which gives Eh directly. At higher temperatures electrochemical cells become very difficult to operate, especially as water is not always present, and although redox conditions (however measured) could still be reported as Eh, in fact it is simpler to use a more direct measure such as f_{O_2}, as we will see. At temperatures roughly in the magmatic range, it is also possible to measure f_{O_2} directly. In this chapter we discuss both parameters, but at the outset it is important to remember that these are simply two equivalent ways of expressing the same property—the state of oxidation or reduction of chemical species within some system.

18.2.1. *Oxidation Potential and Eh–pH Diagrams*

Without doubt, the most widely used measure of oxidation state in lower temperature aqueous and sedimentary systems is the oxidation potential, or Eh. The concept of oxidation potentials has been used by chemists and engineers for close to a century. However, its widespread use in the Earth sciences began only in the early 1950s, primarily with the work of Robert Garrels. Garrels recognized the need for this variable in classifying sedimentary environments and he became interested in the work of the Belgian corrosion engineer Marcel Pourbaix (e.g., Pourbaix, 1949, 1974), who used Eh–pH diagrams to summarize the aqueous chemistry of quite complex systems. The usefulness of Eh–pH diagrams for Earth scientists became immediately evident with publication of a classic paper by Krumbein and Garrels (1952).

The only other intensive variable of comparable significance in aqueous systems is pH. It too is a function of the bulk composition at a given T and P, but both are closely related to a large number of important reactions. Therefore it proves natural to use both as variables in diagrams of systems at fixed T and P, and Eh–pH (or $\log f_{O_2}$–pH) diagrams have become a standard method of displaying and interpreting geochemical data. In the following sections we outline the theoretical basis for calculating these diagrams.

18.2.2. *Cell Voltage and Free Energy of Reaction*

Instead of writing balanced redox reactions such as (18.1) in which both the oxidized and reduced species appear, it will be convenient to split these into two half-reactions or *half-cells* that explicitly include the electron. The reason this will be convenient (in spite of the resulting hassles with conventions) is that we can devise standard voltages for half-cells and then mix and match half-cells to make complete cells, in much the same way that we combine $\Delta_f G^\circ$ values of compounds to make complete reactions. Also, as we will see, the Eh itself is a half-cell voltage. For example, consider the redox reaction

$$\mathrm{Cu} + \frac{1}{2}\mathrm{O_2} = \mathrm{CuO} \qquad (18.1)$$

in which Cu is oxidized to Cu^{2+} and $\frac{1}{2}O_2$ is reduced to O^{2-}. These two aspects of the reaction can be written separately as "half-reactions":

$$\text{Oxidation:} \quad \mathrm{Cu} \rightarrow \mathrm{Cu^{2+}} + 2e \qquad (18.2)$$

$$\text{Reduction:} \quad 2e + \frac{1}{2}\mathrm{O_2} \rightarrow \mathrm{O^{2-}} \qquad (18.3)$$

These half reactions cannot occur by themselves in natural aqueous systems because they would produce negative or positive charge and something else would oxidize or reduce to neutralize this. In Nature half-reactions are always coupled, with one species being reduced while another is oxidized.

For the moment, imagine that half-reactions can occur by themselves. If, as in (18.2), one mole of elemental Cu were oxidized to a mole of Cu^{2+} ions there would be an excess of 2×Avogadro's number worth of electrons in the system. The amount of electrical charge associated with an Avogadro's number of electrons equals 96,485.3 coulombs and is called the *Faraday* ($\mathcal{F}$) of charge. Given the size of standard capacitors, this would require something the size of a small building for storage. Our hypothetical system would also develop a considerable negative voltage due to the accumulation of $2 \times 6.022 \times 10^{23}$ electrons. Stored in a two-farad capacitor this would be $\sim 48,000$ volts, so it should come as no surprise that this process would involve a change in energy. Fortunately we are never obliged to collect electrons like this because there is always another half-reaction that absorbs them. In fact, in measuring redox reactions, we go to great lengths to try to balance half-reactions one against another so that *no* electrons are transferred, i.e., to reach equilibrium. The voltage observed between the two balanced half cells is then related to the amount of work that could be obtained by letting the cell react to completion.

Recall that the electrical work w required to move a charge of $\mathcal{F}$ coulombs through a potential difference $\mathcal{E}$ volts is

$$w = \mathcal{F}\mathcal{E}$$

(joules = coulombs × volts)

where $\mathcal{F}$ is the charge per mole of electrons, so if n is the number of electrons appearing in the reaction as written (e.g. one in (18.2) and (18.3)), there are $n\mathcal{F}$ coulombs of charge, and the work is

$$w = n\mathcal{F}\mathcal{E} \tag{18.4}$$

This electrical work is by definition (Chapter 5) the ΔG associated with the process, as long as the electrical work is the only non-$P\Delta V$ work done. Therefore for any half-cell or complete cell or indeed any electrostatic process in which $n\mathcal{F}$ coulombs are moved through a potential difference $\mathcal{E}$,

$$\Delta G = n\mathcal{F}\mathcal{E} \tag{18.5}$$

or

$$\Delta G^\circ = n\mathcal{F}\mathcal{E}^\circ \tag{18.6}$$

for standard state conditions. (We must check this later for sign convention; see equation (18.15)). As applied to electrochemical cells, these equations are more properly $\Delta\mu = n\mathcal{F}\mathcal{E}$ and $\Delta\mu^\circ = n\mathcal{F}\mathcal{E}^\circ$ because many of the individual free energy terms refer to constituents in solution and hence are partial molar terms.

These equations (18.5, 18.6) connect electrochemistry to the world of thermodynamics. It only remains to devise some way of determining the voltage associated with cells and half-cells, and we should then be able to determine free energies and perhaps other thermodynamic properties of cell reactions.

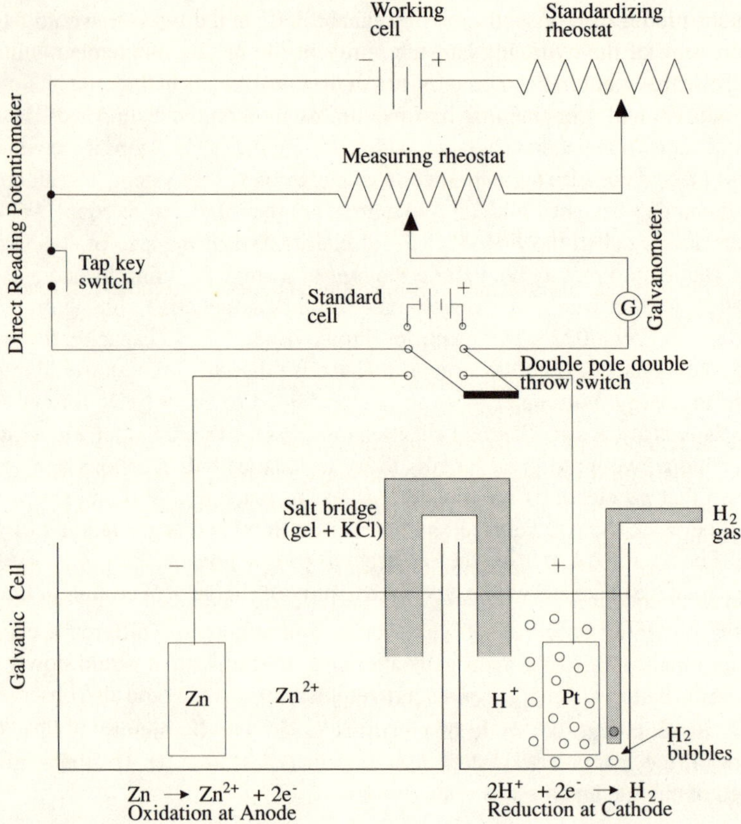

FIG. 18.1. Standard hydrogen electrode connected to a Zn/Zn^{2+} half-cell with a salt bridge and external direct reading potentiometer.

18.2.3. *The Standard Hydrogen Electrode*

In fact, equation (18.6) has been used to determine standard thermodynamic properties by electrochemists throughout much of the twentieth century. We will begin by discussing a cell containing the "hydrogen half-cell," because this will assume a special importance later on. Consider the cell shown in Figure 18.1. On the right is an inert beaker containing the concentration m of an acid such as HCl. Hydrogen gas is bubbled into this solution. A platinum electrode coated with fine-grained "platinum black" crystals (which catalyze reactions) is connected by a wire to another cell that contains a metallic zinc electrode and a Zn^{2+} solution (say $ZnCl_2$). The two beakers are also connected with a "salt bridge," which is a glass tube containing a KCl-impregnated gel.

If the wire between both beakers is connected a reaction begins, just as if all these constituents were mixed together in a single beaker. But as they are separated, a current will begin to flow through the wire as the cell reacts. The amount of current depends mostly on the concentration of zinc and HCl in the two solutions, but also on the pressure of the hydrogen gas, the purity of the metals, the temperature, and anything else that would affect the activities of the constituents in the cell reaction. On the left side, Zn will spontaneously oxidize to Zn^{2+} and on the right, H^+ will reduce to H_2:

$$\text{Oxidation}: \quad \text{Zn} \rightarrow \text{Zn}^{2+} + 2\text{e} \tag{18.7}$$

$$\text{Reduction}: \quad 2\text{H}^+ + 2e \rightarrow \text{H}_2(g) \tag{18.8}$$

These are the half-reactions or half-cell reactions. The total or whole-cell reaction is the sum of these, or

$$\text{Whole cell}: \quad \text{Zn} + 2\text{H}^+ = \text{Zn}^{2+} + \text{H}_2(g) \tag{18.9}$$

As these reactions proceed, electrons flow through the wire from the Zn to the Pt electrode. This would produce a charge imbalance in both beakers if it weren't for the salt bridge, which diffuses Cl^- ions into the right hand beaker and K^+ into the left, and completes the electrical circuit.

This apparatus works as an inefficient battery. In fact, all batteries follow the same principle of connecting one oxidation to one reduction half cell. With an external voltmeter you would observe that whatever the initial potential difference between the two terminals of this zinc-hydrogen battery was to start with, it would slowly drop to 0.0 V as the battery discharges or reacts to equilibrium. You would also observe that the Zn electrode is negative relative to the Pt side. Because the voltage of this whole cell or battery falls as reaction proceeds, voltage *measurements* must be made without permitting any reaction to occur. The direct-reading potentiometer circuit shown in Figure 18.1 can be used to adjust potentials in the potentiometer and battery circuits until they exactly cancel each other. At this point, no current flows (as indicated by the galvanometer), nothing reacts, and the measured voltage is the maximum the cell can deliver. At the null point, the cell is acting reversibly as well, since any slight fluctuation of the potentiometer voltage above or below this setting would cause the cell reaction to run forward or backward.

Let's say that we set up this cell, balance the potential with our potentiometer, and read a cell voltage of $\mathcal{E}$ volts. We could next *suppose* that this voltage is made up of the sum of the two half-cell voltages, i.e., that

$$\mathcal{E}_{\text{cell}} = \mathcal{E}_{\text{zinc half-cell}} + \mathcal{E}_{\text{hydrogen half-cell}} \tag{18.10}$$

Because there is no way to measure the properties of individual half-cells, we are in the same position as we were in Chapter 17 with respect to single ion activities. We must choose a convention; in this case we must choose one half-cell to reference all the others to. For example if we chose the hydrogen half-cell, we could say then that the voltage of the zinc half cell was

$$\mathcal{E}_{\text{zinc half-cell}} = \mathcal{E}_{\text{cell}} - \mathcal{E}_{\text{hydrogen half-cell}} \tag{18.11}$$

Then we could measure a cell containing a copper electrode in a copper solution in the same way, i.e., a copper-hydrogen cell, and say that

$$\mathcal{E}_{\text{copper half-cell}} = \mathcal{E}'_{\text{cell}} - \mathcal{E}_{\text{hydrogen half-cell}} \tag{18.12}$$

We could then calculate the voltage of a cell with the zinc on side and the copper on the other simply by adding the two half-cell voltages, because if we arrange things properly, whatever the voltage of the hydrogen half-cell is, it will cancel out.

Of course it will only cancel out if it was the same in both cases, which means that the pressure of the hydrogen gas and the activity of H^+ in the solution have to be the same. This is accomplished by standardizing on a fugacity of H_2 of one bar, and an activity of H^+ of 1.0, and the hydrogen electrode in this state is called the *standard hydrogen electrode,* or SHE. Similarly, the voltages of the zinc and copper electrodes in this example will change with the zinc and copper ion activities in solution, so they are tabulated for standard conditions of unit activity of the metal and ion at 25°C, 1 bar. With all reactants and products of a half-cell at unit activity, the half-cell voltage becomes the *standard* half cell voltage, $\mathcal{E}^\circ$.

There are some other conventions connected with making sure that the SHE voltage always cancels out, and we consider these next.

18.2.4. *Oxidation and Reduction Sign Conventions*

There is obviously a slight problem with the scheme outlined above, and that is that if we add the voltages in (18.11) and (18.12) we do not get the cell voltage of the zinc-copper cell ($\mathcal{E}_{\text{cell}} + \mathcal{E}'_{\text{cell}}$), because the hydrogen electrode voltage does not cancel but is doubled. That's why we qualified the presentation with "if we arrange things properly".

There are a number of ways to ensure that things "work properly" in combining half-cell potentials, and unfortunately all of them seem to have been used at one time or another. The situation was greatly clarified by the universal acceptance of the conventions adopted by the IUPAC[1] at a meeting in Stockholm in 1953; universal, that is, except for the geochemists. Earth scientists often follow the conventions used in the influential text by Garrels and Christ (1965). In this chapter we use only IUPAC conventions, though in Appendix D we present a complete comparison of the three conventions one is most likely to come across in using *Eh*. Then too at the close of the chapter we try to persuade you not to use *Eh* at all.

The IUPAC recipe for half-cell manipulations is *first,* half-cells are always written and tabulated as reductions when combined with the SHE; and *second,* calculate the cell voltage as the right (reduction) half-cell voltage *minus* the left (oxidation) voltage, whether using the SHE or not. This is the opposite of the way we wrote equations (18.7) and (18.8) for the zinc-hydrogen cell, so we rewrite them (according to the first rule) as

$$\text{Reduction: } \quad Zn^{2+} + 2e \rightarrow Zn$$

$$\text{Oxidation: } \quad H_2(g) \rightarrow 2H^+ + 2e$$

[1]International Union of Pure and Applied Chemistry.

$$\text{Whole cell}: \quad Zn^{2+} + H_2 \rightarrow Zn + 2H^+$$

Considering only standard conditions for now, if $\mathcal{E}^\circ_{Zn}$ is the standard zinc half-cell potential and $\mathcal{E}^\circ_{SHE}$ is the SHE potential, then by the second rule above, the cell potential is

$$\mathcal{E}^\circ_{\text{zinc cell}} = \mathcal{E}^\circ_{Zn} - \mathcal{E}^\circ_{SHE}$$

so that the zinc half-cell potential is

$$\mathcal{E}^\circ_{Zn} = \mathcal{E}^\circ_{SHE} + \mathcal{E}^\circ_{\text{zinc cell}}$$

This cell has a measured potential of −0.763 volts (i.e., the standard zinc electrode is 0.763 volts more negative than the SHE) so

$$\mathcal{E}^\circ_{Zn} = \mathcal{E}^\circ_{SHE} - 0.763 \text{ V}$$

Substituting Cu for Zn we have the copper-hydrogen cell, with

$$\text{Reduction}: \quad Cu^{2+} + 2e \rightarrow Cu; \ \mathcal{E}^\circ_{Cu}$$

$$\text{Oxidation}: \quad H_2(g) \rightarrow 2H^+ + 2e; \ \mathcal{E}^\circ_{SHE}$$

and

$$\text{Whole cell}: \quad Cu^{2+} + H_2 = Cu + 2H^+$$

The potential of this cell is

$$\mathcal{E}^\circ_{\text{copper cell}} = \mathcal{E}^\circ_{Cu} - \mathcal{E}^\circ_{SHE}$$

so that the potential of the copper-cupric ion half-cell is

$$\mathcal{E}^\circ_{Cu} = \mathcal{E}^\circ_{SHE} + \mathcal{E}^\circ_{\text{copper cell}}$$

The measured potential of the copper-hydrogen cell is 0.337 volts (i.e., the standard copper electrode is 0.337 volts more positive than the SHE) so

$$\mathcal{E}^\circ_{Cu} = \mathcal{E}^\circ_{SHE} + 0.337 \text{ V}$$

Now considering the zinc-copper cell, which could be constructed by substituting a copper electrode in a cupric ion solution for the SHE in Figure 18.1, we could write

$$\text{Oxidation}: \quad Zn \rightarrow Zn^{2+} + 2e$$

$$\text{Reduction}: \quad Cu^{2+} + 2e \rightarrow Cu$$

with cell reaction

$$Zn + Cu^{2+} = Zn^{2+} + Cu \tag{18.13}$$

$$\begin{aligned} \mathcal{E}^\circ_{\text{cell}} &= \mathcal{E}^\circ_{Cu} - \mathcal{E}^\circ_{Zn} \\ &= (\mathcal{E}^\circ_{SHE} + 0.337) - (\mathcal{E}^\circ_{SHE} - 0.763) \\ &= 0.337 + 0.763 \\ &= 1.10 \text{ V} \end{aligned}$$

or, we could write

$$\text{Oxidation}: \quad Cu \rightarrow Cu^{2+} + 2e$$
$$\text{Reduction}: \quad Zn^{2+} + 2e \rightarrow Zn$$

with cell reaction

$$Cu + Zn^{2+} = Cu^{2+} + Zn \tag{18.14}$$

$$\begin{aligned} \mathcal{E}^\circ_{\text{cell}} &= \mathcal{E}^\circ_{\text{Zn}} - \mathcal{E}^\circ_{\text{Cu}} \\ &= (\mathcal{E}^\circ_{\text{SHE}} - 0.763) - (\mathcal{E}^\circ_{\text{SHE}} + 0.337) \\ &= -0.337 - 0.763 \\ &= -1.10 \text{ V} \end{aligned}$$

Note that in this convention, the sign of the half-cell voltages does not depend on how the reaction is written (whether as oxidation or reduction), but the sign of the complete cell reaction does depend on how it is written. This corresponds closely to actual lab operations, i.e., in the above cell the zinc electrode would be observed to be negative and the copper positive, however the cell was arranged or written, with 1.10 volts difference between them. But whether this 1.10 V is positive or negative just depends on whether you look at it from the point of view of the zinc electrode "looking up" to the copper electrode, or the copper electrode "looking down" on the zinc electrode.

18.2.5. *The SHE Voltage Equals Zero Convention*

More importantly, note that the convention is successful at having the SHE voltage cancel out, no matter how we consider the cell reaction. Naturally, as it always cancels out, it makes no difference what its voltage is. If we arbitrarily assign a voltage of zero volts to the SHE, then instead of writing

$$\mathcal{E}^\circ_{\text{Cu}} = \mathcal{E}^\circ_{\text{SHE}} + 0.337 \text{ V}$$

for the copper-cupric ion half-cell, we could write

$$\mathcal{E}^\circ_{\text{Cu}} = 0.337 \text{ V}$$

which is a little neater. It is important to realize that this is the only benefit conferred by the $\mathcal{E}^\circ_{\text{SHE}} = 0$ convention. In view of (18.6), a corollary of $\mathcal{E}^\circ_{\text{SHE}} = 0$ is that

$$\Delta G^\circ_{\text{SHE}} = 0$$

i.e., that

$$\frac{1}{2} G^\circ_{H_2} - \bar{G}^\circ_{H^+} - \bar{G}^\circ_e = 0 \tag{17.23}$$

As we saw in the previous chapter, this same convention is the one required to effect an equivalent simplification in the definitions of single-ion properties, and these reasons are sufficient to ensure its universal adoption. Its use is acceptable as long as one realizes that it is quite unnecessary to the workings of thermodynamics.

18.2.6. *Free Energy of Cell Reactions*

In deciding upon conventions, we have various options open when dealing with half-cells, because they can't be measured separately; they are to some extent fictional. But complete cell reactions are different because the changes in various thermodynamic parameters in these reactions *can* be measured or calculated. For example consider reactions (18.13) and (18.14). We can look up the values of $\Delta_f \bar{G}^\circ$ for Cu^{2+} and Zn^{2+} (65.49 and -147.06 kJ mol^{-1} respectively, Wagman et al., 1982), and we find that

$$\Delta_r G^\circ(18.13) = -212.55 \text{ kJ mol}^{-1}$$

and of course

$$\Delta_r G^\circ(18.14) = 212.55 \text{ kJ mol}^{-1}$$

Comparing these with the voltages produced by applying the IUPAC conventions,

$$\mathcal{E}^\circ(18.13) = 1.10 \text{ V}$$

and

$$\mathcal{E}^\circ(18.14) = -1.10 \text{ V}$$

we see that we have to modify (18.5) and (18.6) to read

$$\left.\begin{aligned} \Delta G &= -n\mathcal{F}\mathcal{E} \\ \Delta G^\circ &= -n\mathcal{F}\mathcal{E}^\circ \end{aligned}\right\} \qquad (18.15)$$

to bring the IUPAC conventions into line with standard thermodynamic calculations.

In practice, applications requiring whole-cell calculations are rare in the Earth sciences; however, it is essential to understand both half-cell and whole-cell sign conventions in designing sample and reference electrode combinations for laboratory or field measurements. This is also necessary to understand how pH, Eh, and specific ion electrodes really function.

18.2.7. *Measuring Thermodynamic Properties of Ions from Half-Cell Potentials*

Equations (18.15) allow calculation of the free energy of reaction of cell reactions to be calculated from measured cell voltages, and the use of the SHE reference and the IUPAC conventions then allows calculation of individual ionic free energies. If cell voltages are measured over a range of temperatures, the entropy and enthalpy of the cell reaction and individual ions can also be calculated. Thus

$$\begin{aligned} \Delta_r S^\circ &= -\left(\frac{\partial \Delta_r G^\circ}{\partial T}\right) \\ &= n\mathcal{F}\left(\frac{\partial \mathcal{E}^\circ}{\partial T}\right) \end{aligned}$$

and

$$\begin{aligned}\Delta_r H^\circ &= \Delta_r G^\circ + T\Delta_r S^\circ \\ &= n\mathcal{F}\mathcal{E}^\circ + n\mathcal{F}\left(\frac{\partial \mathcal{E}^\circ}{\partial T}\right)\end{aligned}$$

The heat capacity of reaction could also be calculated from

$$\begin{aligned}\Delta_r C_p^\circ &= \left(\frac{\partial \Delta_r H^\circ}{\partial T}\right) \\ &= n\mathcal{F}\left(\frac{\partial^2 \mathcal{E}^\circ}{\partial T^2}\right)\end{aligned}$$

but as this is a second derivative it is very difficult to get accurate results this way, and calorimetry is normally used. These procedures have served as a major source of the thermodynamic data for aqueous ionic species below 100°C, many of these measurements having been made throughout the first half of this century.

18.2.8. *Electrochemical Cells as Metastable Systems*

A certain amount of confusion may arise from the fact that electrochemical cells can reach equilibrium with their external potentiometer circuits, giving a stable, equilibrium cell voltage of $\mathcal{E}$ volts, implying a non-zero $\Delta_r G$. How does this fit with our knowledge that $\Delta_r G_{T,P}$ is zero at equilibrium? It fits very well, if we realize that the cell at equilibrium is a perfect example of a metastable system according to our definitions in Chapter 3. The external voltage source is a constraint in addition to the minimum constraints of constant T and P, which prevents the metastable system from lowering its G to the minimum possible value. If this constraint is removed by connecting a wire between the two halves of the cell, the cell reaction proceeds until it can go no farther, at which point $\Delta_r G = 0$ and $\mathcal{E} = 0$. Equilibrium cell voltages as measured in the lab or the field are measurements that refer to metastable equilibria. This does not mean that if you measure an Eh in a swamp, the swamp water is necessarily a metastable system. It is the cell reaction that is metastable, not the swamp. This can be so because the cell reaction involves some components (in the reference half-cell) in concentrations completely different from those in the natural system whose Eh is being measured.

18.2.9. *Calculating Eh*

Consider the general cell reaction

$$b\mathrm{B} + c\mathrm{C} = d\mathrm{D} + e\mathrm{E} \tag{18.16}$$

Let's say that this reaction reaches equilibrium with an external measuring system (constraint), giving cell voltage $\mathcal{E}$. If operated under standard conditions, it would give cell voltage $\mathcal{E}^\circ$, and the corresponding free energies of reaction are

$$\Delta_r \mu = -n\mathcal{F}\mathcal{E}$$

and

$$\Delta_r\mu^\circ = -n\mathcal{F}\mathcal{E}^\circ$$

From equation (13.6) we have

$$\Delta_r\mu = \Delta_r\mu^\circ + RT\ln Q \tag{13.6}$$

where

$$Q = \prod_i a_i^{\nu_i}$$
$$= \frac{a_{\mathrm{E}}^e a_{\mathrm{D}}^d}{a_{\mathrm{B}}^b a_{\mathrm{C}}^c}$$

Recall from Chapter 13 that this activity term is referred to as Q rather than K because it refers to a metastable equilibrium. Substitution of equations (18.15) in their partial molar form gives

$$\mathcal{E} = \mathcal{E}^\circ - \frac{RT}{n\mathcal{F}}\ln Q$$

or

$$\mathcal{E} = \mathcal{E}^\circ - 2.3026\frac{RT}{n\mathcal{F}}\log Q \tag{18.17}$$

This is the Nernst equation, after the physical chemist W. Nernst, who derived a similar expression (using concentrations rather than activities) at the end of the last century (Nernst, 1897). As above, n is the number of electrons transferred in the cell reaction (18.16), $\mathcal{F}$ is the Faraday of charge, R is the gas constant and T the temperature (in Kelvins). The constant 2.3026 is added to convert from natural to base 10 logs. At 25°C the quantity 2.3026 $RT/n\mathcal{F}$ has the value 0.05916, which is called the Nernst slope. The importance of (18.17) is that it allows calculation of the potentials of cells having non-standard state concentration (i.e., real cells) from tabulated values of standard half-cell values or tabulated standard free energies.

Equation (18.16) could also be considered to represent a half-cell reaction, except that the electron is not shown. If you have followed our discussion of the single-ion and SHE conventions, you will not be surprised to learn that it does not matter what value the chemical potential of the electron is considered to have because it always cancels out in balanced reactions, and that by convention it is given the value zero. This means that the Nernst equation applies to half-cell reactions as well as cell reactions, as do equations (18.15). And if you have followed all this, you now know what Eh is, because (18.17) when applied to half-cells is the definition of Eh. Thus

$$Eh = \mathcal{E}^\circ - 2.3026\frac{RT}{n\mathcal{F}}\log Q \tag{18.18}$$

In other words Eh of a solution is its half-cell potential. It is the potential that would be observed if that solution formed one side of a cell and the SHE were the other side. For that solution to form such a half-cell, you have to stick an appropriate electrode into it, and arrange a suitable connection between the two half-cells, and we discuss this further on. This is the fundamental equation from which Eh–pH diagrams are calculated. Recalling the convention that we must use reduction half-cells when combined with a SHE, the Nernst equation for the Eh of half-cells will always have the form

$$Eh = \mathcal{E}^\circ - 2.3026\frac{RT}{n\mathcal{F}}\log\left(\frac{\text{reduced form}}{\text{oxidized form}}\right) \tag{18.19}$$

18.3. EXAMPLE CALCULATION OF Eh–pH DIAGRAMS

In this section, we will calculate portions of a simple Eh–pH diagram for the system $Mn - H_2O$ and will show what happens when a third component, CO_2, is added. This illustrates most of the problems encountered in calculating such diagrams. If you wish to add additional components such as SO_4^{2-} or H_2S, the methods are similar and details are provided by Garrels and Christ (1965).

18.3.1. *General Topology of Eh–pH Diagrams*

First, let us examine the completed Eh–pH diagram for $Mn - H_2O$ in Figure 18.2. There are typically four different types of boundaries shown on these diagrams. The top line, labeled O_2/H_2O, represents conditions for water in equilibrium with O_2 gas at 1 atm. Above this line, a P_{O_2} greater than 1 atm is required for water to exist, so that because the diagram is drawn for a pressure of 1 atm, water is not stable above this line. Similarly, the bottom line H_2O/H_2 represents conditions for water in equilibrium with H_2 gas at 1 atm. Below this line, P_{H_2} values greater than 1 atm are required for water to exist, i.e., at 1 atm water is not stable. Therefore the water stability field is between these two lines. The second type of boundary separates the stability fields of minerals or solid phases such as hausmannite (Mn_3O_4) and pyrochroite ($Mn(OH)_2$). These are true phase boundaries: hausmannite is thermodynamically unstable below the hausmannite/pyrochroite boundary and pyrochroite is unstable above it. Thus these first two kinds of boundary represent thermodynamic stability fields for different substances. Notice that on this diagram they all have the same slope (equal to the Nernst slope). The remaining two kinds of line are not stability boundaries at all but refer to concentrations of dissolved ions. For example, the vertical lines within the pyrochroite stability field represent contoured solubilities of pyrochroite as Mn^{2+} concentrations running from 10^{-1} to 10^{-6} m. Finally, the dashed boundaries between aqueous species, such as that between Mn^{2+} and MnO_4^{2-} indicate where the activities of the two species are exactly equal. To the right of this line Mn^{2+} remains present in the solution, but at a lower activity than MnO_4^{2-}, and vice versa.

18.3.2. *Sample Calculations*

It is possible to look up half-cell potentials for many reactions in physical chemistry textbooks and compilations of electrochemical data. However, it is usually a better procedure to choose free energy data, and to use those to calculate Eh and $\mathcal{E}^\circ$ for the reactions of interest.

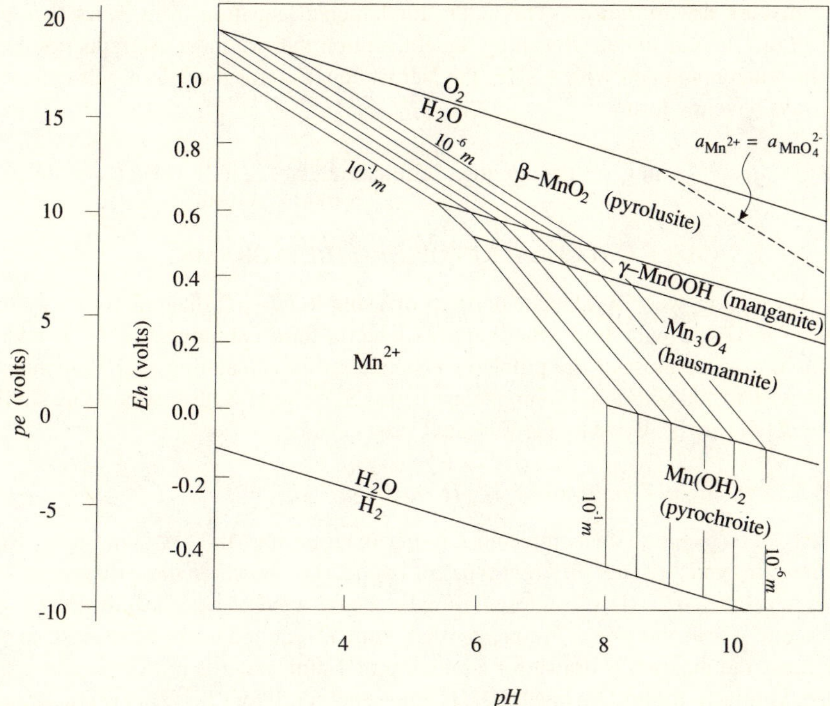

FIG. 18.2. *Eh–pH* and *pe–pH* relations in the system $Mn–H_2O–O_2$ at 25°C, 1 bar. Mn^{2+} activities and stability fields of Mn-oxide minerals are included. The *pe* and *Eh* axes are related by the formula $pe = 5040Eh/T$.

As the first example, we will calculate the two boundaries for the stability field of water.

For the boundary $H_2O(aq) - O_2(g)$, the half-cell reduction reaction is

$$4H^+ + O_2(g) + 4e = 2H_2O(aq) \tag{18.20}$$

for which $n = 4$. Using the tabulated $\Delta_f G^\circ$ for water (all others being zero), we find

$$\begin{aligned} \Delta_r G^\circ &= 2(-237.129) \\ &= -474.258\,\text{kJ} \end{aligned} \tag{18.21}$$

Because

$$\Delta_r G^\circ = -n\mathcal{F}\mathcal{E}^\circ \tag{18.15}$$

then

$$\begin{aligned} \mathcal{E}^\circ &= -(-474258/4 \times 96485.3) \\ &= 1.23\ \text{V} \end{aligned}$$

From the Nernst equation (18.19)

$$Eh = \mathcal{E}^\circ - 2.3026\frac{RT}{n\mathcal{F}}\log[1/(f_{O_2}\cdot a^4_{H^+})]$$

setting $f_{O_2} = 1$ atm gives the equation for the boundary in terms of Eh and pH.

$$Eh = 1.23 - 0.059pH \tag{18.22}$$

For the boundary $H_2(g) - H_2O(aq)$,

$$2H^+ + 2e = H_2(g)$$

$$\begin{aligned}\Delta_r G^\circ &= 0\\ &= -n\mathcal{F}\mathcal{E}^\circ\end{aligned}$$

and

$$Eh = 0 - 2.3026\frac{RT}{n\mathcal{F}}\log(f_{H_2}/a^2_{H^+})$$

or, with $f_{H_2} = 1$ atm, and $n = 2$,

$$Eh = -0.059pH \tag{18.23}$$

For boundary $Mn(OH)_2 - Mn_3O_4$,

$$Mn_3O_4(c) + 2H_2O(aq) + 2H^+ + 2e = 3Mn(OH)_2(c)$$

$$\begin{aligned}\Delta_r G^\circ &= -94140\ \mathrm{J\,mol^{-1}}\\ &= -n\mathcal{F}\mathcal{E}^\circ\end{aligned}$$

$$\begin{aligned}\mathcal{E}^\circ &= -(-94140/2\times 96485)\\ &= 0.488\ \mathrm{V}\end{aligned}$$

and, with $n = 2$,

$$Eh = 0.488 - 0.059pH \tag{18.24}$$

For solubility of Mn_3O_4 as Mn^{2+},

$$Mn_3O_4(c) + 8H^+ + 2e = 3Mn^{2+} + 4H_2O$$

$$\Delta_r G^\circ = -352711\ \mathrm{J}$$

$$\begin{aligned}\mathcal{E}^\circ &= 353711/2 \times 96485 \\ &= 1.828\ \mathrm{V}\end{aligned}$$

$$Eh = \mathcal{E}^\circ - \frac{0.059}{2}\log\left(\frac{a^3_{Mn^{2+}}}{a^8_{H^+}}\right)$$

Or

$$Eh = 1.828 - 0.237pH - 0.0887\log a_{Mn^{2+}} \tag{18.25}$$

This is plotted for selected values of Mn^{2+} activity ranging from 10^{-1} to 10^{-6} in Figure 18.2.

For equal activity contour of Mn^{2+} and MnO_4^{2-},

$$MnO_4^{2-} + 8H^+ + 4e = Mn^{2+} + 4H_2O$$

$$\Delta_r G^\circ = -672369\ \mathrm{J}$$

$$\begin{aligned}\mathcal{E}^\circ &= 672369/4 \times 96485 \\ &= 1.742\ \mathrm{V}\end{aligned}$$

$$Eh = 1.742 - 0.1182pH - 0.0148\log(a_{Mn^{2+}}/a_{MnO_4^{2-}})$$

and where the activities of both aqueous species are equal this reduces to

$$Eh = 1.742 - 0.1182pH \tag{18.26}$$

This boundary lies at high Eh and pH and is illustrated on Figure 18.2.

18.3.3. *Addition of Other Components*

The above examples cover all four types of borders or contour lines found on simple metal–water Eh–pH diagrams. As a next step it is possible to add other components to the system such as CO_2 or H_2S, which produce new aqueous species and carbonate or sulfide minerals. We will illustrate this by adding CO_2 to our $Mn - H_2O$ system. For details on how to include other components such as H_2S/SO_4^{2-} and SiO_2 see Garrels and Christ (1965, Chapter 7).

Figure 18.3 is an Eh–pH diagram calculated for the system Mn–H_2O–CO_2–O_2 at 25°C and $f_{CO_2} = 10^{-3.5}$ atm (the CO_2 partial pressure of normal sea-level air). It differs from Figure 18.2 for the system $Mn - H_2O$ only in the appearance of rhodochrosite, $MnCO_3$, which swamps the fields of $Mn(OH)_2$ and most of Mn_3O_4. The calculations are similar to those above, except that the following carbonate equilibria must be included.

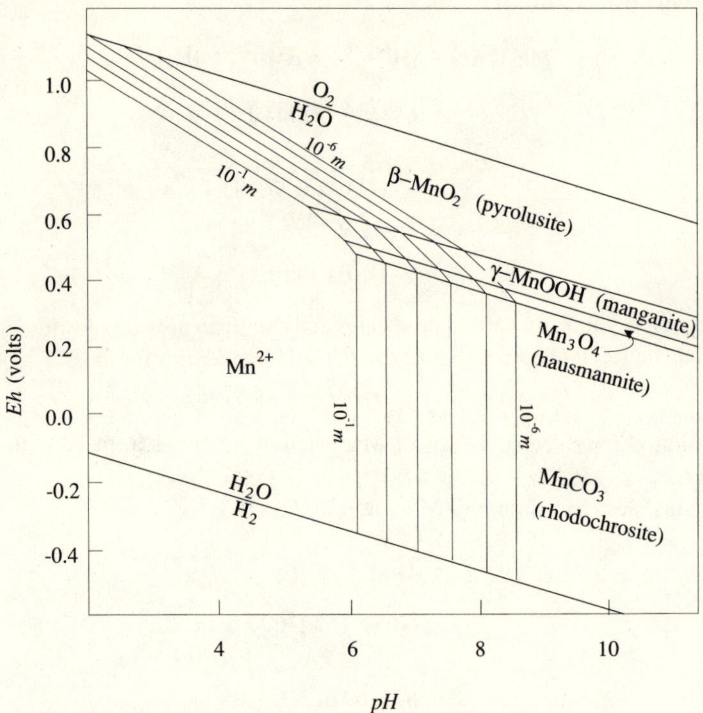

FIG. 18.3. *Eh-pH* relations in the system $Mn - H_2O - CO_2 - O_2$ at 25°C, 1 bar. f_{CO_2} is $10^{-3.5}$ bars.

18.3.4. *Carbonate Equilibria*

Henry's Law:

$$\begin{aligned} a_{H_2CO_3} &= K_H f_{CO_2} \\ &= 10^{-1.5} f_{CO_2} \end{aligned} \tag{18.27}$$

First acid ionization, carbonic acid,

$$\frac{a_{H^+} a_{HCO_3^-}}{a_{H_2CO_3}} = K_{a_1} = 10^{-6.3}$$

so

$$a_{HCO_3^-} = 10^{-6.3} \times 10^{-1.5} \times f_{CO_2}/a_{H^+} \tag{18.28}$$

Second ionization constant, carbonic acid,

$$\frac{a_{H^+} a_{CO_3^{2-}}}{a_{HCO_3^-}} = K_{a_2} = 10^{-10.3}$$

so

$$a_{CO_3^{2-}} = 10^{-10.3} \times 10^{-6.3} \times 10^{-1.5} \times f_{CO_2}/a_{H^+} \tag{18.29}$$

Taking logs (base 10) of (18.27) to (18.29) gives

$$\log a_{H_2CO_3} = -1.5 + \log f_{CO_2} \tag{18.30}$$

$$\log a_{HCO_3^-} = -7.8 + \log f_{CO_2} + pH \tag{18.31}$$

$$\log a_{CO_3^{2-}} = -18.1 + \log f_{CO_2} + 2pH \tag{18.32}$$

These expressions for carbonate species activities can now be substituted into the appropriate half-cell reactions to give the Eh–pH boundaries for the CO_2-containing system.

For the boundary $MnCO_3 - Mn_3O_4$:

$$Mn_3O_4 + 3CO_3^{2-} + 8H^+ + 2e = 3MnCO_3 + 4H_2O \tag{18.33}$$

$$\Delta_r G^\circ = -532623\,\mathrm{J}$$

$$\mathcal{E}^\circ = 532623/2 \times 96485 = 2.76\,\mathrm{V}$$

$$Eh = 2.76 - \frac{0.059}{2} \log\left(1/(a_{H^+}^8 a_{CO_3^{2-}}^3)\right)$$

Substituting (18.32) for $a_{CO_3^{2-}}$ with $f_{CO_2} = 10^{-3.5}$ and simplifying gives:

$$Eh = 0.848 - 0.059pH \tag{18.34}$$

When plotted, this phase boundary swamps the field of pyrochroite and all but the uppermost field of hausmannite, showing that rhodochrosite should be the stable phase at mildly oxidizing to highly reducing conditions even at the low partial pressure of CO_2 in the Earth's atmosphere (see Figure 18.3).

For equal activity contour of Mn^{2+} and $MnHCO_3^+$:

$$Mn^{2+} + HCO_3^- = MnHCO_3^+$$

$$\Delta_r \bar{G}^\circ = -11464\,\mathrm{J}$$

$$\begin{aligned}\log K &= \log\left(a_{MnHCO_3^+}/(a_{Mn^{2+}} a_{HCO_3^-})\right)\\ &= -\Delta_r \bar{G}^\circ/2.303RT\\ &= 11464/(2.303 \times 8.3144 \times 298.15)\\ &= 2.01\end{aligned}$$

For $a_{Mn^{2+}} = a_{MnHCO_3^+}$, $\log a_{HCO_3^-} = -2.01$, and combining with (18.31) gives

$$pH = 9.29 \tag{18.35}$$

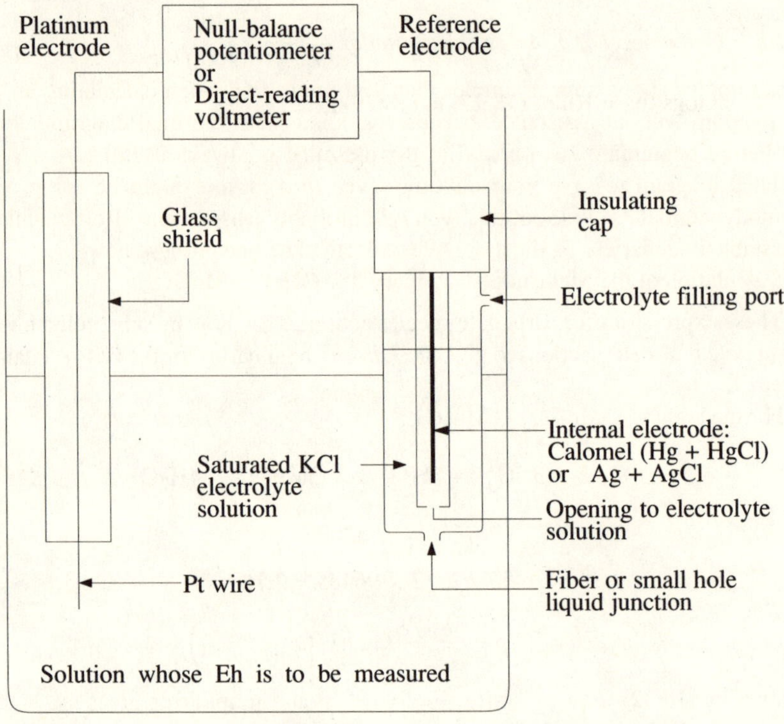

FIG. 18.4. Measurement of Eh showing construction of Pt electrode and of Ag-AgCl or calomel electrode .

For all $pH < 9.29$, the activity of Mn^{2+} will predominate over that of $MnHCO_3^+$ and rhodochrosite solubility can be contoured in terms of Mn^{2+}.

For solubility of $MnCO_3$ as Mn^{2+},

$$MnCO_3 = Mn^{2+} + CO_3^{2-}$$

$$\Delta_r\bar{G}^\circ = 60375 \text{ J}$$

$$\begin{aligned}\log K &= \log a_{Mn^{2+}} + \log a_{CO_3^{2-}}\\ &= -\Delta_r\bar{G}^\circ/2.303RT\\ &= -60375/(2.303 \times 8.3144 \times 298.15)\\ &= -10.58\end{aligned}$$

Substituting (18.32) for $a_{CO_3^{2-}}$, with $f_{CO_2} = 10^{-3.5}$ gives

$$pH = (11.02 - \log a_{Mn^{2+}})/2 \qquad (18.36)$$

This is solved for values of Mn^{2+} activity ranging from 10^{-1} to 10^{-6} and plotted in Figure 18.3 as solubility contours.

18.3.5. *Calculation of a Metastable Boundary*

The examples above appear simpler than you might find when calculating an *Eh–pH* diagram without assistance, because we knew what the final diagrams looked like before beginning. As we said at the outset, it is always useful to try to find a related diagram to save yourself time, even though you might be using newer thermodynamic data. Without this, you will probably waste some effort calculating metastable boundaries. To illustrate, we will calculate one of these now.

Calculation of the Metastable Boundary $Mn(OH)_2$–γ-MnOOH:

$$\gamma\text{–MnOOH} + \text{H}^+ + e = \text{Mn(OH)}_2$$

$$\Delta_r\bar{G}^\circ = -58994\ \text{J}$$

$$\begin{aligned}\mathcal{E}^\circ &= -\Delta_r G^\circ/n\mathcal{F}\\ &= 58994/(1 \times 96485)\\ &= 0.611\ \text{V}\end{aligned}$$

$$Eh = \mathcal{E}^\circ - \frac{RT}{n\mathcal{F}}\log(1/a_{\text{H}^+})$$

or

$$Eh = 0.611 - 0.059pH \tag{18.37}$$

Referring to Figure 18.2, this boundary plots below the phase boundary for Mn_3O_4–MnOOH. If you were to calculate the latter boundary next, it would become apparent that the boundary for $Mn(OH)_2$–MnOOH must be metastable. This kind of trial and error search for stable boundaries can be avoided by listing your aqueous species and mineral phases in order of oxidation state. In this case, we should have observed that the Mn oxides follow the increasing order of oxidation state $Mn(OH)_2 < Mn_3O_4 < MnOOH < MnO_2$. We need not then have calculated the metastable boundary.

18.3.6. *pe – pH Diagrams*

There is a second way of calculating the same kinds of diagrams using the alternative variable *pe* rather than *Eh*. Because this is another way of doing exactly the same thing, it could be argued that the new variable *pe* is unnecessary and redundant. However, the *pe* and *Eh* scales differ numerically, and *pe* calculations are now used about as frequently as *Eh*, so it is worthwhile discussing the use of this second variable.

The concept of *pe* as a means of describing oxidation potential was popularized by the Swedish chemist Lars Gunner Sillén in the 1960s (see review by Truesdell, 1968). The idea was to develop an analogy between *pH*, which refers to hydrated protons, and *pe*, which would refer to hydrated electrons. Like the proton, the electron would be assigned a standard $\Delta_f G^\circ = 0$. Like pH, pe would be defined in terms of activity:

$$pH = -\log a_{\mathrm{H^+}}$$

$$pe = -\log a_e \tag{18.38}$$

To develop the mathematical description of pe, start by considering the half-cell reduction reaction:

$$b\mathrm{B} + c\mathrm{C} + ne = d\mathrm{D} + f\mathrm{F} \tag{18.39}$$

The equilibrium constant for this reaction is

$$K = a_{\mathrm{F}}^f a_{\mathrm{D}}^d / (a_{\mathrm{B}}^b a_{\mathrm{C}}^c a_e^n)$$

and can be calculated in the usual way from $\ln K = -\Delta_r G^\circ / RT$, by assigning a value of zero to $\Delta_f G_e^\circ$. We want to use the a_e^n term separately, so we split up the activity ratio term into two parts,

$$\log K = -n \log a_e + \log \prod_i a_i^{\nu_i} \tag{18.40}$$

where $a_i^{\nu_i}$ now refers to all products and reactants except the electron. As in any equilibrium, the activities a_i and a_e can take on a large range of values, as long as (18.40) is satisfied. If we let Q represent the term $\prod_i a_i^{\nu_i}$, (18.40) becomes

$$\log K = -n \log a_e + \log Q$$

or

$$\log K = n\,pe + \log Q \tag{18.41}$$

If all reactants and products (except the hydrated electron) are in their standard states, $Q = 1$, and (18.41) becomes

$$\log K = n\,pe^\circ$$

or

$$pe^\circ = \frac{1}{n} \log K \tag{18.42}$$

The superscript $^\circ$ on pe° indicates that the activities of all reaction constituents except the electron are in their standard states when (18.42) applies. The energy level of the electron would be different in different half-cells and need not be considered to have a standard state.

Combining (18.41) and (18.42) produces

$$pe = pe^\circ - \frac{1}{n} \log Q \tag{18.43}$$

which is analogous to (18.18).

18.3.7. *Comparison of pe and Eh*

There is a simple, linear relationship between the two variables *pe* and *Eh*, which is evident in comparing the two defining equations:

$$Eh = \mathcal{E}^\circ - 2.3026\frac{RT}{n\mathcal{F}} \log Q \tag{18.18}$$

$$pe = pe^\circ - \frac{1}{n} \log Q \tag{18.43}$$

Hence

$$\begin{aligned} pe &= Eh(\mathcal{F}/2.3026RT) \\ &= 5040 Eh/T \\ &= 16.903\, Eh \text{ at } 298.15 \text{ K} \end{aligned} \tag{18.44}$$

Thus, if you prefer, you can calculate *Eh* and convert to *pe* with equation (18.44), or vice versa. The relationship between the two scales is illustrated on Figure 18.2 where oxidation potential is plotted as both *Eh* and *pe*. An *Eh-pH* diagram looks exactly like a *pe-pH* diagram, except that the Y axis is shifted by the factor $5040/T$.

18.3.8. *Example Calculation, pe–pH*

Because all *pe–pH* calculations are similar, we will only use one example to illustrate the general method.

For boundary $Mn(OH)_2$–Mn_3O_4,

$$Mn_3O_4(c) + 2H_2O(aq) + 2H^+ + 2e = 3Mn(OH)_2$$

$$\begin{aligned} \Delta_r \bar{G}^\circ &= -94140 \text{ J} \\ &= -RT \ln K \end{aligned}$$

Therefore

$$\begin{aligned} \log K &= 94140/(2.303 \times 8.3144 \times 298.15) \\ &= 16.50 \end{aligned}$$

From (18.42)

$$pe^\circ = 16.50/2 = 8.25$$

and from (18.43)

$$pe = 8.25 - \frac{1}{2} \log(1/a_{H^+}^2)$$

or

$$pe = 8.25 - pH \tag{18.45}$$

pe is calculated from *Eh* as follows.

In the above section on Eh, we showed that the boundary $Mn(OH)_2$–Mn_3O_4 can be described by the equation

$$Eh = 0.488 - 0.059pH \tag{18.24}$$

This could be used to calculate an equation for the same boundary in the variables pe and pH from relationship (18.44):

$$pe = 16.903\,Eh \tag{18.44}$$

Substituting (18.44) into (18.24) does reproduce the pe–pH equation (18.45) exactly, showing that the two variables Eh and pe are interchangeable for practical purposes and you can choose whichever you prefer to calculate redox boundaries.

18.3.9. *On the Physical Meaning of pe*

In the above treatment of the variable pe, we began with the definition

$$pe = -\log a_e \tag{18.38}$$

and finished with equation (18.44) relating pe, Eh, and T. We know that Eh, being a half-cell voltage, is not directly measurable, but it is a component of measurable cell voltages, and its link with physical reality is reasonably direct. What about pe, then—is there some specific activity or concentration of aqueous electrons in ionic solutions, and can we measure this activity with an electrode much as we measure H^+ activity with a pH electrode?

It has been amply demonstrated that hydrated electrons do not exist at equilibrium in aqueous solutions (except in the presence of high energy radiation). Therefore in definitions of pe such as (18.38) the electrons are usually referred to as hypothetical and their physical state is not specified. Hostetler (1984) considers two possible physicochemical states electrons could have (even if hypothetical) in aqueous solutions: (i) electrons in an inert solid Pt sample electrode (called $e_{(SE)}$) and (ii) hydrated electrons existing as independent species in aqueous solution ($e_{(aq)}$). He then considers each of these possibilities in detail, and concludes that neither case can be considered to be represented by the pe term in equations such as (18.38). In other words, the quantities pe (in equation 18.44), $pe_{(SE)}$ and $pe_{(aq)}$ are all different quantities. pe in equation (18.44) must refer to some other completely hypothetical, unspecified state that we might call pe_x. Does this mean that pe calculations are in some way physically unjustified or incorrect?

When it comes to physical meaning, parameters in the thermodynamic model sometimes have meanings that are difficult to relate to the real world, as we have seen, particularly in the case of standard states. The example most relevant here is oxygen fugacity, another measure of the state of oxidation of systems, which we will consider in a later section. Oxygen fugacity is often used as a parameter in systems which contain no O_2, just as water contains no free electrons. Still, there are other systems where fugacity approximates partial pressures, and this is a link to reality that *pe* apparently does not have. *pe* is basically a numerical transformation of Eh, but it has an appealing analogy to pH, and there is nothing wrong in using it, whatever it "really" means.

18.4. MEASUREMENT OF Eh

Countless numbers of Eh measurements have been made in natural aqueous and sedimentary systems since the early 1950s by geochemists, biologists, soil scientists and others interested in the surface environment. A comprehensive summary of over 6,000 such measurements is available in an often-cited paper by Baas Becking et al. (1960). Unfortunately, it is possible that many of these measurements have limited physical significance.

18.4.1. *Some Difficulties with Eh Field Measurements*

In more recent years it has been recognized that there are a great many problems involved in measuring physically meaningful oxidation potentials in multi-component natural systems. First of all, natural waters typically contain a great many components, many of which can function as independent redox half-cells. Eh measurements are usually made with a Pt sample electrode and Ag–AgCl or calomel reference electrodes as illustrated in Figure 18.4 and described below. With such electrodes it is essential that all redox processes occur together simultaneously and reversibly at the Pt electrode. In general, this is not the case. Imagine, for example, a lake-bottom water containing many redox couples such as Fe^{2+}/Fe^{3+}, $Mn^{2+}/Mn^{3+}/Mn^{4+}$, H_2S/SO_4^{2-}, NO_3^-/N_2, CO_2/CH_4, and so on. Some of these couples equilibrate rapidly, and others very slowly (particularly the last three). If a stable Eh reading should be obtained in such systems, it is not clear what particular redox couple is involved. Worse still, it is possible for large organic molecules such as humic acids to adsorb to or "poison" the Pt electrode, changing its response. Even molecular oxygen is slow to react at 25°C; for example, Baas Becking et al. (1960, p. 247) observed that measured Eh of seawater was independent of total dissolved O_2. Finally, many natural systems are too dilute or contain redox couples that are so slow to react that it is not possible to obtain stable readings. Systems of this type are said to be poorly poised. There is an analogy here with pH measurements in very pure or weakly buffered systems; for example, it is practically impossible to obtain a stable pH measurement with degassed, distilled water because the system contains almost nothing to which electrodes can respond.

Therefore in general, Eh measurements in natural systems are quite often of only qualitative significance (Morris and Stumm, 1967; Nordstrom and Munoz, 1985; Drever, 1988), although there are systems where Eh can be quantitatively related to independently determined redox couples, such as Fe-rich river waters or acid-mine drainage waters (Crerar et al., 1981; Kleinmann et al., 1981; see also Nordstrom et al., 1979). Field Eh measurements can be very useful in systems where one redox couple (such as Fe^{2+}/Fe^{3+}) predominates, but the chemical composition of the solution should be well characterized before such data are used quantitatively. And in spite of the difficulties of measurement in natural systems, we will always need some way of discussing such systems theoretically, and Eh is much used in this way.

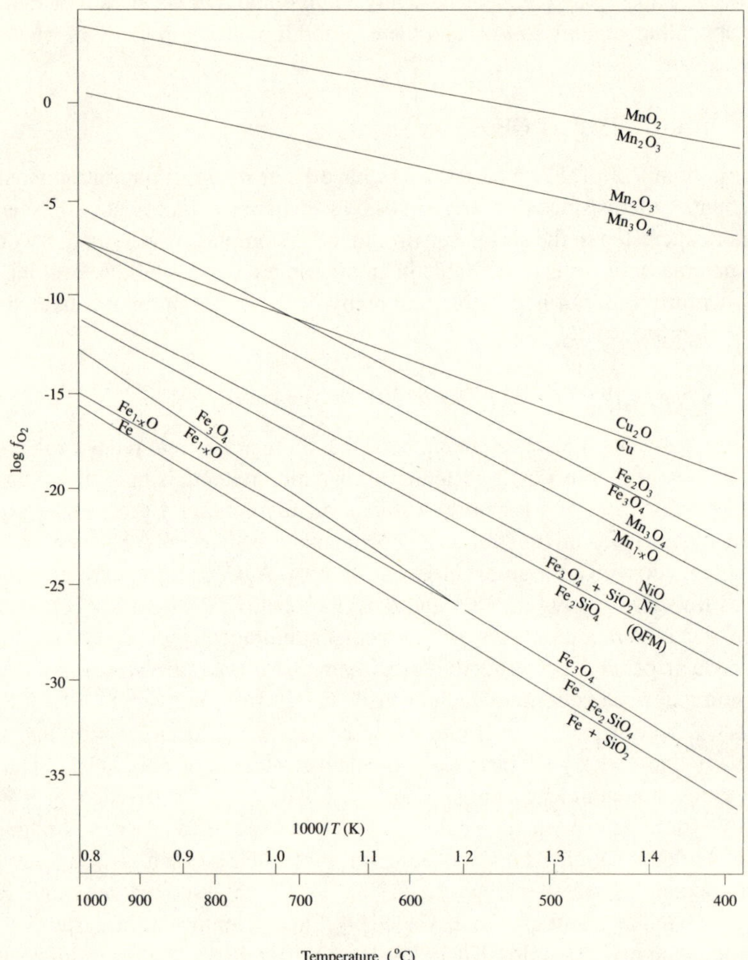

FIG. 18.5. Oxygen fugacities of selected mineral buffer assemblages at 1 bar and 400°C to 1000°C.

18.4.2. *Alternative Measures of Oxidation Potential in Low-Temperature Environments*

We have observed above that Eh measurements in natural systems may be difficult to interpret quantitatively because all possible redox couples might not have equilibrated with each other. In simple systems where dissolved Fe is the principal redox-active component, Eh can sometimes be used to predict the Fe^{3+}/Fe^{2+} activity ratio, but this should not be attempted if other redox couples are present. As an alternative to electrochemical measurements, the activity ratios of interest can often be analyzed directly. For example, there is often a problem with chromium, zinc and arsenic pollution at sites where wood was treated with preservatives. The solubilities of the oxides and hydroxides of these components are highly dependent on valence state, which thus determines the environmental hazard. At such sites you can usually obtain stable Eh readings, but it is not clear to which component they refer. It is therefore safer to chemically analyze ratios of $HAsO_2/H_3AsO_4$ (arsenious to arsenic acid), $Cr^{3+}/HCrO_4^-$ (chromic/bichromate) and so on, and then to check if redox equilibrium has been reached among all species. Also, because molecular oxygen is slow to react at 25°C, environmentalists would never use Eh as a measure of dissolved O_2, but instead measure dissolved oxygen directly, or determine related properties such as biological and chemical oxygen demands (BOD and COD tests). In many such systems, Eh is better used as a kind of qualitative indicator of redox conditions.

The problems at lower temperatures happen because reactions are slow and systems are often far out of equilibrium. (The thermodynamically stable state for all the flora and fauna in any well-aerated lake should be H_2O and CO_2, for example). Reaction rates increase exponentially with temperature however, and at temperatures of 350°C or higher, even reactions involving molecular oxygen can equilibrate in several days time. This is fortunate, because free energy data are sparse for aqueous species at elevated temperatures and it becomes difficult to measure or calculate Eh much above 100°C. Instead, we can use other reactions between redox-sensitive species to calculate redox-related parameters at higher temperatures. This is the subject of the following section.

18.5. OXYGEN FUGACITY

As pointed out in the first section of this chapter, there are many different ways of representing the same fundamental variable—the oxidation state of a system. Of these alternatives, the oxygen fugacity is most commonly used in geochemistry, particularly at higher temperatures.

This is a convenient parameter for a number of reasons. First, many oxidation reactions can be written to include molecular oxygen as a reactant, for example the reduction of hematite to magnetite.

$$6Fe_2O_3 = 4Fe_3O_4 + O_2(g) \tag{18.46}$$

Assuming unit activity for the solids, the equilibrium constant for this reaction is simply

$$\begin{aligned} K &= f_{O_2} \\ &= \exp(-\Delta_r G^\circ/RT) \end{aligned} \tag{18.47}$$

Oxygen fugacities for selected mineral assemblages are plotted versus $1000/T(\mathrm{K})$ and a fixed pressure of 1 bar in Figure 18.5 using data from Huebner (1971) and Chou (1987). Each assemblage is divariant according to the phase rule; restricting the diagram to a constant pressure produces the univariant lines illustrated here.

To find the oxygen fugacity for pressures different from 1 bar, the standard states for the equilibrium constant (18.47) are most conveniently chosen as pure solids at T

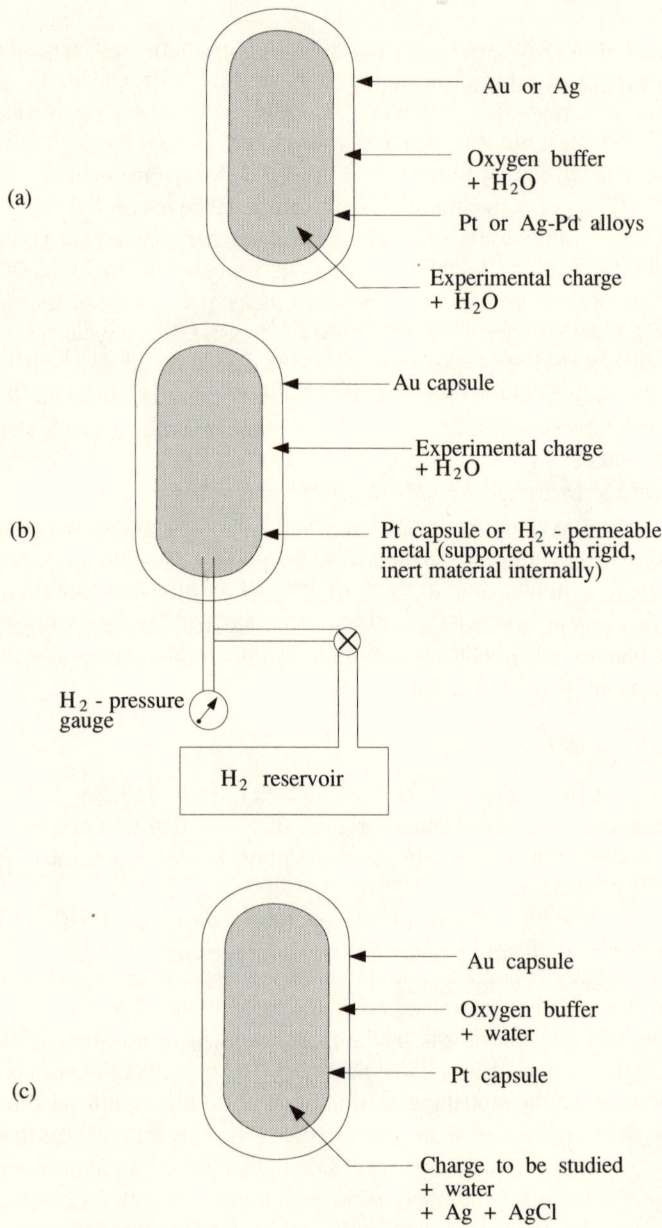

FIG. 18.6. Experimental control and measurement of hydrogen fugacity. (a) Schematic arrangement of an oxygen-buffered experiment. (b) Shaw membrane for external control of hydrogen fugacity. Apparatus can also be used as a hydrogen sensor to measure H_2 pressure externally. (c) Double capsule method of controlling both oxygen fugacity and HCl activity within an inner, Pt capsule.

and P, and ideal $O_2(g)$ at T and 1 bar. From our discussion in Chapter 13 (equation (13.42)) we find that this results in

$$\ln K_{P_2} = \ln K_{P_1} - \Delta_s V^\circ (P_2 - P_1)/RT$$

or in this case

$$\log f_{O_2,P} = \log f_{O_2,1\text{bar}} - \frac{(4V_{magnetite} - 6V_{hematite})(P-1)}{2.303RT} \qquad (18.48)$$

For the magnetite–hematite assemblage (18.46), $\Delta_s V = 0.3548\ \text{J bar}^{-1}$. At $T =$ 1000 K, this works out to a pressure correction to $\log f_{O_2}$ of 0.0185 per kbar, which is small but significant (at 1 bar and 1000 K, $\log f_{O_2}$ is -10.86, so this is a change of +0.17% per kbar).

18.5.1. *On the Physical Significance of Oxygen Fugacity*

With the exception of the Mn oxides, the other mineral assemblages shown in Figure 18.5 coexist at very low oxygen fugacities or pressures. These fugacities also increase exponentially with temperature, so at low T, the calculated equilibrium oxygen fugacities can become absurdly low. As an example, calculate the oxygen fugacity at 25°C for coexistence of hematite and magnetite—a common enough assemblage in iron formations. From (18.47),

$$\begin{aligned}\log f_{O_2} &= \log\ K_{298} \\ &= -(4\Delta_f G^\circ_{mag} - 6\Delta_f G^\circ_{hem})/(2.303RT) \\ &= -72\end{aligned}$$

To indicate how small a number this is, remember that at 1 bar and 0°C, 1 mole (or 6×10^{23} molecules) of gas occupies 22.7 liters. A pressure of 10^{-72} bars would thus be exerted by about 3 molecules of $O_2(g)$ in 10^{50} liters or 10^{47} cubic meters of space—a volume considerably larger than that of a sphere having our Solar System's diameter. What kind of significance can this possibly have?

Asking this question is one of the many ways of demonstrating the confusion between reality and the thermodynamic model. The fugacity of oxygen is a parameter of the thermodynamic model that has no more or no less significance for hematite–magnetite than it has for air or any other system. The only difference is that in air it is a parameter with a numerical value very close to that of the partial pressure of oxygen, while in hematite–magnetite there is no such thing as a partial pressure of oxygen. Nevertheless, the oxygen fugacity is linked by equilibrium constants to all other constituent activities in both systems, many of which *are* physically significant. The oxygen fugacity is approximately equal to a pressure in many systems, but it is also a point on a surface in a mathematical model. As such, its logarithm has no difficulty in sliding down to minus infinity and back if required. It is a parameter that indicates the oxidation state of a system, much like a thermometer indicates the temperature of a system. The fact that the fugacity scale chosen (by the $\lim_{P\to 0} f/P \to 1$ part of our definition) produces some quite low numbers in some cases is immaterial.

Neither does using oxygen fugacities imply anything about the mechanism of oxidation or reduction in real systems. Evidently, systems such as reducing sediments at ordinary temperatures contain essentially no molecular O_2, so that redox couples such as hematite-magnetite will actually react by some other mechanism, such as

$$6Fe_2O_3 + 2H_2 = 4Fe_3O_4 + 2H_2O$$

or

$$12Fe_2O_3 + CH_4 = 8Fe_3O_4 + CO_2 + 2H_2O$$

or any of dozens of other possibilities. But all of these possibilities are linked to the oxygen fugacity by the thermodynamic model, and none is more significant than any other, except perhaps to a scientist using the thermodynamic model to help find out what is "really" happening in nature.

18.5.2. *Oxygen Buffers in Field and Laboratory*

As mentioned in Chapter 14, mineral assemblages are often used in experimental studies to control oxidation states, as first proposed by Hans Eugster (1957). This was used by Eugster and David Wones in their classic studies of the oxidation state of the Fe-rich mica, annite (Eugster and Wones, 1962), and in a great many different studies since then (see review by Chou, 1987). The simplest arrangement is illustrated in Figure 18.6a. An experimental mineral charge is placed with water in a welded Pt or Ag–Pd alloy tube; this is enclosed in a sealed Au tube that contains a mineral f_{O_2} buffer (such as hematite–magnetite) + water. Both capsules are then placed in a high-P autoclave and heated. The inner, Pt capsule is relatively permeable to $H_2(g)$, while the outer Au capsule is not. At equilibrium, f_{O_2} in the inner and outer capsules is identical and f_{O_2} in the inner sample tube is therefore known.

A variation on this idea connects the inner Pt capsule (which is supported with something rigid like quartz crystals) to the exterior with a capillary tube. A known H_2 pressure can then be introduced from an external tank or reservoir; this was originally designed by Shaw (1963) and is called a Shaw membrane. The variation by Gunter et al. (1987) shown in Figure 18.6b permits H_2 pressure generated in the outer sample tube to be monitored with a sensitive pressure transducer or other device; this is called a hydrogen sensor and is a way of measuring f_{H_2} (and therefore f_{O_2}) directly.

Figure 18.6c shows a third double-capsule arrangement that is used to control both f_{H_2} and HCl activity within the inner, Pt capsule. This was developed by Eugster and his students in the 1970s. Within the inner tube, chlorine fugacity f_{Cl_2} is controlled by the equilibrium between Ag and AgCl (which are included along with the sample):

$$AgCl = Ag + \frac{1}{2}Cl_2$$

The buffer assemblage in the outer capsule fixes f_{H_2}; this controls HCl activity in the inner capsule according to the equilibrium

$$\frac{1}{2}H_2 + \frac{1}{2}Cl_2 = HCl$$

Modifications of this method are summarized by Eugster et al. (1987).

Probably the simplest use of f_{O_2} mineral buffers is to include the buffer assemblage as part of the total system being studied. For example, the assemblage pyrite + pyrrhotite + magnetite fixes f_{O_2} at any T and P by the reaction

$$\frac{1}{2}Fe_3O_4 + \frac{3}{2}FeS_2 = 3FeS + O_2(g) \tag{18.49}$$

At the same time, this three-phase assemblage fixes f_{S_2}.

$$FeS + \frac{1}{2}S_2 = FeS_2 \tag{18.50}$$

This double f_{O_2} + f_{S_2} buffer has been used to study metal ion complexing in hydrothermal solutions in the pyrite + pyrrhotite + magnetite system alone (Crerar et al., 1978), as well as in more complicated systems containing the same assemblage plus sphalerite, galena, gold, stibnite, bismuthinite, argentite and molybdenite (Wood et al., 1987). Buffers like this add two more independent intensive variables to those (such as T and P) that can be controlled or measured in laboratory experiments. This extra information is often essential in multicomponent experiments, because according to the phase rule ($f = c - p + 2$) the number of intensive, independent parameters f required to characterize a system increases directly with c, the number of components.

18.5.3. *Solid Electrolyte Fugacity Sensors*

It is also possible to measure oxygen and other gas fugacities with electrochemical sensors. These methods use the same theoretical principles that apply to measurement of Eh. The design and operation of such sensors is described in detail by Sato (1971) and Huebner (1987). We will illustrate the technique briefly, by describing an oxygen fugacity sensor.

The basic principle of operation of a solid electrolyte fugacity sensor is illustrated schematically in Figure 18.7. Two chambers are completely sealed from each other by a solid electrolyte (material through which ions can migrate). The solid electrolyte used for f_{O_2} measurements is most commonly a solid solution of zirconia (ZrO_2) and 15 mole % CaO or 5-10 mole % Y_2O_3. Pure zirconia is an electrical insulator (conductivity $\sim 10^{-8}$ ohm^{-1} cm^{-1}). The Ca or Y doping increases the electrical conductivity to $\sim 10^{-1}$ ohm^{-1} cm^{-1} (at 1000°C) by creating vacant anion sites that permit migration of O^{2-} ions through the crystal structure. Over a certain range of f_{O_2} and T, most electrical current in the electrolyte is ionic, and it exhibits Nernstian behavior. (With zirconia electrolytes this happens for $-18 < \log f_{O_2} < 0$, and 700°C $< T <$ 1200°C.) If the f_{O_2} standard in the cell shown in Figure 18.7 exceeds the unknown f_{O_2}, then $O_2(g)$ will be spontaneously reduced in the standard compartment, and O^{2-} ions will be oxidized in the sample compartment according to the two half-cell reactions.

$$\text{Standard half cell}: \quad O_{2\text{std}} + 4e \rightarrow 2O^{2-} \qquad \mathcal{E}_{\text{red}}$$

$$\text{Sample half cell}: \quad 2O^{2-} \rightarrow 4e + O_{2\text{sample}} \qquad \mathcal{E}_{\text{ox}}$$

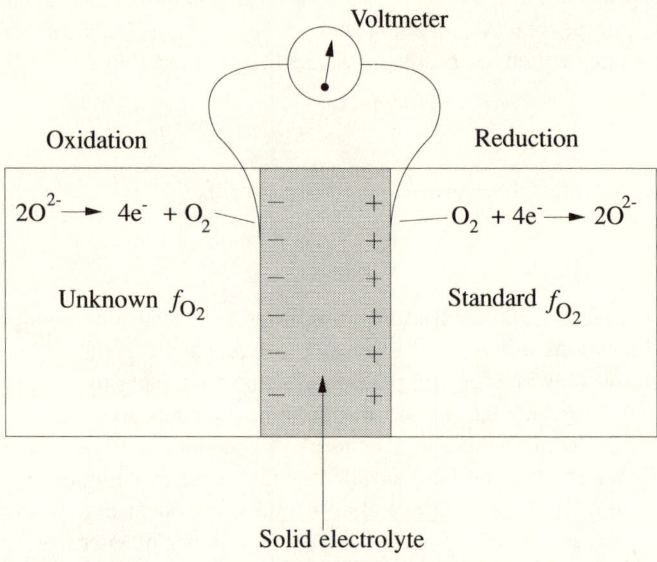

FIG. 18.7. Schematic solid electrolyte oxygen fugacity sensor, where f_{O_2}(standard) > f_{O_2}(unknown).

The whole cell reaction is the sum of these two.

$$\text{Whole cell :}\quad O_{2\,\text{std}} = O_{2\,\text{sample}}$$

In this case $\mathcal{E}_{\text{red}} = \mathcal{E}_{\text{ox}}$ so that $\mathcal{E}_{\text{Whole cell}} = 0$ V. This just means that if pure oxygen at the same T and P is on both sides of the electrolyte, no voltage will be obtained. The cell develops a voltage from the difference in pressure of oxygen on the two sides. Therefore

$$Eh_{\text{cell}} = -\frac{RT}{n\mathcal{F}} \ln \left(\frac{f_{O_2,\text{std}}}{f_{O_2,\text{sample}}} \right) \tag{18.51}$$

Thus the measured EMF of the cell illustrated in Figure 18.7 gives the f_{O_2} of the sample relative to that of a standard (which can be air or pure oxygen). When the circuit is connected in the solid electrolyte cell, O_2 molecules in the standard cell (right side) are reduced at the electrolyte surface to O^{2-} ions and migrate into the solid. At the other sample side of the electrolyte, O^{2-} ions are oxidized to $O_2(g)$; the electrons produced here pass through the external circuit to the right-hand compartment where they are consumed by O_2 reduction. The overall process is analogous to the two-electrode system of Figure 18.4 used to measure Eh; the solid electrolyte serves roughly the same purpose as the liquid junction of a calomel or AgCl reference electrode. Solid electrolyte f_{O_2} sensors have been used to measure f_{O_2} of volcanic gases and lavas and of mineral assemblages in the laboratory; they are also used routinely with blast furnaces and in testing automobile and furnace exhaust systems

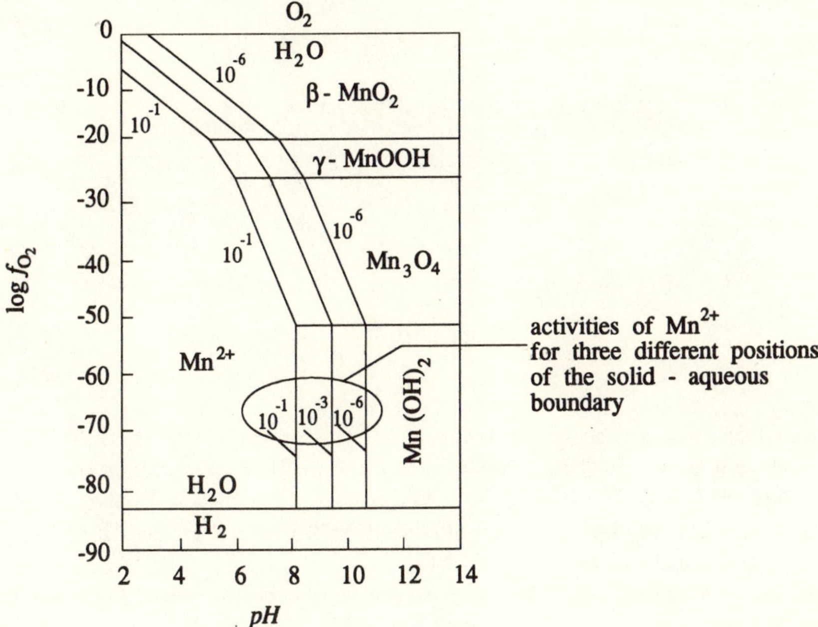

FIG. 18.8. log f_{O_2} – *pH* diagram for the system Mn – H_2O – O_2 at 25°C, 1 atm.

(Huebner, 1987). Solid electrolyte cells have also been designed to measure other gas activities such as those of sulfur and the halogens.

18.5.4. *Calculation of Oxygen Fugacity—pH Diagrams*

Because f_{O_2} and Eh are both indicators of the same thing—oxidation state—it is possible to draw log f_{O_2}–pH diagrams that are analogous to the Eh–pH calculations we have outlined above. To illustrate this we will construct a log f_{O_2}–pH diagram for the same Mn–H_2O–O_2 system at 25°C already described. The completed diagram is shown in Figure 18.8 and should be compared with the analogous Eh–pH diagram of Figure 18.2. The two diagrams are similar except that the phase boundaries on the Eh-pH diagram have the Nernst slope while those on the log f_{O_2}–pH diagram have zero slope. This is because the Nernst equation is not used in any way to calculate the f_{O_2} phase boundaries.

The method of calculating log f_{O_2}–pH boundaries is illustrated with three examples. All other boundaries are derived the same way. Our examples include the boundaries for water stability and for coexisting minerals, and the aqueous solubility contours of a mineral. Notice that half-cell reactions are not involved in these calculations.

For the water stability boundaries, the dissociation reaction of water is

$$2H_2O = O_2(g) + 2H_2(g) \tag{18.52}$$

To calculate the equilibrium constant for this reaction,

$$\begin{aligned}\Delta_r G^\circ &= -2(-237129)\\ &= 474258\ \mathrm{J}\\ &= -RT\ln K\end{aligned}$$

$$\begin{aligned}K &= 10^{-83.1}\\ &= f_{H_2}^2 f_{O_2}\end{aligned} \tag{18.53}$$

The upper boundary occurs at 1 atm $O_2(g)$ pressure or $\log f_{O_2} = 0$. The lower boundary is at 1 atm $H_2(g)$ pressure; from the equilibrium constant (18.53) this corresponds to $\log f_{O_2} = -83.1$. As noted before, water can exist outside of these boundaries, but only if the pressure of oxygen or hydrogen is greater than 1 bar.

For the boundary $Mn(OH)_2 - Mn_3O_4$,

$$0.5\,O_2(g) + 3Mn(OH)_2 = Mn_3O_4 + 3H_2O \tag{18.54}$$

$$\begin{aligned}\Delta_r G^\circ &= -143093\ \mathrm{J}\\ &= -RT\ \ln(1/f_{O_2}^{\frac{1}{2}})\end{aligned}$$

Hence

$$\log f_{O_2} = -50.14 \tag{18.55}$$

For solubility of Mn_3O_4 as Mn^{2+},

$$3H_2O(aq) + 3Mn^{2+} + \frac{1}{2}O_2(g) = Mn_3O_4 + 6H^+ \tag{18.56}$$

$$\begin{aligned}\Delta_r \bar{G}^\circ &= 114223\ \mathrm{J}\\ &= -RT\ln\left(\frac{a_{H^+}^6}{f_{O_2}^{\frac{1}{2}} a_{Mn^{2+}}^3}\right)\end{aligned}$$

Or

$$\log f_{O_2} = 40 - 12pH - 6\log a_{Mn^{2+}} \tag{18.57}$$

This is plotted for selected values of Mn^{2+} activity ranging from 10^{-1} to 10^{-6} in Figure 18.8.

18.5.5. *Interrelating Eh, pH, and Oxygen Fugacity*

The obvious similarity between the Eh–pH and log f_{O_2}–pH diagrams of Figures 18.2 and 18.8 suggests that it should be possible to convert directly from one set of coordinates to the other. This can be done using the half-cell reaction

$$4H^+ + O_2(g) + 4e = 2H_2O(aq) \tag{18.58}$$

and its related Nernst equation,

$$Eh = \mathcal{E}^\circ - \frac{RT}{n\mathcal{F}} \ln[1/(f_{O_2} a^4_{H^+})] \tag{18.59}$$

This equation can be used to interrelate the three variables Eh, pH, and f_{O_2}, as in §18.3.2.

At 25°C, the standard free energy for reaction (18.58) is

$$\begin{aligned} \Delta_r \bar{G}^\circ &= 2(-237129) = -474258\,\text{J} \\ &= -n\mathcal{F}\mathcal{E}^\circ \end{aligned}$$

and therefore $\mathcal{E}^\circ = 1.23$ V and

$$Eh = 1.23 + 0.0148 \log f_{O_2} - 0.0592pH \tag{18.60}$$

Figure 18.9 shows Eh contours calculated from (18.60) drawn on a log $f_{O_2} - pH$ diagram. The fact that we can do this suggests that the true master variable is redox or oxidation state and that f_{O_2}, f_{H_2}, Eh, pe and all other related variables are simply different ways of quantifying the same thing.

18.5.6. *Summary and Comment on Eh versus f_{O_2} as Parameters to Indicate Redox Conditions*

On a broad scale the Earth shows a large range of redox conditions, from the highly reduced Ni–Fe core through various silicate layers up into the zone of free water and eventually into the oxygen-rich atmosphere. Therefore an indicator of the redox state is among the more important of the variables manipulated by geochemists. Like pH, it is an important parameter because it is intimately linked to a large number of reactions of interest to anyone trying to understand the Earth, but like pH it is actually no more or less fundamental than any other intensive parameter. For a closed system at a given T and P, it is completely determined by the bulk composition, as are all intensive parameters (Duhem's Law, §5.4.1), and changes in redox state are accomplished by changing bulk composition. We recommend that students remember Prof. Duhem when they are tempted to "explain" some phenomenon as due to a change in redox conditions. It may be that the phenomenon is linked to a change in redox conditions, but the question is, what changed the redox conditions?

The measurement of redox conditions by means of a cell voltage, where one electrode has a fixed reference potential and the other is expected to react reversibly with natural systems, is attended by a number of problems. The platinum electrode only works well under certain conditions, it is difficult to get the electrode into

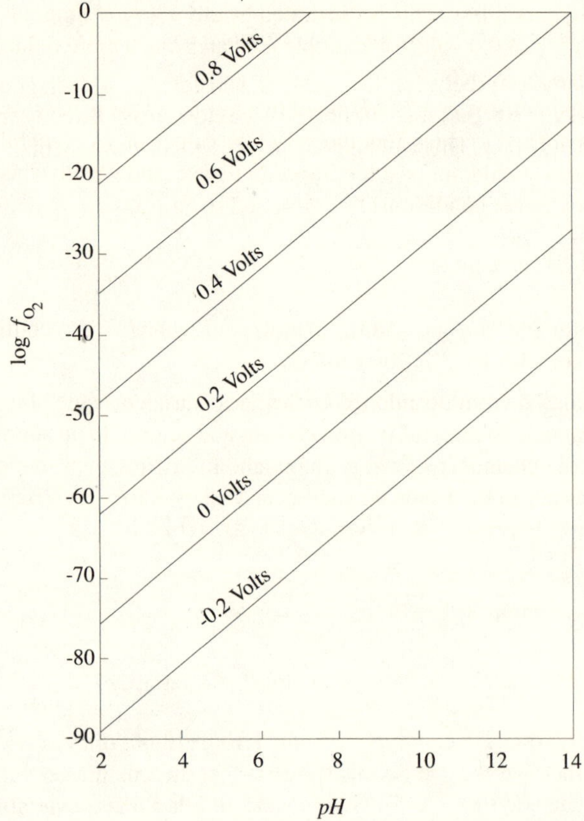

FIG. 18.9. *Eh* contours (in volts) on log $f_{O_2} - pH$ axes, calculated for 25°C, 1 atm.

reducing environments without allowing some oxidation, and the method is restricted to ambient conditions except in research laboratories. We put up with these problems because there is no choice. Oxygen fugacity, on the other hand, although a much simpler concept, can only be directly measured at high temperatures, which seems to rule out its use at Earth-surface conditions where redox measurements are so often required.

Eh has been important in both the measuring and reporting of redox conditions, but an argument can be made that it should be used only in the measurement and not the reporting of redox conditions, i.e., that however measurements are *made*, results should be *reported* as f_{O_2} or log f_{O_2} values. There would be two advantages to this. First, the use of *Eh* entails the use of a relatively complex set of conventions, which are quite difficult to remember unless used continuously (see Appendix D for a complete comparison of conventions). Secondly, and more importantly, *Eh* is less useful than f_{O_2} because it is so commonly linked in reactions with *pH* (giving the "Nernst slope"). This means that a value for *Eh* without an accompanying value for

pH is usually meaningless. This is illustrated by any Eh–pH diagram, in which you can see that an Eh of 0.0 volts for example indicates much more reducing conditions at pH 2 than it does at pH 10.

The conversion from an Eh–pH point to a log f_{O_2} value is very simple (equation 18.60). We have tried to show that the very low values of f_{O_2} generally obtained at low temperatures should not be a hindrance to its use, and we advocate its use in the reporting of any redox conditions.

PROBLEMS

1. If $\Delta_f G°$ for $Pb^{2+}(aq)$ is -5832 cal mol^{-1}, at what Eh is metallic lead in equilibrium with $10^{-6} m$ $Pb^{2+}(aq)$?
2. Water draining from abandoned coal mines causes considerable environmental damage because of its iron content and acidity. In an attempt to understand the chemistry of these waters, the following measurements were made in water in the Loree #2 shaft near Wilkes-Barre, PA. (Barnes et al., USGS Prof. Paper 473-B, 1964).

Sample No.	Depth (ft.)	Eh volts	pH	Fe^{2+} (ppm)	SO_4^{2-} (ppm)
1	224	+0.322	3.41	34	1260
2	292	+0.533	3.55	478	3320
3	467	+0.568	3.36	488	3650
4	751	-0.025	4.0	1073	5800
5	808	-0.103	3.72	1463	6720

Samples 2 and 3 showed a yellow turbidity, the rest were clear. Samples 4 and 5 smelled of rotten eggs, the rest were odorless. Construct an Eh–pH diagram and use it to answer the following questions: 1. Is the yellow turbidity due to fine particles of $Fe(OH)_3$? If so, is the mine water more or less in equilibrium with it or is it actively precipitating? 2. Why is sample 1. clear? 3. Is pyrite stable in any of these samples? 4. In samples 4 and 5 which contain H_2S, are H_2S and SO_4^{2-} apparently at equilibrium, or is sulfur being oxidized or reduced? 5. Above what pH might pyrite and ferric hydroxide coexist? Use -6660 and -166500 cal mol^{-1} for the $\Delta_f G°$ of $H_2S(aq)$ and $Fe(OH)_3$ respectively.

The following half-cell reactions will be useful:

$$Fe^{2+} + 3H_2O = Fe(OH)_3 + 3H^+ + e$$

$$FeS_2 + 8H_2O = Fe^{2+} + 2SO_4^{2-} + 16H^+ + 14e$$

$$H_2S + 4H_2O = 10H^+ + SO_4^{2-} + 8e$$

$$FeS_2 + 11H_2O = Fe(OH)_3 + 2SO_4^{2-} + 19H^+ + 15e$$

Strictly speaking you should draw several diagrams for the different water compositions, or several sets of boundaries on one diagram. You must experiment to determine how many boundaries are necessary to answer the questions. For example, you could safely draw one Fe^{2+} – $Fe(OH)_3$ boundary to consider both samples 2 and 3.

3. Calculate the activity of Fe in coexisting magnetite and hematite at 700°C, 1 atm (the relevant equation is shown in §14.7). Calculate a_{Fe} for the other iron-mineral pairs FeO – Fe_3O_4, FeO – Fe at the same T, P. This should be sufficient to demonstrate the use of a_{Fe} as a redox indicator.
4. Of course, our new a_{Fe} indicator is limited to systems containing iron. But then, f_{O_2} is limited to systems containing oxygen (though not necessarily O_2). True? On the other hand, a_{Fe} can also be used as a sulfidation indicator. Calculate a_{Fe} in coexisting pyrite and pyrrhotite in which a_{FeS} is 0.9, at 500 K, 1 bar.

19

SPECIATION CALCULATIONS

> Recent advances in solution chemistry, thermodynamics and computer technology make it possible to describe quantitatively equilibrium and mass transfer among minerals and aqueous electrolytes in geochemical processes involving large numbers of components, phases, and chemical species at both high and low temperatures and pressures.
>
> Helgeson, H.C., and Kirkham, D.H. (1974)

This chapter focuses on one of the most common questions asked about natural chemical systems: what are the concentrations or activities of the different species present in a system at complete chemical equilibrium? We might be concerned, for example, with oxygen or sulfur fugacities, with the activities of complex ions, or activity ratios of reduced and oxidized species of the same component.

In practice, these calculations range from trivially simple to enormously complex, depending on the number of species (and components) in the system. We will follow roughly this order—from trivial to complex—and outline some of the most common approaches used in performing speciation calculations.

19.1. ROTE METHOD FOR SIMPLE SYSTEMS

This simplest procedure is probably used most often, and works best with systems containing relatively few chemical species. As a general rule of thumb, you might try this if there are fewer than 10 species, but move on to another more sophisticated method for more complicated systems. As an example, we will solve for the equilibrium concentrations of all species in an acetic acid + water solution of a given concentration, m. Specifically, we might be interested in the pH of a 0.1 m HAc solution, but in calculating this we will also get the activities of all other species, whether we need them or not. This is one of the simplest examples imaginable, but the method works exactly the same way with more complicated systems. An excellent reference on this general approach is Butler (1964, Chap. 3). There are six steps to follow:

1. Write all species of relevance or interest. Count the number of unknown species. You will need this many equations.
2. Write all known equilibrium constant equations.
3. If there are charged species, write a charge balance equation.
4. Write all known mass balance equations.
5. You should now have the same number of equations as unknown species. Reduce these by algebraic substitution to one (or two) equations that can be solved for the unknown concentrations. At first, assume all activity coefficients are 1.0.

6. If activity coefficient corrections are required, use the concentrations calculated in step 5 to estimate activity coefficients. Repeat step 5 using these new activity coefficients. Iterate on steps 5 and 6 until successive concentrations and activities stop changing within some acceptable limit.

We will work through our example in detail, remembering that exactly the same procedure applies to more complicated problems. The problem is to calculate all concentrations (or activities) in an m-molal acetic acid solution. Follow the above steps in sequence:

1. Relevant species are: H^+, OH^-, HAc°, and Ac^-. Therefore there are 4 unknowns ($m_{H^+}, m_{OH^-}, m_{HAc^\circ}, m_{Ac^-}$) and we will require 4 equations.
2. Known mass action equations:

$$K_w = m_{H^+} m_{OH^-} \gamma_\pm^2 \tag{19.1}$$

$$K_a = m_{H^+} m_{Ac^-} \gamma_\pm^2 / (m_{HAc^\circ} \gamma_{HAc^\circ}) \tag{19.2}$$

3. Charge balance:

$$m_{H^+} = m_{OH^-} + m_{Ac^-} \tag{19.3}$$

4. Mass balance:

$$m_{HAc,total} = m_{HAc^\circ} + m_{Ac^-} \tag{19.4}$$

Note that to be complete, we should have included H_2O as a fifth species above; in this case equation (19.1) would include water activity as a denominator. Except in extraordinarily concentrated solutions such as those associated with evaporites, this is unnecessary, and water activity can be approximated by 1.0.

5. For the next step, count how many times each unknown occurs in each of the above four equations: m_{H^+} and m_{Ac^-} occur 3 times, and m_{OH^-} and m_{HAc° twice. To reduce these equations to one (or more) equations, start by eliminating those variables that occur the least number of times. We begin by eliminating m_{OH^-}, rearranging (19.1):

$$\boxed{m_{OH^-} = K_w / m_{H^+} \gamma_\pm^2} \tag{19.5}$$

The box is drawn around this equation to remind us that we will return to it to calculate m_{OH^-} at the end.
Next, substitute (19.5) into (19.3) to give

$$m_{H^+} = K_w / (m_{H^+} \gamma_\pm^2) + m_{Ac^-} \tag{19.6}$$

Next, eliminate m_{HAc° from (19.4):

$$\boxed{m_{HAc^\circ} = m_{HAc,total} - m_{Ac^-}} \tag{19.7}$$

Substituting (19.7) into (19.2) produces

$$K_a = \frac{m_{H^+} m_{Ac^-} \gamma_\pm^2}{(m_{HAc,total} - m_{Ac^-})\gamma_{HAc^\circ}} \tag{19.8}$$

Next, eliminate m_{Ac^-} from (19.6):

$$\boxed{m_{Ac^-} = m_{H^+} - K_w/(m_{H^+}\gamma_\pm^2)} \tag{19.9}$$

Finally, plug (19.9) into (19.8) to obtain one final equation in the one unknown, m_{H^+}:

$$K_a = \frac{m_{H^+}(m_{H^+} - K_w/(m_{H^+}\gamma_\pm^2))\gamma_\pm^2}{(m_{HAc,total} - (m_{H^+} - K_w/(m_{H^+}\gamma_\pm^2)))\gamma_{HAc^\circ}} \tag{19.10}$$

Rearranging (19.10) produces the following cubic equation in the variable m_{H^+}:

$$\boxed{\begin{aligned} &m_{H^+}^3\gamma_\pm^4 + (K_a\gamma_{HAc^\circ}\gamma_\pm^2)m_{H^+}^2 \\ &-(K_w\gamma_\pm^2 + K_a\gamma_{HAc^\circ}\gamma_\pm^2 m_{HAc,total})m_{H^+} \\ &-K_wK_a\gamma_{HAc^\circ} \\ &= 0 \end{aligned}} \tag{19.11}$$

This last equation can be solved by Newton-Raphson iteration or some other numerical technique. The first time through, it is simplest to set all activity coefficients equal to 1.0. Equation (19.11) is then solved for m_{H^+}; then (19.9) is solved for m_{Ac^-}, (19.7) for m_{HAc°, and finally, (19.5) for m_{OH^-}. Notice that this takes us back through each of the equations outlined by a box; these were the equations produced each time we first eliminated a variable.

6. These first estimates of the equilibrium concentrations of the species H^+, OH^-, Ac^- and HAc° can now be used to estimate the activity coefficients. If the solution is sufficiently dilute, the Debye-Hückel equation might be used to estimate $\gamma_\pm$; we might estimate γ_{HAc° from the behavior of another neutral-charge species such as CO_2 at the same ionic strength. For more on this, see Chapter 17.
   With these newly estimated activity coefficients, return to equation (19.11), and solve for new concentrations of the four aqueous species. This process is then repeated, or iterated, until successive concentrations and activity coefficients stop changing significantly.

19.1.1. *Simplifying Approximations*

The entire calculation above could have been simplified by making a few chemically reasonable approximations in the original 4 equations (19.1) to (19.4). For example, because HAc is a weak acid, we can safely assume $m_{\mathrm{H^+}} \gg m_{\mathrm{OH^-}}$ and $m_{\mathrm{HAc^\circ}} \gg m_{\mathrm{Ac^-}}$. This simplifies equations (19.3) and (19.4) to

$$m_{\mathrm{H^+}} = m_{\mathrm{Ac^-}} \tag{19.12}$$

and

$$m_{\mathrm{HAc},total} = m_{\mathrm{HAc^\circ}} \tag{19.13}$$

This makes the problem almost trivial. Plug (19.12) and (19.13) back into (19.2) to obtain

$$K_a = m_{\mathrm{Ac^-}}^2 \gamma_\pm^2 / (m_{\mathrm{HAc},total}\gamma_{\mathrm{HAc^\circ}}) \tag{19.14}$$

which may be solved for $m_{\mathrm{Ac^-}}$, and so on. To check your approximations, always substitute the calculated answers back into the original equations (19.1) to (19.4), testing that the left side equals the right side in each case. If it doesn't, then the approximations need to be revised and improved.

19.1.2. *Example: Quartz Solubility in Alkaline Solutions*

To illustrate an equilibrium calculation of the above type, let's choose an example of geothermal interest: the solubility of quartz as a function of pH. Dissolved silica behaves as a very weak acid, H_4SiO_4, and deprotonates to $H_3SiO_4^-$ only at basic pH. If we suppose that basic pH is produced by adding NaOH, then the system of interest is SiO_2(qtz)–H_2O–NaOH. Running through the rote procedure, the species of interest are: $NaOH^\circ$, Na^+, H^+, OH^-, $H_4SiO_4^\circ$, $H_3SiO_4^-$, and $H_2SiO_4^{2-}$. We are presuming that Na–silicate complexing or ion pair formation is insignificant. With 7 unknowns, we require 7 equations which are:

- Mass action equations:

$$K_{\mathrm{NaOH}} = \frac{\gamma_\pm m_{\mathrm{Na^+}} \gamma_\pm m_{\mathrm{OH^-}}}{\gamma^\circ m_{\mathrm{NaOH^\circ}}} \tag{19.15}$$

$$K_{\mathrm{H_4SiO_4}} = \frac{\gamma_\pm m_{\mathrm{H^+}} \gamma_\pm m_{\mathrm{H_3SiO_4^-}}}{\gamma^\circ m_{\mathrm{H_4SiO_4^\circ}}} \tag{19.16}$$

$$K_{\mathrm{H_2O}} = \gamma_\pm m_{\mathrm{H^+}} \gamma_\pm m_{\mathrm{OH^-}} \tag{19.17}$$

$$K_{\mathrm{H_3SiO_4^-}} = \frac{\gamma_\pm m_{\mathrm{H^+}} \gamma_\pm m_{\mathrm{H_2SiO_4^{2-}}}}{\gamma^\circ m_{\mathrm{H_3SiO_4^-}}} \tag{19.18}$$

- Mass balances:

$$\sum \mathrm{Na} = m_{\mathrm{Na^+}} + m_{\mathrm{NaOH^\circ}} \tag{19.19}$$

$$\sum \mathrm{Si} = m_{\mathrm{H_4SiO_4^\circ}} + m_{\mathrm{H_3SiO_4^-}} + m_{\mathrm{H_2SiO_4^{2-}}} \tag{19.20}$$

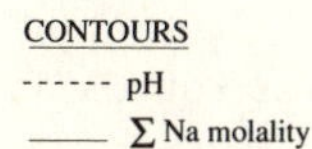

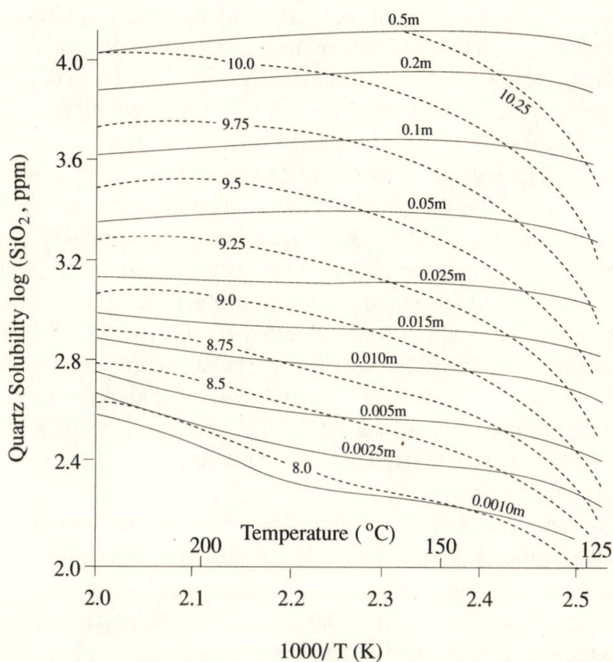

FIG. 19.1. Predicted quartz solubilities as a function of *pH*, temperature and total sodium concentration. After Crerar and Anderson (1971)

- Charge balance:

$$m_{Na^+} + m_{H^+} = m_{H_3SiO_4^-} = m_{OH^-} + m_{H_3SiO_4^-} + 2m_{H_2SiO_4^{2-}} \qquad (19.21)$$

One simplifying approximation can be made: for all non-zero concentrations of NaOH, the *pH* should be basic and we can omit H^+ from the charge balance (19.21). The above 7 equations can then be reduced to 1 non-linear equation in 1 unknown, which can be solved by a numerical technique such as Newton-Raphson iteration. Suitable numerical equation solvers are now available as software for personal computers. The range of solutions to these equations for different NaOH concentrations and temperatures is illustrated in Figure 19.1. At very high NaOH concentrations we would also have to consider the doubly deprotonated species $H_2SiO_4^{2-}$.

While high *pH*s such as those on Figure 19.1 are rare in nature, the diagram suggests a potential engineering application. One of the difficulties in harnessing geothermal power is that silica precipitates within the plumbing systems of power plants. The same thing happens sometimes when spent water is reinjected into subsurface reservoirs. Many schemes have been proposed to stop, or at least slow down, silica precipitation. One rather obvious approach is to add small concentrations of alkali, which raises silica solubility exponentially, as you can see in Figure 19.1. This particular problem has been discussed in some detail by Fleming and Crerar (1982).

The above example produced 7 equations in 7 unknowns and could be solved without much difficulty using a small computer. There is going to be an upper limit to this kind of manual equilibrium calculation, however. That happens when the reduced equations require too much effort to solve numerically. As an example, rephrase the above problem to calculate the solubility of quartz in NaOH + NaHS solutions. The system now becomes SiO_2–H_2O–NaOH–NaHS and the relevant species are: $NaOH^\circ$, Na^+, $NaHS^\circ$, H_2S, HS^-, S^{2-}, H^+, OH^-, $H_4SiO_4^\circ$, $H_3SiO_4^-$, and $H_2SiO_4^{2-}$. We have 11 species and therefore will require 11 equations. This was solved by Crerar and Anderson (1971) using the above method. A similar problem summarized by Helgeson (1964) was to calculate galena (PbS) solubility in NaCl–H_2O–HCl solutions. This required solution of 13 equations, which was also accomplished by the above method. In both cases the equations could only be reduced to two (rather than one) quite complicated expressions in two unknowns. The process of reducing the original equations is often very time-consuming in such cases, and solution of the final equations can also be a challenge.

These last two examples come from a time when computer technology and software were still in their infancy. Today it would be foolish to solve equilibrium problems as complicated as these manually because computer programs are now available to do this quickly and efficiently. In practice, the manual rote method is convenient for quick calculations involving relatively few species and equations. As a general rule of thumb, calculations involving fewer than 7 to 10 species can often be done by hand. With more complicated systems we recommend your nearest computer terminal. The choices available to you at that point are the subject of the rest of this chapter.

19.2. SPECIATION CALCULATIONS BY COMPUTER

19.2.1. *Three Basic Approaches*

There are now hundreds of programs designed to compute chemical equilibrium in complex, multi-component systems. These are too numerous to summarize completely here, but thorough reviews are available in the literature (Van Zeggeren and Storey, 1970; Nordstrom et al., 1979b; Wolery, 1979, 1983; Smith and Missen, 1982; Reed, 1982; Nordstrom and Ball, 1984; Nordstrom and Munoz, 1985, Appendix E; De Capitani and Brown, 1987).

There are three general categories or types of algorithm used: those that use equilibrium constants (as in our rote method above); those that use the free energy of each species and find that set of activities which minimizes the free energy of the entire system; and those that follow reaction progress in small steps towards a state of final equilibrium. We will summarize these methods briefly in the following sections.

19.2.2. *Methods Based on Equilibrium Constants*

There are two types of program within this category. The first is completely general and can be applied to any system for which suitable data are available; the reactions of interest and their equilibrium constants must be supplied by the user for each particular application. The second type of program starts with a built-in set of specific reactions and equilibrium constants, and might be designed, for example, to apply to natural waters ranging from lakes and groundwaters to hydrothermal solutions; this is convenient if all reactions of interest to you are included in the program. If you are working with systems or reactions not included in such programs then it is usually easier to use one of the general algorithms. The relative pros and cons of both approaches should become apparent in the following examples.

19.2.3. *General Algorithms*

These methods essentially follow the rote procedure outlined above. The important difference is that the mass action, mass balance, and charge balance equations are written in generalized mathematical notation. They can then be applied to any chemical system by specifying the reactions and species of interest. The approach we outline here is described in detail by Crerar (1975). However, we include a change in the mass action equations which was not in the original paper; this improves the speed of the method and its chances of success with very complex systems. Consider an arbitrary system of c components containing N chemical species. Equilibrium constants are known for M independent reactions relating some or all of these species.

First, if the system is ionic, write a generalized charge balance:

$$\Psi_i = 0 = \sum_{i=1}^{N} Z_i m_i \tag{19.22}$$

Here Z_i is the valence (including sign) of the ith species, and m_i is its concentration (molality). For example, Z_i would be -2 for SO_4^{2-}, +1 for Na^+, and 0 for $NaCl^\circ$(aq). Next, assume that the total molal concentrations B_e of $N - M - 1$ elements or atomic species are known for this system. Write $N - M - 1$ general mass balances:

$$\Psi_j = 0 = -B_e + \sum_{i=1}^{N} b_{ei} m_i \qquad (j = 2, 3, \ldots, N - M, \text{ as } e = j - 1) \tag{19.23}$$

B_e is the molal concentration of the eth element (such as total Na or total Cl) in the system. Now each of the N chemical species in the system must contain one or more of these elements according to its formula. The parameters b_{ei} refer to the number of atoms of each element in the formula of every species; using the formula H_2SO_4 as an example, $b_H = 2$, $b_S = 1$, and $b_O = 4$.

Finally, we can write the M equilibrium constant or mass action expressions in generalized form as:

$$\Psi_k = 0 = \frac{-K_k}{\prod_{i=1}^{N} \gamma_i^{\nu_{ki}}} + \prod_{i=1}^{N} m_i^{\nu_{ki}} \quad (k = N - M + 1, \ldots, N) \tag{19.24}$$

K_k is the kth equilibrium constant. Each ν_{ki} represents the number of times the ith species occurs in each reaction; it is negative for reactants and positive for products. For example, with the reaction

$$Fe^{2+} + 2Cl^- = FeCl_2^\circ$$

ν_{ki} is -1, -2, and 1 for the species Fe^{2+}, Cl^-, and $FeCl_2^\circ$, respectively, and is 0 for any other species. Activity coefficients, γ_i are included in (19.24) but are not necessary in the mass and charge balances.

We now have 1 charge balance equation (19.22), a total of $N - M - 1$ mass balances (19.23), and M mass action expressions (19.24) for a grand total of N equations in N unknowns. Because of the mass action equations, this is a non-linear system. It could be solved by any of a number of numerical techniques. We will use the Newton-Raphson method in this example.

19.2.4. *Newton-Raphson Solution*

The numerical solution to the above equations can be simplified by taking the logarithm of the expression for the equilibrium constants in equations (19.24). This is helpful because equilibrium constants for reactions in natural systems are often very large or small numbers; using logarithms reduces this possible range and increases the rate at which the iterative method converges. The logarithm (base 10) of (19.24) is:

$$\boxed{\begin{aligned} &\Psi'_k = 0 = -\log K_k + \sum_{i=1}^{N} \nu_{ki} \log \gamma_i + \sum_{i=1}^{N} \nu_{ki} \log m_i \\ &(k = N - M + 1, \ldots, N) \end{aligned}} \tag{19.25}$$

The set of equations to be solved is now (19.22), (19.23), and (19.25). These are marked with boxes above. The Newton-Raphson method requires the complete set of partial derivatives of all of these equations. Taken in order, these are, for the charge balance (19.22)

$$\frac{\partial \Psi_i}{\partial m_i} = Z_i \quad (i = 1, \ldots, N) \tag{19.26}$$

For the mass balances, the partials are

$$\frac{\partial \Psi_j}{\partial m_i} = b_{ei} \quad (i = 1, \ldots, N;\ j = 2, 3, \ldots, N - M) \tag{19.27}$$

Finally, the partials of the mass action equations (19.25) are

$$\frac{\partial \Psi'_k}{\partial m_l}_{m_i \neq m_l} = 0.43429\nu_{jl}/m_l \quad (l = 1, \ldots, N; k = N - M + 1, \ldots, N) \tag{19.28}$$

In the Newton-Raphson method, the system of non-linear equations (19.22), (19.23), and (19.25) is approximated by the linear system made up of the above partial derivatives.

$$\begin{bmatrix} Z_1 & \cdot\cdot & Z_N \\ b_{i,l} & \cdot\cdot & b_{1,N} \\ \vdots & \vdots\vdots & \vdots \\ b_{N-M-1,l} & \vdots\vdots & b_{N-M-1,N} \\ 0.43429\nu_{1,l}/m_l & \cdot\cdot & 0.43429\nu_{1,N}/m_N \\ \vdots & \vdots\vdots & \vdots \\ 0.43429\nu_{M,l}/m_l & \cdot\cdot & 0.43429\nu_{M,N}/m_N \end{bmatrix} \begin{bmatrix} \delta m_1 \\ \delta m_2 \\ \vdots \\ \delta m_{N-M} \\ \delta m_{N-M+1} \\ \vdots \\ \delta m_N \end{bmatrix} = \begin{bmatrix} -\Psi_1 \\ -\Psi_2 \\ \vdots \\ -\Psi_{N-M} \\ -\Psi'_{N-M+1} \\ \vdots \\ -\Psi'_N \end{bmatrix} \tag{19.29}$$

This is the matrix equation

$$\bar{\Psi}_{N\times N} \cdot \bar{\delta}_{1\times N} = \bar{\Psi}_{1\times N}$$

and it can be solved for $\bar{\delta}_{1\times N}$. A first guess is made at the concentrations of all N species, $m_i^{(0)}$. This gives all elements in the matrix $\bar{\Psi}_{N\times N}$, and elements in the vector $\bar{\Psi}_{1\times N}$ are calculated from the boxed equations (19.25), (19.23), and (19.22) above. The next (improved) estimate of the concentrations is given by

$$m_i^{(1)} = m_i^{(0)} + \delta m_i \quad (i = 1, \ldots, N) \tag{19.30}$$

or, in matrix notation,

$$\bar{m}^{(1)}_{1\times N} = \bar{m}^{(0)}_{1\times N} + \bar{\delta}_{1\times N} \tag{19.31}$$

With the new set of guesses $m_i^{(1)}$ the process is repeated iteratively until successive estimations of all concentrations stop changing significantly.

Activity coefficient corrections are treated just as with the rote manual method already discussed. The first time through, all activity coefficients could be set to 1.0 (or some reasonable estimate). Concentrations of species calculated this way are then used to estimate a better set of activity coefficients in each successive iteration.

The above procedure has been coded in FORTRAN as the program EQBRM and a copy suitable for personal computers is included in this book as Appendix E, along with an example showing the proper format for input data. For different applications it is necessary to choose among the available methods of estimating activity coefficients. For example, the Debye-Hückel equation can often be used for dilute systems such as rivers and groundwater, but concentrated brines will require the Pitzer equations or measured coefficients if they are available. For this reason, a subroutine should be written to calculate activity coefficients for your application.

Figure 19.2 illustrates a typical calculation with EQBRM, from Hennet et al. (1988). Here we wished to know the relative importance of soluble inorganic and organic complex ions of Pb in hydrothermal solutions at 100°C. The only organic component considered for this diagram was acetate; the system contained variable concentrations of NaCl, $CaCl_2$, and NaHS, and the pH was allowed to vary. This required considering 25 species: H^+, OH^-, $H_2S(aq)$, HS^-, Na^+, Cl^-, $NaCl(aq)$, $NaHS(aq)$, $HCl(aq)$, Pb^{2+}, $PbCl^+$, $PbCl_2(aq)$, $PbCl_3^-$, $PbCl_4^{2-}$, $PbOH^+$, Ca^{2+}, $CaCl^+$, $CaCl_2(aq)$, $HAc(aq)$, Ac^-, $PbAc_2(aq)$, $PbAc_3^-$, $CaAc^+$, $CaAc_2(aq)$, and $NaAc(aq)$. The only way to solve an equilibrium problem involving this many species is with a computer. The appropriate mass and charge balances and equilibrium constants are tabulated by Hennet et al. (1988). The results compiled in this figure show that Pb(II)–acetate complexes can predominate over inorganic chloride and hydroxide species only at unusually high organic concentrations.

19.2.5. *Programs Designed for Specific Systems*

The program EQBRM and others of its kind are perfectly general in format, and the user defines the reactions and species for each application. For the results shown in Figure 19.2, this kind of program was necessary because the equilibrium constants for the acetate species were not previously available and had not been written into any pre-existing program.

As mentioned in the introduction to this section, there is a second class of programs that solve for chemical equilibrium in pre-defined systems. These contain built-in equilibrium constants, reactions, and sets of species. Many such programs are currently available and a good many of these have been summarized by Nordstrom and Ball (1984). Most of these programs describe chemical equilibria in mineral–water systems.

We will illustrate this class of programs with one of the earliest, and still most used, examples–the program WATEQ. This was described originally by Truesdell

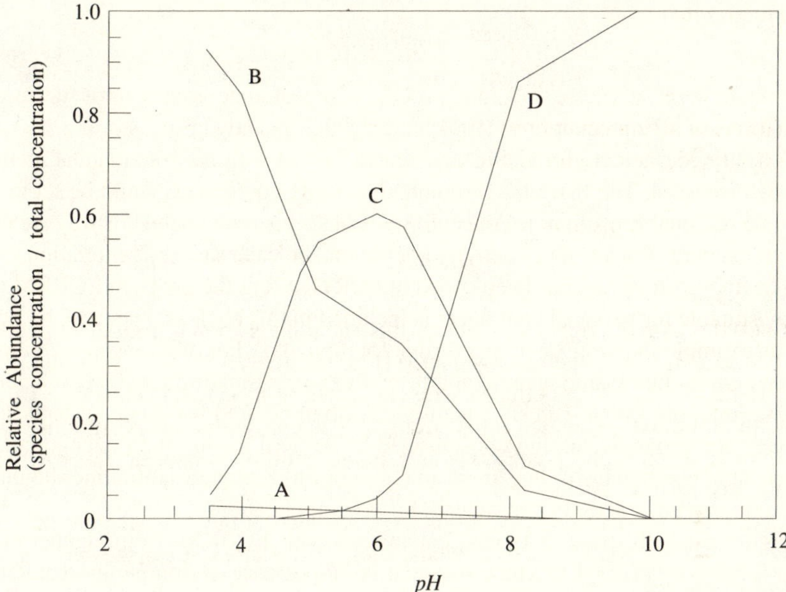

FIG. 19.2. Pb(II) species relative abundance versus pH calculated using EQBRM as described in the text and Appendix E. Conditions: 1.50m Cl; 1.00m Na; 0.25m Ca; 0.20m acetate; $10^{-5}m$ Pb, and 100°C (calculated I=1.50$m \pm 0.05$; neutral pH=6.0). Curve A: Pb(II). Curve B: total Pb(II)–chloro-complexes. Curve C: total Pb(II)–acetate complexes. Curve D: $PbOH^+$. Note that with more Cl-rich brines the relative proportion of metal–organic species (Curve C) decreases.

and Jones (1974) and has gone through many modifications since (e.g., Ball et al., 1980). The program takes as input data total concentrations of all relevant species or elements (Ca, Mg, Na, K, Cl, SO_4^{2-}, HCO_3^-, $SiO_2(aq)$, etc.); field measurements of temperature and pH; and measured Eh and dissolved oxygen, if available.

The general algorithm used by this program was designed by Garrels and Thompson (1962) for one of the first calculations of the equilibrium speciation of seawater. We will summarize the original description by Truesdell and Jones (1974) of the numerical method.

The concentrations or activities of anionic species are calculated first from measured total concentrations and pH and estimated activity coefficients. Using silica as an example, the mass balance on dissolved silica species is

$$m_{Si,total} = m_{H_4SiO_4} + m_{H_3SiO_4^-} + m_{H_2SiO_4^{2-}} \tag{19.32}$$

The mass action or equilibrium constant expressions for ionization of silicic acid, H_4SiO_4, to $H_3SiO_4^-$ and $H_2SiO_4^{2-}$ are

$$K_1 = \frac{m_{H_3SiO_4^-}\gamma_{H_3SiO_4^-}10^{-pH}}{m_{H_4SiO_4}\gamma_{H_4SiO_4}} \tag{19.33}$$

and

$$K_2 = \frac{m_{H_2SiO_4^{2-}}\gamma_{H_2SiO_4^{2-}}10^{-pH}}{m_{H_3SiO_4^-}\gamma_{H_3SiO_4^-}} \tag{19.34}$$

The mass action equations (19.34) and (19.33) can be substituted into the mass balance (19.32) to solve for H_4SiO_4.

$$m_{H_4SiO_4} = \frac{m_{Si,total}}{\left[1 + \gamma_{H_4SiO_4}\left(\frac{K_1 10^{pH}}{\gamma_{H_3SiO_4^-}} + \frac{K_1 K_2 10^{2pH}}{\gamma_{H_2SiO_4^{2-}}}\right)\right]} \tag{19.35}$$

Everything on the right-hand side of this equation is known from the input data. The value of H_4SiO_4 is then plugged into (19.33) to give $H_3SiO_4^-$, and this in (19.34) gives $H_2SiO_4^{2-}$. Activity coefficients are calculated from an extended Debye-Hückel equation and initial estimated ionic strength. The concentrations of all other weak acid anionic species (phosphate, carbonate, borate, sulfide, sulfate, etc.) are calculated by the same method.

Cation and ion-pair concentrations are calculated in the next step based on the above anion concentrations and activities. Taking Ca as an example, the mass balance on total dissolved Ca is:

$$m_{Ca,total} = m_{Ca^{2+}} + m_{CaOH^+} + m_{CaHCO_3^+} + m_{CaCO_3^\circ} + m_{CaSO_4^\circ} \tag{19.36}$$

The mass action equations for formation of the ion pairs $CaOH^+$, $CaHCO_3^+$, $CaCO_3(aq)$, and $CaSO_4(aq)$ are:

$$m_{CaOH^+} = \frac{K_1 a_{OH^-} m_{Ca^{2+}}\gamma_{Ca^{2+}}}{\gamma_{CaOH^+}} \tag{19.37}$$

$$m_{CaHCO_3^+} = \frac{K_2 a_{HCO_3^-} m_{Ca^{2+}}\gamma_{Ca^{2+}}}{\gamma_{CaHCO_3^+}} \tag{19.38}$$

$$m_{CaCO_3^\circ} = \frac{K_3 a_{CO_3^{2-}} m_{Ca^{2+}}\gamma_{Ca^{2+}}}{\gamma_{CaCO_3^\circ}} \tag{19.39}$$

$$m_{CaSO_4^\circ} = \frac{K_4 a_{SO_4^{2-}} m_{Ca^{2+}}\gamma_{Ca^{2+}}}{\gamma_{CaSO_4^\circ}} \tag{19.40}$$

Substituting the above 4 mass action equations into the mass balance (19.36) gives a single expression for the Ca^{2+} free ion concentration:

$$m_{Ca^{2+}} = \frac{m_{Ca,total}}{\left[1 + \gamma_{Ca^{2+}}\left(\frac{K_1 a_{OH^-}}{\gamma_{CaOH^+}} + \frac{K_2 a_{HCO_3^-}}{\gamma_{CaHCO_3^+}} + \frac{K_3 a_{CO_3^{2-}}}{\gamma_{CaCO_3^\circ}} + \frac{K_4 a_{SO_4^{2-}}}{\gamma_{CaSO_4^\circ}}\right)\right]} \tag{19.41}$$

This can be solved for Ca^{2+} molality because all quantities on the right hand side are known. A similar calculation is carried out for all other cations and ion pairs and complexes of these cations (Na, K, Mg, Fe, etc.).

The formation of ion pairs such as $CaHCO_3^+$ and the others considered in equation (19.36) reduces the concentration of free (unassociated) anions such as HCO_3^-, Cl^-, and SO_4^{2-}. The concentrations of all free anions are therefore reduced at this stage by the amount of each tied up in ion pairs and complexes. This changes the ionic strength as well, so new estimates of the activity coefficients must also be made at this point. The calculation of free cation and ion pair concentrations is then repeated with these new estimates of anion molalities and activity coefficients. The iteration is continued until the mass balances for all cations and acid anions (such as equations (19.32) and (19.36) agree satisfactorily with measured quantities.

WATEQ checks for saturation with respect to specific minerals after these calculations are completed by comparing computed ion activity products with solubility products for all minerals listed in the program's data set. For example, the solubility product for $CaSO_4 \cdot 2H_2O$ (gypsum) is

$$K_{sp} = a_{Ca^{2+}} a_{SO_4^{2-}} \tag{19.42}$$

If the predicted ion activity product $IAP = a_{Ca^{2+}} a_{SO_4^{2-}}$ exceeds the solubility product K_{sp} the solution is supersaturated with respect to gypsum. WATEQ provides the concentrations and activities of all species considered, plus a list of all minerals that are supersaturated for these conditions.

As already noted, there are many programs currently available that perform computations similar to WATEQ. One of the more advanced routines is EQ3 (Wolery, 1983). This computes equilibrium speciation using a modified Newton-Raphson algorithm. Recent versions contain data bases applicable at high temperatures, and also include the Pitzer equations for the activity coefficients of concentrated multi-component salt solutions. WATEQ, EQ3, and all similar programs come with built-in data bases for the equilibrium constants of specific reactions. These constants can usually be changed by the user without too much difficulty. In practice, the programs typically consider many more species than needed for specific applications; those species not needed are simply omitted when the program is used. In the user's guide to EQ3, many different examples are considered (Wolery, 1983, pp. 59–81), including the detailed equilibrium speciation of seawater. Recall that this was the problem tackled by Garrels and Thompson (1962) many years before the development of these new, highly sophisticated programs.

19.2.6. *Methods Based on Free Energies*

The second major class of equilibrium-solving routines uses the Gibbs free energies of all species (at the temperature and pressure of interest) rather than equilibrium constants for reactions among the species. Frequently the only other input data required is the total amount, or concentration, of each chemical element in the system (i.e., total moles of Ca, Na, S, Cl, etc.), and perhaps Eh, pH, and temperature.

There is an important fundamental distinction between the free energy based programs and those using equilibrium constants, and this often determines which method to use in specific applications. The free energy programs actually require much more fundamental thermodynamic information, and this can restrict their usefulness.

To illustrate this problem, consider a simple reaction involving minerals and dissolved aqueous species such as

$$Au(s) + 2H_2S(aq) = HAu(HS)_2^\circ + \frac{1}{2}H_2(g) \tag{19.43}$$

The equilibrium constant for such reactions can be measured experimentally, and that gives the standard state free energy of the reaction from

$$\Delta_r G^\circ = -RT \ln K \tag{19.44}$$

There are four species in reaction (19.43) and for the sake of argument we will assume that the standard state free energies of formation $\Delta_f G^\circ$ are well known for Au(c) and $H_2(g)$ but have not been measured for $H_2S(aq)$ and $HAu(HS)_2^\circ$. This is very often a problem with high temperature aqueous species for which free energy data are scarce. If the free energies of all four species were available, then they could be used to calculate the free energy of the reaction:

$$\Delta_r G^\circ = \frac{1}{2}\Delta_f G^\circ_{H_2(g)} + \Delta_f G^\circ_{HAu(HS)_2^\circ} - 2\Delta_f G^\circ_{H_2S(aq)} - \Delta_f G^\circ_{Au(s)} \tag{19.45}$$

Using (19.44) this would give the equilibrium constant for reaction (19.43). The important point here is that the converse is not true: Without the complete set of free energies $\Delta_f G_i^\circ$, it is not possible to calculate either K or $\Delta_r G^\circ$. Equilibrium constants alone do not contain as much thermodynamic information as complete free energy data sets.

With this in mind, you should first check whether free energies are available for all species (at the T and P of interest) before starting a large chemical equilibrium calculation. With luck you will find all necessary free energies and can then select a method based either on equilibrium constants or on free energies. If some free energies are missing, the choice narrows and the free energy methods cannot be used.

Fortunately, thermodynamic data bases are now sufficiently complete that free energy methods can be applied to many systems of geological interest. There are currently many powerful algorithms from which to choose; see Smith and Missen (1982) for summaries and selected computer codes. Several algorithms designed specifically for geochemical applications are outlined by Harvie et al. (1987) and De Capitani and Brown (1988). In the following sections we summarize some of the ways in which these programs have been used.

19.2.7. *Composition of the Early Solar Nebula*

One of the first applications of major geochemical interest was the calculated equilibrium composition of the early solar system. This work is summarized in classic papers by Grossman (1972) and Grossman and Larimer (1974). The method they used is interesting because they calculated equilibrium constants from available free energies; this makes it transitional between the two approaches described above.

This calculation begins by finding the free energies for the 70 most important gaseous species consisting of the 15 elements (H, O, C, N, Mg, Si,..., Cl) thought to have been the most abundant in the solar nebula—the enormous cloud of gas that condensed to form our solar system. Temperatures are presumed to have been 1000 K or higher, and pressures used for the calculation are 10^{-3} to 10^{-4} atm. Hydrogen is by far the most abundant element in such interstellar clouds, and H_2 is the most concentrated H-bearing species. The pressure is so low and temperature so high that the ideal gas law can be invoked and we can approximate total pressure by the partial pressure of H_2:

$$P_{\mathrm{H_2}} \approx P_{total} \tag{19.46}$$

and, because $PV = nRT$,

$$n_{\mathrm{H_2}} = P_{\mathrm{H_2}}/RT \tag{19.47}$$

and

$$n_{\mathrm{H},total} \approx 2n_{\mathrm{H_2}} \tag{19.48}$$

therefore

$$n_{\mathrm{H},total} = 2P_{\mathrm{H_2}}/RT \tag{19.49}$$

Here $n_{\mathrm{H_2}}$ and $n_{\mathrm{H},total}$ are the total number of moles of H_2 molecules and of H atoms per liter, respectively. If A(X)/A(H) is the abundance of any element X relative to that of hydrogen atoms in the solar nebula, then the concentration in moles per liter of element X is simply

$$n_{\mathrm{X},total} = \frac{\mathrm{A(X)}}{\mathrm{A(H)}} n_{\mathrm{H},total} \tag{19.50}$$

We will assume that all species, whether molecules or single atoms, are at complete chemical equilibrium. This means we can write a chemical reaction describing formation of each gas molecule from its constituent gaseous atoms. For water, as an example, this is

$$2\mathrm{H}(g) + \mathrm{O}(g) = \mathrm{H_2O}(g) \tag{19.51}$$

The free energies of all species are known at this T and P, so the equilibrium constants can be calculated for all such reactions (as in equations 19.44 and 19.45 above). We are assuming ideal gas behavior, so activities can be expressed by partial pressures of each gaseous species. For reaction (19.51) the equilibrium constant would be:

$$K = \frac{P_{\mathrm{H_2O}}}{P_{\mathrm{H_2}}^2 \times P_{\mathrm{O}}} \tag{19.52}$$

Using the ideal gas law once again, this can be written as

$$n_{\mathrm{H_2O}} = K\, n_{\mathrm{H}}^2\, n_{\mathrm{O}}\, (RT)^2 \tag{19.53}$$

One equation like (19.53) is written for each of the 70 gaseous molecular species, expressing formation of the species from one or more of the 15 elements considered.

Finally, we know (or have good guesses for) the relative abundances of all 15 elements in the solar nebula. We can therefore write a mass balance for the total moles or gram-atoms per liter of each element. For the element O, this would look like

$$n_{\mathrm{O}} + n_{\mathrm{H_2O}} + n_{\mathrm{CO}} + 2n_{\mathrm{CO_2}} + n_{\mathrm{MgO}} + \cdots = n_{\mathrm{O},total} \tag{19.54}$$

and there are 15 such mass balances, one for each element. Now substitute all expressions such as (19.53) into each of the mass balances. Notice that the only variables on the right side of equations (19.53) are the concentrations of monatomic gas species. Therefore, this gives us a system of 15 simultaneous nonlinear equations in which the sole unknowns are the concentrations of the 15 monatomic gas species:

$$\left.\begin{aligned} n_{\mathrm{H},total} &= f_1(n_{\mathrm{H}}, n_{\mathrm{O}}, n_{\mathrm{C}}, \ldots, n_{\mathrm{Cl}}) \\ n_{\mathrm{O},total} &= f_2(n_{\mathrm{H}}, n_{\mathrm{O}}, n_{\mathrm{C}}, \ldots, n_{\mathrm{Cl}}) \\ n_{\mathrm{C},total} &= f_3(n_{\mathrm{H}}, n_{\mathrm{O}}, n_{\mathrm{C}}, \ldots, n_{\mathrm{Cl}}) \\ \vdots \qquad & \qquad\qquad \vdots \\ n_{\mathrm{Cl},total} &= f_{15}(n_{\mathrm{H}}, n_{\mathrm{O}}, n_{\mathrm{C}}, \ldots, n_{\mathrm{Cl}}) \end{aligned}\right\} \tag{19.55}$$

The above equations can be solved by successive approximation. Initial guesses at the concentrations of the 15 monatomic gas species are plugged into (19.55) to generate improved estimates, and the process is repeated until successive concentrations stop changing significantly. This could be accomplished by rearranging each of the equations (19.55) to forms such as:

$$n_{\mathrm{H}} = n_{\mathrm{H},total} - f_1'(n_{\mathrm{H}}, n_{\mathrm{O}}, n_{\mathrm{C}}, \ldots, n_{\mathrm{Cl}}) \tag{19.56}$$

The above calculation determines the concentrations of gaseous species at temperatures so high that solid phases are not stable. At lower temperatures, solids begin to condense out of this gas and a correction must be made for the amount of each element removed as solids from the gas phase. For example, if the gas is cooled at 10^{-3} atm to 1758 K, corundum condenses according to the reaction

$$\mathrm{Al_2O_3}(s) = 2\mathrm{Al}(g) + 3\mathrm{O}(g) \tag{19.57}$$

This is the first phase to condense in significant quantities from the cooling solar nebula. With the appearance of corundum the concentration of Al and O as this phase must also be included in the mass balances for total O (19.54) and Al. This introduces a new unknown, $n_{\mathrm{Al_2O_3}}(s)$, to the 15 above, leaving us with 15 equations in 16 unknowns. The required 16th equation is simply the equilibrium constant for reaction (19.57), calculated from the free energies of Al(g), O(g), and $\mathrm{Al_2O_3}(s)$ at T and P.

$$\log K_{\text{corundum}} = 2\log P_{\mathrm{Al}} + 3\log P_{\mathrm{O}} - \log a_{\mathrm{Al_2O_3}(s)} \tag{19.58}$$

The solution to this new set of 16 equations therefore gives the concentrations of the 15 monatomic gas species as well as corundum in moles per liter. The concentrations of the remaining 70 gaseous molecular species such as $\mathrm{H_2O}(g)$ are then easily calculated from the set of equations (19.53)—one for each species.

It is necessary to test at each stage for precipitation of solid phases such as corundum. When a solid becomes stable, concentration quotients (calculated assuming no solid) will exceed the equilibrium constant for the gas-solid reaction. For example, corundum is stable when

$$\frac{P_{\mathrm{O}}^3 \times P_{\mathrm{Al}}^2}{a_{\mathrm{Al_2O_3}}} \geq K_{\mathrm{corundum}} \tag{19.59}$$

Every time a new solid phase appears, its concentration must be added to the appropriate mass balance equations and a new equilibrium constant (for formation from its monatomic gaseous constituents) is included in the system of equations to be solved.

Grossman performed this computation as part of his doctoral dissertation, ultimately including about 300 gaseous species and 100 solid phases in a system comprising the 20 most abundant elements in the solar system. The results are summarized in a diagram which is now quite familiar in the fields of cosmochemistry and meteoritics. It shows that the most refractory minerals such as corundum, sphene, melilite, and spinel all condense, as expected, at the highest temperatures. These same minerals are commonly found in the most primitive meteorites, the carbonaceous chondrites. Minerals like the melilites are relatively rare in terrestrial rocks, and it has been suggested that the snowflake-like refractory crystals preserved in such meteorites might actually represent solid material condensed from the solar nebula during the birth of our solar system (Grossman, 1975).

The calculations also perhaps explain the difference in composition between the small, Earth-like inner terrestrial planets and the large outer Jovian planets. The terrestrial planets presumably condensed at much higher temperatures and are thus composed of metals, metal oxides, and silicates. The Jovian planets would have formed at far lower temperatures within the primitive solar nebula and consist predominantly of frozen volatile compounds such as methane, water, ammonia, and so on. Finally, a possible case for early layering of the Earth can be drawn from the calculations: within a cooling nebula metallic Fe and Ni would condense first, followed by spinels, pyroxenes and olivines, with a final lower temperature layer of alkali feldspar, metal oxides, hydrated silicates and, of course, water itself at 0°C.

The Larimer-Grossman condensation model has received its share of criticism in the intervening decade. Much of the debate was raised by the assumption that all solid and gaseous species fully equilibrated, which seems unlikely, particularly at lower temperatures. Despite these concerns, the equilibrium picture fits many of the broad scale chemical features of our solar system remarkably well. It certainly demonstrates the power of thermodynamic equilibrium models, considering the size and complexity of the chemical system that is our niche in the universe.

19.2.8. *Free Energy Minimization*

The above calculation used free energies and the equilibrium constants for specific chemical reactions. For this reason the calculations apply only to the system, species, and reactions considered. It is possible to generalize free energy based equilibrium calculations so that they can be used with systems and species chosen arbitrarily, in a manner analogous to the program EQBRM described above.

We observed in Chapter 3 that a chemical system is at thermodynamic equilibrium when the Gibbs free energy of the system is at a minimum. For a given pressure, temperature, and bulk composition, at equilibrium there will be one or more phases in which the concentrations of all species are fixed. Many different methods have been developed to compute the equilibrium state, as outlined by Van Zeggeren and Storey (1970) and Smith and Missen (1982). As an illustration, we will summarize an algorithm derived by Harvie et al. (1987), which has been applied with great success to geochemical systems.

All generalized equilibrium-solving algorithms based on free energies use some numerical technique to locate the minimum on the free energy-composition surface. These are called optimization methods, and there are many from which to choose (see the summary by Van Zeggeren and Storey, 1970, Ch. 3, for example). The procedure developed by Harvie et al. (1987) uses Lagrangian multipliers.

Let's state the problem in non-mathematical language first. We need to find the unique concentrations of species in a system that minimize the total free energy subject to the following constraints or conditions.

- All mass balance conditions must be met (the concentrations of each component or element contained in all species must equal the total concentration of that component in the system).
- The system must be electrically neutral, so if ions are present, their concentrations must satisfy a charge balance equation.
- No species can have a negative concentration.

To phrase the problem mathematically, the same notation used to describe the EQBRM algorithm above will be used as much as possible. Consider a system containing c components, N chemical species, and ϕ phases. At equilibrium, the total free energy is minimized:

$$\text{minimize} \quad \mathrm{G} = \sum_{i=1}^{N} \mu_i n_i \quad \text{for all species} \tag{19.60}$$

This is subject to the following conditions or constraints:

- Mass balance (c equations, one for each component)

$$\boxed{B_e = \sum\nolimits_{i=1}^{N} b_{ei} n_i \quad \text{(for all components, } e = 1, \ldots c)} \tag{19.61}$$

- Charge balance (ϕ equations, one for each electrolyte phase, p)

$$\boxed{0 = \sum_{all\ i\ in\ phase\ p} Z_{p,i} n_{p,i} \quad \text{(for all phases, } p = 1, \ldots, \phi)} \tag{19.62}$$

- Non-negative concentration constraints (N equations, one for each species, i)

$$n_i \geq 0 \quad \text{(for all species, } i = 1, \ldots, N) \tag{19.63}$$

For numerical purposes, the last constraint is better expressed in terms of a "slack variable" t_i,

$$\boxed{\begin{array}{l} n_i = t_i^2 \quad (i = 1, \ldots, N) \\ where\; t_i = 0\; when\; n_i = 0\; and\; t_i > 0\; when\; n_i > 0 \end{array}} \tag{19.64}$$

In the above equations (19.60) to (19.64), G is the total Gibbs free energy of the system; μ_i is the chemical potential of species i; n_i is the number of moles of species i; N is the total number of species in the system; c is the total number of components (for present purposes considered as the elements); B_e is the number of moles of each component (or element, e) in the system; b_{ei} is the number of moles of component (or element) e contained in one mole of species i; p refers to a separate electrolyte (ionic) solution phase, and ϕ is the total number of these electrolyte phases; $Z_{p,i}$ refers to the valence or charge of the ith species in the pth phase; and t_i is the slack variable defined in (19.64).

The next step is to write the Lagrangian function corresponding to the minimization statement (19.60) and the constraints (19.61), (19.62), and (19.64).

$$\begin{aligned} L = L(\vec{n}, \vec{t}, \vec{\kappa}, \vec{\eta}, \vec{\omega}) = & \sum_{i=1}^{N} \mu_i n_i \\ & - \sum_{i=1}^{c} \kappa_i \sum_{j=1}^{N} (b_{ji} n_j - B_i) \\ & - \sum_{p=1}^{\phi} \eta_p \sum_{j\ in\ p} Z_{p,j} n_{p,j} \\ & - \sum_{i=1}^{N} \omega_j (n_j - t_j^2) \end{aligned} \tag{19.65}$$

In (19.65) we have introduced the Lagrangian multipliers κ_i, η_p, and ω_j, one for each constraint. This increases our total number of unknowns to N of n_i, c of κ_i, ϕ of η_p, and N of ω_j for a total of $2N + c + \phi$ unknowns. We will require the same number of equations to solve for these variables. This is accomplished by taking the partial derivatives of the Lagrangian function (19.65) with respect to each concentration variable, and equating them all to zero.

For example, the partial derivative of (19.65) with respect to the kth species in a non-electrolyte phase is

$$\boxed{\frac{\partial L}{\partial n_k} = 0 = \mu_k - \sum_{i=1}^{c} \kappa_i b_{ki} - \omega_k} \tag{19.66}$$

The same partial derivative for an electrolyte phase p must include the charge balance constraint (19.62).

$$\frac{\partial L}{\partial n_k} = 0 = \mu_k - \sum_{i=1}^{c} \kappa_i b_{ki} - \eta_{p,k} Z_{p,k} - \omega_k \tag{19.67}$$

This gives us a total of N partial derivative equations (19.66) and (19.67), one for each species in the system. We have c mass balance constraints (19.61), ϕ charge balance constraints (19.62), and N non-negative constraints (19.64). Solving these simultaneously with the N partial derivatives above gives us the necessary $2N+c+\phi$ equations in $2N+c+\phi$ unknowns.

This is the basic principle of the Lagrange method for free energy minimization. In practice, the computation can become quite complex and intricate. Part of the problem lies in the non-linear relationship between the chemical potential μ_i and the concentration in equation (19.60).

$$\mu_i = \mu_i^\circ + RT \ln m_i \gamma_i \tag{19.68}$$

Here we have represented concentration by molality, m, (instead of number of moles, n_i) and include an activity coefficient correction, γ_i. As with equilibrium constant-based equilibrium calculations, the activity coefficients can be computed from successive estimates of the concentrations.

As input, this model requires the standard state chemical potentials (or free energies) $\Delta_f G_i^\circ$ of all species and phases, and the bulk composition or total number of moles of each component, B_e. A specific method must always be selected to calculate activity coefficient corrections.

The program of Harvie et al. (1987) has been applied to extremely non-ideal evaporite brines (up to $20m$ concentration), and to equilibria involving silicate melts and solid solutions. The agreement between prediction and observation is impressive. Figure 19.3 shows one series of equilibrium computations for highly nonideal, multicomponent, saturated salt solutions. Here the solubilities of different salts are predicted to high concentrations in the system Na–Cl–SO_4^{2-}–HCO_3^-–CO_3^{2-}–OH^-–H_2O. The Pitzer equations described in Chapter 17 were used to compute activity coefficients in this highly complex system. The excellent agreement with the measured points (dots) demonstrates the power of this method. This also shows how well Pitzer's model for the thermodynamic properties of multicomponent salt solutions actually works.

Many other algorithms have been devised for solving chemical equilibrium problems using free energy data bases for all species and phases. One of the most recent geochemical examples is the program called THERIAK described by De Capitani and Brown (1987). This minimizes free energy and includes non-ideality corrections, but differs from the program of Harvie et al. (1987) in using repeated linear and nonlinear programming steps. The THERIAK algorithm is highly efficient, and is particularly well suited to complex systems in which phase separation (unmixing) can occur in more than one phase. An application of this program to equilibrium conditions in a

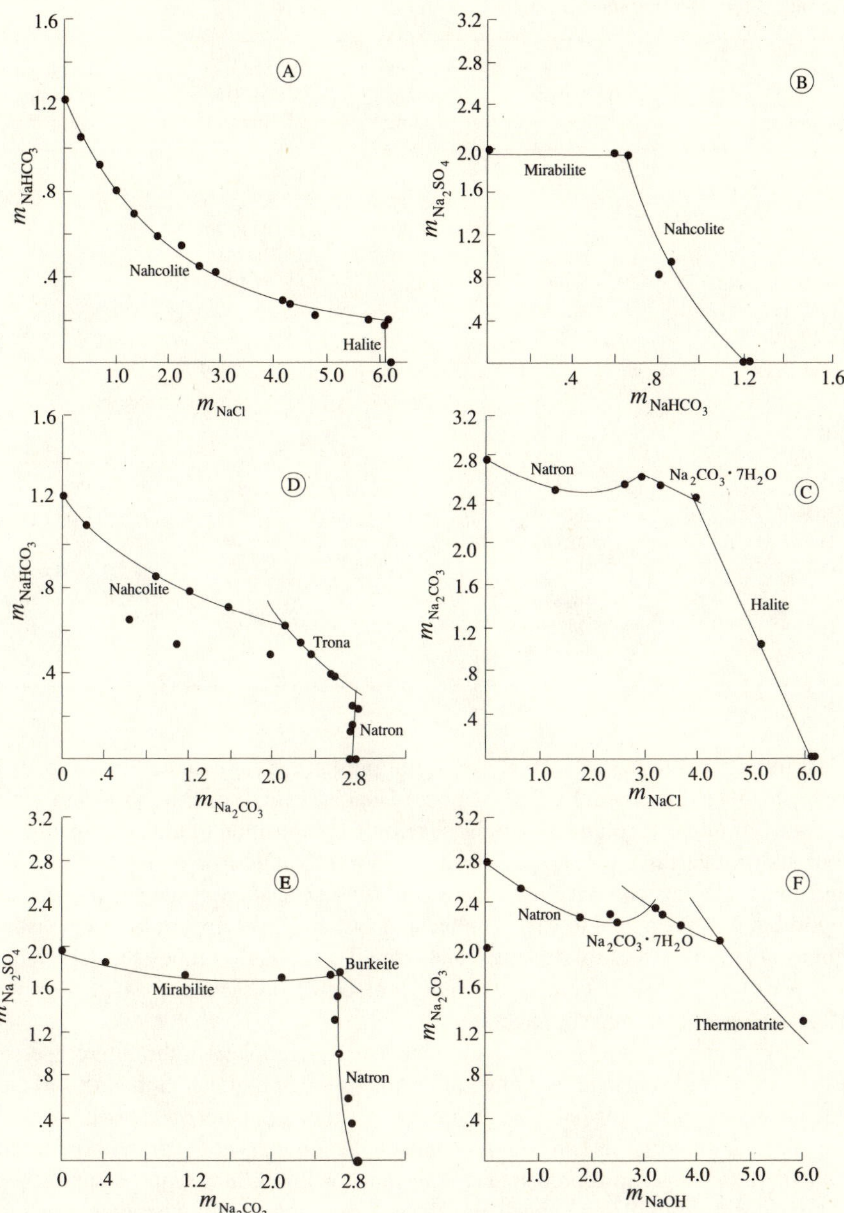

FIG. 19.3. Salt solubilities in the Na-Cl-SO_4-HCO_3-CO_3 -OH-H_2O system. Closed systems. From Harvie et al. (1984)

Table 19.1 Free Energy Minimization Calculation of Chemical Equilibrium in a Blast Furnace. From DeCapitani and Brown (1987). Pressure is 1 atm.

Species	$\Delta_f G°$ (J/mol) at 1050 K	Bulk comp'n (moles)	Stable at 1050 K (moles)	Mole fraction in stable gas phase
Solid Phases				
C(graphite)	0	85.5901		
Fe(iron)	0	3.5270	42.8270	
CaO(lime)	−529190	0.6063	0.7562	
Fe_3O_4(magnetite)	−762660	13.1000		
$CaCO_3$(calcite)	−942450	0.1499		
FeO(wüstite	−193930			
Ideal Gas			282.2639	
O_2	0	20.4600		$6.1303 \cdot 10^{-22}$
H_2	0			$2.2766 \cdot 10^{-2}$
N_2	0	187.1000		$6.6286 \cdot 10^{-1}$
CH_2O	−86110			$8.3543 \cdot 10^{-9}$
CHO	−62550			$3.7258 \cdot 10^{-9}$
OH	22590			$2.8093 \cdot 10^{-13}$
CO_2	−395970			$2.3608 \cdot 10^{-3}$
H_2O	−189870	1.7750		$1.5723 \cdot 10^{-3}$
CH_4	24850	2.5540		$2.3201 \cdot 10^{-5}$
CO	−204640			$2.8918 \cdot 10^{-1}$

blast furnace operating at 1050 K and 1 atm pressure is summarized in Table 19.1. Here the first two columns list free energies of all phases and species considered, at the T and P of interest; column three gives the bulk composition of the system; column four shows that three phases coexist at equilibrium: solid iron and calcium oxide, and a gas. The last column shows the equilibrium speciation of the gas phase. This algorithm can be applied to very complicated systems involving multiple coexisting immiscible liquids and solid phases, and overall computation times are quite rapid. The blast furnace calculation, for example, took only 0.43 seconds on a 48 megabyte 1987-vintage mainframe computer.

As free energy data bases become increasingly complete in the future, we can expect to see much heavier use of equilibrium-solving programs. Software will also become more widely available as the power of small computers increases; many of the necessary algorithms can be run on personal computers at the present time. This kind of quick, fingertip approach to equilibrium conditions in the most complex geochemical systems will certainly change the way we handle thermodynamic questions in geochemistry, and should ultimately change the nature of the questions we ask.

19.3. REACTION PATH CALCULATIONS

In natural systems, chemical reactions often start far removed from equilibrium, then progress along towards a final stable (or metastable) state. Chemical weathering is a good example: surface waters and freshly exposed rock are normally completely out of equilibrium when first brought together. Given time, however, they will begin to react and change composition. These changes will always proceed towards chemical equilibrium, but many different paths may be followed. In multi-mineral systems many reactions might occur simultaneously and the overall chemical behavior can become quite complex. Many processes occur in parallel, for example, when a magma of one composition is injected into chemically different country rock and begins to react with pre-existing minerals. Large mineral deposits such as the porphyry coppers are produced when hydrothermal solutions pass through wallrocks of completely different chemistry, and so on.

Rather than taking on problems of this magnitude at the outset, let's illustrate the principles involved with a very simple example—the oxidation of iron to hematite:

$$\boxed{2Fe + \tfrac{3}{2}O_2 \rightarrow Fe_2O_3} \qquad (19.69)$$

Assume this process begins in a chemically closed box containing a random amount of fresh iron at T and 1 bar total oxygen pressure. Of course, iron and oxygen are completely out of equilibrium when first exposed to each other. Reaction will progress spontaneously in the direction that reduces the free energy of the system, as indicated by the arrow.

In fact, several different processes are possible here, depending on initial amounts of reactants and the rates of certain reactions. This illustrates, right at the outset, one of the greatest difficulties in applying these types of calculations to real systems: additional non-thermodynamic information is often required before choosing geologically realistic reactions. By writing reaction (19.69) we are presuming this process occurs in preference to all other possibilities. Elemental iron, when exposed to oxygen, might react to form wüstite, magnetite, or other possible oxides rather than hematite. For this example, assume that hematite nucleates more rapidly than the other oxides. This kind of additional kinetic information is necessary before we can even choose an appropriate reaction.

Presuming then that reaction (19.69) is what actually occurs (or perhaps the reaction we wish to model in spite of what actually occurs) we can consider two different cases:

1. If the initial oxygen/iron ratio is well above the stoichiometric value (3/2)/2 = 3/4 dictated by reaction (19.69), all original iron will be converted to hematite. In this case the final state would consist of hematite and gaseous oxygen.

2. If the initial oxygen/iron ratio is below the stoichiometric value, then some elemental iron will remain, in association with hematite and gaseous oxygen. The f_{O_2} for this assemblage would be $10^{-86.7}$ bars (as calculated from the equilibrium constant for reaction (19.69) at 25°C). This is an interesting situation, because hematite and iron are thermodynamically metastable under these conditions with respect to magnetite.

If we drop the restriction that magnetite cannot nucleate, then two additional processes might occur in case 2 above:

$$3Fe + 2O_2 \rightarrow Fe_3O_4$$

$$3Fe_2O_3 \rightarrow 2Fe_3O_4 + \frac{1}{2}O_2$$

This could ultimately produce magnetite plus residual oxygen gas as the final assemblage. Because the end result is so different, we see how important outside kinetic information is to this kind of reaction path model.

19.3.1. *The Extent of Reaction or Progress Variable*

For the sake of argument, assume that case 1 above actually occurs. We can speak of the extent to which this reaction has progressed from the initial nonequilibrium conditions towards its final state; for example, when half the initial Fe is oxidized, the reaction has progressed halfway. You can regard some of this material as a concrete example of the theory discussed in Chapter 14. For any balanced chemical reaction,

$$\sum_i \nu_i M_i = 0 \tag{19.70}$$

where ν_i are the stoichiometric coefficients in the chemical reaction and are positive for products, negative for reactants (e.g., for reaction (19.69), $\nu_{Fe} = -2$, $\nu_{O_2} = -3/2$, and $\nu_{Fe_2O_3} = +1$). M_i represents the molecular weight of the ith species in the reaction.

Assume that the chemical system is completely closed so that no material can enter or leave. As our initial system of Fe and O_2 begins to react it will form a small amount of Fe_2O_3; we can express the change in mass of each species for this small amount of reaction by the equation

$$dm_i = \nu_i M_i d\xi \tag{19.71}$$

where ξ is called the "degree of advancement of the reaction", "the extent of reaction" or "the progress variable," and is described in units of moles. The concept of the progress variable was introduced by Th. De Donder in the 1920s. If reaction (19.69) progresses by some small amount, say $d\xi = 0.001$ moles, then (19.71) tells us that the mass of Fe will decrease by $dm_i = -2(55.8)(0.001)$ or $-0.112\,g$ throughout the course of this small reaction step. For a generalized reaction such as (19.70) the differential changes in mass are related to the progress variable by

$$\frac{dm_1}{\nu_1 M_1} = \frac{dm_2}{\nu_2 M_2} = \cdots = \frac{dm_N}{\nu_N M_N} = d\xi \tag{19.72}$$

For reaction (19.69) this becomes

$$\frac{dm_{Fe}}{-2M_{Fe}} = \frac{dm_{O_2}}{-\frac{3}{2}M_{O_2}} = \frac{dm_{Fe_2O_3}}{M_{Fe_2O_3}} = d\xi \tag{19.73}$$

You can get a better intuitive feeling for the physical significance of the progress variable by rearranging (19.73) to express explicitly the change in mass of any species with extent of reaction:

$$\left.\begin{aligned} dm_{O_2} &= -\tfrac{3}{2}M_{O_2}d\xi \\ dm_{Fe_2O_3} &= M_{Fe_2O_3}d\xi \\ dm_{Fe} &= -2M_{Fe}d\xi \end{aligned}\right\} \tag{19.74}$$

It is apparent here that the progress variable determines how much reactant is destroyed or product created during a step in a chemical reaction. The larger the step (or $d\xi$) the greater these changes in mass. The relative changes in amounts of reactant and product species are controlled simply by the stoichiometry of the reaction itself (as expressed by equation (19.73)). Note that like all differentials, $d\xi$ can be large or small, and positive or negative (in which case the reaction runs backwards).

It is often more convenient to deal with moles (and concentrations) rather than masses in typical applications. For this, simply divide masses of all species i by their molecular weights M_i.

$$n_i = \frac{m_i}{M_i} \tag{19.75}$$

Then (19.71) becomes

$$dn_i = \nu_i d\xi \tag{19.76}$$

where n_i is the number of moles of the ith species.

The relationship between reaction progress, change in moles of different species, and the overall reaction stoichiometry as in (19.72) now simplifies to

$$\frac{dn_i}{\nu_i} = d\xi \tag{19.77}$$

or, using reaction (19.69) again as an example,

$$-\frac{dn_i}{2} = -\frac{dn_{O_2}}{3/2} = \frac{dn_{Fe_2O_3}}{1} = d\xi \tag{19.78}$$

Finally, integrating (19.77) gives

$$n_i = n_i^\circ + \nu_i \Delta\xi \tag{19.79}$$

We can use this equation to plot the "progress" of reaction (19.69). This is illustrated in Figure 19.4a,b for both case 1 (with excess initial O_2) and case 2 (with excess Fe), as discussed above. Using case 1 as an example, consider a closed system initially containing 10 moles each of Fe and O_2. After a reaction step $\Delta\xi = 1$, the system will contain 8 moles of Fe, 8.5 moles of O_2, and 1 mole of freshly formed

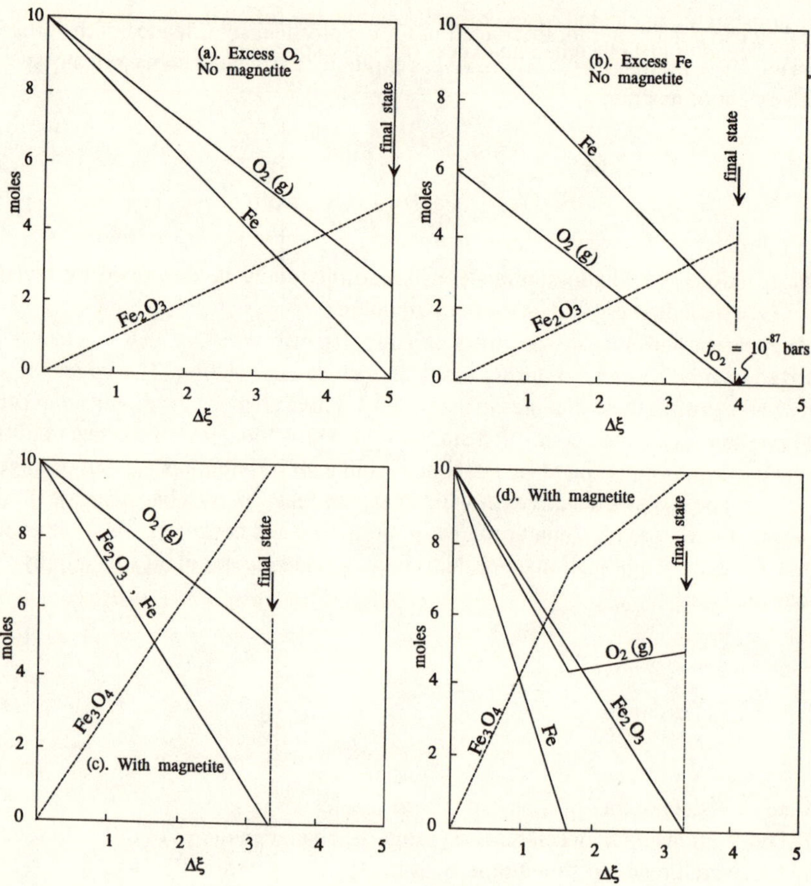

FIG. 19.4. Progress of reaction for $2Fe + \frac{3}{2}O_2(g) = Fe_2O_3$. (a) Excess initial O_2. Closed system consisting of 10 moles each of Fe and $O_2(g)$, initially. Reaction stops at $\Delta\xi = 5$ with 5 moles of hematite and 2.5 moles oxygen. (b) Excess initial Fe. Closed system containing 10 moles Fe and 6 moles O_2, initially. Reaction stops at $\Delta\xi = 4$ with 2 mole Fe and $\log f_{O_2} = -87$. (c) Simultaneous oxidation of Fe and reduction of hematite to magnetite, where rates of both reactions are equal. Initial moles of O_2, Fe, and Fe_2O_3=10, and of Fe_3O_4=0. (d) Same as (c), except rate of Fe oxidation is twice that of Fe_2O_3 reduction.

Fe_2O_3. The reaction ends when all initial Fe is consumed, which occurs for a reaction progress $\Delta\xi = 5$ (found by solving equation 19.79 for $n_{Fe} = 0$).

19.3.2. *Multiple Reactions and Relative Rates*

The example in Figure 19.4 is simple because only one reaction occurs at a time. In nature, many reactions can occur at once, and this situation can be treated by extending the above reasoning to a series of simultaneous reactions. We have already mentioned that if magnetite nucleated quicker than hematite, the following two reactions might occur instead of (19.69) when an excess of Fe is exposed to O_2.

$$3Fe + 2O_2 \rightarrow Fe_3O_4 \quad (rxn\ a)$$

$$3Fe_2O_3 \rightarrow 2Fe_3O_4 + \tfrac{1}{2}O_2 \quad (rxn\ b)$$

Both reactions would run simultaneously and this could be described by revising (19.77) to include these and any other parallel reactions.

$$dn_i = \sum_{rxn=1}^{R} \nu_{i,rxn} \quad (rxn = 1, \ldots, N) \tag{19.80}$$

where the subscript rxn refers to one of a total of R simultaneous reactions, and $\nu_{i,rxn}$ is the stoichiometric coefficient of the ith species in each reaction. In our example there are two simultaneous reactions, a and b.

If we assume that the above two reactions run at exactly the same rate, then the simultaneous changes in all four species with reaction progress can be calculated in the following manner.

$$\left.\begin{aligned} dn_{\mathrm{Fe}} &= -3d\xi_a \\ dn_{\mathrm{O_2}} &= -2d\xi_a + \tfrac{1}{2}d\xi_b \\ dn_{\mathrm{Fe_3O_4}} &= d\xi_a + 2d\xi_b \\ dn_{\mathrm{Fe_2O_3}} &= -3d\xi_b \end{aligned}\right\} \tag{19.81}$$

where $d\xi_a = d\xi_b$.

This appears fairly simple and would be, except that the same problem of requiring non-thermodynamic information presents itself again. We encountered this first in choosing hematite rather than magnetite as the most likely product in reaction (19.69). In fact, we cannot realistically presume both reactions a and b will run at precisely the same rate. If the rates are different, then the net reaction path (or changes in species concentrations) could differ considerably from this simplified case. This means we need to introduce into the calculation some way of defining reaction rates.

The rate, r, at which a reaction progresses is simply

$$r = \frac{d\xi}{dt} \tag{19.82}$$

The reaction rate can also be expressed by the derivative with respect to time of equation (19.76).

$$r = \frac{1}{\nu_i}\frac{dn_i}{dt} = \frac{d\xi}{dt} \tag{19.83}$$

or

$$\frac{dn_i}{dt} = \nu_i \cdot r \tag{19.84}$$

Taking the different reaction rates r_a and r_b into account, the calculation (19.81) can then be rephrased in terms of changes with time or relative rates. For example, if reaction a progresses faster than b, then ξ_a must increase faster than ξ_b. The relative rates can be calculated from

$$\frac{r_a}{r_b} = \frac{d\xi_a/dt}{d\xi_b/dt} \tag{19.85}$$

or

$$d\xi_a = \frac{d\xi_b r_a}{r_b} \tag{19.86}$$

The progress or reaction path of our example can then be rewritten to include relative rates of the two reactions a and b by substituting (19.86) into equations (19.81).

$$\left.\begin{aligned} dn_{\mathrm{Fe}} &= -3\frac{d\xi_b r_a}{r_b} \\ dn_{\mathrm{O_2}} &= -2\frac{d\xi_b r_a}{r_b} + \frac{1}{2}d\xi_b \\ dn_{\mathrm{Fe_3O_4}} &= \frac{d\xi_b r_a}{r_b} + 2d\xi_b \\ dn_{\mathrm{Fe_2O_3}} &= -3d\xi_b \end{aligned}\right\} \tag{19.87}$$

To test the importance of relative rates, we can draw two different reaction path diagrams: one for the case where rates of reactions a and b are equal, and one for which the rate of a is twice that of b ($d\xi_a/dt = 2d\xi_b/dt$, and $r_a = 2r_b$). To do this, equations (19.81) and (19.87) are integrated (to give expressions similar to equation 19.79). The results are plotted in Figure 19.4c,d, for a starting composition of 10 moles each of O_2, Fe, and Fe_2O_3 and 0 starting moles of Fe_3O_4. Comparing the two figures, you can see that the end result in both cases is the same (9.9 moles Fe_3O_4; 5.0 moles O_2; $\Delta\xi_{final} = 3.3$). However, the paths taken for the two cases are considerably different. In the second case, where the rates are different, Fe disappears early on in the progress of the reaction, while it remains to just before the final state when the rates are equal.

For practical applications this requires, of course, that reaction rates be known and therefore takes us well outside the realm of thermodynamics into the sister field of chemical kinetics. In fact, rigorous reaction path computations are an ingenious blend of kinetics and thermodynamics, as we shall see in the following section.

19.3.3. *Application to Mineral–Water Reactions*

Reaction path computations of the above general type were first applied to mineral systems by Harold Helgeson and his colleagues in the 1960s. Much of the early work is summarized by Helgeson (1979). The program EQ6 developed for the U.S. Dept. of Energy by Tom Wolery (1979) is a widely used modern version of the original

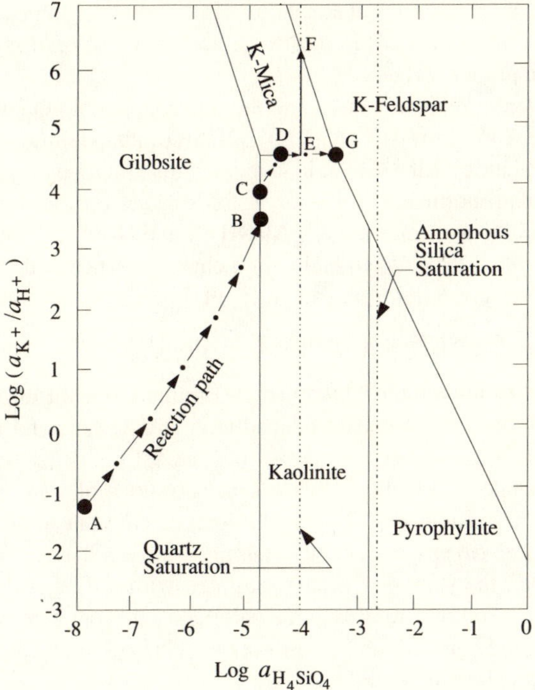

FIG. 19.5. Equilibrium activity diagram for the system K_2O–Al_2O_3–SiO_2–H_2O at 25°C, 1 bar (after Helgeson, 1979). The reaction path ABCDG represents successive stages in the hydrolysis of K-feldspar, which correspond to stages ABCDG in Figure 19.6.

programs, which now includes the Pitzer equations for the activity coefficients of mixed electrolyte solutions (see Chapter 17). A reaction path model is also included in the program PHREEQE, which was developed to interpret groundwater chemistry (Plummer et al., 1983). For a discussion of these and other related programs, see Nordstrom and Ball (1984).

To illustrate applications of reaction path calculations to natural systems, we will draw heavily from the papers of Helgeson and his co-authors. The notation used below is consistent with Helgeson's as much as possible, to facilitate reference to his articles.

We will begin as Helgeson did, by considering what happens when a sample of pure K-feldspar ($KAlSi_3O_8$) reacts with initially pure water. At the outset, we know that fresh K-feldspar and water are completely out of equilibrium. Figure 19.5 is a phase diagram for the analogous system $K_2O - Al_2O_3 - SiO_2 - H_2O$. This shows K-feldspar separated from gibbsite $Al(OH)_3$ by stability fields of K-mica and kaolinite. The model assumes that K-feldspar dissolves congruently when placed in pure water (i.e., the solute produced by K-feldspar dissolution has the same stoichiometry as the feldspar), and that incongruent solubility is the result of (eventual) precipitation of a different phase or phases from that solution. For the model to be accurate, we must of course know the nature of the solute species produced on dissolution, and their thermodynamic properties.

Helgeson et al. (1969) assumed that the only important aqueous species in this system are K^+, Al^{3+}, $Al(OH)^{2+}$, $Al(OH)_4^-$, H_4SiO_4, and $H_3SiO_4^-$. If you perform a speciation calculation of the kind we have been discussing on a solution having K, Al and Si in the proportions 1:1:3 (as in K-feldspar) at very low concentration, you find that the dominant species are K^+, $Al(OH)_4^-$, and H_4SiO_4. Essentially, $Al(OH)_4^-$ must dominate the Al species to maintain a charge balance with K^+. Therefore the dissolution reaction of K-feldspar can be approximated by

$$KAlSi_3O_8 + 8H_2O \rightarrow K^+ + Al(OH)_4^- + 3H_4SiO_4 \tag{19.88}$$

However, this does not tell the whole story. Also increasing during dissolution of the feldspar are all the other species produced (Al^{3+}, $Al(OH)^{2+}$, and $H_3SiO_4^-$), though at lower concentrations. These are shown in Figure 19.6, rising from the lower left corner with a slope of 1.0. Initially, the concentrations of all these species are so small that the concentrations of H^+ and OH^- remain constant at 10^{-7}. The species all increase from zero ($\log m = -\infty$) maintaining the overall 1:1:3 stoichiometry, and would stop when the solution became saturated with K-feldspar. However, *before* that happens, the solution may become saturated with other minerals, which will precipitate as K-feldspar continues to dissolve. This is the essential element of the reaction path calculations.

To find out whether the solution has become saturated with another mineral, the solubility products of all minerals in the system considered (i.e., all minerals which contain any combination of the elements in the system) must be compared against the corresponding Ion Activity Product (IAP) in the solution after each increment of dissolution. This can be literally hundreds of minerals in large model systems. In the relatively simple K-feldspar case, there are only a few minerals that could possibly form. The first of these, according to the data used by Helgeson et al. (1969), is gibbsite. The solubility product for gibbsite is

$$Al(OH)_3(s) = Al^{3+} + 3OH^-; \quad K_{sp} = a_{Al^{3+}} a_{OH^-}^3 \tag{19.89}$$

Saturation in gibbsite occurs when its activity product exceeds the solubility product for gibbsite.

$$a_{Al^{3+}} a_{OH^-}^3 > K_{sp} \tag{19.90}$$

Under the model conditions, this happens at point A on Figure 19.5.

Now if K-feldspar continues to dissolve (or, if you prefer, we continue to perform speciation calculations for solutions in which K, Al, and Si continue to increase in the ratio 1:1:3), gibbsite will continue to precipitate, and the remaining solution will have compositions that follow the path A→B on Figure 19.5. During this process, the silica content of the solution continues to increase until at point B, a silica-bearing mineral (kaolinite) becomes stable. The coexistence of gibbsite and kaolinite buffers the activity of silica according to

$$2Al(OH)_3(s) + 2SiO_2(aq) = Al_2Si_2O_5(OH)_4(s) + H_2O$$

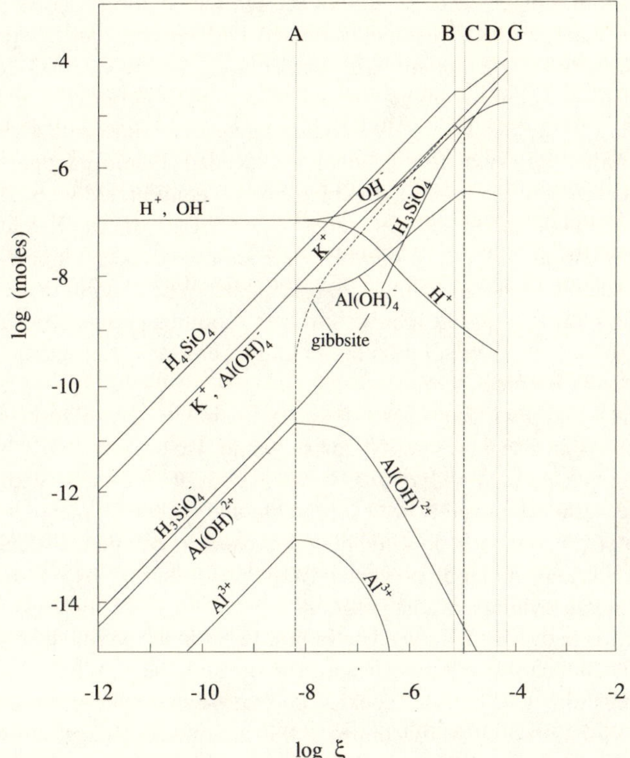

FIG. 19.6. Number of moles of species in the aqueous phase (solid lines) and minerals (only gibbsite is shown) produced and destroyed (dashed lines) per kg of water during the hydrolysis of K-feldspar (after Helgeson et al., 1969, Helgeson, 1979). The increase in kaolinite between B and G and of muscovite between D and G is omitted for clarity. Letters ABCDG correspond to Figure 19.5. ξ is the progress variable.

Therefore, as K-feldspar continues to dissolve, aqueous SiO_2 does not increase, but is used to convert previously precipitated gibbsite into kaolinite. K^+ continues to increase, and the net result is the path B→C. At C, all gibbsite is used up, and the solution composition can resume increasing in silica content, and follows a path C→D roughly parallel to its original path (A→B), only this time precipitating kaolinite rather than gibbsite. Along C→D, K, Al, and Si are all increasing in solution, until at D, K-mica (muscovite) begins to precipitate, because its solubility product is exceeded. Again, coexistence of minerals buffers a solution parameter, this time (a_{K^+}/a_{H^+}), through the relation

$$KAl_3Si_3O_{10}(OH)_2 + \frac{3}{2}H_2O + H^+ = \frac{3}{2}Al_2Si_2O_5(OH)_4 + K^+$$

Because the ratio (a_{K^+}/a_{H^+}) is fixed, but SiO_2 continues to increase as K-feldspar dissolves, kaolinite reacts to form muscovite, and the solution follows path D→E, at which point the solution becomes saturated with quartz, and if equilibrium is maintained, quartz will precipitate. With four components (K_2O, Al_2O_3, SiO_2, H_2O, a maximum of four phases can coexist at our arbitrarily chosen T and P (25°C, 1 atm) according to the phase rule. With quartz, muscovite, kaolinite and water, this number has now been reached, and cannot be exceeded (K-feldspar doesn't count; it is being used as a source of solutes, and has not yet equilibrated with the solution). Therefore if we continue to add K_2O, Al_2O_3, and SiO_2 from the K-feldspar to the solution, the solution will stay at point E while kaolinite reacts with the solution to form muscovite, and quartz continues to precipitate. When kaolinite is all used up, additional dissolution of K-feldspar will drive the solution composition along E→F, with the SiO_2 content of the solution buffered by the presence of quartz. At point F, K-feldspar finally becomes stable.

Alternatively, at point E, if quartz does not precipitate, the solution composition could continue from E to G, where K-feldspar would also become stable, but this time in a solution oversaturated with quartz. It would coexist metastably with muscovite and kaolinite, rather than stably with muscovite and quartz. In nature, quartz quite often does not precipitate at low temperatures, and in computer calculations that simulate equilibrium, it can be prevented from "precipitating" by removing it from the list of minerals available to the program.

The reaction path ABCDE traced on Figure 19.5 shows intermediate phases such as gibbsite first appearing and then disappearing. Along the path BC all three phases gibbsite + kaolinite + K-feldspar coexist, despite the fact that this is a metastable assemblage. If the reaction were stopped at this intermediate stage, this three-phase assemblage would be left behind as the only remnant of an incomplete process. This kind of arrested reaction is very common in nature, particularly at lower temperatures where reaction rates are slower. In such situations, reaction path calculations can provide important insight into the chemical evolution of geological systems that seem nonsensical in terms of classical thermodynamics.

Quasistatic Reactions

Before going on to just how this kind of calculation is performed, we want to reiterate here that what we have just described is a perfect example of a *quasistatic reaction*, first mentioned in §3.7.2. K-feldspar plus water is an unstable assemblage at 25°C, 1 atm. Unless constrained from reacting either by separating them or simply by slow kinetics, they will react irreversibly. The process we have discussed above simulates what would happen if they were allowed to react very briefly (i.e., 0.000001 moles or some similarly small $d\xi$ of K-feldspar is dissolved in water), then the constraint preventing dissolution is reapplied, and the dissolved constituents achieve equilibrium. Then a second small increment of irreversible dissolution is allowed, followed by equilibration of the solution, etc. After each step, when the constraint is applied, we have a metastable system, and each time the constraint is released, the resulting reaction is irreversible, no matter how small in extent. The result is a large succession of metastable equilibrium states extending from the initial metastable state to some final stable or metastable equilibrium state. Each intermediate state is often referred to as in "partial equilibrium", because the solution constituents equilibrate with each other, but not with K-feldspar. We prefer to think of the system after each step as in complete metastable equilibrium because of the extra constraint, not in partial equilibrium.

Naturally, real reactions may not follow the quasistatic path outlined here. Nevertheless, calculating that path has resulted in increased understanding of natural processes, which is usually the goal in constructing models.

19.3.4. *Summary of the Numerical Model*

In this section we briefly outline the numerical method used in a PATH program to compute reaction paths for mineral-water systems. For complete details we recommend the articles by Helgeson (1968) and (1979), and the notation in these articles is used below.

Continuing with our example of K-feldspar, we can write a reaction representing dissolution of feldspar and precipitation of gibbsite; this occurs along the first path A→B in Figure 19.5. Gibbsite is considered to be in equilibrium with the aqueous solution, and all dissolved aqueous species are equilibrated.

$$\begin{aligned}&\bar{n}_{\mathrm{KAlSi_3O_8}}\mathrm{KAlSi_3O_8} + \bar{n}_{\mathrm{H_2O}}\mathrm{H_2O} \rightarrow \bar{n}_{\mathrm{Al(OH)_3}}\mathrm{Al(OH)_3}(gibbsite)\\ &+\bar{n}_{\mathrm{K^+}}\mathrm{K^+} + \bar{n}_{\mathrm{Al^{3+}}}\mathrm{Al^{3+}} + \bar{n}_{\mathrm{Al(OH)^{2+}}}\mathrm{Al(OH)^{2+}} + \bar{n}_{\mathrm{Al(OH)_4^-}}\mathrm{Al(OH)_4^-}\\ &+\bar{n}_{\mathrm{H_4SiO_4}}\mathrm{H_4SiO_4} + \bar{n}_{\mathrm{H_3SiO_4^-}}\mathrm{H_3SiO_4^-} + \bar{n}_{\mathrm{H^+}}\mathrm{H^+} + \bar{n}_{\mathrm{OH^-}}\mathrm{OH^-}\end{aligned} \tag{19.91}$$

The quantities $\bar{n}$ above are the reaction coefficients of each species s. For convenience, we will require that the system contain 1000 g water, and assign $\bar{n}_{\mathrm{KAlSi_3O_8}} = -1$. Each $\bar{n}_s$ then represents the change in the number of moles of the subscripted species s with a reaction step $d\xi$:

$$\frac{dm_s}{d\xi} = \bar{n}_s \tag{19.92}$$

which is analogous to equation (19.77).

Reaction (19.91), to be completely rigorous, should include all species of significance, but some are ignored here for brevity. We might even have written some species that are really products as reactants, but this will ultimately become evident from the signs of each $\bar{n}_s$ (negative for reactants, positive for products). The exact values of the 10 reaction coefficients $\bar{n}_s$ in (19.92) depend on the extent of the reaction, and are therefore unknown at this stage. To solve for each individual $\bar{n}_s$ we will require 10 equations in the 10 unknowns $\bar{n}_s$.

We can begin assembling these equations by writing equilibrium constants for 5 independent reactions between the species in (19.91).

$$a_{\mathrm{OH^-}}a_{\mathrm{H^+}} = K_w \tag{19.93}$$

$$a_{\mathrm{Al^{3+}}}a^3_{\mathrm{OH^-}} = K_{sp}(gibbsite) \tag{19.94}$$

$$\frac{a_{\mathrm{Al^{3+}}}a^4_{\mathrm{OH^-}}}{a_{\mathrm{Al(OH)_4^-}}} = K_{\mathrm{Al(OH)_4^-}} \tag{19.95}$$

$$\frac{a_{\mathrm{Al^{3+}}}a_{\mathrm{OH^-}}}{a_{\mathrm{Al(OH)^{2-}}}} = K_{\mathrm{Al(OH)^{2-}}} \tag{19.96}$$

$$\frac{a_{\mathrm{H^+}}a_{\mathrm{H_3SiO_4^-}}}{a_{\mathrm{H_4SiO_4}}} = K_{\mathrm{H_4SiO_4}} \tag{19.97}$$

The next step is to take the derivative of each of the above 5 equations with respect to the progress variable and rearrange. We will illustrate this just for the first equation (19.93) since the method is identical for the remaining four. The derivative of (19.93) with respect to ξ is

$$a_{\mathrm{OH^-}}\left(\frac{da_{\mathrm{H^+}}}{d\xi}\right) + a_{\mathrm{H^+}}\left(\frac{da_{\mathrm{OH^-}}}{d\xi}\right) = 0 \tag{19.98}$$

Substituting from (19.93) gives

$$\frac{K_w}{a_{\mathrm{H^+}}}\left(\frac{da_{\mathrm{H^+}}}{d\xi}\right) + \frac{K_w}{a_{\mathrm{OH^-}}}\left(\frac{da_{\mathrm{OH^-}}}{d\xi}\right) = 0 \tag{19.99}$$

which simplifies to

$$\frac{1}{a_{\mathrm{H^+}}}\left(\frac{da_{\mathrm{H^+}}}{d\xi}\right) + \frac{1}{a_{\mathrm{OH^-}}}\left(\frac{da_{\mathrm{OH^-}}}{d\xi}\right) = 0 \tag{19.100}$$

Now because $a_s = \gamma_s m_s$, equation (19.92) can be rewritten in the form

$$\bar{n}_s = \frac{dm_s}{d\xi} = \frac{1}{\gamma_s}\left(\frac{da_s}{d\xi}\right) \tag{19.101}$$

Substituting from this for $(da_s/d\xi)$ in (19.100) gives the desired equation.

$$\boxed{\frac{\bar{n}_{\mathrm{H^+}}}{m_{\mathrm{H^+}}} + \frac{\bar{n}_{\mathrm{OH^-}}}{m_{\mathrm{OH^-}}} = 0} \tag{19.102}$$

Similar manipulation of the remaining four mass action equations gives similar expressions.

$$\boxed{\frac{\bar{n}_{\mathrm{Al^{3+}}}}{m_{\mathrm{Al^{3+}}}} + \frac{3\bar{n}_{\mathrm{OH^-}}}{m_{\mathrm{OH^-}}} = 0} \tag{19.103}$$

$$\boxed{\frac{\bar{n}_{\mathrm{Al^{3+}}}}{m_{\mathrm{Al^{3+}}}} + \frac{4\bar{n}_{\mathrm{OH^-}}}{m_{\mathrm{OH^-}}} - \frac{\bar{n}_{\mathrm{Al(OH)_4^-}}}{m_{\mathrm{Al(OH)_4^-}}} = 0} \tag{19.104}$$

$$\boxed{\frac{\bar{n}_{\mathrm{Al^{3+}}}}{m_{\mathrm{Al^{3+}}}} + \frac{\bar{n}_{\mathrm{OH^-}}}{m_{\mathrm{OH^-}}} - \frac{\bar{n}_{\mathrm{Al(OH)^{2-}}}}{m_{\mathrm{Al(OH)^{2-}}}} = 0} \tag{19.105}$$

$$\frac{\bar{n}_{H^+}}{m_{H^+}} + \frac{\bar{n}_{H_3SiO_4^-}}{m_{H_3SiO_4^-}} - \frac{\bar{n}_{H_4SiO_4}}{m_{H_4SiO_4}} = 0 \tag{19.106}$$

In the 5 boxed equations above, the concentrations m_s of each species are known, giving 5 equations in the 10 unknowns n_s. The remaining 5 equations are simply mass balances expressing the transfer of the 5 elements in reaction (19.91) relative to one mole of K-feldspar in a closed system.

Mass Balances

Al :
$$\bar{n}_{Al^{3+}} + \bar{n}_{Al(OH)^{2-}} + \bar{n}_{Al(OH)_4^-} = -\bar{n}_{KAlSi_3O_8} \tag{19.107}$$

Si :
$$\bar{n}_{H_4SiO_4} + \bar{n}_{H_3SiO_4^-} = -3\bar{n}_{KAlSi_3O_8} \tag{19.108}$$

O :
$$\begin{aligned}&\bar{n}_{Al(OH)^{2-}} + 4\bar{n}_{Al(OH)_4^-} + 4\bar{n}_{H_4SiO_4}\\ &+4\bar{n}_{H_3SiO_4^-} + \bar{n}_{H_2O} + \bar{n}_{OH^-} = -8\bar{n}_{KAlSi_3O_8}\end{aligned} \tag{19.109}$$

H :
$$\begin{aligned}&\bar{n}_{Al(OH)^{2-}} + 4\bar{n}_{Al(OH)_4^-} + 4\bar{n}_{H_4SiO_4} + 3\bar{n}_{H_3SiO_4^-}\\ &+2\bar{n}_{H_2O} + \bar{n}_{H^+} + \bar{n}_{OH^-} = 0\end{aligned} \tag{19.110}$$

K :
$$\bar{n}_{K^+} = -\bar{n}_{KAlSi_3O_8} \tag{19.111}$$

The boxed equations (19.102) to (19.106) and (19.107) to (19.111) constitute a set of 10 linear equations in the 10 unknowns $\bar{n}_s$, and can be solved by matrix algebra. The reaction path calculation is started by assigning a set of (very small) initial concentrations m_s to the 10 species in the original reaction (19.91). The set of 10 linear equations above is then solved for the 10 reaction coefficients $\bar{n}_s$. These coefficients are used to calculate the new concentrations of all species after a small reaction step $\Delta\xi$. This is accomplished with equation (19.79), which we used with the example of oxides forming on iron (Figure 19.4a,b,c,d).

$$m_s = m_s^\circ + \bar{n}_s \cdot \Delta\xi \tag{19.112}$$

Here, m_s refers to molality of aqueous species, and moles of solid per kg water for any precipitated minerals. The activity coefficients of all species are computed at each step; these are used to check whether any solid phases have become supersaturated by comparing ion activity products with solubility products (as in 19.90). When a phase precipitates, the set of equations is adjusted accordingly, incorporating the new product or reactant.

In practice, the reaction coefficients change with reaction progress. For this reason, small increments of ξ on the order of 10^{-5} are advisable. More rigorous path calculations provide for these changes by modifying (19.112) to include the derivative $\bar{n}'_s = (d\bar{n}_s/d\xi)$.

$$\Delta m_s = \bar{n}_s \Delta\xi + \bar{n}'_s \frac{(\Delta\xi)^2}{2!} \tag{19.113}$$

This is simply a truncated Taylor's series on m_s. The first derivatives $\bar{n}'_s$ are computed by expanding the matrix equation above and solving for them as well as for the reaction coefficients $\bar{n}_s$. Expressions for the first derivatives are provided by Helgeson (1979).

The path calculation proceeds iteratively, using concentrations m_s from each step to compute new reaction coefficients $\bar{n}_s$. These in turn give new concentrations for the next step.

A reaction path calculation gives the concentrations of all aqueous species and the number of moles of all minerals produced or destroyed at each successive step of the reaction. Results for the hydrolysis of K-feldspar based on the species and equations considered above are illustrated in Figure 19.6 (from Helgeson et al, 1969, and Helgeson, 1979). The letters ABCDE on this figure correspond exactly to the reaction path shown on Figure 19.5. It is clear that concentrations of aqueous species and solids alike change by many orders of magnitude throughout the reaction. The appearance and disappearance of intermediate phases such as gibbsite is modeled quantitatively. Final equilibrium is attained when the total free energy of the system reaches a minimum.

$$G = \sum_{i=1}^{N} \mu_i n_i \quad \text{minimized} \tag{19.114}$$

This can be tested at each step, and for the K-feldspar example in Figure 19.6 occurs at the extent of reaction $\log \xi = -4.1$ This is, in fact, another way (albeit, more time consuming) of performing a free energy-based equilibrium calculation. The extra information obtained makes it worthwhile.

The most recent computations of chemical reaction paths couple chemical kinetics, path calculations, and fluid flow models. This can be accomplished by alternating between fluid flow and reaction path calculations in small time steps, with reaction kinetics included as we have described above. Several examples of this type are summarized by Brimhall and Crerar (1987, pp. 302–306). With this kind of approach it should ultimately become possible to model the detailed physical and chemical evolution of quite complex natural mineral systems. With inclusion of three-dimensional space as well as temperature and pressure gradients, there are challenges for the foreseeable future.

APPENDIX A

CONSTANTS AND NUMERICAL VALUES

The SI (Système International) Units

Base (fundamental) Units		
Physical Quantity	**SI unit**	**Symbol**
Length	meter	m
Mass	kilogram	kg
Time	second	s
Electric current	ampere	A
Temperature	kelvin	K
Amount of substance	mole	mol

Derived SI Units

Physical quantity	SI unit	Symbol for SI unit	Unit in terms of base units	Unit in terms of other SI units
Velocity (speed)			m/s	
Acceleration			m/s^2	N/kg
Force	newton	N	$kg\,m/s^2$	J/m
Pressure	pascal	Pa	$kg/(m\,s^2)$	N/m^2
Energy	joule	J	$kg\,m^2/s^2$	$N\,m$
Entropy	joule per kelvin	S	$kg\,m^2/(s^2\,K)$	J/K
Power	watt	W	$kg\,m^2/s^3$	J/s
Momentum			$kg\,m/s$	
Frequency	hertz	Hz	s^{-1}	
Electric charge	coulomb	C	$A\,s$	$V\,F$
Voltage (emf)	volt	V	$kg\,m^2/(A\,s^3)$	W/A; C/F
Electric resistance	ohm	Ω	$kg\,m^2/(A^2\,s^3)$	V/A
Capacitance	farad	F	$A^2\,s^4/(kg\,m^2)$	C/V

Fundamental Physical Constants[a]

Quantity	Symbol in this text	Value	Units
constant of gravitation	g	6.67259	10^{-11} m^3 kg^{-1} s^{-2}
elementary charge	e	1.60217733	10^{-19} C
Avogadro constant	N_a	6.022136	10^{23} mol^{-1}
Faraday constant	$\mathcal{F}$	96485.309	C mol^{-1}
molar gas constant	R	8.314510	J mol^{-1}K^{-1}
Boltzmann constant, R/N_a	k	1.380658	10^{-23}J K^{-1}
molar volume[b]	V_m	0.02241410	m^3 mol^{-1}

[a] Cohen, E.R., and Taylor, B.N., 1988, The 1986 CODATA recommended values of the fundamental physical constants: Jour. Phys. Chem. Ref. Data, v. 17, pp. 1795–1803.

[b] The volume per mole of ideal gas at 101325 Pa and 273.15 K

Miscellaneous Useful Conversions and Older Units

ln 10	2.302585
ln x	ln 10 × $\log_{10} x$
1 cal	4.184 J
R	1.987216 cal K^{-1} mol^{-1}
$\mathcal{F}$	96485.309 J volt^{-1} mol^{-1}
	23060.542 cal volt^{-1}mol^{-1}
$RT/\mathcal{F}$	0.02569273 volts (T = 298.15 K)
2.302585 $RT/\mathcal{F}$	0.0591597 volts (T = 298.15 K)
1 bar	10^5 pascal
	14.504 psi
	0.10 J cm^{-3}
	0.0239006 cal cm^{-3}
1 atm	1.01325 bar
	101325 pascal
	14.696 psi
1 cm^3	0.10 J bar^{-1}
	0.0239006 cal bar^{-1}
1 Å	1 angstrom = 10^{-8} cm

APPENDIX B

EXPRESSIONS FOR ΔG°, ΔS°, ΔH°, $\ln K$ AS FUNCTION OF TEMPERATURE FOR VARIOUS VERSIONS OF THE HEAT CAPACITY FUNCTION

$$\boxed{\Delta C_p^\circ = 0}$$

$$\Delta H_T^\circ = \Delta S_{T_r}^\circ$$

$$\Delta S_T^\circ = \Delta S_{T_r}^\circ$$

$$\Delta G_T^\circ = \Delta H_{T_r}^\circ - T\Delta S_{T_r}^\circ$$

$$\ln K_T = \ln K_{T_r} - \frac{\Delta H_{T_r}^\circ}{R}\left(\frac{1}{T} - \frac{1}{T_r}\right)$$

$$\boxed{\Delta C_p^\circ = \text{constant}}$$

$$\Delta H_T^\circ = \Delta H_{T_r}^\circ + \Delta C_p^\circ(T - T_r)$$

$$\Delta S_T^\circ = \Delta S_{T_r}^\circ + \Delta C_p^\circ \ln(T/T_r)$$

$$\Delta G_T^\circ = \Delta G_{T_r}^\circ - \Delta S_{T_r}^\circ(T - T_r) + \Delta C_p^\circ(T - T_r - T \cdot \ln\frac{T}{T_r})$$

$$\ln K_T = \ln K_{T_r} - \frac{\Delta H_{T_r}^\circ}{R}\left(\frac{1}{T} - \frac{1}{T_r}\right) + \frac{\Delta C_p^\circ}{R}\left(\ln\frac{T}{T_r} + \frac{T_r}{T} - 1\right)$$

$$\boxed{\Delta C_p^\circ = \Delta a + \Delta b T}$$

$$\Delta H_T^\circ = \Delta H_{T_r}^\circ + \Delta a(T - T_r) + \frac{\Delta b}{2}(T^2 - T_r^2)$$

$$\Delta S_T^\circ = \Delta S_{T_r}^\circ + \Delta a \ln\frac{T}{T_r} + \Delta b(T - T_r)$$

$$\Delta G_T^\circ = \Delta G_{T_r}^\circ - \Delta S_{T_r}^\circ(T - T_r) + \Delta a(T - T_r - T\ln\frac{T}{T_r}) + \frac{\Delta b}{2}(-T^2 - T_r^2 + 2TT_r)$$

$$\ln K_T = \ln K_{T_r} - \frac{\Delta H_{T_r}^\circ}{R}\left(\frac{1}{T} - \frac{1}{T_r}\right) + \frac{\Delta a}{R}\left(\ln\frac{T}{T_r} + \frac{T_r}{T} - 1\right) + \frac{\Delta b}{2R}\left(T + \frac{T_r^2}{T} - 2T_r\right)$$

$$\boxed{\Delta C_p^\circ = \Delta a + \Delta b T + \Delta c T^2}$$

$$\Delta H_T^\circ = \Delta H_{T_r}^\circ + \Delta a(T - T_r) + \frac{\Delta b}{2}(T^2 - T_r^2) + \frac{\Delta c}{3}(T^3 - T_r^3)$$

$$\Delta S_T^\circ = \Delta S_{T_r}^\circ + \Delta a \ln\frac{T}{T_r} + \Delta b(T - T_r) + \frac{\Delta c}{2}(T^2 - T_r^2)$$

$$\Delta G_T^\circ = \Delta G_{T_r}^\circ - \Delta S_{T_r}^\circ(T - T_r) + \Delta a(T - T_r - T\ln\frac{T}{T_r})$$
$$+\frac{\Delta b}{2}(-T^2 - T_r^2 + 2TT_r) + \frac{\Delta c}{6}(-T^3 - 2T_r^3 + 3TT_r^2)$$

$$\ln K_T = \ln K_{T_r} - \frac{\Delta H^\circ_{T_r}}{R}\left(\frac{1}{T} - \frac{1}{T_r}\right) + \frac{\Delta a}{R}\left(\ln\frac{T}{T_r} + \frac{T_r}{T} - 1\right)$$
$$+\frac{\Delta b}{2R}\left(T + \frac{T_r^2}{T} - 2T_r\right) + \frac{\Delta c}{6R}\left(T^2 + \frac{2T_r^3}{T} - 3T_r^2\right)$$

$$\boxed{\Delta C_p^\circ = \Delta a + \Delta b T - \Delta c T^{-2}}$$

(Maier and Kelley, 1932)

$$\Delta H^\circ_T = \Delta H^\circ_{T_r} + \Delta a(T - T_r) + \frac{\Delta b}{2}(T^2 - T_r^2) + \Delta c\left(\frac{1}{T} - \frac{1}{T_r}\right)$$

$$\Delta S^\circ_T = \Delta S^\circ_{T_r} + \Delta a \ln\frac{T}{T_r} + \Delta b(T - T_r) + \frac{\Delta c}{2}\left(\frac{1}{T^2} - \frac{1}{T_r^2}\right)$$

$$\Delta G^\circ_T = \Delta G^\circ_{T_r} - \Delta S^\circ_{T_r}(T - T_r) + \Delta a(T - T_r - T\ln\frac{T}{T_r})$$
$$+\frac{\Delta b}{2}(-T^2 - T_r^2 + 2TT_r) + \Delta c\left(\frac{T^2 + T_r^2 - 2TT_r}{2TT_r^2}\right)$$

$$\ln K_T = \ln K_{T_r} - \frac{\Delta H^\circ_{T_r}}{R}\left(\frac{1}{T} - \frac{1}{T_r}\right) + \frac{\Delta a}{R}\left(\ln\frac{T}{T_r} + \frac{T_r}{T} - 1\right)$$
$$+\frac{\Delta b}{2R}\left(T + \frac{T_r^2}{T} - 2T_r\right) + \frac{\Delta c}{R}\left(\frac{-T^2 - T_r^2 + 2TT_r}{2T^2T_r^2}\right)$$

$$\boxed{\Delta C_p^\circ = \Delta a + \Delta b T + \Delta c T^{-2} + \Delta f T^2 + \Delta g T^{-\frac{1}{2}}}$$

(Haas and Fisher, 1976)

$$\Delta H^\circ_T = \Delta H^\circ_{T_r} + \Delta a(T - T_r) + \frac{\Delta b}{2}(T^2 - T_r^2) - \Delta c\left(\frac{1}{T} - \frac{1}{T_r}\right) + \frac{\Delta f}{3}(T^3 - T_r^3) + 2\Delta g(T^{\frac{1}{2}} - T_r^{\frac{1}{2}})$$

$$\Delta S^\circ_T = \Delta S^\circ_{T_r} + \Delta a \ln\frac{T}{T_r} + \Delta b(T - T_r) + \frac{\Delta c}{2}\left(\frac{1}{T^2} - \frac{1}{T_r^2}\right) + \frac{\Delta f}{2}(T^2 - T_r^2) - 2\Delta g\left(\frac{1}{T^{\frac{1}{2}}} - \frac{1}{T_r^{\frac{1}{2}}}\right)$$

$$\Delta G^\circ_T = \Delta G^\circ_{T_r} - \Delta S^\circ_{T_r}(T - T_r) + \Delta a(T - T_r - T\ln\frac{T}{T_r})$$
$$+\frac{\Delta b}{2}(-T^2 - T_r^2 + 2TT_r) + \Delta c\left(\frac{-T^2 - T_r^2 + 2TT_r}{2TT_r^2}\right)$$
$$+\frac{\Delta f}{6}(-T^3 - 2T_r + 3TT_r^2) + 2\Delta g\left(2T^{\frac{1}{2}} - T_r^{\frac{1}{2}} - \frac{T}{T_r^{\frac{1}{2}}}\right)$$

$$\ln K_T = \ln K_{T_r} - \frac{\Delta H^\circ_{T_r}}{R}\left(\frac{1}{T} - \frac{1}{T_r}\right) + \frac{\Delta a}{R}\left(\ln\frac{T}{T_r} + \frac{T_r}{T} - 1\right)$$
$$+\frac{\Delta b}{2R}\left(T + \frac{T_r^2}{T} - 2T_r\right) + \frac{\Delta c}{R}\left(\frac{T^2 + T_r^2 - 2TT_r}{2TT_r^2}\right)$$
$$+\frac{\Delta f}{6R}\left(T^2 + \frac{2T_r^3}{T} - 3T_r^2\right) + \frac{2\Delta g}{R}\left(\frac{-2}{T^{\frac{1}{2}}} + \frac{T_r^{\frac{1}{2}}}{T} + \frac{1}{T_r^{\frac{1}{2}}}\right)$$

N.B.: The original Haas and Fisher formula has a 2 in the b term which is here absorbed in our b term, $i.e.$ Δb here is twice the Haas and Fisher Δb.

$$\boxed{\Delta C_p^\circ = \Delta a + \Delta b T^{-\frac{1}{2}} + \Delta c T^{-2} + \Delta d T^{-3}}$$

(Berman and Brown, 1985)

$$\Delta H^\circ_T = \Delta H^\circ_{T_r} + \Delta a(T - T_r) + 2\Delta b(T^{\frac{1}{2}} - T_r^{\frac{1}{2}}) - \Delta c\left(\frac{1}{T} - \frac{1}{T_r}\right) - \frac{\Delta d}{2}\left(\frac{1}{T^2} - \frac{1}{T_r^2}\right)$$

$$\Delta S^\circ_T = \Delta S^\circ_{T_r} + \Delta a \ln\frac{T}{T_r} - 2\Delta b\left(\frac{1}{T^{\frac{1}{2}}} - \frac{1}{T_r^{\frac{1}{2}}}\right) - \frac{\Delta c}{2}\left(\frac{1}{T^2} - \frac{1}{T_r^2}\right) - \frac{\Delta d}{3}\left(\frac{1}{T^3} - \frac{1}{T_r^3}\right)$$

$$\Delta G^\circ_T = \Delta G^\circ_{T_r} - \Delta S^\circ_{T_r}(T - T_r) + \Delta a(T - T_r - T\ln\frac{T}{T_r}) + 2\Delta b\left(2T^{\frac{1}{2}} - T_r^{\frac{1}{2}} - \frac{T}{T_r^{\frac{1}{2}}}\right)$$
$$+\Delta c\left(\frac{-T^2 - T_r^2 + 2TT_r}{2TT_r^2}\right) + \frac{\Delta d}{6}\left(-\frac{1}{T^2} + \frac{3}{T_r^2} - \frac{2T}{T_r^3}\right)$$

$$\ln K_T = \ln K_{T_r} - \frac{\Delta H^\circ_{T_r}}{R}\left(\frac{1}{T} - \frac{1}{T_r}\right) + \frac{\Delta a}{R}\left(\ln\frac{T}{T_r} + \frac{T_r}{T} - 1\right)$$
$$+\frac{2\Delta b}{R}\left(\frac{T_r^{\frac{1}{2}}}{T} - \frac{2}{T^{\frac{1}{2}}} + \frac{1}{T_r^{\frac{1}{2}}}\right) + \frac{\Delta c}{R}\left(\frac{T^2 + T_r^2 - 2TT_r}{2T^2T_r^2}\right)$$
$$+\frac{\Delta d}{6R}\left(\frac{1}{T^3} - \frac{3}{TT_r^{\frac{1}{2}}} - \frac{2}{T_r^3}\right)$$

$$\boxed{\Delta C_p^\circ = \Delta C^\circ_{p_r} \cdot \frac{T(\partial\alpha/\partial T)_P}{T_r(\partial\alpha/\partial T)_{P_r}}}$$

for $\Delta C^\circ_{p_r}$ positive
Mesmer (1985)

$$a = \ln K_r + \frac{\Delta H^\circ_r}{RT_r} - \frac{\Delta C^\circ_{p_r} \cdot \alpha_r}{RT_r(\partial\alpha/\partial T)_{P_r}}$$

$$b = -\frac{\Delta H^\circ_r}{R} + \frac{(T_r\alpha_r + \ln\rho_r)\Delta C^\circ_{p_r}}{RT_r(\partial\alpha/\partial T)_{P_r}}$$

$$c = \frac{-\Delta C^\circ_{p_r}}{RT_r(\partial\alpha/\partial T)_{P_r}}$$

$$\begin{aligned}\Delta H^\circ &= -R\,(b + c(T\alpha + \ln\rho)) \\ &= \Delta H^\circ_r + \frac{\Delta C^\circ_{p_r}}{T_r(\partial\alpha/\partial T)_{P_r}}\left(T\alpha - T_r\alpha_r + \ln\frac{\rho}{\rho_r}\right)\end{aligned}$$

$$\begin{aligned}\Delta S^\circ &= R(a - c\alpha) \\ &= \Delta S^\circ_r + \frac{\Delta C^\circ_{p_r}}{T_r(\partial\alpha/\partial T)_{P_r}}\,(\alpha - \alpha_r)\end{aligned}$$

$$\begin{aligned}\Delta C^\circ_p &= -RTc(\partial\alpha/\partial T)_P \\ &= \Delta C^\circ_{p_r}\cdot\frac{T(\partial\alpha/\partial T)_P}{T_r(\partial\alpha/\partial T)_{P_r}}\end{aligned}$$

$$\begin{aligned}\Delta V^\circ &= -cR\beta \\ &= \frac{\Delta C^\circ_{p_r}\cdot\beta}{T_r(\partial\alpha/\partial T)_{P_r}}\end{aligned}$$

$$\begin{aligned}\Delta G^\circ &= -R(aT + b + c\cdot\ln\rho) \\ &= \Delta G^\circ_r + \Delta H^\circ_r\left(1 - \frac{T}{T_r}\right) + \frac{\Delta C^\circ_{p_r}}{RT_r(\partial\alpha/\partial T)_{P_r}}\left(\alpha_r(T - T_r) + \ln\frac{\rho}{\rho_r}\right)\end{aligned}$$

$$\begin{aligned}\ln K &= a + b/T + (c\cdot\ln\rho)/T \\ &= \ln K_r - \frac{\Delta H^\circ_r}{R}\left(\frac{1}{T} - \frac{1}{T_r}\right) + \frac{\Delta C^\circ_{p_r}}{RT_r(\partial\alpha/\partial T)_{P_r}}\left(\frac{1}{T}\ln\frac{\rho_r}{\rho} - \frac{\alpha_r}{T}(T - T_r)\right)\end{aligned}$$

APPENDIX C

STANDARD STATE THERMODYNAMIC PROPERTIES OF SELECTED MINERALS AND OTHER COMPOUNDS

These data are included primarily for use in problem sets, such as those in this book or others set by course instructors. Transition state and other data have been omitted for simplicity.

The data are mostly from the program SUPCRT92 (Johnson, Oelkers, and Helgeson, 1992). Entries with an asterisk (*) are from other sources, and have been added only to broaden the range of problems which may be formulated. No attempt has been made to make the data self-consistent. Many of the gram formula weights are slightly inconsistent with the latest values of the atomic weights (DeLaeter, 1988).

A note to the unwary: A column heading of $b \times 10^3$ means that the column entries (e.g. 2.82) are equal to $b \times 10^3$, not that the entry is to be multiplied by 10^3. Thus in this case $b \times 10^3 = 2.82$, or $b = 2.82 \times 10^{-3}$.

Name	Formula	Gram Formula Weight	¢ f G⁻ cal mol¡ 1	¢ f H⁻ cal mol¡ 1	S⁻ cal mol¡ 1K¡ 1	V⁻ cm^3 mol¡ 1	C_p^- coefficients: a cal mol¡ 1	b × 10^3 cal mol¡ 1K¡ 2	c × 10¡ 5 cal mol¡ 1K
Elements									
*aluminum	Al	26.982	0	0	6:776	9 999	4 94	2 96	
*carbon (graphite)	C	12.011	0	0	5:74	5 298	4 03	1:14	2 04
*carbon (diamond)	C	12.011	2900	1895	2 38	3 417	2 27	3 06	1 54
copper	Cu	63.546	0	0	7 923	7:113	5 41	1 50	
oxygen	O_2	31.9988	0	0	49 029	24789 2	7:16	1 0	0 40
*silicon	Si	28.086	0	0	4 496	12 056	5:70	0:70	1 04
sulfur	S_2(g)	64.120	18953	30681	54 536	24789 2	8:72	0:16	0 90
*sulfur (rhombic)	S(s)	32.060	0	0	7 60	15 511	3 58	6 24	
iron (fi)	Fe	55.8470	0	0	6 52	7 092	3 04	7 58	-0 60
Oxides									
corundum	fi − Al_2O_3	101.961	−374824	−397145	12:18	25 575	27 49	2 82	8 38
ferrous oxide	FeO	71.846	−60097	−65020	14 52	12 00	12:122	2 072	0:750
hematite	Fe_2O_3	159.692	−178155	−197720	20 94	30:274	23 49	18 60	3 55
magnetite	Fe_3O_4	231.539	−242574	−267250	34 83	44 524	21 88	48 20	0 00
lime	CaO	56.079	−144366	−151790	9 50	16:764	11 67	1 08	1 56
periclase	MgO	40.311	−136086	−143800	6 44	11 248	10:18	1:74	1 48
K oxide	K_2O	94.203	−77056	−86800	22 5	40 38	18 51	8 65	0 88
Na oxide	Na_2O	61.979	−89883	−99140	17 935	25 00	18 25	4 89	2 89
fi-quartz	SiO_2	60.085	−204646	−217650	9 88	22 688	11 22	8 20	2:70
fl-quartz	SiO_2	60.085					14 41	1 94	
cristobalite	SiO_2	60.085	−203895	−216755	10 372	25:74	13 98	3 34	3 81
coesite	SiO_2	60.085	−203541	−216616	9 65	20 641	11 0	8 2	2:70
chalcedony	SiO_2	60.085	−204276	−217282	9 88	22 688	11 22	8 20	2:70
amorphous silica	$SiO_2 \cdot nH_2O$	60.085	−202892	−214568	14 34	29 0	5 93	47 20	22:7
spinel	$MgAl_2O_4$	142.273	−517006	−546847	19 27	39:71	36:773	6 415	9:709

water	$H_2O(l)$	18.015	−56688	−68317	16:712	18:1	18.04		
*steam	$H_2O(g)$	18.015	-54525	-57935	44:763	24789.2	12:665	10:404	0:013
cuprite	Cu_2O	143.0794	−35384	−40830	22.08	23.437	14.08	5.88	0:76
tenorite	CuO	79.5394	−30568	−37200	10:18	12.22	11.53	1.88	1:76
Sulfates									
anhydrite	$CaSO_4$	136.138	−315925	−342760	25.50	45.94	16:78	23.60	
*gypsum	$CaSO_4 \cdot 2H_2O$	172.168	−429540	−483420	46.40	74.69			
barite	$BaSO_4$	233.398	−325563	−352100	31.60	52:1	33.80		8.43
Hydroxides									
boehmite	$AlO(OH)$	59.988	−217250	−235078	11.58	19.535	14.435	4.20	0.00
diaspore	$AlO(OH)$	59.988	−218402	−237170	8.43	17:760	14.435	4.20	0.00
gibbsite	$Al(OH)_3$	78.004	−276168	−309065	16:75	31.956	8.65	45.6	0.0
brucite	$Mg(OH)_2$	58.327	−199646	−221390	15.09	24.63	24:147	4.012	6:11
*goethite	$FeO(OH)$	88.854	−116766	−133683	14.43	20.82			
Carbonates									
calcite	$CaCO_3$	100.089	−269880	−288552	22:15	36.934	24.98	5.24	6.20
aragonite	$CaCO_3$	100.089	−269683	−288531	21.56	34:15	20:13	10.24	3.34
dolomite	$CaMg(CO_3)_2$	184.411	−517760	−556631	37.09	64.365	41.557	23.952	9.88
magnesite	$MgCO_3$	84.321	−245658	−265630	15:70	28.018	19:731	12.539	4:748
Sulfides									
chalcopyrite	$CuFeS_2$	183.515	−44900	−44453	31:15	42.83	20:79	12.80	1.34
bornite	Cu_5FeS_4	501.803	−86704	−79922	99.29	98.6	49:76	35.08	1.35
troilite	FeS	87.907	−24084	−24000	14.41	18.2	5:19	26.40	
pyrite	FeS_2	119.967	−38293	−41000	12.65	23.94	17.88	1.32	3.05

Name	Formula	Gram Formula Weight	$\Delta_f G^\circ$ cal mol^{-1}	$\Delta_f H^\circ$ cal mol^{-1}	S° cal mol^{-1}K^{-1}	V° cm^3 mol^{-1}	C_p° coefficients: a cal mol^{-1}	$b \times 10^3$ cal mol^{-1}K^{-2}	$c \times 10^{-5}$ cal mol^{-1}K
Ortho and Ring Silicates									
kyanite	Al_2SiO_5	162.046	−580956	−616897	20.00	44.09	41.3931	6.8165	12.8821
andalusite	Al_2SiO_5	162.046	−580587	−615866	22.20	51.53	41.3108	6.2926	12.3921
sillimanite	Al_2SiO_5	162.046	−580091	−615099	23.13	49.90	40.0240	7.3905	11.6741
gehlenite	$Ca_2Al_2SiO_7$	274.205	−903148	−951225	48.1	90.24	63.74	8.00	15.12
grossular	$Ca_3Al_2Si_3O_{12}$	450.454	−1496307	−1582737	60.87	125.3	104.017	17.013	27.318
andradite	$CaFe_2Si_3O_{12}$	508.184	−1296819	−1380345	70.13	131.85	113.532	15.636	30.889
monticellite	$CaMgSiO_4$	156.476	−512829	−540800	26.4	51.47	36.82	5.34	8.00
merwinite	$Ca_3Mg(SiO_4)_2$	328.719	−1036526	−1090796	60.5	104.4	72.97	11.96	14.44
akermanite	$CaMgSi_2O_7$	272.640	−879362	−926497	50.03	92.81	60.09	11.40	11.40
fayalite	Fe_2SiO_4	203.778	−330233	−354119	35.45	46.39	36.51	9.36	6.70
forsterite	Mg_2SiO_4	140.708	−491564	−520000	22.75	43.79	35.81	6.54	8.52
cordierite	$Mg_2Al_3(AlSi_5O_{18})$	584.969	−2061279	−2183199	97.33	233.22	143.83	25.80	38.60
hydrous cordierite	$Mg_2Al_3(AlSi_5O_{18}) \cdot H_2O$	602.984	−2121350	−2255676	111.43	241.22	155.23	25.80	38.60
clinozoisite	$Ca_2Al_3Si_3O_{12}(OH)$	622.882	−1549240	−1643781	70.64	136.2	106.118	25.214	27.145
zoisite	$Ca_2Al_3Si_3O_{12}(OH)$	622.882	−1549179	−1643691	70.747	135.9	106.118	25.214	27.145
epidote	$Ca_2FeAl_2Si_3O_{12}(OH)$	651.747	−1450906	−1543992	75.28	139.2	117.622	12.816	31.864
lawsonite	$CaAl_2Si_2O_7(OH)_2 \cdot H_2O$	314.2	−1073628	−1158324	55.8	7101.32	81.80	23.36	16.26
Chain and Band Silicates									
wollastonite	$CaSiO_3$	116.164	−369225	−389590	19.60	39.93	26.64	3.60	6.52
Ca-Al pyroxene	$CaAl_2SiO_6$	218.125	−742067	−783793	35.0	63.5	54.13	6.42	14.9
jadeite	$NaAl(SiO_3)_2$	202.140	−679445	−722116	31.9	60.4	48.16	11.42	11.87
enstatite	$MgSiO_3$	100.396	−348930	−369686	16.2	31.276	24.55	4.74	6.28
ferrosilite	$FeSiO_3$	131.931	−267160	−285625	22.6	32.592	26.49	5.07	5.55
diopside	$CaMg(SiO_3)_2$	216.560	−723780	−765378	34.2	66.09	52.87	7.84	15.74
hedenbergite	$CaFe(SiO_3)_2$	248.106	−638998	−678276	40.7	68.27	54.81	8.17	15.01

anthophyllite	$Mg_7Si_8O_{22}(OH)_2$	780.872	−2715430	−2888749	128.6	264.4	180.682	60.574	38.462
tremolite	$Ca_2Mg_5Si_8O_{22}(OH)_2$	812.410	−2770245	−2944038	131.19	272.92	188.222	57.294	44.822
Framework Silicates									
analcime	$NaAlSi_2O_6 \cdot H_2O$	220.155	−738098	−790193	56.0	97.1	53.49	24.14	8.88
low albite	$NaAlSi_3O_8$	262.244	−886308	−939680	49.51	100.07	61.70	13.90	15.01
anorthite	$CaAl_2Si_2O_8$	278.210	−954078	−1007552	49.1	100.79	63.311	14.794	15.44
maximum microcline	$KAlSi_3O_8$	278.337	−895374	−949188	51.13	108.741	63.83	12.90	17.05
nepheline	$NaAlSiO_4$	145.227	−472872	−500241	29.72	54.16	35.908	6.458	7.328
kalsilite	$KAlSiO_4$	158.167	−481750	−509408	31.85	59.89	29.43	17.36	5.32
wairakite	$CaAl_2Si_4O_{12} \cdot 2H_2O$	434.411	−1477432	−1579333	105.1	186.87	100.40	44.47	16.43
laumontite	$CaAl_2Si_4O_{12} \cdot 4H_2O$	470.441	−1596823	−1728664	116.1	207.55	123.20	44.47	16.43
Sheet Silicates									
kaolinite	$Al_2Si_2O_5(OH)_4$	258.161	−905614	−982221	48.53	99.52	72.77	29.20	21.52
pyrophyllite	$Al_2Si_4O_{10}(OH)_2$	360.616	−1255997	−1345313	57.2	126.6	79.432	39.214	17.282
chrysotile	$Mg_3Si_2O_5(OH)_4$	277.134	−964871	−1043123	52.9	108.5	75.82	31.60	17.58
antigorite	$Mg_{48}Si_{34}O_{85}(OH)_{62}$	4536.299	−15808020	−17070891	861.36	1749.13	1228.45	513.76	286.68
talc	$Mg_3Si_4O_{10}(OH)_2$	379.289	−1320188	−1410920	62.34	136.25	82.48	41.61	13.34
annite	$KFe_3(AlSi_3O_{10})(OH)_2$	511.890	−1147156	−1232195	95.2	154.32	106.43	29.77	19.31
phlogopite	$KMg_3(AlSi_3O_{10})(OH)_2$	417.286	−1396187	−1488067	76.1	149.66	100.61	28.78	21.50
muscovite	$KAl_2(AlSi_3O_{10})(OH)_2$	398.313	−1336301	−1427408	68.8	140.71	97.56	26.38	25.44
paragonite	$NaAl_2(AlSi_3O_{10})(OH)_2$	382.201	−1326012	−1416963	66.4	132.53	97.43	24.50	26.44
7-Å chlinochlore	$Mg_5Al(AlSi_3O_{10})(OH)_8$	555.832	−1957101	−2113197	106.5	211.5	162.82	50.62	40.88
14-Å chlinochlore	$Mg_5Al(AlSi_3O_{10})(OH)_8$	555.832	−1961703	−2116964	111.2	207.11	166.50	42.10	37.47
prehnite	$Ca_2Al(AlSi_3O_{10})(OH)_2$	412.389	−1390537	−1482089	65.07	140.33	91.60	37.82	19.60
margarite	$CaAl_2(Al_2Si_2O_{10})(OH)_2$	398.187	−1394370	−1486023	63.8	129.4	102.50	16.35	28.05

Conventional standard partial molar properties of aqueous species at 25°C and 1 bar, plus some of the coefficients required to calculate these properties at high T and P using the revised HKF model, and using θ = 228 K and Ψ = 2600 bars.

Ion	$\Delta_f \bar{G}_j^\circ$	$\Delta_f \bar{H}_j^\circ$	$\bar{S}_j^\circ$	$\bar{V}_j^\circ$	$10a_1$	$10^{-2}a_2$	a_0	$10^{-4}a_4$	c_1	$10^{-4}c_2$	$r_{x,j}$
H^+	0	0	0	0	0	0	0	0	0	0	0
Li^+	−69933	−66552	2.7	0.40	−0.0237	−0.0690	11.5800	−2.7761	19.2	−0.24	0.68
Na^+	−62591	−57433	13.96	−1.2	1.8390	−2.2850	3.2560	−2.726	18.18	−2.981	0.97
K^+	−67527	−60270	24.2	9.06	3.5590	−1.4730	5.435	−2.712	7.40	−1.791	1.33
Mg^{2+}	−108505	−111367	−33.0	−17.7	−0.8217	−8.5990	8.3900	−2.390	20.80	−5.892	0.66
Ca^{2+}	−132120	−129800	−13.5	−15.0	−0.1947	−7.2520	5.2966	−2.4792	9.00	−2.522	0.99
Sr^{2+}	−134760	−131670	−7.53	−17.8	0.7071	−10.1508	7.0027	−2.3594	10.7452	−5.0818	1.12
Ba^{2+}	−134030	−128500	2.3	−10.10	2.7383	−10.0565	−0.0470	−2.3633	3.80	−3.450	1.34
F^-	−67340	−80150	−3.15	3.1	0.6870	1.3588	7.6033	−2.8352	4.46	−7.488	1.33
Cl^-	−31379	−39933	13.56	21.4	4.0320	4.8010	5.5630	−2.847	−4.40	−5.714	1.81
Br^-	−24870	−29040	19.8	28.3	5.2690	6.5940	4.7450	−3.143	−3.80	−6.811	1.96
OH^-	−37595	−54977	−2.56	−4.7	1.2527	0.0738	1.8423	−2.7821	4.15	−10.346	1.40
NO_3^-	−26504	−49429	35.1	29.00	6.6740	8.4530	−0.1563	−4.204	7.70	−6.725	2.81
HCO_3^-	−140277	−164898	23.5	24.60	7.3760	1.6350	2.5770	−3.143			2.17
HS^-	2878	−3848	16.3	20.21	5.0860	4.7860	2.2070	−2.847	3.41	−6.046	1.84
SO_4^{2-}	−177930	−217400	4.5	12.9	8.3014	−1.9846	−6.2122	−2.697	1.64	−17.998	3.14
$NaCl^\circ$	−92910	−96120	28.0	24.0	5.0363	4.7365	3.4154	−2.9748	10.8	−1.30	
SiO_2 (aq)	−199190	−209775	18.0	16.1	1.9	1.7	20.0	−2.70	29.10	−51.20	
CO_2 (aq)	−92250	−98900	28.10	33.4	6.2466	7.4711	2.8136	−3.0879	40.0325	8.8004	
CO_3^{2-}	−126191	−161385	−11.95	4.0	2.8524	−3.9844	6.4142	−2.6143	−3.3206	−17.1917	

Units:

$\Delta_f \bar{G}_j^\circ$	$\Delta_f \bar{H}_j^\circ$	$\bar{S}_j^\circ$	$\bar{V}_j^\circ$	a_1	a_2	a_3	a_4	c_1	c_2	$r_{x,j}$
cal mol^{-1}	cal mol^{-1}	cal mol^{-1}K^{-1}	cm^{3}mol^{-1}	cal mol^{-1}bar^{-1}	cal mol^{-1}	cal K mol^{-1}bar^{-1}	cal K mol^{-1}	cal mol^{-1}K^{-1}	cal K mol^{-1}	Å

APPENDIX D

ELECTROCHEMICAL CELL CONVENTIONS

IUPAC Convention	Garrels & Christ, etc.
1. The cell is represented by a schematic diagram, e.g., $Zn\|Zn^{2+}\|\|Cu^{2+}\|Cu$	same
2. The cell schematic indicates the direction in which the cell reaction is to be written. The left-hand electrode is the reactant and the right-hand electrode is the product, e.g., $Zn\|Zn^{2+}\|\|Cu^{2+}\|Cu$ $Zn + Cu^{2+} \rightarrow Zn^{2+} + Cu$ In cases where an electrode cannot be identified with a reactant or product: The left-hand electrode supplies electrons to the outer circuit (i.e., oxidation takes place). The right hand electrode accepts electrons from the outer circuit (i.e., reduction takes place). e.g., $Pt\|Fe^{2+}, Fe^{3+}\|\|Mn^{2+}, Mn^{3+}\|Pt$ $Fe^{2+} + Mn^{3+} \rightarrow Fe^{3+} + Mn^{2+}$	same
3. The e.m.f. of the cell is given by : $\mathcal{E} = \mathcal{E}_{\text{right electrode}} - \mathcal{E}_{\text{left electrode}}$	$\mathcal{E} = \mathcal{E}_{\text{left}} - \mathcal{E}_{\text{right}}$
4. The e.m.f. of a cell is related to the Gibbs free energy of the cell reaction by: $\Delta G = -n\mathcal{F}\mathcal{E}$	$\Delta G = +n\mathcal{F}\mathcal{E}$
5. The electrode potential of a half-cell is equal in magnitude and sign to the e.m.f. of a cell formed with the electrode in question on the right and the standard hydrogen electrode ($\mathcal{E}^\circ = 0$) on the left.	SHE on right
6. The standard electrode potentials are tabulated as reductions, e.g. $Zn^{2+} + 2e \rightarrow Zn;\ \mathcal{E}^\circ = -0.763$ V	tabulated as oxidations
However, the electrode potential is a sign-invariant quantity, i.e., $Zn \rightarrow Zn^{2+} + 2e;\ \mathcal{E}^\circ = -0.763$ V	same
7. For the reaction $a\text{A} + b\text{B} + \cdots \rightleftharpoons c\text{C} + d\text{D} + \cdots$ the Nernst expression is $\mathcal{E} = \mathcal{E}^\circ - \frac{RT}{n\mathcal{F}} \ln \left[\frac{a_C^c a_D^d \cdots}{a_A^a a_B^b \cdots}\right]$	$\mathcal{E} = \mathcal{E}^\circ + \frac{RT}{n\mathcal{F}} \ln \left[\frac{a_C^c a_D^d \cdots}{a_A^a a_B^b \cdots}\right]$
8. In view of 6., the Nernst expression for a half-cell is given by $\mathcal{E} = \mathcal{E}^\circ - \frac{RT}{n\mathcal{F}} \ln \left[\frac{\text{reduced form}}{\text{oxidized form}}\right]$	$\mathcal{E} = \mathcal{E}^\circ + \frac{RT}{n\mathcal{F}} \ln \left[\frac{\text{oxidized form}}{\text{reduced form}}\right]$

The difference between these conventions is basically that G&C and most other geological authors write half-cell potentials as oxidations (no. 8) whereas the international convention is to write them as reductions (no. 6). Since the sign of the potential is the same either way, the ΔG must be of opposite sign in the two conventions (no. 4), and this affects nos. 7 and 8.
There is an older convention (e.g. Latimer, 1952) in which $\mathcal{E}^\circ$ is not sign invariant. This convention has disappeared, but be careful if using older literature.

An Example

IUPAC	$2H_2O + 2e = H_2 + 2OH^-$	reduction; $\mathcal{E}^\circ = -0.828V$
	$H_2 = 2H^+ + 2e$	oxidation; $\mathcal{E}^\circ = 0$
	$2H_2O = 2H^+ + 2OH^-$	
G & C	$H_2 + 2OH^- = 2H_2O + 2e$	oxidation; $\mathcal{E}^\circ = -0.828V$
	$2H^+ + 2e = H_2$	reduction; $\mathcal{E}^\circ = 0$
	$2H^+ + 2OH^- = 2H_2O$	
		⇑
		N.B.: $\mathcal{E}$ is sign invariant

IUPAC Convention	Garrels & Christ, etc.
SHE on left (oxidation)	SHE on right (reduction)
Thus	Thus
$2H_2O = 2H^+ + 2OH^-$	$2H^+ + 2OH^- = 2H_2O$
$\mathcal{E}^\circ_{\text{cell}} = \mathcal{E}^\circ_{\text{right}} - \mathcal{E}^\circ_{\text{left}}$	$\mathcal{E}^\circ_{\text{cell}} = \mathcal{E}^\circ_{\text{left}} - \mathcal{E}^\circ_{\text{right}}$
$= -0.828v - 0$	$= -0.828V - 0$
$= -0.828v$	$= -0.828V$
$\Delta G^\circ = -n\mathcal{F}\mathcal{E}^\circ$	$\Delta G^\circ = +n\mathcal{F}\mathcal{E}^\circ$
$= -2 \times 23.06 \times (-0.828)$	$2 \times 23.06 \times (-0.828)$
$= +38.187$ kcal mol^{-1}	$= -38.187$ kcal mol^{-1}
$= -RT \ln K$	$= -RT \ln K$
$\log K = -38187/(4.576 \times 298.15)$	$\log K = 38187/(4.576 \times 298.15)$
$= -28.0$	$= +28.0$
$K = (a^2_{H^+} \cdot a^2_{OH^-})\big/a^2_{H_2O}$	$K = a^2_{H_2O}\big/(a^2_{H^+} \cdot a^2_{OH^-})$
$a_{H^+} \cdot a_{OH^-} = K^{\frac{1}{2}}$	$a_{H^+} \cdot a_{OH^-} = K^{-\frac{1}{2}}$
$= 10^{-14}$	$= 10^{-14}$

N.B.: In both conventions $\mathcal{E}^\circ_{\text{left}}$ refers to $\mathcal{E}^\circ_{\text{oxidation half-cell}}$ and $\mathcal{E}^\circ_{\text{right}}$ refers to $\mathcal{E}^\circ_{\text{reduction half-cell}}$. SHE is not necessarily one of the half-cells.

APPENDIX E

EQBRM—A FORTRAN SPECIATION PROGRAM

This appendix contains a computer program called EQBRM written in FORTRAN77 which calculates the concentrations of all aqueous species in a system at equilibrium. The algorithm and some of the theory are outlined in Chapter 19. We describe here how to use the program, using a simple calculation as an example.

The program calculates activity coefficients of ionic species by calling a subroutine called ACTCF which calculates activity coefficients using the Davies revision of the Debye-Hückel equation:

$$\log \gamma_i = \frac{-Z_i^2 A\sqrt{\bar{I}}}{1+\sqrt{\bar{I}}} + 0.2AZ_i^2\bar{I}$$

where $\bar{I}$ is the true ionic strength, representing concentrations of all species as corrected for ion-pair and complex formation:

$$\bar{I} = \frac{1}{2}\sum_{i=1}^{N} m_i Z_i^2$$

Subroutine ACTF can be rewritten using other equations if desired.

How to set up data for the program EQBRM

EQBRM was designed to be completely general, letting you define the species, reactions and component concentrations of interest to you. To do this systematically, go through the following steps:

1. Write all N species of interest and assign each a number from 1 to N. The concentrations of these species are the N unknowns.
2. Write out M mass balance or equilibrium constant equations ($j = 1, 2, \ldots, M$). These should be written so the lowest activities are in the denominator to reduce the chances of matrix singularity.
3. Assign the reaction coefficients ν_{ij} based on the number of times the ith species appears in the jth reaction; these coefficients are negative for reactants and positive for products for each reaction as written.
4. Write $N - M - 1$ mass balances, giving the total concentration of each element or component B_e, (such as total Na, total Cl, or total NH_3 in the system.
5. Assign the parameters $b_{e,i}$ which refer to the number of atoms of each element or component in the formula of each species. For example, from the formula for the species H_2SO_4 we can set $b_H = 1$, $b_S = 1$, $b_O = 4$, and $b_{e,i} = 0$ for all other elements.

6. Write valences, Z_i, of all species.
7. Make reasonable first guesses at concentrations of all species.
8. Choose a value for the A term in the activity coefficient expression.

Example worksheet for data input to EQBRM

Here we will solve for the equilibrium concentrations of all species in the 4-component, 10-species system 0.25m NH_4Cl–0.25m NaCl–0.25m KCl–H_2O at 300°C and vapor-saturated pressure. Set up a data worksheet following the six steps outlined above.

Step 1: There are 10 species of interest in this system (knowing that it is acidic and that m_{OH^-} is negligible). We assign these species numbers, as follows:

Species	NH_4	NH_4OH°	H^+	HCl°	NH_4Cl°	Cl^-	Na^+	$NaCl^\circ$	K^+	KCl°
Assigned Number	1	2	3	4	5	6	7	8	9	10

Step 2: There are 5 known mass action equations and equilibrium constants relating these species:

$$\frac{a_{NH_4^+}}{a_{NH_4OH^\circ}a_{H^+}} = K_1; \qquad \log K_1 = 4.677$$

$$\frac{a_{NH_4^+}a_{Cl^-}}{a_{NH_4Cl^\circ}} = K_2; \qquad \log K_2 = -0.82$$

$$\frac{a_{Na^+}a_{Cl^-}}{a_{NaCl^\circ}} = K_3; \qquad \log K_3 = -1.013$$

$$\frac{a_{HCl^\circ}}{a_{H^+}a_{Cl^-}} = K_4; \qquad \log K_4 = 1.24$$

$$\frac{a_{K^+}a_{Cl^-}}{a_{KCl^\circ}} = K_5; \qquad \log K_5 = -0.735$$

Step 3: Assign reaction coefficients to each species in each of the above five reactions:

					Reaction Coefficients					
Species	NH_4^+	NH_4OH°	H^+	HCl°	NH_4Cl°	Cl^-	Na^+	$NaCl^\circ$	K^+	KCl°
Rxn 1	1	−1	−1	0	0	0	0	0	0	0
Rxn 2	1	0	0	0	−1	1	0	0	0	0
Rxn 3	0	0	0	0	0	1	1	−1	0	0
Rxn 4	0	0	−1	1	0	−1	0	0	0	0
Rxn 5	0	0	0	0	0	1	0	0	1	−1

Steps 4 and 5: Write the 4 available mass balances and assign the parameters $b_{e,i}$, the number of times each component or element (e) occurs in the formula of each species (i) in each mass balance:

$$B_1 = \sum NH_3 = 0.25m = m_{NH_4^+} + m_{NH_4OH^\circ} + m_{NH_4Cl}$$
$$B_2 = \sum Cl = 0.75m = m_{HCl} + m_{NH_4Cl} + m_{Cl^-} + m_{NaCl^\circ} + m_{KCl^\circ}$$
$$B_3 = \sum Na = 0.25m = m_{Na^+} + m_{NaCl^\circ}$$
$$B_4 = \sum K = 0.25m = m_{K^+} + m_{KCl^\circ}$$

Species (i)	NH_4	NH_4OH°	H^+	HCl°	NH_4Cl°	Cl^-	Na^+	$NaCl^\circ$	K^+	KCl°
$b_{NH_3,i}$	1	1	0	0	1	0	0	0	0	0
$b_{Cl,i}$	0	0	0	1	1	1	0	1	0	1
$b_{Na,i}$	0	0	0	0	0	0	1	1	0	0
$b_{K,i}$	0	0	0	0	0	0	0	0	1	1

Step 6: Write the charge balance for ionic species, and assign valences, Z_i:

$$m_{NH_4^+} + m_{H^+} + m_{Na^+} + m_{K^+} = m_{Cl^-}$$

Species	NH_4	NH_4OH°	H^+	HCl°	NH_4Cl°	Cl^-	Na^+	$NaCl^\circ$	K^+	KCl°
Z_i	1	0	1	0	0	-1	1	0	1	0

Step 7: First guesses at concentrations of all species:

Species	NH_4	NH_4OH°	H^+	HCl°	NH_4Cl°	Cl^-	Na^+	$NaCl^\circ$	K^+	KCl°
$\sim m$	0.1	0.002	0.002	0.005	0.15	0.2	0.1	0.15	0.1	0.15

Step 8: Choose the A parameter (0.51, 0.53, 0.60, 0.69, 0.81, 0.98, 1.256 at 25, 50, 100, 150, 200, 250, and 300°C, respectively). See Helgeson et al. (1981) for a more complete list.

Creating the input file

This completes the definition of our system. The above information is then written in a data file named EQBRM.DAT for use by the program. Enter N and M, the number of species and equilibrium constants, on the first line. Enter the reaction coefficients for each species from 1 to 10, taking one line for each of the 5 reactions. Enter the total molality of NH_3 on the next line, and the $b_{NH_3,i}$ parameters for species 1 to 10 on the next line. Repeat this sequence of two data lines for each of the mass balances on Cl, Na and K. Write the log of the equilibrium constants in order on the next line or lines, with no more than 8 entries per line. Enter the valences of each species on the following line. Write the initial guesses at species concentrations on the following lines, again with no more than 8 entries per line. Finally, on the last line, enter a value for the Debye-Hückel A parameter. Here is what the listing looks like, with comments:

Contents of the file EQBRM.DAT

`10,5`	N,M (No. species; no. K's)
`1,-1,-1,0,0,0,0,0,0,0`	
`1,0,0,0,-1,1,0,0,0,0`	signed reaction
`0,0,0,0,0,1,1,-1,0,0`	coefficients
`0,0,-1,1,0,-1,0,0,0,0`	for species 1,2,. . . ,10
`0,0,0,0,0,1,0,0,1,-1`	
`0.25`	total NH_3 (B_1)
`1,1,0,0,1,0,0,0,0,0`	NH_3 mass balance parameters ($b_{NH_3,i}$)
`0.75`	total Cl (B_2)
`0,0,0,1,1,1,0,1,0,1`	Cl mass balance parameters ($b_{Cl,i}$)
`0.25`	total Na (B_3)
`0,0,0,0,0,0,1,1,0,0`	Na mass balance parameters ($b_{Na,i}$)
`0.25`	total K (B_4)
`0,0,0,0,0,0,0,0,1,1`	K mass balance parameters ($b_{K,i}$)
`4.677,-0.82,-1.013,1.24,-0.735`	$\log(K_1, K_2, K_3, K_4, K_5)$
`1,0,1,0,0,-1,1,0,1,0`	valences (Z_i)
`0.1,0.002,0.002,0.005,0.15,0.2,0.1,0.15`	
`0.105,0.146`	two lines of first guesses at m_i.
`1.256`	Debye-Hückel A parameter.

Shown below is a listing of the FORTRAN77 source code for the main program EQBRM and for subroutines ACTCF (for the activity coefficients) and LINSOL, DDECOM and DSOLVE, (which perform the necessary matrix algebra). These should be compiled and linked, and can then be run provided the data file EQBRM.DAT is also present.

As written, the program can handle up to 30 species. This can be changed by altering statement 017 and recompiling. This program is useful for relatively small model systems, where you want complete control over the species and data involved.[1] For natural systems, where you might be dealing with dozens or hundreds of species, programs such as EQ3, WATEQ, SOLMINEQ, etc., may be more appropriate.

Program output for the example given is included after the program listings.

[1]For a recent example of its use, see Anderson and Cermignani, Can. Mineralogist, v. 29, p. 965–980, 1991.

```
      PROGRAM EQBRM
C ----------------------------------------------------------------------
C PROGRAM TO COMPUTE CHEMICAL EQUILIBRIUM IN GENERALIZED SYSTEMS
C GIVEN N CHEMICAL SPECIES RELATED BY M MASS ACTION EXPRESSIONS,
C N-M-1 MASS BALANCES AND A CHARGE BALANCE.
C
C THIS VERSION USES LOG(K) IN MAIN EQUATIONS AND CORRECTS FOR ACTIVITY
C COEFFICIENTS USING SUBROUTINE ACTCF.
C REVISED FOR IBM PC, MICROSOFT FORTRAN 77, VERSION 5.0, 29 MAY 1990.

C DATA MUST BE ENTERED IN THE FILE EQBRM.DAT AND MAY BE ENTERED
C SEPARATED BY COMMAS IN FIELDS NOT EXCEEDING FORMATTED FIELD
C LENGTH.  LINK WITH SUBROUTINES LINSOL, DDECOM, DSOLVE AND ACTCF.
C WRITTEN BY DAVID CRERAR.
C ----------------------------------------------------------------------
      IMPLICIT REAL*8(A-H,O-Z)
C  NDIM IS MAXIMUM DIMENSION OF ALL ARRAYS AND IS PASSED TO ALL SUBROUTINES

      PARAMETER(NDIM=30)
      EXTERNAL LINSOL
      INTEGER V(NDIM,NDIM),G(NDIM,NDIM),Z(NDIM),IPVT(NDIM)
      REAL*8 LOGK(NDIM),A(NDIM),C(NDIM,NDIM),Y(NDIM),AOLD(NDIM),
     1 DELTAA(NDIM),CHOLD(NDIM,NDIM),YHOLD(NDIM),TEST(NDIM),B(NDIM),
     2 RLV(NDIM,NDIM),WORK(NDIM),ANS(NDIM),BLEEN(NDIM,NDIM),IS,
     3 GAMMA(NDIM),GAMOLD(NDIM)
      OPEN(6,FILE='EQBRM.DAT')
      OPEN(2,FILE='eqbrm.out', status='unknown')
C ----------------------------------------------------------------------
C  READS ALL INPUT DATA
C ----------------------------------------------------------------------
C  READS NUMBER OF CHEMICAL SPECIES OR UNKNOWNS N, AND THE NUMBER OF
C  MASS ACTION EXPRESSIONS M.
C  ENTER N, THEN M SEPARATED BY BLANK OR COMMA ON THE SAME LINE.
      READ(6,300)N,M
  300 FORMAT(2I5)
C  READS STOICHIOMETRIC COEFFICIENTS V(J,I) OF EACH SPECIES IN
C  MASS ACTION EXPRESSIONS. I IS THE SPECIES (I=1,...,N), AND J THE
C  MASS ACTION EXPRESSION (J=1,...,M). THE INTEGER V(J,I) IS NEGATIVE
C  FOR REACTANT AND POSITIVE FOR PRODUCT SPECIES.
C  ENTER ALL V(1,J) ON FIRST LINE SEPARATED BY COMMAS, THEN ALL V(2,J)
C  ON NEXT LINE AND SO ON TO ALL V(J,J) ON THE JTH LINE.
      DO 2 J=1,M
      READ(6,301)(V(J,I),I=1,N)
  301 FORMAT(20I4)
    2 CONTINUE
C  READS THE MASS BALANCE COMPOSITION COEFFICIENTS G(J,I) AND MASSES
C  B(J) IN MOLAL UNITS. I REFERS TO SPECIES (I=1,...N) AND J TO THE
C  NUMBER OF MASS BALANCES (J=1,...,N-M-1).
C  ENTER B(1) ON ONE LINE FOLLOWED BY ALL G(1,I) SEPARATED BY COMMAS ON
C  THE NEXT LINE AND SO ON TO B(J) AND ALL G(J,I) ON 2XJTH LINE.
      J1=N-M-1
      DO 4 J=1,J1
      READ(6,392)B(J)
  392 FORMAT(F10.0)
      READ(6,302)(G(J,I),I=1,N)
  302 FORMAT(20I4)
    4 CONTINUE
C  READS LOG OF ALL EQUILIBRIUM CONSTANTS, LOGK(J), (J=1,...,M).
C  ENTER ALL LOGK(J) IN ASCENDING ORDER SEPARATED BY COMMAS OR BLANKS
C  ON ONE OR MORE CONSECUTIVE LINE(S) WITH NO MORE THAN 8 PER LINE.
C  USE EITHER E,D, OR F FORMAT.
      READ(6,303)(LOGK(J),J=1,M)
  303 FORMAT(8F10.0)
C  READS SPECIES CHARGES FOR CHARGE BALANCE.  INTEGER Z(I) IS THE
C  SIGNED VALENCE OF SPECIES I.
C  ENTER ALL Z(I) SEPARATED BY COMMAS ON ONE LINE.
```

```
      READ(6,301)(Z(I),I=1,N)
C
C  READS FIRST GUESSES AT ALL SPECIES CONCENTRATION, A(I)
C  ENTER ALL A(I) SEPARATED BY COMMAS ON ONE OR MORE CONSECUTIVE LINES
C  WITH NO MORE THAN 8 PER LINE.
C  DATA CAN BE IN F OR D FORMAT.
      READ(6,304)(A(I),I=1,N)
  304 FORMAT(8F10.0)
C
C read the Debye-Huckel "A" parameter from file
C
 READ(6, 305)dbha
305     FORMAT(F10.4)
C  ---------------------------------------------------------------------
C  PRINTS ALL INPUT DATA
      WRITE(2,701)N,M
  701 FORMAT(' ',9X,'RESULTS',///,10X,'NUMBER OF UNKNOWNS=N=',I3,/,
     110X,'NUMBER OF MASS ACTION EXPRESSIONS=M=',I3,///)
      WRITE(2,702)
  702 FORMAT(' ',9X,'MASS ACTION EQUILIBRIUM CONSTANTS LOGK(J)',/
     1 10X,' AND STOICHIOMETRIC REACTION COEFFICIENTS V(J,I)'//,
     2 3X,'LOGK(J)',15X,'V(J,I)'/)
      DO 703 J=1,M
      WRITE(2,704)LOGK(J),(V(J,I),I=1,N)
  704 FORMAT(1H ,1PD11.3,5X,30I3)
  703 CONTINUE
      WRITE(2,705)
  705 FORMAT(' ',///,1X,'MASS BALANCE MOLALITIES B(J), AND COMPOSITION
     1 COEFFICIENTS b(J,I)',//,3X,'B(J)',18X,'b(J,I)',/)
      DO 706 J=1,J1
      WRITE(2,704)B(J),(G(J,I),I=1,N)
  706 CONTINUE
      WRITE(2,707)(Z(I),I=1,N)
  707 FORMAT(' ',///,26X,'VALENCES Z(I)',//,16X,30I3)
      WRITE(2,714)
      WRITE(2,708)(A(I),I=1,N)
  714 FORMAT(1H ,///,10X,'FIRST GUESSES AT SPECIES CONCENTRATIONS A(I)'
     1 //)
  708 FORMAT(1H ,1P20D11.3//)
*
* now the Debye-Huckel A parameter
*
 WRITE(2, 719)dbha
719     FORMAT(///' DEBYE-HUCKEL "A" PARAMETER:  '1pd12.4//)
C ----------------------------------------------------------------------
C  SETS COUNTER FOR ACTIVITY COEFF ITERATIONS AND SETS ALL INITIAL
C  ACTIVITY COEFF'S TO ZERO
      NUMGAM=O
      DO 400 I=1,N
      GAMMA(I)=1.0D0
  400 CONTINUE


C ----------------------------------------------------------------------
C  FILLS OUT THE COEFFICIENT MATRIX C(N,N).
C  NOTE THAT LINEAR CHARGE AND MASS BALANCE EQUATIONS ARE ASSIGNED FIRST ROW

C  POSITIONS. THIS IS ESSENTIAL IF BROWN'S GAUSSIAN ELIMINATION METHOD
C  IS USED IN PLACE OF NEWTON-RAPHSON.
C  FOR CHARGE BALANCE SUBMATRIX.
      DO 8 I=1,N
      C(1,I)=Z(I)
    8 CONTINUE
C  FOR MASS BALANCES SUBMATRIX
      M2=N-M
      DO 7 J=2,M2
```

```
      DO 7 I=1,N
      C(J,I)=G(J-1,I)
    7 CONTINUE
C ----------------------------------------------------------------------
C  ACTIVITY COEFFICIENT DETERMINATION - FROM SUBROUTINE ACTCF
  410 CONTINUE
      DO 404 I=1,N
      GAMOLD(I)=GAMMA(I)
  404 CONTINUE
C ------------------------------------------------------------------------
C  NDIM IS MAXIMUM DIMENSION OF ARRAYS (DEFINED IN OPENING PARAMETER
C  STATEMENT). A AND Z ARE ARRAYS OF SPECIES CONCENTRATIONS AND VALENCES,
C  N IS NUMBER OF SPECIES, AND GAMMA IS THE CALULATED ARRAY OF ACT. COEFF'S.

C ------------------------------------------------------------------------
      IF(NUMGAM.GT.0)CALL ACTCF(NDIM,A,Z,N,GAMMA, dbha)
      NUMGAM=NUMGAM+1
      IF(NUMGAM.EQ.1000)GO TO 990
*     WRITE(2,401)NUMGAM
* 401 FORMAT('1',///,5X,'GAMMA ITERATION NUMBER =',I3/)
*     DO 402 I=1,N
*      WRITE(2,403)I,GAMMA(I)
*  403 FORMAT(' ',5X,'GAMMA(',I3,')=',1PD10.2)
*  402 CONTINUE
C ------------------------------------------------------------------------
C  ONLY THE SUBMATRIX OF MASS ACTION EXPRESSIONS CHANGES WITH EACH ITERATION
,
C  SO ITERATION ON CONCENTRATIONS BEGINS HERE
C ------------------------------------------------------------------------
      NUM=0
  900 CONTINUE
      NUM=NUM+1
      IF(NUM.EQ.100)GO TO 990
C  FOR SUBMATRIX OF MASS ACTION EXPRESSIONS
      DO 600 J=1,M
      DO 600 I=1,N
      RLV(J,I)=V(J,I)
  600 CONTINUE
      M1=M2+1
      DO 6 J=M1,N
      DO 6 I=1,N
      JHOLD=J-M2
      C(J,I)=0.43429*RLV(JHOLD,I)/A(I)
    6 CONTINUE
C ------------------------------------------------------------------------
C  FILLS OUT THE VECTOR Y(N) OF MASS ACTION, AND OF MASS AND CHARGE
C  BALANCE FUNCTIONS
C -------------------------------------------------------------------------
C  FOR CHARGE BALANCE IN Y VECTOR
      HOLD=0.0D0
      DO 13 I=1,N
      RLZ=Z(I)
      HOLD=HOLD+RLZ*A(I)
   13 CONTINUE
      Y(1)=-HOLD
C  FOR MASS BALANCE FUNCTIONS IN Y VECTOR
      DO 12 J=2,M2
      HOLD=0.0D0
      DO 11 I=1,N
      RLG=G(J-1,I)
      HOLD=HOLD+RLG*A(I)
   11 CONTINUE
      Y(J)=B(J-1)-HOLD
   12 CONTINUE
C  FOR MASS ACTION FUNCTIONS IN Y VECTOR.
      DO 10 J=M1,N
```

```
      JHOLD=J-M2
      FHOLD=0.0D0
      DO 9 I=1,N
      FHOLD=FHOLD+RLV(JHOLD,I)*(DLOG10(A(I))+DLOG10(GAMMA(I)))
    9 CONTINUE
      Y(J)=LOGK(JHOLD)-FHOLD
   10 CONTINUE
C  ------------------------------------------------------------------------
C     USES MATLIB SUBROUTINE LINSOL TO SOLVE THE EQUATION:
C     C(N,N)*DELTAA(N)=Y(N). THE DELTAA VALUES APPEAR IN THE
C     DELTAA MATRIX, THE DETERMINANT APPEARS IN DET.
C  ------------------------------------------------------------------------
C     MAKE COPY OF COEFFICIENT MATRIX C.
      DO 3001 I=1,N
      DO 3001 J=1,N
      BLEEN(I,J)=C(I,J)
 3001 CONTINUE
      CALL LINSOL(BLEEN,Y,N,NDIM,DELTAA,DET,COND,WORK,IPVT)
C  ------------------------------------------------------------------------
C  FOR NEW ESTIMATE OF CONCENTRATIONS A(I).
      DO 14 I=1,N
      AOLD(I)=A(I)
      A(I)=DELTAA(I)+AOLD(I)
C  TO PREVENT NEGATIVE CONCENTRATIONS
      IF(A(I).LE.0.0)A(I)=1.0D-10
      TEST(I)=DABS((A(I)-AOLD(I))/AOLD(I))*100.0
   14 CONTINUE
C  ========================================================================
C  PRINTS, ON SCREEN, THE Y(N) VECTOR, SOLUTION VECTOR DELTAA(N),
C  AND THE SET A(N) OF NEW CONCENTRATION ESTIMATES, FOR EACH ITERATION.
      WRITE(*,709)NUM
  709 FORMAT(1H ,////,9X,'ITERATION NUMBER',I3,/)
      WRITE(*,715)
  715 FORMAT(1H ,//,4X,'DELTAA(J)',4X,'Y(J)',8X,'A(J)'/)
      DO 716 J=1,N
      WRITE(*,996)DELTAA(J),Y(J),A(J)
  996 FORMAT(1H ,1X,1P20D11.3/)
  716 CONTINUE
C  ------------------------------------------------------------------------
C     WRITES OUT JACOBIAN DETERMINANT OF C(N,N), CALCULATED IN
C     LINSOL (ABOVE). IF THIS APPROACHES ZERO, ITERATION MAY NOT CONVERGE
C  ------------------------------------------------------------------------
      WRITE(*,713)DET
  713 FORMAT(1H ,///,9X,'JACOBIAN DETERMINANT OF C(N,N)=',1PD12.5)
C   PRINTS ERROR MESSAGES FROM LINEAR EQUATION SOLVING SUBROUTINE 'LINSOL'.
      WRITE(*,723)COND
  723 FORMAT(1H ,4X,'SUBROUTINE LINSOL CONDITION PARAMETER=',
     1D12.5,///,)
C  ------------------------------------------------------------------------
C  IF COND=1, MATRIX INVERTED BY LINSOL WAS PERFECTLY NONSINGULAR;
C  IF COND=1E32, MATRIX WAS SINGULAR AND ANSWER IS SUSPECT.
C  ------------------------------------------------------------------------
C  ITERATES ON A(I) IF ANY ONE OR MORE VALUES OF TEST(I) EXCEED CHOSEN
C  CRITICAL ERROR.
      DO 15 I=1,N
      IF(TEST(I).GT.0.1)GO TO 900
   15 CONTINUE
*      WRITE(2,722)
*  722 FORMAT(1H ,//,5X,'FINAL CONCENTRATIONS')
*      IS=0.0D0
*      DO 720 I=1,N
*      IS=Z(I)*Z(I)*A(I)+IS
*      WRITE(2,721)I,A(I)
*  720 CONTINUE
*      IS=IS/2.0
*  721 FORMAT(1H ,5X,'A(',I3,')=',1PD11.3)
```

```
*      WRITE(2,395)IS
*  395 FORMAT(' ',/,5X,'IONIC STRENGTH=',1PD11.3,1X,'MOLAL'///)
       IF(NUMGAM.EQ.1)GO TO 410
       DO 421 I=1,N
       TEST1=(DABS(GAMMA(I)-GAMOLD(I)))/GAMOLD(I)*100.0
       IF(TEST1.GT.0.1)GO TO 410
  421 CONTINUE
       WRITE(2,411)
  411 FORMAT(' ',/,5X,'GAMMA ITERATIONS SUCCESSFULLY CONVERGED')
       GO TO 999
  990 WRITE(2,700)
  700 FORMAT(1H ,9X,'ERROR. ITERATED 100 TIMES WITHOUT CONVERGENCE.',
     1//)
999      WRITE(2,401)NUMGAM
401      FORMAT('1',///,5X,'GAMMA ITERATION NUMBER =',I3/)
 DO 402 I=1,N
 WRITE(2,403)I,GAMMA(I)
403      FORMAT(' ',5X,'GAMMA(',I3,')=',1PD11.3)
402      CONTINUE

 WRITE(2,722)
722      FORMAT(1H ,//,5X,'FINAL CONCENTRATIONS')
 IS=0.0D0
 DO 720 I=1,N
 IS=Z(I)*Z(I)*A(I)+IS
 WRITE(2,721)I,A(I)
720      CONTINUE
 IS=IS/2.0
721      FORMAT(1H ,5X,'A(',I3,')=',1PD11.3)
 WRITE(2,395)IS
395      FORMAT(' ',/,5X,'IONIC STRENGTH=',1PD11.3,1X,'MOLAL'///)
 WRITE(2,666)
  666    FORMAT('1'//)
 STOP
 END
C=====================================================================
C
       SUBROUTINE ACTCF(NDIM,A,Z,N,GAMMA, dbha)
C  FOR ACTIVITY COEFF FROM DAVIES EXTENSION OF THE DEBYE-HUCKEL EQUATION
       IMPLICIT REAL*8 (A-H,O-Z)
       REAL*8 IS
       INTEGER Z(NDIM)
       DIMENSION A(NDIM), GAMMA(NDIM), GAMLOG(100)

C  FOR IONIC STRENGTH
       IS=0.0D0
       DO 1 I=1,N
       IS=IS + Z(I)**2*A(I)
1      CONTINUE
       IS=IS/2.0
       DO 2 I=1,N
       GAMLOG(I) = -(Z(I)**2*dbha*DSQRT(IS))/(1.0+DSQRT(IS))
      *   +0.2*dbha*Z(I)**2*IS
       GAMMA(I) = 10.0**GAMLOG(I)
2      CONTINUE
       RETURN
       END
C=====================================================================
C
       SUBROUTINE DDECOM(NDIM,N,A,COND,DET,IPVT,WORK)
C  LU-DECOMPOSITION OF A MATRIX
       INTEGER NDIM,N
       DOUBLE PRECISION A(NDIM,NDIM),COND,WORK(NDIM),DET
       INTEGER IPVT(NDIM)
       DOUBLE PRECISION EK,T,ANORM,YNORM,ZNORM
       INTEGER NM1,I,J,K,KP1,KB,KM1,M
```

```
      IPVT(N)=1
      IF(N.EQ.1)GO TO 80
      NM1=N-1
C
C   COMPUTE 1-NORM OF A
C
      ANORM=0.0D0
      DO 10 J=1,N
      T=0.0D0
      DO 5 I=1,N
      T=T+DABS(A(I,J))
    5 CONTINUE
      IF(T.GT.ANORM)ANORM=T
   10 CONTINUE
C
C    GAUSSIAN ELIMINATION WITH PARTIAL PIVOTING
C
      DO 35 K=1,NM1
      KP1=K+1
C
C  FIND PIVOT
C
      M=K
      DO 15 I=KP1,N
      IF(DABS(A(I,K)).GT.DABS(A(M,K)))M=I
   15 CONTINUE
      IPVT(K)=M
      IF(M.NE.K)IPVT(N)=-IPVT(N)
      T=A(M,K)
      A(M,K)=A(K,K)
      A(K,K)=T
C
C  SKIP STEP IF PIVOT IS ZERO
C
      IF(T.EQ.0.0D0)GO TO 35
C
C  COMPUTE MULTIPLIERS
C
      DO 20 I=KP1,N
      A(I,K)=-A(I,K)/T
   20 CONTINUE
C
C  INTERCHANGE AND ELIMINATE BY COLUMNS
C
      DO 30 J=KP1,N
      T=A(M,J)
      A(M,J)=A(K,J)
      A(K,J)=T
      IF(T.EQ.0.0D0)GO TO 30
      DO 25 I=KP1,N
      A(I,J)=A(I,J)+A(I,K)*T
   25 CONTINUE
   30 CONTINUE
   35 CONTINUE
C
C   SOLVE(A-TRANSPOSE)*Y=E
      DO 50 K=1,N
      T=0.0D0
      IF(K.EQ.1)GO TO 45
      KM1=K-1
      DO 40 I=1,KM1
      T=T+A(I,K)*WORK(I)
   40 CONTINUE
   45 EK=1.0D0
      IF(T.LT.0.0D0)EK=-1.0D0
      IF(A(K,K).EQ.0.0D0)GO TO 90
```

```
      WORK(K)=-(EK+T)/A(K,K)
   50 CONTINUE
      DO 60 KB=1,NM1
      K=N-KB
      T=0.0D0
      KP1=K+1
      DO 55 I=KP1,N
      T=T+A(I,K)*WORK(K)
   55 CONTINUE
      WORK(K)=T
      M=IPVT(K)
      IF(M.EQ.K)GO TO 60
      T=WORK(M)
      WORK(M)=WORK(K)
      WORK(K)=T
   60 CONTINUE
C
      YNORM=0.0D0
      DO 65 I=1,N
      YNORM=YNORM+DABS(WORK(I))
   65 CONTINUE
C
C    SOLVE A*Z=Y
C
      CALL DSOLVE(NDIM,N,A,WORK,IPVT)
C
      ZNORM=0.0D0
      DO 70 I=1,N
      ZNORM=ZNORM+DABS(WORK(I))
   70 CONTINUE
C
C   ESTIMATE CONDITION
C
      COND=ANORM*ZNORM/YNORM
      IF(COND.LT.1.0D0)COND=1 0D0
C
C    COMPUTE DETERMINANT
C
      DET=IPVT(N)*A(1,1)
      DO 75 I=2,N
      DET=DET*A(I,I)
   75 CONTINUE
C
      RETURN
C
C 1 BY 1 MATRIX
C
   80 COND=1.0D0
      IF(A(1,1).NE.0.0D0) RETURN
C
C   EXACT SINGULARITY
C
   90 COND=1.0D+32
      RETURN
      END
C=====================================================================
C
      SUBROUTINE DSOLVE(NDIM,N,A,B,IPVT)
C   SOLUTION OF DECOMPOSED LINEAR EQUATION MATRIX
C   ---------------------------------------------
      INTEGER NDIM,N,IPVT(NDIM)
      DOUBLE PRECISION A(NDIM,NDIM),B(NDIM)
      INTEGER KB,KM1,NM1,KP1,I,K,M
      DOUBLE PRECISION T
C
C   FORWARD ELIMINATION
```

```
C
      IF(N.EQ.1)GO TO 50
      NM1=N-1
      DO 20 K=1,NM1
      KP1=K+1
      M=IPVT(K)
      T=B(M)
      B(M)=B(K)
      B(K)=T
      DO 10 I=KP1,N
      B(I)=B(I)+A(I,K)*T
   10 CONTINUE
   20 CONTINUE
C
C  BACK SUBSTITUTION
C
      DO 40 KB=1,NM1
      KM1=N-KB
      K=KM1+1
      B(K)=B(K)/A(K,K)
      T=-B(K)
      DO 30 I=1,KM1
      B(I)=B(I)+A(I,K)*T
   30 CONTINUE
   40 CONTINUE
   50 B(1)=B(1)/A(1,1)
      RETURN
      END
C==========================================================================
C
      SUBROUTINE LINSOL(A,B,N,NDIM,C,DET,COND,WORK,IPVT)
C  SUBROUTINE TO SOLVE SIMULTANEOUS LINEAR EQUATIONS
C  ADAPTED FROM "COMPUTER METHODS FOR MATHEMATICAL COMPUTATIONS"
C  BY G.E. FORSYTHE ET AL.(1977)
C  ---------------------------------------------------------------------
C
      INTEGER N,NDIM,IPVT(NDIM),I,J,RCODE
      DOUBLE PRECISION A(NDIM,NDIM),C(NDIM),B(NDIM),WORK(NDIM),DET,
     *  COND,CONDP1
C
C  COPY MATRIX B (RIGHT HAND SIDE VECTOR) INTO MATRIX C.
C  B REMAINS UNTOUCHED, C IS DESTROYED AND REPLACED WITH
C  THE ANSWERS.
      DO 5 M=1,N
      C(M)=B(M)
    5 CONTINUE
C
C  DECOMPOSITION OF THE COEFFICIENT MATRIX, USING SUBROUTINE DDECOM.
C  NOTE-THIS CALLS AND MUST BE LINKED WITH SUBROUTINES DDECOM AND DSOLVE.
C
      CALL DDECOM(NDIM,N,A,COND,DET,IPVT,WORK)
C  IF COND=1, MATRIX WAS PERFECTLY NONSINGULAR; IF COND=1E32, MATRIX WAS
C  SINGULAR AND ANSWER IS SUSPECT.
C  IF THE MATRIX IS SINGULAR TO WORKING PRECISION THEN EXIT,OTHERWISE CONTIN
UE.
C
      CONDP1=COND+1
      IF(CONDP1.EQ.COND)RETURN
C
C   SOLUTION OF THE COEFFICIENT MATRIX
C
      CALL DSOLVE(NDIM,N,A,C,IPVT)
      RETURN
      END
```

Program Output

The results of each iteration leading to the final result are sent to the screen but scroll by too fast to read. They are usually of little interest. The final results are written to a file called EQBRM.OUT. Remember to rename this file if you wish to save it, as it will be overwritten next time the program is run. The final results in this case are:

GAMMA ITERATION NUMBER = 5

GAMMA(1)= 4.04D-01	that is, $\gamma_{NH_4} = 0.404$
GAMMA(2)= 1.00D+00	$\gamma_{NH_4OH} = 1.0$
GAMMA(3)= 4.04D-01	$\gamma_{H^+} = 0.404$
GAMMA(4)= 1.00D+00	$\gamma_{HCl^\circ} = 1.0$
GAMMA(5)= 1.00D+00	$\gamma_{NH_4Cl^\circ} = 1.0$
GAMMA(6)= 4.04D-01	$\gamma_{Cl^-} = 0.404$
GAMMA(7)= 4.04D-01	$\gamma_{Na^+} = 0.404$
GAMMA(8)= 1.00D+00	$\gamma_{NaCl^\circ} = 1.0$
GAMMA(9)= 4.04D-01	$\gamma_{K^+} = 0.404$
GAMMA(10)= 1.00D+00	$\gamma_{KCl^\circ} = 1.0$

In this simple scheme, all singly charged ions have the same γ, and all doubly charged ions would have another γ.

FINAL CONCENTRATIONS

A(1)= 1.630D-01	that is, $m_{NH_4} = 0.1830$
A(2)= 2.844D-03	$m_{NH_4OH} = 0.002844$
A(3)= 1.206D-03	$m_{H^+} = 0.001206$
A(4)= 1.638D-03	$m_{HCl^\circ} = 0.001638$
A(5)= 8.416D-02	$m_{NH_4Cl^\circ} = 0.08416$
A(6)= 4.782D-01	$m_{Cl^-} = 0.4782$
A(7)= 1.385D-01	$m_{Na^+} = 0.1385$
A(8)= 1.118D-01	$m_{NaCl^\circ} = 0.1115$
A(9)= 1.755D-01	$m_{K^+} = 0.1755$
A(10)= 7.450D-02	$m_{KCl} = 0.07450$

As a check, you might calculate one or two K values, and compare with the input. For example

$$
\begin{aligned}
K_6 &= \frac{a_{HCl^\circ}}{a_{H^+}a_{Cl^-}} \\
&= \frac{0.001638 \times 1.0}{(0.001206 \times 0.404)(0.4782 \times 0.404)} \\
&= 17.40
\end{aligned}
$$

$$\log K_6 = 1.24 \quad \text{which is the input value.}$$

APPENDIX F

ANSWERS TO PROBLEMS

Chapter 2

2. (a). $du = \frac{yz(y+z)}{(x+y+z)^2}dx + \frac{xz(x+z)}{(x+y+z)^2}dy + \frac{xy(x+y)}{(x+y+z)^2}dz$
   (b). $du = (\frac{1}{y} - \frac{z}{x^2})dx + (\frac{1}{z} - \frac{x}{y^2})dy + (\frac{1}{x} - \frac{y}{z^2})dz$
   (c). $du = e^{(x+y^2)}dx + 2y\,e^{(x+y^2)}dy$
3. (a). homogeneous 2nd degree in x, y, z. (b). homogeneous 3rd degree in x, y, z. (c). homogeneous 5th degree in x, y. (d),(e). not homogeneous.
6. (a). Not exact. Integrating factor is $1/xy$. (b). Not exact. Integrating factor is $1/x^3y$. (c). Not exact. Integrating factor is y^2/x. Other integrating factors are also possible.
7. $w = y^2 + p^2/4$
8. $f = -2(p/3)^{3/2} + q^2/8)$

Chapter 5

1. 903.3°C. Yes. 2. 133.1 degrees. 3. $\Delta_r H° = -4139$ cal mol^{-1}; $\Delta_r G° = -4144$ cal mol^{-1}; $\Delta_r U° = -4138.99$ cal mol^{-1}. 5. -74.9155 cal K^{-1} mol^{-1} 6. gibbsite stable. 7. gibbsite.

Chapter 7

2. -407593.5 J mol^{-1}. 3. -371383 J mol^{-1}. 4. -319135 J mol^{-1}. 5. -460709 J mol^{-1}. 6. -407576 J mol^{-1}. 7. -1466 cal mol^{-1}. 8. -616456 cal mol^{-1}.

Chapter 8

1. Isentropic work between A and B $w = 3558$ J mol^{-1}. 4. Triple point at (a). 3932 bars, 483°C. (b). 2902 bars, 430°C. 5. 64504 bars. 6. hematite stable. About 151 K, or -122°C. 7. $\Delta_f H^\circ_{\beta,600\text{K}} = -18890$ cal mol^{-1}.

Chapter 9

3. (a). slope 159.8 cal mol^{-1}. (b). $X_B = 0.337$. (c). $\mu_A = -203$ cal mol^{-1} and $\mu_B = -43$ cal mol^{-1}

Chapter 10

1. 1.0191 cal mol^{-1} K^{-1}

Chapter 11

1. $f^\circ_{H_2O} = 1054$ bars; $f^\circ_{CO_2} = 666$ bars

Chapter 12

1. 1. (a). 1.0 (b). 1.4928 (c). 1.4933 (d). $10^{-14.946}$
   2. (a). 1.0 (b). 1.787 (c). 5.408 (d). Estimate a value for f_{ice} at T by any means, such as extrapolating fugacities from lower T.
   3. (a). 0.001 (b). 0.000472
   4. (a). At system equilibrium. (b). When the same standard state is used for both — system equilibrium irrelevant.
   5. -11533 cal mol^{-1}
   6. 7.070 (Same as $a^{qtz}_{\text{SiO}_2}$ using this standard state).
2. In equation 12.1, the value of μ° varies with T. To use a constant T standard state, you must include the difference between μ° in this equation and the μ° in the state you have chosen.
3. 0.02654; 92.36 bars.
4. Pure CO_2 at 600°C, 2000 bars.
5. $a_{\text{NaCl}} = 10^{-4.248}$.

Chapter 13

1. (a). $10^{-12.45}$ bars (b). $10^{-12.53}$ bars
2. 0.0310 bars; 1.098 bars
3. $1.0017 \times 10^{-4}\, m$; $1.934 \times 10^{-3}\, m$.
4. (a.) $\Delta_r G^\circ = -5255$ cal mol^{-1} (b). Pyrrhotite is stoichiometric FeS, *not* $Fe_{0.92}S$, because the calculated ΔG° is between standard states. (c). $\Delta_f G^\circ_{\text{FeS}_2} = -30283$ cal mol^{-1}
5. $f_{\text{CO}_2} = 135$ bars. Need f of pure CO_2 at 749 K, 2 kbar, and activity coefficient of CO_2.
6. $a_{\text{SiO}_2} = 0.321$, pure quartz at T, P standard state. Therefore silica concentration is $0.321 \times 0.208 = 0.066\, m$.
7. 9912 bars, if $\Delta C^\circ_p = 0$.
8. $\Delta_f G^\circ_{\text{H}_4\text{SiO}_4} = -310882$ cal mol^{-1}.

Chapter 14

1. (a). Traditional, 1130 units mol^{-1}; B-H, 780 units mol^{-1}.
   (b). Compound and elements at T, P; -80 units mol^{-1}.
   (d). Traditional and B-H $\Delta_f G^\circ$ values are always negative, showing that our elemental G° values are too small, relative to G° for ABC.
   (e). $K = 14.6$
   (f). $P = 2161$ bars; $S^\circ = 0.61$ units K^{-1} mol^{-1}.
   (g). In both cases the phase boundary has a negative slope with α below and β above. However in (i), the boundary passes above and to the right of 25°C, 1 bar, and in (ii) it passes below and to the left.

(h). (i). 850 units mol^{-1}. (ii). 850, 400 units mol^{-1}. (iii) A+B+C react to form ABC until one of them is used up. (iv). 320 units mol^{-1} (equation 14.25).

(i). $a_B = 0.355$ (std. state pure B at T, P). $a_B = 9.014$ (std. state pure B at T, 1 bar).

Chapter 17

1. For $\alpha = 2$; $g(x) = 0.1419$; $g'(x) = -0.1113$
   For $\alpha_1 = 1.4$; $g(x) = 0.23509$; $g'(x) = -0.14801$
   For $\alpha_2 = 12$; $g(x) = 0.0045687$; $g'(x) = -0.0045687$
   $B'_{Na,Cl} = -0.009681$; $B'_{Na,SO_4} = -0.04075$; $B'_{Ca,Cl} = -0.059097$; $B'_{Ca,SO_4} = -0.074156$;
   $B_{Ca,Cl} = 0.5449$; $B_{Ca,SO_4} = 0.70384$; $C_{Na,Cl} = 0.0006350$;
   $C_{Na,SO_4} = 0.001757$; $C_{Ca,Cl} = -0.0001202$; $C_{Ca,SO_4} = 0.0$;
   $\Phi_{Na,Ca} = -0.01384$; $\Phi_{Cl,SO_4} = -0.05384$;
   $\Phi'_{Na,Ca} = \Phi'_{Cl,SO_4} = 0.01355$; $F = -1.047296$.
2. $\omega_{Ca^{2+},25} = 1.2366 \times 10^5$; $\omega_{Ca^{2+},300} = 1.2441 \times 10^5$; $\omega_{SO_4^{2-},25} = 3.1924 \times 10^5$; $\omega_{SO_4^{2-},300} = 3.1987 \times 10^5$;
3. $\Delta_a G^\circ{}_{anhydrite,300} = -325408$ cal mol^{-1}; $\Delta_a G^\circ{}_{Ca^{2+},300} = -127082$ cal mol^{-1}; $\Delta_a G^\circ{}_{SO_4^{2-},300} = -172062$ cal mol^{-1};
4. $\log K_{sp,25} = -4.306$; $\log K_{sp,300} = -9.960$
5. $\log K_{NaCl,25} = 0.777$; $\log K_{NaCl,500,1} = -2.262$; $\log K_{NaCl,500,2} = -1.287$

Chapter 18

1. −0.304 V.
2. 1. Yes. Near equilibrium, because the samples plot almost on the boundary for their concentration. 2. outside $Fe(OH)_3$ stability range (actually $Fe(OH)_3$ is metastable, but can be treated as stable here). 3. yes, 4 and 5. 4. sulfate is being reduced. 5. a little over $pH7$.
3. $a_{Fe} = 10^{-6.20}; 10^{-2.91}$; 1.0 for $Fe_3O_4 - Fe_2O_3$, $Fe_3O_4 - FeO$, $Fe - FeO$ respectively.
4. True. $a_{Fe} = 10^{-5.95}$.

BIBLIOGRAPHY

Adcock, S.W., 1988, Computer calculation of the thermodynamic and electrostatic properties of water at elevated temperatures and pressures: Unpublished ms. and computer program H2O. Available from Dr. S.W. Adcock, Geological Survey of Canada, 601 Booth St., Ottawa, Canada, K1A 0E8.

Anderson, G.M., 1970, Some thermodynamics of dehydration equilibria: Amer. Jour. Science, v. 269, pp. 392–401.

Anderson, G.M., 1976, The accuracy and precision of calculated mineral dehydration equilibria, *in* D.G. Fraser, ed., Thermodynamics in Geology: Dordrecht, Holland, D.Reidel Publishing Co., pp. 115–136.

Anderson, G.M., Castet, S., Schott, J., and Mesmer, R.E., 1991, The density model for estimation of thermodynamic parameters of reactions at high temperatures and pressures: Geochim. Cosmochim. Acta, v. 55, pp. 1769–1779.

Baas Becking, L.G.M., Kaplan, I.R., and Moore, D., 1960, Limits of the natural environment in terms of *pH* and oxidation-reduction potentials: Jour. Geol., v. 68, pp. 243–284.

Badger, P.H., 1967, Equilibrium Thermodynamics: Boston, Allyn and Bacon, Inc., 626 pp.

Ball, J., Nordstrom, D.K., and Jenne, E.A., 1980, WATEQ3—Additional and revised thermochemical data and computer coding for WATEQ2.: U.S. Geol. Survey Water Resources Invest., 78-116, 109 pp.

Barnes, H.L., and Ernst, W.G., 1963, Ideality and ionization in hydrothermal fluids: The system $MgO–H_2O–NaOH$: Amer. Jour. Sci., v. 261, pp. 129–150.

Barrett, T.J., Anderson, G.M., and Lugowski, J., 1988, The solubility of hydrogen sulfide in 0–5m NaCl solutions at 25°C and one atmosphere: Geochim. Cosmochim. Acta, v. 52, pp. 807–811.

Barrow, G.H., 1966, Physical Chemistry, 2nd ed.: New York, McGraw-Hill, 843 pp.

Barton, P.B. Jr., 1972, Equilibrium: article *in* The Encyclopedia of Geochemistry and Environmental Sciences. Vol. IVA, pp. 725–728, Encyclopedia of Earth Sciences Series, R.W. Fairbridge, ed., New York, Van Nostrand Rheinhold Co.

Barton, P.B. Jr., Bethke, P.M., and Toulmin, P. IIIrd., 1963, Equilibrium in ore deposits: Mineral. Soc. Amer. Spec. Paper 1, pp. 171–185.

Beattie, J., and Stockmayer, W., 1942, Chapter 2, *in* H.S. Taylor and S. Gladstone, ed., Treatise on Physical Chemistry: Princeton, NJ, D. Van Nostrand Co., 812 pp.

Beattie, J.A., 1955, Chapter 3, part C, *in* F.D. Rossini, ed., Thermodynamics and Physics of Matter: Princeton, NJ, Princeton University Press.

Benson, S.W., 1968, Thermochemical Kinetics: New York, Wiley and Sons, 223 pp.

Bent, Henry A., 1965, The Second Law: New York, Oxford University Press, 429 pp.

Bent, Henry A., 1972, A note on the notation and terminology of thermodynamics: Jour. Chem. Education, v. 49, pp. 44–46.

Berman, R.G., and Brown, T.H., 1984, A thermodynamic model for multicomponent melts, with application to the system $CaO-Al_2O_3-SiO_2$: Geochim. Cosmochim. Acta, v. 48, pp. 661–678.

Berman, R.G., Brown, T.H., and Greenwood, H.J., 1985, An internally consistent thermodynamic data base for minerals in the system: Atomic Energy of Canada Ltd., v. TR-377, Whiteshell Nuclear Res. Establishment, Pinawa, Canada, 62 pp.

Berman, R.G., Engi, M., Greenwood, H.J., and Brown, T.H., 1986, Derivation of internally-consistent thermodynamic data by thetechnique of mathematical programming: a review with application to the system $MgO-SiO_2-H_2O$: Jour. Petrology, v. 27, pp. 1331–1364.

Berman, R.G., 1988, Internally consistent thermodynamic data for minerals in the system $Na_2O-K_2O-CaO-MgO-FeO-Fe_2O_3-Al_2O_3-SiO_2-TiO_2-H_2O-CO_2$: Jour. Petrology, v. 29, pp. 445–522.

Boas, M.L., 1966, Mathematical Methods in the Physical Sciences: New York, Wiley and Sons, 778 pp.

Bockris, J.O'M., and Reddy, A.K.N., 1970, Modern Electrochemistry, 1: New York, Plenum, 622 pp.

Born, M., 1920, Volumen und Hydrationswarme der Ionen: Ziet. Physik., v. 1, pp. 45–48.

Bowers, T.S., and Helgeson, H.C., 1983, Calculation of the thermodynamic and geochemical consequences of nonideal mixing in the system H_2O-CO_2-NaCl on phase relations in geologic systems: Geochim. Cosmochim. Acta., v. 47, pp. 1247–1275.

Bowers, T.S., Jackson, K.J., and Helgeson, H.C., 1984, Equilibrium Activity Diagrams: Berlin, Springer Verlag, 397 pp.

Bradley, D.J., and Pitzer, K.S., 1979, Thermodynamics of electrolytes 12. Dielectric properties of water and Debye-Hückel parameters to 350°C and 1 Kbar: Jour. Phys. Chem., v. 83, pp. 1599–1603.

Brantley, S.L., Crerar, D.A., Møller, N.E., and Weare, J.H., 1984, Geochemistry of a modern marine evaporite: Bocana de Virrila, Peru: Jour. Sed. Petrology, v. 54, pp. 477–462.

Bridgman, P.W., 1914, A complete collection of thermodynamic formulas: Phys. Rev., v. 3, pp. 273–281.

Bridgman, P.W., 1926, A Condensed Collection of Thermodynamic Formulas: Cambridge, Harvard University Press, 34 pp.

Bridgman, P.W., 1941, The Nature of Thermodynamics:, Harvard University Press, 239 pp.

Brimhall, G.H., and Crerar, D.A., 1987, Ore fluids: magmatic to supergene: *in* I.S.E. Carmichael and H.P. Eugster, ed., Thermodynamic Modeling of Geological Materials: Minerals, Fluids and Melts, v. 17, Reviews in Mineralogy, Mineral. Soc. Amer., pp. 235–322.

Brown, T.H., 1977, Introduction to non-ideal and complex solutions: Chapter 9 *in* H.J. Greenwood, ed., Short Course in Application of Thermodynamics to Petrology and Ore Deposits, Toronto, Mineral. Assoc. Canada, pp. 126–135.

Butler, J.N., 1964, Ionic Equilibrium, a Mathematical Approach: Reading, MA, Addison-Wesley, 547 pp.

Callen, H.B., 1960, Thermodynamics: New York, Wiley and Sons, 376 pp.

Canagaratna, S.G., 1969, A critique of the definitions of heat: Amer. Jour. Physics, v. 37, pp. 679–683.

Carmichael, I.S.E., Turner, F.J., and Verhoogen, J., 1974, Igneous Petrology: New York, McGraw-Hill, 739 pp.

Cermignani, C., and Anderson, G.M., 1973, Origin of a diopside-tremolite assemblage near Tweed, Ontario: Can. Jour. Earth Sci., v. 10, pp. 84–90.

Chou, I-M., 1987, Oxygen buffer and hydrogen sensor techniques at elevated pressures and temperatures, *in* G.C. Ulmer and H.L. Barnes, ed., Hydrothermal Experimental Techniques: New York, Wiley-Interscience, pp. 61–99.

Crerar, D.A., Susak, N., and Schwartz, S., 1978, Solubility of the buffer assemblage pyrite and pyrrhotite and magnetite in NaCl solutions from 200 to 350°C: Geochim. Cosmochim. Acta., v. 42, pp. 1427–1437.

Crerar, D.A., 1975, A method for computing multicomponent chemical equilibria based on equilibrium constants: Geochim. Cosmochim. Acta., v. 39, pp. 1375–1384.

Crerar, D.A., and Anderson, G.M., 1971, Solubility and solvation reactions of quartz in dilute hydrothermal solutions: Chem. Geol., v. 8, pp. 107–122.

Crerar, D.A., Means, J.L., Yuretich, R., Borcsik, M., Amster, J., and Hastings, D., 1981, Hydrochemistry of the NJ Coastal Plain II: transport and deposition of iron, aluminum, dissolved organic matter: Chem. Geol., v. 31, pp. 23–44.

Davies, P.C.W., 1977, Thermodynamics and cosmology , *in* The Physics of Time Asymmetry: Berkeley, CA, University of California Press, pp. 80–111.

DeCapitani, C., and Brown, T.H., 1987, The computation of chemical equilibrium in complex systems containing non-ideal solutions: Geochim. Cosmochim. Acta., v. 51, pp. 2639–2652.

Debye, P.J., 1954, The Collected Papers of P.J. Debye: New York, Interscience, 700 pp.

Debye, P., and Hückel, E., 1923, The theory of electrolytes. I. Lowering the freezing point and related phenomena: Physik. Z. Leipzig, v. 24, pp. 185–206.

Debye, P., and Hückel, E., 1923, Theory of electrolytes II. The limiting law of electrical conductivity: Physik. Z. Leipzig, v. 24, pp. 305–325.

DeLaeter, J.R., 1988, Atomic weights of the elements 1987: Jour. Phys. Chem. Ref. Data, v. 17, pp. 1791–1793.

Denbigh, K., 1966, The Principles of Chemical Equilibrium: Cambridge, Cambridge University Press, 494 pp.

Denbigh, K.G. and Denbigh, J.S. 1985, Entropy in Relation to Incomplete Knowledge: Cambridge University Press, 164 pp.

Dence, J.B., 1975, Mathematical Techniques in Chemistry: New York, John Wiley and Sons, 442 pp.

Dickerson, R.E., 1969, Molecular Thermodynamics:, New York, W.A. Benjamin, 452 pp.

Douglas, T.B., and King, E.G., 1968, High-temperature drop calorimetry, *in* J.P. McCullough and D.W. Scott, ed., Experimental Thermodynamics v. 1: London, Butterworths, pp. 293–332.

Drever, J.I., 1988, The Geochemistry of Natural waters, 2nd ed.: Englewood Cliffs, NJ, Prentice Hall, 437 pp.

Dyson, F., 1979, Time without end: physics and biology in an open universe: Rev. Mod. Phys., v. 51, pp. 447–460.

Eugster, H.P., 1957, Heterogeneous reactions involving oxidation and reduction at high pressures and temperatures: Jour. Phys. Chem., v. 26, pp. 1760–1761.

Eugster, H.P., 1959, Reduction and oxidation in metamorphism, *in* P.H. Ableson, ed., Researches in Geochemistry v. 1: New York, Wiley and Sons, pp. 397–426.

Eugster, H.P., 1986, Minerals in hot water: Amer. Mineral., v. 71, pp. 655–673.

Eugster, H.P., Chou, I.-M., and Wilson, G.A., 1987, Mineral solubility and speciation in supercritical chloride fluids, *in* G.C. Ulmer and H.L. Barnes, ed., Hydrothermal Experimental Techniques: New York, Wiley-Interscience, pp. 1–19.

Eugster, H., Albee, A., Bence, A., Thompson, J.B., Jr., and Waldbaum, D., 1972, The two-phase region and excess mixing properties of paragonite-muscovite crystalline solutions: Jour. Petrology, v. 13, pp. 147–179.

Eugster, H.P., Harvie, C.E., and Weare, J.H., 1980, Mineral equilibria in a 6–7 component sea-water system, Na–K–Mg–Ca–SO_4–Cl–H_2O, at 25°C: Geochim. Cosmochim. Acta, v. 44, pp. 1335–1347.

Eugster, H.P., and Wones, D., 1962, Stability relations of the ferruginous biotite, annite: Jour. Petrology, v. 3, pp. 82–125.

Feynman, R.P., Leighton, R.B., and Sands, M., 1963, The Feynman Lectures on Physics, Vol. 1, pp. 4–2: Addison Wesley Publishing Co., New York.

Fleming, B.A., and Crerar, D.A., 1982, Silicic acid ionization and calculation of silica solubility at elevated temperature and *pH*. Application to geothermal fluid processing and reinjection: Geothermics, v. 11, pp. 15–29.

Flowers, G., 1979, Correction of Holloway's (1977) adaption of the modified Redlich-Kwong equation of state for calculation of the fugacities of molecular species in supercritical fluids of geologic interest: Contrib. Mineral. Petrol., v. 69, pp. 315–318.

Fowler, R.H., and Guggenheim, E.A., 1939, Statistical Thermodynamics: Cambridge, Cambridge University Press, 693 pp.

Franks, F., 1972, Introduction-Water, the unique chemical, *in* F. Franks, ed., Water. A Comprehensive Treatise New York, Plenum, pp. 1–20.

Frautschi, S., 1982, Entropy in an expanding universe: Sci., v. 217, pp. 593–599.

Friedman, H.L., 1962, Ionic Solution Theory: New York, Interscience, 265 pp.

Garrels, R.M., and Christ, C.L., 1965, Solutions, Minerals, and Equilibria: New York, Harper and Row, 450 pp.

Garrels, R.M., and Thompson, M.E., 1962, A chemical model for sea water at 25°C and one atmosphere total pressure: Amer. Jour. Sci., v. 260, pp. 57–66.

Gasser, R.P.H., and Richards, W.G., 1974, Entropy and Energy Levels: Oxford, Oxford University Press, 134 pp.

Gibbs, J.W., 1875, On the equilibrium of heterogeneous substances. Reprinted in: "The Scientific Papers of J.Willard Gibbs" Vol.1 Thermodynamics, pp. 55–349, Dover Publications Inc., New York, 1961.

Gill, P.E., Murray, W., and Wright, M.H., 1981, Practical Optimization: New York, Academic Press, 401 pp.

Goranson, R.W., 1930, Thermodynamic Relations in Multi-component Systems: Washington D.C., Carnegie Institution of Washington, 329 pp.

Gordon, T.M., 1977, Derivation of internally consistent thermochemical data from phase equilibrium experiments using linear programming: Chapter 13 *in* H.J. Greenwood, ed., Short Course in Application of Thermodynamics to Petrology and Ore Deposits, Toronto, Mineral. Assoc. Canada, pp. 185–198.

Green, E.J., 1970, On the perils of thermodynamic modelling: Geochim. Cosmochim. Acta, v. 34, pp. 1029–1033.

Greenwood, H.J., 1989, On models and modeling: Can. Mineralogist, v. 27, pp. 392–401.

Grossman, L., 1972, Condensation in the primitive solar nebula: Geochim. Cosmochim. Acta., v. 36, pp. 597–619.

Grossman, L., 1975, The most primitive objects in the solar system: Sci. Amer., v. 232,pp. 30–38.

Grossman, L., and Larimer, J., 1974, Early chemical history of the solar system: Rev. Geophy. Space Phys., v. 12, pp. 71–101.

Grover, J., 1977, Chemical mixing in multicomponent systems:an introduction to the use of Margules and other thermodynamic excess functions to represent non-ideal behavior, *in* D.G. Fraser, ed., Thermodynamics in Geology Boston, Reidel, pp. 67–97.

Guggenheim, E.A., 1935, The specific thermodynamic properties of aqueous solutions of strong electrolytes: Philos. Mag., v. 19, pp. 588–643.

Guggenheim, E.A., 1952, Mixtures: Oxford, Clarendon Press, 270 pp.

Guggenheim, E.A., 1959, Thermodynamics: Amsterdam, North-Holland Publishing Co., 476 pp.

Gunter, W.D., Myers, J., and Girsperger, S., 1987, Hydrogen: metal membranes, *in* G.C. Ulmer and H.L. Barnes, ed., Hydrothermal Experimental Techniques: New York, Wiley-Interscience, pp. 100–120.

Haar, L., Gallagher, J.S., and Kell, G.S., 1984, NBS/NRC Steam Tables. Thermodynamic and Transport Properties and Computer Programs for Vapor and Liquid States of Water in SI Units: Washington, Hemisphere Publishing Corporation, 320 pp.

Haas, J.L. Jr., and Fisher, J.R., 1966, Simultaneous evaluations and correlation of thermodynamic data: Amer. Jour. Sci., v. 276, pp. 525–545.

Harbaugh, J.W., and Bonham-Carter, G., 1970, Computer Simulation in Geology: New York, Wiley and Sons, 575 pp.

Harned, H.S., and Owen, B.B., 1958, The Physical Chemistry of Electrolyte Solutions, 3rd ed.: New York, Reinhold, 803 pp.

Harvie, C.E., and Weare, J.H., 1980, The prediction of mineral solubilities in natural waters: the Na–K–Mg–Ca–Cl–SO_4–H_2O system from zero to high concentration at 25°C: Geochim. Cosmochim. Acta, v. 44, pp. 981–987.

Harvie, C.E., Eugster, H.P., and Weare, J.H., 1982, Mineral equilibria in the six component seawater system, Na–K–Mg–Ca–SO_4–Cl–H_2O at 25°C. II. Compositions of the saturated solutions: Geochim. Cosmochim. Acta, v. 46, pp. 1603–1618.

Harvie, C.E., Greenberg, J.P., and Weare, J.H., 1987, A chemical equilibrium algorithm for highly non-ideal multiphase systems: free energy minimization: Geochim. Cosmochim. Acta., v. 51, pp. 1045–1057.

Harvie, C.E., Møller, N., and Weare, J.H., 1984, The prediction of mineral solubilities in natural waters: The Na–K–Mg–Ca–H-Cl-SO_4–OH–HCO_3–CO_3–CO_2–H_2O system to high ionic strengths at 25°C: Geochim. Cosmochim. Acta., v. 48, pp. 723–751.

Helgeson, H.C., 1964, Complexing and Hydrothermal Ore Deposition: New York, Macmillan, 128 pp.

Helgeson, H.C., 1968, Evaluation of irreversible reactions in geochemical processes involving minerals and aqueous solutions. I. thermodynamic relations: Geochim. Cosmochim. Acta., v. 32, pp. 853–877.

Helgeson, H.C., 1969, Thermodynamics of hydrothermal systems at elevated temperatures and pressures: Amer. Jour. Science, v. 267, pp. 729–804.

Helgeson, H.C., 1979, Mass transfer among minerals and hydrothermal solutions, *in* H.L. Barnes, ed., Geochemistry of Hydrothermal Ore Deposits, 2nd ed.: New York, Wiley-Interscience, pp. 568–610.

Helgeson, H.C., Brown, T.H., Nigrini, A., and Jones, T.A., 1970, Calculation of mass transfer in geochemical processes involving aqueous solutions: Geochim. Cosmochim. Acta, v. 34, pp. 569–592.

Helgeson, H.C., Delany, J.M., Nesbitt, H.W., and Bird, D.K., 1978, Summary and critique of the thermodynamic properties of rock-forming minerals: Amer. Jour. Sci., v. 278A, pp. 1–229.

Helgeson, H.C., Garrels, R.M., and Mackenzie, F.T., 1969, Evaluation of irreversible reactions in geochemical processes involving minerals and aqueous solutions II. applications: Geochim. Cosmochim. Acta., v. 33, pp. 455–481.

Helgeson, H.C., and Kirkham, D.H., 1974a, Theoretical prediction of the thermodynamic behavior of aqueous electrolytes at high temperatures and pressures: I. Summary of the thermodynamic/electrostatic properties of the solvent: Amer. Jour. Sci., v. 274, pp. 1089–1198.

Helgeson, H.C., and Kirkham, D.H., 1974b, Theoretical prediction of the thermodynamic behavior of aqueous electrolytes at high pressures and temperatures: II. Debye-Hückel parameters for activity.: Amer. Jour. Sci. , v. 274, pp. 1199–1261.

Helgeson, H.C., and Kirkham, D.H., 1976, Theoretical prediction of the thermodynamic properties of aqueous electrolytes at high pressures and temperatures: III. Equation of state for aqueous species at infinite dilution: Amer. Jour. Sci., v. 276, pp. 97–240.

Helgeson, H.C., Kirkham, D.H., and Flowers, G.C., 1981, Theoretical prediction of the thermodynamic properties of aqueous electrolytes at high pressures and temperatures: IV. Calculation of activity coefficients, osmotic coefficients, and apparent molal and standard and relative partial molal properties to 600°C and 5 kb: Amer. Jour. Sci., v. 281, pp. 1249–1493.

Hemingway, B.S., Haas, J.L. Jr., and Robinson, G.R. Jr., 1982, Thermodynamic Properties of Selected Minerals in the system Al_2O_3–CaO–SiO_2–H_2O at 298.15 K and 1 bar (10^5 Pascals) Pressure and at Higher Temperatures: U.S. Geol. Survey, Bull. 1544, 70 pp.

Henley, R.W., Truesdell, A.H., Barton, Jr., P.B., and Whitney, J.A., 1984, Reviews in Economic Geology, Vol. 1: Fluid-Mineral Equilibria in Hydrothermal Systems: , Society of Economic Geologists, 267 pp.

Hennet, R., Crerar, D., and Schwartz, J., 1988, Organic complexes in hydrothermal systems: Econ. Geol., v. 83, pp. 742–764.

Herzfeld, C.M., 1962, The thermodynamic temperature scale, its definition and realization: Chapter 6, pp. 41–50 *in* Temperature, Its Measurement and Control in Science and Industry, C.M. Herzfeld, ed. Vol. 1, Basic Standards and methods, F.G. Brickwedde, ed. New York, Rheinhold Publishing Co., 848 pp.

Hill, T.L., 1960, An Introduction to Statistical Thermodynamics: Reading, MA, Addison-Wesley, 508 pp.

Holloway, J., 1977, Fugacity and activity of molecular species in supercritical fluids, *in* D.G. Fraser, ed., Thermodynamics in Geology: Dordrecht-Holland, Reidel, pp. 161–181.

Hostetler, J.D., 1984, Electrode reactions, aqueous electrons, and redox potentials in natural waters: Amer. Jour. Sci., v. 284, pp. 734–759.

Huebner, J.S., 1971, Buffering techniques for hydrostatic systems at elevated pressures, *in* G.C. Ulmer, ed., Research Techniques for High Pressure and High Temperature: New York, Springer-Verlag, pp. 123–177.

Huebner, J.S., 1987, Use of gas mixtures at low pressure to specify oxygen and other fugacities of furnace atmospheres: *in* G.C. Ulmer and H.L. Barnes, ed., Hydrothermal Experimental Techniques, Wiley-Interscience, pp. 20–60.

Islam, J.N., 1979, The ultimate fate of the universe: Sky and Telescope, v. 57, pp. 13–18.

Jacobs, G.K., and Kerrick, D.M., 1981, APL and FORTRAN programs for a new equation of state for H_2O, CO_2, and their mixtures at supercritical conditions: Computers & Geosciences, v. 7, pp. 131–143.

Johnson, J.W., Oelkers, E.H., and Helgeson, H.C., 1992, SUPCRT92: A software package for calculating the standard molal thermodynamic properties of minerals, gases, aqueous species, and reactions from 1 to 5000 bars and 0° to 1000°C: Computers and Geosciences (in press).

Kac, M., 1982, Marginalia, dehydrated elephants revisited: Amer. Scient., v. 70,pp. 633–634.

Kerrick, D., and Jacobs, G., 1981, A modified Redlich-Kwong equation for H_2O, CO_2, and H_2O–CO_2 mixtures at elevated pressures and temperatures: Amer. Jour. Sci., v. 281, pp. 735–767.

Kielland, J., 1937, Individual activity coefficients of ions in aqueous solutions: Jour. Amer. Chem. Soc., v. 59, pp. 1675–1678.

Kivelson, D., and Oppenheim, I., 1966, Work in irreversible expansions: Jour. Chem. Education, v. 43, pp. 233–235.

Kleinmann, R.I.P. , Crerar, D., and Pacelli, R.R., 1981, The biochemistry of acid mine drainage and a method to control acid formation: Trans. AIME, v. 270, pp. 300–305.

Klotz, I.M., 1964, Chemical Thermodynamics: New York, W.A. Benjamin, 468 pp.

Klotz, I.M., and Rosenberg, R.M., 1972, Chemical Thermodynamics, 3rd ed.: Menlo Park, CA, W.A. Benjamin, 444 pp.

Korzhinskii, D.S., 1959, Physicochemical Basis of the Paragenesis of Minerals: New York, Consultant's Bureau, Inc., 142 pp.

Krumbein, W.C., and Garrels, R.M., 1952, Origin and classification of chemical sediments in terms of *pH* and oxidation-reduction potentials: Jour. Geol., v. 60, pp. 1–33.

Latimer, W.M., 1952, Oxidation Potentials, 2nd ed.: Englewood Cliffs, NJ, Prentice Hall, 392 pp.

Lewins, J., 1985, Teaching Thermodynamics: New York, Plenum Press, 517 pp.

Lewis, G.N., and Randall, M., 1923, Thermodynamics: New York, McGraw-Hill.

Lietzke, M.H., and Stoughton, R.W., 1962, The calculation of activity coefficients from osmotic coefficient data: Jour. Phys. Chem., v. 66, pp. 508–509.

Liu, C.T., and Lindsay, W.T., 1972, Thermodynamics of sodium chloride solutions at high temperatures: Jour. Solution Chem., v. 1, pp. 45–69.

Long, F.A., and McDevit, W.F., 1952, Activity coefficients of nonelectrolytes in aqueous electrolyte solutions: Chem. Rev., v. 51, pp. 119–169.

Maier, C.G., and Kelly, K.K., 1932, An equation for the representation of high temperature heat content data: Amer. Chem. Soc. Jour., v. 54, pp. 3243–3246.

Margenau, H., and Murphy, G.M., 1956, The Mathematics of Physics and Chemistry, 2nd ed.: New York, D. Van Nostrand, 604 pp.

Margules, M., 1895, Uber die Zusammen setzung der gesattigten Dampfe von Mischugen: Sitzungber: Wien Akad., v. 104, pp. 1243–1278.

Mayer, J.E., 1950, The Theory of Ionic Solutions: Jour. Chem. Phys., v. 18, pp. 1426–1436.

Mayer, J.E., and Mayer, M., 1940, Statistical Mechanics: New York, Wiley and Sons, 495pp.

Mazo, R.M., and Mou, C.Y., 1979, Introduction to the statistical mechanics of solutions, *in* R.M. Pytkowicz, ed., Activity Coefficients in Electrolyte Solutions v. 1: Boca Raton, FL, CRC Press, pp. 29–64.

McCullough, J.P., and Scott, D.W., eds., 1968, Experimental Thermodynamics, v1. Calorimetry of Non-reacting Systems: London, Butterworths, 606 pp.

McGlashan, M.L., 1966, The use and misuse of the laws of thermodynamics: Jour. Chem. Education, v. 43, pp. 226–232.

McGlashan, M.L., 1979, Chemical Thermodynamics: Academic Press, London, 345 pp.

McMillan, W.G., and Mayer, J.E., 1945, The statistical thermodynamics of multicomponent systems: Jour. Chem. Phys., v. 13, pp.276–303.

Mesmer, R.E., 1985, A model for estimation of thermodynamic quantities for reactions—uncertainties from such predictions: Paper presented at Second International Symposium on Hydrothermal Reactions at The Pennsylvania State University.

Monnin, C., 1989, An ion interaction model for the volumetric properties of natural waters: Density of the solution and partial molar volumes of electrolytes to high concentrations at 25°C: Geochim. et Cosmochim. Acta, v. 53, pp. 1177–1188.

Monnin, C., 1990, The influence of pressure on the activity coefficients of the solutes and on the solubility of minerals in the system $Na–Ca–Cl–SO_4–H_2O$ to 200°C and 1 kbar, and to high NaCl concentration: Geochim. Cosmochim. Acta, v. 54, pp. 3265–3282.

Møller, N., 1988, The prediction of mineral solubilities in natural waters: A chemical equilibrium model for the $CaSO_4–NaCl–CaCl_2–H_2O$ system to high temperatures and concentration: Geochim. Cosmochim. Acta, v. 52, pp. 821–838.

Monk, C.B., 1961, Electrolytic Dissociation: London, Academic Press, 320 pp.

Morris J.C., and Stumm, W., 1967, Redox equilibria and measurements of potentials in the aquatic environment: *in* Equilibrium Concepts in Natural Water Systems. Adv. in Chem. Series, No. 67, pp. 270–285. Amer. Chem. Soc., Washington, D.C.

Nafziger, R.H., and Muan, A., 1967, Equilibrium phase compositions and thermodynamic properties of olivines and pyroxenes in the system MgO–"FeO"–SiO_2: Amer. Mineral., v. 52, pp. 1364–1385.

Nash, L.K., 1972, Chemthermo: A Statistical Approach to Classical Chemical Thermodynamics: Reading, Addison-Wesley, 207 pp.

Nash, L.K., 1971, Elements of Statistical Thermodynamics: Reading, Addison-Wesley, 138 pp.

Navrotsky, A., 1977, Progress and new directions in high temperature calorimetry: Phys. Chem. Mineral., v. 2, pp. 89–104.

Nernst, W., 1897, Die elektrolytische Zersetzung Wassriger Losungen: Ber., v. 30,pp. 1547–1563.

Nordstrom, D.K., and Munoz, J.L., 1985, Geochemical Thermodynamics: Menlo Park, CA, Benjamin/Cummings, 477 pp.

Nordstrom, D.K., and Ball, J.W., 1984, Chemical models, computer programs and metal complexation in natural waters, *in* C.J.M. Kramer and J.C. Duinker, ed., Complexation of Trace Metals in Natural Waters The Hague, Nijhoff and Junk, pp. 149–164.

Nordstrom, D.K., Jenne, E.A., and Ball, J.W., 1979, Redox equilibria of iron in acid mine waters, *in* E.A. Jenne, ed., Chemical Modelling in Aqueous Systems v. Sympos. Ser. 93: Amer. Chem. Soc., pp. 51–79.

Nordstrom, D.K., Plummer, L.N., Wigley, T.M.L., Wolery, T.J., and Ball, J.W., 1979, A comparison of computerized chemical models for equilibrium calculations in aqueous systems, *in* E.A. Jenne, ed., Chemical Modelling in Aqueous Solutions v. Sympos. Ser. 93: , Amer. Chem. Soc., pp. 857–892.

Oelkers, E.H., and Helgeson, H.C., 1991, Calculation of activity coefficients and degrees of formation of neutral ion pairs in supercritical electrolyte solutions: Geochim. Cosmochim. Acta, v. 55, pp. 1235–1251.

Partington, J.R., 1957, A Short History of Chemistry, 3rd. ed.: New York, Harper and Row.

Perrin, C.L., 1970, Mathematics for Chemists: New York, Wiley-Interscience, 453pp.

Pippard, A.B., 1966, Elements of Classical Thermodynamics: Cambridge University Press, 165 pp.

Pitzer, K.S., 1975, Thermodynamics of electrolytes. V. Effects of higher-order electrostatic terms: Jour. Solution Chem., v. 4, pp. 249–265.

Pitzer, K.S., 1979, Theory: ion interaction approach, *in* R.M. Pytkowicz, ed., Activity Coefficients in Electrolyte Solutions v. 1: Boca Raton, CRC Press, pp. 157–208.

Pitzer, K.S., 1981, Characteristics of very concentrated solutions: *in* Proceedings, Nobel Symposium on the Chemistry and Geochemistry of Solutions at High Temperatures and Pressures (Stockholm, Royal Swedish Acad. Sci.), F. Wickmann and D. Rickard, eds. Vol. 13/14 of Phys. and Chem. of the Earth. New York, Pergamon Press.

Pitzer, K.S., 1987, A thermodynamic model for aqueous solutions of liquid-like density: *in* I.S.E. Carmichael and H.P. Eugster, eds., Thermodynamic Modeling of Geological Materials: Minerals, Fluids and Melts, v. 17, Reviews in Mineralogy, Mineral. Soc. Amer., pp. 97–142.

Pitzer, K.S., and Brewer, L., 1961, Revised Edition of "Thermodynamics" by G.N. Lewis and M. Randall, 2nd ed.: New York, McGraw-Hill, 723 pp.

Plummer, L.N., Parkhurst, D.L., and Thorstenson, D.C., 1983, Development of reaction models for groundwater systems: Geochim. Cosmochim. Acta., v. 47, pp. 665–685.

Pourbaix, M., 1949, Thermodynamics of Dilute Aqueous Solutions: London, E. Arnold, 136 pp.

Pourbaix, M., 1974, Atlas of Electrolytical Equilibria in Aqueous Solutions: Houston, TX, Nat. Ass. Corrosion Engineers, 644 pp.

Powell, R., 1977, Activity-composition relationships for crystalline solutions, *in* D.G. Fraser, ed., Thermodynamics in Geology Boston, Reidel, pp. 57–65.

Prausnitz, J.M., 1969, Molecular Thermodynamics of Fluid Phase Equilibria: Englewood Cliffs, NJ, Prentice Hall, 523 pp.

Prausnitz, J.M., Lichtenthaler, R.N., and Azevedo, E.G. de, 1986, Molecular Thermodynamics of Fluid Phase Equilibria, 2nd ed.: Englewood Cliffs, NJ, Prentice Hall, 600 pp. Prigogine, I., and Defay, R., 1954, Chemical Thermodynamics: London, Longmans Green, 543 pp.

Prigogine, I., and Defay, R., 1954, Chemical Thermodynamics: London, Longmans Green, 543 pp.

Pytkowicz, R.M., ed., 1979, Activity Coefficients in Electrolyte Solutions, 162: Boca Raton, FL, CRC Press, v. 1 288 pp., v.2 330 pp.

Quist, A.S., and Marshall, W.L., 1968, Electrical conductances of aqueous sodium chloride solutions from 0 to 800°C and at pressures to 4000 bars: Jour. Phys. Chem., v. 72, pp. 684–703.

Ramberg, H., 1971, Temperature changes associated with adiabatic decompression in geological processes: Nature, v. 234, pp. 539–540.

Randall, M., and Failey, 1927a, The activity coefficients of gasses in aqueous salt solutions: Chem. Rev., v. 4, pp. 271–284.

Randall, M., and Failey, 1927b, The activity coefficients of non-electrolytes in aqueous salt solutions from solubility measurements: Chem. Rev., v. 4, pp. 285–290.

Randall, M., and Failey, 1927c, The activity coefficients of the undissociated part of weak electrolytes: Chem. Rev., v. 4, pp. 291–318.

Redlich, O., 1976, Thermodynamics: Fundamentals, Applications: New York, Elsevier, 277 pp.

Redlich, O., and Kwong, J., 1949, The thermodynamics of solutions. V. An equation of state. Fugacities of gaseous solutions.: Chem. Rev., v. 44, pp. 233–244.

Reed, M.H., 1982, Calculation of multicomponent chemical equilibria and reaction processes in systems involving minerals, gases, and an aqueous phase: Geochim. Cosmochim. Acta., v. 46, pp. 513–528.

Reiss, H., 1965, Methods of Thermodynamics: New York, Blaisdell, 217 pp.

Richardson, S.W., Gilbert, M.C., and Bell, P.M., 1969, Experimental determination of kyanite-andalusite and andalusite-sillimanite equilibria:the aluminum silicate triple point: Amer. Jour. Sci., v. 267, pp. 259–272.

Robie, R.A., Hemingway, B.S., and Fisher, J.R., 1978, Thermodynamic Properties of Minerals and Related Substances at 298.15 K and 1 Bar (10^5 Pascals) Pressure and at Higher Temperatures: U.S. Geological Survey Bull. 1452, 456 pp.

Robie, R.A., and Hemingway, B.S., 1972, Calorimeters for Heat of Solution and Low-Temperature Heat Capacity Measurements: U.S. Geological Survey Prof. Paper 755, 32 pp.

Robinson, R.A., and Stokes, R.H., 1968, Electrolyte Solutions: London, Butterworths, 571 pp.

Robinson, G.R., Haas, J.L. Jr., Schafer, C.M., and Haselton, H.T. Jr., 1983, Thermodynamic and Thermophysical Properties of Selected Phases in the $MgO–SiO_2–H_2O–CO_2$, $CaO–Al_2O_3–SiO_2–H_2O–CO_2$ and $Fe–FeO–Fe_2O_3–SiO_2$: Open-File Rpt. 83-79:, U.S. Geol. Surv., 429 pp.

Runciman, W., Sengupta, D., and Gourley, J.T., 1973, The polarized spectra of iron and silicates. II. Olivine: Amer. Mineral, v. 58, pp. 451–456.

Sato, M., 1971, Electrochemical measurements and control of oxygen fugacity and other gaseous fugacities with solid electrolyte sensors, *in* G.C. Ulmer, ed., Research Techniques for High Pressure and High Temperature: New York, Springer-Verlag, pp. 43–99.

Saxena, S.K., 1973, Thermodynamics of Rock-forming Crystalline Solutions: New York, Springer-Verlag, 188 pp.

Schumann, R. Jr., 1955, Application of Gibbs-Duhem equations to ternary systems: Acta Metallurgica, v. 3, pp. 219–226.

Shaw, A.N., 1935, The derivation of thermodynamical relations for a simple system: Roy. Soc. London, Phil. Trans., v. A243, pp. 299–328.

Shaw, H.R., 1963, Hydrogen–water vapor mixtures: control of hydrothermal atmospheres by hydrogen osmosis: Science, v. 139, pp. 1220–1222.

Sillén, L.G., and Martell, A.E, 1964, Stability Constants of Metal-Ion Complexes. Special Publication 17: London, The Chemical Society, 754 pp.

Smith, W.R., and Missen, R.W., 1982, Chemical Reaction Equilibrium Analysis: Theory and Algorithms: New York, Wiley and Sons, 364 pp.

Stokes, R.H., and Robinson, R.A., 1948, Ionic hydration and activity in electrolyte solutions: Jour. Amer. Chem. Soc., v. 70, pp. 1870–1878.

Stull, D.R., and Prophet, H., 1971, JANAF thermochemical tables: 2nd ed., Nat. Stand. Ref. Data Ser., Nat. Bur. Stand. (U.S.), vol. 37, 1141 pp.

Sweeton, F.H., Mesmer, R.E., and Baes, Jr., C.F., 1974, Acidity measurements at elevated temperatures. VII. Dissociation of water: Jour. Solution Chemistry, v. 3, pp. 191–214.

Tanger, J.C. IV, and Helgeson, H.C., 1988, Calculation of the thermodynamics and transport properties of aqueous species at high pressures and temperatures: Revised equations of state for the standard partial molal properties of ions and electrolytes: Amer. Jour. Science, v. 288, pp. 19–98.

Thompson, J.B. Jr., 1959, Local equilibrium in metasomatic processes, *in* P.H. Ableson, ed., Researches in Geochemistry v. 1: New York, Wiley and Sons, pp. 427–457.

Thompson, J.B. Jr., 1967, Thermodynamic properties of simple solutions, *in* P.H. Ableson, ed., Researches in Geochemistry v. 2: New York, Wiley and Sons, pp. 340–361.

Thompson, J.B. Jr., 1970, Geochemical reaction and open systems: Geochim. Cosmochim. Acta, v. 34, pp. 529–551.

Thompson, J.B. Jr., 1982, Composition space: an algebraic and geometric approach: *in* John M. Ferry, ed., Characterization of Metamorphism through Mineral Equilibria v. 10, Mineralogical Society of America, pp. 1–32.

Thompson, J.B. Jr., and Waldbaum, D.R, 1968, Mixing properties of sanidine crystalline solutions: I. Calculations based on ion-exchange data: Amer. Mineral., v. 53, pp. 1965–1999.

Thompson, J.B. Jr., and Waldbaum, D.R., 1969, Mixing properties of sanidine crystalline solutions:III. Calculations based on two-phase data.: Amer. Mineral., v. 54, pp. 811–838.

Tinoco, I Jr., Sauer, K., and Wang, J.C., 1985, Physical Chemistry, Principles and Applications in the Biological Sciences, 2nd ed.: Englewood Cliffs, NJ, Prentice Hall, 706 pp.

Tremaine, P.R., Sway, K., and Barbero, J.A., 1986, The apparent molar heat capacity of aqueous hydrochloric acid from 10 to 40°C: Jour. Solution Chem., v. 15, pp. 1–22.

Truesdell, A.H., 1968, The advantage of using pE rather than Eh in redox equilibrium calculations: Jour. Geol. Education, v. 16, pp. 17–20.

Truesdell, A.H., and Jones, B.F., 1974, WATEQ, a computer program for calculating chemical equilibria of natural waters: Jour. Res. U.S. Geol. Survey, v. 2, pp. 233–248.

Truesdell, C., 1980, The Tragicomical History of Thermodynamics 1822–1854: New York, Springer Verlag, 372 pp.

Tunell, G., 1931, The definition and evaluation of the fugacity of an element or compound in the gaseous state: Jour. Phys. Chem., v. 35, pp. 2885–2913.

Tunell, G., 1960, Relations Between Intensive Thermodynamic Quantities and Their First Derivatives in a Binary System of One Phase: San Francisco, W.H. Freeman, 47 pp.

Ulbrich, H., and Waldbaum, D., 1976, Structural and other contributions to the third law entropies of silicates: Geochim. Cosmochim. Acta., v. 40, pp. 1–24.

Van Zeggeren, F., and Storey, S.H., 1970, The Computation of Chemical Equilibria: Cambridge, Cambridge University Press, 176 pp.

Vera, J., and Prausnitz, J., 1972, Generalized van der Waal's theory for dense fluids: Chem. Eng. Jour., v. 3, pp. 1–13.

Wagman, D.D., Evans, W.H., Parker, V.B., Schumm, R.H., Halow, I., and Bailey, S.M., 1982, The NBS tables of chemical thermodynamic properties: Jour. Phys. Chem. Ref. Data, Vol. II, Supplement no. 2, Amer. Chem. Soc., Washington, D.C.

Waldbaum, D.R., 1971, Temperature changes associated with adiabatic decompression in geological processes: Nature, v. 232, pp. 545–547.

Waldbaum, D.R., and Thompson, J.B., Jr., 1969, Mixing properties of sanidine crystalline solutions: IV. Phase diagrams from equations of state: Amer. Mineral., v. 54, pp. 1274–1298.

Waldbaum, D.R., and Thompson, J.B., Jr., 1968, Mixing properties of sanidine crystalline solutions: II. Calculations based on volume data: Amer. Mineral., v. 53, pp. 200–2017.

Wall, F.T., 1958, Chemical Thermodynamics: San Francisco, CA, Freeman, 422 pp.

Walther, J.V., and Helgeson, H.C., 1977, Calculation of the properties of aqueous silica and the solubility of quartz and its polymorphs at high pressures and temperatures: Amer. Jour. Sci., v. 277, pp. 1315–1351.

Weill, D.F., and Fyfe, W.S., 1964, A discussion of the Korzhinskii and Thompson treatment of thermodynamic equilibrium in open systems: Geochim. Cosmochim. Acta, v. 28, pp. 565–576.

Weare, J.H., 1987, Models of mineral solubility in concentrated brines with application to field observations: Chapter 5 *in* I.S.E. Carmichael and H.P. Eugster, eds., Thermodynamic Modeling of Geological Materials: Minerals, Fluids and Melts, v. 17, Reviews in Mineralogy, Mineral. Soc. Amer., pp. 143–176.

Weinberg, S., 1977, The First Three Minutes: New York, Basic Books, 188 pp.

Wolery, T.J., 1979, Calculation of chemical equilibrium between aqueous solution and minerals: the EQ3/6 software package: Livermore, CA, Lawrence Livermore Nat'l. Lab.

Wolery, T.J., 1983, EQ3NR, a computer program for geochemical aqueous speciation-solubility calculations: user's guide and documentation: Livermore, CA, Lawrence Livermore Nat'l. Lab.

Wood, S.A., Crerar, D., Brantley, S.L., and Borcsik, M., 1984, Mean molal stoichiometric activity coefficients of alkali halides and related electrolytes in hydrothermal solutions: Amer. Jour. Sci., v. 284, pp. 668–705.

Wood, S.A., Crerar, D., and Borcsik, M., 1987, Solubility of the assemblage pyrite-pyrrhotite-magnetite-sphalerite-galena-gold-stibnite-bismuthinite- argentite-molybdenite in H_2O–$NaCl$–CO_2 solutions from 200 to 350°C: Econ. Geol., v. 82, pp. 1864–1887.

Wood, S.A., and Spera, F.J., 1984, Adiabatic decompression of aqueous solutions: Applications to hydrothermal fluid migration in the crust: Geology, v. 12, pp. 707–710.

Yoder, H.S., 1952, The MgO–Al_2O_3–SiO_2–H_2O system and the related metamorphic facies: Amer. Jour. Sci., Bowen Vol., pp. 569–627.

Zemaitis, J.F., D.M. Clark, M. Rafal, and Scrivner, N.C., 1986, Handbook of aqueous electrolyte thermodynamics: Amer. Instit. Chem. Engineers, New York, 852 pp.

Zen, E-an, 1961, Mineralogy and petrology of the system Al_2O_3–SiO_2–H_2O in some pyrophyllite deposits of North Carolina: Amer. Mineralogist, v. 46, pp. 52–66.

Zen, E-an, 1962, Problem of the thermodynamic status of the mixed-layer minerals: Geochim. Cosmochim. Acta, v. 26, pp. 1055–1067.

Zen, E-an, 1963, Components, phases, and criteria of chemical equilibrium in rocks: Amer. Jour. Sci., v. 261, pp. 929–942.

INDEX